Environmental Science and Engineering

The ultimate goal of this series is to contribute to the protection of our environment, which calls for both profound research and the ongoing development of solutions and measurements by experts in the field. Accordingly, the series promotes not only a deeper understanding of environmental processes and the evaluation of management strategies, but also design and technology aimed at improving environmental quality. Books focusing on the former are published in the subseries Environmental Science, those focusing on the latter in the subseries Environmental Engineering.

Junwen Zhang · Roger Ruan ·
Mohammed J. K. Bashir
Editors

Environmental Pollution Governance and Ecological Remediation Technology

Springer

Editors
Junwen Zhang
School of Energy and Mining Engineering
China University of Mining
and Technology (Beijing)
Beijing, China

Roger Ruan
University of Minnesota
St. Paul, MN, USA

Mohammed J. K. Bashir
University Tunku Abdul Rahman
Kampar, Malaysia

ISSN 1863-5520 ISSN 1863-5539 (electronic)
Environmental Science and Engineering
ISBN 978-3-031-25283-9 ISBN 978-3-031-25284-6 (eBook)
https://doi.org/10.1007/978-3-031-25284-6

This Springer imprint is published by the registered company Springer Nature Switzerland AG
The registered company address is: Gewerbestrasse 11, 6330 Cham, Switzerland

Committee Members

Conference General Chairs

Prof. Roger Ruan (ASABE Fellow, IFT Fellow), University of Minnesota, USA
Prof. Xiaozhi Wang (Dean), Yangzhou University, China
Prof. Tianpeng Gao (Dean), Xi'an University, China

Technical Program Committee Chairs

Prof. Abdeltif Amrane, University of Rennes 1, France
Prof. Fawang Zhang, China Geological Survey, China

Publication Chairs

Prof. Junwen Zhang, China university of Mining and Technology (Beijing), China
Prof. Roger Ruan (ASABE Fellow, IFT Fellow), University of Minnesota, USA
Prof. Mohammed J. K. Bashir, University Tunku Abdul Rahman, Malaysia

Local Organizing Committees

Prof. Wenbin Zhang, Honghe University, China
Prof. Shouyu Zhang, University of Shanghai for Science and Technology, China
Prof. Jianfeng Zhang, Institute of subtropical Forestry, Chinese Academy of Forestry, China

Prof. Xiaoqing Wang, Chongqing Jiaotong University, China
Prof. Junwen Zhang, China university of Mining and Technology (Beijing), China

Technical Program Committees

Assoc. Prof. Junyuan Guo, Chengdu University of Information Technology, China
Prof. Tao Zhu, China University of mining and technology (Beijing), China
Prof. Shoubin Zhang, University of Jinan, China
Prof. Haibo Li, Northeastern University, China
Prof. Li Zhang, Beijing University of Technology, China
Prof. Xiao Tan, Hohai University, China
Prof. Ping Wang, Central South University of Forestry and Technology, China
Prof. Siyue Li, Wuhan Institute of Technology, China
Prof. Shengsen Wang, Yangzhou University, China
Prof. Yu Gao, Shenyang University of Technology, China
Prof. Xinbo Wang, Shandong University, China
Prof. Fang Deng, Nanchang Hangkong University, China
Prof. Zhibao Huo, Shanghai Ocean University, China
Prof. Yufeng Jiang, Lanzhou Jiaotong University, China
Prof. Yu Hong, Beijing Forestry University, China
Prof. Yu Shen, Nanjing Forestry University, China
Prof. Xueqing Shi, Qingdao University of Technology, China
Prof. Lingjun Kong, Guangzhou University, China
Prof. Xiaofeng Wang, Dalian Maritime University, China
Prof. Hualin Jiang, Nanchang Hangkong University, China
Prof. Yongming Zhang, Shanghai Normal University, China
Prof. Xiaojian Zhou, Yangzhou University, China
Prof. Min Li, College of Environmental Science and Engineering, Beijing Forestry University, China
Prof. Tiecheng Wang, Northwest A&F University, China
Prof. Yuqi Wang, School of Environmental Science and Engineering, Guangzhou University, China
Prof. Panyue Zhang, College of Environmental Science and Engineering, Beijing Forestry University, China
Prof. Wanjun Wang, Guangdong University of Technology, China
Prof. Jin Wang, School of Environment, Beijing Jiaotong University, China
Prof. Nianzhi Jiao, Xiamen University, China
Prof. WenfaTan, South University, China
Prof. Tao Zhang, China Agricultural University, China
Prof. Abdeltif Amrane, University of Rennes 1, France
Prof. George A. O'Doherty, Northeastern University, USA
Prof. Carlos Caceres, The University of Queensland, Australia
Prof. Lam Kin Che, The Chinese University of Hongkong, HK

Preface

The 2022 2nd International Conference on Environmental Pollution and Governance (ICEPG 2022) was successfully held on September 23–25, 2022, in Shanghai, China (virtual conference). ICEPG 2022 presented the latest research results in related fields and provided opportunities for delegates from all over the world to exchange experience and new ideas, establish business or research relations as well as find global partners for future collaborations.

We were delighted to invite three distinguished professors to serve as our conference general chairs. They are Prof. Roger Ruan from the University of Minnesota, USA; Prof. Xiaozhi Wang from Yangzhou University, China; and Prof. Tianpeng Gao from Xi'an University, China, respectively. Over 80 individuals attended the conference. Divided into three parts, the conference agenda covered keynote speeches, oral presentations, and online Q&A discussion. Firstly, keynote speakers were each allocated 30–45 minutes to hold their speeches. Then in the oral presentations, the excellent papers we had selected were presented by their authors by turns.

During the conference, five sophisticated professors were invited to address keynote speeches. Among them, Prof. Roger Ruan, one of our conference general chairs, delivered a keynote speech on: complete gas, liquid, and solid waste utilization for sustainable development. Aiming to ensure a sustainable development through complete recycling and re-utilizing solid, liquid, and gas wastes, his research discussed a series of advanced technologies or systems to extract value out of the wastes while treating them at the same time. Additionally, Prof. Fawang Zhang from China Geological Survey, China, made a report on the Essence of Mine Groundwater Pollution "Source Prevention" and Its Relationship with "Terminal Governance." This paper started with the root cause of the mine groundwater pollution and discussed the relationship between governance problems of the combination of "source prevention" and "terminal governance" in order to provide theoretical basis for the remediation and treatment of groundwater pollution in mines. The wonderful and dramatic speeches of each keynote speaker had triggered heated discussion. Moreover, every participant praised this conference for disseminating useful and insightful knowledge.

After months of well preparation and hard work, the proceedings of ICEPG 2022 covering a bunch of papers are smoothly published. These papers feature but are not limited to the following areas: sewage treatment, environmental monitoring, water pollution control project, soil pollution control, etc. All the papers have been checked through rigorous review and processes to meet the requirements of publication.

On behalf of the organizing committee, we would like to express our heartfelt appreciation to all the keynote speakers, peer reviewers, and all the participants. Particularly, we would like to acknowledge the Springer, for the endeavor of all its colleagues in publishing this paper volume. We firmly believe that all the attendees have had fruitful discussions and gained valuable knowledge and will enjoy the opportunity for future collaborations.

The Committee of ICEPG 2022

Beijing, China — Junwen Zhang
St. Paul, USA — Roger Ruan
Kampar, Malaysia — Mohammed J. K. Bashir

Contents

Part I
Environment Monitoring and Causes of Pollution Analysis

Chapter 1
The Water Quality Survey of Underground and Surface Water in Songmuzhai

Jiazhi Liang, Honglei Wei, Shaohong You, and Pingping Jiang

Abstract The study area was adjacent to the industrial area, the water was polluted by industrial activity in several years ago. In order to study the water quality of this area whether got better, we chose three spring water monitoring points, three drilling monitoring points and two surface water sampling points were selected. The pH, temperature, conductivity, and major ions (K^+, Na^+, Ca^{2+}, Mg^{2+}, Cl^-, HCO_3^-) were measured for a long time. Based on the correlation and seasonal variation characteristics of ions, the pollution characteristics of major irons were discussed. The results show that surface water is easily affected by human activities, while the background spring (S2) was not affected by human activities significatively. Thus, the water quality in this area was not affected by industrial activity.

Keywords Surface water · Spring · Drilling · Water quality

1.1 Introduction

As a typical karst area in southern China, the southwest region has strong karst development. There are 3066 karst underground rivers. The karst groundwater reserves account for 51% of the total groundwater. The karst water resources in Guizhou and Chongqing even account for the total groundwater resources of their provinces and

---

J. Liang · H. Wei · S. You · P. Jiang (✉)
Guilin University of Technology, Guilin 541004, China
e-mail: Jiangpp@glut.edu.cn

J. Liang
Guangxi Zhongpinzhi Environmental Monitoring Co., Ltd., Guilin 541004, China

H. Wei
Guilin Jiahua Environmental Technology Co., Ltd., Guilin 541004, China

S. You
Technical Innovation Center of Mine Geological Environment Restoration Engineering in Southern Stony Hill Area, Nanning, China

J. Zhang et al. (eds.), *Environmental Pollution Governance and Ecological Remediation Technology*, Environmental Science and Engineering,
https://doi.org/10.1007/978-3-031-25284-6_1

municipalities. 80 and 73% of the total amount (Pu et al. 2010; Yuan 2014). The quality of karst groundwater varies greatly among provinces, municipalities directly under the Central Government and autonomous regions due to different geographical locations, geological conditions and human activities (Kačaroğlu 1999). According to the source of pollution, it is divided into agricultural pollution, industrial pollution and urban pollution, etc.

The study area was located in the karst area. The water was polluted by industrial activity in several years ago. At that time, the surface water was not drunk, and the soil was not planted. The water quality improved after several years whether or not. Thus, in order to know the state of water quality, the springs, drillings and surface waters were chosen. The purpose of this paper: (1) to understand the characteristics of water quality; (2) the main source of ions in water.

1.2 Materials and Methods

1.2.1 Study Area

The study area is located in the county of Songmuzhai within Hezhou city of the Guangxi Province. The main sources of pollution of water were agricultural pollution, domestic waste water and industrial waste water. The annual rainfall in the study area is abundant, but the seasonal distribution is uneven, mainly in May–July and October-December. The landform of the study area is the Gufeng Plain, which is located on the wing of the syncline, with faults developed at about 84° to the west of the north. The strata around the study area include the Lower Carboniferous Datang Stage (C_1d), the Lower Carboniferous Yanguan Stage (C_1y), the Lower Jurassic Daling Formation (J_1d), and the lower Jurassic Shiti Formation (J_1s_1), Lower Tertiary (Ea), Quaternary Update (Q_{1-2}) and Holocene (Q_{3-4}). (The hydrogeological and distribution of samplings keep the same with Jiang et al. (2018).

1.2.2 Hydrochemical Parameters

Samplings were collected from eight points, including three springs (S1, S2, and S3), three drillings and two surface water (Table 1.1). Field measurements of pH, temperature, and electric conductivity (EC) were collected with a portable multi-parameter water quality analyzer (WTW multi 3430, Germany). The analytical precisions for these measurements were 0.01, 1 μs/cm, and 0.1 °C, for pH, EC, and temperature, respectively. The pH and EC values were compensated to 25 °C.

Table 1.1 Water sample points

Number	Type	Current situation
S1	Spring	Pollution
S2	Spring	Background value
S3	Spring	Water quality was influenced by rain
ZK1	Drilling	Pollution
ZK2	Drilling	Pollution
ZK3	Drilling	Pollution
F	Surface water	Water quality was influenced by rain
XH-1	Surface water	Pollution

1.2.3 Analytical Methods

Water samples were filtered using 0.45-μm acetate filter membranes and collected in 50-mL clear polyethylene bottles. Cations and anions were analyzed using two samples each. Cation samples were acidified (pH) using 0.2 mL 1:1 HNO_3, while the other water samples were preserved at 4 °C. All of the geochemical analyses were conducted at the Environmental and Geochemical Analysis Laboratory at the Institute of Karst Geology of the Chinese Academy of Geological Science. Cations and heavy metals (K^+, Na^+, Ca^{2+}, Mg^{2+}) as well as anions (Cl^-, SO_4^{2-}, and NO_3^-) were measured on an ICP-OES spectrometer (IRIS Intrepid II XSP, Termo Fisher Scientifc, USA) and Ion Chromatograph (861 Advanced Compact IC Metrohm, Swiss), respectively, with the analytical precision of 0.01 mg/L for both. HCO_3^- was measured in triplicate via hydrochloric acid titration, and it exhibited an average error of less than 5% (Jiang et al. 2020).

1.3 Results and Discussion

1.3.1 Physicochemical Parameters and Major Irons

The results of p, temperature and EC were presented in Table 1.2. The pH values of ZK2 and F were 4.545–6.922, while other samplings of pH values were higher (6.897–7.611). The EC value in S2 was lowest (319–368 μs/cm), which was not influenced by seasons. However, most of the EC values in other samplings were over 1000 μs/cm. Meanwhile, the EC values were influenced by human activities.

Table 1.2 Parameters at site and common ions in water

	Time	pH	T	EC	K^+	Na^+	Ca^{2+}	Mg^{2+}	Cl^-	SO_4^{2-}	HCO_3^-	TDS
			°C	μs/cm	mmol/L							mg/L
S1	Dec	6.467	21.2	2930	0.099	5.726	6.050	2.629	20.172	0.160	4.000	1476
	Jan	6.514	21.2	3020	0.102	6.743	5.950	2.638	22.014	0.055	4.197	1560
	Mar	6.474	21.0	2360	0.104	4.417	5.350	2.048	13.217	0.054	3.803	1115
	Apr	6.537	21.9	2020	0.091	3.775	4.100	1.775	13.884	0.036	3.607	1046
	May	6.956	21.3	1016	0.098	1.706	2.600	0.955	5.762	0.034	2.803	562
	June	6.690	21.0	1030	0.088	1.583	2.525	0.846	5.259	0.052	3.197	556
	July	6.709	21.4	978	0.088	1.713	2.592	0.870	5.051	0.054	2.391	506
	Sep	7.063	21.4	1167	0.089	2.348	2.896	0.969	6.612	0.054	2.328	578
	Oct	6.838	21.4	1436	0.092	4.187	3.123	0.988	8.642	0.057	2.817	733
S2	Dec	7.588	12.4	319	0.015	0.066	1.750	0.224	0.139	0.070	3.246	288
	Jan	7.527	16.0	323	0.016	0.182	1.725	0.238	0.425	0.075	2.885	278
	Mar	7.102	21.3	330	0.008	0.053	1.575	0.222	0.126	0.055	3.328	283
	Apr	7.188	23.4	322	0.007	0.059	1.500	0.221	0.142	0.038	3.213	272
	May	7.496	24.6	341	0.054	0.083	1.550	0.307	0.158	0.040	3.410	291
	June	7.160	24.5	368	0.010	0.053	1.600	0.270	0.092	0.070	3.328	285
	July	7.171	24.5	339	0.011	0.051	1.635	0.262	0.092	0.070	3.303	285
	Sep	7.063	24.5	335	0.009	0.054	1.619	0.232	0.097	0.068	3.303	283
	Oct	7.528	21.3	340	0.014	0.069	1.644	0.238	0.152	0.071	3.493	298
S3	Dec	7.611	12.2	3360	0.151	11.252	6.375	2.354	24.451	0.242	2.705	1667
	Jan	7.582	16.5	3310	0.148	12.878	5.725	2.117	26.944	0.186	2.705	1744

(continued)

Table 1.2 (continued)

	Time	pH	T	EC	K^+	Na^+	Ca^{2+}	Mg^{2+}	Cl^-	$SO_4{}^{2-}$	$HCO_3{}^-$	TDS
			°C	μs/cm	mmol/L							mg/L
	Mar	7.162	19.6	2740	0.141	9.073	5.025	1.539	18.522	0.128	2.607	1291
	Apr	7.370	25.5	1657	0.203	6.127	3.025	0.948	11.535	0.403	2.803	915
	May	7.579	24.2	368	0.016	0.231	1.500	0.302	0.309	0.063	3.508	304
	June	7.430	26.0	348	0.033	0.361	1.475	0.306	0.403	0.099	3.803	332
	July	7.213	23.5	383	0.039	0.383	1.583	0.325	0.352	0.114	3.429	318
	Sep	7.596	24.2	415	0.040	0.548	1.640	0.334	0.859	0.097	3.146	320
	Oct	7.321	23.0	1490	0.099	6.307	3.104	0.919	11.059	0.131	2.310	842
ZK1	Jan	6.897	21.0	2670	0.075	3.267	8.450	2.667	20.014	0.166	3.262	1411
	Mar	7.091	17.8	1681	0.058	3.581	4.850	0.977	12.316	0.181	1.902	880
	Apr	7.473	23.9	670	0.037	1.682	1.825	0.408	4.367	0.163	1.902	410
ZK2	Jan	6.247	18.8	10,180	0.295	25.670	20.325	9.400	93.358	0.048	4.607	5448
	Mar	6.620	19.4	6250	0.179	16.696	13.600	3.658	61.127	0.108	4.508	3529
	Apr	6.651	20.1	6660	0.194	18.712	15.125	4.866	56.400	0.135	4.902	3535
	May	6.441	22.0	8500	0.226	19.422	14.250	7.046	57.014	0.122	5.607	3637
	June	6.400	23.4	9250	0.244	26.769	18.900	9.242	78.986	0.050	6.607	4906
	July	6.536	23.1	9260	0.249	27.765	19.830	9.642	80.975	0.359	7.236	6013
	Sep	6.849	23.5	9040	0.245	28.956	18.895	8.885	81.515	0.086	5.789	4900
	Oct	6.530	22.5	9320	0.314	31.852	20.193	8.617	83.558	0.093	6.000	5101
ZK3	Jan	7.181	18.8	1461	0.123	2.361	4.050	1.808	9.363	0.098	3.803	845

(continued)

Table 1.2 (continued)

	Time	pH	T	EC	K^+	Na^+	Ca^{2+}	Mg^{2+}	Cl^-	SO_4^{2-}	HCO_3^-	TDS
			°C	μs/cm	mmol/L							mg/L
	Mar	7.127	19.6	1108	0.062	1.126	3.700	1.154	9.549	0.107	4.098	821
	Apr	7.092	20.1	1053	0.067	1.050	3.175	1.173	6.106	0.111	4.000	656
	May	7.471	21.4	948	0.177	1.027	2.850	1.183	4.861	0.165	3.803	598
	June	7.370	23.7	950	0.068	0.731	2.950	1.277	4.429	0.106	3.803	571
	July	7.406	23.4	920	0.069	0.699	2.828	1.243	4.358	0.107	3.744	559
	Sep	7.862	24.4	854	0.065	0.673	2.862	1.068	3.770	0.100	3.760	531
	Oct	7.538	22.0	794	0.066	0.535	2.970	0.785	3.326	0.093	4.056	527
XH-1	Mar	6.937	19.3	3030	–	–	–	–	–	–	2.393	160
	Apr	7.014	23.1	1246	0.060	3.585	2.778	0.795	8.515	0.114	2.705	697
	May	7.313	26.2	812	0.099	2.311	2.100	0.507	5.129	0.087	2.393	504
	June	7.190	29.3	553	0.065	1.167	1.800	0.385	2.271	0.096	2.607	360
	July	7.187	25.0	920	0.061	2.087	2.059	0.516	4.275	0.094	2.328	454
	Sep	7.584	25.0	756	0.092	4.055	3.041	0.777	8.213	0.088	2.517	691
	Oct	7.298	20.6	1623	0.121	6.322	3.226	0.948	12.003	0.093	2.282	876
F	Apr	6.922	22.5	781	0.097	2.019	2.350	0.458	5.085	0.161	2.607	510
	May	4.454	23.2	2620	0.048	10.899	5.350	0.478	23.106	0.040	0.000	1302
	June	4.600	27.4	3320	0.058	13.982	6.875	0.583	28.811	0.052	0.000	1646
	July	4.747	25.2	3330	0.016	4.374	2.394	0.210	9.338	0.038	0.252	553
	Oct	7.107	22.7	895	0.143	2.347	2.478	0.473	5.501	0.182	2.648	545

Table 1.3 Correlation analysis of ions in water

	K^+	Na^+	Ca^{2+}	Mg^{2+}	Cl^-	HCO_3^-
K^+	1					
Na^+	0.805**	1			*	
Ca^{2+}	0.803**	0.941**	1			
Mg^{2+}	0.832**	0.877**	0.952**	1		
Cl^-	0.819**	0.968**	0.988**	0.937**	1	
SO_4^{2-}	0.311	0.034	−0.028	−0.039	−0.023	
HCO_3^-	0.503**	0.340*	0.525**	0.670**	0.467**	1

* Significantly correlated at 0.05 two-sided
** Significantly correlated at 0.01 two-sided

1.3.2 The Source of Major Ions

The sources of major ions are influenced by geographical environment, climatic conditions, rock types and human activities, etc. (Karmegam et al. 2011). The cation and anion millimolar concentration of S2 were $Ca^{2+} > Mg^{2+} > K^+, Na^+$, and $HCO_3^- > Cl^- > SO_4^{2-}$, respectively. The cation and anion millimolar concentration of drillings, S1 and S3 were $Ca^{2+} > Na^+ > Mg^{2+} > K^+$, and $Cl^- > HCO_3^- > SO_4^{2-}$, respectively. The cation and anion millimolar concentration of surface water were $Na^+ > Ca^{2+} > Mg^{2+} > K^+$, and $Cl^- > HCO^{3-} > SO_4^{2-}$, respectively.

The TDS of S2 was not influenced by season, which was 272–298 mg/L. However, the TDS in other samplings was significant different in time and space, which was 304–5448 mg/L. The Gibbs was used to analysis the ions source (Ju et al. 2010; Wang et al. 2014; Liu et al. 2018). According to the Gibbs, the major ion source of S2 was water- rock interaction. The major ion source of drillings, water surfaces, S1 and S3 were discharge of industrial wastewater. The correlation between K^+, Na^+, Ca^{2+}, Mg^{2+}, Cl^- and HCO_3^- in water is significant (Table 1.3). Because ions from the same source are correlated, thus the sources of K^+, Na^+, Ca^{2+}, Mg^{2+}, Cl^- and HCO_3^- were the same. Ye et al. (2019) and Wang et al. (2013) had the same results. They found that the K^+, Na^+, Ca^{2+}, Mg^{2+}, Cl^- and HCO_3^- in water come from the discharge of industrial water (Fig. 1.1).

1.3.3 Seasonal Variation Characteristics of Major Ions

From December to July, and then to October, the rainfall in the study area first decreased and then increased, and the conventional ion concentrations of S1, S2, S3, ZK1, ZK3 and XH-1 also decreased first and then increased, indicating that each Point regular ions are affected by dilution. Several studies have shown the effect of dilution, however, with the increase of rainfall, the concentration of conventional

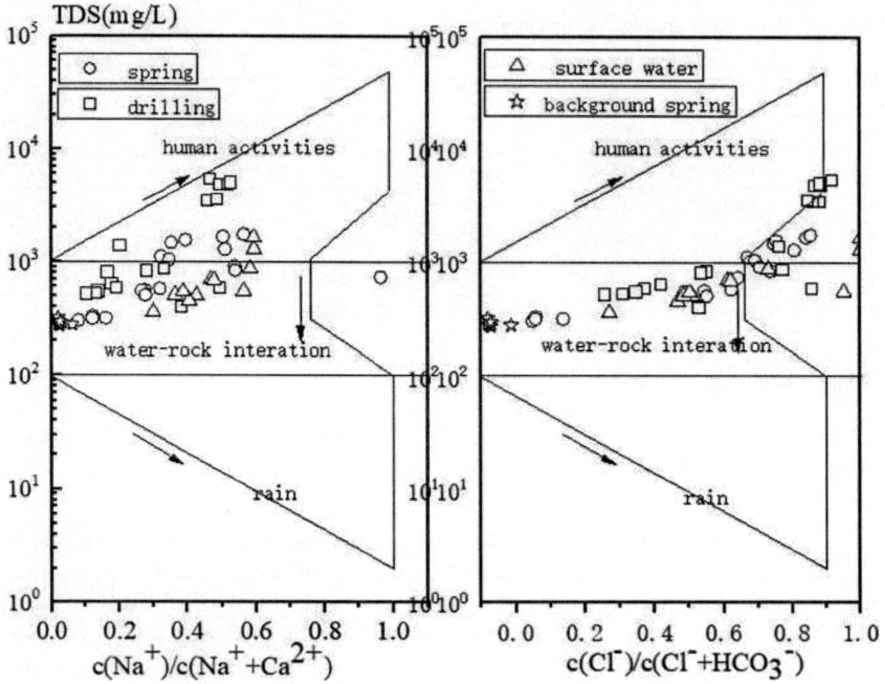

Fig. 1.1 Gibbs figure which reflects the characters of ions in water

ions at ZK2 and F increases, indicating that they are less diluted by rain, and heavy metal elements are also this feature appeared. ZK2 is close to the industrial pollution source, and F is the sewage discharge outlet, so the water quality of ZK2 and F points is directly affected by industrial wastewater. The increase in rainfall may exacerbate the seepage of industrial wastewater, causing the ion concentrations at ZK2 and F to increase instead during the wet season. However, the other sampling sites were less directly affected by industrial wastewater, so in the presence of increased rainfall, the dilution effect was mainly manifested. In the downstream of F, XH-1 is affected by the multiple effects of leakage at point F, water supply and rainfall of the tributaries of the East Main Canal, and the ion concentration also shows a decreasing trend in the wet season, that is, dilution is the main factor.

1.4 Conclusions

In this paper, to know the water quality of study area, three springs, three drillings and two surface waters were chosen. The main conclusions can be summarized a follows:

(1) The water–rock interaction in the background spring (S2) is the main source of ions; the chemical types of water at other points are complex, and human activities are the main source of ions. The K^+, Na^+, Ca^{2+}, Mg^{2+}, Cl^- and HCO_3^- in water come from industrial wastewater discharge.
(2) The major iron concentrations of S1, S2, S3, ZK1, ZK3 and XH-1 were affected by the dilution effect of rainfall, and the relationship between ion concentration and rainfall was positive.

References

Jiang PP, Zou YE, Zhang Q et al (2018) Natural degradation of ammonia nitrogen in groundwater in Songmu village of Hezhou, Guangxi. Environ Eng 36(07):62–66, 26
Jiang PP, Yu G, Zhang Q et al (2020) Chemical weathering and CO_2 consumption rates of rocks in the Bishuiyan subterranean basin of Guangxi, China. Sci Rep 10:11677
Ju JT, Wang Y, Xie MP et al (2010) Composition, spatial distribution, and environmental significance of water irons in Pumayun Co catchment, southern Tibet. J Geog Sci 20(01):109–120
Kačaroğlu F (1999) Review of groundwater pollution and protection in Karst Areas. Water Air Soil Pollut 113:337–356
Karmegam U, Chidambaram S, Prasanna MV et al (2011) A study on the mixing proportion in groundwater samples by using Piper diagram and Phreeqc model. Chin J Geochem 30(4):490–495
Liu JJ, Zhao YS, Huang X et al (2018) Spatiotemporal variations of hydrochemistry and its controlling factors in the Yarlung Tsangpo River. China Environ Sci 38(11):4289–4297
Pu JB, Yuan DX, Jiang YJ et al (2010) Hydrogeochemistry and environmental meaning of Chongqing subterranean karst streams in China. Adv Water Sci China 21(05):628–636
Wang JZ, Wu JL, Zeng HA et al (2013) Characteristics of water isotope and hydrochemistry in Hetao plain of inner Mongolia. J Earth Sci Environ 35(4):104–112
Wang XY, Li ZQ, Zhou P et al (2014) River water chemical characteristics and controls during the spring flood period in Yushugou Bbasin, Hami, Eastern Tianshan Mountains, China. Arid Land Geog 37(5):922–930
Ye HJ, Zhang RX, Wu P et al (2019) Characteristics and driving factor of hydrochemical evolution in karst water in the critical zone of Liupanshui mining area. Earth Sci 44:2887–2898
Yuan DX (2014) Research on major environmental geological problems and countermeasures in Karst Shishan Area in Southwest China

Chapter 2
Analysis of Environmental Factors of TC Intensification: A Case Study of Three Continuous Super Typhoons in November 2006

Kefeng Shi

Abstract Tropical cyclones (TCs) are one of the natural disasters that draw significant damage to human society. This article aims to discover the potential environmental factors that lead to one of the most catastrophic tropical cyclone disasters ever recorded: the three continuous supertyphoon landfall in the Philippines in November 2006. The event is so unique because each year around 30 tropical cyclones are forming in the Northwestern Pacific Ocean Basin on average, but among which there were few precedent cases of 3 continuous Supertyphoons made their landfall one by one at the same location around a month, and environmental factors that drive such phenomenon require the combined interaction of many different environmental factors. Hence to discover these potential driving factors, the analysis will be conducted through discussing the observation in diversified aspects incorporating satellite data and previous researcher's idea. It is found out that in the oceanic aspect, both the El Niño Southern Oscillation (ENSO) phenomenon and the recovery effect from the salinity barrier can help sustain a relatively high sea surface temperature that energizes the development of the continuously occurred 3 STYs. On the other hand, the El Niño in 3CSTY's case, can increase both the intensity and the sustaining time of the TCs and shift the position of TCs southeastward. Considering the atmospheric conditions, it is revealed that the meso-scaled weather systems can affect TC's development and generation location, and the CISK effect explains the role of low vertical wind shear plays in thermodynamic perspective.

Keywords Typhoon · ENSO · Wind shear · Sea surface temperature

---

K. Shi (✉)
Department of Atmospheric and Oceanic Sciences, McGill University, Montreal, QC H3A 0B9, Canada
e-mail: kefeng.shi@mail.mcgill.ca

J. Zhang et al. (eds.), *Environmental Pollution Governance and Ecological Remediation Technology*, Environmental Science and Engineering,
https://doi.org/10.1007/978-3-031-25284-6_2

2.1 Introduction

A tropical cyclone (TC), as aforementioned, is a complicated weather system that will draw tremendous loss to human society in case of landfalls. Speaking of scientists' discovery on TCs, the admitted truth is that great progress has been made to find out the fundamental mechanisms of TC generation and development. Nonetheless, like many other atmospheric phenomenons, there are too much uncertainty and elements to be taken into consideration before the true colors of TCs are revealed because the Earth's atmospheric system, like all other fluid systems, is so chaotic and intricate to simulate or evaluate. In this article, the abbreviate "3CSTY" stands for "three continuous super typhoons". In the 3CSTY's case in this research, the environmental factors that lead to the 3 super typhoons will be investigated. Constrained by the limited theoretical base and the complexity of the meteorological model (TC) itself, the analysis of this unique case is not able to go far beyond fundamental properties. As a result, this article synthesizes the result from legible satellite data and the accumulated experience in the related field by scientists. The fundamental mechanism of TC formation will be discussed firstly. The environmental factors, in the later parts, will be categorized into two parts: oceanic factors, and atmospheric factors, because TC's basic property is a product of atmosphere-sea interaction. It is hoped for this research to be a helpful resource for TC study as it is derived from a case study including multi-time TC generation and landfall in a short period, therefore the research on similar meteorological cases in the future can have a precedent comparison.

2.2 The Fundamental Environmental Factors of TC Formation

2.2.1 The Basic Information Related to Tropical Cyclones and the TCs in the 3CSTY Case

Tropical cyclone is a kind of powerful (meso-scaled) weather system. The structure of TC contains a vertically developed cloud system presented in a vortex shape. A classic TC is composed of a cyclone eye, eyewall, and spiral rainbands. A mature TC usually has a central lowest pressure around 960 hPa and a highest windspeed above 60 kt. TC, as a kind of weather system with a diameter around 100–2000 km, can release enormous energy and can do significant harm to the people in or around its passing route through intense wind, heavy rain, and storm surge.

In the 3CSTY case, 3 super typhoons (STY) made their landfall in the Philippines one by one around November of 2006, and they caused unprecedented damage to the local economy and residents. The three STYs all reached the highest level of typhoon intensity category according to the National Meteorological Center of CMA:

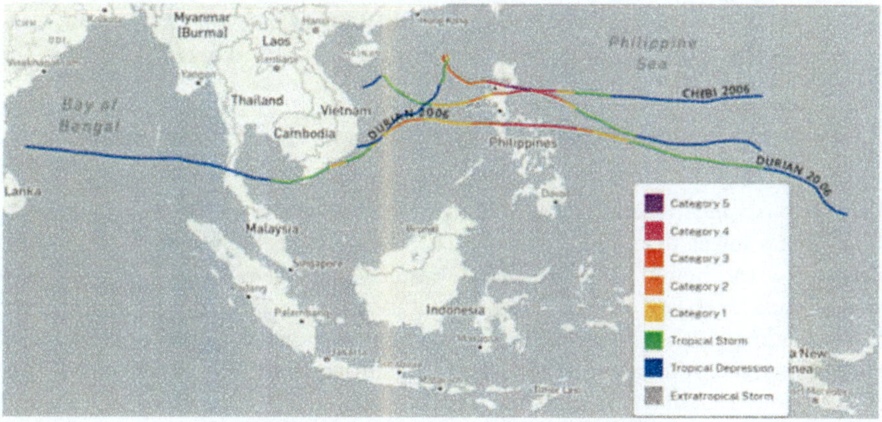

Fig. 2.1 The tracks of the 3CSTY, with the level of intensity of the TCs

the supertyphoon (STY) level, with typhoon CIMARON (0619) reached the wind speed of 140 kt, typhoon CHEBI (0620) reached 125 kt and typhoon DURIAN (0621) reached 135 kt. The estimated economical loss in the landfall areas of 3CSTY reached 560 million dollars. The continuous production of STYs in the same ocean basin is an extremely infrequent phenomenon, the reason of which has an indispensable relationship to several environmental factors, which will be discussed in the latter part of the article (Fig. 2.1).

2.2.2 *Basic Influence Factors in the Developing Stage of TCs*

First, it is necessary to get into the production mechanisms of an ordinary typhoon, because STYs are developed from embryos that originated in the tropical depression disturbance in the intertropical convergence zone (ITCZ). After the disturbance forms, strict environmental factors are required so that the disturbance can create and sustain a rotating conversion form: Warm SST (Sea Surface Temperature) which is no lower than 26.5 °C, 5 unit of latitude difference from the equator, and for the environmental wind situation, small wind shear in the vertical direction is required. All these factors provide TC embryos with a thermodynamically favorable condition to develop. For an embryo which has already evolved into a tropical storm (TS, with wind speed at the lower altitude of its central area reached 17.2–24.4 m/s), it still needs the support of many other kinds of environmental factors so that it can burst to reach the intensity of a super typhoon (STY, with wind speed at the lower altitude of its central area reached 51 m/s).

2.3 Relationship Between TC Development and the Oceanic Factors

2.3.1 *Sea Surface temperature's Influence on STY Formation*

It is previously mentioned that the forming of TC requires a sea surface temperature (SST) of no lower than 26.5 °C since the SST must be high enough to provide plenty of warm, humid water vaper. The water vapor, then, ascends to the lifting condensation level (LCL) to release latent heat and cause precipitation. Consequently, the circulation of this process (will be discussed in later parts), with the contribution of Coriolis force, form a conversing form that subsequently leads to the embryo of TC. The role SST plays, in this case, is the supporter of water vapor and energy. As a result, the positive relationship between the SST of the ocean basin and the intensity of the TC produced by this basin is observed by Gabriel et. al. (2007). What is special in the case of 3CSTY is the continuity of the formation of 3 STYs, because after one TC passes through a certain area of the ocean, it increases the ocean's mixed layer depth due to the surging waves and turbulence produced by the wind wave. Such phenomenon will decrease the SST around the area where TC had passed through. The research led by Jiayi Pan and Yujuan Sun reveals that the Satellite SST images show that on 3 November 2006, the maximum temperature cooling reached 4.48 C around the location 19.508 N, 116.268 E (Pan and Sun 2013). The reason for this, according to Jiayi et. al, is that the sub-critical condition of the upper-ocean dynamics when the typhoon translation decelerated. Given an ocean basin with unfavorable SST temperature, the TC generation after the passing of TC CIMARON requires a recovery of the oceanic temperature condition to develop. It is shown that stratification due to salinity can induce a barrier layer to build up higher SST (Wang et al. 2011). Besides, the barrier layer can affect currents by trapping momentum input from the wind in a shallow surface layer, thereby producing strong flows in the mixed layer (Vialard and Delecluse 1998). With the suppression on the mixed layer of the ocean, the dynamics of the upper ocean then have a stronger ability to recover to higher SST that can energize later TCs (CHEBI and DURIAN) in the subtropical area of the Northwest Pacific Ocean.

2.3.2 *TC's Formation Frequency and Intensity Related to ENSO Phenomenon*

When the relationship between TCs and SST is discussed, the El Niño–Southern Oscillation (ENSO) phenomenon is an indispensable aspect to be taken into consideration. The ENSO phenomenon can be categorized into two aspects: the El Niño and the La Niña. Fall 2006 was the developing period of a weak El Niño event, in which the southeast trades in the eastern Pacific Ocean are much weaker than usual,

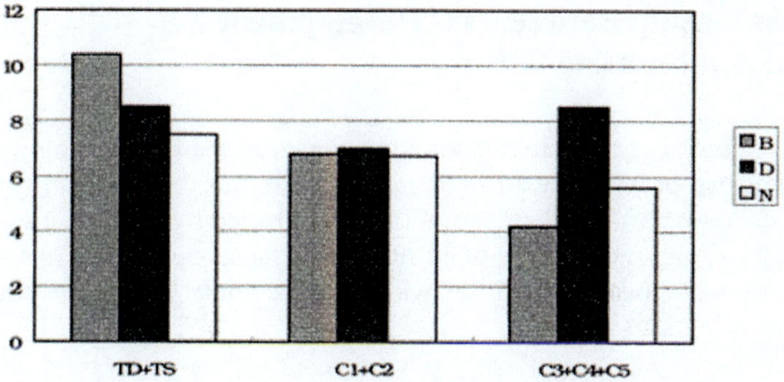

Fig. 2.2 The average number of TCs (1979–99) based on the Saffir–Simpson scale generation (Chia 2002)

and thus exerts less offshore drifting from the western coast of South America. This makes the warm water in the western Pacific Ocean to flow back to the east and increases both the sea level and SST in the eastern Pacific Ocean. When it comes to how ENSO affects TCs, the intensity of the TCs, the location of generation, and the sustaining time are the three major aspects to focus on. For the intensity of TC, the derivation from the figure discovered by HSIN HSING CHIA below shows that more intense typhoon occurs in the El Niño events (Chia 2002). (B refers to the TCs generated in La Niña years, D refers to which in El Niño years, and N refers to normal years) The peak intensity level of CIMARON (level C5), CHEBI (level C4), and DURIAN (level C4) in this case, then, can significantly prove the result shown in Fig. 2.2.

Regarding how ENSO affects the generation location of TCs, meteorologists have observed a tendency that a southeastward displacement of the mean tropical storm and typhoon genesis region usually occur in El Niño years. This is a piece of confounding evidence to illustrate why the tracks of STYs are both converged to the Philippine area in the 3CSTY case: 3 driving factors lead to this result, firstly, the ENSO effect made the southeastern part a more favorable environment for TC genesis. Secondly, the subtropical high extended its body further to the west due to the stable westerly system at that time provides a barrier for the TCs in the north. Finally, the cold air mass suppression from Eastern Asia converges typhoon's track from NWP to the South China Sea.

Consequently, the southeastern displacement that is previously mentioned induced by the El Niño events elongated the sustaining time of TC, because when a TC is generated in lower latitude (no lower than 5 unit of latitude difference from the equator), it will take more time on traveling in tropical-subtropical ocean basin before heading to the northeastern or northwestern direction and thus perish in strong westerlies or cold seawater. Considering the presence of such a tendency in the 3CSTY case, both CIMARON and DURIAN had an impressively long lifetime of 14 and 16 days.

2.4 Relationship Between TC Development and Atmospheric Factors

Air mass is a broad concept in meteorology, however, there are mainly two kinds of air masses that occur in the air mass movement in this article: the α type meso-scaled weather systems, the diameter of the area covered by which can range from 2000 to 200 km, and another type of air mass movement is presented in a relatively micro-scaled and vertical way: it is the wind shear activities in the inner structure of TC.

2.4.1 TC Development and the Influence of Meso-Scaled Weather System

When it comes to the α type meso-scaled weather systems, the feature may vary with location and time. For the 3CSTY's case, the northern hemisphere has already entered boreal fall. Therefore, at this time, the two major factors that took place in eastern Asia and northwestern Pacific Ocean are the cold front advanced from the northeastern part of Eurasia and the southeastern airflow originated from the Indian Ocean. The main effect of the cold front is not only to suppress and shift the TC path westward, but also to form a large temperature gradient between the front and the TC path. The research from Yuei-An Liou et. al exquisitely revealed the exact mean angle of the southern movement of TC track in the presence of cold front in the north: the inclination angle θ1 of the winter super-typhoons had smaller values and a narrower range (10°–26°) than those (13°–40°) of the summer super-typhoons due to the effects of the cold front, where θ1 is the angle of deflection of TC into higher latitudes with respect to the equator (Liou 2018). In both CHEBI's and DURIAN's case, the northerly and the easterly wind create a suppression from the northeast and consequently decrease θ1 (Figs. 2.3 and 2.4).

Besides, another worth-mentioning phenomenon caused by the α type meso-scaled air mass movement is that a large number of convective cloud systems will form when the cold front encounters the warm, humid air mass on the northern side of the TC. Because when the cold, dry air encounters the secondary rainband and the distant rainband of the TC, the large temperature gradient will trigger the development of convection and finally it will lead to the burst of the outer mesoscale convective system (OMCS). In TC CIMARON's case, this phenomenon can be observed: a row of the mesoscale convective system is formed around the colliding surface between the cold air mass and the northern side of the TC.

On the other hand, the southwesterly airflow can play the role of a water vapor extractor and carrier. Plenty of warm water vapor is transferred by such airflow from the Indian Ocean and the low-level jet streams then carries these water vapor to supply the horizontal conversion of the TC. Such effect on typhoons can span through the entire TC season from summer to winter, but the strength of it varies due to the

Fig. 2.3 Easterly wind at the edge of the subtropical high and the cold air mass in the north limit TC DURIAN's track

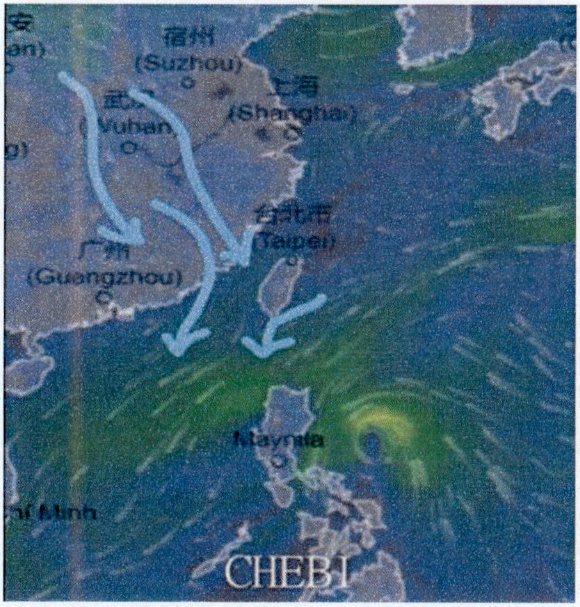

Fig. 2.4 Cold air mass movement from China generated a northeasterly which suppress TC CHEBI to the southwest

intensity of Indian monsoon difference in summer and winter. The CISK effect in the latter part of the article indicates the warm, humid air as the energy resource for the structural circulation of TCs. In TC CIMARON's case, the role of southwesterly can be observed more clearly since the generation time of TC CIMARON is relatively earlier than its descendants, so it can receive the beneficial effect from southwesterly when the summer monsoon season in the Indian Ocean had not gone to an end (Fig. 2.5).

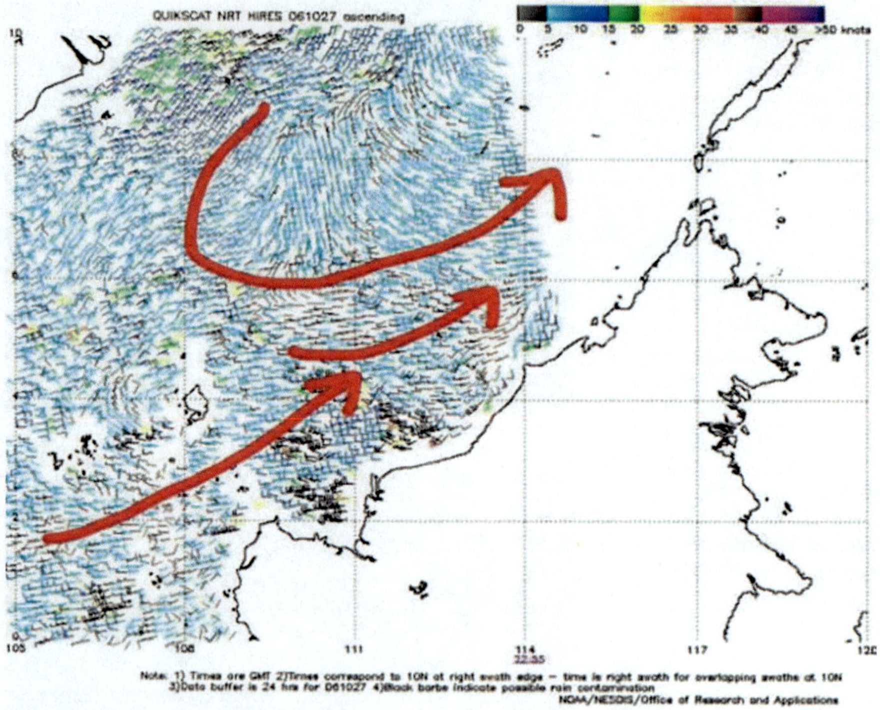

Fig. 2.5 Weak southwest airflow is observed in the area southwest to Philippine area where TC CIMARON is bursting (QuikSCAT 2005)

2.5 TC Development and Vertical Wind Shear

Cold front and southwestern airflow are factors that provide TCs with a further boost in their developing period from TS to STY. Nevertheless, one factor that is essential for TC throughout their entire lifetime is inside the structure of TC itself: the low environmental vertical wind shear. The role of wind shear was also mentioned in part two as the several most important elements to take place for TC generation. Vertical wind shear is usually detrimental to TC formation and development because when the vertical wind shear is too strong, the link between lower-level circulation and high-level convective cloud formation will be weakened by the shear effect in the vertical direction. Such link is necessary across the entire lifespan of TCs because the sustaining of TC needs its structure at different altitude to collaborate. The essence of the link is a mutual effect between the heat released by the cumulus cloud in high altitude and the humid air conversion induced by low-level disturbance. The processing mechanism of such effect can be explained by the property of the two sides of the link: the disturbance requires thermal energy to develop, and the formation of cumulus/convective clouds can fruit thermal energy in the form of latent heat released from the condensation of warm, humid water vapor; conversely, the disturbance continuously leads to the convergence of humid air which produces cumulus clouds. The mutual effect will ultimately drive to divergence in high altitude and the convergence in the low altitude (due to the pressure gradient). Such effect is known as the conditional instability of the second kind (CISK) and the continuous processing of such effect leads to TC development from embryos. The perspective of MARK DEMARIA shows a different analysis of the mutual effect in TC structure: instead of the more commonly known "ventilation" form of the CISK effect, MARK DEMARIA chose to use "tilt-like" to describe what happens when strong vertical wind shear occurs, and it is further conveyed in her research that such a tilt-like mechanism leads to mid-level warming which is unfavorable for convective activity and thus suppresses TC development (Demaria 1996).

2.6 Conclusion

In this article, the case of the 3CSTY (TC CIMARON, TC CHEBI, and TC DURIAN) is taken as a platform to illustrate the potential cause of the generation of such an unprecedented, continuous, and powerful weather system. The discussion is conducted by discovering oceanic factors and atmospheric factors which lead to the disaster. The result can be derived from this case: for the oceanic condition in the 3CSTY case, on one hand, the sustained high SST due to the recovery for the SST by the salinity barrier after a TC pass can energize the development of the continuously occurred 3 STYs. On the other hand, the ENSO, or El Niño in 3CSTY's case, can shift the position of TC to southeast and also increase both the intensity and the sustaining time of the TC. When it comes to the atmospheric conditions, the cold front in the north and the southwestern airflow from the Indian Ocean are found to be the two major atmospheric factors that can influence TCs' development by determining their position of genesis and further tracks. In another aspect, the points related to the CISK effect indicate environmental conditions with low vertical wind shear as a more favorable environmental condition for TC generation and development.

Meanwhile, the conclusion derived by this study and the case itself is merely the tip of the iceberg, and it still requires further discovery through approaches like numerical weather analysis with computational models before it can be practically used In this article the information relating to TC and equatorial waves is not completely presented, nevertheless, the influence of which on TC development is tremendous, and further correlative research can discuss such influence and apply the result in operating weather forecasting. Future study is needed for a more comprehensive picture.

Acknowledgements In this research, I appreciate the academic support from the atmospheric and oceanic department of McGill University. Besides, the data from NOAA historical hurricane track website and WUNDERMAP (wunderground.com) which support the research by providing necessary TC information like wind speed, lifespan, and their tracks. Without which it is impossible to conduct a study on the environmental situation based on the atmospheric data at the time of 3CSTY presence. Besides, the research receives a contribution from QuikSCAT/SeaWinds Scatterometer as an indispensable satellite observed data resource in this study for analyzing the influence of meso-scaled air mass movement. This research also receives academic support from the atmospheric and oceanic department of McGill University.

References

Chia HH (2002) The interannual variability in the genesis location of tropical cyclones in the Northwest Pacific

Demaria M (1996) the effect of vertical shear on tropical cyclone intensity change, Hurricane Research Division, NOAA/AOML

Gabriel AV, Brian JS (2007) Effect of remote sea surface temperature change on tropical cyclone potential intensity

Liou YA (2018) Season-dependent distributions and profiles of seven super-typhoons (2014) in the Northwestern Pacific ocean from satellite cloud images senior member. IEEE, Ji-Chyun Liu, Chung Ping Liu, and Chung-Chih Liu

Pan J, Sun Y (2013) Estimate of ocean mixed layer deepening after a typhoon passage over the South China Sea by using satellite data

QuikSCAT/SeaWinds Scatterometer, National Environmental Satellite, Data, and Information Service, STAR Center for Satellite Applications and Research, 2005

Vialard J, Delecluse P (1998a) An OGCM study for the TOGA decade. Part I. Role of salinity in the physics of the western Pacific fresh pool. J Phys Oceanogr 28:1071–1088. Vialard J, Delecluse P (1998b) An OGCM study for the TOGA decade. Part II. Barrier-layer formation and variability. J Phys Oceanogr 28:1089–1105

Wang X, Hanc G, Qi Y, Li W (2011) Impact of barrier layer on typhoon-induced sea surface cooling. Dyn Atmos Oceans

Chapter 3
Study on the Dissolution Law of the Main Precipitates Produced in the Treatment of Cyanide Containing Wastewater

Yanbo Chen, Guangsheng Li, Xingfu Zhu, and Qing Xu

Abstract During the treatment of cyanide containing wastewater, a variety of precipitates will be generated, including $Fe_2Fe(CN)_6$, $Fe_4[Fe(CN)_6]_3$, $Cu_2Fe(CN)_6$, $Zn_2Fe(CN)_6$, CuCN, $Zn(CN)_2$. The solid–liquid separation of these precipitates is realized through pressure filtration, so as to realize the removal of cyanide in wastewater. Therefore, the stable existence of these precipitates is a necessary condition for the effective treatment of cyanide containing wastewater. Through theoretical calculation and experimental research, this paper analyzes whether these precipitates are dissolved and the degree of dissolution under acidic (ph = 2) and alkaline (ph = 11) conditions. Thus, the precipitation dissolution under acid and alkaline conditions can be judged according to the precipitation type generated by cyanide containing wastewater, and then the efficient and safe implementation of cyanide containing wastewater treatment, pressure filtration, filter cake washing and other operations can be realized by controlling the acid and alkaline of wastewater.

Keywords Cyanide containing wastewater · Precipitation · Acidic · Alkaline · Dissolution characteristics

3.1 Introduction

During cyanide leaching of gold bearing ores, copper, zinc, iron and other metal minerals associated with gold ores directly or indirectly react with CN^-, O_2, OH^- in the leaching solution to produce a variety of metal cyanide complex ions, mainly including copper cyanide complex ion $Cu(CN)_3{}^{2-}$ (the main form of copper cyanide complex ion), zinc cyanide complex ion $Zn(CN)_4{}^{2-}$, iron cyanide complex ion

Y. Chen (✉) · G. Li · X. Zhu
Selection and Metallurgy Laboratory of Shandong Gold Mining Technology Co., Ltd., Yantai 261441, Shandong, China
e-mail: 595907901@163.com

Q. Xu
Shandong Academy of Scientific and Technological Information, Jinan 250101, Shandong, China

J. Zhang et al. (eds.), *Environmental Pollution Governance and Ecological Remediation Technology*, Environmental Science and Engineering,
https://doi.org/10.1007/978-3-031-25284-6_3

$Fe(CN)_6^{4-}$, etc. These metal cyanide complex ions are the main component of cyanide in cyanide containing wastewater, and also the main object of cyanide containing wastewater treatment (Beneficiation test research method 2017; Jun et al. 2016).

The treatment methods of cyanide containing wastewater can be divided into destruction method and comprehensive recovery method. Destruction methods mainly include oxidation method, electrolysis method, high temperature decomposition method, microbial decomposition method and natural purification method; Comprehensive recovery methods mainly include acidification method, ion exchange method, chemical precipitation method, etc. Among them, oxidation method, acidification method and chemical precipitation method are widely used, and precipitation will be generated, including $Fe_2Fe(CN)_6$, $Cu_2Fe(CN)_6$, $Zn_2Fe(CN)_6$, CuCN, $Zn(CN)_2$. The stable existence of precipitation is the prerequisite for cyanogen fixation and removal. In addition, in the process of experimental research and actual operation, considering safety and environmental protection, precipitation pressure filtration, filter cake washing and other operations, it is necessary to adjust the feeding and washing water of the filter press to alkaline (Theory and method of cyanide wastewater treatment for gold extraction 2015). Therefore, the dissolution characteristics of these precipitates under acidic or alkaline conditions have important practical significance.

3.2 Theoretical Calculation

Theoretically calculate whether $Fe_2Fe(CN)_6$, $Fe_4[Fe(CN)_6]_3$, $Cu_2Fe(CN)_6$, $Zn_2Fe(CN)_6$, CuCN, $Zn(CN)_2$will dissolve and the degree of dissolution under the conditions of strong acidity (pH = 2) and strong alkalinity (pH = 11), so as to guide the subsequent experimental research.

3.2.1 *Dissolution Characteristics Under Acidic Conditions (pH = 2)*

Dissolution characteristics of $Fe_2Fe(CN)_6$

$$Fe_2Fe(CN)_6 \rightleftharpoons 2Fe^{2+} + Fe(CN)_6^{4-} \quad Fe(CN)_6^{4-} \rightleftharpoons Fe^{2+} + 6CN^-$$

$$CN^- + H^+ \rightleftharpoons HCN \uparrow$$

$$K^{\theta}_{s\,(Fe2Fe(CN)6)} = 10^{-35},\ C_{(Fe(CN)64-)} = 6.3 \times 10^{-13}\ mol/L$$

$$pH = 2,\ C_{(H^+)} = 10^{-2}\ mol/L$$

$$C_{(CN^-)} * C_{(H^+)} = 6.3 \times 10^{-15} < K^{\theta}_{a\,(HCN)} = 6.2 \times 10^{-10}$$

$$C_{(CN^-)} * C_{(H^+)}/K^{\theta}_{a\,(HCN)} = 10^{-5},$$

The ratio is small. So, theoretically, under acidic conditions (pH = 2~7), $Fe_2Fe(CN)_6$will not be dissolved.

Dissolution characteristics of Fe4[Fe(CN)6]3

$$Fe_4\left[Fe(CN)_6\right]_3 \rightleftharpoons 4Fe^{3+} + 3Fe(CN)_6^{4-} \quad Fe(CN)_6^{4-} \rightleftharpoons Fe^{2+} + 6CN^-$$
$$CN^- + H^+ \rightleftharpoons HCN \uparrow$$
$$K^{\theta}_{s\,(Fe4[Fe(CN)6]3)} = 10^{-42}, C_{(Fe(CN)_6^{4-})} = 8.67 \times 10^{-22}\ mol/L$$
$$pH = 2C_{(H^+)} = 10^{-2}\ mol/L$$
$$C_{(CN^-)} * C_{(H^+)} = 8.67 \times 10^{-24} < K^{\theta}_{a\,(HCN)} = 6.2 \times 10^{-10}$$
$$C_{(CN^-)} * C_{(H^+)}/K^{\theta}_{a\,(HCN)} = 1.4 \times 10^{-14},$$

The ratio is small. Therefore, theoretically, under acidic conditions (ph = 2~7), $Fe_4[Fe(CN)_6]_3$will not cause dissolution problems.

Dissolution characteristics of Cu2Fe(CN)6

Similar to the dissociation of $Fe_2Fe(CN)_6$, at pH = 2, $Cu_2Fe(CN)_6$ is dissociated into Cu^{2+} and $Fe(CN)_6{}^{4-}$, and $Fe(CN)_6{}^{4-}$will not be further dissociated at pH = 2. Therefore, in theory, $Cu_2Fe(CN)_6$ will not dissolve under acidic conditions (pH = 2~7).

Dissolution characteristics of Zn2Fe(CN)6

Similar to the dissolution of $Fe_2Fe(CN)_6$, at pH = 2, $Zn_2Fe(CN)_6$ dissociates into Zn^{2+} and $Fe(CN)_6{}^{4-}$, and $Fe(CN)_6{}^{4-}$ will not dissolve at pH = 2. Therefore, in theory, $Zn_2Fe(CN)_6$ will not dissociate under acidic conditions (pH = 2~7).

CuCN dissolution characteristics

$$CuCN \rightleftharpoons Cu^+ + CN^- \quad CN^- + H^+ \rightleftharpoons HCN \uparrow$$
$$K^{\theta}_{s\,(CuCN)} = 3.47 \times 10^{-20}, C_{(CN^-)} = 1.86 \times 10^{-10}\ mol/L$$
$$pH = 2C_{(H^+)} = 10^{-2}\ mol/L$$
$$C_{(CN^-)} * C_{(H^+)} = 1.86 \times 10^{-12} < K^{\theta}_{a\,(HCN)} = 6.2 \times 10^{-10}$$
$$C_{(CN^-)} * C_{(H^+)}/K^{\theta}_{a\,(HCN)} = 3 \times 10^{-1},$$

The ratio is close to 1. Therefore, theoretically, a small amount of CuCN precipitation will dissolve under acidic conditions (pH = 2~7).

Dissolution characteristics of Zn(CN)2

$$Zn(CN)_2 \rightleftharpoons Zn^{2+} + 2CN^- \quad CN^- + 2H^+ \rightleftharpoons HCN \uparrow$$

$$K_{s\,(Zn(CN)2)}^{\theta} = 10^{-12}, C_{(CN^-)} = 2 \times 10^{-4}\ \text{mol/L}$$
$$pH = 2, C_{(H^+)} = 10^{-2}\ \text{mol/L}$$
$$C_{(CN^-)} * C_{(H^+)} = 1.86 \times 10^{-6} > K_{a\,(HCN)}^{\theta} = 6.2 \times 10^{-10}$$
$$C_{(CN^-)} * C_{(H^+)}/K_{a\,(HCN)}^{\theta} = 3 \times 10^{3},$$

The ratio is large, so theoretically, under acidic conditions (ph = 2~7), $Zn(CN)_2$ will dissolve completely.

3.2.2 *Dissolution Characteristics Under Alkaline Conditions (Ph = 11)*

Dissolution characteristics of Fe2Fe(CN)6

$$Fe_2Fe(CN)_6 \rightleftharpoons 2Fe^{2+} + Fe(CN)_6^{4-} \quad 2Fe^{2+} + 4OH^- \rightleftharpoons 2Fe(OH)_2 \downarrow$$
$$2Fe(OH)_2 + 1/2O_2 + H_2O = 2Fe(OH)_3 \downarrow$$
$$K_{s\,(Fe2Fe(CN)6)}^{\theta} = 1.0 \times 10^{-35}, C_{(Fe^{2+})} = 1.26 \times 10^{-12}\ \text{mol/L}$$
$$pH = 11, C_{(OH^-)} = 10^{-3}\ \text{mol/L}$$
$$C_{(Fe^{2+})} * C^2_{(OH^-)} = 1.26 \times 10^{-18} < K_{s\,(Fe(OH)2)}^{\theta} = 4.87 \times 10^{-17}$$

There are $Fe_2Fe(CN)_6$, $Fe(OH)_2$ and $Fe(OH)_3$ precipitates in the reaction system, including $Fe_2Fe(CN)_6$ and $Fe(OH)_2$ precipitation conversion reaction, as well as oxidation–reduction reaction. $C_{(Fe^{2+})}*C^2_{(OH^-)}/K_s{}^{\theta}{}_{(Fe(OH)2)} = 2.59 \times 10^{-1}$, the ratio is small. Theoretically, ferrous blue will not have obvious dissolution problem under the condition of pH = 11. However, $Fe(OH)_2$ will be rapidly oxidized to $Fe(OH)_3$, and the solubility product of $Fe(OH)_3$ is KS θ $K_s{}^{\theta}{}_{(Fe(OH)3)} = 1.1 \times 10^{-39}$, promote the dissociation balance of ferrous blue to move towards the dissociation direction. Therefore, $Fe_2Fe(CN)_6$ will be dissolved under alkaline conditions (pH = 7~11), especially when the dissociated oxygen increases after increasing the stirring intensity, the dissolution rate is significantly accelerated (Selected current gold standards 2017).

Dissolution characteristics of Fe4[Fe(CN)6]3

$$Fe_4\left[Fe(CN)_6\right]_3 \rightleftharpoons 4Fe3^+ + 3Fe(CN)_6^{4-} \quad Fe^{3+} + 12OH^- \rightleftharpoons 4Fe(OH)_3$$
$$K_{s\,(Fe4[Fe(CN)6]3)}^{\theta} = 10^{-42}, C_{(Fe^{3+})} = 11.56 \times 10^{-22}\ \text{mol/L}$$
$$pH = 11,\ C_{(OH^-)} = 10^{-3}\ \text{mol/L}$$
$$C_{(Fe^{3+})} * C^3_{(OH^-)} = 11.56 \times 10^{-31} > K_{s\,(Fe(OH)3)}^{\theta} = 1.1 \times 10^{-39}$$

Under alkaline conditions, there are $Fe_4[Fe(CN)_6]_3$, $Fe(OH)_2$ and $Fe(OH)_3$ precipitates in $Fe_2Fe(CN)_6$ dissolution reaction system, including $Fe_2Fe(CN)_6$ and $Fe(OH)_2$ precipitation conversion reaction, as well as oxidation–reduction reaction. After the dissolution of $Fe_4[Fe(CN)_6]_3$, Fe^{3+} is produced, and Fe^{3+} reacts directly with OH^- to produce $Fe(OH)_3$ with small solubility product, which makes the dissociation equilibrium of Prussian blue move towards the dissociation direction. Therefore, under alkaline conditions (pH = 7~11), $Fe_4[Fe(CN)_6]_3$ will have obvious dissolution problems (Peng et al. 2018).

Dissolution characteristics of Cu2Fe(CN)6 and Zn2Fe(CN)6

The solubility product of $Cu_2Fe(CN)_6$ and $Zn_2Fe(CN)_6$ is not found, and the dissolution under the condition of ph = 11 needs to be tested.

Dissolution characteristics of CuCN

$$CuCN \rightleftharpoons Cu^+ + CN^- \quad Cu^+ + OH^- \rightleftharpoons CuOH \downarrow$$
$$K^{\theta}_{s\,(CuCN)} = 3.47 \times 10^{-20}, C_{(Cu^+)} = 1.86 \times 10^{-10} mol/L$$
$$pH = 11, C_{(OH^-)} = 10^{-3}\ mol/L$$
$$C_{(Cu^+)} * C_{(OH^-)} = 1.86 \times 10^{-13} > K^{\theta}_{s\,(CuOH)} = 1.0 \times 10^{-14}$$
$$C_{(Cu^+)} * C_{(OH^-)}/K^{\theta}_{s\,(CuOH)} = 18.6,$$

The two values are close, so when pH = 11, the dissolution of CuCN precipitation will not be too large.

Dissolution characteristics of Zn(CN)2

$$Zn(CN)_2 \rightleftharpoons Zn^{2+} + 2CN^- \quad Zn^{2+} + 2OH^- \rightleftharpoons Zn(OH)_2$$
$$K^{\theta}_{s\,(Zn(CN)2)} = 10^{-12}, C_{(Zn^{2+})} = 1 \times 10^{-4}\ mol/L$$
$$pH = 11, C_{(OH^-)} = 10^{-3}\ mol/L$$
$$C_{(Zn^{2+})} * C^2_{(OH^-)} = 1.0 \times 10^{-10} > K^{\theta}_{s\,(Zn(OH)2)} = 1.2 \times 10^{-17}$$

$Zn(CN)_2$ and $Zn(OH)_2$ coexist in precipitation dissolution equilibrium, because $C_{(Zn^{2+})} * C^2_{(OH^-)}/K_s^{\theta}{}_{(Zn(OH)2)} = 10^7$, the ratio is large, so when pH = 11, CuCN precipitation will be completely transformed into $Zn(OH)_2$ precipitation.

Table 3.1 Results of $Fe_2Fe(CN)_6$ precipitation dissolution test

Test no.	pH	Whether to join SCN^-	Detection result, mg/L	
			Fe	TCN
1#	2	No	370.80	29.12
2#		Yes	369.96	29.12
3#	11	No	136.20	422.25
4#		Yes	112.16	374.73
Stock solution			367.40	29.12

3.3 Experimental Study

3.3.1 $Fe_2Fe(CN)_6$ Dissolution Test

$Fe_2Fe(CN)_6$ precipitation suspension was prepared with $K_4Fe(CN)_6{\cdot}3H_2O$ and $FeSO_4{\cdot}7H_2O$. Take four groups of equal suspension and number them 1#, 2#, 3#, 4#. Adjust the pH of 1 #, 2 # to 2, add 22 g/l SCN^-, adjust the pH of 3 #, 4 # to 11, and add 22 g/l SCN^-, stir for 0.5 h and then filter (Table 3.1).

If $Fe_2Fe(CN)_6$ precipitation dissolves obviously, $Fe(CN)_6{}^{4-}$ will enter the liquid again, and the Fe concentration will increase significantly. According to the test results in the table, compared with the stock solution, Fe did not increase significantly at pH = 2. It shows that $Fe_2Fe(CN)_6$ precipitation is stable under the condition of pH = 2.

Under the condition of pH = 11, it was observed that $Fe_2Fe(CN)_6$ precipitation was significantly dissolved. Compared with the stock solution, 3#, 4# total cyanide increased significantly, indicating that $Fe_2Fe(CN)_6$ precipitation will be significantly dissolved under the condition of pH = 11.

3.3.2 Fe4[Fe(CN)6]3 Dissolution Test

$Fe_2Fe(CN)_6$ precipitation suspension was prepared with $K_4Fe(CN)_6{\cdot}3H_2O$ and $FeSO_4{\cdot}7H_2O$., and $Fe_4[Fe(CN)_6]_3$ suspension was prepared after strong stirring and oxidation for a period of time. Take four groups of equal suspension and number them 1#, 2#, 3#, 4#. Adjust the pH of 1 #, 2 # to 2, add 22 g/l SCN^-, adjust the pH of 3 #, 4 # to 11, and add 22 g/l SCN^-, stir for 0.5 h and then filter.

If the $Fe_4[Fe(CN)_6]_3$ precipitation dissolves significantly, $Fe(CN)_6{}^{4-}$ will enter the liquid again, and the Fe concentration will increase significantly. It can be seen from Table 3.2 that compared with the stock solution, Fe does not increase significantly at pH = 2. It shows that $Fe_4[Fe(CN)_6]_3$ precipitation is stable under the condition of pH = 2.

Table 3.2 Results of $Fe_4[Fe(CN)_6]_3$ precipitation dissolution test

Test no.	pH	Whether to join SCN^-	Detection result, mg/L	
			Fe	TCN
1#	2	No	62.36	0.1
2#		Yes	60.14	0.1
3#	11	No	583.60	1623.90
4#		Yes	578.24	1632.42
Stock solution			60.14	0.1

During the test, it was found that when the pH was adjusted to about 10.5, $Fe_4[Fe(CN)_6]_3$ began to dissolve, and under the condition of $pH = 11$, $Fe_4[Fe(CN)_6]_3$ precipitation significantly dissolved. Compared with the stock solution, 3#, 4# total cyanide increased significantly. It shows that under the condition of $pH = 11$, $Fe_4[Fe(CN)_6]_3$ precipitation will dissolve significantly, and the dissolution degree is greater than that of $Fe_2Fe(CN)_6$.

Under two pH conditions, when scn- is added, the concentration of Fe and CN^- in the treated liquid has no obvious change. It shows that SCN^- has no obvious solubilization effect on $Fe_4[Fe(CN)_6]_3$ precipitation under both acidic and alkaline conditions (Tao 1985).

3.3.3 *Cu2Fe(CN)6 Dissolution Test*

$Cu_2Fe(CN)_6$ precipitation suspension was prepared with $K_4Fe(CN)_6 \cdot 3H_2O$ and $CuSO_4 \cdot 5H_2O$. Take four groups of equal suspension and number them 1#, 2#, 3#, 4#. Adjust the pH of 1 #, 2 # to 2, add 22 g/l SCN^-, adjust the pH of 3 #, 4 # to 11, and add 22 g/l SCN^-, stir for 0.5 h and then filter.

If the precipitation of $Cu_2Fe(CN)_6$ dissolves obviously, the concentration of Fe and CN^- will increase significantly. It can be seen from Table 3.3 that, compared with the original solution, there is no significant change in the concentration of Fe and CN^- in the treated liquid under the condition of $pH = 2$. It shows that $Cu_2Fe(CN)_6$ precipitation is stable under acidic conditions. Under the condition of $pH = 11$, the concentration of Fe and CN^- in the treated liquid increased, indicating that $Cu_2Fe(CN)_6$ precipitation has the problem of dissolution under alkaline conditions.

Under two pH conditions, when SCN^- is added, the concentration of Fe and CN^- in the treated liquid has no obvious change, indicating that SCN^- has no obvious solubilization effect on $Cu_2Fe(CN)_6$ precipitation under both acidic and alkaline conditions (Tao 1990).

Table 3.3 Results of $Cu_2Fe(CN)_6$ precipitation dissolution test

Test no.	pH	Whether to join SCN^-	Detection result, mg/L		
			Cu	Fe	TCN
1#	2	No	103.20	13.30	11.35
2#		Yes	119.50	19.85	21.27
3#	11	No	8.25	64.23	234.041
4#		Yes	7.86	57.42	241.10
Stock solution			97.66	5.84	11.35

Table 3.4 Results of $Zn_2Fe(CN)_6$ precipitation dissolution test

Test no.	pH	Whether to join SCN^-	Detection result, mg/L	
			Zn	Fe
1#	2	no	18.44	7.88
2#		Yes	20.28	8.45
3#	11	no	2.97	8.84
4#		Yes	1.41	5.65
Stock solution			24.26	8.15

3.3.4 Zn2Fe(CN)6 Dissolution Test

$K_4Fe(CN)_6{\cdot}3H_2O$ and $ZnSO_4{\cdot}7H_2O$ were used to prepare $Zn_2Fe(CN)_6$ precipitation suspension. Take four groups of equal suspension and number them 1#, 2#, 3#, 4#. Adjust the pH of 1 #, 2 # to 2, add 22 g/l SCN^-, adjust the pH of 3 #, 4 # to 11, and add 22 g/l SCN^-, stir for 0.5 h and filter respectively.

If the precipitation of $Zn_2Fe(CN)_6$ dissolves obviously, $Fe(CN)_6{}^{4-}$ will enter the liquid again, and the Fe concentration will increase significantly. It can be seen from Table 3.4 that Fe does not increase significantly under the conditions of $pH = 2$ and 11, indicating that $Zn_2Fe(CN)_6$ precipitation can exist stably in the range of $pH = 2{\sim}11$.

It can be seen from Table 3.4 that SCN^- has no significant effect on the precipitation and dissolution of $Zn_2Fe(CN)_6$ at $pH = 2$ and 11.

3.3.5 CuCN Dissolution Test

Prepare suspension with CuCN, and take four groups of equal suspension, numbered 1#, 2#, 3#, 4#, respectively. Adjust the pH of 1 #, 2 # to 2, add 22 g/l SCN^-, adjust the pH of 3 #, 4 # to 11, and add 22 g/l SCN^-, stir for 0.5 h and then filter.

If CuCN precipitation dissolves obviously, the concentrations of Cu and CN^- will increase significantly compared with the original solution. It can be seen from Table

Table 3.5 Results of CuCN precipitation dissolution test

Test no.	pH	Whether to join SCN^-	Detection result, mg/L	
			Cu	TCN
1#	2	no	51.95	24.30
2#		Yes	20.60	713.46
3#	11	no	101.70	72.80
4#		Yes	1052.19	822.67
Stock solution			7.75	18.40

3.5 that under the conditions of pH = 2 and 11, the concentrations of Cu and CN^- increased. It shows that CuCN precipitation will dissolve under two pH conditions.

Under the condition of pH = 2, the concentration of CN^- increases significantly after adding SCN^- (the dissociated Cu^+ combines with SCN^- to form $Cu(SCN)_2$ or CuSCN, and the pH increases rapidly after adding thiocyanate), indicating that SCN^- will promote the dissociation of CuCN precipitation under acidic conditions. Under the condition of pH = 11, the concentration of Cu^+ and CN^- increases significantly after adding SCN^- (the dissociated Cu^+ generates $Cu(SCN)_2$, CuSCN, $Cu(OH)_2$ precipitation, the concentration of Cu^+ decreases, forming a CN^- rich environment, and Cu^+ and CN^- form $CuCN_3{}^{2-}$ dominated complex ions, which greatly increases both Cu and TCN), indicating that SCN^- also promotes the dissociation of CuCN precipitation under alkaline conditions.

3.3.6 Zn(CN)2 Dissolution Test

$Zn(CN)_2$ Precipitation suspension was prepared with ZnSO4·7H2O and 30%NaCN solution. Take three groups of equal suspension and number them 1#, 2#, 3#. Adjust its pH to 2, 9 and 11 respectively. Mix each for 0.5 h and then filter.

If the prepared $Zn(CN)_2$ precipitation dissolves significantly under acidic conditions, CN^- will combine with H^+ to form HCN overflow, and the concentration of Zn^{2+} will increase significantly. It can be seen from Table 3.6 that after the pH of the prepared $Zn(CN)_2$ precipitation suspension is adjusted to 2, the concentration of Zn^{2+} in the filtrate increases from 55.61 to 697.6 mg/l. At the same time, during the test, it was found that after the pH was adjusted to 2, $Zn(CN)_2$ Precipitation dissociated rapidly, and the suspension became transparent.

It can be seen from Table 3.6 that after the pH of the prepared $Zn(CN)_2$ Precipitation suspension was adjusted to 11, the total cyanide increased significantly. At the same time, during the test, it was found that after the pH was adjusted to 11, the precipitated particles became significantly thicker. It indicates that the precipitation type has changed, and it is confirmed that $Zn(CN)_2$ precipitation is dissolved and converted to $Zn(OH)_2$ at pH = 11.

Table 3.6 Results of $Zn(CN)_2$ precipitation dissolution test

Test no.	pH	Whether to join SCN^-	Detection result, mg/L	
			Zn	TCN
1#	2	No	697.6	–
2#	11		–	136.23
Stock solution			55.61	0.1

3.4 Conclusion

Through theoretical calculation and experimental research, whether $Fe_2Fe(CN)_6$, $Fe_4[Fe(CN)_6]_3$, $Cu_2Fe(CN)_6$, $Zn_2Fe(CN)_6$, CuCN, $Zn(CN)_2$ precipitates dissolve under acidic (pH = 2) and alkaline (pH = 11) conditions and the degree of dissolution are analyzed. In the process of cyanide containing wastewater treatment, the precipitation dissolution under acid and alkaline conditions can be judged according to the precipitation type generated by cyanide containing wastewater, and then the efficient and safe implementation of cyanide containing wastewater treatment, pressure filtration, filter cake washing and other operations can be realized by controlling the acid and alkaline of wastewater.

References

Lu J et al (2016) Brief description of treatment methods of cyanide containing waste liquid. Indust Energy Saving Technol 01:67

Peng X et al (2018) Experimental study on ozone oxidation treatment of cyanide containing wastewater. Mining& Metallurgy 01: 69–72

Secretariat of the National Gold Standardization Technical Committee (2017) Selected current gold standards. Metallurgical Industry Press, Beijing China, P 262

Tao Y (1985) Reagent consumption and concentration control of gold cyanide leaching operation. Nonferr Met (Benef) 05:37–42

Tao Y (1990) Ways to reduce sodium cyanide consumption. Gold 11:31–36

Theory and method of cyanide wastewater treatment for gold extraction. Metallurgical Industry Press, Beijing, 2015, pp 144–172

ZHANG X L (2017) Beneficiation test research method. Chemical Industry Press, Beijing China, p 151

Chapter 4
Correlation Analysis Between GDP, Sewage and Heavy Metal Pollution, and Human Health in the Past 14 years—A Case Study of Gansu Province

Bin Yue, Zhuoxin Yin, Guohua Chang, Yuanyuan Shen, Youyi Zhao, and Tianpeng Gao

Abstract In order to understand the effect of the sewage and pollution in recent years on the local economy and people's health in Gansu Province, the discharge of sewage and wastewater contained heavy metal, GDP, health expenditure and mortality were statistically analyzed from 2004 to 2017. The results showed that the discharge of sewage totally presented an increasing trend rapidly between 2004 and 2015, and reached the maximum value in 2015. At the same time, there was a peak in the total discharge of five kinds of heavy metal pollutants (Pb, Cr(VI), Cd, Hg and As) in 2005, and then decreased gradually, later stabilized. Discharge of Cd, Pb and As were dominant in the total discharge of heavy metals. GDP almost increased linearly, and was positively correlated with the emission of wastewater (0.975). The mortality of local population was positively correlated with the discharge of heavy metal in sewage.

Keywords Sewage discharge · Heavy metal pollution · GDP · Mortality

4.1 Introduction

Gansu Province played an important role in the construction of ecological security barrier in China, and its water environmental status had an important influence on maintaining and stabilizing the local ecosystem and human health. Gansu Province

B. Yue · Z. Yin · G. Chang · Y. Zhao · T. Gao (✉)
College of Urban Environment, The Engineering Research Center of Mining Pollution Treatment and Ecological Restoration of Gansu Province, Lanzhou City University, Gansu 730070, China
e-mail: zkgtp@163.com

Y. Shen · T. Gao
College of Biological and Environmental Engineering, Xi'an University, Xi'an 710065, China

Academy of Eco Xi'an, Xi'an University, Xi'an 710065, China

J. Zhang et al. (eds.), *Environmental Pollution Governance and Ecological Remediation Technology*, Environmental Science and Engineering,
https://doi.org/10.1007/978-3-031-25284-6_4

is an important water conservation area located in the upper reaches of the Yellow River and the Yangtze River. It is also an important part of the Qinghai Tibet plateau ecological barrier, the Loess Plateau Sichuan Yunnan ecological barrier, and the Northsern sand prevention belt determined in the 12th Five Year Plan in China. According the national main functional zone planning, nearly 90% of the land area in Gansu Province is restricted and prohibited development zone, which plays an important ecological regulation role in water and soil supply, resources protection, water and soil loss prevention, and ecological security in the Yellow River, and Yangtze River basins and many other areas (Zhang and Gao 2015). It is also one of the most complex and vulnerable areas of natural ecological types in China, which is of the harsh natural conditions, the fragile ecological environment and the weak carrying capacity of the ecological system. Hence, it is of great significance to protect ecological environment in Gansu not only for its own sustainable development, but also for the northwest and even for the maintenance of the ecological security of the whole country.

Gansu is currently facing a serious water shortage, with less per capita water resources. In recent years, with the development of economy and the improvement of people's living standards in Gansu Province, the water security crisis characterized by water resources shortage and water environment pollution has not only seriously affected the local industrial and agricultural production, but also seriously threatened the ecological environment security and human health. Gansu Province is an important base of nonferrous metals and petrochemical industry in China, and its industry obvious characteristics of heavy industry (Teng et al. 2016). Therefore, there are many heavy metals existed in the production wastewater discharge. Due to heavy metals' characteristics of high toxicity, persistence, accumulation and difficult degradation, they are widely concerned and listed as the priority control pollutant in China's water (Jiang et al. 2014). This manuscript makes a statistical analysis of sewage discharge in Gansu Province from 2004 to 2017, including the discharge amount of five heavy metals, GDP, health cost and mortality rate of five heavy metals order to indicate the impact of sewage and heavy metal pollution on local economy and people's health in Gansu Province, and to remind the local government departments to strengthen water pollution control and achieve standard discharge.

4.2 Experimental Details

4.2.1 Sources

The data of wastewater discharge, the discharge of five heavy metal pollutants (Pb, Cr(VI), Cd, Hg and As) in sewage in Gansu Province are from China Environmental Statistics Yearbook (2005) and Gansu Development Yearbook (2005). GDP data are from China Statistical Yearbook (2005), total health cost and mortality data are from Gansu Development Yearbook (2005) and China Health Statistics Yearbook (2005).

4.2.2 *Methods*

The correlation analysis among the urban sewage discharge, the total discharge of five heavy metals, mortality and GDP was carried out by SPSS 21.0.

4.3 Results

4.3.1 *Wastewater Emission Status from 2004 to 2017*

The change in wastewater emission in Gansu Province between 2004 and 2017 is shown in Fig. 4.1. The wastewater emission fluctuated little from 2004 to 2007, and it increased year by year after 2008. From 2010 to 2013, the wastewater emission increased rapidly. In these years, the wastewater emission in 2005 was the smallest, 437.28 million tons; it reached the maximum value of 670.72 million tons in 2015, and from 2004 to 2015, the wastewater discharge increased by 219.01 million tons. In 2016 and 2017, the wastewater emission showed a downward trend. According to the fitting results between the wastewater emission of per unit GDP and time, it was found that the wastewater emission of per unit GDP gradually decreases exponentially with time (top right of Fig. 4.1). The fitting Eq. (4.1) between wastewater emission of per unit GDP and time is:

$$y = 8.73 + 20.45 \exp(563.45 - 0.2812\,t),\ R^2 = 0.9896 \tag{4.1}$$

where y is wastewater emission of per unit GDP and t is time.

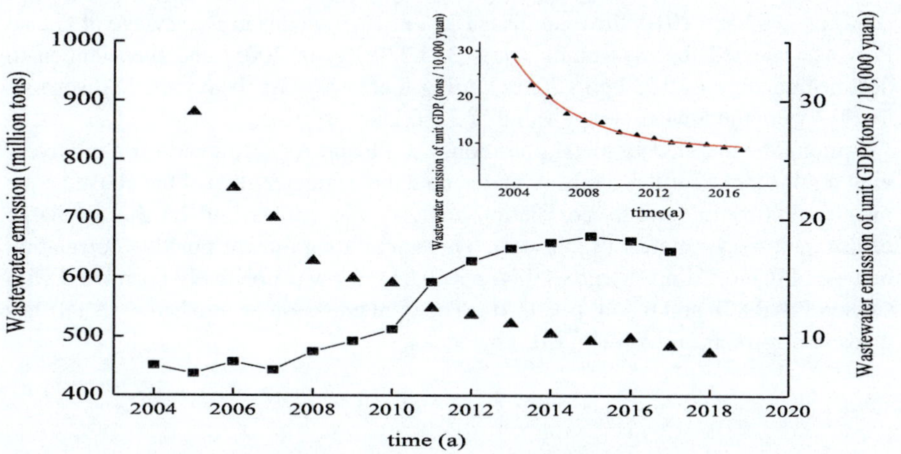

Fig. 4.1 Variation of wastewater emission (black square) and wastewater emission of per unit GDP (black up-pointing triangle) in Gansu Province from 2004 to 2017

4.3.2 *Emission of Metal Pollutants in Wastewater from 2004 to 2017*

During 2004–2017, the discharge of heavy metal pollutants such as Pb, Cr(VI), Cd, Hg and As from wastewater in Gansu Province is shown in Table 4.1. The discharge of heavy metal pollutants in industrial wastewater was provided in China Environmental Statistics Yearbook before 2010, and the discharge of heavy metal pollutants in wastewater was given after 2010. From 2004 to 2009, the discharge of Hg in industrial wastewater basically maintained at about 0.3 tons, and in 2010, it dropped to 0.006 tons. From 2011 to 2015, the discharge of Hg in wastewater was also relatively stable, basically maintained at about 0.1 t/a. After 2016, it dropped to 0.02 tons. From 2004 to 2010, the discharge of Cd in industrial wastewater showed a general downward trend. Before 2006, the discharge was relatively high, and reached the largest value of 19.75 tons in 2005. Since then, it has gradually declined to a minimum of 0.53 tons in 2017. From 2004 to 2010, the discharge of Cr(VI) fluctuated between 0.5 and 2.0 tons. In 2006, the discharge reached the peak value of 1.961 tons. From 2011 to 2017, the trend of Cr(VI) in wastewater was relatively gentle, and the discharge of Cr(VI) was below 1.0 t/a. From 2004 to 2010, the amount of Pb in industrial wastewater fluctuated from 34 to 82 tons, and reached a peak of 81.96 tons in 2005. From 2011 to 2017, the discharge trend of Pb in wastewater is relatively gentle, all of which are below 10 tons. From 2004 to 2010, the emission of As from industrial wastewater fluctuated sharply, among which the maximum was 250.76 tons in 2005. After 2011, the annual emission of As decreased slowly. From 2004 to 2010, the order of annual discharge of five heavy metals is as As (90.35 tons) > Pb (54.47 tons) > Cd (9.72 tons) > Cr(VI) (1.09 tons) > Hg (0.25 tons). After 2011, the order of annual discharge of the five heavy metals is Pb (6.10 tons) > As (3.48 tons) > Cd (1.16 tons) > Cr(VI) (0.39 tons) > Hg (0.07 tons).

The change trend of total discharge of heavy metals in sewage is in Fig. 4.2. Between 2004 and 2010, the total discharge of heavy metals in the sewage of Gansu Province reached the maximum value (353,720 kg) in 2005, and then tended to decline, reaching 44,193 kg in 2010. During these 14 years, heavy metal discharge in 2017 were the lowest, with a value of 5260 kg.

Among the total heavy metal pollution, Cd, Pb and As accounted for a relatively high proportion. In order to further understand the characteristics of the above heavy metal pollution in sewage, correlation analysis was carried out on the discharge of different heavy metals (Table 4.3). There was a significant positive correlation between Pb and Cd discharge (0.934, $p = 0.01$). As was positively correlated with Cd and Pb (0.878 and 0.854, $p = 0.01$). The total emission of five heavy metals has a positive correlation with Hg, Cd, Pb and As.

Table 4.1 Discharge of heavy metals in industrial wastewater (2004–2010) and in wastewater (2011–2017) from Gansu

Year	Metal emission (ton)				
	Hg	Cd	Cr(VI)	Pb	As
2004	0.3	13.61	1.31	56.93	75.63
2005	0.27	19.75	0.98	81.96	250.76
2006	0.314	15.272	1.961	74.205	94.56
2007	0.183	7.202	1.5	48.964	44.771
2008	0.323	4.274	0.829	46.811	80.52
2009	0.336	4.209	0.573	37.997	80.562
2010	0.006	3.688	0.452	34.42	5.627
2011	0.072	1.425	0.24	6.885	5.736
2012	0.09	1.304	0.381	6.793	3.751
2013	0.095	1.289	0.415	8.068	4.286
2014	0.088	0.824	0.499	4.382	3.266
2015	0.107	1.161	0.722	5.84	4.696
2016	0.02	1.57	0.13	7.32	1.7
2017	0.02	0.53	0.36	3.42	0.93

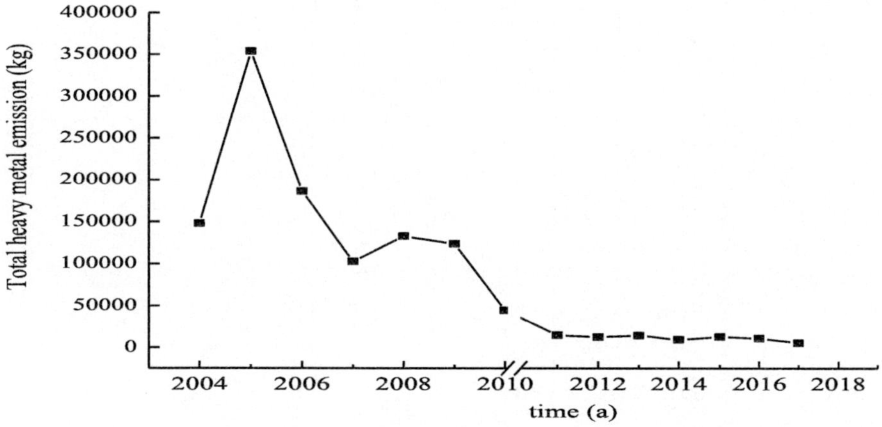

Fig. 4.2 Variation of total emission of five heavy metals from wastewater discharge from 2004 to 2017

4.3.3 GDP, Total Health Expenditure, and Population Mortality Rate of Gansu Province

The data of GDP, total health expenditure and the ratio of total health expenditure to GDP and population mortality in Gansu Province from 2004 to 2017 were shown in

Table 4.3 Pollution of heavy metals in industrial wastewater (2004–2010) and in wastewater (2011–2017) from Gansu Correlation between Wastewater Emission, Heavy Metal Discharge, Total Health Expenditure (THE), Health Expenditure Per Capita (HEPC), etc.

	Wastewater emission	Hg	Cd	Cr	Pb	As	Total emission of metals	GDP	THE	HEPC	Mortality rate
Wastewater emission	1										
Hg	−0.767*	1									
Cd	−0.791**	0.677**	1								
Cr	−0.730**	0.688**	0.753**	1							
Pb	−0.934**	0.782**	0.934**	0.797**	1						
As	−0.720**	0.709**	0.878**	0.501	0.854**	1					
Total emission of metals	−0.810**	0.753**	0.932**	0.620*	0.930**	0.985**	1				
GDP	0.975**	−0.806**	-0.829**	−0.746**	−0.928**	−0.729**	−0.817**	1			
THE	0.938**	−0.767**	-0.763**	−0.715**	−0.875**	−0.669*	−0.757**	0.977**	1		
HEPC	0.942**	−0.775**	-0.768**	−0.724**	−0.882**	−0.673**	−0.762**	0.980**	1.000**	1	
Mortality rate	−0.720**	0.797**	0.571*	0.642*	0.705**	0.622*	0.665**	−0.639*	−0.525	−0.539**	1

Table 4.2 Data of GDP, the total health expenditure and mortality in Gansu Province in 2004–2017

Year	GDP (billion yuan)	Total health expenditure (billion yuan)	Health expenditure per capita (yuan)	Ratio of total health expenditure to GDP (%)	Mortality rate (‰)
2004	155.893	8.992	343.45	5.77	6.52
2005	193.398	10.661	410.91	5.51	6.57
2006	227.67	12.441	447.36	5.46	6.62
2007	270.24	14.715	562.27	5.45	6.65
2008	317.611	21.389	813.85	6.73	6.68
2009	338.756	26.398	1001.65	7.79	6.71
2010	412.075	29.538	1153.86	7.17	6.02
2011	412.075	39.360	1543.99	7.84	6.03
2012	565.02	44.472	1725.36	7.87	6.05
2013	633.069	51.821	2006.89	8.27	6.08
2014	683.682	56.975	2199.13	8.34	6.11
2015	679.032	65.407	2516.09	9.63	6.15
2016	720.037	75.406	2889.18	10.47	6.18
2017	745.99	81.27	3095.17	10.59	6.52

Table 4.2. From 2004 to 2017, the GDP of Gansu Province gradually increased from 155.893 billion Yuan to 745.99 billion Yuan. The total health expenditure and the health expenditure per capita almost showed a linear growth trend. Before 2008, the total health expenditure was under 15 billion Yuan, and it was over 20 billion Yuan after 2008. After 2013, the total health expenditure is over 50 billion Yuan. The ratio of total health expenditure to GDP shows a slow growth trend, reaching the maximum value (10.59%) in 2017. From 2004 to 2009, the change of population mortality was relatively stable. From 2010 to 2017, the population mortality showed a slow growth trend.

4.3.4 Correlation Analysis of Indicator

There was a negative correlation between the discharge of wastewater and the discharge of Hg, Cd, Cr, Pb, As and the total amount of the five heavy metals, among which the discharge of wastewater was a significant negative correlation with the discharge of Pb ($-0.934, p = 0.01$) (Table 4.3). The correlation coefficient between wastewater discharge and GDP, total health expenditure, and total health expenditure per capita was more than 0.930. The wastewater discharge was negatively correlated with local mortality. GDP was negatively correlated with the discharge of Hg, Cd, Cr, Pb and As. There was a significant positive correlation between GDP and the total health expenditure and the health expenditure per capita; the pollution of the above five heavy metals were positively correlated with population mortality, among which the correlation coefficient between Hg and mortality is the highest (0.797, $p = 0.01$); while the correlation coefficient between mortality and GDP, the total health expenditure and the health expenditure per capita were all negative, which indicts that the heavy metal emissions have an impact on the health of the local people.

4.4 Discussion

Water is one of the important basic substances for human survival, and its quality plays an important role in keeping human health. If the heavy metals discharged through the water enter the environment, they would enter the environments of water, soil and atmospheric through adsorption, migration, volatilization, etc. or enter the biological or human body through the food chain. As heavy metals are not biodegradable and have the characteristics of bioaccumulation, once they enter organisms, long-term accumulation will pose a direct threat to the human health (Chen et al. 2017).

Therefore, it is great significance for maintaining human health and improving the quality of ecological environment to prevent, control and eliminate water pollutants in water environment. Among the five heavy metals involved in this study, mercury is a highly toxic heavy metal substance. Mercury and its compounds can invade human body through many ways, such as respiratory tract, skin or digestive tract. When inorganic mercury and its compounds enter the human body, they will cause damage to liver, kidney, heart, thyroid, brain and other organs, and even lead to nervous system disorder and chronic mercury poisoning (Wei et al. 2014).

From 2004 to 2017, the average proportion of the first, second and third industries in the economic structure of Gansu Province were 12.79%, 43.86% and 43.35% respectively. In the second industry, the contribution rate of industrial output value is close to 77% (National Bureau of Statistics of the People's Republic of China 2005). The second industry is in the leading position in Gansu Province. In the second industry, the pillar industries are mainly metallurgy, petrochemical industry, non-ferrous metals and other industries relying on raw materials and energy development. High energy consumption, high pollution and resource-based industries have a large proportion in the economic structure of Gansu. Miao et al. reported that there were 170 enterprises mainly engaged in non-ferrous metal mining and processing, and among which there was 67 enterprises involve in non-ferrous metal smelting and rolling processing industry (Liu 2009; Miao et al. 2010).

In the wastewater discharge of Gansu Province, the contents of heavy metals Pb, Cd, Cr and As was higher, which also showed the high proportion of secondary industry in the industrial structure of Gansu Province. Therefore, in the future economic development, the government of Gansu needs to strengthen the sewage treatment efforts, minimize the discharge of heavy metals, increase the sewage emission reduction work and actively promote the recycling of water resources. In recent years, the problem of heavy metal pollution of soil in Gansu Province has become increasingly prominent, which is obviously closely related to heavy metal emission enterprises. According to the soil pollution survey in Gansu Province, the farmland soil on both sides of Dongdagou River Basin in Baiyin City and the soil around Baiyin Company Industrial Park have been seriously polluted by heavy metal wastewater. The main pollutants in the soil were Cd, As, Hg and Pb, and the area of the polluted area was 13.58 km^2 (Science et al. 2010). The quality and safety of local agricultural products are also greatly threatened. The heavy metals Cd and Pb in crops exceed the standard seriously, which affects food safety and human and animal health (Tian et al. 2012). In 2006, in the Incident of Blood Lead Exceeding" in Huixian County of Gansu Province, more than 300 local people blood lead exceeded the standard, and all the land within 400 m of the enterprise's lead emission was polluted. (Ma and Mu 2006).

Therefore, the environmental pollution caused by heavy metals has seriously affected the local human health and the sustainable development of social economy, the pollution control is urgent.

In the future, the provincial government of Gansu should strengthen the prevention and treatment of heavy metal pollution, establish and improve the relevant legislation on the prevention and control of heavy metal pollution, further adjust and optimize the industrial structure, promote clean production, and gradually eliminate enterprises with backward technology, equipment and serious pollution.

4.5 Conclusion

From 2004 to 2017, the sewage discharge of Gansu Province increased first and then decreased, reaching a peak in 2015. During this period, the pollution of heavy metals Cd, Pb and As in the wastewater were relatively high. With the growth of GDP, the sewage discharge of per unit GDP gradually decreased from 2004 to 2017. There was a positive correlation between the death rate and the emission of all metals in Gansu Province. Therefore, in the future economic development of Gansu Province, it is necessary to strengthen the control of heavy metal discharge from sewage, which will be of great significance to protect the health of local residents and promote the sustainable development of local economy, society and environment.

Acknowledgements The research was funded by the National Natural Science Foundation of China (Grant No. 31860176), the Research Project of Universities in Gansu (2020A-124), Key Research and Development Program of Gansu Province (No. 20YF3FA037).

References

Chen G, Zhao YJ, Hu BQ, Yang JM, Huang YY (2017) Spatial and temporal distribution of heavy metals in riparian zone and soil environmental quality analysis. Environ Pollut Cont 39(9):981–986

Editorial board of Gansu Development Yearbook, Gansu Development Yearbook. China Statistics Press, Beijing, 2005–2018

Environmental Science and Design Institute of Gansu Province, Comprehensive prevention and control plan of heavy metal pollution in Gansu Province, 2010

Jiang DS, Li M, Cui YB (2014) Acute toxic effects of heavy metals and chlorophenols on *Limnodrilus hoffmeisteri* and safety assessment of the aquatic environment. China Environ Sci 34(6):1572–1578

Liu BX (2009) Problems in the industrial structure of Gansu Province and the countermeasures. Res Develop 6:76–79

Ma J, Mu CY (2006) Administrative omission of environmental protection in China—from "the Incident of Blood Lead Exceeding" in Huixian County. Gansu Province, Gansu Agric 12:173

Miao XH, He LP, Tao HY, Quan HB (2010) Current situation and countermeasures of heavy metal pollution control in Gansu Province. Gansu Sci Technol 26(23):36–38

National Bureau of Environmental Statistics of the People's Republic of China, China Environmental Statistics Yearbook. China Statistics Press, Beijing, 2005–2017

National Bureau of Statistics of the People's Republic of China, China Statistical Yearbook. China Statistics Press, Beijing, 2005–2018

National Health and Family Planning Commission, China Health Statistics Yearbook. China Union Medical University Press, Beijing, 2005–2018

Teng YQ, Zhang Q, Zhang AN (2016) Transformation of Gansu's industrial structure and its benefit analysis. Gansu Sci Technol 32(16):16–19

Tian QC, Yang TB, Shi PH, Zeng CS, Wang TX (2012) Analysis of heavy metal sources and measures for preventing and controlling them in Baiyin city. Environ Monit China 28(6):40–45

Wei YH, Guo JQ, Chen ZM, Mo ZY, Xie H, Mao JY, Liang GY, Liu HL (2014) Effects on human health of mercury pollution in the environment and its prevention measures. Popular Sci Technol 16(3):59–61

Zhang Y, Gao F (2015) The synthetic evaluation on the ecological defense construction in Gansu and the influence factors. J Arid Land Resour Environ 29(11):93–98

Chapter 5
Vegetation Dynamic Monitoring and Driving Factors Analysis in Chengdu in Recent 20 Years Based on Google Earth Engine

Peijuan Wu, Zihan Jiang, and Hong Fu

Abstract Vegetation dynamic monitoring and driving factor analysis are important contents of regional ecological environment assessment. Google Earth Engine (GEE), the MODIS NDVI remote sensing images of Chengdu from 2000 to 2021 were collected, and the maximum synthesis method, trend Finally, temperature, precipitation and artificial impervious area were selected to analyze the vegetation at the pixel scale. The results showed that: (1) Chengdu was mainly composed of medium and high coverage areas The results showed that: (1) Chengdu was mainly composed of medium and high coverage areas, accounting for 57.6% of the total area. Among them, high coverage areas were mainly distributed in the west of Chengdu, while low coverage areas and medium and low coverage areas were mainly distributed in the central urban area of Chengdu; (2) The (2) The vegetation coverage in Chengdu had not changed significantly in the past 20 years ($P > 0.05$). The area with improved vegetation accounted for 40% and the area with degraded vegetation accounted for 39%. Among them, the areas with significantly improved vegetation (Slope > 0.01) were mainly in Jinniu District, Chenghua District, Wuhou District, Among them, the areas with significantly improved vegetation (Slope > 0.01) were mainly in Jinniu District, Chenghua District, Wuhou District, Qingyang District and Jinjiang District, while the severely degraded areas (Slope < −0.01) were mainly in Pidu District and Xindu District, Qingbaijiang District, Shuangliu District and Wenjiang District; (3) At the regional scale, the central urban area of Chengdu showed a significant (3) At the regional scale, the central urban area of Chengdu showed a significant trend of degradation ($P < 0.05$, Slope = −0.0027/year), while Dayi County showed a significant trend of improvement ($P < 0.05$, Slope = 0.0011/year); (4) The overall temperature in Chengdu did not change significantly ($P > 0.05$), and the precipitation increased significantly in each pixel ($P < 0.05$, Slope = 10.63 mm/year), and the area with a

P. Wu · H. Fu (✉)
College of Architecture and Environment, Sichuan University, Chengdu 610065, China
e-mail: fh196657@163.com

Z. Jiang
School of Advavced Technology Xi'an Jiaotong-Liverpool University, Suzhou 215000, China

J. Zhang et al. (eds.), *Environmental Pollution Governance and Ecological Remediation Technology*, Environmental Science and Engineering,
https://doi.org/10.1007/978-3-031-25284-6_5

larger increase was in the eastern part of Chengdu. In addition, the artificial impervious area showed a significant increasing trend ($P < 0.05$, Slope = 68.30 km^2/year). The research results can provide a scientific basis for the formulation of ecological protection and restoration policies in Chengdu.

Keywords Vegetation · Dynamic monitoring · Driving factors · GEE · Chengdu

5.1 Introduction

Vegetation ecosystems can provide a variety of functions and services to humans, such as water conservation, atmospheric purification and climate regulation (Migliavacca et al. 2021; Xueling 2004; Pan et al. 2021). However, with global warming, frequent climate extremes and anthropogenic disturbances, regional vegetation ecosystems are at risk of degradation, which will affect biodiversity conservation and human sustainable development goals (Yilun et al. 2022; Huang et al. 2018). Especially in the active areas of first-tier cities, with the rapid economic development and population increase in recent years, it has led to more and more engineering construction in various areas and the expansion of urban planning areas. These human activities will eventually lead to the reduction of vegetation area and fragmentation of urban landscape pattern, which will lead to the risk of urban flooding and heat island effect, and will cause a series of more serious ecological and environmental problems (Li et al. 2014; Zongxue et al. 2020) The risk of urban flooding and heat island effect will lead to a series of more serious ecological problems. For this reason, it is necessary to understand the monitoring of regional urban vegetation dynamics.

In recent years, domestic and foreign scholars have mainly used Landsat and MODIS NDVI series remote sensing images for vegetation dynamics monitoring studies (Jun-Nan et al. 2018; Ma Na and Yunfeng 2012). For example, Junnan et al. (2018) based on MODIS NDVI data, used a vegetation cover model to dynamically monitor the vegetation change in Yunnan Province in the past 16 years; Baoquan (2013) Jia Baoquan analyzed the dynamic changes of vegetation in Beijing based on Landsat TM satellite images and discussed the reasons for the changes; Chen et al. (2010) analyzed the spatial and temporal changes of vegetation cover in the Bailong River basin in the past decade based on MODIS NDVI and Landsat data using the hybrid image element decomposition method. However, the data of the above studies were collected and processed on local computers, which led to the whole process being very tedious and consumed computer memory. Especially with the advent of the era of big data remote sensing, processing remote sensing data locally can no longer meet the needs of researchers. For this reason, Google has launched a cloud platform for online processing of big data remote sensing images called Google Earth Engine (GEE) (Tamiminia et al. 2020). GEE cloud platform integrates Landsat, MODIS and other common remote sensing data and various analysis algorithms, which provides a platform for long time large scale remote sensing data processing and reduces the workload (Hao et al. 2018; Gorelick et al. 2017).

As a new first-tier city and the core city of the Chengdu-Chongqing Economic Circle, Chengdu has always enjoyed the reputation of being the "Land of Heaven". In the past two decades, urban construction in Chengdu has been in full swing and a large amount of land area has been occupied, resulting in significant changes in vegetation cover. With people's aspiration for a better life, vegetation cover has become an important index for urban environment evaluation. Therefore, it is important to study the spatial and temporal characteristics of the dynamic changes of vegetation cover in Chengdu in recent years. In addition, besides human activities affect the change of urban vegetation, climate change also affects the growth of vegetation (Xin Zhongbao and Jiongxin 2007; Hui-Xia et al. 2011). Therefore, it is also crucial to study the human and climate drivers of vegetation cover, which will provide some scientific basis for urban ecological planning.

In summary, in this study, to solve the above problems, MODIS NDVI remote sensing data will be selected based on the GEE cloud platform, and annual NDVI will be calculated using the maximum synthesis method, and trend analysis and correlation analysis methods will be selected to analyze the spatial and temporal patterns of vegetation and the driving factors from 2000 to 2021 in Chengdu City. The selected driving factors include temperature, precipitation and artificial impervious surface factor. The main answers are the improvement and degradation of vegetation in each region of Chengdu city in the last two decades, and the spatial and temporal patterns of the drivers.

5.2 Data and Methods

5.2.1 Overview of the Study Area

Chengdu (102° 54′ E–104° 53′ E, 30° 05′ N–31° 26′ N) is one of the 15 sub-provincial cities in China, located in southwestern China, geographically (see Fig. 5.1). The total area is 14,335 km^2, accounting for 2.9% of the total area of Sichuan, with an average altitude of about 770 m and an altitude difference of 4800 m. Due to the huge vertical height difference, the terrain is mainly plains, hills and mountains, and the region is rich in soil resources, forest resources, biological resources and water resources (Chenghu et al. 2009; Dazhong et al. 2009). The region is rich in soil resources, forest resources, biological resources and water resources. Chengdu is located in a subtropical monsoon climate zone, with an average annual temperature of about 16 °C and an annual rainfall of about 1000 mm (Wang et al. 2011). The annual rainfall is about 1000 mm. Among them, there is more rainfall in summer and less drought in winter and spring (Yiqi and Tiangui 2021; Luying et al. 2020). The annual rainfall is about 1000 mm. At present, there are 12 urban areas, 5 county-level cities and 3 counties in Chengdu.

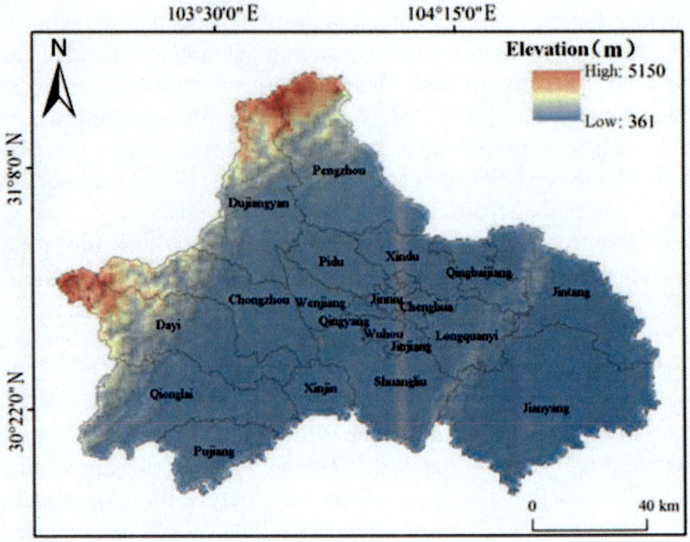

Fig. 5.1 Geographical location of the study area

5.2.2 *Google Earth Engine Platform*

Google Earth Engine (https://earthengine.google.com), or GEE for short, is a cloud processing platform with powerful satellite image analysis capabilities, integrating a variety of advanced geographic information algorithms, which can solve many of the user's geospatial analysis problems (Gorelick et al. 2017; Tao and Hong 2018). In this cloud platform, users can access a large amount of open remote sensing product data and analyze remote sensing data online for free based on a variety of algorithms, and finally they can visualize and export the results. The platform mainly uses JavaScript or Python language for parallel processing of large-scale remote sensing data, which greatly improves the data processing efficiency and is widely used in multi-scale spatial analysis research such as vegetation dynamics monitoring, land use classification, climate change simulation and ecological environment evaluation (Tamiminia et al. 2020; Hao et al. 2018).

5.2.3 *Data Sources*

Vegetation normalized index (Third Level). The vegetation normalized index (NDVI) used in this study was selected from the MODIS MOD13Q1 V6 product, which has a spatial resolution of 250 m and a temporal resolution of 16 days (Jun-Nan et al. 2018). The spatial resolution of the data is 250 m and the temporal resolution is 16 days. In this study, the NDVI data of each year from 2000 to 2021 were selected for fusion based on the maximum synthesis method on the GEE cloud processing platform and cropped using the vector boundary of Chengdu city, and a total of 22 raster data of NDVI with year-by-year time series were obtained to reflect the best vegetation growth of Chengdu city in each of the last 20 years.

Meteorological data (Third Level). The temperature and precipitation data used in this study were obtained from the National Tibetan Plateau Science Data Center (https://data.tpdc.ac.cn), which is a remote sensing product with month-by-month data, spatial resolution of 1 km, and data format of .nc format. The product is based on the Delta spatial downscaling method to downscale the CRU TS v4.02 dataset and WorldClim v2.0 dataset (Peng et al. 2019). In this study, the monthly weather data from 2001 to 2020 were extracted based on the brick() function inside the R language raster package, and the annual mean temperature raster data were obtained by averaging the monthly temperature for each year individually based on the image element scale, and the annual total precipitation raster data were obtained by summing the precipitation data for each year.

Anthropogenic activity data (Third Level). Man-made impervious surface data were chosen to represent the anthropogenic activity data in this study. The artificial impervious surface data used are from Finer Resolution Observation and Monitoring—Global Land Cover (http://data.ess.tsinghua.edu.cn/), a remote sensing product with year-by-year data and a spatial resolution of 30 m. This dataset is based on Landsat remote sensing imagery on the GEE cloud platform. The data set is based on Landsat remote sensing images on the GEE cloud platform, and the spatial mask and feature evaluation algorithms are used to map the increasing artificial impervious surface year by year (Gong et al. 2020). The data set is based on Landsat remote sensing images on the GEE cloud platform. In this study, year-by-year impervious surface data from 2000 to 2018 were obtained by this dataset and ARCGIS 10.8 software to reflect the impact of urbanization on vegetation in Chengdu city.

5.2.4 Research Methodology

Trend analysis (Third Level). In this study, the trend analysis method was used to investigate the trends of vegetation cover changes at multiple time scales in Chengdu city at the image element scale. The method uses a one-dimensional linear regression model to analyze the trend of NDVI image data from 2000 to 2021, and the slope of NDVI change is fitted at the image element scale using the least squares method. The advantage of this method is that it eliminates the influence of NDVI anomalies in individual years on the overall results and can truly reflect the vegetation change in the region (Yi and Mingguo 2007). The expressions are as follows The expression of the calculation is as follows (Jun-Nan et al. 2018) The expression is as follows

$$Slope = \frac{n \times \sum_i^n (i \times NDVI_i) - \left(\sum_i^n i\right) \times \left(\sum_i^n NDVI_i\right)}{n \times \sum_i^n i^2 - \left(\sum_i^2 i\right)^2} \tag{5.15.1}$$

In Eq. (5.1), i is the year (i = 2000,2001,...,2021); NDVIi denotes the NDVI image data in year i; n is the NDVI time series span, which is 22 in this study; and Slope denotes the slope of the fit. In this study, the trend of NDVI in Chengdu city within the time series was obtained based on GEE cloud platform ee.Reducer.linearFit() at the image element scale. If the Slope is positive, it means that NDVI is increasing and vegetation is improving in this time series. The larger the absolute value of Slope, the more obvious the change of vegetation cover.

Correlation analysis (Third Level). Pearson correlation analysis is a classical mathematical method to measure the degree of correlation between two sets of variables (Ma et al. 2021). In this study, the Pearson correlation coefficient was used to analyze the correlation degree of NDVI over time at different regional scales as a whole, and to analyze the correlation degree of the driving factors temperature, precipitation and impervious surface over time in the Chengdu metropolitan area, which is calculated as follows (Ma et al. 2021): The correlation coefficient is calculated as follows

$$r = \frac{\sum_{i=1}^{n}(x_i - \overline{x})(y_i - \overline{y})}{\sqrt{\sum_{i=1}^{n}(x_i - \overline{x})^2 \sum_{i=1}^{n}(y_i - \overline{y})^2}} \quad (5.25.2)$$

In Eq. (5.2), x and y are the two groups of variables; n is the number of variables in each group; r is the Pearson correlation coefficient, which ranges from −1 to 1. If r is positive, the two groups of variables are positively correlated with each other. The larger the absolute value of r, the stronger the correlation between the two groups of variables. This analysis was performed using the corrplot package in the R software.

5.3 Results

5.3.1 Spatial and Temporal Patterns of Annual Mean NDVI

The mean value of multi-year NDVI from 2000 to 2021 was 0.76, and the mean maximum and minimum values were 0.96 and 0.10, respectively. five vegetation cover classes were classified according to NDVI values from low to high (see Fig. 5.2), and the area and percentage of each class were calculated. Table 5.1 The area and proportion of each class were calculated. It was found that Chengdu was mainly a medium–high vegetation cover area, accounting for 57.6% of the total area, followed by a high vegetation cover area, accounting for 35.2% of the total area. From the spatial pattern of NDVI in Fig. 5.2 and the topography in Fig. 5.1, it can be found that the vegetation cover in Chengdu shows obvious spatial heterogeneity, the NDVI value decreases with the increase of longitude, and the topographic characteristics of vegetation cover show a spatial pattern of higher altitude areas than plains. Among them, the high vegetation cover areas are mainly distributed in the high altitude areas

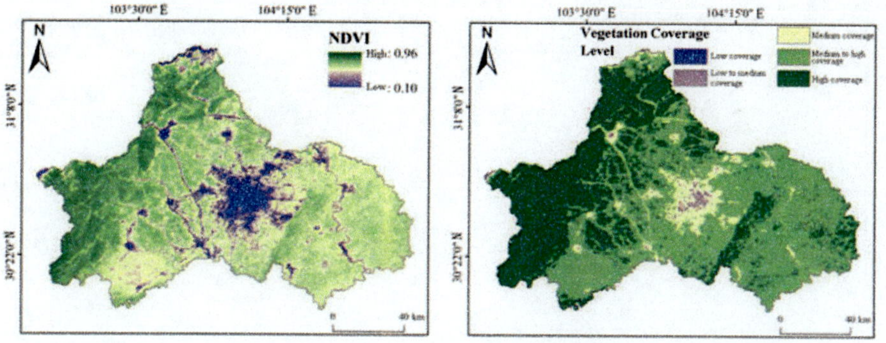

Fig. 5.2 The pattern of average vegetation coverage in Chengdu in recent 20 years

Table 5.1 Statistics of average vegetation coverage grades in Chengdu in the past 20 years

Vegetation coverage	Vegetation coverage level	Area (km^2)	Proportion (%)
0.0–0.2	Low coverage	1	0
0.2 to 0.4	Low to medium coverage	155	0.009
0.4–0.6	Medium coverage	1052	0.063
0.6–0.8	Medium to high coverage	9628	0.576
0.8–1.0	High coverage	5883	0.352

in the western part of Chengdu, while the low cover areas, medium–low cover areas and medium cover areas are mainly distributed in the central urban area of Chengdu.

The average NDVI in 2006 was 0.787, and the average NDVI in 2020 was 0.744. Seven classes of change were classified from low to high according to Slope values (see Fig. 5.3), and the area and percentage of each class were counted. Table 5.1 The area and percentage of each class were calculated. It was found that Chengdu was mainly a slightly improved area, accounting for 32.7% of the total area, followed by a slightly degraded area, accounting for 24.4% of the total area. The annual mean NDVI improved in 40% of the area, degraded in 39%, and basically unchanged in 21%. The least distributed areas with obvious improvement accounted for 0.8%, mainly located in Jinniu District, Chenghua District, Wuhou District, Qingyang District and Jinjiang District. Seriously degraded areas accounted for 5.9%, mainly located in Pidu District, Xindu District, Qingbaijiang District, Shuangliu District and Wenjiang District. 39.6% of the areas with significant changes in NDVI and 60.4% of the areas without significant changes (Table 5.2).

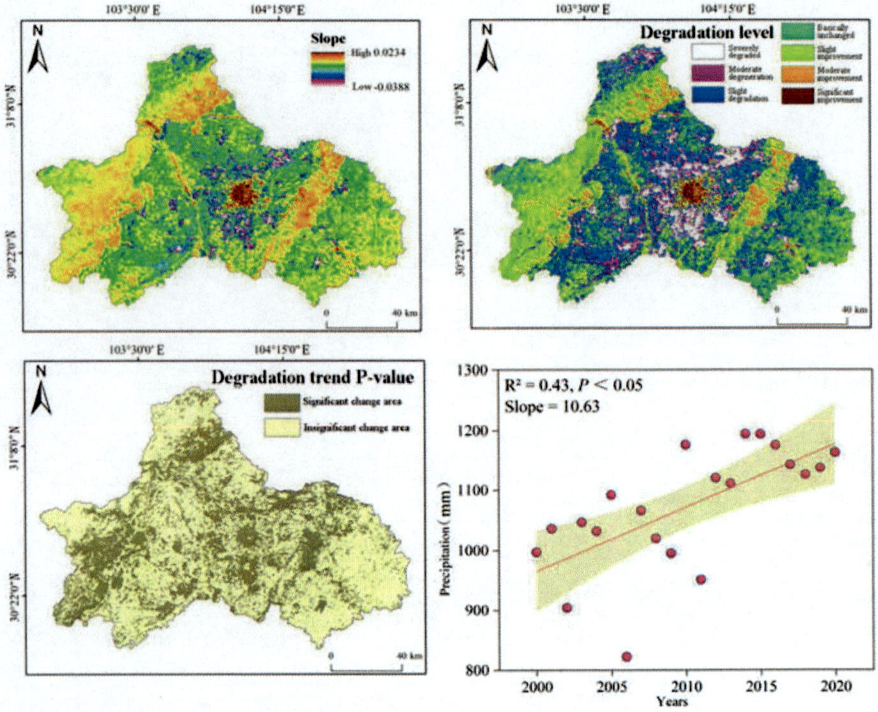

Fig. 5.3 Vegetation change trend in Chengdu in recent twenty years

Table 5.2 Vegetation change trend grade statistics in Chengdu in recent 20 years

NDVI change trend (Slope)	Degradation level	Area (km^2)	Proportion (%)
< −0.0100	Severely degraded	989	0.059
−0.0100 to −0.0050	Moderate degeneration	1472	0.088
−0.0050 to −0.0010	Slight degradation	4072	0.244
−0.0010–0.0010	Basically unchanged	3512	0.210
0.0010–0.0050	Slight improvement	5463	0.327
0.0050–0.0100	Moderate improvement	1072	0.064
> 0.0100	Significant improvement	139	0.008

5.3.2 *Regional-Scale NDVI Trend Analysis*

In the regional scale analysis, considering that the 12 urban areas are more influenced by human activities due to the economic development, the 12 urban areas are combined into Chengdu central city for trend analysis. The multi-year NDVI trends of each region in Chengdu (see Fig. 5.4). It was found that only the vegetation trends

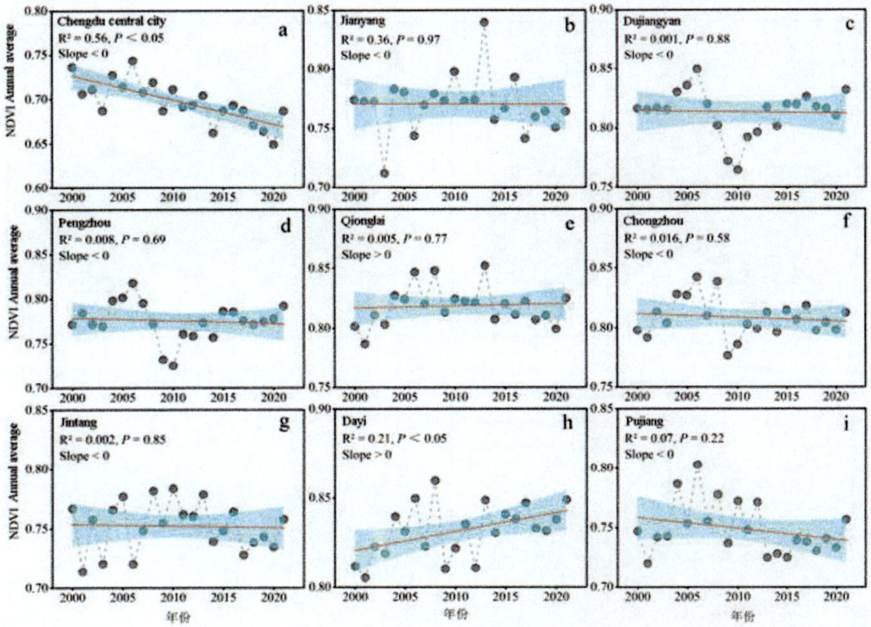

Fig. 5.4 Vegetation change trend of Chengdu at the regional scale in the past two decades

in Chengdu Central Urban Area ($R^2 = 0.56$, $P < 0.05$) and Dayi County ($R^2 = 0.21$, $P < 0.05$) were significant, while the other regions did not change significantly. The vegetation in Chengdu Central City was significantly degraded in the last 20 years with a Slope of −0.0027/year, while Dayi County was significantly improved with a Slope of 0.0011/year.

5.3.3 Analysis of Driving Factors

Temperature factors (Third Level). Spatial patterns and trends of temperature factors in Chengdu (see Fig. 5.3). The average multi-year temperature in Chengdu from 2000 to 2020 is 16.1 °C, and the average maximum and minimum values are 17.7 °C and −5.9 °C, respectively. Among them, and Longquanshan region in the western part of Chengdu are at a lower temperature, while all other regions basically fall into the high value range. In terms of temporal trend, the overall change in temperature in Chengdu is not significant ($P = 0.12$), and the Slope is 0.01 °C/year, showing a weak increasing trend. The area with decreasing temperature was mainly located in the western high-altitude region of Chengdu, accounting for 8.9%. the lowest mean temperature was 15.5 °C in 2000, and the highest mean temperature was 16.4 °C in 2006. The regions with significant changes in temperature accounted for 32.5% and those with no significant changes accounted for 67.5% (Fig. 5.5).

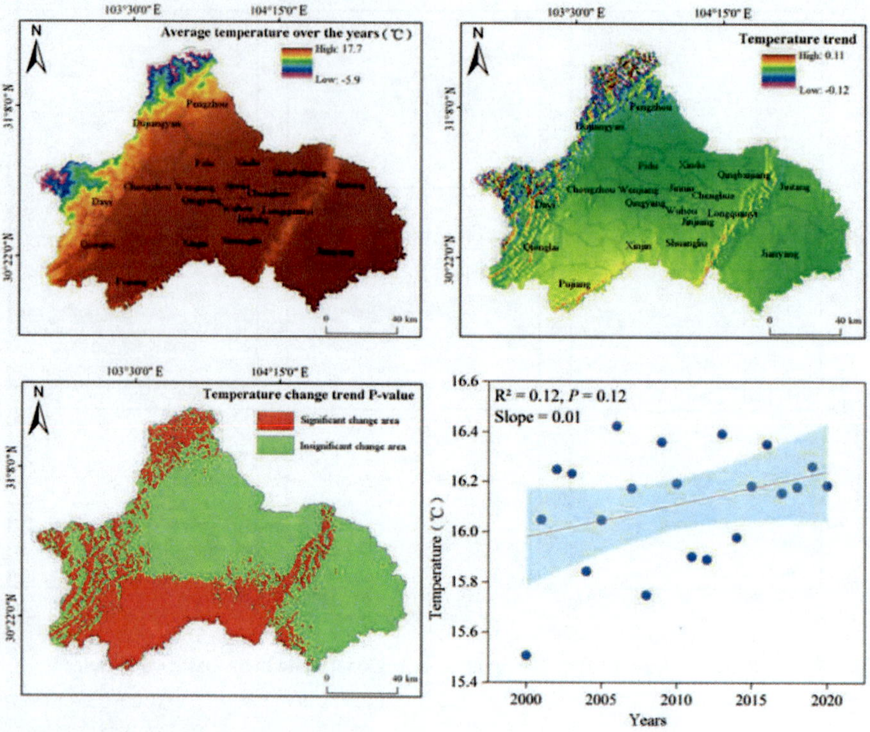

Fig. 5.5 Spatial pattern and variation trend of temperature factors in Chengdu

Precipitation factors (Third Level). Figure 5.6 shows the spatial patterns and trends of precipitation factors in Chengdu. The average multi-year precipitation in Chengdu from 2000 to 2020 is 1072.5 mm, and the average maximum and minimum values are 1578.7 mm and 800.2 mm, respectively, with more precipitation in the southwestern part of Chengdu. In terms of temporal trends, precipitation increases significantly ($R^2 = 0.43$, $P < 0.05$) for each metropolis with a Slope of 10.63 mm/year. the regions where precipitation increases significantly are mainly located in the eastern part of the metropolis. the average precipitation in 2006 was the least, 822.5 mm. the average precipitation in 2014 was the most, 1194.6 mm. The precipitation in 97.8% of the regions with significant changes and 2.2% of the regions with insignificant changes.

Artificial impervious surface (Third Level). Spatial patterns and trends of man-made impervious surfaces in Chengdu (see Fig. 5.7). It was found that the man-made impervious surface in Chengdu City showed a significant increasing trend from 2000 to 2018 ($R^2 = 0.99$, $P < 0.05$) with a Slope of 68.30 km^2/year. among them, the impervious surface increased the most in 2017, with 185.20 km^2. This was followed by 2016 with an increase of 102.65 km^2.

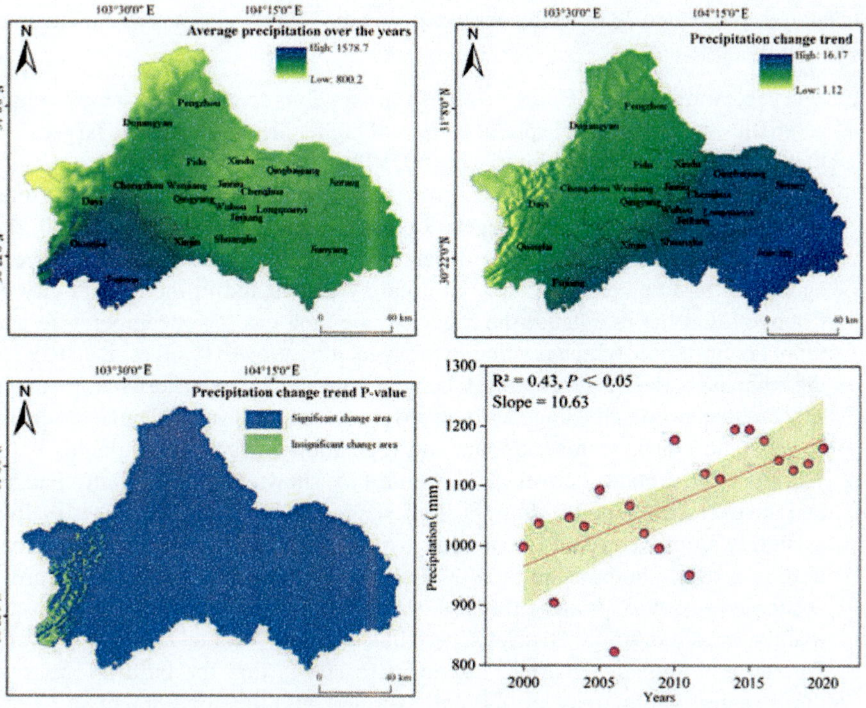

Fig. 5.6 Spatial pattern and variation trend of precipitation factors in Chengdu

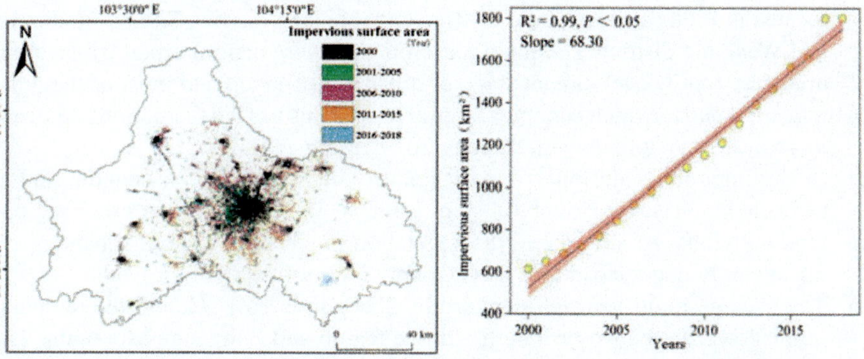

Fig. 5.7 Spatial pattern and changing trend of artificial impervious surface in Chengdu

5.4 Conclusion

Based on the GEE cloud platform, MODIS NDVI remote sensing data were selected for this study, and the dynamic monitoring and driver analysis of vegetation cover in

Chengdu from 2000 to 2021 were conducted, and the following research conclusions were mainly obtained.

(1) The vegetation cover of Chengdu city shows obvious spatial heterogeneity characteristics, and the overall spatial pattern of higher elevation areas is larger than that of plains. The average multi-year NDVI in Chengdu is 0.76, and it is mainly in the middle and high vegetation cover areas, accounting for 57.6% of the total area. Among them, the high vegetation cover areas are mainly located in the high altitude areas in the western part of Chengdu. The low-coverage areas and medium–low-coverage areas are mainly distributed in the central city of Chengdu. On the one hand, the high altitude area has lower temperature and more precipitation, which is suitable for vegetation growth (Goli et al. 2010). On the other hand, the western area of Chengdu is located in the eastern margin of the Tibetan Plateau, and the huge altitude difference and rich vertical band spectrum of vegetation make water abundant and vegetation types more available.
(2) The overall vegetation cover of Chengdu city shows non-significant changes and the overall vegetation shows a slight degradation trend. Chengdu is mainly a slightly improved area, accounting for 32.7% of the total area. The areas with improved annual average NDVI accounted for 40% and degraded areas accounted for 39%. Among them, the areas with significant improvement are mainly located in Jinniu District, Chenghua District, Wuhou District, Qingyang District and Jinjiang District. The main reason is that the built-up areas of these central urban areas are basically formed, and the government started to focus on urban green space coverage area and create urban ecological parks and road vegetation landscapes at multiple points to improve urban vegetation coverage (Fibing 2008; Zhanfu et al. 2020). Severely degraded areas are mainly located in Pidu district, Xindu district, Qingbaijiang district, Shuangliu district and Wenjiang district. The main reason is that these central urban fringe areas are in the rapid development stage of urban construction and most of them are industrial parks, which occupy a large area, and this has led to a serious decrease in surface vegetation cover (Xudong 2019; Wang et al. 2002).
(3) In the regional scale, only the vegetation change trend in Chengdu central urban area and Dayi County is significant, while the change in other regions is not obvious. Among them, the vegetation in Chengdu central urban area is significantly degraded, while Dayi County is significantly improved.
(4) The average multi-year temperature in Chengdu is 16.1 °C, and the temperature is lower in the western high-altitude region and Longquan Mountain. The overall change of temperature in Chengdu is not significant and shows a weak increasing trend. The regions with decreasing temperature are mainly located in the western high-altitude region of Chengdu; the average multi-year precipitation in Chengdu is 1072.5 mm, with more precipitation in the southwestern region. The precipitation of each image element in Chengdu City showed a significant increasing trend. Among them, the regions with greater increase in precipitation are mainly located in the eastern part of Chengdu; the man-made impervious surface in Chengdu shows a significant increasing trend. Among

them, the impervious surface increased the most in 2017, with 185.20 km^2. This will lead to a gradual warming and humidification of Chengdu, which will benefit the vegetation growth in natural ecological areas. However, with the significant increase in urban impervious surface, it also poses a challenge to the ecological environment of the central city of Chengdu. For this reason, it is also crucial to plan urban planning sites scientifically and to include urban ecological health in the assessment of economic development.

References

Baoquan J (2013) Analysis of vegetation change and its causes in Beijing based on TM satellite image data. J Ecol 33(5):1654–1666

Chen E, Zhao XJ (2022) Spatial and temporal variation of vegetation cover in the Bailong River basin from 2010 to 2020. Green Technol 24(10):1–7

Chenghu Z, Weiming C, Jinkai Q et al (2009) Study on the 1:1 million digital landform classification system of China. J Earth Inf Sci 11(6):707–724

Dazhong D, Keming C, Shiqian W et al (2009) Shale gas resource evaluation method and its application in Sichuan basin. Nat Gas Ind 29(5):33–39

Fibing C (2008) Analysis and optimization of landscape patterns of green space systems in the Central City of Chengdu. Southwest University

Goli Z, Ouyang H, Zhang X et al (2010) Vegetation cover change on the Tibetan Plateau and its response to climate change based on ecogeographic zoning. Geogr Studies, 2010, 29(11): 2004–2016.

Gong P, Li X, Wang J et al (2020) Annual maps of global artificial impervious area (GAIA) between 1985 and 2018. Rem Sens Environ 236:111510

Gorelick N, Hancher M, Dixon M et al (2017) Google earth engine: planetary-scale geospatial analysis for everyone. Rem Sens Environ 202(1):18–27

Hao BF, Han XJ, Ma MG et al (2018) Research progress on the application of Google Earth Engine in earth and environmental sciences. Rem Sens Technol Appl 33(4):600–611

Baorong H, Baoyin L, Zhisheng H et al (2018) A comprehensive study on the inflection point of ecological and environmental quality in China. Proc Chin Acad Sci 33(10):1072–1082

Li H-X, Liu G-H, Fu B-J (2011) Response of vegetation growth to climate change and human activities in the Sanjiangyuan area based on NDVI. J Ecol 31(19):5495–5504

Jun-Nan X, Chao P, Wei-Ming C et al (2018) Analysis of vegetation cover change in Yunnan Province based on MODIS-NDVI. J Earth Inf Sci 20(12):1830–1840

Li F, Wang RS, Zhao D (2014) Urban eco-infrastructure based on ecosystem services: status, problems and prospects. J Ecol 34(1):190–200

Luying Z, Wen M, Bo P (2020) Characteristics of climate change in Chengdu Plain. J Chengdu Univ Inf Eng 35(2):179–187

Ma BX, Jing J, Xu Y et al (2021) Spatial and temporal variation of vegetation NPP in Yunnan-Guizhou-Gui karst region from 2000 to 2019 and its relationship with climate change. J Ecol Environ 30(12):2285–2293

Na M, Yunfeng H, Dafang Z et al (2012) Vegetation cover pattern and dynamics in Zhenglan Banner, Inner Mongolia based on remote sensing and image dichotomous model. Geoscience 32(2):251–256

Migliavacca M, Musavi T, Mahecha MD et al (2021) The three major axes of terrestrial ecosystem function. Nature 598(7881):468–472

Pan Q, Zheng H, Wang ZH et al (2021) Advances in the study of plant functional traits on ecosystem services. J Plant Ecol 45(10):1140–1153

Peng S, Ding Y, Liu W et al (2019) 1 km monthly temperature and precipitation dataset for China from 1901 to 2017. Syst Sci Data 11(4):1931–1946

Tamiminia H, Salehi B, Mahdianpari M et al (2020) Google earth engine for geo-big data applications: a meta-analysis and systematic review. ISPRS J Photogram Rem Sens Environ 164:152–170

Tao Z, Hong T (2018) A study of vegetation cover change and urban expansion in Beijing-Tianjin Hebei from 2001–2015 based on google earth engine. Rem Sens Technol Appl 33(4):593–599

Wang BE, Fan GZ, Dong YP et al (2011) Characterization of climate and phenology in Sichuan and Chongqing. Geoscience 31(6):674–681

Wang Z-H, Wang G-H, Wu C-Y (2002) Eco-industrial parks: a strategic choice for sustainable industrial development in China. Sci Technol Herald 20(0212):29–32

Zhongbao X, Jiongxin X, Wei Z (2007) Impacts of climate change and human activities on vegetation cover changes on the Loess Plateau. Chin Sci Ser D 37(11):1504–1514

Xudong S (2019) Practice and innovation of urban master planning in Chengdu. Planner 20:67–74

Xueling G (2004) Study on the service function of vegetation ecosystem and its spatial characteristics in the Qinling Mountains. Northwestern University, Taiyuan

Yi S, Mingguo M (2007) Analysis of vegetation cover changes in northwest China based on SPOT VEGETATION data. China Desert 27(1):89–93

Yilun H, Guoxu J, Jihong L (2022) Interpretation of IPCC AR6 report: terrestrial and freshwater ecosystems and changes in their services. Adv Clim Chang Res 18(4):395–404

Yiqi Y, Tiangui X (2021) Analysis of climate characteristics of Chengdu Plain Economic Zone. Highland Mount Meteorol Res 41(2):143–149

Zhanfu L, Jin Z, Yating L et al (2020) Spatial and temporal evolution of urban greening level in China from 2000 to 2017 and its influencing factors. Geogr Arid Regions 43(2):481–490

Zongxue X, Hao C, Meifang R et al (2020) Progress of urban flooding mechanism and risk assessment in China. Adv Water Sci 31(5):713–724

Chapter 6
Research on High Concentration Sulfate Treatment in Cyanide-Containing Liquid of a Gold Smelting Company

Guanglin Qin, Guangsheng Li, and Xingfu Zhu

Abstract In the lean liquid of a gold smelting company, the concentration of sulfate radicals keeps rising. When the temperature is lowered, sodium sulfate crystals are produced, which affects normal production. In this paper, based on the analysis of the increase of sulfate concentration and the cause of crystallization, through the comparison between the ettringite method and the calcium chloride method, the calcium chloride method can reduce the sulfate concentration in the cyanide-containing liquid to below 30 g/L. To ensure that no crystallization occurs during production in winter, the method is technically feasible and economically reasonable.

Keywords Gold · Sulfate · Smelting · Lime-aluminum method

6.1 Research Status of Sulfate Treatment

At present, there are many methods to remove sulfate radical, such as physical method, chemical method, biological method, ion exchange method, membrane separation method, adsorption method and so on (Siliva et al. 2003; Lv and Huang 2003). Among them, the physical method can be divided into freezing method and brine exclusion method. The basic principle of chemical precipitation method is to add substances that can react with sulfate radicals to form sulfate precipitation, and then achieve the purpose of removing sulfate radicals through separation (Guangsheng et al. 2001; Baumgartner et al. 2006; Michali et al. 2002; Ericsson and Hallmans 1994). There are mainly calcium chloride method, barium chloride method, barium carbonate method, ultra-high lime aluminum method, etc. In nature, sulfur exists in three forms: elemental sulfur, sulfide and (sulfite) sulfate (Hani Bouhabila et al.

G. Qin (✉) · G. Li · X. Zhu
Beneficiation and Mentallurgy Laboratory Branch Company, Shandong Gold Mining Technology Co., Ltd., Jinan, China
e-mail: 49260892@qq.com

J. Zhang et al. (eds.), *Environmental Pollution Governance and Ecological Remediation Technology*, Environmental Science and Engineering,
https://doi.org/10.1007/978-3-031-25284-6_6

2001). The existing biological process for treating high-concentration sulfate wastewater has problems such as long start-up time, slow treatment speed, low efficiency, and large consumption of organic matter. The methods of treating sulfate radicals include ion exchange method, membrane separation technology, adsorption method, etc. (Luan et al. 1992).

6.2 Analysis of the Causes of Elevated Sulfate Radicals

The analysis of the reasons for the increase of sulfate concentration in the cyanide-containing water system is mainly divided into the following three aspects.

6.2.1 Acidification Stage of Neutralizing Liquid

In production, the gold leaching poor solution is an alkaline liquid. In the process of recovering free cyanide, copper, zinc and other ions in the poor solution, the pH value of the liquid needs to be adjusted to about 2.0. In the acidification process of the lean liquid, sulfuric acid is added to the lean liquid, and the hydroxide ions in the neutralized liquid react with the hydrogen ions in the sulfuric acid to form water, and the sulfate radicals are freed in the solution, which increases the sulfate radical concentration.

6.2.2 Acidizing Liquid Neutralization Stage

After the free cyanide, copper, zinc and other ions in the cyanide-containing liquid are recovered, the pH value of the liquid is acidic, and the pH value needs to be adjusted to 11–12 before returning to the production system. In the neutralization stage of the acidified solution, lime is added, hydroxide and hydrogen ions react to form water, and calcium ions react with sulfate to form calcium sulfate precipitation. In the production analysis, it was found that after adding lime to the acidified solution, the sulfate concentration was higher than that in the lean solution before acidification. The main reason is that the generated calcium sulfate is densely attached to the lime surface, preventing the reaction from continuing, so that the sulfate concentration in the liquid cannot continue to decrease.

Table 6.1 Changes in the solubility of sodium sulfate at different temperatures

Temperature/°C	0	10	20	30	40	50	60	70
Solubility/g	4.9	9.1	19.5	40.8	48.8	46.2	45.3	44.3

6.2.3 Alkali Oxidation During Cyanation

In the incoming gold concentrate of the smelting company, both pyrite and a small amount of pyrrhotite will undergo oxidation under alkaline conditions. Among them, the sulfur in pyrite is oxidized to thiosulfate under certain conditions, and further oxidized to become sulfate, which increases the sulfate concentration in the system. On the other hand, orpimentite is rapidly oxidized in the presence of oxygen and water, and microorganisms in the environment accelerate the entire oxidation process due to catalysis, and the sulfur in orpiment is eventually oxidized to sulfate radicals.

6.3 Analysis of the Causes of Sodium Sulfate Crystallization

The solubility changes of sodium sulfate at different temperatures are shown in Table 6.1.

It can be seen from Table 6.1 that the solubility of sodium sulfate varies greatly under different temperature conditions. In the standard state, the solubility of sodium sulfate at 0 °C is 4.9 g, and the concentration of sulfate radicals that produce crystals is 33.13 g/L. Therefore, when the sulfate concentration is reduced to below 33.13 g/L, sodium sulfate crystals will not be generated in the liquid whose temperature is higher than 0 °C.

6.4 Experimental Studies

6.4.1 Acidizing Solution Neutralization Test

When adding lime to the acidified solution to saturation, the amount of lime is 16 kg/m^3. After 20 min of reaction time, the suspension is filtered to measure the sulfate radical content in the filtrate. The test results are shown in Table 6.2.

Table 6.2 Variation of sulfate concentration in acidification solution neutralization test

Sample	Acidifying solution	Neutralizer
Concentration/(g L^{-1})	76.89	47.08

As can be seen from Table 6.2, after the acidification solution was neutralized with lime, the concentration of sulfate radicals decreased from 76.89 g/L to 47.08 g/L, a decrease of 29.81 g/L.

6.4.2 *Experiment on Sulfate Radical Treatment by Ettringite Method*

The test was carried out with the diluted neutralization solution (10 g/L), and the factors such as reaction time and dosage of pharmaceutical agents were explored.

Response time effect test

Ca^{2+}, $Al^{3+}$and $SO_4{}^{2-}$ were added in a molar ratio of 5:3:1, calcium oxide and sodium metaaluminate were added according to the theoretical requirements, and the reaction temperature was 25 °C. The effect of the reaction time on the removal of sulfate radicals was investigated (Fig. 6.1).

It can be seen from Table 6.1 that when the reaction time is greater than 15 min, with the increase of time, the mass concentration of sulfate in the solution after the reaction has no obvious change, and the reaction time is determined to be 15 min.

Effect of lime addition on sulfate removal

Under the condition that the reaction time is 15 min, the reaction temperature is 25 °C, and the molar ratio of aluminum ion and sulfate ion is 3:1, the molar ratio of calcium ion and sulfate group is 1, 2, 3, 4, and 5, respectively. Different lime dosage conditions are investigated. Effect on sulfate removal (Fig. 6.2).

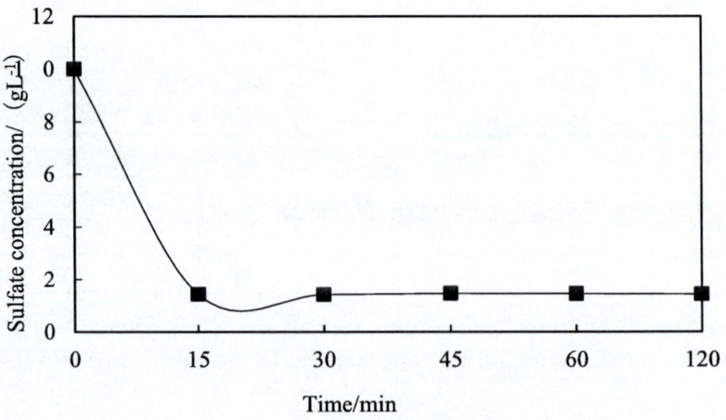

Fig. 6.1 Influence of reaction time on sulfate removal efficiency

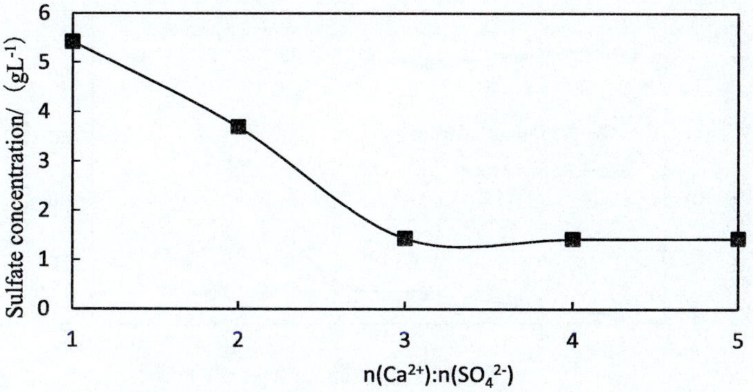

Fig. 6.2 Effect of lime addition on sulfate removal

The effect of sodium metaaluminate addition on the removal effect

Under the above optimal conditions, corresponding amounts of sodium metaaluminate were added according to the molar ratio of aluminum ions to sulfate radicals of 1.0, 1.5, 2.0, 2.5, and 3.0, respectively, and the effects of different dosages on the removal of sulfate radicals were investigated.

As can be seen from Fig. 6.3, when the molar ratio of aluminum ions to sulfate ions is 1, the remaining sulfate concentration in the solution is the lowest.

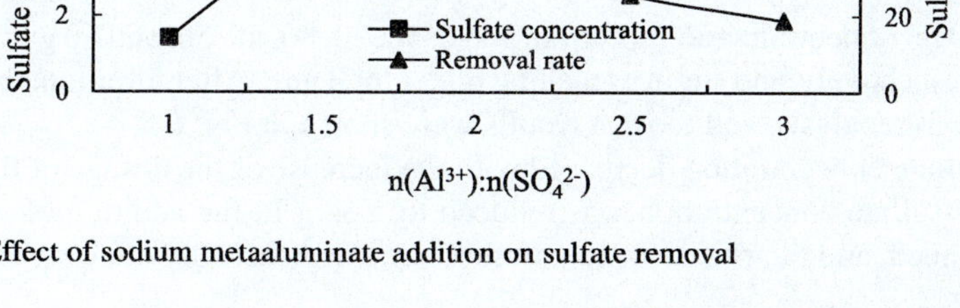

Fig. 6.3 Effect of sodium metaaluminate addition on sulfate removal

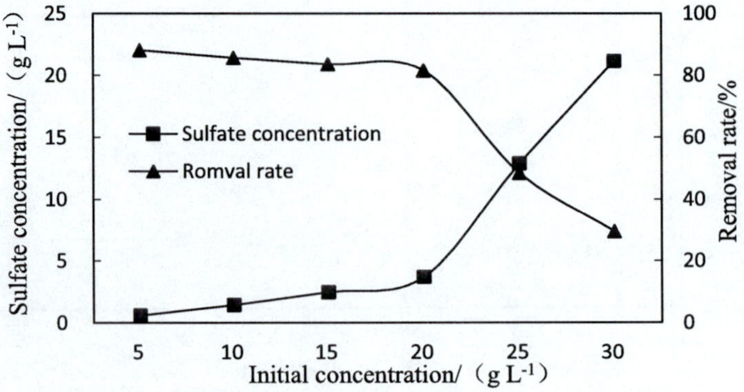

Fig. 6.4 Changes of sulfate concentration and removal rate under different initial concentrations

6.4.3 Calcium Chloride Treatment of Sulfate Radical Test

Influence of sulfate concentration on removal effect

Under optimized conditions, the effect of different initial sulfate concentrations on sulfate removal was investigated. The test results are shown in Fig. 6.4.

It can be seen from Fig. 6.4 that when the sulfate concentration in the liquid is greater than 20 g/L, the removal rate drops significantly. Therefore, the ettringite method is not suitable for treating liquids with sulfate concentration higher than 20 g/L.

6.4.4 Calcium Chloride Treatment of Sulfate Radical Test

Calcium chloride dosage test

Take 1000 ml of neutralization solution, add 0, 10, 20, 30, 40, 50, and 60 g of calcium chloride respectively, and stir in a leaching mixer for 3 min. After filtration, the filtrate was taken for analysis, and the test results were shown in Fig. 6.5.

The sulfate concentration decreased with the increase of the dosage of the agent. When the sulfate concentration was reduced to 5.62 g/L, the addition of medicine was continued, and the sulfate concentration remained basically unchanged.

Response time test

50 g of calcium chloride was added to 1000 ml of neutralization solution, and the reaction time was 1, 3, 5, 7, and 10 min, respectively, to determine the optimal reaction time. The test results are shown in Fig. 6.6.

After adding calcium chloride to the neutralization solution, the reaction speed was faster. Within 1 min reaction time, the sulfate concentration decreased to 8.09 g/L, and the reaction basically stopped after 5 min. Continuing to prolong the reaction

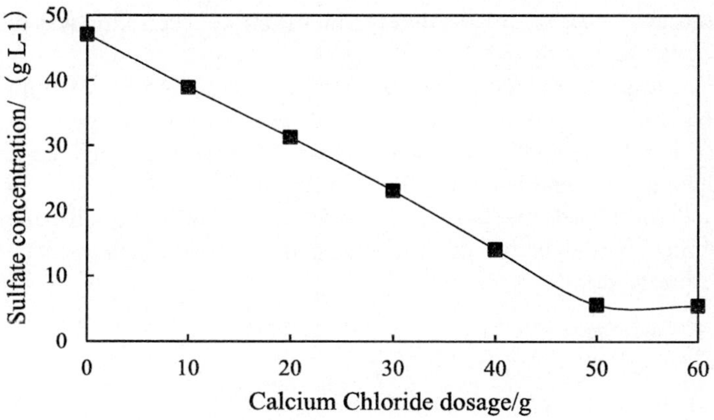

Fig. 6.5 Variation of sulfate concentration with calcium chloride dosage

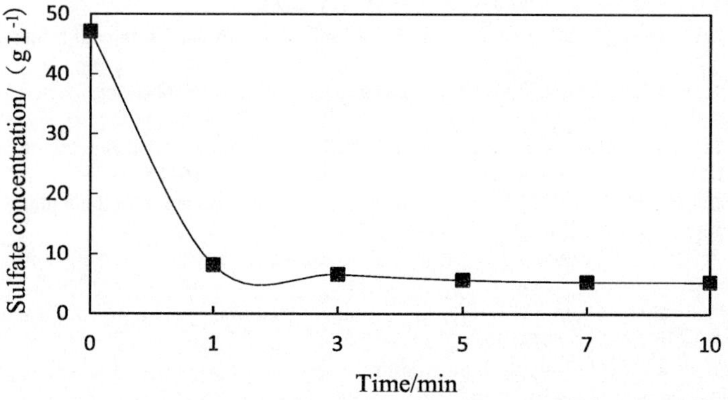

Fig. 6.6 Response time results

time, the sulfate concentration changed little. Therefore, the test reaction time was determined to be 5 min.

6.5 Conclusion

(1) The main product of winter crystallization in the lean liquid of a smelting company is sodium sulfate. The main reason for the crystallization of the product is that the concentration of sulfate in the system is too high. Since the solubility of sodium sulfate decreases with the decrease of temperature, when the temperature decreases in winter, the sulfate radical in the liquid changes from an unsaturated state to a supersaturated state, and sodium sulfate crystallizes out.

(2) By neutralizing the acidified solution, the sulfate concentration was reduced by 29.81 g/L.
(3) In the neutralization liquid treatment test, by comparing the calcium chloride method and the ettringite method, it is determined that the calcium chloride method is feasible and economically reasonable to treat the high concentration of sulfate in the cyanide-containing lean liquid.
(4) The calcium chloride method can reduce the sulfate concentration in the cyanide-containing liquid to below 30 g/L to ensure that no crystallization occurs during winter production.

References

Baumgartner L, Reid R, Dupraz C et al (2006) Sulfate reducing bacteria inmicrbialmats: Changing prradigms, new discoveries. Sediment Geol 185:131–145

Bernt E, Bengt H (1994) Trentment and disposal of saline wastewater from coal mines in Poland, (98):239–248

Bouhabila EH, Aim RB, Here B (2001) Fouling characterization in menbrane bioreactor, 22(5):123–132

Li G, Deng H, Bian X et al (2001) Experimental study and application of freezing crystallization of a gold cyanide liquid sodium sulfate. China Metal Bull (9):156–157

Luan ZK, Tang HX et al (1992) Chemical process of acid mine drainage in the aquatic system of copper mine area. J Environ Sci (China) 4(3):42–48

Lv CY, Huang MY (2003) Study of batch difluoride with model-La cat-ion exchange resin. J Nanhua Univ Med Edn 31(4):386–388

Michali B, Stanislaw K, Klaudia W (2002) Application of membrane techniques in a water softening process. Desalination 145(1/3):321–327

Siliva AJ, Varesche F et al (2003) Sulphate removal from industrial wastewater using apacked-bed anaerobic reactor. Process Biochem

Chapter 7
Wind Climate Analysis in the Yellow Sea and Bohai Sea Based on ERA5 Reanalysis Data

Yue Yu, Yi-Tao Wang, Rui-Ping Yang, Chun-Xin Li, Hai-Ying Liu, Ding-Jiang Wei, and Wei Li

Abstract Sea surface wind is closely related to human life. It is of great significance to fully grasp the characteristics of wind phenology (climate state of wind field) for navigation safety, disaster prevention and reduction and other maritime activities. Based on the ERA5 wind field reanalysis data provided by the European Center for Medium-Range Weather Forecasts (ECMWF), this paper studies the sea surface wind field in Yellow Sea and Bo Hai region. The main analysis includes the spatial and temporal distribution characteristics of monthly mean wind direction and annual mean wind speed, and the spatial and temporal distribution characteristics of monthly mean and annual mean wind frequency (above level 6). It is hoped that it can provide reference for maritime activities such as navigation safety and disaster prevention and mitigation.

Keywords Yellow Sea · Bohai Sea · ERA5 wind field · Wind climate

7.1 Introduction

Yellow sea, Bo Hai is located in the east coast of Eurasia, adjacent to the Korean Peninsula, as the northernmost sea of China, due to its special geographical environment, are prone to produce severe wind weather all year round. The sea surface wind is inseparable from human activities, and the casualties and property losses caused by wind and waves account for a large proportion of all Marine disasters every year. From 1982 to 1990, 14,345 ships of various sizes were capsized due to typhoon and waves in China's coastal waters, with an average annual loss of more than 2600 ships and over 520 people killed or missing. In November 1979, China's bohai Ii sank due to a cold wave. Therefore, it is of great significance to fully grasp

Y. Yu · Y.-T. Wang (✉) · C.-X. Li · H.-Y. Liu · D.-J. Wei · W. Li
Dalian Naval Academy, Liaoning, China
e-mail: wytbmdx@163.com

R.-P. Yang
No. 91776 of PLA, Beijing, China

J. Zhang et al. (eds.), *Environmental Pollution Governance and Ecological Remediation Technology*, Environmental Science and Engineering,
https://doi.org/10.1007/978-3-031-25284-6_7

the characteristics of wind conditions for maritime navigation safety and disaster prevention and mitigation.

Predecessors have done research on the characteristics of the wind in the surrounding sea areas of China, which has played a great role in promoting. The monthly mean spatial and temporal distribution characteristics of sea surface wind speed over the South China Sea are analyzed by using the altimeter data from 1987 to 1988, and compared with the ship data (Qi 1999). The monsoon transition rule of the South China Sea is found. Meanwhile, the mean wind speed of the South China Sea is smaller in the monsoon transition period and larger in the northeast monsoon period (Zhao 2006). Based on the ERS21, ERS22 and QuikSCAT scatterometer sea surface wind field data from 1992 to 2005, the vector empirical orthogonal function decomposition of the data shows that the first mode of anomalous wind field (VEF-1) shows a winter-summer oscillation pattern, which reflects the monsoon characteristics of the Sea surface wind field. The second mode (VEOF-2) shows a spring–autumn oscillation pattern, which mainly reflects the characteristics of sea surface wind field during the transition period between winter and summer monsoon. Judy etc. (Liu and He 2003). Based on the QuikSCAT vector wind data from 1999 to 2002, the monthly variation and spatial distribution characteristics of gale frequency over the South China Sea are analyzed statistically. The results show that the QuikSCAT vector wind data have high reliability after comparing with the data from the South China Sea Island observation station. It is also found that the maximum center of northeast monsoon is located in bashi Channel and Taiwan Strait, and there are the submajor center of northeast monsoon and the maximum center of southwest monsoon in the central and southern part of the South China Sea. Based on CCMP wind field data from July 1987 to December 2009, the seasonal characteristics, wind frequency and extreme wind speed characteristics of sea surface wind field in the surrounding sea areas of China are analyzed (Liu and Zheng 2013). It is found that there are large wind speed regions in spring and summer in the offshore areas of China. The distribution characteristics of gale frequency and annual mean wind speed are roughly consistent, and the large values are mainly distributed in the waters around Taiwan Strait, Luzon Strait and Ryukyu Islands. Zheng (2020, 2021) created the wave climate dataset and wave energy dataset of the Maritime Silk Road for the first time globally, which will make positive contribution to disaster reduction and industrialization of renewable energy.

The resolution of data used in previous studies is low, which is difficult to meet the accuracy required by sea navigation. The time series of used data is short, which is not accurate enough to reflect the characteristics of wind climate state. And the analysis of sea surface wind speed is mostly quarterly, which has little reference significance for the safety of sea navigation in the short and medium term. Based on the ERA5 wind field reanalysis data from January 1991 to December 2020, the spatial and temporal distribution characteristics of wind direction, wind speed and wind frequency of sea surface wind field in Yellow Sea and Bo Hai regions are analyzed, hoping to provide reference for navigation safety and disaster prevention and mitigation activities.

7.2 Data and Methodology

7.2.1 Data Introduction

ERA5 reanalysis data is the fifth generation of atmospheric reanalysis data for global climate by ECMWF. Compared with previous ERA series data, ERA5 has better applicability and higher correlation coefficient (Lu and Li 2021) At the same time, the wind direction and speed data have good matching under typhoon condition (Li and Liu 2021), has been widely used at home and abroad (Zhang 2011). The spatial range of ERA5 data is 90 °S–90 °N, 180 °W–180 °E; The spatial resolution is 0.25° × 0.25°, and the time series is 00:00 from January 1, 1950 to the present, which is constantly updated. The data used in this paper is ERA5 10 m wind field on the sea surface (U and V directions), with a spatial resolution of 0.25° × 0.25°.The time resolution is 6 h by 6 h. The time series is From January 1991 to December 2020.

7.2.2 Methodology

The processing method of wind direction and speed is monthly mean wind speed superimposed UV vector map and annual mean wind speed. Due to the high resolution of UV wind field data, the density of the vector map leads to poor display effect. Therefore, the data resolution of the UV vector map used in this paper is 0.5° × 0.5°. The processing method of gale frequency is to count the frequency of monthly mean wind speed and annual mean wind speed over level 6.

7.3 Wind Climate Analysis

7.3.1 Monthly Mean Wind Direction and Wind Speed

The monthly wind direction and speed of the sea surface in Yellow Sea and Bo Hai were averaged for many years, and the distribution characteristics of the monthly wind direction and speed were analyzed according to the seasons.

Spring (March, April, May) is the transition season of the monsoon, the wind speed is smaller. The prevailing wind in Bo Hai is mainly SW wind, and the average wind speed is less than 3 m/s. The average wind speed showed an increasing trend, and the large value of the average wind speed was concentrated in Liaodong Bay, which was caused by the topographic effect. The prevailing wind direction changed from W-WN wind to SW wind in the northern part of Yellow Sea, and from N-WN wind to southerly wind in the middle and southern part of Yellow Sea. The average wind speed is within 3 m/s, and the relatively low value area of the average wind speed gradually moves from northwest to southeast.

In March (Fig. 1a), the wind direction in Bo Hai was unstable and the wind speed was small, with the average wind speed within 1 m/s. W-WN wind was dominant in the northern part of Yellow Sea, and N-WN wind was dominant in the central and southern part of Yellow Sea, with an average wind speed of 0.5–3 m/s and an increasing trend from west to east. April (Fig. 1b) is the representative month of spring. SWSW wind is dominant in The Bo Hai, and the average wind speed is between 1 and 2.5 m/s. The maximum average wind speed is in Liaodong Bay, Followed by Bohai Bay and Laizhou Bay. SW wind prevails in the northern part of Yellow Sea, with an average wind speed of 1.5–2.5 m /s. There is a relatively large value area in the sea area east of chengshan Mountain. S-SW wind prevailed in the central and southern part of Yellow Sea, and the average wind speed showed a relatively low value area, with a decreasing trend from west to east. The average wind speed was smaller than that in the northern Part of Yellow Sea, ranging from 0 to 2.5 m/s. In May (Fig. 1c), SW wind prevails in Bo Hai, and the liaodong Bay is the area with high wind speed, with the average wind speed up to 3 m/s. The prevailing south wind over Yellow Sea, the average wind speed is 2–3 m/s. There is a large area with average wind speed of 2.5–3 m /s on the west side of the southern Yellow Sea.

In summer (June, July and August), s-SE winds prevail over Yellow Sea and Bo Hai. In summer, the wind speed was small and the average wind speed changed little. The average wind speed in Bo Hai was 0.5–3.5 m/s, and the large values were concentrated in Liaodong Bay, Bohai Bay and Laizhou Bay. The mean wind speed of Yellow Sea is concentrated in the west of the south of Yellow Sea. The contour distribution changes from east to west to south to north, which is related to the movement trend of the subtropical high system controlling Yellow Sea and Bo Hai.

In June (Fig. 2a), Yellow Sea and Bo Hai are affected by summer monsoon, mainly southerly wind. The average wind speed in the Bo Hai was 2–2.5 m/s, and in Yellow Sea was 0.5–3.5 m/s, and the wind speed decreased from west to east.

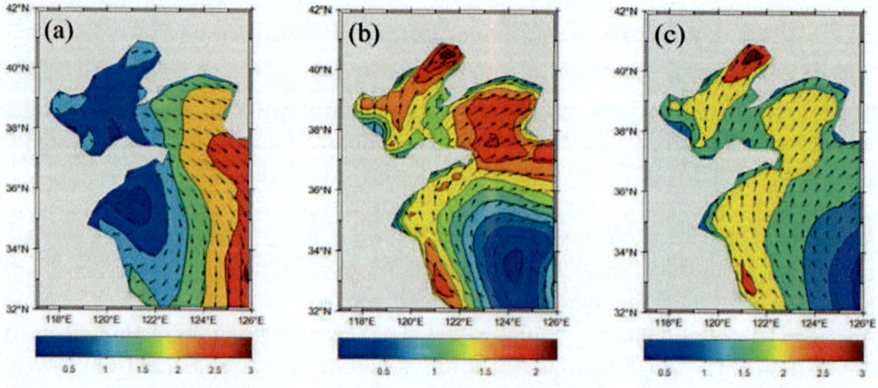

Fig. 7.1 Average wind direction and speed in March, April and May in Yellow Sea and Bo Hai in recent 30 years, unit: m/s

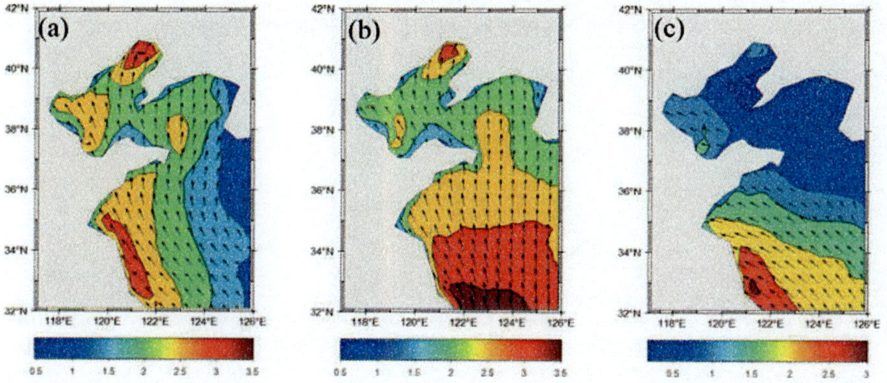

Fig. 7.2 Average wind direction and speed in June, July and August in Yellow Sea and Bo Hai in recent 30 years, unit: m/s

July (Fig. 2b) is the representative month of summer, which is strongly influenced by summer monsoon. S-wind prevails in Yellow Sea and Bo Hai, and the average wind speed is the highest in summer, ranging from 2–3.5 m/S. The decreasing trend of mean wind speed over Yellow Sea changed from east–west to north–south. In August (Fig. 2c), S–SE winds prevail in Yellow Sea and Bo Hai due to the influence of monsoon. The average wind speed in the Bo Hai and the northern part of Yellow Sea decreased significantly, and was less than 1 m/s, while it was 1–3 m /s in the central and southern part of Yellow Sea.

In autumn (September, October and November), the Siberian high gradually recovered, and the wind direction began to change to winter wind. N-EN wind prevailed over Yellow Sea and Bo Hai, and the wind speed gradually increased. The average wind speed of the Bo Hai is less than 2.5 m/s, and the wind speed of the northerly wind in the Bo Hai area is often less than that of the southerly wind because of the topography. The average wind speed over Yellow Sea ranged from 1 to 4.5 m/s, and the maximum value of wind speed was concentrated in the middle and south of Yellow Sea. The contour lines gradually shift from north to south to east to west.

In September (Fig. 3a), the wind direction in Bo Hai is uncertain, and the average wind speed is less than 1 m/s. After the wind direction turns to the north, the large value area of wind speed corresponding to summer wind disappears. N-EN wind prevailed over Yellow Sea, and the wind speed in the northern part of Yellow Sea was roughly the same as that in the Bo Hai. Compared with the northern part, the wind speed in the central and southern parts was 1.5–3.5 m /s higher than that in the northern part of Yellow Sea. The wind speed showed a decreasing trend from south to north. October (Fig. 3b) is the representative month of autumn. WN wind prevails in Bohai Sea, and the average wind speed changes little, less than 1.5 m/s. Compared with September, the distribution of wind direction and speed in the Yellow Sea did not change much, and the average wind speed increased slightly in the range of 1.5–4.5 m/s. In November (Fig. 3c), WN wind prevailed over The Bo Hai, and the average

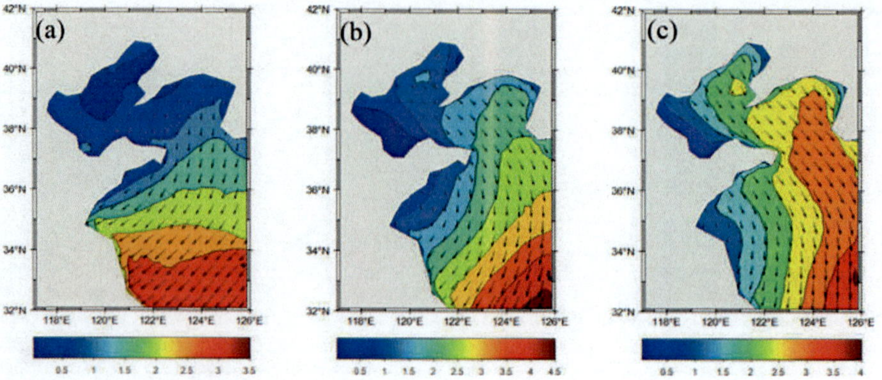

Fig. 7.3 Average wind direction and speed in September, October and November in Yellow Sea and Bo Hai in recent 30 years, unit: m/s

wind speed increased to 0.5–2.5 m/s. There was a large area of relative wind speed in the south of Liaodong Bay. WN wind prevailed in the northern part of Yellow Sea, and the distribution of average wind speed did not change, and the wind speed increased to 2.5–3 m/s. The N wind prevailed in the central and southern part of Yellow Sea, and the maximum area of wind speed gradually moved southward, and the distribution of wind speed gradually changed from north to south to east to west.

Winter (December, January and February) is the monsoon season, with frequent cold air and n-WN wind prevailing in the Yellow and Bo Hai. The distribution of contour lines is east–west, and the maximum wind speed is concentrated in the eastern part of Yellow Sea, with the average wind speed up to 6 m/s.

In December (Fig. 4a), WN wind prevails over Bo Hai, with an average wind speed of 1–3 m /s. The wind direction in Yellow Sea is the same as that in the Bo Hai, and the wind speed is larger, ranging from 2 to 5.5 m/s. The overall distribution is east–west, the maximum value is located in the southeast of Yellow Sea, and the wind speed can reach 5.5 m/s. January (Fig. 4b) is the representative month in winter. The WN wind continues to prevail in the Yellow and Bo Hai under the influence of the cold air moving southward. The variation of wind speed in Bo Hai is small, ranging from 1 to 3 m/s. The wind speed in Yellow Sea increased slightly, from 3 to 6 m/s. In February (Fig. 4c) WE wind continues to prevail in The Yellow and Bo Hais. The wind speed in the Bo Hai decreased to 1–2 m /s due to the gradual weakening of the cold air southward, and the wind speed in Yellow Sea decreased slightly to 3–5 m/s.

7.3.2 Annual Mean Wind Speed

The annual average sea surface wind speed in Yellow Sea and Bo Hai area was processed for many years, and the distribution characteristics of annual average sea surface wind speed were analyzed.

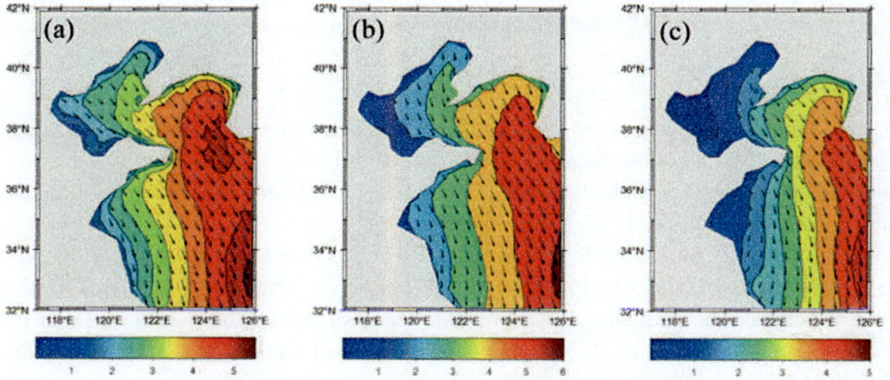

Fig. 7.4 Average wind direction and speed in December, January and February in Yellow Sea and Bo Hai in recent 30 years, unit: m/s

As shown in Fig. 7.5, the annual average wind speed in Bo Hai is less than 1 m/s. The maximum average wind speed is distributed in Liaodong Bay, and the sub-maximum area is Laizhou Bay and Bohai Bay, which is caused by strong southwest wind due to topography. The maximum annual mean wind speed of Yellow Sea is distributed in the southeast of Yellow Sea, and the average annual wind speed of the maximum center can reach more than 2 m/s. The sub-maximum area was in the middle and southwest of Yellow Sea, and the annual average wind speed was above 1.5 m/s. The mean wind speed shows a zonal distribution from west to east and increases gradually.

Fig. 7.5 Average sea surface wind speed in Yellow Sea and Bo Hai in recent 30 years, unit: m/s

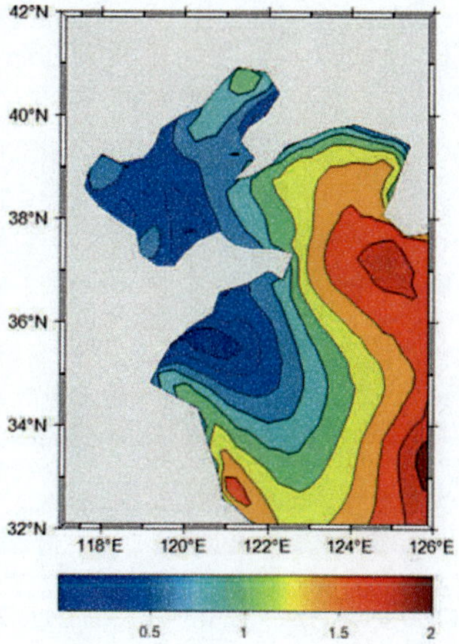

7.4 Gale Frequency Characteristics

The frequency of strong wind in yellow sea and Bo Hai is 5–15% in spring. Bo Hai is 5–8%, and liaodong Bay is the largest region. The sub-large values are Laizhou Bay and Bohai Bay. In Yellow Sea, the wind frequency is 5–15%. The wind frequency basically shows east–west belt distribution, and the maximum value is in the southeast of Yellow Sea, which can reach about 18%.

In March (Fig. 6a), the gale frequency in Bo Hai is 2–7%, and that in Yellow Sea is 5–18%. Figure 6b of April has roughly the same distribution pattern of isolines, with a decrease in frequency. The gale frequency of Bo Hai is 1–6%, and that of Yellow Sea is 4–10%.The frequency of may (Fig. 6c) continued to decrease, with the Bo Hai gale frequency being 1–5%, and Yellow Sea gale frequency being 2–5%.

The frequency of strong winds in Yellow Sea and Bo Hai decreased to 0.5–7% in summer. The frequency of strong wind in Bo Hai is 0.5–3%, and that in Yellow Sea is 1–7%. Compared with spring, the large value area is enlarged, and the contour distribution gradually changes from east to west to south to north.

In June (Fig. 7a), the wind frequency of Bo Hai is within 1.5%, and that of Yellow Sea is 1–3.5%. In July (Fig. 7.7b), the wind frequency of Bo Hai is within 2%, and that of Yellow Sea is 1.5–5%. In August (Fig. 7c), the gale frequency increased to 1–3% in Bo Hai and 2–7% in Yellow Sea.

The frequency of strong winds in the yellow and Bo Hais increases in autumn. The gale frequency in Bo Hai is 2–17%.The frequency of strong wind in yellow Sea is 3–17%, and the contour gradually changes from north to south to east to west, and the maximum value area gradually moves west, up to 17%.

In September (Fig. 8a), the gale frequency of Bo Hai is 1–8%, and that of Yellow Sea is 2–10%.In October (Fig. 8b), Bo Hai was affected by autumn gale, and the frequency of gale increased significantly, up to 12%.In Yellow Sea, the frequency of strong wind is 4–11%. In November (Fig. 8c), the Bo Hai continued to be affected

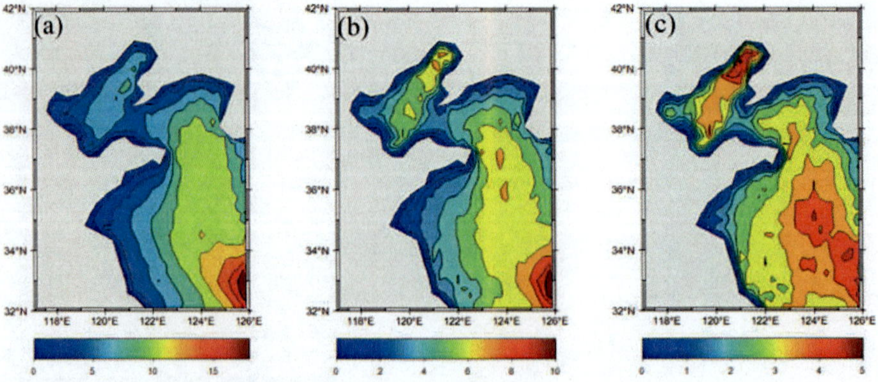

Fig. 7.6 Average frequency of winds above Force 6 in March, April and May in Yellow Sea and Bo Hai in recent 30 years, unit: %

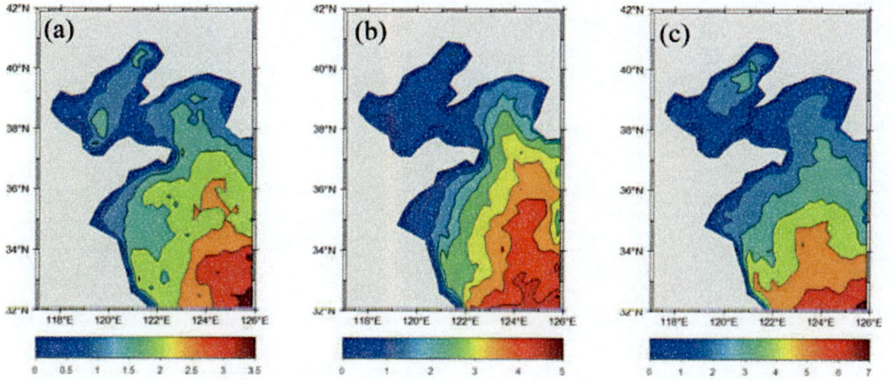

Fig. 7.7 Average frequency of winds above Force 6 in June, July and August in Yellow Sea and Bo Hai in recent 30 years, unit: %

by the autumn gale, and the gale frequency was up to 18%. The gale frequency in Yellow Sea also continued to rise, which was 6–18%.

Affected by cold air, the frequency of strong winds in Yellow Sea and Bo Hai increases significantly in winter. Bo Hai gale frequency is 2–15%, yellow Sea gale frequency is 5–25%.The large value region is basically unchanged, and the contour line continues to be distributed east–west.

In December (Fig. 9a), the frequency of strong wind in Bo Hai is 2–15%, and that in Yellow Sea is 5–25%. In January (Fig. 9b), the gale frequency in Bo Hai decreased to 2–10%, while the gale frequency in Yellow Sea barely changed. Compared with last month, the frequency of Bo Hai gale in February (Fig. 9c) was basically unchanged at 2–10%, and the frequency range of Yellow Sea gale was basically unchanged, but the gradient of isoline distribution was more obvious.

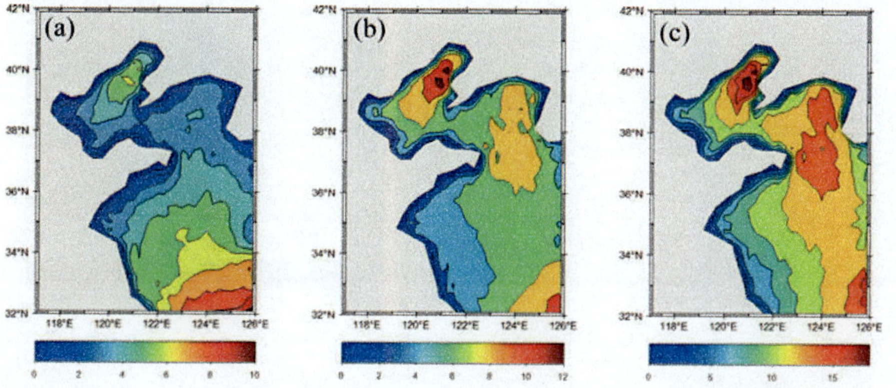

Fig. 7.8 Average frequency of winds above Force 6 in September, October and November in Yellow Sea and Bo Hai in recent 30 years, unit: %

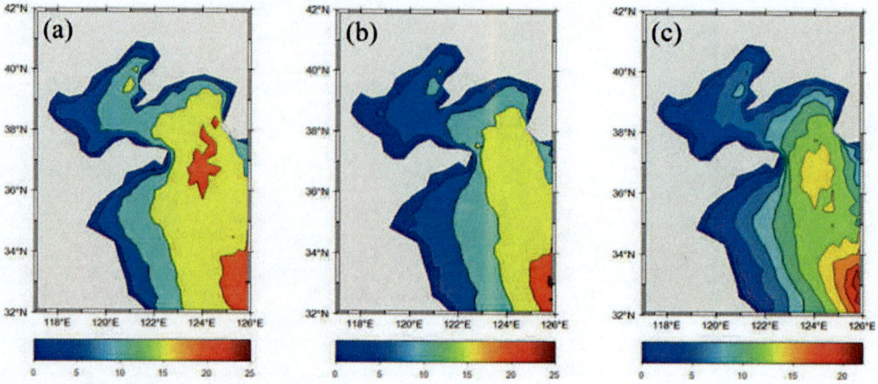

Fig. 7.9 Average frequency of winds above Force 6 in December, January and February in Yellow Sea and Bo Hai in recent 30 years, unit: %

As shown in Fig. 7.10, the distribution of gale frequency is roughly the same as that of annual mean wind speed. The frequency of gales in Bo Hai is 1–8%, and the maximum value area is Liaodong Bay. The frequency of strong wind in Yellow Sea is 4–12%, and the maximum value is in the southeast of Yellow Sea. The contour lines are distributed east–west.

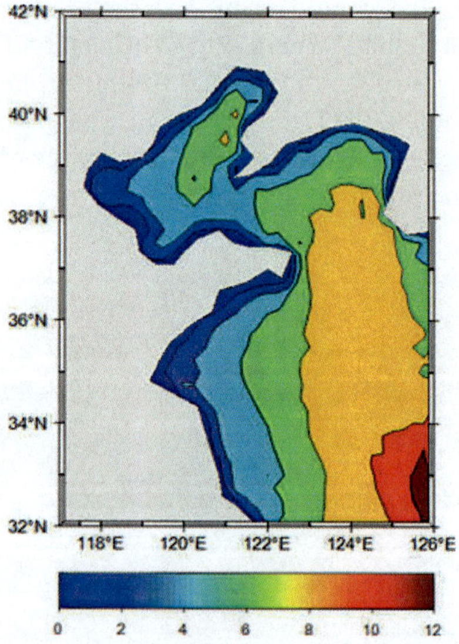

Fig. 7.10 Average frequency of winds above Force 6 in Yellow Sea and Bo Hai in recent 30 years, unit: %

7.5 Conclusions

7.5.1 Wind Climate Characteristics

Wind speed is smaller in March, April and May. The prevailing wind direction in Bo Hai is mainly SW wind, and the average wind speed is less than 3 m/s. The average wind speed showed an increasing trend, with the maximum value of 3 m/s in Liaodong Bay and the second maximum value in Laizhou Bay and Bohai Bay. The prevailing wind direction changed from W-WN wind to SW wind in the northern part of Yellow Sea, and from N–WN wind to southerly wind in the middle and southern part of Yellow Sea. The average wind speed is within 3 m/s, and the relatively low value area of the average wind speed gradually moves from northwest to southeast.

S-SE winds prevail in June, July and August over Yellow Sea and Bo Hai, and the wind speed is small and the average wind speed has little change. The large value area of wind speed in Bo Hai has no change. The maximum area of wind speed in Yellow Sea is concentrated in the south west of Yellow Sea. The contour distribution gradually changes from east to west to south to north.

In September, October and November, N-EN winds prevail over Yellow Sea and Bo Hai, and the wind speed increases gradually. The average wind speed in Bo Hai is less than 2.5 m/s. The average wind speed over Yellow Sea ranges from 1 to 4.5 m/s, and the maximum value of wind speed is concentrated in the middle and south of Yellow Sea, which can reach 4.5 m/s. The contour lines gradually shift from north to south to east to west.

N-WN winds prevail in December, January and February in Yellow Sea and Bo Hai, and the maximum average wind speed in January is 1–6 m/s. The distribution of contour lines is east–west, and the maximum wind speed is concentrated in the eastern part of Yellow Sea, with the average wind speed up to 6 m/s.

The average annual wind speed was less than 1 m/s in The Bo Hai. The maximum average wind speed was distributed in Liaodong Bay with an average wind speed of about 1 m/s, and the second maximum was located in Laizhou Bay and Bohai Bay. The maximum annual mean wind speed of Yellow Sea is distributed in the southeast of Yellow Sea, and the average annual wind speed of the maximum center can reach more than 2 m/s. The sub-maximum area was in the middle and southwest of Yellow Sea, and the annual average wind speed was above 1.5 m/s. Contour lines are distributed in east–west bands and gradually increase from west to east.

7.5.2 Gale Frequency Characteristics

The frequency of strong wind in March, April and May is 5–15% in yellow sea and Bo Hai. Bo Hai is 5% ~ 8%, and liaodong Bay is the largest region. The sub-large values are Laizhou Bay and Bohai Bay. In Yellow Sea, the wind frequency is 5–15%.

The wind frequency basically shows east–west belt distribution, and the maximum value is in the southeast of Yellow Sea, which can reach about 18%.

In June, July and August, the frequency of strong winds in yellow Sea and Bo Hai decreased to 0.5–7%.The frequency of strong wind in Bo Hai is 0.5–3%, and that in Yellow Sea is 1–7%. Compared with spring, the large value area is enlarged, and the contour distribution gradually changes from east to west to south to north.

In September, October and November, the frequency of strong winds in yellow sea and Bo Hai increased. The gale frequency in Bo Hai is 2–17%.The frequency of strong wind in yellow Sea is 3–17%, and the contour gradually changes from north to south to east to west, and the maximum value area gradually moves west, up to 17%.

The frequency of strong winds in December, January and February increased significantly. Bo Hai gale frequency is 2–15%, yellow Sea gale frequency is 5–25%.The large value region is basically unchanged, and the contour line continues to be distributed east–west.

The wind frequency distribution is roughly the same as the annual mean wind speed distribution. The frequency of gales in Bo Hai is 1–8%, and the maximum value area is Liaodong Bay. The frequency of strong wind in Yellow Sea is 4–12%, and the maximum value is in the southeast of Yellow Sea. The contour lines are distributed east–west.

7.5.3 *Significance*

The paper mainly analyzes the distribution characteristics of wind direction, speed and gale frequency in Yellow Sea and Bo Hai, and obtains specific conclusions, which have certain reference significance for navigation safety, disaster prevention and reduction and other maritime activities.

References

Liu C, He X (2003) Statistical characteristics of QuikSCAT vector Wind and remote sensing analysis of strong winds in the South China Sea. J Trop Meteorol (S1):107–117

Li A, Ze L (2021) Applicability of the ERA5 reanalysis data to China adjacent Sea under typhoon condition. Marine Sci (in Chinese) 45(10):71–80

Liu T, Zheng C (2013) Analysis of seasonal characteristics, Wind Frequency and Extreme Wind Speed characteristics of Sea Surface Wind field around China. J Yanbian Univ (Nat Sci Edn) 33(02)6:148–152. https://doi.org/10.16379/j.carolcarrollnki. ISSN 1004-4353.2013.02.017

Lu R, Li X (2021) Comparative study on the applicability of ERA-Interim and ERA5 reanalysis data in Jiangsu Province. Marine Forecast 38(04):27–37.

Qi Y, Shi P (1999) Analysis of monthly mean characteristics of wind and wave in south China Sea using satellite altimeter data. Trop Oceanogr (02):90–96

Zhang D, Zheng C, Chen X (2011) Study on the wind field over China from 1999 to 2009 based on QuikSCAT/NCEP wind field. Marineforecast 28(04):58–64

Zhao X (2006) Study on wind and wave algorithm of China Sea Scatterometer and analysis of spatio-temporal characteristics of sea surface wind field and effective wave height. Graduate University of Chinese Academy of Sciences (Institute of Oceanology)

Zheng CW (2021) Global oceanic wave energy resource dataset—with the Maritime Silk Road as a case study. Renew Energy 169:843–854

Zheng CW, Zhang G (2020) Establishment of wave climate datasets: case study for the Maritime Silk Road. In: 2020 6th international conference on big data and information analytics (BigDIA), pp 451–456. https://doi.org/10.1109/BigDIA51454.00080

Chapter 8
Effects of Different Irrigation Methods on Vegetation Restoration of Mine Abandoned Land in Semi-Arid Grassland

Hao Rong, Abias, Nan Ge, and Henglu Miao

Abstract In the process of vegetation restoration in grassland wasteland, the changes of total plant community coverage, above-ground biomass, soil moisture and water use efficiency on the slopes of dumps under different irrigation treatments were studied through comparative experiments, in order to improve vegetation restoration in grassland mining wasteland. Irrigation technology provides scientific basis. The research results show that with the increase of water deficit in the process of vegetation restoration on the slopes of abandoned land in grassland mining areas, plant growth and development are limited, the coverage decreased significantly ($P < 0.05$). The soil moisture content of 0–30 cm in different periods of the mixed forage growing season basically showed a variation law that increased with the increase of rainfall and irrigation water. The average soil moisture content of RDI3, RDI4, RDI5 treatments were 23.1, 32.3 and 41.2% lower than that of the fully watered treatment (CK). Compared with each irrigation treatment level, the water use efficiency of mild and moderate deficient irrigation (when the soil moisture content reaches 55% and 45% of the soil field water holding capacity, respectively) was higher.

Keywords Abandoned land · Irrigation amount · Water use efficiency

8.1 Introduction

The arid and semi-arid grassland area is an important ecological security barrier in northern my country, and it is also one of the key areas for the main distribution and output of coal resources in my country (Feng et al. 2020).Under the background of climate change, coupled with the disturbance of grazing and other human activities, the vegetation in this area is seriously degraded, and the ecosystem is relatively fragile (Riley 1995; Zhou et al. 1982).With the continuous increase in the scale and intensity

H. Rong (✉) · Abias · N. Ge · H. Miao
Institute of Water Resources for Pastoral Area Ministry of Water Resources, Hohhot 010020, Inner Mongolia, China
e-mail: mksrh@126.com

J. Zhang et al. (eds.), *Environmental Pollution Governance and Ecological Remediation Technology*, Environmental Science and Engineering,
https://doi.org/10.1007/978-3-031-25284-6_8

of mineral resource mining in the region, the area of loose accumulation wasteland formed by resource mining has also continued to increase, and the contradiction between mineral resource development and grassland environmental protection has become increasingly acute. Due to its unique formation method, the soil structure of mine wasteland is different from that of ordinary soil. It has the characteristics of loose soil structure, non-layered soil, low effective water content, and poor erosion resistance (Gang et al. 2018). The average annual precipitation in arid and semi-arid regions is low and the effectiveness of natural rainfall is poor. Soil moisture is the biggest limiting factor for land reclamation and ecological restoration in the region (Xiaoyun et al. 2018). During the mining and utilization of large-scale coal mines in the grassland area, the water resources available for ecological restoration of the mining area are very limited, and the lack of effective water supply in the vegetation restoration of the abandoned mine land makes it extremely difficult to restore the vegetation in the abandoned land, especially its slopes. The scientific method of water-saving irrigation on the slope is an effective way to solve the shortage of soil moisture on the slope of the abandoned land in the mining area. Compared with flat-land irrigation, slope-surface irrigation can easily lead to uneven distribution of water, and excessive irrigation can also cause soil erosion on slopes and waste of water resources (). In this study, aiming at the ecological construction problems of difficult vegetation restoration and serious soil erosion on the slopes of abandoned land in grassland mining areas, based on the water balance equation, the growth status and water use efficiency of mixed planting of leguminous and gramineous herbs on the slopes of abandoned land under different water replenishment treatments were set up. The comparative experiment of this paper provides a scientific basis for the sustainable utilization of water resources and the improvement of ecological restoration technology in the construction of green mines in grassland areas.

8.2 Materials and Methods

8.2.1 Overview of the Test Area

The test site is located in the dump site in the southeast of the open-pit coal mining area of Datang International Shengli East No. 2, Xilinhot City, Inner Mongolia. The dump site is an artificially reshaped pile topography formed by the overburden (the overburden thickness is about 40 cm) after the coal gangue is discharged. It is composed of platforms and slopes alternately in the shape of steps, and the height of each step is about 25 m. Its soil texture is mainly sandy soil, the average soil bulk density of 0–40 cm soil layer is 1.45 g/cm^3, and the field water holding capacity is 23.9% (volume water content). The test area belongs to the arid and semi-arid continental climate in the middle temperate zone. The average annual precipitation is 294.9 mm, and the precipitation is concentrated in June to September. The average temperature for many years is 1.7 °C, and the average annual evaporation is 2350.2 mm.

8.2.2 Experimental Design

In this study, two commonly cultivated herbs in semi-arid areas, *Medicago sativa* and *Agropyron cristatum* were selected as test materials. The experimental plots were arranged on the slope of the dump (33° slope). The width of each plot was 3 m, and the length of the plot was 15 m. The irrigation adopted the ground drip irrigation method controlled by the intelligent irrigation system. Legumes and grasses were mixed according to the local planting mode. The ratio is 1:1, the row spacing is 25-–30 cm, and the sowing depth is 2 cm. Experiments were conducted in the second year of the mixed forage growth.

The experiment was set up with 6 lower irrigation treatment levels (Table 8.1), and each treatment was repeated 3 times, with a total of 18 experimental plots. Irrigation time chooses three key stages for the growth of mixed grass, namely the initial growth stage (greening stage to branching stage, April 20–May 10), rapid growth stage (branching stage to budding stage, May 25—6 has 20 days), vigorous growth period (budding period to flowering period, July 10–August May). During the test, when the soil water content of each treatment dropped to the lower design water limit of the soil water content in each 30 cm soil layer, irrigation was carried out until the relative soil water content reached the upper limit (75%) of the field water holding capacity.

Table 8.1 The upper limit of irrigation water for the trial of adjusted deficit irrigation for vegetation restoration in abandoned land (unit: %)

Name	Water treatment (% of field water capacity)			Explanation
	Early stage of growth	Rapid growth period	Vigorous growth period	
CK	75	75	75	Contrast, fully irrigated
RDI1	65	65	65	Treatment 1, micro-adjustment deficit irrigation
RDI2	55	55	55	Treatment 2, slight deficit irrigation
RDI 3	45	45	45	Treatment 3, moderate deficit irrigation
RDI 4	35	35	35	Treatment 4, heavy deficit irrigation
RDI 5	25	25	25	Treatment 5, Intensity deficit irrigation

8.2.3 Test Method

Total vegetation coverage

In 2021, during the period of vigorous plant growth (early August), the coverage frame method was used to randomly select 3 plots (1 m × 1 m) in each experimental plot to determine the total vegetation coverage in each plot.

The dry matter mass of vegetation on the ground was measured by cutting

At the end of each growing season, 3 (1 m × 1 m) quad-rats were randomly selected, cut on the ground, and the dead branches and leaves were removed, and weighed after drying.

Soil moisture content

The soil moisture uses cloud smart soil moisture sensor to monitor the soil moisture content of the 0–30 cm soil layer every day, and the data is measured every 10 cm.

Calculation of water consumption

The water consumption of plants in each growth stage is calculated using the water balance equation (Xiaoqian et al. 2018), and the calculation formula is:

$$ET = W_0 - W_t + W_T + P_0 + K + M$$

In the formula: ET is the water consumption in a certain period of time (mm); W_0 and W_t are the water storage amount in the planned soil layer in a certain period of time (mm); W_T is the increased water amount due to the change of the planned wet layer (mm); P_0 is the effective Rainfall (mm); K is groundwater recharge (mm), there is no influence of groundwater in this experiment, so $K = 0$; M is the amount of irrigation in a certain period (mm).

Water Use Efficiency

Water Use Efficiency (WUE) = Plant Yield (Y)/Water Consumption (ET).

8.3 Results and Analysis

8.3.1 Effects of Different Regulated Deficit Irrigation Treatments on Vegetation Characteristics

Comparing the changes of aboveground biomass of vegetation under fully irrigated (CK) and five regulated deficit irrigation, it can be seen from Fig. 8.1 that the order of aboveground biomass from large to small is CK > RDI1 > RDI2 > RDI3 > RDI4

> RDI5. Significant difference was found between CK, RDI1, RDI2, RDI3 and RDI4, RDI5 ($p < 0.05$). With the increase of water deficit, the aboveground biomass of vegetation showed a downward trend, and when the soil moisture was lower than 45% of the field capacity, the aboveground biomass decreased significantly, indicating that the soil available water content decreased at this time, and the plants began to suffer from drought stress., in order to absorb soil water and improve plant drought resistance, the proportion of photosynthetic products allocated to plant roots increased, resulting in a decrease in the proportion of aboveground parts.

The results of analyzing the changes in the total coverage of the plant community in the experimental plots under different regulated deficit irrigation treatments showed that (Fig. 8.2), with the increase of the water deficit, the total coverage of the plant community gradually decreased, and the total coverage of the plant community from large to small was CK > RDI1 > RDI2 > RDI3 > RDI4 > RDI5, RDI3, RDI4, RDI5 were reduced by 4.0%, 24.7%, 32.8% compared with full irrigation (CK), respectively. The results of variance analysis showed that CK, RDI1, RDI2, RDI3 and RDI4, RDI5 The difference in the total coverage of the plant community between the two groups reached a significant level ($p < 0.05$), which was consistent with the changes in the aboveground biomass of vegetation under different irrigation treatments, indicating that the moderate or more water deficit had affected the absorption and utilization of water by the mixed forages. This affects the normal growth of plants.

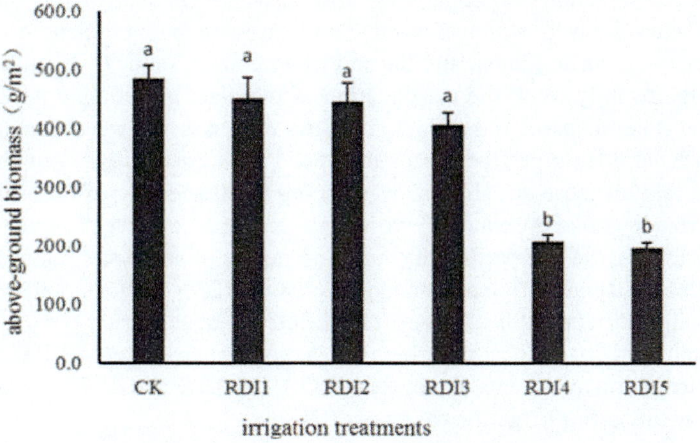

Fig. 8.1 Variation of above-ground biomass vegetation in different treatments

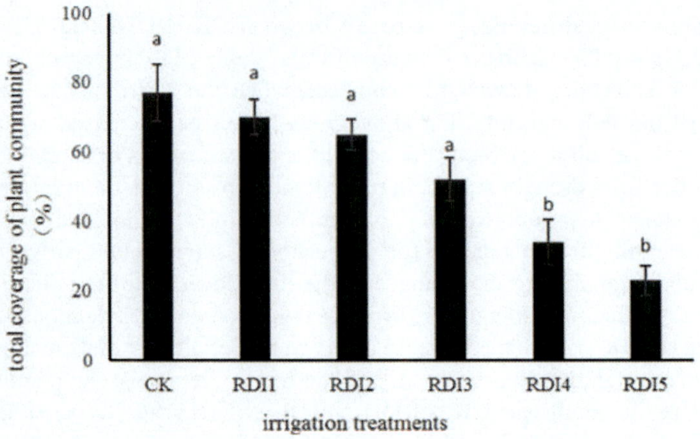

Fig. 8.2 The total coverage of plant community of under different treatments

8.3.2 Changes of Soil Moisture Content in Different Regulated Deficit Irrigation Treatments

Figure 8.3 shows the changes of soil moisture in the initial growth, rapid growth and vigorous growth stages of mixed forage under different water deficit adjustment treatments. In the second year of planting, alfalfa and ice grass are mainly distributed in the 0–30 cm soil layer., which showed that plants were more efficient in using soil water in the 0–30 cm soil layer, and the soil water content in the 0–30 cm soil layer changed significantly. With the continuation of the plant growth and development process, the water demand of the plant gradually increases, the water demand of the rapid growth period is larger, the fluctuation of soil moisture decreases obviously, and the natural rainfall increases significantly during the flowering period of the plant, and the herb enters the reproductive growth stage. Plants' need for irrigation begins to decline. During the test period, the variation ranges of soil water content in each regulated deficit irrigation treatment ranged from 6.6 to 22.5 cm^3/cm^3, and the change range of soil water content was larger. The difference of soil water content reached a significant level ($p < 0.05$), and the average soil water content of RDI3, RDI4, and RDI5 irrigation treatments decreased by 23.1%, 32.3% and 41.2%, respectively, compared with the full irrigation treatment (CK).

8.3.3 Influence of Regulated Deficit Irrigation on Water Use Efficiency

In 2021, the natural precipitation in the early and rapid growth period of plants is relatively low. The precipitation in April and May is only 11.8 and 17.0 mm. In the arid

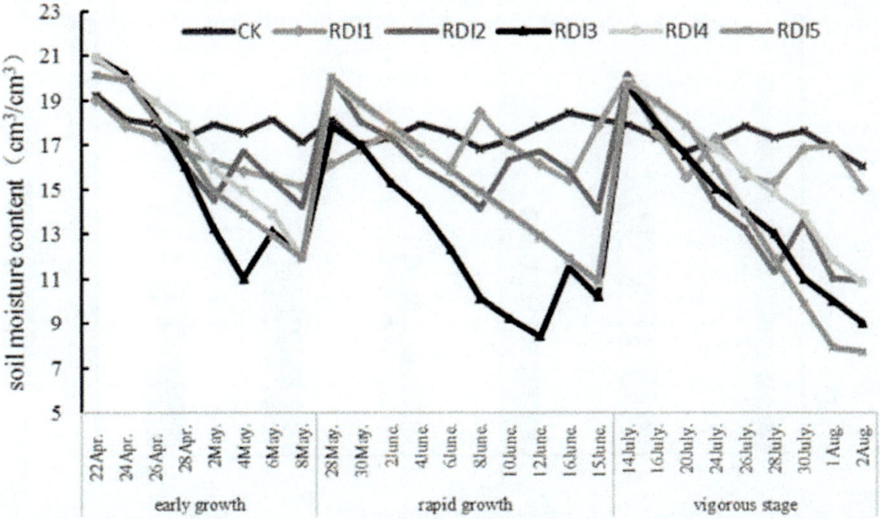

Fig. 8.3 Changes of soil moisture content under different regulated deficit irrigation treatments

and water-deficient environment, the amount of irrigation water is the main factor affecting the water consumption of vegetation. The results of plant water consumption under the regulated deficit irrigation treatment showed (Fig. 8.4) that the water consumption of plants increased with the increase of the irrigation amount, and the water consumption of full irrigation (CK) was significantly higher than that of the regulated deficit irrigation treatment ($p < 0.05$). Among the irrigation treatments, the difference in water consumption between slightly-regulated deficit irrigation (RDI1) and moderate deficit-regulated (RDI3), severe deficit-regulated (RDI4), and intensive deficit-regulated (RDI5) irrigation treatments reached a significant level ($p < 0.05$). The water consumption and water use efficiency of plants were significantly affected by the amount of irrigation. With the increase of the amount of irrigation, the water consumption gradually increased, while the water use efficiency gradually decreased. Severe (RDI4) and intensive (RDI5) treatments had lower water use efficiency and were significantly lower than other treatments ($p < 0.05$). The water use efficiency of the two treatments with mild deficit adjustment (RDI2) and moderate deficit adjustment (RDI3) was significantly higher than that of other irrigation treatments ($p < 0.05$), and both reached more than 10 g/kg. These two irrigation methods have higher water use efficiency.

8.4 Conclusions

(1) In the process of vegetation restoration on the slopes of abandoned land in arid and semi-arid grassland mining areas, when the soil moisture content is

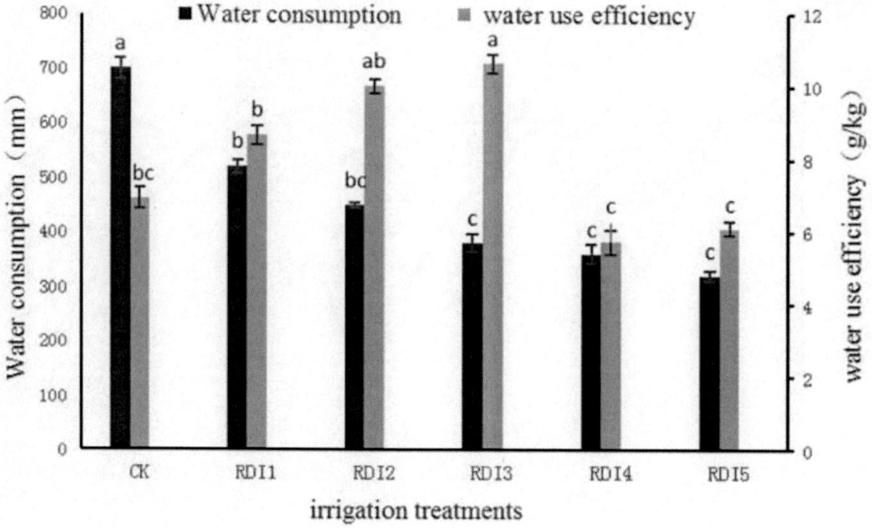

Fig. 8.4 Water consumption and water use efficiency under different regulated deficit irrigation treatments

lower than 45% of the field water holding capacity, the total coverage of plant communities and aboveground biomass decrease significantly, and plant growth and development are significantly affected by drought stress.

(2) There was no significant difference in total plant coverage and aboveground biomass between slight, mild, moderate regulated deficit irrigation and full irrigation. The mild and moderate regulated deficit irrigation could improve the water use efficiency of mixed forage.

Acknowledgements This work was supported by the Inner Mongolia Natural Science Foundation (2021MS04021) and the Inner Mongolia Autonomous Region Science and Technology Plan Project (2019GG023).

References

Ahirwal J, Maiti SK et al (2016) Assessment of soil properties of different land uses generated due to surface coal mining activities in tropical Sal (Shorea robusta) forest, India. Catena: An Interdisciplin J Soil Sci Hydrol Geomorphol Focus Geoecol Lands Evol 140:155–163

House GL, Bever JD (2020) Biochar soil amendments in prairie restorations do not interfere with benefits from inoculation with native arbuscular mycorrhizal fungi. Restor Ecol 28(4):785–795

Hou X, Liu S, Zhao S et al (2018) Interaction mechanism between floristic quality and environmental factors during ecological restoration in a mine area based on structural equation modeling. Ecol Eng 124

Lv G, Liu Y, Li Y et al (2018) Spatial variation characteristics of physical properties of surface soil moisture in dump slope of Haizhou open-pit coal mine. Chin J Soil Sci 49(1):69–77
Meng F, Zhang Z, Chen S (2020) Ecological environment impact assessment of coal resources development in Baorixile minging area. China Coal 44(5):553–564
Riley SJ (1995) Aspects of the differences in the erodibility of the waste rock dump and natural surfaces, Ranger Uranium Mine, Northern Territory, Australia. Appl Geogr 15(4):309–323
Wu X, Xia G, Li Y et al (2018) Effect of regulated deficited irrigation on growth, photosynthetic characteristics and water use efficiency of black peanut. J Shenyang Agric Univ 49(1):57–64
Zhou W, Yang H, Huang L et al (2017) Grassland degradation remote sensing monitoring and driving factors quantitative assessment in China from 1982 to 2010. Ecol Indicators 83:303–313

Chapter 9
Characteristics and Patterns of Microplastic Distribution in Zhoushan

Wei Chen, Lu Cao, Zhiyuan Jin, Sen Li, and Yan Li

Abstract As an emerging pollutant, microplastics has attracted more and more attention, especially microplastics in the ocean. In this paper, the abundance of microplastics in the sediments and their influencing factors were studied in the waters around Shengsi County and Liuheng Town in Zhoushan City, and the results showed that the average abundance of microplastics in Zhoushan was 25.59 n kg^{-1}. The abundance of microplastics at L3 reached 74.90 n kg^{-1}, and the abundance of microplastics at L4 was the lowest, only 4.28 n kg^{-1}. The sediment is mainly dominated by blue fibrous microplastics, most of which are CE or RY, and the size is mostly > 0.5 mm.

Keywords Microplastic · Abundance · Distribution

9.1 Introduction

With the rapid development of society, plastic products are becoming more and more inseparable in our lives. Global production of plastic products has risen sharply since the 1950s, and according to current trends, it is expected that more than 25 billion tons of plastic will be produced worldwide by 2050 (Geyer et al. 2017). The vast majority of marine litter in the world's oceans is plastic, accounting for about 60–80% of all waste (Derraik 2002). These plastics in the environment after the repeated action of photochemical degradation, thermal oxidation and biological corrosion and other processes, will be decomposed into small particle size plastic particles, when the plastic particle size is less than 5 mm, it is defined as microplastics (Thompson et al. 2004). Microplastics are everywhere, and microplastics can be found in the washing and care products that people use every day, tap water and air. Microplastics have the characteristics of small volume and large specific surface area, so their ability to adsorb pollutants is very strong, especially some persistent organic pollutants (Rios et al. 2007) and heavy metals (Holmes et al. 2012), which are more likely to be

W. Chen (✉) · L. Cao · Z. Jin · S. Li · Y. Li
School of Marine Engineering Equipment, Zhejiang Ocean University, Zhoushan 316022, China
e-mail: 956477794@qq.com

J. Zhang et al. (eds.), *Environmental Pollution Governance and Ecological Remediation Technology*, Environmental Science and Engineering,
https://doi.org/10.1007/978-3-031-25284-6_9

adsorbed by microplastics. Microplastics are also easily transmitted in the food chain (Farrell and Nelson 2013), and the heavy metals and other pollutants they adsorb will be concentrated in humans due to the enrichment of the food chain, causing harm to the human body. Some animals will cause their own eating difficulties after ingesting microplastics, and will also have a certain impact on their viability (Fossi et al. 2012).

Many scholars have conducted research on marine microplastics. Zhang et al. analyzed the sediments in the deep sea of the western Pacific Ocean and found that the average abundance of microplastics was 240 n kg^{-1} (Zhang et al. 2020). Jiang et al. summarized the abundance of microplastics in offshore China with a richness of 0.13–545 items/m^3 and a concentration of 20–7900 items/kg in estuarine sediments (Jiang et al. 2022). Microplastic pollution in the ocean is already very serious. Zhoushan is rich in fishery resources, and microplastics will have a bad impact on the fishery, so it is necessary to determine the abundance of microplastics in Zhoushan waters, understand the current situation of microplastic pollution in Zhoushan waters, and provide reference for follow-up research.

9.2 Material and Methods

9.2.1 The Study Area and Sample Point Settings

Zhoushan City is located in the East China Sea on the south side of the Yangtze River estuary and the outer side of Hangzhou Bay, consisting of 1390 large and small islands, with an area of 22,200 square kilometers, of which the land area is 1440 square kilometers. Zhoushan is rich in fishery resources and is one of the four famous fishing grounds in the world. In this study, a total of 10 sampling points were set up in the sea near Shengsi County and Liuheng Town in Zhoushan City, of which the sampling points in the sea near Shengsi County were numbered S1–S4, and the sampling points in the sea near Liuheng Town were numbered L1–L6, and the specific sampling point distribution was shown in Fig. 9.1.

9.2.2 Sample Collection and Processing

The sampling of this study was carried out in March 2022 in the waters around Shengsi County and Liuheng Town, Zhoushan City, and the main work was sediment sample collection. The sediment sample collection and method are the same as the methods of Cao et al. to treat the soil in the middle and lower reaches of the Yangtze River (Cao et al. 2021). The sediment samples are collected with Peterson mud collectors, and the samples are transferred to the aluminum box with a stainless steel shovel, and after drying, different particle size soil samples (<0.5 mm and >0.5 mm) are screened with a stainless steel sieve, and then the soil samples are weighed for

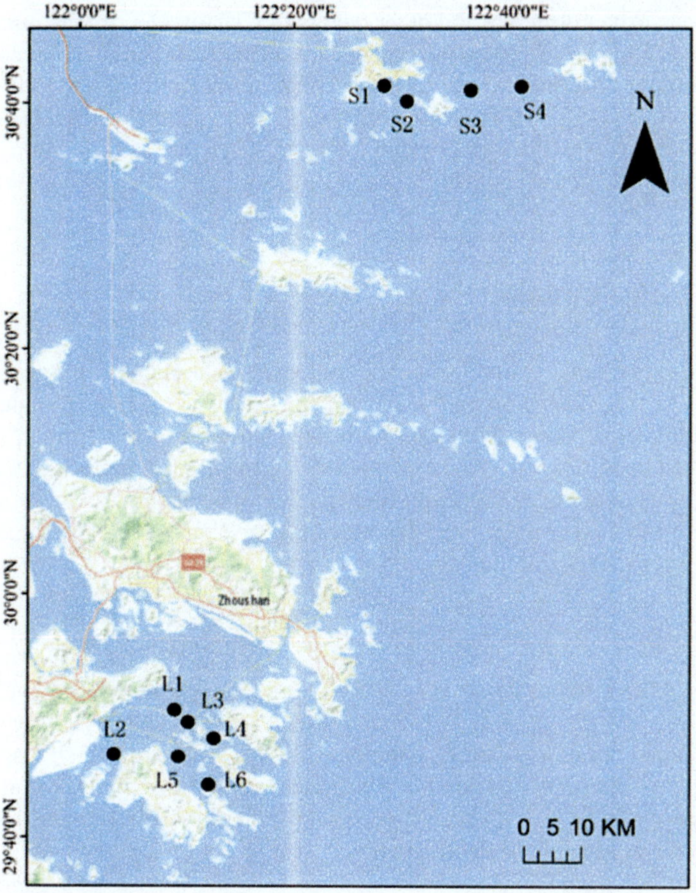

Fig. 9.1 Sample point distribution plot

abundance calculations. The separation of microplastics in soil samples is based on the different densities of microplastics and solutions, first put 10 g of soil samples into a 45 ml centrifuge tube, followed by saturated NaCl solution (1.2 g/cm^3) and shake well to prevent soil particle adhesion from affecting the experimental results. After shaking thoroughly, centrifuge the tube at a rate of 4000 rpm for 5 min to recover the supernatant, then repeat the experiment three times until there is no noticeable float in the supernatant. Subsequently, 30% hydrogen peroxide solution was added to the supernatant to digest for more than 12 h to exclude the interference of plant fibers on subsequent microscopy. Finally, the supernatant is suction (3 μm of filtration membrane) using a filtration device, and the membrane is carefully stored for the next step of detection.

The identification of microplastics is mainly divided into two steps, the first is the preliminary identification, the filter membrane under the microscope (SMZ25, Nikon Corporation, Japan) in a "zigzag" way from left to right, according to the shape, color, luster, surface structure and other factors to visually identify the microplastic particles, and then the suspected microplastics with antistatic surgical forceps to pick out, in the Fourier transform micro-infrared spectrometer (Nicolet iN10, Thermo Fisher Scientific, USA) under spectral testing. The obtained spectrum is compared to the standard spectral library of the polymer, and when the similarity is higher than 60%, the polymer type is confirmed.

9.2.3 *Quality Control*

During the experiment, cotton test clothes and masks should be worn to reduce the impact of clothing fiber drop on the experiment. At the same time, ultrapure water is used in all steps. When pumping, the beaker needs to be rinsed with ultrapure water at least three times to reduce errors. During the digestion process, the beaker is covered with aluminum foil to prevent the microplastics in the air from having an impact on the experiment. All experiments are performed by the same person, reducing experimental contingency.

9.2.4 *Data Processing*

Plot the sample point distribution using ArcGis 10.2 software, which is done with Origin 2019b. The units of microplastic abundance in the sediments in this study are denoted by "$n \cdot kg^{-1}$", where n represents the number of microplastics and kg represents the dry weight of the sediment.

9.3 Results and Discussion

9.3.1 *Abundance Distribution of Microplastics*

The results show that microplastics are found in all sampling points, and the plastic abundance of different sampling points is shown in Fig. 9.2. The average abundance of microplastics in Zhoushan is 25.59 n kg^{-1}. Among all the sampling points, the abundance of microplastics at L3 reached 74.90 n kg^{-1}, and the abundance of microplastics at L4 was the lowest, only 4.28 n kg^{-1}. In the sea near Liuheng Town, the abundance of microplastics in L1, L2 and L3 is significantly higher than that of L4, L5 and L6, which may be due to the rapid flow of water at L4, L5 and L6, and the sediment on the seafloor is very easy to be washed, so the abundance of microplastics in these places is low.

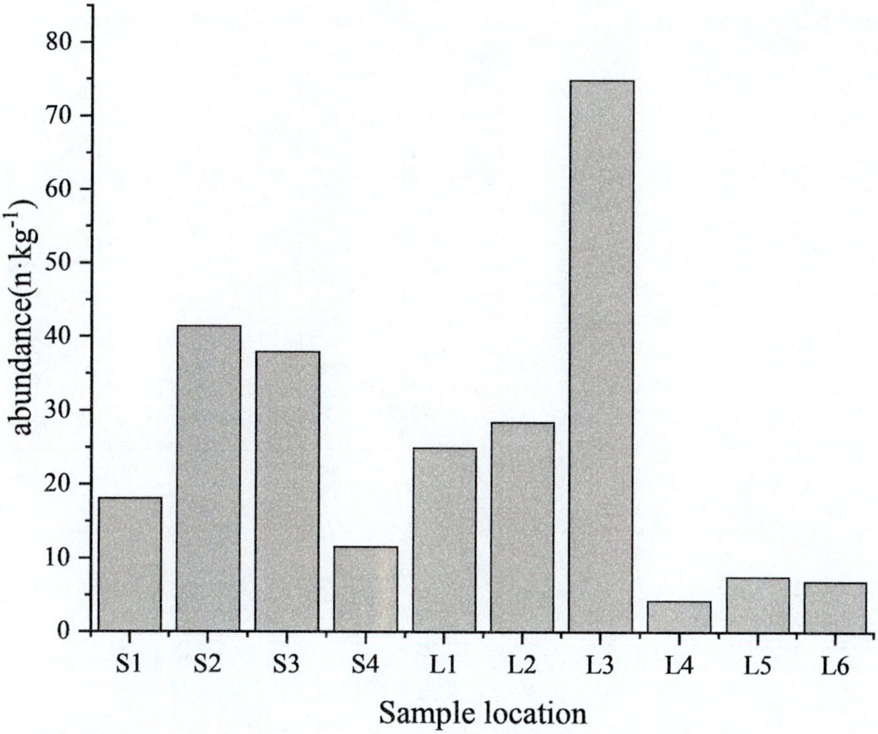

Fig. 9.2 Distribution characteristics of microplastics abundance at different sampling points

9.3.2 Morphological Characteristics of Microplastics

According to the results of the microscope visual inspection, the shape of the microplastics is divided into three categories: fibers, fragments and films. The various shapes of microplastics are counted, and the results are shown in Fig. 9.3. In the two sea areas, the proportion of fiber microplastics is the largest, of which the proportion of fiber microplastics in the sea near Shengsi is 83.33%, and the proportion of fiber microplastics in the sea near Liuheng is 93.22%. The high content of fibrous microplastics in sediments is largely due to the abundance of fishery resources in Zhoushan, so there are more fishing nets used when fishing for seafood at sea, and when discarded fishing nets are discarded in the ocean, they will decompose into fibrous microplastics after a long period of physical and chemical changes. No film microplastics have been found in the sea near Shengsi, and the proportion of film microplastics in the sea near Liuheng is only 1.69%.

A total of six colors of microplastics were found in all samples, accounting for the specific proportion as shown in Fig. 9.4, of which blue microplastics accounted for the highest, accounting for about 70.13% of the total, and black microplastics accounted for the least, accounting for only 1.30%.

Fig. 9.3 The proportion of microplastic shapes in different sampling points

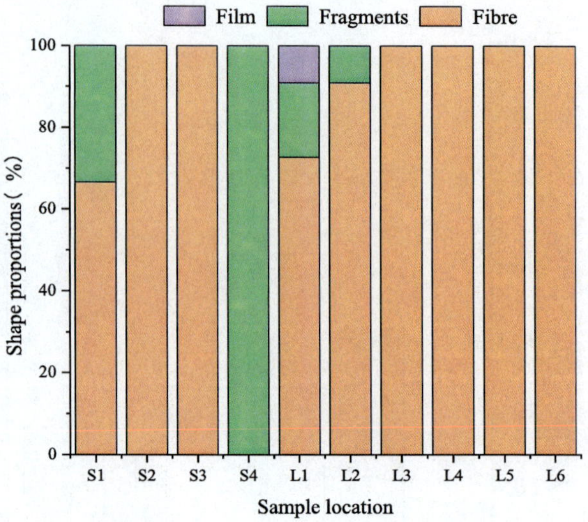

Fig. 9.4 The proportion of color of microplastics in different sampling points

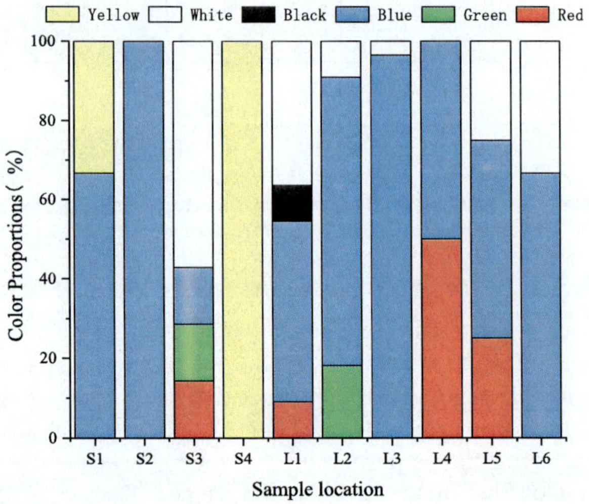

This study divided qualitative microplastic sizes into two ranges: 0.1–0.5 mm and > 0.5 mm. Among them, > 0.5 mm of microplastics are more, accounting for about 94.81% of all microplastics. It can be seen that the microplastics near these two seas are mainly dominated by larger microplastics.

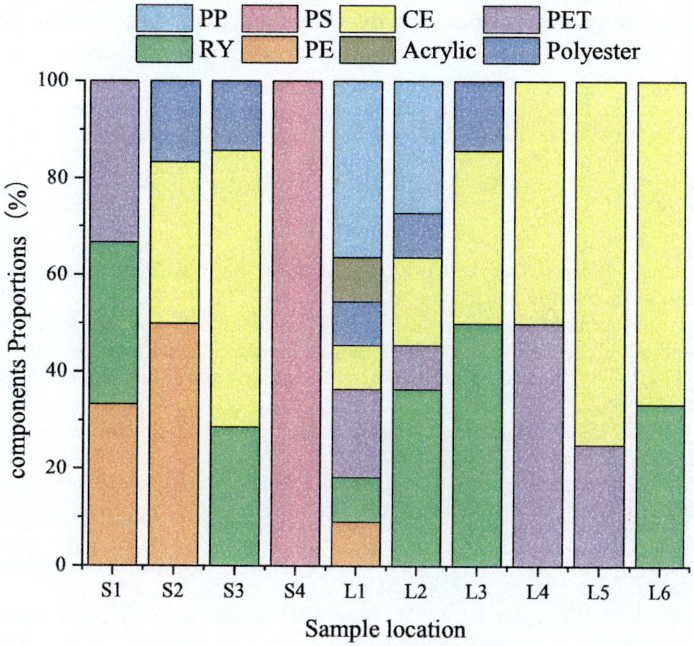

Fig. 9.5 The proportion of microplastic components in different sampling points

9.3.3 The Main Components of Microplastics

The main microplastics in Zhoushan are polypropylene (PP), polystyrene (PS), cellulose (CE), polyethylene terephthalate (PET), rayon (RY), polyethylene (PE), Crylic and Polyester. This is shown in Fig. 9.5. Among them, CE and RY accounted for the highest proportion, 32.47% and 29.87%, respectively, and most of them appeared in fibers. CE and RY are mainly derived from human clothing. Browne et al. found through experiments that a piece of clothing produces more than 1900 fibers after each wash (Browne et al. 2011). Napper et al. estimate that an average of 6 kg of acrylic fabric releases more than 728,000 fibers after washing (Napper and Thompson 2016). It can be seen that most of the microplastics in the sea near Zhoushan still come from humans.

9.4 Conclusion

The average abundance of microplastics in the sea near Zhoushan was 25.59 n kg^{-1}, the highest abundance of microplastics in L3, reaching 74.90 n kg^{-1}, and the lowest abundance of microplastics in L4, only 4.28 n kg^{-1}. Most of the samples were blue fibrous microplastics, mainly CE or RY, which were speculated to be mainly from nearby residents. Thus, The human impact on the marine environment is also enormous.

References

Browne MA, Crump P, Niven SJ et al (2011) Accumulation of microplastic on shorelines woldwide: sources and sinks. Environ Sci Technol 45(21):9175–9179

Cao L, Wu D, Liu P et al (2021) Occurrence, distribution and affecting factors of microplastics in agricultural soils along the lower reaches of Yangtze River China. Sci Total Environ 794:148694

Derraik JGB (2002) The pollution of the marine environment by plastic debris: a review. Mar Pollut Bull 44(9):842–852

Farrell P, Nelson K (2013) Trophic level transfer of microplastic: Mytilus edulis (L.) to Carcinus maenas (L.). Environ Pollut 177:1–3

Fossi MC, Panti C, Guerranti C et al (2012) Are baleen whales exposed to the threat of microplastics? A case study of the Mediterranean fin whale (Balaenoptera physalus). Mar Pollut Bull 64(11):2374–2379

Geyer R, Jambeck JR, Law KL (2017) Production, use, and fate of all plastics ever made. Sci Adv 3(7):e1700782

Holmes LA, Turner A, Thompson RC (2012) Adsorption of trace metals to plastic resin pellets in the marine environment. Environ Pollut 160:42–48

Jiang Y, Yang F, Kazmi SSUH et al (2022) A review of microplastic pollution in seawater, sediments and organisms of the Chinese coastal and marginal seas. Chemosphere 286:131677

Napper IE, Thompson RC (2016) Release of synthetic microplastic plastic fibres from domestic washing machines: effects of fabric type and washing conditions. Mar Pollut Bull 112(1–2):39–45

Rios LM, Moore C, Jones PR (2007) Persistent organic pollutants carried by synthetic polymers in the ocean environment. Mar Pollut Bull 54(8):1230–1237

Thompson RC, Olsen Y, Mitchell RP et al (2004) Lost at sea: where is all the plastic? Science 304(5672):838–838

Zhang D, Liu X, Huang W et al (2020) Microplastic pollution in deep-sea sediments and organisms of the Western Pacific Ocean. Environ Pollut 259:113948

Chapter 10
Construction Waste Production Prediction and Correlation Analysis Based on ARIMA-BP Combination Model

Heng Yang, Yang Liu, Jingjue Jia, and Xiaoyu Liu

Abstract Reliable construction waste production data are prerequisites for construction waste management and resource utilization. However, since there is no relevant construction waste production data in various cities in China, it is necessary to calculate and predict construction waste production in different cities. This paper aims to verify a reliable prediction method of construction waste production based on the data in the statistical yearbook. Through the study, it is found that ARIMA-BP combination forecasting model has shown ideal accuracy in the prediction of construction waste production in Tianjin, Kunming and Xining, and the average relative error has reached about 0.01. This paper shows that in the next few years, the production of construction waste in the three cities has increased to varying degrees, and the highest production of construction waste in Tianjin in the next few years will reach more than 35 million tons. At the same time, by analyzing the composition proportion and influencing factors of construction waste, it is hoped that this paper will provide help for the management and resource utilization of construction waste.

Keywords Construction waste · ARIMA-BP combination model · Production prediction · Analysis and evaluation

---

H. Yang (✉) · Y. Liu · J. Jia · X. Liu
School of Geomatics and Urban Spatial Informatics, Beijing University of Civil Engineering and Architecture, No. 15 Yongyuan Rd., Daxing District, Beijing 102616, China
e-mail: 2108160320005@stu.bucea.edu.cn

Y. Liu
Beijing Advanced Innovation Center for Future Urban Design, Beijing University of Civil Engineering and Architecture, No. 1 Zhanlanguan Rd., Xicheng District, Beijing 100044, China

Beijing Key Laboratory of Urban Spatial Information Engineering, No. 15 Yangfangdian Rd., Haidian District, Beijing 100038, China

J. Zhang et al. (eds.), *Environmental Pollution Governance and Ecological Remediation Technology*, Environmental Science and Engineering,
https://doi.org/10.1007/978-3-031-25284-6_10

10.1 Introduction

With the rapid development of urbanization, the construction industry has brought new opportunities for development, and the construction market has also shown a prosperous scene. But at the same time, with the continuous progress of construction activities, frequent construction and demolition activities have produced a lot of construction waste, which has brought great pressure to the urban environment. In China, it is estimated that more than 2 billion tons of construction waste are generated each year, accounting for 40% of total urban waste generation.

At present, China, the United States, Japan and other countries have paid attention to the evaluation and prediction of construction waste, and the subsequent improvement of the utilization rate of these wastes. Many scholars have done a lot of research on garbage classification, yield prediction, resource utilization and so on (Jialong 2012; Huang et al. 2018). In order to formulate scientific and reasonable construction waste management and treatment schemes, it is necessary to study the change trend of construction waste production. The existing methods for predicting the annual production of construction waste mainly include multiple linear regression method, time series method, material flow prediction method, and model construction (Zhang et al. 2021; Qiao et al. 2020). Based on the research ideas of many scholars, this paper uses autoregressive integrated moving average model (ARIMA) prediction model, back propagation (BP) neural network model and ARIMA-BP combination model to predict the production of construction waste, and makes corresponding comparison, so as to select a better prediction method.

10.2 Related Theory and Foundation

10.2.1 Estimate Index

At present, there are many kinds of construction waste, which cannot be accurately calculated and predicted. According to the statistics of relevant personnel, the influence of construction area, demolition area and decoration area on the production of construction waste accounts for more than 95%. Therefore, this paper mainly studies the construction waste at the construction stage, demolition stage and decoration stage (Yan 2008).

This paper mainly uses the building area estimation method to estimate. For the generation of construction waste in the construction stage, 500–600 tons of construction waste is generated on average for every 10,000 square meters of building construction (Chao et al. 2007). Through multiple consultations and field investigations on the construction site, this paper takes the generation rate of construction waste as 540 tons/ 10,000 m^2. At the same time in the demolition process, according to the relevant statistical data show that the production of garbage removal per square meter is about 1–1.3 tons. In this paper, the production rate of demolition waste is mainly determined as 1.15 tons/m^2. In the decoration stage, according to the 'construction waste quantity calculation standard', we take each square meter of decoration area to produce 0.1 ton construction waste. The estimation formula of annual construction waste production in different stages is as in

$$Q_1 = S_1 * Y_1 \tag{10.1}$$

$$Q_2 = S_2 * Y_2 \tag{10.2}$$

$$Q_3 = S_3 * Y_3 \tag{10.3}$$

where Q_1 is the amount of garbage generated in the construction stage, Q_2 is the amount of garbage generated in the demolition stage, Q_3 is the amount of garbage generated in the decoration stage, S_1 is the construction area, S_2 is the construction demolition area, S_3 is the construction decoration area, Y_1 is the construction garbage generation rate, Y_2 is the demolition garbage generation rate, and Y_3 is the decoration garbage generation rate.

Therefore, the estimation formula of annual construction waste production is as

$$Q = Q_1 + Q_2 + Q_3 \tag{10.4}$$

10.2.2 *Model Introducing*

ARIMA Prediction Model.

ARIMA prediction model, also known as ARIMA (p, d, q), is the most common statistical model used for time series prediction model. In the ARIMA (p, d, q) model, p is the autoregressive order, d is the difference number needed to make the non-stationary time series become a stationary time series, and q is the moving average order. ARIMA modeling is to stabilize the original sequence, determine the optimal model through the process of model identification and test, and carry out prediction analysis. The model is mathematically described as

$$\Delta^d y_t = \theta_0 + \sum_{i=1}^{p} \varphi_i \Delta^d y_{t-1} + \sum_{j=1}^{q} \theta_j \varepsilon_{t-1} \tag{10.5}$$

where y_t is the original time series, $\Delta^d y_t$ denotes the stationary sequence of y_t after d sub-difference, ε_t denotes the residual term, φ_i (i = 1, 2, ..., p) and θ_j (j = 1, 2, ..., q) are self-regression coefficients and moving average coefficients to be estimated, respectively.

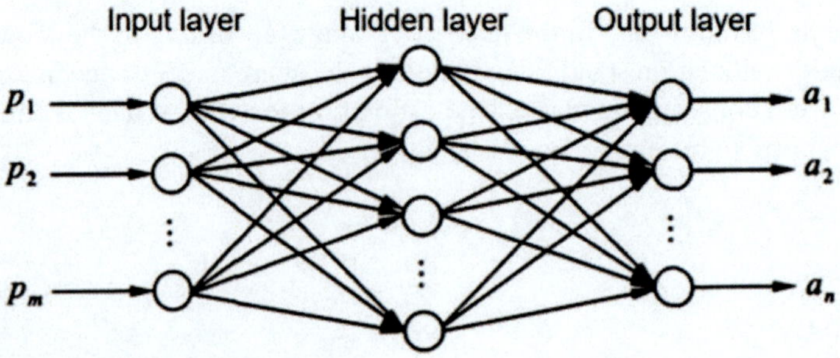

Fig. 10.1 Schematic diagram of BP neural network

BP Neural Network

BP neural network is a multilayer feedforward neural network trained by error back propagation algorithm. The learning process is mainly divided into two parts. The first part is the forward transmission part of the input signal. After network training, the weights and thresholds of the reverse correction model are obtained by comparing the output results with the real values. BP neural network consists of three parts: the input layer, the middle layer and the output layer. The neurons in the former layer only input information to the neurons in the latter layer, and there is no feedback phenomenon. There is no information transmission between the networks in the same layer. The middle layer, also known as the hidden layer, can directly affect the changes in the weight coefficients of the other two layers, thereby affecting the performance of the entire network. Figure 10.1 is a typical BP neural network diagram.

10.3 Research Scheme

10.3.1 Research Area

In this paper, the research area is China's Tianjin, Kunming and Xining three cities, these three cities are located in different regions of China, and there are great differences in urban economic level.

Tianjin

Tianjin is located in north China, downstream Haihe River Basin, is the largest port city in northern China. By the end of 2021, the total output value of Tianjin's construction industry was 465.305 billion yuan, with a total resident population of 13.73 million and the regional GDP was 156.9505 billion yuan.

Kunming

Kunming is located in southwest China, central Yunnan-Guizhou Plateau. By the end of 2021, the total output value of Kunming's construction industry was more than 350 billion yuan, with 8.502 million permanent residents and the regional GDP was 72.225 billion yuan.

Xining

Xining is located in the northwest of China, is the eastern portal of the Qinghai-Tibet Plateau. In 2021, the total output value of the construction industry in Xining City was 47.402 billion yuan, and the resident population was 2.4756 million and the regional GDP was 15.488 billion yuan.

The research area of this paper is shown in Fig. 10.2.

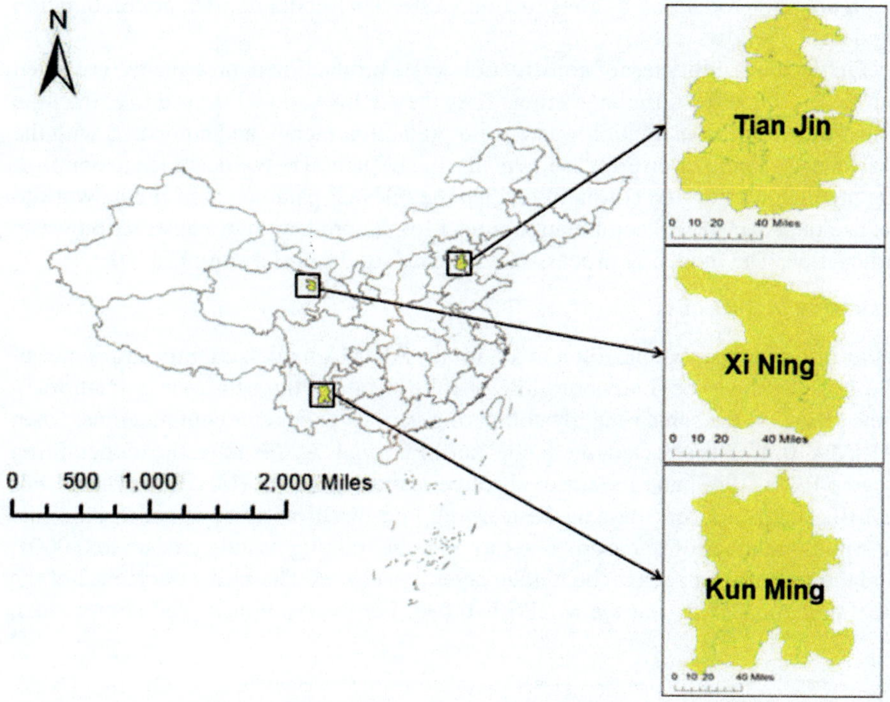

Fig. 10.2 Research area

10.3.2 Related Experiments

Data Source

Based on the statistical data of 'China Statistical Yearbook', this paper respectively inquires the construction area data and completion area data of Tianjin, Kunming and Xining in the past two decades. According to relevant data, the demolition area of old buildings is about 10% of the new building area in the year. And the decoration area is based on the actual completed area. The total output of construction waste is estimated by Formulas 10.1, 10.2, 10.3 and 10.4.

Experimental Process

In order to ensure the accuracy of the prediction results, this paper uses ARIMA prediction model, BP neural network model and ARIMA-BP combination model to predict the production of construction waste. The modeling idea of combination model is as follows.

Firstly, the stability test of construction waste production data sequence is carried out to determine the difference order. Then the ARIMA model is used to extract the linear part of the construction waste yield prediction model, and compared with the real value, the prediction error is e_t. Finally, the BP neural network is used to optimize the prediction error and constantly adjust the relevant parameters of the network to realize the ARIMA-BP combination model for the prediction of construction waste production. The modeling process of combined model is shown in Fig. 10.3.

Experimental Results

Take Tianjin as an example. First of all, for the ARIMA model, after the difference of the model, it shows obvious aboriginality at the level, and this sequence is a stationary time series. At the same time, by comparing various parameter combinations, when ARIMA (0, 1, 0) is selected, the model performs well. At this time, the model fitting degree R2 is 0.948, and the mean absolute percentage error (MAPE) is 0.052. For the BP neural network model, the learning rate of BP network is set to 0.02, the maximum number of iterations is set to 500, the training accuracy is set to 0.0001, and the number of nodes in the hidden layer is set to 10. The fitting degree R2 of the final model is 0.972, and the MAPE is 0.038. For the combined prediction model,

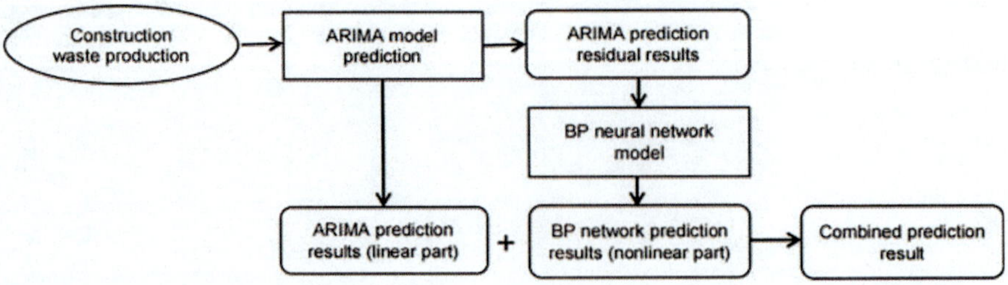

Fig. 10.3 Combination modeling process

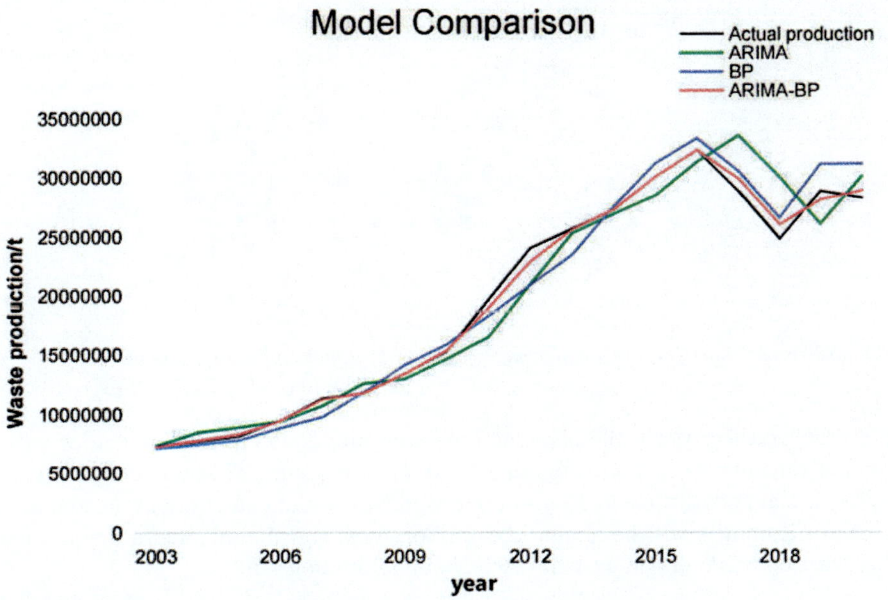

Fig. 10.4 Comparison of predicted and real values of different models

the number of hidden layer nodes is set to 8, and other parameters are basically the same. The fitting degree R2 of the final model is 0.993, and the MAPE is 0.010. Figure 10.4 shows the comparison between the predicted values and the real values of the three models. Through comparative analysis, it can be found that the accuracy of ARIMA-BP combination model is greatly improved compared with a single model.

At the same time, the three models were used to compare and analyze the construction waste production in Kunming and Xining. It was found that the ARIMA-BP combination model showed advantages in the prediction accuracy of construction waste production in other two cities. For Kunming, the fitting degree R2 of the model was 0.997, and the MAPE was 0.007. For Xining, the fitting degree R2 of the model was 0.99, and the MAPE was 0.012. Therefore, ARIMA-BP model is used to predict the construction waste in the three cities. The final prediction results are shown in Table 10.1.

Table 10.1 Prediction results of construction waste production

City	Years					
	2021	2022	2023	2024	2025	2026
Tianjin/10,000 tons	2934.5	3029.0	3168.5	3304.9	3429.5	3578.6
Kunming/10,000 tons	1898.8	1809.5	1918.7	2048.8	2147.5	2347.5
Xining/10,000 tons	331.8	346.0	368.2	350.6	360.4	388.5

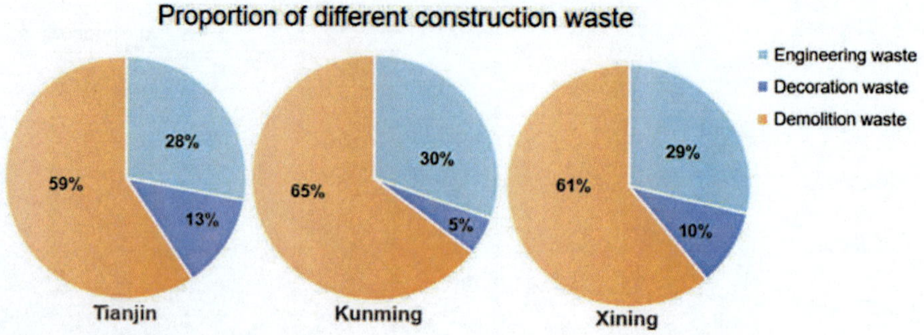

Fig. 10.5 Proportion of different construction waste in Tianjin, Kunming and Xining

In the next few years, the construction waste production in the three cities will generally maintain a slow growth trend trend. The maximum amount of construction waste in Tianjin will exceed 35 million tons, the maximum amount of construction waste in Kunming will be about 23.5 million tons, and the maximum amount of construction waste in Xining will reach about 3.9 million tons.

10.4 Analysis and Evaluation

10.4.1 Component Analysis

By counting the annual averages of construction waste generation in three cities over the period 2001–2020, we can get the results as shown in the Fig. 10.5.

Through the proportion of construction waste in three cities, we can clearly find that the proportion of demolition waste is the largest, accounting for 59%, 65% and 61% respectively. Decorative waste accounts for the smallest proportion of the three kinds of waste production. From the perspective of historical data, we should focus on strengthening the control of the amount of demolition waste. At the same time, we should pay attention to the change of the amount of construction waste and decoration waste, and increase the resource management of construction waste.

10.4.2 Correlation Analysis

There are many factors that affect the amount of construction waste. According to the statistics of different scholars, we select three main factors: population, economy, gross product of construction industry (Xianhua et al. 2016). Taking Tianjin as an example, by querying the data of Tianjin, we selected Pearson correlation coefficient

Table 10.2 Impact values of different factors

Method	Influencing factor		
	Population	Economy	Gross product of construction industry
Pearson	0.967	0.959	0.995
Spearman	0.976	0.943	0.988

and Spearman correlation coefficient to analyze. The results of the analysis are shown in Table 10.2.

Through correlation analysis, we can find that the influence values of the three factors are all close to 1, indicating that they have a great impact on the production of construction waste. Among them, the correlation between the gross product of construction industry and the production of construction waste is the highest.

10.5 Conclusions

Based on ARIMA model, BP neural network model and ARIMA-BP combination model, this paper uses the historical data of 'China Statistical Yearbook' to predict the annual production of construction waste in three representative cities of different levels and regions in China. The results show that the output of construction waste in Tianjin, Kunming and Xining will increase in the next few years. Relevant departments need to increase construction waste resource treatment. By analyzing the composition and influencing factors of construction waste, it is found that demolition waste occupies a large proportion in different cities, and it is necessary to strengthen the control of demolition waste production. The population, economy, gross product of construction industry have great influence on the production of construction waste. Through prediction and analysis, we hope to provide auxiliary data support for urban construction waste disposal decisions.

Acknowledgements This work was financially supported by The National Key Research and Development Program of China (2018YFC0706003) and the Pyramid Talent Training Project of Beijing University of Civil Engineering and Architecture (21082717008), and funded by the Beijing Key Laboratory of Urban Spatial Information Engineering (2020216).

References

Chao L, Xiaozhao X, Li Z (2007) Thoughts and suggestions on the current situation of construction waste treatment. Sci Technol Inform 386

Huang B, Wang X, KuaH et al (2018) Construction and demolition waste management in China through the 3R principle. Resour Conserv Recycl 36–44

Jialong C et al (2012) Thinking on utilization status of domestic construction waste resources. Compr Utiliz Resour China 47–50

Qiao L, Liu D, YuanX et al (2020) Generation and prediction of construction and demolition waste using exponential smoothing method: a case study of Shandong province, China. Sustainability 5094

Xianhua Z, Mingchen Q, Tingchen W (2016) Analysis of influencing factors of construction waste output based on grey correlation degree. J Shenyang Arch Univ 46–52

Yan X (2008) Present situation analysis and comprehensive utilization of construction waste in Xi'an. Masteral thesis, Chang'an University

Zhang M, Li D, Jingyang Let al (2021) Prediction of construction and demolition waste production based on material flow analysis. J Environ Eng Technol 869–878

Chapter 11
Seasonal Dynamics of Reservoir Environmental Factors and Planktonic Algae

Weihan Xi, Peng Chen, and Xiufang Hu

Abstract Reservoir algae outbreaks are related to many factors such as water quality, local weather and hydrological conditions. Water quality conditions are the basis for algal growth and development, and the large amount of nutrients such as nitrogen and phosphorus in the water provide a direct carbon source for algal growth and reproduction. To investigate the relationship between environmental factors and the seasonal dynamics of algae in Shuang Xi Kou Reservoir, water bodies at different points of the reservoir were sampled and monitored. The correlation analysis between environmental factors such as water temperature, total nitrogen (TN), total phosphorus (TP) and permanganate index (CODMn) and planktonic algae was selected. The results showed that the important causes of the gradual increase of planktonic algae in Shuang Xi Kou Reservoir were related environmental factors such as total nitrogen and total phosphorus, and the planktonic algae mainly consisted of cyanobacteria, green algae, diatoms and cryptophytes, which provided reference for the analysis of nutrient sources in Shuang Xi Kou Reservoir and the monitoring and early warning of planktonic algae breeding in Shuang Xi Kou Reservoir.

Keywords Planktonic algae · Environmental factors · Dynamic analysis · Relevance

11.1 Introduction

Planktonic algae are primary producers in aquatic ecosystems and are a major food source for aquatic plants and animals (Chen et al. 2021; Gu et al. 2021; Yanhong and Chunhua 2022). Their biomass and community composition are externalizations of environmental changes in aquatic ecosystems and changes in trophic status

W. Xi (✉) · P. Chen (✉) · X. Hu
Zhejiang University of Water Resources and Electric Power 310081, Hangzhou, China
e-mail: 1823234574@qq.com

P. Chen
e-mail: chenp@zjweu.edu.cn

J. Zhang et al. (eds.), *Environmental Pollution Governance and Ecological Remediation Technology*, Environmental Science and Engineering,
https://doi.org/10.1007/978-3-031-25284-6_11

within the water column. As such, planktonic algae have been widely used to assess ecosystem environments and assess eutrophication in water bodies (Yanhong and Chunhua 2022; Jiancheng et al. 2022). SPSS correlation analysis is now used to study the correlation between planktonic algal communities and environmental factors in Shuang Xi Kou Reservoir from September 2020 to August 2021, with the aim of providing primary data for the management of eutrophication in this reservoir and laying the foundation for data analysis for subsequent series of studies.

11.2 Materials and Methods

11.2.1 Regional Overview

Shuang Xi Kou Reservoir is a comprehensive medium-sized reservoir for water supply, flood control and irrigation, located in Zhangshan Village, Dayin Town, Yuyao City, Zhejiang Province. The river where Shuang Xi Kou Reservoir is located, Dayin Creek, is a tributary of the Yao River in the Yongjiang River system. The reservoir area has a typical subtropical monsoon climate. Significant changes in precipitation over the years, the average precipitation is 1783.4 mm and the catchment area above the dam site is 40.01 km. According to 《Zhejiang Province Water Function, Water Environment Functional Zone Classification Scheme》, the target water quality of the Shuang Xi Kou Reservoir implements the Class II standard of the Surface Water Environmental Quality Standard (GB3838-2002).

11.2.2 Samling Methods

Seven monitoring points were set up in the reservoir at the stagnant area, outlet, lake core, mixing area and inlet, and one monitoring point for nutrient indicators was set up near the outlet. Fifteen conventional indicators are monitored. They are Water temperature (WT), pH, dissolved oxygen (DO), turbidity, conductivity, permanganate index, ammonia nitrogen, total phosphorus, total nitrogen, nitrate, total chlorophyll, cyanobacteria, diatoms, cryptophytes, biotoxicity (luminous bacteria). The monitoring section monitors water samples six times a day from September 2020 to August 2021.

11.3 Data Analysis

Application of Excel 2019, SPSS 22.0 software for statistical and analysis of data.

11.3.1 Environmental Factors

The seasonal dynamics of environmental factors in the Shuang Xi Kou Reservoir from September 2020 to August 2021 are shown in Table 11.1. in which, the annual average water temperature is 21.07 °C, with the highest in summer (June–August) and the lowest in winter (December–January); pH value is 8.12 annually, with alkaline water quality; dissolved oxygen is 9.47 mg/L annually, with the highest dissolved oxygen level in spring (March–May), followed by summer and autumn (September–November) lowest. Turbidity averaged 2.41 NTU per year, with the highest turbidity in spring and the lowest in autumn; conductivity averaged 71.51 us/cm per year, with the higher conductivity in spring and summer and the more turbid waters, and the lowest conductivity in winter and the purer waters; ammonia nitrogen mass concentration averaged 0.04 mg/L per year, with the lowest in summer and about one-fifth of the mass concentration in autumn; nitrate nitrogen averaged 1.27 mg/L per year, with the highest in summer and the lowest in autumn. 1.27 mg/L, with the lowest mass concentration in summer and the highest in spring, 6.25 times that in summer; permanganate index 1.83 mg/L per year, with the highest index in spring, the water body is more polluted by organic matter, the rest of the season is more balanced; total phosphorus content 0.02 mg/L per year, more stable throughout the year; total nitrogen content 1 mg/L per year, with the higher content in spring and the second highest in autumn.

11.3.2 Seasonal Dynamics of Planktonic Algae in Shuang Xi Kou Reservoir

The seasonal dynamics of planktonic algae in the Shuang Xi Kou Reservoir from September 2020 to August 2021 are shown in Table 11.2. Longitudinally, the highest annual average levels of cryptophytes and the lowest levels of cyanobacteria were observed. Horizontally, cyanobacteria levels were highest in summer, 9.7 times higher than the lowest winter levels. Green algae levels were highest in summer, followed by autumn and lowest in spring. Diatoms were highest in spring, 5.3 times higher than the lowest winter levels. Cryptophytes had the highest spring levels and the lowest winter levels.

11.3.3 Environmental Factors in Relation to Planktonic Algae

The correlations between planktonic algae and each major nutrient, pH, etc. in Shuang Xi Kou Reservoir from September 2020 to August 2021 are shown in Table 11.3 below. The data in Tables 11.2 and 11.3 are based on synchronous dynamic seasonal

Table 11.1 Seasonal dynamics of environmental factors in Shuang Xi Kou reservoir from September 2020 to August 2021

Environmental factors	Spring 2021	Summer 2021	Autumn 2020	Winter 2020	Annual average
Water temperature (°C)	18.73 ± 10.47	27.63 ± 5.87	23.19 ± 7.21	12.89 ± 7.81	21.07 ± 12.63
pH	8.76 ± 1.05	9.01 ± 0.94	7.49 ± 2.24	7.12 ± 0.76	8.12 ± 1.83
Dissolved oxygen (DO) mg/L	10.87 ± 2.33	9.94 ± 3.96	7.69 ± 2.11	9.77 ± 1.73	9.47 ± 4.43
Turbidity (NTU)	4.55 ± 8.55	3.73 ± 6.27	0.64 ± 6.26	1.92 ± 15.28	2.41 ± 14.79
Electrical conductivity (us/cm)	76.13 ± 22.77	85.99 ± 20.11	68.14 ± 18.96	58.56 ± 10.14	71.51 ± 34.59
Ammonia nitrogen (mg/L)	0.03 ± 0.07	0.01 ± 0.02	0.06 ± 2.24	0.05 ± 2.59	0.04 ± 2.6
Nitrate nitrogen (mg/L)	2 ± 2.26	0.32 ± 0.53	1.36 ± 2.64	1.24 ± 2.35	1.27 ± 2.73
Permanganate index (CODMn) mg/L	2.33 ± 1.27	1.53 ± 0.33	1.9 ± 0.3	1.53 ± 0.13	1.83 ± 1.78
Total phosphorus (TP) mg/L	0.02 ± 0	0.02 ± 0.01	0.02 ± 0.01	0.02 ± 0.01	0.02 ± 0.01
Total nitrogen (TN) mg/L	1.17 ± 0.9	0.88 ± 0.28	1 ± 0.38	0.93 ± 0.37	1 ± 0.73
Nitrogen to phosphorus ratio (TN/TP)	58.67 ± 45.17	60.33 ± 36.67	68.83 ± 44.17	68.83 ± 41.17	62.92 ± 113.00

Table 11.2 Seasonal dynamics of planktonic algae in Shuang Xi Kou reservoir from September 2020 to August 2021

Planktonic algae	Average				
	Spring 2021	Summer 2021	Autumn 2020	Winter 2020	Annual average
Total chlorophyll (μg/L)	38.68 ± 79.65	34.46 ± 48.04	15.29 ± 19.41	11.23 ± 15.54	24.64 ± 93.69
Cyanobacteria (ug/L)	2.3 (0–34.55)	4.76 (0–23.02)	0.52 (0–2.75)	0.49 (0–3.70)	1.8 (0–34.55)
Green algae (ug/L)	5.3 (0–25.35)	7.15 (0–20.57)	6.6 (0–27.25)	6.3 (0–12.23)	7.53 (0–100.20)
Diatomaceae (ug/L)	11.9 (0–98.54)	8.23 (0–34.11)	3.81 (0–11.49)	2.26 (0–7.73)	6.28 (0–38.54)
Cryptophyte (ug/L)	19.04 (0–68.20)	14.32 (0–62.38)	4.3 (0–11.61)	2.19 (0–8.77)	9.01 (0–68.20)

variation information at the monthly mean range scale. The correlation between total nitrogen and green algae was significant ($p < 0.01$), with a moderate negative correlation; cyanobacteria and diatoms were all significantly correlated with cryptophytes ($p < 0.01$), with moderate positive correlation and high positive correlation, with correlation coefficients of 0.752 and 0.850 respectively; cryptophytes were significantly correlated with total chlorophyll ($p < 0.01$), with a high positive correlation, with a correlation coefficient of 0.957, and permanganate index was significantly correlated with total nitrogen ($p < 0.05$), with a moderate positive correlation coefficient of 0.567; pH was significantly correlated with water temperature ($p < 0.01$), cyanobacteria ($p < 0.01$) and diatoms ($p < 0.01$), all with a moderate positive correlation. This suggests that total chlorophyll was mainly provided by cryptophytes. Permanganate index and nitrogen and phosphorus nutrients are key factors influencing the structure of the planktonic algal community in Shuang Xi Kou Reservoir, and if supervision and control of Shuang Xi Kou Reservoir is not strengthened, further risk of water blooms will occur.

11.4 Discussion

Combined with Table 11.3, it can be seen that pH is a direct environmental factor for algal growth and reproduction, with values fluctuating little throughout the year and the water body being weakly alkaline. As the index of dissolved oxygen rises, the number of cryptophytes also rises, with the highest levels of dissolved oxygen in spring and the number of cryptophytes increasing from the same period to the average highest value for the monitoring period. The increase in total nitrogen is a decisive indicator of the increase in green algae (Ting et al. 2017; Xia et al. 2012).

Table 11.3 Correlation analysis of environmental factors and planktonic algae

	pH	WT	DO	NH_4-N	Nitrate nitrogen	Permanganate Index	TP	TN	Cyanobacteria	Green algae	Diatomaceae	Cryptophyte	Total chlorophyll
pH	1												
WT	0.678*	1											
DO	0.442	−0.298	1										
NH_4-N	0.337	−0.014	−0.575*	1									
Nitrate nitrogen	0.113	−0.429	0.135	0.259	1								
Permanganate Index	0.281	0.188	0.006	−0.079	0.043	1							
TP	0.008	−0.12	0.097	−0.424	0.225	−0.002	1						
TN	0.071	−0.066	0.032	−0.27	−0.036	0.567*	0.105	1					
Cyanobacteria	0.582*	0.272	0.405	−0.532	−0.228	−0.022	−0.169	0.297	1				
Green algae	0.183	0.246	−0.041	0.517	0.085	−0.221	−0.444	−0.769**	−0.107	1			
Diatomaceae	0.630*	0.223	0.437	−0.473	0.106	0.282	0.341	0.562*	0.456	−0.544	1		
Cryptophyte	0.709**	0.187	0.620*	−0.644*	0.157	0.245	0.178	0.453	0.752**	−0.351	0.850**	1	
Total chlorophyll	0.844**	0.334	0.606*	−0.495	0.126	0.182	0.066	0.291	0.774**	−0.111	0.811**	0.957**	1

* Correlations were significant at the 0.05 level (two-tailed)
** Correlations were significant at the 0.01 level (two-tailed)

From the long-term monitoring data, spring and summer are the periods of greatest seasonal fluctuations in the mass concentration of planktonic algae in the Shuang Xi Kou Reservoir, and are the periods of highest algal blooms. Combined with the nutrient data in Table 11.1, it can be seen that the nitrate nitrogen content, high manganese index, total nitrogen and total phosphorus content are higher in spring than in the remaining three seasons, which is the time of concentration of nutrient salts, so that nutrient salts are an important factor in the surge of planktonic algae.

11.5 Conclusion

(1) The planktonic algae in the surface water of Shuang Xi Kou Reservoir mainly consist of Phylum Green Algae, Phylum Cyanobacteria and Phylum Diatoms and Cryptophyta, and in terms of the number of species Phylum Diatoms > Phylum Cryptophyta > Phylum Cyanobacteria > Phylum Green Algae, with Phylum Diatoms being the main dominant species group in Shuang Xi Kou Reservoir.

(2) The composition of the planktonic algae in Shuang Xi Kou Reservoir from 2020 to 2021 is dominated by diatoms and cryptophytes, with high nutrient content and high manganate index in spring and summer, indicating that the seasonal dynamics of nutrients is the main factor causing the surge of planktonic algae.

(3) The pH and total nitrogen mass concentration are the main environmental factors affecting the seasonal variation of planktonic algae in Shuang Xi Kou Reservoir, control of the reservoir should mainly focus on nitrogen control, the focus should be on the high incidence of planktonic algae in spring and summer.

Acknowledgements This study is supported by National College Student's Innovation and Entrepreneurship Training Program (No. 202211481012).

References

Chen X, Li X, Du G, Liu B (2021) Dynamic characteristics of phytoplankton community structure in Gaobeidian lake in relation to water environment factors [J]. Anhui Agric Sci 49(23):60–66

Gu Y, Lanling Y, Jianliang Y, Mingliang L, Mingliang L, Hailan X, Quanyuan W (2021) Seasonal dynamics and environmental factors of planktonic algae in the Qiantang River (Hangzhou section)[J]. South-North Water Transf Water Conservancy Sci Technol (in English) 19(02):334–341

Jiancheng H, Zhibin S, Jiacheng Z, Tan L, Tao T (2022) A comparative study of the community patterns of planktonic and facultative algae in the upper Jinsha river and their relationships with environmental factors[J]. J Ecol 42(02):590–599

Ting G, Qianli MA, Zhenxing W, Li W, Ling-ai Y, Zhencheng X, Xuemin Z, Rongchang L, Yu L (2017) Summer eutrophication conditions and characteristics of cyanobacterial water bloom outbreaks in Turtle rock reservoir [J]. Environ Sci 38(10):4141–4150

Xia Y, Xing'e H, Lei C, Qizhou W, Zhiyuan L (2012) A review of factors influencing algal blooms[C]. In: Proceedings of the 2012 annual academic conference of the chinese society of environmental sciences vol 2, pp 603–609

Yanhong L, Chunhua H (2022) Characteristics of planktonic algal communities in Poyang lake and their correspondence with environmental factor paradigms[J]. Environ Monit Manage Technol 34(01):33–37+48

Chapter 12
Effect of Immobilized Bacteria and *Bidens Pilosa* L on the Remediation of Heavy Metal Contaminated Soil

Haoming Wang, Yuan Liu, Yueqing He, Yuanyuan Shen, Xicheng Tai, Jing Xu, Xiaoxiao Li, Bin Yue, Guohua Chang, and Tianpeng Gao

Abstract In the study, three tolerant bacteria were isolated, and the immobilized bacteria was constructed with corn stalk as the carrier. The immobilized inoculants combined with alfalfa were used in pot experiments to explore the combined remediation of heavy metal-contaminated soil. The quality of heavy metal contaminated soil after remediation was comprehensively evaluated by measuring the changes of heavy metal content in *Bidens pilosa* L roots, stems and leaves, soil enzyme activity and soil heavy metal content. The results show the inoculation with immobilized bacteria can improve soil enzyme activity, and the immobilized inoculants increased the catalase and ascorbate peroxidase activity, and decreased the contents of reduced Glutathione and soluble sugars. So, the result show that may provide a new microbial enhanced-phytoremediation of metal-polluted soils. Combined remediation of immobilized bacteria and *Bidens pilosa* L improved the quality of heavy metal-contaminated soil. The results of the study will help us further understand the mechanism of microbial inoculation to reduce the toxicity of heavy metals, and provide guidance for forage cultivation in heavy metal contaminated soil.

Keywords Heavy metal · *Bidens pilosa* L · Remediation · Immobilization bacteria

---

Haoming Wang and Yuan Liu these authors have contributed equally to this study.

H. Wang · Y. Liu · Y. He
School of Biological and Pharmaceutical Engineering, Lanzhou Jiaotong University, Lanzhou 730070, China

X. Tai · B. Yue · G. Chang · T. Gao (✉)
Engineering Center for Pollution Control and Ecological Restoration in Mining of Gansu Province, Lanzhou City University, Lanzhou 730070, China
e-mail: zkgtp@163.com

Y. Shen · J. Xu · X. Li · T. Gao
Academy of Eco Xi'an, Xi'an University, 710065, Xi'an, China

School of Biology and Environmental Engineering, Xi'an University, Xi'an 710000, China

J. Zhang et al. (eds.), *Environmental Pollution Governance and Ecological Remediation Technology*, Environmental Science and Engineering,
https://doi.org/10.1007/978-3-031-25284-6_12

12.1 Introduction

Soil heavy metal pollution is an environmental problem facing the world that needs to be solved (Yu et al. 2021). The abundant heavy metal resources have laid the material foundation for the healthy and stable development of the national economy (Hu et al. 2014). However, the long-term mining process, lack of technology, capital and poor management have caused serious impacts on the soil and environment around the mine, and the accumulation of large amounts of tailings and waste slag has caused adverse impacts on the natural ecological environment and social and economic life (Lin 2009). The tailings contain heavy metals, which are released and migrated to the surrounding soil and groundwater through surface biogeochemistry and enter the human body through the food chain, thus posing a serious threat to the health and ecological environment of the residents around the mine site (Wuana and Okieimen 2011). The treatment of heavy metal pollution in lead and zinc mines has always been a hot spot and a difficult area of research in China (Li et al. 2018). Due to the relative stability of heavy metal elements in the environment and the difficulty of degradation, no treatment method has been found for the remediation of contaminated soil with heavy metals that can be applied on a large scale (Sprynskyy et al. 2006).

To reduce the total contaminant mass in soil, a number of remediation methods have been practiced, including soil replacement, electro kinetic removal, thermal treatment, soil washing, and phytoextraction (Ahmed et al. 2022; Xu et al. 2021). On the other hand, to mitigate bioavailability or bioaccumulation, various chemical immobilization techniques are often employed using a range of chemical compounds, such as lime, phosphate compounds, and organic compounds. In addition, verification, solidification, and phyto-stabilization have also been investigated at the field (Chen et al. 2010). However, most of the field cleanup actions are often practiced by remediation industries. As a result, information is often less systematic or long-term monitoring of completed site is lacking (Ayangbenro and Babalola 2017).

Bioremediation is a technique used to remove environmental contaminants from ecosystem. It utilizes the biological mechanisms inherent in microbes and plants to eradicate hazardous pollutants and restore the ecosystem to its original condition (Ali et al. 2013). Heavy metal is known to dislodge components in biological molecules, hindering the functions of the molecules and changing enzymes, protein or membrane transporter structure of function thereby becoming toxic to plants (Gupta et al. 2016). Microorganisms are essential in remediation of heavy metal contaminated environments as they have a variety of ways to endure metal toxicity. The exploitation of microorganisms to sequester, precipitate, or change the oxidation state of numerous heavy metals has been widely studied (Abdelkrim et al. 2020). Plant plays an important role in bioremediation, which are five different types of major processes involved in the phytoremediation and include: (1) phyto-extraction, (2) rhizofiltration, (3) phyto-volatization, (4) phyto-degradation, and (5) phyto-stabilization (Kang et al. 2016).

In this study, biochar was employed as microbial material to make immobilized bacteria, which and *Bidens pilosa* L remediate soil contaminated with heavy metals

(Pb, Zn and Cu). The objective of this study was evaluated the immobilized bacteria on the potential of *Bidens pilosa* L to alleviate heavy metal stress. The effect of Chlorophyll (Chl), malondialdehyde (MDA), soil enzyme, and plant enzyme as well as the heavy metals (Pb, Zn and Cu) bio-concentration and transfer factors of the radish were also analyzed to obtain a more thorough understanding for the plant growth and heavy metal uptake in the *Bidens pilosa* L.

12.2 Materials and Methods

12.2.1 Sample Collection and Bacterial Isolation

Bacteria were isolated from the soil collected from a Pb–Zn mining areas with Cu contamination (Longnan, Gansu, China). The center of the Bijiashan Pb–Zn mining area is at 33°47′30″, 105°42′00″E. And the 0–20 cm tailings soil was collected by the five-point sampling method. Three groups were taken in parallel, and the test soils were stored in sterile plastic bags at 4 °C for subsequent experiments.

12.2.2 Plant and Immobilized Bacteria

The Immobilized three strains of bacteria used from Lanzhou Jiaotong University (*Acinetobacter* L1, *Exiguobacterium acetylicum* L2 and *Klebsiella oxytoca* L3), which were isolated from Bijiashan Pb–Zn mining area in autumn 2020. After conducting the antagonistic experiment on the three strains, it was determined that they had no antagonistic effect, and were prepared according to the following method.

The strain construction material was corn stalk. L1, L2 and L3 were adsorbed and immobilized. The three strains were activated for 8–10 h respectively. When the OD600 of the bacterial liquid was about 0.8, the volume ratio of L1:L2:L3 was 2:1:1. Transferring the mixed bacterial solution into the sterilized LB medium with 2% inoculum, culturing for 48 h, collecting the composite bacterial cells by centrifugation, inoculating the bacterial cells in LB medium, and making a certain concentration of bacteria seed fluid. Weigh 2.0 g of corn stalk into a 200 mL conical flask, sterilize it and cool it to room temperature, add 100 mL of seed solution to it, place it at 30 °C, 140 r/min constant temperature shaking incubator, and fix it for 48 h. Take out, centrifuge at 4000 rpm for 10 min to obtain immobilized bacteria, wash the bacteria with sterilized 0.9% physiological saline solution, repeat 3 times, the obtained solid is the immobilized bacteria.

12.2.3 Pot Experiments

The pot-culture experiment was carried on in a greenhouse at Xian University, Shanxi Province, China from May 19 to October 22. The effect of immobilized bacteria inoculation on growth and heavy metal uptake in *Bidens pilosa* L plants were investigated in pots (d × h: 20.5 × 18.5 cm) containing 3 kg of 2 mm sieved mine soil. The contaminated mine soil was irrigated with 200 ml of water every two days and incubated for 14 days. There was a pallet to prevent water losses and to protect the soil from metal leaching. Seeds were disinfected with 10% H_2O_2 for 10 min and then rinsed with water three times. Each pot was sown with 50 seeds. After the plants had grown for 30 days, 25 embryonic plants were left that were vigorous, free of pests and diseases and of similar size and maturity.

The experiment design was as follows: (1) No heavy metal contaminated soil in 30 km around the mine (CK0). (2) No heavy metal contaminated soil with *Bidens pilosa* L (CK1). (3) No heavy metal contaminated soil with immobilized bacteria (S0). (4) No heavy metal contaminated soil with immobilized bacteria and *Bidens pilosa* L (S1). (5) Mine soil with *Bidens pilosa* L. (T1). (6) Mine soil (T0). (7) Mine soil with immobilized bacteria (H0). (8) Mine soil with *Bidens pilosa* L and immobilized bacteria (H1).

Each pot was watered with 100 ml every two days and placed in outdoors for 100 days on 19 May 2021. For the climatic conditions during the planting period, the temperature was 25–40 °C during the day, 15–30 °C at night, and the humidity was about 70%.

12.2.4 Plant Analysis

Total chlorophyll content was determined by using the method of Li et al. (2018). Reduced glutathione content was determined according to the method of Winterbourn and Metodiewa (1994). Malondialdehyde and soluble sugar content were determined by using the method of Zou (2000). Superoxide dismutase (SOD) activity was determined by NBT method (Wang et al. 2001). The activity of peroxidase (POD) was determined by guaiacol method (Fehrmann and Dimond 1967). Catalase (CAT) activity was determined by ultraviolet absorption method (Abei 1984). The ascorbate peroxidase activity was determined via reduction of the absorption of ascorbate at 290 nm with the modified method (Alen'kina and Nikitina 2020).

12.2.5 Soil and Plant Heavy Metal Analysis

According to heavy metal content of soil and plant, we calculated bio-concentration factor (BCF) and translocation factor (TF) with following formulas: (Takarina and Pin 2017).

BCF = the heavy metal (Pb, Zn and Cu) concentration in plants/the heavy metal (Pb, Zn and Cu) concentration in soil × 100.

TF = the heavy metal (Pb, Zn and Cu) concentration in aboveground parts of plants/the heavy metal (Pb, Zn and Cu) concentration in plant roots × 100.

12.2.6 Determination of Soil Enzymatic Activity

Urease, phosphatase, sucrase, and catalase activity were selected to be the typical indicators of soil health to evaluate the effect of heavy metal on soil health. Meanwhile, urease, phosphatase, sucrase and catalase were used simultaneously as indicators for the evaluation of soil remediation effects by *Bidens pilosa* L and immobilized bacteria. And these selected enzyme activities were measured spectrophotometrically. Urease activity was determined by the method described by Guan and described by Fernández-Calviño (Guan 1986; Fernández-Calviño et al. 2010). Phosphatase activity was measured by Sun method (Sun et al. 2017). Catalase activity was measured by back titrating residual H_2O_2 with $KMNO_4$ (Johnson and Temple 1964).

Soil sucrase activity was determined according to Hu et al. (2014) method. Sucrase activity was expressed by the glucose released by soil. The urease activity was express as milligrams of NH_4^+-N released by per kilogram of dry soil 24 h. The phosphatase activity was expressed as micrograms pNP produced by per kilogram of dry soil (Yang et al. 2021). Catalase activity may be related to the metabolic activity of aerobic organisms and has been used as an indicator of soil fertility (Trasar-Cepeda et al. 2000).

12.2.7 Statistical Analysis

All statistical analyses were performed by analysis of variance (ANOVA) and compared by Duncan test ($p < 0.05$) using the softer ware Statistical Package for Social Sciences (SPSS 22.0). Sampling analyses was repeated at least three times under the same conditions to decrease experimental errors. Data were shown as mean ± S.D. of three replicates.

12.3 Result

12.3.1 Changes in Plant Physiological Indicators in Soil Remediation Experiments

In the study, the trend of chlorophyll change is completely opposite to that of Malondialdehyde. The rank of chlorophyll is CK1 > T1 > H1 > S1 (Fig. 12.1a), but the rank of MDA is CK1 < T1 < H1 < S1 (Fig. 12.1b). The change of Chlorophyll and MDA indicated compared with the non-inoculated control (CK1), heavy metal (T1 and H1) and immobilized bacteria (S1) both caused the decrease of chlorophyll content in *Bidens pilosa* L and affect the photosynthesis of plants. In addition, heavy metals and immobilized bacteria both lead to the peroxidation of cell membrane lipids, resulting in the increase of MDA content.

Compared with the control (CK1), we find the content of soluble sugar (SS) and reduced glutathione (GSH) increased in T1 treatment, but SS and GSH content in L1 lower than T1. It indicated that the inoculation of immobilized bacteria alleviated the osmotic imbalance of plant cells caused by heavy metal ions.

Compared with the un-inoculated group (CK1), HM treatment (T1) caused antioxidant including APX, POD and SOD activity decreased. In addition, the inoculated of immobilized bacteria (S1) reduced the activities of four antioxidant enzymes (including POD, CAT, APX and SOD) in plants (Fig. 12.1e–h). Compared with the un-inoculated group under HM treatment (T1), the inoculated of immobilized bacteria under contaminated (H1) reduced POD and SOD activity, but promoted APX and CAT activity. In our study, suggesting inoculated of immobilized bacteria

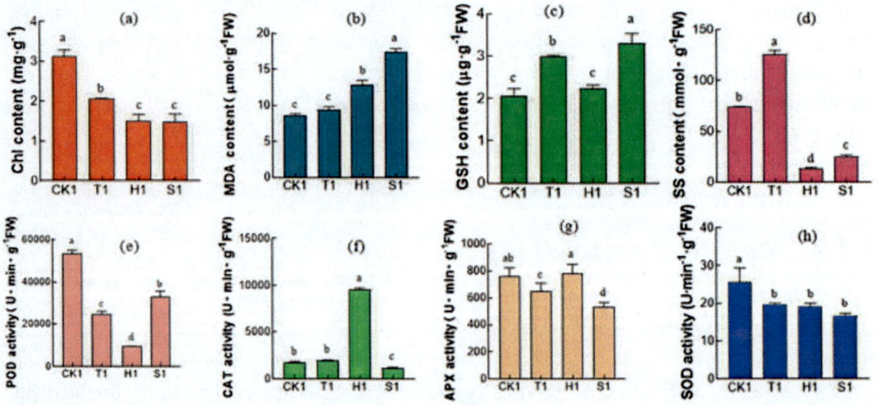

Fig. 12.1 Restoration experiment—changes in plant physiological indicators among treatment groups. **a** Chl(Chlorophyll), **b** MDA(Malondialdehyde), **c** GSH(Reduced glutathione), **d** SS(Soluble sugar), **e** POD(Peroxidase), **f** CAT(Catalase), **g** APX(Ascorbic acid peroxidase), and **h** SOD(Superoxide dismutase). Error bars represent the standard deviations of the means of plant physiological indicators

could effectively improve the ability of defense system against oxidative damage induced by combined toxicity of heavy metal.

In the process of remediation, the addition of immobilized bacteria led to *Bidens pilosa* L turning on antioxidant enzymes to resist stress, and the osmotic substances did not play an important role. The changes including the activities of four antioxidant enzymes and the contents of two osmotic substances indicated that *Bidens pilosa* L formed different coping mechanisms when they are stressed by heavy metals.

12.3.2 Changes in Soil Enzyme in Soil Remediation Experiments

The four soil enzyme activities of T1 were higher than T0, and T0 was lower than CK0 (Fig. 12.2). It indicated heavy metal pollution will reduce soil enzyme activity, and phytoremediation will improve heavy metal soil enzyme activity and improve heavy metal soil quality.

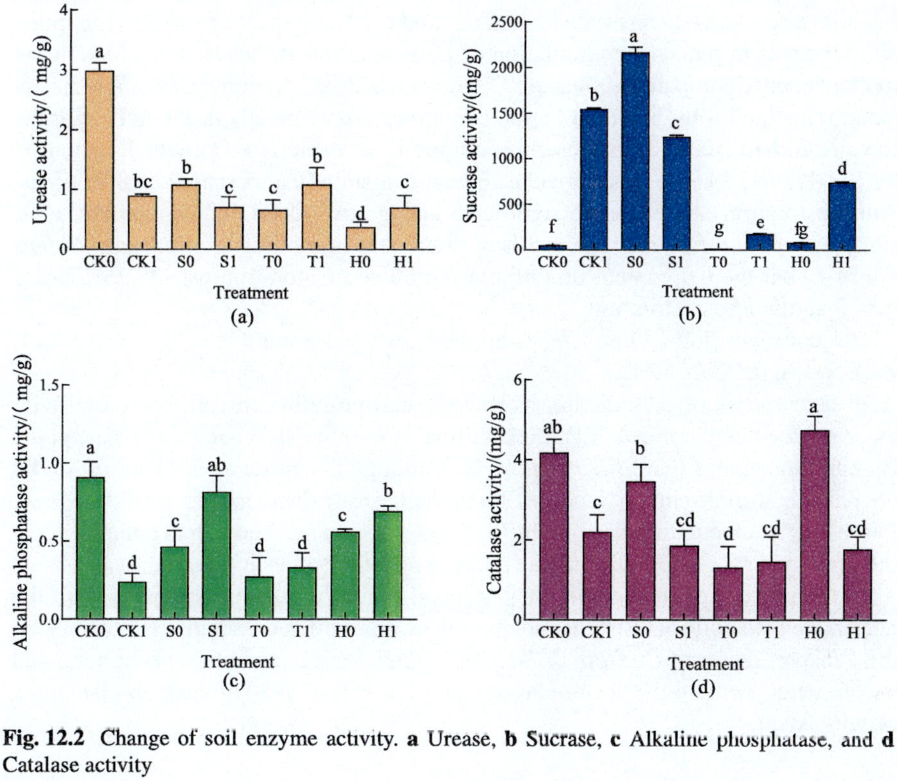

Fig. 12.2 Change of soil enzyme activity. **a** Urease, **b** Sucrase, **c** Alkaline phosphatase, and **d** Catalase activity

Among all groups, urease activity and alkaline phosphatase activity in CK1 were highest. Compared with CK1, the urease activity between S1 and H1 decreased by 16.3% and 18.7% respectively, but T1 increased by 6.18%. *Bidens pilosa* L increased soil nitrogen levels. The phosphatase activities of S1 and H1 increased significantly by 226.9% and 204.3% compared with CK1, and T1 increased by 30.9%.

This study shows that the activities of urease, phosphatase and sucrase can be improved by single phytoremediation, single immobilized bacteria remediation or combined remediation in heavy metal contaminated soil. The reason is that some functional groups in the bacteria blocked the combination of heavy metals and enzymes, resulting in the enhancement of related enzyme activities in the soil (Abdelkrim et al. 2020). This is consistent with the results of this study.

12.3.3 Heavy Metal Content in Plants and Soil

The content of heavy metal under different treatments were measured after repaired and were shown in Table 12.1. The highest contents of Pb, Zn and Cu were located in plant roots (Pb: 22.8–170.8 mg/kg, Zn: 204.8–370.6 mg/kg, Cu: 10.3–92.0 mg/kg). For non-essential elements such Pb^{2+}, Zn^{2+} and Cu^{2+} ion, there are no specific transport channels in plants. Accumulation of large amounts of heavy metal ions in the roots of plants compared to leaves. Compared with the treatment of single heavy metal remediation by plants (T1), the content of heavy metals in Pb and Zn in the roots of *Bidens pilosa* L increased by immobilized bacteria (H1), which increased by 39.08% and 80.95%, respectively, and the content of Cu increased by 8.01 times. After restoration, there was no significant change in soil Pb and Zn contents after single restoration of *Bidens pilosa* L or combined immobilization bacteria *Bidens pilosa* L, but the Cu content of soil after combined restoration was 30.25% lower than that of single restoration.

Bio-concentration and transfer factors among Pb, Zn and Cu of *Bidens pilosa* L were shown in Table 12.2.

In the process of remediation of heavy metal pollution in soil, compared with the non-inoculated control (T1), the addition of immobilized bacteria enhanced the Cu-enrichment ability of *Bidens pilosa* L. Although the enrichment factor of Zn did not change, the enrichment factor of Pb in the L group decreased from 1.39 to 1.24. Interestingly, inoculation of immobilized bacteria enhanced copper accumulation in the shoot and root of *Bidens pilosa* L providing BCF values higher than 1.0.

In the process of remediating heavy metal pollution in soil, compared with the non-inoculated control (T1), the inoculation of immobilized bacteria (H1) enhanced the transport factor of Cu from 0.17 to 0.36. Therefore, the addition of immobilized bacteria can promote the accumulation and transport of heavy metals in plants to a certain extent.

Table 12.1 Contents of heavy metals in various parts and soils of *Bidens pilosa* L

Treatment	*Bidens pilosa* L		
	Pb	Zn	Cu
T1 (Leaf)	138.8 ± 12.1c	127.1 ± 5.6e	7.6 ± 0.4de
H1 (Leaf)	76.4 ± 4.8f	154.7 ± 3.2f	33.4 ± 5.0c
T1 (Stem)	133.6 ± 4.7 cd	324.0 ± 5.1b	2.7 ± 0.4f
H1 (Stem)	119.3 ± 5.3e	160.9 ± 2.7f	9.9 ± 1.0d
T1 (Root)	122.8 ± 5.0de	204.8 ± 3.6c	10.2 ± 2.0d
H1 (Root)	170.8 ± 2.7b	370.6 ± 1.5a	92.0 ± 3.6a
T1 (Soil)	288.8 ± 7.6a	189.1 ± 2.2e	52.4 ± 5.5b
H1 (Soil)	296.4 ± 5.2a	197.8 ± 5.1d	29.6 ± 4.2c

Note Data are means (n = 3) ± SE. The same letters in each column indicate no significant differences according the Duncan test ($p < 0.05$). Control means that no treatments have been applied (neither heavy metals nor bacterial inoculation)

Table 12.2 Bio-concentration and transfer factors of *Bidens pilosa* L

	Treatment	*Bidens pilosa* L		
		Pb	Zn	Cu
Enrichment factor (BCF)	T1	1.39	3.47	0.39
	H 1	1.24	3.47	4.57
Transfer coefficient (TF)	T1	0.66	1.15	0.17
	H 1	0.42	0.55	0.36

12.4 Discussion

In our pot experiments, we inoculated *Bidens pilosa* L and immobilized bacteria to assess their phyto-remediation performances on soil contaminated with heavy metal.

Heavy metals at harmful levels obstruct normal plant functioning and act as an obstacle to metabolic processes in various ways. The functional groups of important molecules in the cell are hindered by metal toxicity and the normal functioning of enzymes and pigments in the disrupted biomolecules, which interferes with the structure of the cytoplasmic membrane (Gupta et al. 2016; Alen'kina and Nikitina 2020). In this study, the results demonstrated that heavy metal stress and inoculation led to the decrease Chl content and the increase of GSH and MDA in *Bidens pilosa* L. Heavy metal stress led to a significant increase in SS content, but the inoculation caused a significant decrease in SS content.

Active antioxidant systems occur naturally in plants which remove the toxicity produce by ROS. Plant cells lessen the undesirable effects of free radicals by generating enzymatic antioxidants such as superoxide dismutase, catalase and glutathione reductase, and non-enzymatic antioxidants such as ascorbate, glutathione, alkaloids, tocopherols that remove the free radicals (Sharma et al. 2012). Compared with the non-inoculated control (CK1), the inoculated of immobilized bacteria (S1) reduced the activities of four antioxidant enzymes (including POD, CAT, APX and SOD) in plants. In addition, when soil polluted, compared with the non-inoculated control (T1), the access of immobilized bacteria (H1) reduced the activities of two antioxidant enzymes (SOD and POD). Therefore, we can infer that combined inoculation of immobilized bacteria was more effective than the non- inoculation in increasing the plant's resistance to oxidative damage caused by excessive accumulation of ROS in heavy metal (Xun et al. 2015). Overall, the Chl content of *Bidens pilosa* L in polluted soil decreased compared to *Bidens pilosa* L in unpolluted soil. On the contrary, the content of MDA, GSH and SS increased. In addition, the changes in the activities of the four antioxidant enzymes in *Bidens pilosa* L were inconsistent. Except for the increase in CAT enzyme activity, the activities of the other three antioxidant enzymes (POD, APX, and SOD) decreased. However, the immobilized inoculants increased the activities of CAT and APX, and decreased the contents of GSH and SS.

The study showed that urease activity, alkaline phosphatase and catalase activity was decreased among contaminated soil. Balyaeva et al. found the Pb decreased the activities of urease, catalase, invertase, and acid phosphatase significantly (Belyaeva et al. 2005). The activities of sucrase, catalase and alkaline phosphatase enzymes in the contaminated soil higher than no-inoculation of immobilized bacteria. It indicated immobilized bacteria effective than no-inoculation to increase the activities of these soil enzymes. Xun et al. found the inoculation with AMF and PGPR might contribute to enhanced soil microbial activity to promote oil pollutant decomposition and transformation into nontoxic substances (Xun et al. 2015). Therefore, this study shows that inoculation with immobilized bacteria can improve soil enzyme activity.

The results TF and BCF show that the inoculation of immobilized bacteria improved Cu uptake. Although inoculation of immobilized bacteria slightly decreased Pb and Zn form soil into *Bidens pilosa* L, increased Pb, Zn and Cu uptake into the root of *Bidens pilosa* L. Jiang et al. found that Pb and Cd contents varied from 38 to 192% and from 5 to 191% in inoculated plants growing in heavy metal-contaminated soils compared to the un-inoculated control, respectively (Jiang et al. 2021). So, the result show that may provide a new microbial enhanced-phytoremediation of metal-polluted soils.

In general, application of immobilized bacteria has positive effect in enhancing phyto-remediation of contaminated soil. Inoculation with immobilized bacteria can relieves the negative impact caused by stress on the *Bidens pilosa* L, and leads to more rapid rates of remediation in contaminated soil. Although the mechanisms are not well understood, the study supports the hypothesis that immobilized bacteria create favorable micro environments that enhance the activities of soil enzymes and remediation efficiency.

12.5 Conclusions

In this study, we found that heavy metal stress reduced the enzymatic activities of APX, POD and SOD in plants, while inoculation of immobilized bacteria under heavy metal stress reduced POD and SOD activities but promoted APX and CAT activities, suggesting that inoculation of immobilized bacteria could effectively scavenge some of the reactive oxygen radicals. As for other physiological indicators, the chlorophyll content and MDA, GSH, and SS contents of plants decreased under heavy metal stress, while inoculation of immobilized bacteria significantly increased the contents of GSH and SS, which assisted in the repair of oxidative damage by participating in plant non-enzymatic systems.

In terms of soil enzyme activity and heavy metal content, heavy metal stress reduced urease, alkaline phosphatase, and catalase activities, whereas inoculation with immobilized bacteria effectively helped to restore soil enzyme activity. The results of TF and BCF showed that inoculation with immobilized bacteria enhanced copper uptake by plants under complex heavy metal stress, although the mechanism of combined plant-microbial repair under environmental abiotic stress is still unclear, but Our results support that immobilized bacteria can effectively mitigate the damage caused by environmental stresses to plants and soils.

Acknowledgements The authors thank Key Laboratory for ecological restoration and high-quality development of Qinling Mountains in the upper and middle reaches of the Yellow River, Xi'an Key Laboratory of plant stress physiology and ecological restoration technology.

Funding This work was funded by National Natural Science Foundation of China (31860176, 32101415), Key Research & Development Program of Shaanxi (2020ZDLSF06-06, 2021JQ-791, 2021JQ-792, 2021JM-510), Key Research & Development Program of Gansu (20YF3FA037), and XAWLKYTD012.

References

Abdelkrim S, Jebara SH, Saadani O et al (2020) In situ effects of Lathyrus sativus-PGPR to remediate and restore quality and fertility of Pb and Cd polluted soils. Ecotoxicol Environ Saf 192:110260

Abei H (1984) Catalase in vitro. Methods Enzymol 105:121–126

Ahmed SF, Kumar PS, Rozbu MR et al (2022) Heavy metal toxicity, sources, and remediation techniques for contaminated water and soil. Environ Technol Innov 25:102114

Alen'kina SA, Nikitina VE (2020) Effect of Azospirillum lectins on the ascorbate peroxidase activity and ascorbic acid content in wheat seedling roots exposed to abiotic stresses. Appl Biochem Microbiol 56(2):211–218

Ali H, Khan E, Sajad MA (2013) Phytoremediation of heavy metals—concepts and applications. Chemosphere 91(7):869–881

Ayangbenro AS, Babalola OO (2017) A new strategy for heavy metal polluted environments: a review of microbial biosorbents. Int J Environ Res Public Health 14(1):94

Belyaeva ON, Haynes RJ, Birukova OA (2005) Barley yield and soil microbial and enzyme activities as affected by contamination of two soils with lead, zinc or copper. Biol Fertil Soils 41(2):85–94

Chen J, Shi Z, Hu L B et al (2010) Site assessment, long-term monitoring and regulatory concerns for application of phytoremediation technology for remediation of heavy metal/metalloid-contaminated soils. In: Faerber T, Herzog J (ed) Solid waste management and environmental remediation. Nova Science Publishers Inc

Fehrmann H, Dimond AE (1967) Peroxidase activity and phytophthora resistance in different organs of potato plan. Phytopathology 57(1):69

Fernández-Calviño D, Soler-Rovira P, Polo A et al (2010) Enzyme activities in vineyard soils long-term treated with copper-based fungicides. Soil Biol Biochem 42(12):2119–2127

Guan SM (1986) Studies on soil enzyme and its methods. China Agriculture Press, Beijing

Gupta A, Joia J, Sood A et al (2016) Microbes as potential tool for remediation of heavy metals: a review. J Microb Biochem Technol 8(4):364–372

Hu H, Jin Q, Kavan P (2014) A study of heavy metal pollution in China: current status, pollution-control policies and countermeasures. Sustainability 6(9):5820–5838

Jiang S, Duan L, Dai G et al (2021) Immobilization of heavy metal (loid) s in acid paddy soil by soil replacement-biochar amendment technology under normal wet condition. Environ Sci Pollut Res 28(48):68886–68896

Johnson JL, Temple KL (1964) Some variables affecting the measurement of "catalase activity" in soil. Soil Sci Soc Am J 28(2):207–209

Kang CH, Kwon YJ, So JS (2016) Bioremediation of heavy metals by using bacterial mixtures. Ecol Eng 89:64–69

Li Y, Liu C, Zhang J et al (2018) Variation in leaf chlorophyll concentration from tropical to cold-temperate forests: association with gross primary productivity. Ecol Ind 85:383–389

Lin L (2009) The status of heavy metal pollution and its control in the Pb and Zn mining districts of China. J Anhui Agric Sci 37(30):14837–14838+14870

Sharma P, Jha AB, Dubey RS et al (2012) Reactive oxygen species, oxidative damage, and ant oxidative defense mechanism in plants under stressful condition. J Bot

Sprynskyy M, Buszewski B, Terzyk AP et al (2006) Study of the selection mechanism of heavy metal (Pb^{2+}, Cu^{2+}, Ni^{2+}, and Cd^{2+}) adsorption on clinoptilolite. J Colloid Interface Sci 304(1):21–28

Sun J, Mei H, Gao F (2017) Ratiometric detection of copper ions and alkaline phosphatase activity based on semiconducting polymer dots assembled with rhodamine B hydrazide. Biosens Bioelectron 91:70–75

Takarina ND, Pin TG (2017) Bio concentration factor (BCF) and translocation factor (TF) of heavy metals in mangrove trees of Blanakan fish farm. Makara J Sci 77–81

Trasar-Cepeda C, Leiros MC, Seoane S et al (2000) Limitations of soil enzymes as indicators of soil pollution. Soil Biol Biochem 32(13):1867–1875

Wang Y, Oberley LW, Murhammer DW (2001) Antioxidant defense systems of two lipidopteran insect cell line. Free Radical Biol Med 30(11):1254–1262

Winterbourn CC, Metodiewa D (1994) The reaction of superoxide with reduced glutathione. Arch Biochem Biophys 314(2):284–290

Wuana RA, Okieimen FE (2011) Heavy metals in contaminated soils: a review of sources, chemistry, risks and best available strategies for remediation. Int Sch Res Not

Xu DM, Fu RB, Wang JX et al (2021) Chemical stabilization remediation for heavy metals in contaminated soils on the latest decade: available stabilizing materials and associated evaluation methods-A critical revie. J Clean Prod 321:128730

Xun F, Xie B, Liu S et al (2015) Effect of plant growth-promoting bacteria (PGPR) and arbuscular mycorrhizal fungi (AMF) inoculation on oats in saline-alkali soil contaminated by petroleum to enhance phytoremediation. Environ Sci Pollut Res 22(1):598–608

Yang R, Wang J, Zhu L et al (2021) Effects of interaction between enrofloxacin and copper on soil enzyme activity and evaluation of comprehensive toxicity. Chemosphere 268:129208

Yu T, Jing T Y, Liu X, Ma XD, Yang ZF, Hou QY, Xia XQ, Li FY (2021) Research progress in current status of soil heavy metal pollution and analysis technology. Geol China 48(2):460–476 (in Chinese with English abstract)

Zou Q (2000) Experimental guidance of plant physiology. China Agricultural Press, Beijing

Chapter 13
The Micro-RNA Profile of a Native Fish *Anabas Testudineus* (Bloch, 1792) Under in Situ Exposure of Pollution from Electronic-Waste Dismantling

Wanglong Zhang

Abstract Electronic waste (e-waste) now has become a global concern. Unsafe dismantling of e-waste in some regions has caused serious environmental pollution by various sorts of contaminants. Under such complex pollution, the true status of wild species, especially the fishes under in situ exposure is recruiting increasing interests. More than the genetic changes, what has happened on the epigenetic aspect is still largely unknown. We sampled the *Anabas testudineus* from two polluted and one reference site, de novo characterized the miRNA sequences and annotated them, and finally found 183 miRNAs (108 is conserved and 75 is novel) in the liver. Further miRNA expression profile comparisons are indicated to be site-specific with a number of miRNAs being altered. Pathway enrichment indicated the pollution may modulate a series of metabolism- and stress-related processes. The current study reveals miRNA status and highlights the miRNA sequencing is a promising tool for toxicology research and environment monitoring.

Keywords *Anabas testudineus* · Electronic waste · Epigenetic · Micro-RNA · Molecular pathway

13.1 Introduction

Electronic waste (e-waste) now has become a global concern. The production has been increasing rapidly (Forti et al. 2020). Numerous reports have figured out its production, fate, and impacts to the environment, especially, in regions where take the recycling of e-waste as an important industry. In some low- or middle-income countries, the e-waste has usually been recycled in unsafe procedures and thus hurt the ecosystem as well as humans (Zeng et al. 2020). For instance, the most toxic compounds released from rough recycling procedures under concern is heavy metals and organic flame retardants, including Pb, Cu, Ag, polychlorinated biphenyls

W. Zhang (✉)
College of Life Sciences, Qufu Normal University, Qufu 273165, Shandong, China
e-mail: zhangwanglong@qfnu.edu.cn

J. Zhang et al. (eds.), *Environmental Pollution Governance and Ecological Remediation Technology*, Environmental Science and Engineering,
https://doi.org/10.1007/978-3-031-25284-6_13

(PCBs), chlorinated paraffins, etc. It is now clear that pollution from rough e-waste dismantling can induce bioaccumulation of toxic compounds in wild organisms and humans. The concentrations of toxicants and their impacts on many physiological parameters and biomarkers of human health have been usually evaluated and reviewed (Matovu et al. 2019). In addition, pollution from e-waste dismantling has been reported to pose adverse effects on wild organisms (Qin et al. 2009). However, many of the studies are performed in the lab, the true status of wild species under such pollution needs more attention.

Until now, various molecular aspects of ecotoxicity of organisms exposed to environmental pollution have been reported, including on the mRNA, protein, and enzyme activity. More than those genetic levels, the aspect of epigenetic alterations is largely unknown, including on both wild organisms and on humans. The miRNA plays an important role in shaping various aspects of biological processes. Fish, especially the model species zebrafish, could serve as a good model in the epigenetic impacts by the pollution or certain contaminants (Tai and Freeman 2020). More than that, with the development of omics, a series of non-model wild species can be determined, and which also enables the exploration of the species living natively in situ in the polluted region and reflects their true status and the alterations in them.

As the key investigation target and animal model, *Anabas testudineus* has substantial significance since it can better adapt to the seriously e-waste pollution. Resolving its biological alterations especially on the miRNA level would provide a good angle to fully understand the e-waste pollution-induced alteration. Thus, this study aims both to de novo characterize the miRNAs of this species and to find the alterations of miRNA. The results suggest a non-model species can be used in ecotoxicology studies as a promising tool for better mechanistic exploration and species conservation.

13.2 Material and Methods

13.2.1 Fish Sampling and RNA Extraction

Anabas testudineus individuals of similar size were sampled using fishing nets from two e-waste-polluted sites, st1 ($n = 4$) and st2 ($n = 5$), in an e-waste dismantling town–Guiyu in China. At a local lab, after sedation with MS222, the liver was immediately excised and stored in RNAlater at −80 °C until RNA extraction. As the reference, fish from a fishing farm within similar size (named hn, $n = 4$) serve as the background and enable comparison between polluted and reference sites. The animal study procedures complied with the ARRIVE guidelines, UK Animals (Scientific Procedures) Act 1986, and the associated guidelines.

13.2.2 MiRNA Sequencing and Bioinformatics

Micro RNA characterization was performed using the liver total RNA. After total RNA separation in PAGE gel, the 18-30nt RNA were striped and recycled. Then the recycled RNA was prepared in the following protocol: 5' end and 3' end adaptor ligation, cDNA synthesis, PCR and library fragment selection, library quantification, and pooling cyclization. Then the library was sequenced on BGISEQ-500.

Raw data were filtered by removing reads in low quality, with primer contamination, with poly A, and shorter than 18nt. Bowtie is used to map clean reads to the transcriptomic reference and to other sRNA databases. The miRNA targets were computationally predicted using RNAhybrid and miRnada. To identify the differentially expressed miRNAs (DEMs), the miRNA expressions were normalized to obtain the expression of transcript per million (TPM). The miRNAs with q-value ≤ 0.001 and |log2 foldchange)| ≥ 1 were identified as DEMs. Target genes under DEMs' regulation were kept performing KEGG pathway-based analysis to help further understand DEMs' target genes' biological functions. The expression profile of miRNAs of each group has undergone principal component analysis (PCA).

13.3 Results and Discussion

13.3.1 Contamination Background

It is well known that unsafe e-waste dismantling could cause serious environmental pollution by metal elements and organic compounds. Moreover, between the two polluted sites, the st1 is more seriously polluted than the st2 based on our previous observation of contamination, both on the heavy metal and persistent organic pollutants respects. Based on this background, this study expects to see the differences of miRNA profile among the individuals live in reference (hn), polluted (st2), and the more polluted site (st1).

13.3.2 MiRNA Overview

As a none-model species, the de novo characterization of the miRNA would add biological knowledge to this species and benefit the environment science on the aspect of revealing the true status and the changes happened under such serious pollution from e-waste dismantling. Generally, the miRNA sequencing is performed in SE50 mode on BGISEQ 500 platform, the clean tag for each sample is ranged from 26.38 to 29.33 M. Clean tags were mapped to the de novo assembled reference transcriptomic sequences, and RNA databases such as miRbase and Rfam. Figure 13.1a shows the average distribution of tags of all the samples, including the mature (37%),

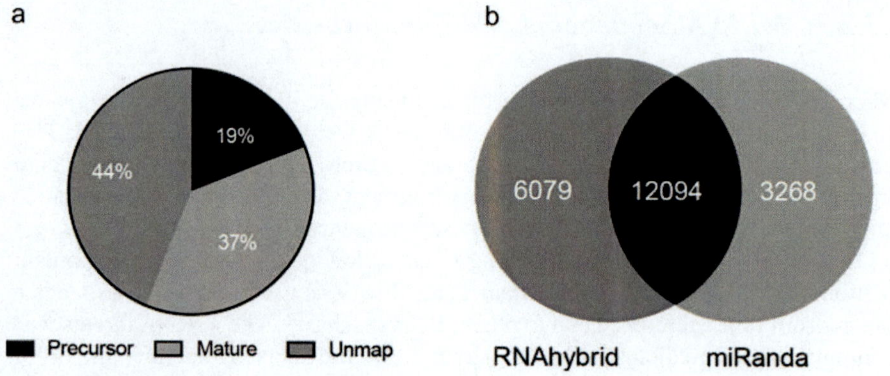

Fig. 13.1 **a** Statistics of the annotation of the miRNA sequences. **b** Amount of the target prediction using RNAhybrid and miRanda. (Author produced this plot.)

precursor (19%), and unmap (44%). As a result, around 94–101 miRNAs were identified as known conserved ones in all the samples, 14–29 miRNAs in each sample, and merged to be 75 miRNAs, were predicted to be novel ones. In summary, in total there are 183 miRNAs identified in all liver samples of *Anabas testudineus*. The environment contaminants induce the toxic effects in which miRNAs play an important role and can be altered, such as metals, organic contaminants, or other xenobiotics (Balasubramanian et al. 2020). Therefore, we determined the miRNAs profile of the liver of *Anabas testudineus* living under the e-waste pollution and the reference to reveal their potential differences. Figure 13.1b shows the predicted targets of all the miRNAs, about 12,094 genes can be predicted by both of the two databases.

13.3.3 Pollution-Induced Variations

As shown in Fig. 13.2, a PCA analysis clearly shows the fish from polluted sites (st1 and st2) and the clean site is separately located in the plot, indicating the variance among samples from distinct sites. Moreover, the miRNA expression profiles of samples from e-waste polluted sites are more closely located but still have a slight difference. The 66 DEMs, of hn vs. st1 comparison, are predicted to target 6032 genes of the assembled mRNA gene dataset with 4790 genes could be annotated in KEGG database. The 72 DEMs, of hn vs. st2 comparison, are predicted to target 6741 genes of the assembled mRNA gene dataset with 5351 genes could be annotated in KEGG database. By performing the KEGG pathway enrichment of the different DEMs' target genes, the hn vs. st1, indicating the potentially affected/differential pathways in the fish under st1's pollution situation.

As shown in Fig. 13.3, a series of metabolism- (lysine degradation, glycerolipid metabolism, purine metabolism, histidine metabolism, glycerophospholipid metabolism, etc.), stress- (ubiquitin-mediated proteolysis, protein processing in

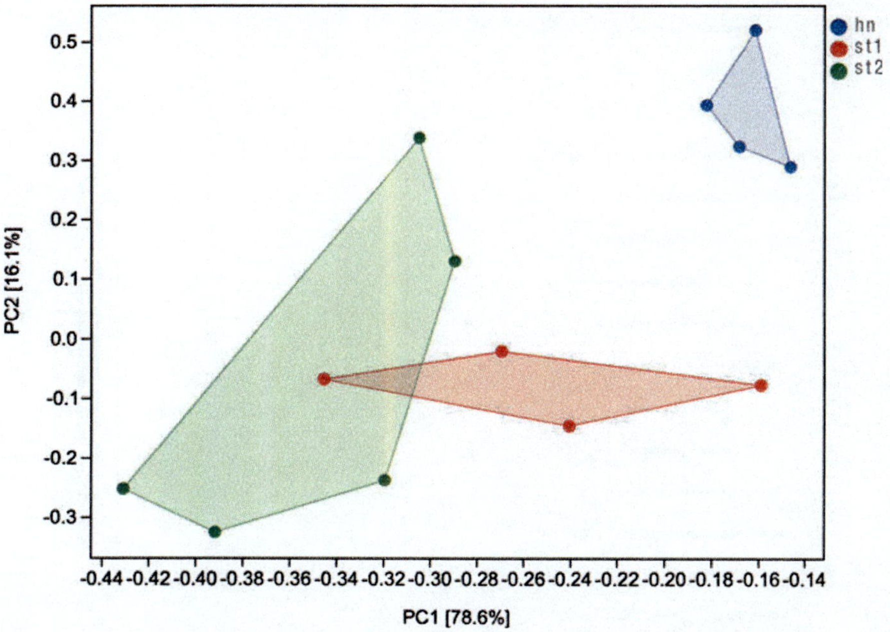

Fig. 13.2 PCA analysis of the samples from hn, st1, and st2. (Author produced this plot.)

endoplasmic reticulum, nucleotide excision repair, apoptosis, etc.), and endocrine- (thyroid hormone signaling pathway, estrogen signaling pathway, etc.) related pathways were revealed.

13.3.4 Distinction Between Various Polluted Sites (St1 versus St2)

To discriminate the site-specific effects, which might be caused by their distinct pollution burden, results indicate that even a majority of DEMs of hn versus st1 and hn versus st2 are the same, there are still site-specific miRNAs indicating the site-specific miRNA profile in st1 (17 miRNAs, 47% is predicted novel miRNA in *Anabas testudineus*) and st2 (23 miRNAs, 61% is predicted novel miRNA in *Anabas testudineus*) versus hn, which may result from the distinct pollution burden. By comparing the st1 versus st2, there are 46 DEMs predicted to target 3926 genes of the assembled mRNA gene dataset with 3369 genes could be annotated in KEGG database.

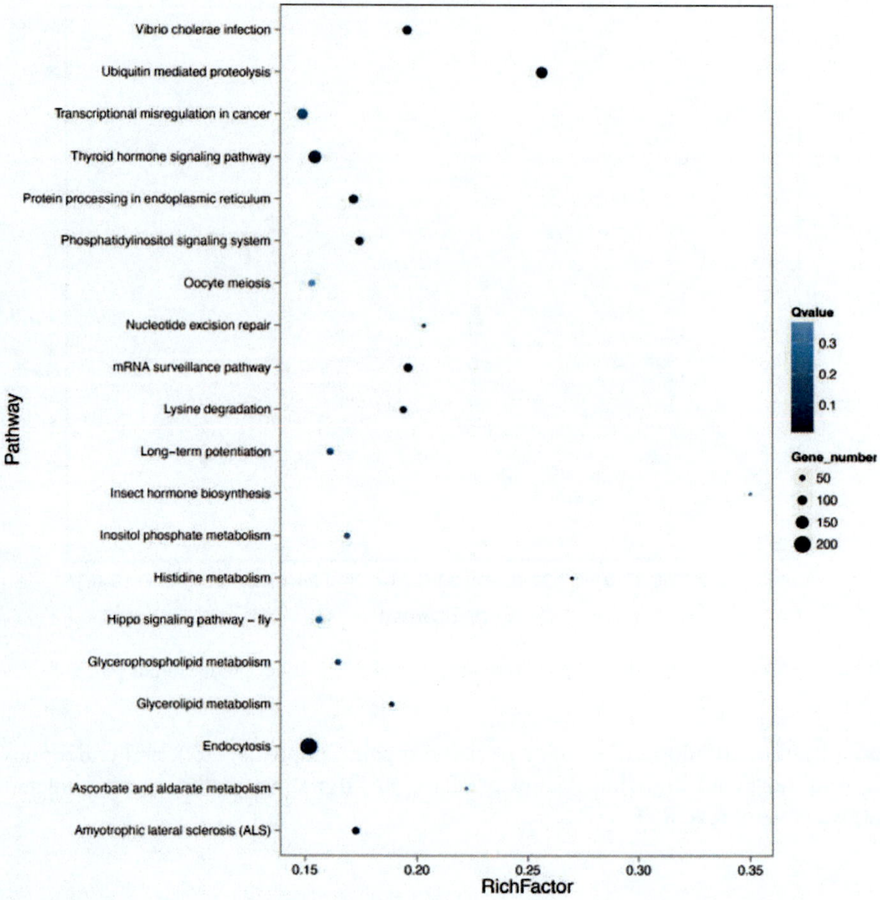

Fig. 13.3 KEGG pathway (level 3) analysis of the target genes of DEMs between hn versus st1. (Author produced this plot.)

13.3.5 Key Pathways Altered by e-waste Pollution

Using the pathway enrichment analysis, we have identified many miRNAs and their target genes-involved pathways. Considering the exposure from serious pollution from e-waste dismantling, we have seen some toxicity related pathways being enriched, e.g., the stress-related and endocrine-disturbance-related ones. In the result, the thyroid hormone signaling pathway is significantly enriched. Currently, many environmental contaminants observed in e-waste dismantling and polluted regions were found to have thyroid hormone receptor disturbing potential as the mimics of natural hormones T3 and T4, such as TBBPA. A recent report compared the TBBPA and other flame retardants and revealed its miRNA alterations contrast with its derivative TBBPA-DBPE (Dasgupta et al. 2021). This is in accordance with certain contaminants found in the e-waste polluted site which usually have the endocrine-disturbance potential.

13.4 Conclusion

In this study, we have focused on a true environment science concern about what are the true status and the alterations of a native fish species under the globally concerning problem of pollution from electronic-waste dismantling. This study first de novo characterized the miRNAs of the non-model species, *Anabas testudineus*, with many miRNAs being novel ones. Moreover, by quantifying the miRNA profile, the polluted individuals do have dissimilar miRNA profiles to the ones from reference site. Even of those from the two polluted sites (st1 and st2) are slightly different as indicated by the PCA result and DEMs. It is found that the DEMs in the liver target a series of metabolism- and stress-related biological processes and pathways. In summary, environmental pollution is toxic to wild species, a native species under in situ exposure could be an indicator of the pollution. The miRNA profiles may give predictive markers for environment monitoring. However, we should also admit that the miRNA profiles give a spectrum of target genes that might be altered, which may need additional examination and mechanistic exploration, luckily, the multi-omics strategy, such as integrating RNA-seq data can help future mechanistic research and conservation.

Acknowledgements This work was supported by the Key Research and Development Project of Jining City (No. 2021JNZC076).

References

Balasubramanian S, Gunasekaran K, Sasidharan S, Jeyamanickavel Mathan V, Perumal E (2020) MicroRNAs and xenobiotic toxicity: an overview. Toxicol Rep 7:583–595

Dasgupta S, Dunham CL, Truong L, Simonich MT, Sullivan CM, Tanguay RL (2021) Phenotypically anchored mRNA and miRNA expression profiling in zebrafish reveals flame retardant chemical toxicity networks. Front Cell Dev Biol 9:663032

Forti V, Balde CP, Kuehr R, Bel G (2020) The global e-waste monitor 2020: quantities, flows and the circular economy potential

Matovu H, Sillanpää M, Ssebugere P (2019) Polybrominated diphenyl ethers in mothers breast milk and associated health risk to nursing infants in Uganda. Sci Total Environ 692:1106–1115

Qin X, Xia X, Li Y, Zhao Y, Yang Z, Fu S, Tian M, Zhao X, Qin Z, Xu X, Yang Y (2009) Ecotoxicological effects of mixed pollutants resulted from e-wastes recycling and bioaccumulation of polybrominated diphenyl ethers in Chinese loach (Misgurnus anguillicaudatus). J Environ Sci (china) 21(12):1695–1701

Tai JKAC, Freeman JL (2020) Zebrafish as an integrative vertebrate model to identify miRNA mechanisms regulating toxicity. Toxicol Rep 7:559–570

Zeng X, Huo X, Xu X, Liu D, Wu W (2020) E-waste lead exposure and children's health in China. Sci Total Environ 734:139286

Chapter 14
A Study on Influencing Factors of Carbon Emissions Based on 14 Typical Countries

Shanshan Wu, Lily Zhang, Xin Zhao, Ying Mou, and Wanlei Xue

Abstract This paper reviews relevant studies at home and abroad, and sorts out carbon emission influencing factors from six aspects: economic growth, demographic factors, industrial structure, energy factors, international trade, and foreign direct investment. The research takes 14 typical countries as samples, and classifies all countries into five categories according to the direction of carbon emission influencing factors, according to promoting factors, and inhibiting factors. The paper is expected to provide a theoretical basis for provinces and cities to compare carbon emission paths horizontally in a more targeted manner and learn from the experience of policy and measures.

Keywords Carbon emissions · Influencing factors · Overview

14.1 Introduction

The climate crisis has become a major global challenge for human society today. As an important force in global climate governance, China has implemented a series of strategies, measures, and actions to address climate change. At present, *the Action Plan for Reaching Carbon Dioxide Peak Before 2030* has mapped a path to carbon peak, neutrality, and a $1 + N$ policy system with a clear timetable, roadmap, and construction plan. It demonstrates China's determination to achieve the "dual carbon goals". Due to the significant differences among Chinese provinces in the economic base, industrial structure, and energy structure, it is necessary to consider our nation jointly to achieve carbon peak and neutrality. However, many provinces have proceeded without cooperation throughout implementation, and have

S. Wu (✉) · L. Zhang
State Grid Energy Research Institute Co. Ltd., Beijing, China
e-mail: wushanshan@sgeri.sgcc.com.cn

X. Zhao · Y. Mou · W. Xue
Economic and Technology Research Institute, State Grid Shandong Electric Power Company, Jinan, China

J. Zhang et al. (eds.), *Environmental Pollution Governance and Ecological Remediation Technology*, Environmental Science and Engineering,
https://doi.org/10.1007/978-3-031-25284-6_14

not pursued the path of carbon peak and neutrality based on provincial conditions. Therefore, studying the factors affecting carbon emissions in typical countries and the magnitude of their role in the past literature would be beneficial for provinces to develop carbon emission reduction strategies and deploy their own goals scientifically, and be beneficial for the central government to coordinate provinces to achieve carbon neutrality milestones in sequence. This research will serve as a theoretical foundation for promoting general synergy and the thorough implementation of the carbon-neutrality target in China.

14.2 Factors Influencing Carbon Emissions

There is a rich body of research on the factors influencing carbon emissions. Initially, scholars focused on the relationship between economic growth and the environment. As the research progresses and with more international carbon emission reduction practices, scholars have discussed mainly the economic development, industrial structure, technological development, demographic factors, energy factors, international trade, and foreign direct investment aspects. However, there are differences in the effects of these factors on carbon emissions. Most studies show that economic growth, population size, population consumption, and industrial structure have positive effects on carbon emissions, while energy efficiency and structure negatively affect carbon emissions. Also, the transmission paths of international trade, foreign direct investment, and technological development are complex, and there is no uniform conclusion on their effects on carbon emissions. These effects are shown in Table 14.1.

14.3 Study and Measurement of Carbon Emission Influencing Factors in Typical Countries

Reviewing the existing studies, it is further found that the influence of each factor on carbon emissions is also different in different regions and at different times. Also, the factors are not independent of each other but affect each other. The influence of each factor on carbon emissions is complex.

Therefore, this paper reviews the relevant literature and classifies all countries into five categories according to the facilitating and inhibiting factors: (1) India is the country where all four factors, economic, demographic, and energy, are facilitating factors. (2) Australia and Italy are the countries where economic, demographic, and energy structure are the facilitating factors and energy intensity is the inhibiting factor. (3) Countries where economy and population are facilitating factors and energy intensity and energy structure are inhibiting factors include the United States, the United Kingdom, France, Germany, China, Japan, and Canada. (4) Brazil and South

Table 14.1 Description of factors and indicators affecting carbon emission (Made by myself)

Factors	Tier 1 Indicators	Characteristics	Direction of effects
Economic development	Economic development	GDP (Halliru et al. 2021; Odugbesan and Adebayo 2020), Per capita GDP (Han and Liu 2020), GDP growth rate (Batool et al. 2022)	+
	Financial development	Domestic credit to the private sector as the ratio of GDP (Batool et al. 2022; Weili et al. 2022)	–
	Urbanization	The urban population as a percentage of the total population (Ahmed et al. 2022; Chang et al. 2021)	Uncertain
Industrial structure	Industrial structure	Proportion of value added of tertiary industry to Regional GDP (Ahmed et al. 2022) Industrial added value as a percentage of GDP (Wang and Wang 2021)	Uncertain
	Advanced Industrialization	Ratio of tertiary industry output value to secondary industry output value (Zhao et al. 2022)	–
	Industrial Rationalization	$TL_{it} = \sum_{m=1}^{n} \left(\frac{Y_m}{Y}\right) \ln \left(\frac{Y_m/L_m}{Y/L}\right)$[1] (Zhao et al. 2022)	Uncertain
Technological development	Technology level	Patent application (Abid et al. 2022; Liu et al. 2021) Number of patent applications granted (Ahmed et al. 2022)	–
	Communication technology	Internet consumers as a percentage of the populace (Batool et al. 2022) Fixed telephone users and mobile cellular users (Weili et al. 2022)	Uncertain
Demographic factors	Population size	Population total (Halliru et al. 2021; Pan et al. 2021), Number of permanent residents (Chang et al. 2021)	+
	Age structure	The proportion of working-age population (Pan et al. 2021) The percentage of the population aged 65 and over (Wang and Wang 2021)	+

(continued)

[1] Y indicates total industrial output value, L denotes total employment in industry, Yi represents the output value of the i industry, and Li indicates the number of employees in the i industry.

Table 14.1 (continued)

Factors	Tier 1 Indicators	Characteristics	Direction of effects
	Consumption structure	The Engel coefficient (Pan et al. 2021)	–
Energy factors	Energy consumption	Energy consumption (Halliru et al. 2021; Odugbesan and Adebayo 2020)	+
	Energy structure	The share of coal consumption in total energy consumption (Ahmed et al. 2022; Chang et al. 2021)	+
	Energy efficiency	GDP per unit of energy use (Irfan et al. 2021)	–
	Energy intensity	Total energy use divided by GDP (Ahmed et al. 2022; Chang et al. 2021)	+
	Renewable energy	Share of RE use to total final energy use (Irfan et al. 2021) Electricity production from renewable sources (Ahmed et al. 2022)	–
International trade	International trade level	Ratio of total trade to the GDP (Irfan et al. 2021) The sum of merchandise exports and imports divided by the value of GDP (Wang and Wang 2021)	Uncertain
	Freedom of trade	KOF globalization (Lu et al. 2022)	+
Foreign direct investment	Foreign direct investment	Net inflows of GDP% (Abid et al. 2022), Total inflow of FDI (Liu et al. 2021) Actual utilization of foreign direct investment (Ahmed et al. 2022)	Uncertain

Africa are the countries where economy and population are the facilitating factors and energy intensity is the inhibiting factor. (5) The countries where the economy is a contributing factor and population and energy intensity are inhibiting factors are Russia and South Korea. The specific classification is shown in Tables 14.2, 14.3, 14.4, 14.5 and 14.6.

14.4 Conclusions

Existing studies on the factors influencing carbon emissions in various countries provide a good basis for further exploring suitable pathways to achieve carbon peak under different regional development bases, but the following deficiencies exist:

Table 14.2 Category 1: countries and the magnitude of the influencing factors (Made by myself)

Facilitating factors: economic growth, population factors, energy intensity and energy structure				
	Economic growth	Population factors	Energy intensity	Energy structure
India	Coefficient per unit of GDP 729.4 (Mengyao et al. 2017) GDP decoupling index 0.8 (Yingming and Jiandong 2019) Percentage of CO_2 90% (Inglesi-Lotz 2018) Coefficient per unit of GDP 0.453832 (Xiaohui et al. 2012) Coefficient per unit of GDP 2.853 (Xinyue 2018) Economic growth CO_2 change 600 Mt (Mengyao et al. 2017) Economic growth coefficient 58.43 (Mengyao et al. 2017)	Urbanization rate coefficient − 1.97E + 04 (Mengyao et al. 2017) Decoupling index for population size 0.2 (Yingming and Jiandong 2019) Percentage of CO_2 − 40%(Inglesi-Lotz 2018) Contribution of demographic factors 4.8 Gigatons CO_2 Total population coefficient 0.457001 (Xiaohui et al. 2012) Coefficient per unit of GDP 2.315 (Xinyue 2018) Urbanization rate coefficient 0.467 (Xinyue 2018) Change in population size CO_2 300 Mt (Mengyao et al. 2017)	Decoupling index of energy intensity −0.07 (Yingming and Jiandong 2019) Percentage of CO_2 170% (Inglesi-Lotz 2018) Energy intensity coefficient 0.103175 (Xiaohui et al. 2012) Energy intensity CO_2 change 20 Mt (Mengyao et al. 2017) Energy intensity coefficient 58.43 (Mengyao et al. 2017)	Renewable energy share coefficient −4.31E + 04 (Mengyao et al. 2017) Decoupling index of the energy structure 0.03 (Yingming and Jiandong 2019) Energy structure CO_2 change 20 Mt (Mengyao et al. 2017)

Table 14.3 Classification 2 countries and the magnitude of the role of influencing factors (Made by myself)

Facilitating factors: economic growth, population factors, energy intensity; inhibiting factor: energy structure

	Economic growth	Population factors	Energy intensity	Energy structure
Australia	Coefficient per unit of GDP 13.12 (Mengyao et al. 2017)	Urbanization rate coefficient -3.84E + 04 (Mengyao et al. 2017) Contribution of demographic factors 0.02 Gigatons CO_2 (Arto and Dietzenbacher 2014)		Renewable energy share coefficient 3.00E + 02 (Mengyao et al. 2017)
Italy	Coefficient per unit of GDP 5.3 (Mengyao et al. 2017) Economic growth contribution 0.54 Gigatons CO_2 (Odugbesan and Adebayo 2020)	Urbanization rate coefficient -6.73E + 05 (Mengyao et al. 2017) Total population contribution 0.43 Gigatons CO_2 (Arto and Dietzenbacher 2014)	Energy intensity contribution – 0.12 Gigatons CO_2 (Odugbesan and Adebayo 2020)	Renewable energy share coefficient 5.3 (Mengyao et al. 2017)

Table 14.4 Classification 3 countries and the magnitude of the role of influencing factors (Made by myself)

Facilitating factors: economic growth, population size; inhibiting factor: energy intensity, energy structure				
	Economic growth	Population factors	Energy intensity	Energy structure
The United States	Coefficient per unit of GDP 145.10 (Mengyao et al. 2017) GDP decoupling index 0.81 (Yingming and Jiandong 2019) Coefficient per unit of GDP 0.383753 (Xiaohui et al. 2012) Economic growth contribution 6.7 Gigatons CO_2 (Odugbesan and Adebayo 2020) Coefficient per unit of GDP 6.754 (Xinyue 2018) Economic growth CO_2 change 3200 Mt (Mengyao et al. 2017)	Urbanization rate coefficient -3.27E + 04 (Mengyao et al. 2017) Decoupling index for population size 0.19 (Yingming and Jiandong 2019) Contribution of demographic factors 2.5 Gigatons CO_2 (Arto and Dietzenbacher 2014) Total population coefficient 0.410776 (Xiaohui et al. 2012) Total population contribution 11 Gigatons CO_2 (Odugbesan and Adebayo 2020) Coefficient of population size −2.845 (Xinyue 2018) Urbanization level coefficient 0.17588 (Xinyue 2018) Change in population size CO_2 1000 Mt (Mengyao et al. 2017)	Decoupling index of energy intensity −1.02 (Yingming and Jiandong 2019) Energy intensity coefficient 0.410776 (Xiaohui et al. 2012) Energy intensity contribution -6 Gigatons CO_2 (Odugbesan and Adebayo 2020) Energy intensity CO_2 change −4000 Mt (Mengyao et al. 2017)	Renewable energy share coefficient −1.01E + 05 (Mengyao et al. 2017) Decoupling index of the energy structure −0.23 (Yingming and Jiandong 2019)

(continued)

Table 14.4 (continued)

Facilitating factors: economic growth, population size; inhibiting factor: energy intensity, energy structure

	Economic growth	Population factors	Energy intensity	Energy structure
Canada	Coefficient per unit of GDP 14.47 (Mengyao et al. 2017) Economic growth contribution 1 Gigatons CO_2 (Odugbesan and Adebayo 2020) Coefficient per unit of GDP 0.3499 (Rüstemoğlu and Andrés 2016)	Urbanization rate coefficient –3.81E + 04 (Mengyao et al. 2017) Contribution of demographic factors 0.5 Gigatons CO_2 (Arto and Dietzenbacher 2014) Total population contribution 1 Gigatons CO_2 (Odugbesan and Adebayo 2020) Urbanization level coefficient 1.0178 (Rüstemoğlu and Andrés 2016)	Energy intensity contribution – 0.5 Gigatons CO_2 (Odugbesan and Adebayo 2020)	Renewable energy share coefficient −6.00E + 02 (Mengyao et al. 2017) Energy structure coefficient – 0.0504 (Rüstemoğlu and Andrés 2016)
United Kingdom	Coefficient per unit of GDP 3.97 (Mengyao et al. 2017) Economic growth contribution 1 Gigatons CO_2 (Odugbesan and Adebayo 2020)	Urbanization rate coefficient 1.01E + 04 (Mengyao et al. 2017) Total population contribution 1.3 Gigatons CO_2 (Odugbesan and Adebayo 2020)	Energy intensity contribution – 1.1 Gigatons CO_2 (Odugbesan and Adebayo 2020) Impact size-0.37	Renewable energy share coefficient −3.02E + 04 (Mengyao et al. 2017)
France	Coefficient per unit of GDP 6.69 (Mengyao et al. 2017) Economic growth contribution 0.9 Gigatons CO_2 (Odugbesan and Adebayo 2020)	Urbanization rate coefficient – 3.55E + 04 (Mengyao et al. 2017) Total population contribution 1 Gigatons CO_2 (Odugbesan and Adebayo 2020)	Energy intensity contribution – 0.7 Gigatons CO_2 (Odugbesan and Adebayo 2020)	Renewable energy share coefficient −5.80E + 03 (Mengyao et al. 2017)

(continued)

Table 14.4 (continued)

Facilitating factors: economic growth, population size; inhibiting factor: energy intensity, energy structure				
	Economic growth	Population factors	Energy intensity	Energy structure
Germany	Coefficient per unit of GDP 0.489929 (Xiaohui et al. 2012) Economic growth contribution 2 Gigatons CO_2 (Odugbesan and Adebayo 2020) Economic growth CO_2 change 600 Mt (Mengyao et al. 2017)	Total population coefficient – 0.060283 (Xiaohui et al. 2012) Total population contribution 2.3 Gigatons CO_2 (Odugbesan and Adebayo 2020) Change in population size CO_2 20 Mt (Mengyao et al. 2017)	Energy intensity coefficient 0.381661 (Xiaohui et al. 2012) Energy intensity contribution – 1.2 Gigatons CO_2 (Odugbesan and Adebayo 2020) Energy intensity CO_2 change – 659 Mt (Mengyao et al. 2017)	Energy structure CO_2 change – 200 Mt (Mengyao et al. 2017)
China	Coefficient per unit of GDP 2115.51 (Mengyao et al. 2017) GDP decoupling index 1.06 (Yingming and Jiandong 2019) Proportion of CO_2 152% (Inglesi-Lotz 2018) Coefficient per unit of GDP 0.359480 (Xiaohui et al. 2012) Impact size 4.76 Coefficient per unit of GDP 1.743 (Xinyue 2018) Economic growth CO_2 change 2000 Mt (Mengyao et al. 2017)	Urbanization rate coefficient -1.56E + 05 (Mengyao et al. 2017) Decoupling index for population size 0.04 (Yingming and Jiandong 2019) Proportion of CO_2 −110% (Inglesi-Lotz 2018) Contribution of demographic factors 5.8 Gigatons CO_2 (Odugbesan and Adebayo 2020) Total population coefficient 0.409451 (Xiaohui et al. 2012) Impact size 0.18 Coefficient per unit of GDP 2.315 (Xinyue 2018) Urbanization level coefficient 0.67614 (Xinyue 2018) Change in population size CO_2 80 Mt (Mengyao et al. 2017)	Decoupling index of energy intensity −0.98 (Yingming and Jiandong 2019) Proportion of CO_2 −60% (Inglesi-Lotz 2018) Energy intensity coefficient – 0.170076 (Xiaohui et al. 2012) Impact size −0.48 Energy intensity CO_2 change – 1000 Mt (Mengyao et al. 2017)	Renewable energy share coefficient −1.26E + 05 (Mengyao et al. 2017) Decoupling index of the energy structure 0.03 (Yingming and Jiandong 2019) Impact size −0.01 Energy structure CO_2 change – 100 Mt (Mengyao et al. 2017)

(continued)

Table 14.4 (continued)

Facilitating factors: economic growth, population size; inhibiting factor: energy intensity, energy structure

	Economic growth	Population factors	Energy intensity	Energy structure
Japan	Coefficient per unit of GDP 42.42 (Mengyao et al. 2017) GDP decoupling index 0.94 (Yingming and Jiandong 2019) Coefficient per unit of GDP 0.336083 (Xinyue 2018) Economic growth contribution 1.4 Gigatons CO_2 (Odugbesan and Adebayo 2020) Economic growth CO_2 change 800 Mt (Mengyao et al. 2017)	Urbanization rate coefficient –1.68E + 04 (Mengyao et al. 2017) Decoupling index for population size 0.06 (Yingming and Jiandong 2019) Total population coefficient 0.623781 (Xinyue 2018) Total population contribution 1 Gigatons CO_2 (Odugbesan and Adebayo 2020) Change in population size CO_2 100 Mt (Mengyao et al. 2017)	Decoupling index of energy intensity −0.22 (Yingming and Jiandong 2019) Energy intensity coefficient 0.178707 (Xiaohui et al. 2012) Energy intensity contribution –0.5 Gigatons CO_2 (Odugbesan and Adebayo 2020) Energy intensity CO_2 change –500 Mt (Mengyao et al. 2017)	Renewable energy share coefficient −1.08E + 04 (Mengyao et al. 2017) Decoupling index of the energy structure −0.75 (Yingming and Jiandong 2019) Energy structure CO_2 change –100 Mt (Mengyao et al. 2017)

Table 14.5 Classification 4 countries and the magnitude of the role of influencing factors (Made by myself)

Facilitating factors: economic growth, demographic factors; inhibiting factors: energy intensity				
	Economic growth	Demographic factors	Energy intensity	Energy Structure
Brazil	Proportion of CO_2 48%(Rüstemoğlu and Andrés 2016) Proportion of CO_2 58.9%(Rüstemoğlu and Andrés 2016) Coefficient per unit of GDP 2.8562 (Rüstemoğlu and Andrés 2016)	Proportion of CO_2 30% (Rüstemoğlu and Andrés 2016) Contribution of demographic factors 0.25Gigatons CO_2 (Arto and Dietzenbacher 2014) Impact size 0.30 Proportion of CO_2 37% (Rüstemoğlu and Andrés 2016) Urbanization level coefficient −2.3069 (Rüstemoğlu and Andrés 2016)	Proportion of CO_2 8% (Rüstemoğlu and Andrés 2016) Impact size 0.03 Proportion of CO_2 −3.5% (Rüstemoğlu and Andrés 2016)	Impact size 0.06 Energy structure coefficient −0.7281 (Rüstemoğlu and Andrés 2016)
South Africa	Proportion of CO_2 47%(Inglesi-Lotz 2018)	Proportion of CO_2 110% (Inglesi-Lotz 2018)	Proportion of CO_2 −5%(Inglesi-Lotz 2018)	

Table 14.6 Classification 5 countries and the magnitude of the role of influencing factors (Made by myself)

Facilitating factors: economic growth; inhibiting factors: demographic factors, energy intensity				
	Economic growth	Demographic factors	Energy intensity	Energy Structure
Russia	Proportion of CO_2 −50% (Inglesi-Lotz 2018) Proportion of CO_2 191.5% (Rüstemoğlu and Andrés 2016)	Proportion of CO_2 −110% (Inglesi-Lotz 2018) Contribution of demographic factors −1.1 Gigatons CO_2 (Arto and Dietzenbacher 2014) Proportion of CO_2 22.9% (Rüstemoğlu and Andrés 2016)	Proportion of CO_2 43% (Inglesi-Lotz 2018) Proportion of CO_2 −15.5% (Rüstemoğlu and Andrés 2016)	
Korea	Coefficient per unit of GDP 35.06 (Mengyao et al. 2017)	Urbanization rate coefficient −4.98E + 04 (Mengyao et al. 2017)		Renewable energy share coefficient −1.66E + 04 (Mengyao et al. 2017)

(1) Global research on the factors influencing carbon emissions emerged in Europe and the United States in the 1990s and culminated in China in 2012. This means that many of the literature findings use data from before 2010. Currently, the economic and political context of the country has changed dramatically, and attention needs to be paid to timeliness when referring to existing studies.
(2) Most of the relevant comparative studies focus on the comparison of the world's major economies or major organization members, such as BRIC and OECD internal comparisons. Few studies have been conducted for countries such as Argentina, Mexico, and South Africa. Thus, the comprehensiveness of research needs to continue to be improved.
(3) In the existing literature, the factors affecting carbon emissions include economic growth, demographic factors, industrial structure, energy intensity and energy mix, international trade, and foreign direct investment. However, when comparing countries, most studies focus more on economic growth, demographic factors, energy intensity, and energy structure. On the one hand, it indicates that the above factors are universal for the study of carbon emissions; on the other hand, the impact of industrial structure, international trade, and foreign direct investment on carbon emissions could be further studied.

Acknowledgements This work was supported by Research on Shandong Province's Economic and Electricity Development Adapting to Dual-Carbon Strategy and Dual-cycle New Pattern.

References

Abid A, Mehmood U, Tariq S, Haq ZU (2022) The effect of technological innovation, FDI, and financial development on CO_2 emission: evidence from the G8 countries. Environ Sci Pollut Res 29:11654–11662
Ahmed M, Shuai C, Ahmed M (2022) Influencing factors of carbon emissions and their trends in China and India: a machine learning method. Environ Sci Pollut Res
Arto I, Dietzenbacher E (2014) Drivers of the growth in global greenhouse gas emissions. Environ Sci Technol 48:5388–5394
Batool Z, Raza SMF, Ali S, Abidin SZU (2022) ICT, renewable energy, financial development, and CO_2 emissions in developing countries of East and South Asia. Environ Sci Pollut Res 29:35025–35035
Chang K, Du Z, Chen G, Zhang Y, Sui L (2021) Panel estimation for the impact factors on carbon dioxide emissions: a new regional classification perspective in China. J Clean Prod 279:123637
Halliru AM, Loganathan N, Golam Hassan AA (2021) Does FDI and economic growth harm environment? Evidence from selected West African countries. Transnational Corporations Rev 13:237–251
Han Y, Liu Y (2020) Influence of economic growth and energy consumption on carbon emissions in Yangtze river economic belt. IOP Conf Ser: Earth Environ Sci 601:012007
Inglesi-Lotz R (2018) Decomposing the South African CO_2 emissions within a BRICS countries context: signalling potential energy rebound effects. Energy 147:648–654
Irfan M, Mahapatra B, Ojha RK (2021) Examining the effectiveness of low-carbon strategies in South Asian countries: the case of energy efficiency and renewable energy. Environ Dev Sustain 23:11936–11952

Liu X, Wahab S, Hussain M, Sun Y, Kirikkaleli D (2021) China carbon neutrality target: revisiting FDI-trade-innovation nexus with carbon emissions. J Environ Manage 294:113043

Lu Z, Mahalik MK, Mahalik H, Zhao R (2022) The moderating effects of democracy and technology adoption on the relationship between trade liberalisation and carbon emissions. Technol Forecast Soc Chang 180:121712

Mengyao H, Weidong L, Zhipeng T, Yan X (2017) Analysis of factors affecting carbon emissions in major countries in the world—based on a variable coefficient panel model. Resour Sci 39:2420–2429

Odugbesan JA, Adebayo TS (2020) The symmetrical and asymmetrical effects of foreign direct investment and financial development on carbon emission: evidence from Nigeria. SN Appl Sci 2:1982

Pan C, Wang H, Guo H, Pan H (2021) How do the population structure changes of China affect carbon emissions? An empirical study based on ridge regression analysis. Sustainability 13:3319

Rüstemoğlu H, Andrés AR (2016) Determinants of CO_2 emissions in Brazil and Russia between 1992 and 2011: a decomposition analysis. Environ Sci Policy 58:95–106

Wang Q, Wang L (2021) The nonlinear effects of population aging, industrial structure, and urbanization on carbon emissions: a panel threshold regression analysis of 137 countries. J Clean Prod 287:125381

Weili L, Khan H, Khan I, Han L (2022) The impact of information and communication technology, financial development, and energy consumption on carbon dioxide emission: evidence from the belt and road countries. Environ Sci Pollut Res 29:27703–27718

Xiaohui S, Yufen Z, Yimei W, Yinchang F (2012) Analysis of the impact of population factors on carbon emissions based on IPAT extended model. Environ Sci Res 25:109–115

Xinyue Z (2018) Comparative study on influencing factors of carbon emissions between China, America and India based on STIRPAT model (Master). Shanxi University of Finance and Economics

Yingming Y, Jiandong S (2019) Research on the decoupling and driving factors of carbon emissions from energy consumption in major countries in the world. Coal Eng 51:173–177

Zhao J, Jiang Q, Dong X, Dong K, Jiang H (2022) How does industrial structure adjustment reduce CO_2 emissions? Spatial and mediation effects analysis for China. Energy Econ 105:105704

Chapter 15
Study on the Determination of 14 Heavy Metals in the Soil by ICP-MS with Microwave Digestion

Xiaohui Wu, Qing Xu, Qiqiang Niu, Wanxia Xing, Cong Hou, and Hongqi Shao

Abstract In order to establish an efficient sample preparation method for heavy metal elements analysis in the soil samples, the effect of acid system, microwave digestion time and temperature were studied. 14 heavy metals, such as Be, V, Cr, Mn, Co, Ni, Cu, Zn, Mo, Cd, Sb, Tl, Pb, U, were analyzed by inductively-coupled plasma mass spectrometry (ICP-MS) under inductively coupled plasma collision mode. The results showed that the calibration curve of 14 heavy metals was linear and the limit of detection was in the range of 0.01–1.00 mg/kg. The precision was verified by certified reference soil samples, with relative standard deviation of 0.90–5.88%, which were both satisfactory.

Keywords Microwave digestion · Heavy metals · Soil · Preparation method · Inductively-coupled plasma mass spectrometry

15.1 Introduction

With the development of society and economy, heavy metals pollution in the soil was mainly caused by human activities (Sun and Li 2015). Heavy metals pollution in the soil would eventually endanger the health of mankind because of the characteristics of long-term accumulation, and irreversible (Mbaa et al. 2022). Therefore, how to analyze the heavy metals in the soil efficiently, quickly and accurately was of great significance to soil environmental quality supervision and protection. According to

X. Wu · Q. Xu (✉) · Q. Niu · H. Shao
Shandong Institute of Scientific and Technical Information, Jinan 250101, China
e-mail: sgzxxq@163.com

W. Xing
Shandong College of Highway Technician, Jinan 250101, China

C. Hou (✉)
Shandong Provincial Eco-Environment Monitoring Center, Jinan 250101, China
e-mail: huayao_hc@163.com

J. Zhang et al. (eds.), *Environmental Pollution Governance and Ecological Remediation Technology*, Environmental Science and Engineering,
https://doi.org/10.1007/978-3-031-25284-6_15

the sample preparation and analysis methods studied before (Liu et al. 2020), ICP-MS was a technique used for elemental and isotopic analysis, which combines the remarkable characteristics of ICP-MS for atomising and ionising samples with the advantages of excellent sensitivity, low interference, low Limit of Detection (LOD) and simultaneous analysis of multiple elements (Zhao et al. 2018). In this paper, the acid system, microwave digestion time and temperature were investigated to analyse heavy metals in the soil to provide a new method. Method validation, such as LOD, linearity, recovery, and precision were determined.

15.2 Experimental

15.2.1 Instrumentation and Reagents

Instrumentation. The inductively-coupled plasma mass spectrometer used in this work was an Aglient 7700 × ICP-MS (Aglientt Technologies, America); Microwave digestion system and Milli-Q system were from Milston and Merck respectively.

Reagents. Multi-elements standard solution, ICP-MS tuning solution and internal standard solution(10 mg/L) were all purchased from Aglient Technologies; All certified reference material for soil、HNO_3、HF and H_2O_2 were analytical reagent.

15.2.2 Operating Conditions

Before each experiment, the ICP-MS was tuned by an aqueous multi-elements standard solution (0.1 μg/L) to ensure the instrument meeting the test requirements. Operating conditions were shown in Table 15.1.

Table 15.1 Operating conditions for ICP-MS spectrometer

Operating conditions	Values	Operating conditions	Values
RF Power/W	1550	Auxiliary gas flow rate/L min^{-1}	0.9
Sampling depth/mm	8.0	Integral time/s	1
Carrier gas flow rate/L min^{-1}	1.08	Oxide ion yield/%	1.10
Plasma gas flow rate/L min^{-1}	15.0	Double-charged ion yield/%	1.44

Table 15.2 Operating programme for the microwave digestion

Step	Heating Time(min)	Final temperature(°C)	Dwell time(min)
1	5	120	2
2	5	150	5
3	5	190	20

15.2.3 Sample Preparation and Digestion

Before ICP-MS analysis techniques could be performed on solid materials, the sample need further preparation into soluble liquid. The soil sample (0.1 g) was weighed and transferred directly into the microwave digestion oven, Subsequently, 4 mL of HNO_3, 2 mL of H_2O_2 and 2 mL of HF were added successively to digest the soil sample, digestion programme was shown in Table 15.2. After cooling at room temperature, all the digestion liquors were quantitatively diluted to 50 mL with ultra-pure water.

15.3 Results and Discussion

15.3.1 Preparation Conditions Investigation

Selection of sample weight. According to EPA3051 (EPA Method 3051A, Method 3051A (SW-846) 2007), EPA3052 (EPA Method 3052, SW-846 Test Method 3052 1996) 和 EPA6020 (EPA Method 6020B, Method 6020B (SW-846) 2014), the weight of soil samples was determined to be 0.10 g.

Selection of acid system and amounts. Nitric acid was a strong oxidising agent and was widely used in ICP-MS as acid medium because it behaves ideally under microwave conditions. Since it was known that hydrogen peroxide had clarifying effect, the mixture of nitric acid and hydrogen peroxide had already been used as digestion reagents to improve oxidizability (Wang et al. 2006). HF could destroy soil lattice to extract heavy metal elements which was existed in the soil lattice. During ICP-MS analysis, ion interference was mainly caused by the addition of Ar, O, H, Cl, and N, mainly by the interference element with the mass–charge ratio less than 82 (Zhao and Zhang 2014), so Cl^-could cause ion interference. The specific conditions were shown in Table 15.3.

Three kinds of acid system (A:HNO_3–HCl–HF; B:HNO_3–HCl–HF–H_2O_2; C:HNO_3–HF–H_2O_2) were used as digestion solution. The results showed that, for system A, without H_2O_2, the digestion degree of organic matter in the soil was low, the release of metal ions was incomplete, and the recovery was low; for system B, the recovery of element Pb was low which could be induced by the reaction of Cl^- with Pb^{2+} to form precipitation of $PbCl_2$; for system C, after microwave digestion, the

Table 15.3 Common interference ions

Mass number	Analyted elements	Interference ions
51	V	$^{35}Cl^{16}O$, $^{37}Cl^{14}N$
52	Cr	$^{36}Ar^{16}O$, $^{40}Ar^{12}C$, $^{35}Cl^{16}OH$, $^{37}Cl^{14}NH$
55	Mn	$^{37}Cl^{18}O$, $^{23}Na^{32}S$
60	Ni	$^{44}Ca^{16}O$, $^{23}Na^{37}Cl$
63	Cu	$^{40}Ar^{23}Na$, $^{12}C^{16}O^{35}Cl$, $^{12}C^{14}N^{37}Cl$

digestion solution was clear, indicating that this acid system could digest soil sample well. Therefore, HNO_3–HF–H_2O_2 acid system was selected as digestion solution.

Selection of digestion time and temperature. Digestion degree would be affected by temperature and time, so microwave digestion temperature and time were optimized in this part. The results were shown in Fig. 15.1 and Fig. 15.2.

Temperature of 170 °C, 180 °C and 190 °C were selected as the final digestion temperature to be studied in this experiment. As shown in Fig. 15.1, the higher final temperature, the better recovery, so 190 °C was the best final digestion temperature.

Digestion time was maintained at 10 min, 15 min, 20 min and 25 min. As shown in Fig. 15.2, the soil digested completely with prolonging the digestion time, the digestion curve would be mild after 20 min, so 20 min was selected as maintaining time.

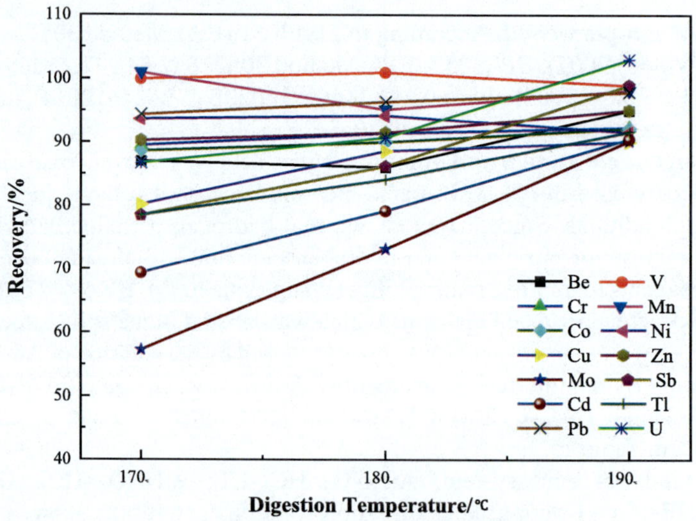

Fig. 15.1 The relation between recovery rate and digestion temperature

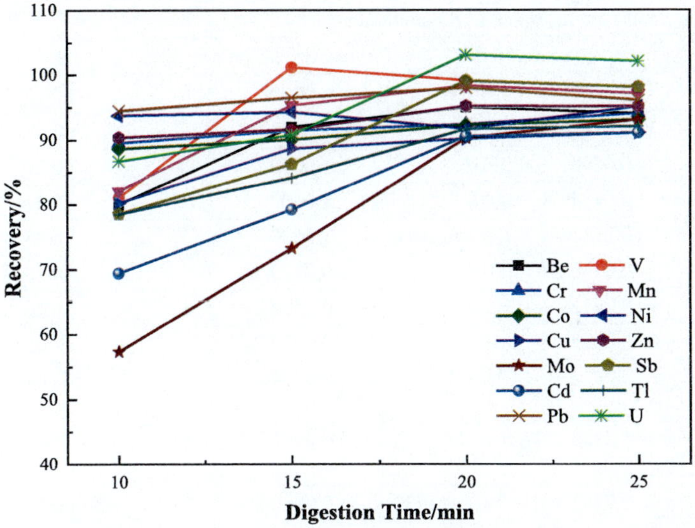

Fig. 15.2 The relation between recovery rate and digestion time

15.3.2 *Selection of Internal Standard Element*

The elements, such as Sc, Ge, Rh, In, Re, Bi, were selected as internal standard elements to analyse heavy metal elements. There would be interference of $^{28}Si^{16}O^{1}H$ and $^{12}C^{16}O_2{}^{1}H$ when element Sc was selected as internal standard element, and Sc, In, Re, and Bi might be presented in the soil samples, all these could affect the recovery of internal standard element, so element Rh was selected as internal standard element.

15.3.3 *Limit of Detection*

According to concentration gradient in Table 15.4, the calibration curve was established with the concentration of elements as the horizontal and the ratio as the longitudinal.

Table 15.4 Recommended range of the standard curve

Elements		Recommended range (μg/L)
Low concentration element	Cd, Tl, Be, Mo, U, Sb	0, 0.10, 0.20, 0.50, 1.00, 5.00, 10.00
High concentration element	Cr, Ni, Cu, Zn, Pb, Co, V, Mn	0, 1.00, 10.0, 50.0, 100, 200, 500

Table 15.5 Linear parameter and LOD of elements

Elements	Linear equation	Correlation coefficient R^2	LOD (mg/kg)
Be	$y = 0.0040x + 6.5516 \times 10^{-5}$	1.0000	0.12
V	$y = 0.0242x + 3.4034 \times 10^{-4}$	1.0000	0.02
Cr	$y = 0.0298x + 0.0110$	1.0000	0.21
Mn	$y = 0.0184x + 0.0028$	1.0000	0.21
Co	$y = 0.0541x + 2.1238 \times 10^{-4}$	1.0000	0.03
Ni	$y = 0.0148x + 0.0013$	0.9999	0.20
Cu	$y = 0.0412x + 0.0088$	0.9999	0.20
Zn	$y = 0.0073x + 0.0099$	0.9999	1.00
Sb	$y = 7.0972 \times 10^{-4}x + 2.2255 \times 10^{-4}$	0.9999	0.04
Mo	$y = 6.3383 \times 10^{-4}x + 1.0775 \times 10^{-5}$	0.9999	0.03
Cd	$y = 2.9341 \times 10^{-4}x + 2.1739 \times 10^{-5}$	0.9999	0.02
Tl	$y = 0.0022x + 1.0291 \times 10^{-5}$	0.9999	0.02
Pb	$y = 0.0030x + 0.0026$	0.9999	0.17
U	$y = 0.0035x + 5.4684 \times 10^{-6}$	0.9999	0.01

Limit of Detection was tested and calculated according to HJ168-2020, the results were shown in Table 15.5. All the calibration curves yielded the R^2 higher than 0.9999 using linear regression.

15.3.4 Precision

The precision of analytical method were researched by two kinds of certified reference soil samples (GSS-4, GSS-18) according to preparation method mentioned before. The measured values and certified values were depicted in Table 15.4, and relative standard deviations ($n = 6$) were lower than 5.88% in all cases (Table 15.6).

15.4 Conclusions

In this paper, the determination of 14 heavy metal elements in the soil using microwave digestion followed by ICP-MS analysis has been performed. The effect of different acid system, microwave digestion time and temperature were investigated to ensure the appropriate digestion conditions, and HNO_3–HF–H_2O_2 acid system was selected as digestion system. The introduction of HNO_3–HF–H_2O_2 acid system

Table 15.6 Precision and accuracy test results

	GSS-4			GSS-18		
	Measured values (mg/kg)	Certified values (mg/kg)	RSD%	Measured values (mg/kg)	Certified values (mg/kg)	RSD%
Be	1.62	1.85	4.69	1.67	1.70	2.51
V	240	247	1.30	63.1	66	1.02
Cr	373	370	1.24	55.3	55	2.82
Mn	1410	1420	3.11	529	529	1.47
Co	22.0	22.0	3.63	10.0	10.2	0.90
Ni	65.8	64	2.37	25.2	25	1.97
Cu	42.3	40	1.26	19.7	19.5	0.95
Zn	211	210	3.01	63.2	63	2.23
Sb	6.88	6.30	2.03	0.88	0.88	4.15
Mo	2.82	2.60	2.21	0.58	0.61	4.11
Cd	0.33	0.35	2.69	0.15	0.15	3.58
Tl	0.92	0.94	4.90	0.56	0.55	1.38
Pb	61.1	58.0	3.02	19.9	20	1.97
U	6.31	6.70	5.88	2.17	2.3	3.01

could reduce the types of acids and avoid the interference induced by the use of HCl on the premise of ensuring accuracy. According to the index of the LOD, correlation coefficient, recovery, and precision, it was found that the sample preparation method mentioned before was highly sensitive for simultaneous quantification of 14 heavy metal elements. It could provide a new method for the analysis of heavy metals in the soil to support soil environmental quality supervision and protection.

References

EPA Method 3051A, Method 3051A (SW-846) (2007) Microwave assisted acid digestion of sediments, sludges, and oils [S]. Washington, U.S. EPA

EPA Method 3052, SW-846 Test Method 3052 (1996) Microwave assisted acid digestion of siliceous and organically based matrices [S]. Washington, U.S. EPA

EPA Method 6020B, Method 6020B (SW-846) (2014) Inductively coupled plasma-mass spectrometry [S]. Washington, U.S. EPA

Liu X, Sun W, Wen T (2020) Determination of 23 metal elements in detailed soil survey samples by inductively coupled plasma-mass spectrometry with three acid stepwise digestion. Rock Min Anal 39(5):793–800

Mbaa B, Ppm B, Sp C, Itu A (2022) Health risk assessment of heavy metals concentration from soil; a case study of the mayo-dallah in southern area of chad. J Radiat Res Appl Sci 15(1):130–138

Sun J, Li M (2015) Determination of heavy metal elements in various soil by microwave assisted extraction with acid mixture and ICP-MS. Environ Chem 34(4)

Wang J, Nakazato T, Sakanishi K (2006) Single step microwave digestion with HNO_3 alone for determination of trace elements in coal by ICP spectrometry. Talanta 68:1584–1590

Zhao X, Zhang L, Zhang J (2014) Spectral interference and elimination of environmental analysis with ICP-MS. Environ Monit China 30(3):101–106

Zhao S, Cao S, Luo L et al (2018) A preliminary investigation of metal element profiles in the serum of patients with bloodstream infections using inductively-coupled plasma mass spectrometry (ICP-MS). Clin Chim Acta 485:323–332

Chapter 16
Effects of Vegetation Recovery Modes on Soil Nutrients in Mine Wasteland of Steppe

Dan Shan, Tiegang Zhang, Hao Rong, Kairan Xu, and Fei Zhang

Abstract The community structure of varying artificial vegetation, main soil nutrients were investigated in a typical dump of a large-scale prairie open-pit coalmine. The results showed that the artificial vegetation significantly increased the soil nitrogen (N) and organic carbon (OC) content. The soil total nitrogen (TN) and available nitrogen (AN) content in vegetated areas containing *Medicago sativa* and *Melilotus suaveolens* were significantly higher than in other vegetated areas. Moreover, the soil organic carbon (OC) content under *Populus simonii*, *Pinus tabulaeformis*, and *Caragana Korshinskii* communities was significantly higher than that under other vegetation types. Compared with natural restoration vegetation, artificial vegetation was found to have an insignificant impact on soil total phosphorus (TP), available phosphorus (AP) and total potassium (TK) content. The soil fertility of the dump platform was significantly higher than that of the slope, and the correlation between the content of soil available nitrogen (AN), available phosphorus (AP), and available potassium (AK) and the reference values was higher, suggesting the content of available nutrients was a better indicator for soil fertility in the dumpsite of the prairie mining area. The results presented herein provide scientific support for the optimization of vegetation restoration models in the dumps of open-pit coalmines.

Keywords Vegetation modes · Soil nutrient · Mine wasteland

D. Shan (✉) · T. Zhang
Yinshanbeilu Grassland Eco-Hydrology National Observation and Research Station, China Institute of Water Resources and Hydropower Research, Beijing 100038, China
e-mail: mksshd@126.com

D. Shan · T. Zhang · H. Rong · K. Xu · F. Zhang
Institute of Water Resources for Pastoral Area Ministry of Water Resources, Hohhot 010020, Inner Mongolia, China

J. Zhang et al. (eds.), *Environmental Pollution Governance and Ecological Remediation Technology*, Environmental Science and Engineering,
https://doi.org/10.1007/978-3-031-25284-6_16

16.1 Introduction

Artificial vegetation is the most commonly used and most effective method of ecological restoration of mine dumps. This method can effectively increase the coverage of plant communities and species diversity. The enhancement of plant roots can improve the soil environment, increase soil fertility for production, and improve ecosystem diversity and stability (Henry and Houerou 2000; Houerou 2000; Li et al. 2015). Artificial reclamation is an important part of recovering and protecting the environments of mining areas during modern coal production (Li et al. 1997). Soil is the site of many ecological processes in the ecosystem and determines the relationship between the species composition of vegetation communities and primary productivity in arid and semi-arid ecosystems. Quantifying the relationship between soil characteristics and vegetation landscapes would facilitate an understanding of ecosystem processes and vegetation management (Asensio et al. 2013; Zhao et al. 2016; Zhang et al. 2018). Ecological restoration of mine dumps located in arid and semi-arid areas is challenging because of the complexity of the material composition, the soil, the low precipitation, the short frost-free period, and the relatively fragile ecological environment. It needs is essential for the ecological restoration of mine dumps for selecting artificial vegetation types according to natural conditions, soil characteristics, and ecological construction. This study investigated soil nutrients under vegetation recovery modes in coalmine dumps of steppe and plant communities were explored for the restoration of mine wasteland.

16.2 Materials and Method

16.2.1 Study Area

The open-pit coalmine investigated in this study is located in Xilinhot, Xilin Gol League, Inner Mongolia, China. This area has a mid-temperate arid and semi-arid climate, with average annual precipitation of 293 mm, average annual evaporation of 1835 mm, and an average annual temperature of 1.7 °C. The soil in the area is typical chestnut soil and the vegetation coverage of 40%. The slope height of each step is about 20 m. The stepped slope is covered with soil after coal gangue and soil mixed to form a covered slope to depths of 50–80 cm (Liu and Xing 2009). The vegetation restoration of the Southern Dump successively conducted in 2012.

16.2.2 Method

Vegetation recovery modes were selected for the vegetation restoration area, including arbor forests (*Ulmus pumila*, *Pinus tabulaeformis*, *Populus simonii* and

Platycladus orientalis), bushes (*Caragana korshinskii* and *Artemisia desertorum*), and artificial grassland (*Melilotus suaveolens* and *Medicago sativa*). Selected information regarding each type of vegetation community is shown in Table 16.1.

At the end of plant growth (early September 2021), nine samples of soil were collected from random locations within plots of each vegetation type. Soil was collected from depths of 0–20 cm and 20–40 cm from each sample site using soil drills. It was then placed in aluminum boxes and transported to the laboratory for analysis. The soil measurement indicators included pH, soil total nitrogen (TN), available nitrogen (AN), total phosphorus (TP), available phosphorus (AP), total potassium (TK), available potassium (AK) and soil organic carbon (OC).

Table 16.1 Selected plant community information

No	Vegetation type	Restoration duration (year)	Distribution	Height (cm)	Coverage (%)	Canopy density
I	*Ulmus pumila* 1	9	Platform	172	–	0.20
II	*Ulmus pumila* 2	8	Slope	78	–	0.26
III	*Pinus tabulaeformis*	9	platform	211	–	0.14
IV	*Populus simonii*	8	Platform	194	–	0.17
V	*Platycladus orientalis*	10	Slope	65	54	–
VI	*Caragana Korshinskii* 1	8	Platform	71	48	–
VII	*Caragana Korshinskii* 2	9	Slope	64	30	–
VIII	*Artemisia desertorum* 1	9	Platform	45	57	–
IX	*Artemisia desertorum* 2	9	Slope	42	36	–
X	*Melilotus suaveolens*	8	Platform	47	43	–
XI	*Medicago sativa*	8	Platform	34	45	–
CK	Natural restoration	11	Platform	25	38	–

16.3 Results

16.3.1 *Soil pH and Nitrogen of Vegetation Recovery Modes*

The large dumpsite formed by open-site coal mining in the grassland area has changed the original mechanical composition and structure of the soil, resulting a barren condition, poor water retention capacity, and low microbial activity (Shan et al. 2017). The soil pH value of each test plot was > 8.5, suggesting an alkaline characteristic, which is not conducive to the growth and development of plants. This also contributes to difficulty restoring the vegetation in abandoned mining areas. No significant difference was observed between control and artificial vegetation areas ($p > 0.05$). Changes in the main soil nutrients in the 0–20 cm and 20–40 cm soil layers of different artificial vegetation types were identified for comparison analysis. Soil TN and AN in the grassland open-pit mine dump, suggesting the significant influence of artificial vegetation on N cycling. The TN in the soil to the depth of 0–20 cm was significantly different between the 11 artificially vegetated areas, which were all significantly higher than in the control area ($p < 0.05$). Among all vegetation areas, the soil TN in the *Pinus tabulaeformis* community (786.9 mg/kg) was highest, followed by the *Medicago sativa* and *Melilotus suaveolens* communities. Similarly, the TN in the soil to the depth of 20–40 cm was also significantly different between different communities ($p < 0.05$), with the highest TN being observed in *Medicago sativa* community, followed by *Melilotus suaveolens*. However, *Artemisia desertorum* community on the dump slope had the lowest soil TN to the depth of 20–40 cm, which was significantly lower than the control areas. The soil AN and TN had similar changes among investigated plots, with the highest content being observed to the depth of 0–20 cm in the *Medicago sativa* community, followed by *Caragana korshinskii* and *Pinus tabulaeformis* communities on the dump platform. For the soils to the depth of 20–40 cm, the AN content was highest in *Ulmus pumila* community, followed by *Melilotus suaveolens*, with the lowest content observed in the control (50.14 mg kg^{-1}).

16.3.2 *Phosphorus and Potassium of Vegetation Recovery Modes*

With the exception that the average soil TP content to the depth of 20–40 cm is higher than that to the depth of 0–20 cm, the mean values of TN, AN, AP, TK, AK, and OC content in the soil from each artificial vegetation type were higher to the soil depth of 0–20 cm than 20–40 cm. As shown in Tables 16.2 and 16.3, vegetation restoration practices changed the main nutrient elements. When compared with control areas, the TP content in artificially vegetated areas did not show a significant increase while the soil TP content in the *Artemisia desertorum* and *Caragana korshinskii* communities were relatively lower. The changes in the soil AP content of artificially vegetated areas

Table 16.2 Soil nutrient in the 0–20 cm soil layer in different vegetated areas (mg kg^{-1}/%)

No	TN	AN	TP	AP	TK	AK	OC%	pH
I	480.4e	69.09ab	147.1ab	0.9a	31.9a	144.8c	0.72d	8.8a
II	404.5b	86.14bc	150.5ab	1.6a	29.8a	103.4a	0.58c	8.8a
III	786.9i	110.61d	232.2e	2.0ab	26.3a	135.6bc	0.71d	8.9a
IV	508.1f	68.91ab	238.2e	1.1a	29.8a	112.6ab	0.46b	8.8a
V	478.3e	61.31a	155.6ab	1.3a	27.6a	93.2a	0.34a	8.7a
VI	420.6c	111.83d	163.9bc	1.4a	32.0a	105.7a	0.45b	8.8a
VII	412.6bc	85.30bc	135.4a	1.3a	31.1a	113.5ab	0.35a	8.6a
VIII	458.5d	98.17d	144.3ab	1.8sb	30.8a	136.1bc	0.54bc	8.6a
IX	398.0b	98.56c	157.4ab	1.6ab	28.4a	126.8b	0.52bc	8.5a
X	598.3g	67.60ab	209.3d	2.8b	28.1a	126.0b	0.70d	8.8a
XI	606.4g	127.17e	183.1c	2.9b	30.4a	110.9ab	0.58c	8.7a
CK	275.2a	82.34b	168.4bc	1.7ab	25.9a	99.2a	0.34a	9.0a
Mean	485.65	88.92	173.78	1.7	29.34	117.32	0.52	8.8

Note Significant at $p < 0.05$. Lowercase letters indicate comparison between treatments

were not obvious, suggesting little effect on soil P by different vegetation restoration practices. The soil TK content in each investigated plot was between 25.6 g kg^{-1} and 31.9 g kg^{-1}, with insignificant differences between the naturally restored control plots and the artificially vegetated treatments, and insignificant changes between vegetation types. The AK varied from 73.1 g kg^{-1} to 144.8 g kg^{-1}, with the lowest value observed at soil depths of 20–40 cm in the control plot. The magnitude of changes of AK was higher than that of TK between treatments. Soil AK to the depth of 0–20 cm in *Ulmus pumila* community and to 20–40 cm in *Caragana korshinskii* community was significantly higher than in others. The TK in the soil of the dumpsite was high, with a moderate AK content, suggesting sufficient K in the dumpsite available to plant uptake.

16.3.3 Soil Organic Carbon of Vegetation Recovery Modes

Comparative analysis was performed to investigate changes in soil OC in varying vegetated areas. The soil OC to the depth of 0–20 cm in the control areas, *Platycladus orientalis* and *Caragana Korshinskii* communities on the dump slope was significantly lower, while it was significantly higher in the *Ulmus pumila, Pinus tabulaeformis*, and *Melilotus suaveolens* communities on the dump platform than in others ($p < 0.05$). A lower soil OC content was observed in the 20–40 cm soil layer of the naturally restored control and the *Artemisia desertorum* communities (VIII, IX). The soil OC content to the depth of 20–40 cm in the *Pinus tabulaeformis, Populus*

Table 16.3 Soil nutrient in the 20–40 cm soil layer in different vegetated areas (mg kg^{-1}/%)

No	TN	AN	TP	AP	TK	AK	OC%	pH
I	576.8g	104.62d	263.8f	1.9a	29.4a	94.3bc	0.49bc	9.0a
II	450.2de	50.52a	173.1d	0.8a	28.3a	81.0ab	0.48bc	8.8a
III	542.0f	78.20bc	213.7e	1.2a	27.1a	116.1cd	0.63c	9.0a
IV	420.7c	85.66cd	267.4f	1.7a	28.4a	135.6d	0.52c	8.9a
V	444.1d	87.57cd	132.6bc	1.3a	29.9a	92.9bc	0.49bc	8.7a
VI	467.7e	77.84bc	176.7d	1.0a	29.5a	98.6c	0.57c	8.8a
VII	406.8c	56.71a	143.0c	0.8a	27.7a	79.5ab	0.40ab	8.8a
VIII	324.7b	54.82a	116.0b	1.6a	31.6a	87.4b	0.35a	8.9a
IX	304.4a	64.53ab	98.3a	1.5a	26.2a	76.1a	0.33a	8.7a
X	616.1h	102.94d	253.8f	2.2a	27.8a	138.2d	0.47bc	8.8a
XI	600.5h	86.94cd	170.7d	2.0a	28.6a	86.4b	0.39ab	8.9a
CK	336.7b	50.14a	123.5bc	1.4a	25.6a	73.1a	0.32a	9.0a
Mean	457.56	75.04	177.72	1.5	28.34	96.60	0.45	8.9

Note Significant at $p < 0.05$. Lowercase letters indicate comparison between treatments

simonii, and *Caragana Korshinskii* communities were significantly higher than that in other vegetated areas.

16.4 Conclusions

The soil N and OC under vegetation recovery modes increased significantly, and the TN and AN in the *Medicago sativa* community and *Melilotus suaveolens* community were significantly higher than in other artificial vegetation types. The soil OC content in the areas vegetated with *Pinus tabulaeformis, Populus simonii,* and *Caragana korshinskii* was significantly higher than that of other vegetation types. No significant changes in soil TP, AP, and TK were observed in artificially vegetated areas relative to the control. However, artificial vegetation practices had a significantly higher impact on AK than TK, with AK content significantly higher in *Ulmus pumila* community and *Caragana korshinskii* community in the dumpsite platform than other vegetation types.

When compared with the surrounding natural grassland, the TN, AN, OC, and P in the soil increased slightly, but remained at a low level after vegetation restoration. However, there was sufficient K in the soil, with no significant K changes observed between natural restoration and artificial vegetation. Therefore, the lack of nutrients such as soil N and P were likely the limiting factor for vegetation restoration.

Acknowledgements This work was supported by the Inner Mongolia Autonomous Region Science and Technology Plan Project (2019GG023), Ordos City Science and Technology Cooperation Major Project (2021EEDSCXQDFZ2016).

References

Asensio V, Covelo EF, Kandeler E (2013) Soil management of copper mine tailing soils—sludge amendment and tree vegetation could improve biological soil quality [J]. Sci Total Environ 456–457:82–90

Henry N, Houerou L (2000) Restoration and rehabilitation of arid and semiarid Mediterranean ecosystems in North Africa and West Asia: a review [J]. Arid Soil Res Rehabil 14(1):3–14

Houerou HL (2000) Restoration and rehabilitation of arid and semiarid Mediterranean ecosystems in North Africa and West Asia: a review [J]. Arid Soil Res Rehabil 14(1):3–14

Li QF, Cao JY, Zhang SL et al (1997) Vegetation restoration in Zungeer open-cut coal mine—natural recovery in soil refusal site [J]. Grassland China 2:23–25

Li JC, Wang X, Yue JY et al (2015) Evaluation on soil ecologic fertility during vegetation succession in Antaibao open pit [J]. Res Soil Water Conserv 22(1):66–71

Liu WM, Xing WY (2009) Research on dump process in Shengli Donger surface mine [J]. Opencast Mining Technol 4:14–16

Shan D, He JL, Xing ED et al (2017) Effects of microbial fertilizer on microorganism and soil enzyme activity in coal mine dump of typical steppe [J]. Bull Soil Water Conserv 37(3):81–85

Zhang ZT, Wang JM, Zhang JR et al (2018) Interaction between reclaimed soil and vegetation in mining area: a review [J]. Soil 50(02):239–247

Zhao Y, Zhang P, Hu Y et al (2016) Effects of re-vegetation on herbaceous species composition and biological soil crusts development in a coal mine dumping site[J]. Environ Manage 57(2):298–307

Chapter 17
Study on the Dissociation Characteristics of Main Ions in Cyanide Containing Wastewater

Yanbo Chen, Guangsheng Li, and Xingfu Zhu

Abstract According to the cyanide containing wastewater produced by cyanide leaching in gold smelting enterprises, through theoretical analysis and Experimental Research on the characteristics of ion dissociation, oxidation and precipitation dissolution, the dissociation and oxidation characteristics of the main metal complex ions and the dissolution characteristics of the main sediment generated in the treatment process are preliminarily mastered. Based on this, several relatively feasible methods and processes of cyanide containing wastewater with different ion composition are proposed, which can effectively avoid the blind attempt of various treatment methods in the process of experimental research, and improve the efficiency of experimental research.

Keywords Cyanide containing wastewater · Dissociation of metal complex ions · Ion oxidation · Precipitation dissolution

17.1 Introduction

Cyanide containing wastewater refers to industrial wastewater containing cyanide, which will be produced by gold smelting, coke oven steelmaking, electroplating, etc. Cyanide in cyanide containing wastewater can be divided into free cyanide and complex cyanide, which mainly include Fe $(CN)_6{}^{4-}$, Cu $(CN)^{2-}$, Cu $(CN)_3{}^{2-}$, Cu $(CN)_4{}^{3-}$, Zn $(CN)_4{}^{2-}$, Pb $(CN)_4{}^{2-}$, Ag $(CN)_2{}^{-}$, Au $(CN)_2{}^{-}$, Ni $(CN)_4{}^{2-}$, Hg $(CN)_4{}^{2-}$. Cyanide containing wastewater of gold smelting enterprises has the characteristics of high cyanide content and complex composition (Gao 2000). Once the treated ore changes, the ionic composition of cyanide containing wastewater will also change. Even if the cyanide containing wastewater of the same ore is treated, the ionic composition will also change with the change of production conditions.

Y. Chen (✉) · G. Li · X. Zhu
Selection and Metallurgy Laboratory of Shandong Gold Mining Technology Co. Ltd., Yantai 261441, Shandong, China
e-mail: 595907901@163.com

J. Zhang et al. (eds.), *Environmental Pollution Governance and Ecological Remediation Technology*, Environmental Science and Engineering,
https://doi.org/10.1007/978-3-031-25284-6_17

The unstable ion composition of cyanide containing wastewater increases the difficulty of treatment, which requires better pertinence and adaptability of treatment methods (General chemistry 2011; Li et al. 2017). The treatment methods of cyanide containing wastewater can be divided into destruction method and comprehensive recovery method. The destruction method mainly includes oxidation method, electrolysis method, high temperature decomposition method, microbial decomposition method and natural purification method. The comprehensive recovery method mainly includes acidification method, chemical precipitation method, etc.. The essence of destruction method is ion dissociation, and the degradation and precipitation of ions in comprehensive recovery method are also related to ion dissociation. Therefore, it is of great significance to study the dissociation characteristics of ions for the treatment of cyanide containing wastewater (Jun et al. 2016).

17.2 Dissociation Characteristics of Main Metal Complex Ions

Since the metal elements in gold bearing minerals are mainly Fe, Cu and Zn, the main metal complex cyanogens in cyanide containing wastewater of gold smelting enterprises are Fe $(CN)_6{}^{4-}$, Cu $(CN)_3{}^{2-}$ (copper cyanide complex ion mainly exists in this form), Zn $(CN)_4{}^{2-}$. Therefore, the dissociation characteristics of these three metal complex ions were mainly studied.

17.2.1 Theoretical Analysis

The metal complex cyanogen dissociates under different pH conditions, which may produce ferrocyanide, cyanide, hydroxide, thiocyanide precipitation and free cyanogen, and then realize the removal of cyanogen through filtration, blowing off and other measures. The reaction mechanism of commonly used acidification method, oxidation method and chemical precipitation method are closely related to the dissociation characteristics of metal complex ions. The ionic stability of metal elements and cn- is the main factor affecting the dissociation characteristics. In addition, the change of ion concentration will also have an important impact on the dissociation (Peng et al. 2018).

(1) The dissociation equilibrium equation of $Fe(CN)_6{}^{4-}$: $Fe(CN)_6{}^{4-} \rightleftharpoons Fe^{2+} + 6CN^-$. Under acidic conditions, CN^- and H^+ form HCN, which reduces the concentration of CN^-,and causes the dissociation equilibrium of $Fe(CN)_6{}^{4-}$ to shift to the right. Under alkaline conditions, Fe^{2+} and OH^- form $Fe(OH)_2$ precipitation, which is further oxidized to $Fe(OH)_3$, which reduces the concentration of Fe^{2+} and causes the dissociation equilibrium of $Fe(CN)_6{}^{4-}$ to shift

to the right. Therefore, theoretically, $Fe(CN)_6^{4-}$ may be dissociated under acid and alkaline conditions.

(2) The dissociation equilibrium equation of $Cu(CN)_3^{2-}$ is:$Cu(CN)_3^{2-} \rightleftharpoons Cu(CN)_2^- + CN^-$; $Cu(CN)_2^- \rightleftharpoons CuCN\downarrow + CN^-$; $CuCN \rightleftharpoons Cu^+ + CN^-$.Under acidic conditions, $CN^-$and H^+ form HCN, which reduces the concentration of CN^- and causes the dissociation equilibrium of $Cu(CN)_3^{2-}$, $Cu(CN)_2^-$, CuCN to shift to the right. In alkaline solution, Cu^+ and OH^- form CuOH, which is very unstable and will be oxidized by O_2 to $Cu(OH)_2$, which reduces the concentration of Cu^+ and causes the dissociation equilibrium of CuCN to shift to the right. Therefore, theoretically, $Fe(CN)_6^{4-}$ may be dissociated under acid and alkaline conditions.

(3) The dissociation equilibrium equation of $Zn(CN)_4^{2-}$ is: $Zn(CN)_4^{2-} \rightleftharpoons Zn(CN)_3^- + CN^-$; $Zn(CN)_3^- \rightleftharpoons Zn(CN)_2\downarrow + CN^-$; $Zn(CN)_2 \rightleftharpoons Zn^{2+} + 2CN^-$. Under acidic conditions, CN^- and H^+ form HCN, which reduces the concentration of CN^- and causes the dissociation equilibrium of $Zn(CN)_4^{2-}$, $Zn(CN)_3^-$, $Zn(CN)_2$ to shift to the right. Under alkaline conditions, Zn^{2+} and OH^- form $Zn(OH)_2$ precipitation, which reduces the concentration of Zn^{2+} and causes the dissociation equilibrium of $Zn(CN)_2$ to shift to the right. Therefore, theoretically, under acid and alkaline conditions, $Zn(CN)_4^{2-}$ may be dissociated.

In conclusion, $Fe(CN)_6^{4-}$, $Cu(CN)_3^{2-}$ and $Zn(CN)_4^{2-}$ theoretically dissociate under acidic and alkaline conditions, but the degree of dissociation is difficult to calculate quantitatively. It is a relatively simple and effective method to quantitatively study the degree of dissociation of ions through experiments under set conditions.

17.2.2 Experimental Study

Weigh a certain amount of $K_4Fe(CN)_6\cdot 3H_2O$ to prepare aqueous solution, and take a certain amount of $CuSO_4\cdot 5H_2O$, $ZnSO_4\cdot 7H_2O$ and 30% NaCN solution to prepare $Na_2Cu(CN)_3$ solution and $Na_2Zn(CN)_4$ solution respectively. Take out part of the solution as the stock solution for test, divide the remaining solution equally, adjust the pH to 2 and 11 respectively, magnetic stirring and observe the test phenomenon, and filter the test after 30 min.

(1) After the pH is adjusted to 2, $K_4Fe(CN)_6\cdot 3H_2O$ aqueous solution changes from colorless and transparent to light green. The test results in Table 17.1 show that under the condition of pH = 2, the total cyanide is lower than that of the original solution. It shows that $Fe(CN)_6^{4-}$ ion will dissociate under the condition of ph = 2. Under the condition of pH = 11, the color of the solution does not change, and the total cyanide does not change significantly compared with the original solution, indicating that $Fe(CN)_6^{4-}$ exists stably under this condition.

(2) After the pH is adjusted to 2, the aqueous solution of $Na_2Cu(CN)_3$ changes from colorless and transparent to turbid, and a large number of white precipitates

Table 17.1 Effect of acidity and alkalinity on dissociation characteristics

Test items	$Fe(CN)_6^{4-}$			$Cu(CN)_3^{2-}$			$Zn(CN)_4^{2-}$		
	Stock solution	pH = 2	pH = 11	Stock solution	pH = 2	pH = 11	Stock solution	pH = 2	pH = 11
Fe	555.7	553.1	556.4	–	–	–	–	–	–
Cu	–	–	–	620.09	10.32	616.82	–	–	–
Zn	–	–	–	–	–	–	588.36	585.52	575.56
FCN	–	–	–	–	–	–	239.76	211.13	242.04
TCN	1587.09	1476.43	1576.71	627.44	101.92	614.95	1106.59	232.97	1106.59

are formed. Table 17.1 shows that under the condition of pH = 2, the copper concentration is significantly lower than that of the stock solution. It shows that under the condition of pH = 2, $Cu(CN)_3^{2-}$ dissociates and generates CuCN precipitation. Under the condition of pH = 11, the concentration of Cu and total cyanide have no obvious change compared with the original solution, indicating that $Cu(CN)_3^{2-}$ exists stably under this condition.

(3) In the process of pH adjustment to 2, the aqueous solution of $Na_2Zn(CN)_4$ first formed white precipitates, and then quickly dissolved and became colorless and transparent. Table 17.1 shows that under the condition of pH = 2, the concentration of Zn changes little, and the total cyanide is significantly lower than that of the stock solution. It shows that under the condition of pH = 2, $Zn(CN)_4^{2-}$ dissociated Zn exists in the form of ions, while CN^- overflows in the form of HCN. When the pH is adjusted to 11, the solution is colorless and transparent. Under the condition of pH = 11, the concentration of Zn and total cyanide have no obvious changes compared with the original solution, indicating that $Zn(CN)_4^{2-}$ exists stably under this condition.

17.3 Oxidation and Dissociation Characteristics of Main Ions

17.3.1 Theoretical Analysis

According to information, under alkaline conditions, oxides can oxidizeCN^-, $Fe(CN)_6^{4-}$, $Cu(CN)_3^{2-}$, $Zn(CN)_4^{2-}$, SCN^-into cyanate ($M(CNO)_n$). Theoretically, the stability, reducibility and catalysis of metal complex ions have varying degrees of influence on the oxidation and dissociation of these ions, and the specific oxidation effect of these ions needs to be tested (Theory and method of gold extraction cyanidation wastewater treatment 2015).

17.3.2 Experimental Study

Weigh a certain amount of 30%NaCN solution, $K_4Fe(CN)_6 \cdot 3H_2O$ and KSCN to prepare aqueous solution, and take a certain amount of $CuSO_4 \cdot 5H_2O$, $ZnSO_4 \cdot 7H_2O$ and 30% NaCN solution to prepare $Na_2Cu(CN)_3$ solution and $Na_2Zn(CN)_4$ solution respectively. Take out part of the solution as the stock solution for test, adjust the pH of the remaining solution to 11, add the same amount of hydrogen peroxide, magnetic stirring and observe the test phenomenon, and filter the test after 30 min.

It can be seen from Table 17.2 that the concentration of CN^- before and after the oxidation treatment of the prepared 30% NaCN solution is greatly reduced, and the oxidation effect is good, indicating that CN^- can be treated by alkaline oxidation

Table 17.2 Results of ion oxidation test under alkaline conditions

Ion category	TCN, mg/L		Oxidation rate, %
	Stock solution	After oxidation treatment	
CN^-	2600	226.92	91.27
$Fe(CN)_6^{4-}$	15,600	13,877.71	11.04
$Cu(CN)_3^{2-}$	7800	629.70	91.93
$Zn(CN)_4^{2-}$	10,400	8736.42	16.00
SCN^-	58,000	0.1	100.00

method. Due to the catalytic oxidation of Cu, $Cu(CN)_3^{2-}$ also has a good oxidation treatment effect, while $Fe(CN)_6^{4-}$, $Zn(CN)_4^{2-}$ have a poor oxidation treatment effect, indicating that $Cu(CN)_3^{2-}$ is more suitable for alkaline oxidation treatment than $Fe(CN)_6^{4-}$, $Zn(CN)_4^{2-}$. Compared with the other four ions, SCN^- oxidation treatment has the best effect. From this test, it can be preliminarily judged that cyanide containing wastewater or treated liquid dominated by CN^-, $Cu(CN)_3^{2-}$, SCN^- can be treated by alkaline oxidation method or further treatment.

17.4 Conclusion

(1) $Cu(CN)_3^{2-}$ dissociates under acidic conditions, releases CN^-, and generates CuCN precipitation (in the presence of SCN^-, it will be partially converted to CuSCN precipitation). Due to the catalysis of copper ions, $Cu(CN)_3^{2-}$ is easy to oxidize and decompose under alkaline conditions to form $Cu(OH)_2$ Precipitation and CNO^-. Therefore, acidification method or alkaline oxidation method should be considered for the removal of $Cu(CN)_3^{2-}$. For the wastewater with high content of $Cu(CN)_3^{2-}$, acidification method or oxidation method should be adopted for the treatment process.

(2) $Fe(CN)_6^{4-}$ is relatively stable under acidic and alkaline conditions. For the treatment objects with low content of copper, zinc and other elements (unable to completely form $Cu_2Fe(CN)_6$, $Zn_2Fe(CN)_6$ precipitation with $Fe(CN)_6^{4-}$), the acidification treatment effect is poor. It is not easy to oxidize and decompose under alkaline conditions. Therefore, the removal of $Fe(CN)_6^{4-}$ should not be treated by acidification or oxidation alone. An appropriate amount of metal cations such as iron, copper and zinc should be added under acidic or alkaline conditions, and the chemical precipitation method should be used to remove it.

(3) Under acidic conditions, $Zn(CN)_4^{2-}$ will be dissociated step by step, first converted to $Zn(CN)_2$ precipitation. With the increase of acidity, it will be completely dissociated, releasing CN^- and Zn^{2+}. It is relatively stable under alkaline conditions and is not easy to dissociate or be oxidized. Therefore, in order to avoid the complete dissociation of $Zn(CN)_4^{2-}$ and the release of CN^- under acidic conditions, and increase the release of HCN and the difficulty of

treatment, it is advisable to remove it in the form of $Zn(CN)_2$, $Zn_2Fe(CN)_6$ precipitation under weak acidic conditions.

(4) Due to the catalytic oxidation of Cu,$Cu(CN)_3^{2-}$ also has a good oxidation treatment effect, while $Fe(CN)_6^{4-}$, $Zn(CN)_4^{2-}$ have a poor oxidation treatment effect, indicating that $Cu(CN)_3^{2-}$ is more suitable for alkaline oxidation treatment than $Fe(CN)_6^{4-}$, $Zn(CN)_4^{2-}$. Compared with the other four ions, SCN^- oxidation treatment has the best effect. From this test, it can be preliminarily judged that cyanide containing wastewater or treated liquid dominated by CN^-, $Cu(CN)_3^{2-}$, SCN^- can be treated by alkaline oxidation method or further treatment.

In addition to the common CN^-, $Cu(CN)_3^{2-}$, $Fe(CN)_6^{4-}$ and $Zn(CN)_4^{2-}$, cyanide containing wastewater may also contain SCN^-, $Pb(CN)_4^{2-}$, $Ag(CN)_2^-$, $Au(CN)_2^-$, $Ni(CN)_4^{2-}$, $Hg(CN)_4^{2-}$ plasma. The composition of these ions in the wastewater undoubtedly increases the difficulty of treating cyanide containing wastewater, and its dissociation characteristics need to be further studied.

References

Gao D (2000) 20 year review of cyanide containing wastewater treatment technology [J]. Gold 01:46–49

General chemistry [M] (2011) Beijing, Higher Education Press, pp 90–103

Jun L et al (2016) Brief description of treatment methods of cyanide containing waste liquid [J]. Ind Energy Saving Technol 01:67

Li C et al (2017) Review on treatment methods of cyanide containing wastewater [J]. Shanxi Chem Ind 06:149–151

Peng X et al (2018) Experimental study on ozone oxidation treatment of cyanide containing wastewater [J]. Min Metall 01:69–72

Theory and method of gold extraction cyanidation wastewater treatment [M] (2015) Beijing, Metallurgical Industry Press, pp. 43–44, 146–147

Chapter 18
Comparison of the Flocculation Flow Field Measurements in Taylor-Couette Reactor

Yuhong Mao, Chao Chen, Yue Wang, and Xiangpeng Huang

Abstract Particle image velocimetry (PIV) was used to measure the velocity field of the Taylor-Couette flocculation flow with different flocs as the tracer. It can be seen that these different flocs can be used as trace particles in PIV for flow field measurement. The results indicated that the velocity vector maps obtained by PIV with different flocs can express the variation characteristics of the flocculation flow field structure of the vortex. At the same time, the velocity vector field obtained by PAC flocs generated in real time as trace particle is almost similar to that obtained by $FeCl_3$. The major changes of the flow field in the coagulation process are mainly due to the change of rotation speed rather than the change of flocculants type or dosage.

Keywords Flocculation · PIV · Flow field measurement · Velocity vector

18.1 Introduction

Turbulent flocculation dynamics may be closely related to the turbulent motion. Therefore, an understanding of the velocity field of flocculation process is crucial for the engineering applications of flow such as Taylor-Couette and other mixing device. The flow field structure has an important influence on the operational performance of the flocculation process. However, the flocculation process is hard to expound clearly and make a veracious measure because it is very complicated as well as various. And it is extremely easy to change as long as there is a slight disturbance in the measurement process, so the monitoring method is very important (Chang 2011; Mao et al. 2018).

Taylor-Couette device consists of two concentric cylinders, a rotating inner cylinder and a fixed concentric outer cylinder. Working fluid pivot around the inner cylinder in the annulus gap between the two concentric cylinders. When the angular

Y. Mao (✉) · C. Chen · Y. Wang · X. Huang
School of Environmental and Municipal Engineering, Lanzhou Jiaotong University, Lanzhou 730070, Gansu, China
e-mail: maoyuhong@126.com

J. Zhang et al. (eds.), *Environmental Pollution Governance and Ecological Remediation Technology*, Environmental Science and Engineering,
https://doi.org/10.1007/978-3-031-25284-6_18

velocity is in the lower range, it can be seen that the flow has stability and orientation by the Reynolds number (Re). The fundamental state of this device is called a circular Couette flow, it become the cornerstone of the development of hydraulic stability theory since Couette's study to use this device to measure viscosity, and Taylor's experiment of the stability of Couette flow.

Coagu-flocculation experiment was conducted in Taylor-Couette device with kinds of flocculants, synchronously, PIV was applied to measure the flow field during flocculation process without any other special seed particles. Therefore, this is undoubtedly logical that PIV can be used as a useful synchronous measurement tool in the in situ observation during flocculation processes. Later, one can study the influence and the relationship between vortex centrifugal field and the flocculation efficiency according to the measurement of vortex velocity vector field of the flocculation process.

18.2 Experiment

18.2.1 Equipment

For the convenience of vortex flow field reconstruction and control, the Taylor-Couette flow apparatus was used as the coagu-flocculation experimental device, which was made in our own laboratory, and PIV system was applied to the flocculation experimental device. It consists of two concentric cylinders. The rotating inner cylinder is made of stainless steel, and has diameter $2r_i = 75$ mm. The fixed outer cylinder is made of plexiglas, and has diameter $2r_O = 100$ mm. The resulting gap width is $d = r_O - r_i = 12.5$ mm, the radius ration is $\eta = r_O/r_i = 0.75$. The entire cylinder height L = 440 mm, the cylinder aspect ratio, $\Gamma = L/(r_O - r_i) = 35.2$. The entire cylinder was enclosed in a square plexiglas box, which was filled with some working fluid having a refractive index matched to that of Plexiglas, to eliminate the optical distortions.

The PIV system used in this experiment came from Dantec Dynamics of Denmark. It consists of doubled pulsed Nd:YAG laser, high speed flowsense 2 M CCD camera wich resolution is 1200 × 1600 pixels and the pixel pitch is 7.3854 μm, flow map system HUB, host computer and flowmanager 4.71 software. The Flowmanager software was used to both keep track of the recorded data, the set-ups and the experimental configuration, image processing and data analysis. The PIV system data acquisition rate is 0 ~15 Hz per second. The PIV system was applied to coagu-flocculation experiment in the Taylor-Couette flow apparatus. The schematic diagram of the experimental setup is shown in Fig. 18.1. For this apparatus, the critical Reynolds number $Re_c = 83$, when laminar Taylor vortices appear. It is nearly the same as that reported (Mao et al. 2015; Wang 2022) which corresponding the inner cylinder rotation speed n is less than 2 rpm.

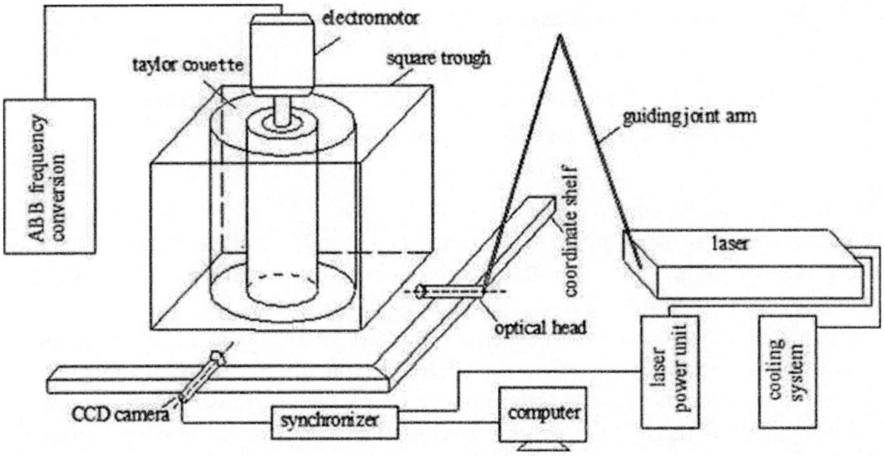

Fig. 18.1 Schematic diagram of the measuring system for Taylor-Couette flow field

18.2.2 Method

Different flocculants (Include: polyaluminium chloride, PAC; $FeCl_3$; polymeric ferric sulfate, PFS) with different dosage(10, 20 mg/L) are used in the coagulation process at different rotation speed. The flocculation experimental procedure is as follows:

Tap water 1500 ml, 2% suspension of kaolin 7.5 ml were added into the Taylor-Couette reactor. The kaolin particles are uniformly dispersed in mixture water filled in the annulus space by strong mixing at the high rotating speed of the inner cylinder (500 r/min) for 2 min. The turbidity of this mixture water equals 100 NTU after the strongly mixing. Subsequently, 1% kinds of flocculants stock solutions were added at the target dosage into the Taylor-Couette reactor samples, then strongly stirring again by the high rotating speed (500 r/min) for 2 min to mix them well. Later, Start ABB frequency converter system to control the Taylor-Couette apparatus to rotate at the required rotation, stir slowly at different target rotating speeds given for 10 min to fully flocculate react in the water sample in the annulus so as to the flocculate reaction is complete in the target vortex field.

Then the flow field measurement with kaolin flocs during flocculation processes can start. The velocity vector field is obtained by PIV, the trace particles are the flocs generated in the measurement process rather than dedicated tracer particles. The trace particles were not added in advance but the flocs generated in real time during the flocculation process. Start PIV system during the flocculation reaction in first 10 min at the target rotation speed to collect data of flow field. All flow field measurement experiments were performed using the same protocol.

18.3 Results and Discussion

18.3.1 *The Instantaneous Velocity Vector Maps Obtained with PAC Flocs as the Trace Particles*

The instantaneous velocity vector maps obtained with PAC flocs as the tracer during the flocculation process in different rotation speed at any time. Instantaneous velocity vector maps of some typical flow field were selected to representative features of the vortex flow, and combine them into patterns as shown in Fig. 18.2.

One can observe that the vortices on the velocity vector image are changed along with the rotating speed. It can be seen that below 10 rpm, there are closed, independent and similar to each other even at different speeds. Yet their size distribution is uniform on velocity vector map. In the range between 10 and 40 rpm, the vortices gradually from loose, open to combined with each other, the size of vortex is gradually nonuniform, that is, vortices appear in pairs with one big and one small. With the rotating speed of the inner cylinder is increased, the vortices are combined more compact with each other, the change of vortices sizes is regular and the vortex distribution become uniform again when the roating speed is between 40 and 65 rpm.

But after 65 rpm, the vortices gradually become independent, integrity and closed entirely, the vortex volume size is uniform and there are no mainstream liquid movement between 60 and 100 rpm. Later, the vortex distribution gradually interferes with each other, the integrity of vortices is gradually destroyed and the neighboring vortices connect with each other again, and there maybe produce open or non-closed vortices, the internal vortices may be contains several irregular small vortices when n is greater than 100 rpm.

The velocity maps with the dosage at 10 and 20 mg/L are compared, which are slightly different from those obtained under pure hydraulic flow field with dedicated trace particles. Because the floc particles produced during the flocculation flow field are different in size and may be unevenly distributed, which will lead to more long vectors and green vectors in the velocity vector maps (the green vector is a substitution vector). Therefore, the presence of green vector in the vector map indicates that the flocs are irregular, imperfect and cannot be captured by CCD. The velocity vector obtained under the dose 10 mg/L is closer to that of pure hydraulic flow field, and the green vector is relatively less than that of 20 mg/L.

18.3.2 *The Instantaneous Velocity Vector Maps Obtained with $FeCl_3$ Flocs as the Trace Particles*

Just like the situation of using PAC flocculants above, the instantaneous velocity vector maps were obtained with $FeCl_3$ flocs as the tracer during the flocculation process in different rotation speed at any time. Several instantaneous velocity vector

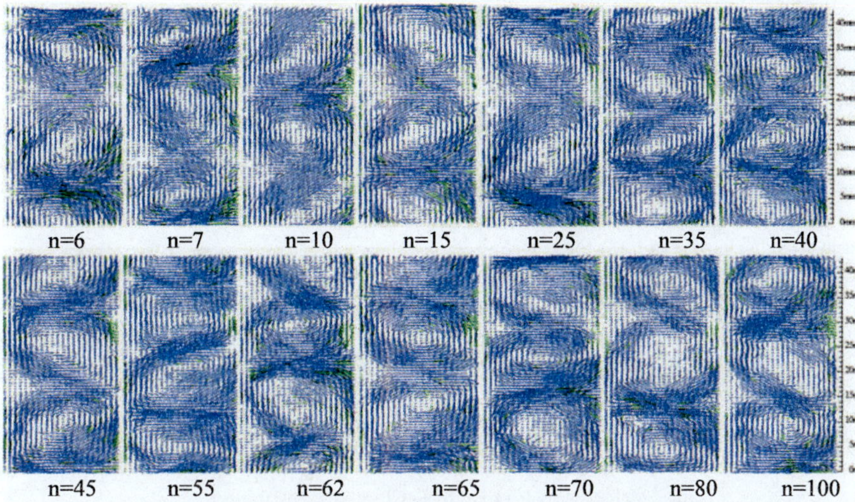

a. Velocity vector map of PAC at dosage 10 mg/L

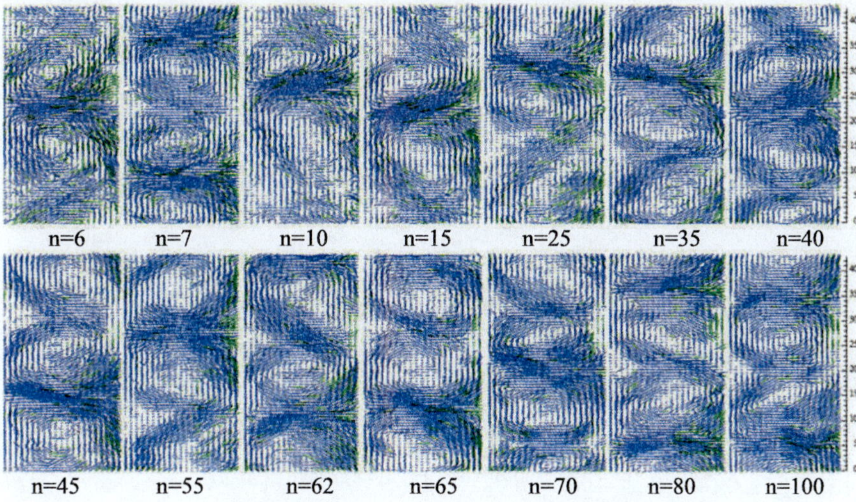

b. Velocity vector map of PAC at dosage 20 mg/L

Fig. 18.2 The velocity vector maps obtained with PAC flocs as the tracer, n is the rotation speed of the inner cylinder (r/min)

maps of some typical flow field were selected to representative features of the vortex flow, and combine them into patterns as shown in Fig. 18.3.

It can be seen that the morphology of vortices are also changed along with the rotation speed. Like the results of PAC, the flow field is changing with the rotation speed during the coagulation process, and the dosage or type of flocculants have little influence on the velocity vector of the flocculation flow field.

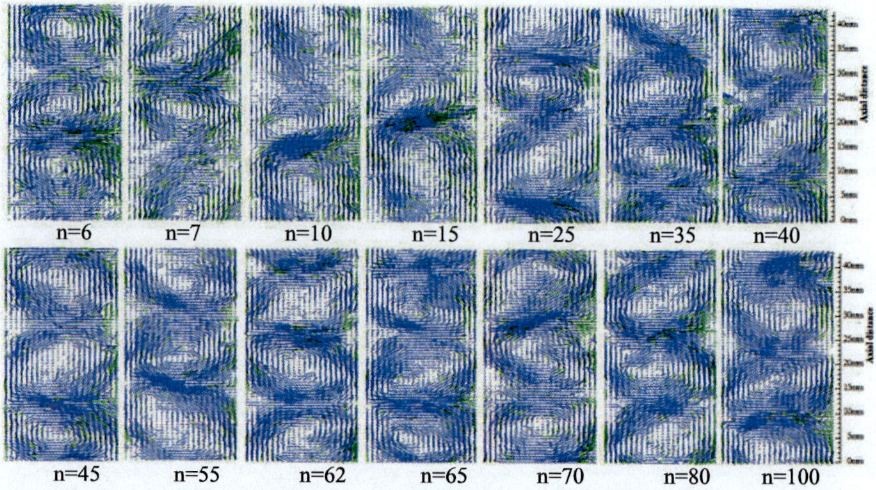

a. Velocity vector map of $FeCl_3$ at dosage 10 mg/L

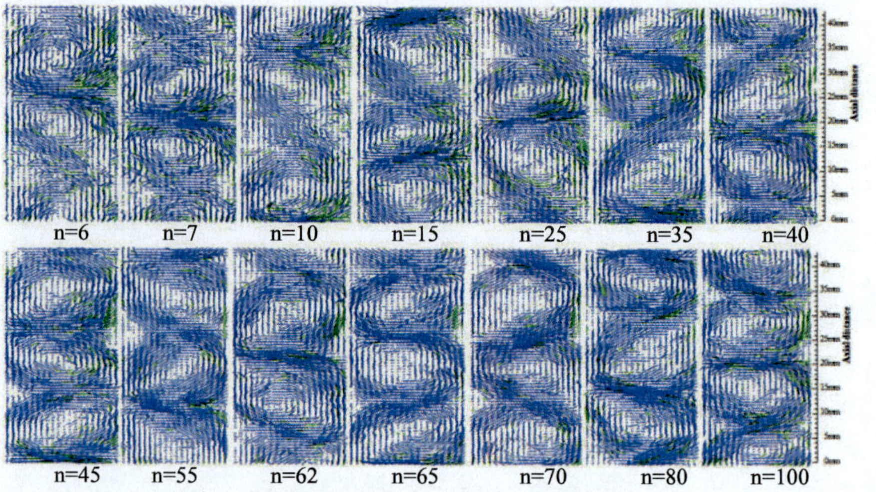

b. Velocity vector map of $FeCl_3$ at dosage 20 mg/L

Fig. 18.3 The velocity vector maps obtained with $FeCl_3$ flocs as the tracer, n is the rotation speed of the inner cylinder (r/min)

Unlike the velocity vector field of PAC, the velocity vector of $FeCl_3$ obtained under the dose 20 mg/L is closer to that of pure hydraulic flow field, and the green vector is relatively less than that of 10 mg/L. This also means that the vortex flow field of the flocculation process will be affected by adding flocculants.

No matter what the type and dosage of flocculants there are, as long as the dosage is appropriate with any flocculants, it will weaken the influence of the flocculation on vortex flow field, making it closer to the velocity vector diagram of pure hydraulic flow field. At the same time, with more closer to the value of the optimal dosage of any flocculants, this weakening effect will be stronger. The optimal dosage of various flocculants to obtain the optimal velocity field is different, This shows that different flocculants and their dosage have different effects on the flow field. This means that different flocculants and their different dosage have different effects on the flow field. The measurement results of kinds of flocculants can also reflect the major changes process of the flow field in the process of coagulation is mainly due to the change of rotation speed rather than the change of flocculants type or dosage. This also indicates that the vortex flow field will be affected after adding flocculants.

18.4 Conclusions

These different flocs can be used as trace particles in PIV for flow field measurement. The instantaneous velocity vector maps obtained with different flocs can express the variation characteristics of the flocculation flow field. The velocity vector obtained by PAC flocs is almost similar to that obtained by $FeCl_3$.The major changes of the flow field in the coagulation process are mainly due to the change of rotation speed rather than the change of flocculants type or dosage.

Acknowledgements We thank the financial support from the National Natural Science Foundation of China (No. 51968038).

References

Chang Q (2011) The flocculation of water treatment [M]. China Chemical Industry Press, pp 33–39, 174–175

Mao YH, Chang Q, Zeng LY (2015) Flocculation effects of submicroscopic vortices [J]. China Environ Sci 32(3):467–472

Mao YH, Ye Q, Deng JW (2018) Large eddy PIV study of Taylor-Couette flow and its influence on the coagulation process [J]. Acta Sci Circum 38(6):2374–2381

Wang SM (2022) Numerical simulation of the effect of geometric size on the characteristics of Taylor-Couette vortex [D]. Lanzhou Jiaotong University

Chapter 19
Air Pollution in a Typical City of Beijing-Tianjin-Hebei Pollution Channel Under the COVID-19 Epidemic

Dayong Tian, Shiyu Liu, Shuai Liu, Rui Xu, and Youjun Liu

Abstract Lockdowns during the novel coronavirus (COVID-19) reduced human activity and vehicle use, providing opportunities for air pollution research. This study selected Anyang City, a typical air pollution city, as an example, and revealed the variation of air pollution factors before and after the lockdown. The results showed that $PM_{2.5}$ and PM_{10} exceed the standards with the frequencies of 69.23% and 38.46%, respectively. Most pollutants decreased after the lockdown compared with those before the lockdown. The decline rates of pollutants before and after the epidemic were 37.93% ($PM_{2.5}$), 37.76% (PM_{10}), 18.14% (SO_2), 38.58% (CO), 56.69% (NO_2). Conversely, O_3 increased by 137.47%. It shows that human activities, especially the reduction of vehicle use, have a greater impact on NO_2, O_3 than others pollutants.

Keywords COVID-19 · Lockdown · Air pollution · O_3

19.1 Introduction

Northern China is one of the world's most densely populated regions (Lu et al. 2019; Wang et al. 2019; Xu et al. 2022). The area has been suffering from severe air pollution in recent decades, especially on the south of Beijing, namely the Beijing-Tianjin-Hebei (BTH) air pollution transmission channel (Fig. 19.1) (Tian et al. 2020). In order to reduce the air pollution in the area, the Chinese government implemented the Air Pollution Prevention and Control Law in 2015 (Jie et al. 2018; Liu et al. 2010; Rohde and Muller 2015), and the air quality has been improved during 2016 to 2020. However, the novel coronavirus (COVID-19) suddenly broke out in Wuhan city of Hubei Province before the Chinese lunar new year of 2020 (also known as Spring Festival). The government quickly took measures to put the cities under lockdown. During the lockdown (about 30 days after the first day of the Spring Festival, Anyang

D. Tian (✉) · S. Liu · S. Liu · R. Xu · Y. Liu
School of Chemical and Environmental Engineering, Anyang Institute of Technology, Anyang 455000, China
e-mail: astdy@126.com

J. Zhang et al. (eds.), *Environmental Pollution Governance and Ecological Remediation Technology*, Environmental Science and Engineering,
https://doi.org/10.1007/978-3-031-25284-6_19

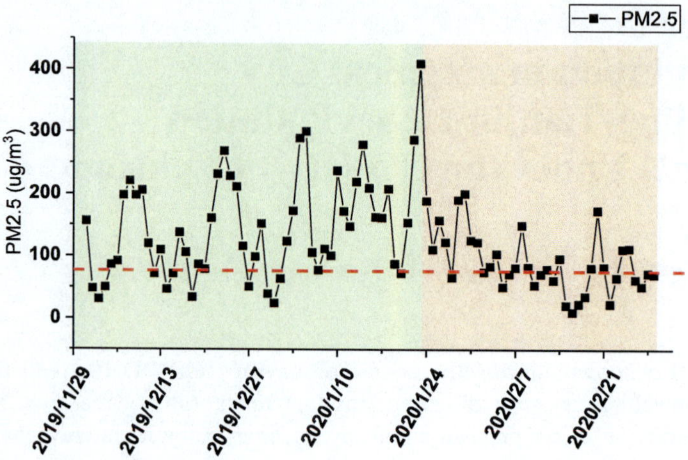

Fig. 19.1 Variation of daily concentrations of $PM_{2.5}$ before and after lockdown in Anyang city (from December 1, 2019 to February 23, 2020). The city is put under lockdown on January 24, 2020. The solid black line denotes daily concentrations and the red dotted line represents the national secondary standard of air quality in China

is put under lockdown on January 24, 2020), more than 70% of vehicles and 30% of factories were suspended, which resulted in the significantly reduction of human activities as well as the use of civilian vehicles. This provides an opportunity for the study of the variation of air pollution in China.

Coal, oil and natural gas are main fuels of industry and daily lives of urban residents in China. The combustion of coal and fuel oil has resulted in an increasing atmospheric concentration of PM, SO_2, NO_2 and CO. Anyang City is also an important industrial city along the BTH pollution transmission corridor. The industrial structure is dominated by heavy industry and chemical industry, such as steel manufacturing, coking plants and pharmaceutical plants. The city has been listed as the top 10 cities with the most serious air pollution in China (Wang et al. 2010), and ranked the last among 168 cities in China in September 2020 (O'Dell et al. 2019). The city, one of the eight ancient capitals of China, is located in the northern of Henan Province. Similar to other cities on the Beijing-Tianjin-Hebei transmission channel, the west of the city is the Taihang Mountain and the east is the North China Plain. The special terrain makes the region prone to adverse atmospheric diffusion conditions and temperature inversion in winter, resulting in the accumulation of air pollutants (O'Dell et al. 2019).

Many previous studies have focused on how air pollution varied with different seasons and locations. However, it still remains unknown how air pollution will change under significant reduced human activities due to the lockdown of a city.

The aim of this study is therefore to take Anyang city as an example and investigate daily concentrations of six air pollutants, including $PM_{2.5}$, PM_{10}, SO_2 NO_2,

CO_2, and O_3, before and after the COVID-19 lockdown, and identify main pollutants by comparing the concentrations with national air quality standards, and reveal the variation of air pollution factors before and after the lockdown, which might provide an insight to prevention and control of air pollution along BTH air pollution transmission channels in China.

19.2 Materials and Methods

Daily concentrations of six factors, including PM_{10}, $PM_{2.5}$, SO_2, NO_2, CO, O_3, in Anyang city before and after the COVID-19 lockdown were collected from the Data Center of the Ministry of Ecology and Environment of China. The sample size during pre-lockdown were 54 days, and the data was collected from December 1, 2019 to January 23, 2020. Daily concentrations during post-lockdown were collected from January 24 to February 29, 2020, a total of 37 days.

The arithmetic mean method was employed to calculate the averages of daily concentrations of the six pollutants before and after lockdown (Eq. 19.1). The decline rates of the six pollutants before and after lockdown are calculated using Eq. (19.2).

$$c_{\text{averages}} = \left(\sum_{1}^{n} c_{\text{daily mean concentrations}}\right)/n \tag{19.1}$$

$$x\%(\text{rate of decline}) = \left(\frac{c_{\text{average - pre - lockdown}} - c_{\text{average - post - lockdown}}}{c_{\text{average - pre - lockdown}}}\right) \times 100\% \tag{19.2}$$

$C_{\text{average-post lockdown}}$ and $C_{\text{average-post lockdown}}$ denote the averages of daily concentrations before and after lockdown, respectively. The sample sizes, *n*, are 54 and 37 for $C_{\text{average-post lockdown}}$ and $C_{\text{average-post lockdown}}$, respectively. $x\%$ represents the decline rates of pollutant concentrations during lockdown period compared with that during pre-lockdown period.

19.3 Results and Discussion

19.3.1 Variation of Daily Concentrations of $PM_{2.5}$

The variation of $PM_{2.5}$ before and after lockdown in Anyang city are shown in Fig. 19.1. It can be seen that, from November 29, 2019 to February 23, 2021, $PM_{2.5}$ in Anyang city have a frequency of 68.13% exceeding the national secondary standard. Among them, the frequency of exceeding the national secondary standard during pre-lockdown was 77.78%, and after the lockdown of the city, the

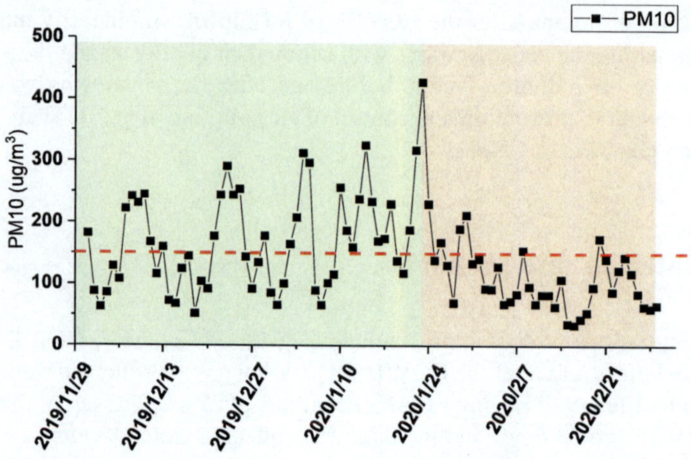

Fig. 19.2 Variation of daily concentrations of PM_{10} before and after lockdown in Anyang city (from December 1, 2019 to February 23, 2020). The city is put under lockdown on January 24, 2020

frequency of exceeding the national secondary standard was 54.05%. The averages of daily concentrations before and after the lockdown are 146.31 ± 83.79 and 90.81 ± 48.32 ug/m^3, decreased by 37.8% after the lockdown, which indicating $PM_{2.5}$ significantly decreased due to the reduced human activity, especially civilian vehicle emission ($p < 0.01$).

19.3.2 Variation of Daily Concentrations of PM_{10}

The variation of PM_{10} before and after lockdown in Anyang city is shown in Fig. 19.2. It can be seen that PM_{10} in Anyang city have a frequency of 38.46% exceeding the national secondary standard (53.70% during pre-lockdown and 16.22% after the lockdown). The average of daily concentrations was 171 ± 82.56 ug/m^3 before the lockdown, which is 37.76% higher than that after the lockdown (106.43 ± 48.26 ug/m^3), indicating PM_{10} significantly decreased due to the reduced human activity ($p < 0.01$).

19.3.3 Variation of Daily Concentrations of SO_2

Figure 19.3. shows that SO_2 in Anyang city did not exceed the national secondary standard. The average of daily concentrations was 17.33 ± 5.26 ug/m^3 before the lockdown, which is 18.14% higher than that after the lockdown (14.19 ± 5.71 ug/m^3), indicating the decrease of SO_2 during the lockdown is less than $PM_{2.5}$ and PM_{10}.

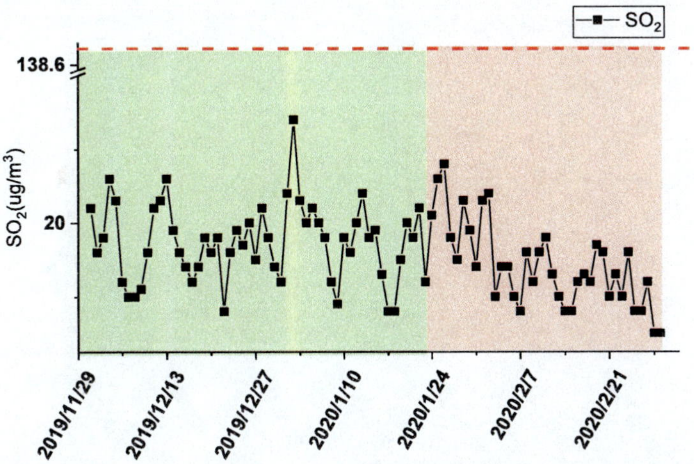

Fig. 19.3 Variation of daily concentrations of SO_2 before and after lockdown in Anyang city (from December 1, 2019 to February 23, 2020). The city is put under lockdown on January 24, 2020

19.3.4 Variation of Daily Concentrations of CO

As shown in Fig. 19.4, the daily concentrations of CO do not exceed the national secondary standard, except on January 23, 2020. The averages of daily concentrations before and after the lockdown are 2.06 ± 0.72 and 1.26 ± 0.43 ug/m^3, decreased by 38.58%, indicating CO significantly decreased due to the reduced human activity ($p < 0.01$).

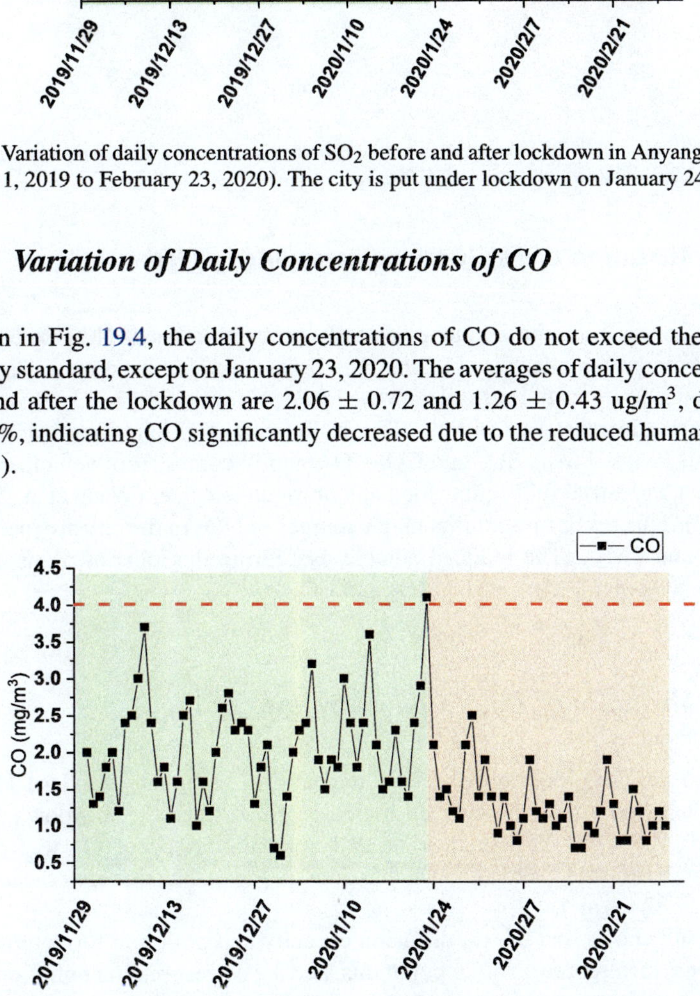

Fig. 19.4 Variation of daily concentrations of CO before and after lockdown in Anyang city. The city is put under lockdown on January 24, 2020

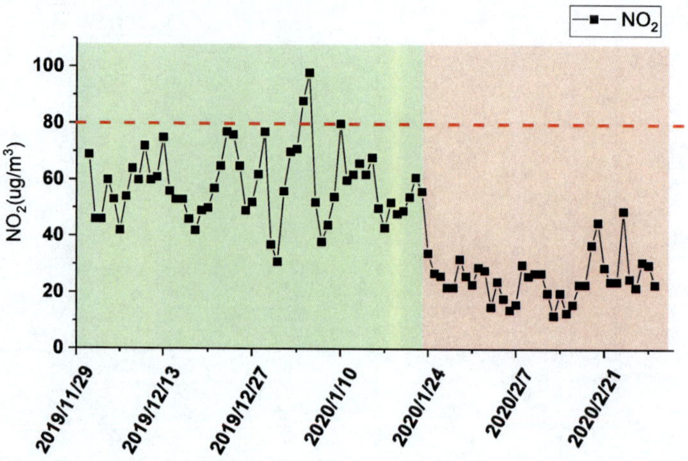

Fig. 19.5 Variation of daily concentrations of NO_2 before and after lockdown in Anyang city. The city is put under lockdown on January 24, 2020

19.3.5 Variation of Daily Concentrations of NO_2

As shown in Fig. 19.5, the frequency of daily concentrations of NO_2 exceeding the national secondary standard are 11.11% and 0 before and after lockdown. The average of daily concentrations before and after the lockdown were 58.17 ± 13.01 and 25.19 ± 7.87 ug/m^3, decreased by 56.69%. The decline rate of NO_2 is significantly higher than that of $PM_{2.5}$, PM_{10}, SO_2 and CO. NO_2 mainly comes from vehicle emission, power plant, industrial fuel combustion and nitrogen fertilizer (Wang et al. 2019; Xu et al. 2022). This result implies that major sources of NO_2 in the city are inconsistent with SO_2 and $PM_{2.5}$. The reduced vehicle use during the lockdown has a greater impact on NO_2 than $PM_{2.5}$, PM_{10}, SO_2 and CO.

19.3.6 Variation of Daily Concentrations of O_3

Figure 19.6 shows O_3 does not exceeded the national secondary standard. However, O_3 is the only pollutant showing an increased trend after the lockdown, with an increase of 137.47%. The average of daily concentrations of O_3 was 34.81 ± 17.18 ug/m^3 before the lockdown, which is 56.67% lower than that after the lockdown (82.68 ± 21.82 ug/m^3).

This result shows that the O_3 diffusion capacity and pollution formation mechanism are both different from other pollutants, and the corresponding pollution control measures should not be the same. Although NO_2 and volatile organic compounds (VOCs) emissions decreased, previous studies show that ozone concentrations in 2018 were also significantly higher than that in 2016 (Tian et al. 2020). The increased

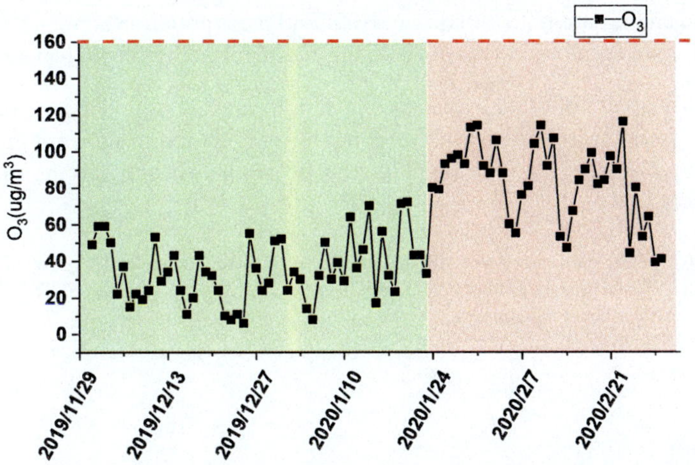

Fig. 19.6 Variation of daily concentrations of O_3 before and after lockdown in Anyang city. The city is put under lockdown on January 24, 2020

O_3 concentrations might be due to the weakened titration effect of NO (Hashim et al. 2021; Yin et al. 2021). Similar phenomena were found in Wuhan, China and Baghdad Iraq (Hashim et al. 2021; Yin et al. 2021). In addition, the decreased $PM_{2.5}$ and PM_{10} lead to increased solar transmittance and enhance photochemical reactions, which may also be a contributing factor.

Considering that ozone control mechanisms are more complex than other pollutants, it is necessary to take coordinated measures to control VOCs and NOx, preventing titration effects. Based on the frequencies of exceeding the standard and increasing trends, Ozone, along with $PM_{2.5}$ and PM_{10} were identified as key factors in future air pollution control.

19.4 Conclusion

(1) The frequencies of daily concentrations of $PM_{2.5}$ and PM_{10} exceeding the national secondary standards are 69.23% and 38.46%, respectively. SO_2 and O_3 do not exceed the standards, and CO and NO_2 rarely exceed the standards.
(2) Five pollutants, except for O_3, decreased after COVID-19 lockdown compared with pre-lockdown, and decline rates were 37.93% ($PM_{2.5}$), 37.76% (PM_{10}), 18.14% (SO_2), 38.58% (CO) and 56.69% (NO_2), respectively. The average of daily concentrations of O_3 increased by 137.47% after the lockdown.
(3) The reduced human activities, especially reduced vehicle use, have a greater impact on NO_2 and O_3 than $PM_{2.5}$, PM_{10}, NO_2 and CO. The mechanism of O_3 is different from other five factors.

(4) Based on the national standards and trends of these pollution factors, PM_{25} PM_{10} and O_3 are identified as main factors in future air pollution control strategies.

The study of air pollution in this special period can help us understand the impact of vehicle emissions and industrial emissions on air pollution, which is of great social and economic significance for the protection of people's health and the sustainable development.

Acknowledgements This work was funded by Scientific and Technological Project of Henan province (222102320103) and Huan Shui Scholar Project of Anyang Institute of Technology. We are grateful for their financial support.

References

Hashim BM, Al-Naseri SK, Al-Maliki A, Al-Ansari N (2021) Impact of COVID-19 lockdown on NO_2, O_3, $PM_{2.5}$ and PM_{10} concentrations and assessing air quality changes in Baghdad, Iraq. Sci Total Environ 754(8):141978

Jie W, Xu-rong S, Ying-ze T, Jiao X, Guo-liang S, Yin-chang F (2018) Analysis of chemical composition of the fine particulate matter in summer in Tianjin city via a single particle aerosol mass spectrometer (SPAMS). Environ Sci 39(8):3492–3501

Liu XH, Zhang Y, Cheng SH, Xing J, Zhang Q, Streets DG, Jang C, Wang WX, Hao JM (2010) Understanding of regional air pollution over China using CMAQ, part I performance evaluation and seasonal variation. Atmos Environ 44(20):2415–2426

Lu X, Vogt RD, Li H, Han S, Mo X, Zhang Y, Ullah S, Chen C, Han X. Li H (2019) China's ineffective plastic solution to haze. Science (New York, NY) 364(6446):1145

O'Dell K, Ford B, Fischer EV, Pierce JR (2019) Contribution of wildland-fire smoke to US $PM_{2.5}$ and its influence on recent trends. Environ Sci Technol 53(4):1797–1804

Rohde R, Muller RA (2015) Air pollution in China: mapping of concentrations and sources. PLoS ONE 10(8):0135749

Tian D, Fan J, Jin H, Mao H, Geng D, Hou S, Zhang P, Zhang Y (2020) Characteristic and spatiotemporal variation of air pollution in Northern China based on correlation analysis and clustering analysis of five air pollutants. J Geophys Res: Atmos 125(8):e2019JD031931

Wang S, Song T, Shiraiwa M, Song J, Ren H, Ren L, Wei L, Sun Y, Zhang Y, Fu P (2019) Occurrence of aerosol proteinaceous matter in urban Beijing: an investigation on composition, sources and atmospheric processes during the "APEC Blue" period. Environ Sci Technol 53(13):7380–7390

Wang JL, Li SM, Liu XL, Wu XJ (2010) An analysis of the fog distribution in Beijing for the 2001–2005 period using NOAA and FY data. Atmos Res 96(4):575–589

Xu C, Zhang Z, Ling G, Wang G, Wang M (2022) Air pollutant spatiotemporal evolution characteristics and effects on human health in North China. Chemosphere 294(12):133814

Yin H, Liu C, Hu Q, Liu T, Wang S, Gao M, Xu S, Zhang C, Su W (2021) Opposite impact of emission reduction during the COVID-19 lockdown period on the surface concentrations of $PM_{2.5}$ and O_3 in Wuhan, China. Environ Pollut 289(3):117899

Chapter 20
Effects of Rhizobia and Pseudomonas Fluorescens Complex Microbial Community on the Soil Fertility of Sandy Loess

Yongmei Zhao and Yirou Lu

Abstract ***Objective*** To detect the effects of the compound bacterial agent composed of Rhizobia WP1-3 and Pseudomonas RQ3-01 on soil fertility in sandy loess. ***Methods*** The symbiosis ratio of WP1-3 and RQ3-01 bacteria was analyzed by antagonism test. The sandy loess soil of cherry orchard and wheat field was selected as the samples, and the compound bacterial solution was added to the soil for treatment. The contents of available nitrogen, phosphorus and potassium in the soil and the contents of soil enzymes were detected at 3d, 6d and 9d of culture. ***Results*** Antagonism test showed that when the volume co-occurrence ratio of WP1-3 and RQ3-01 bacteria was 1:1, the growth number of composite bacteria was the highest, reaching 3.96 × 106 CFU/mL. On the 9th day after application of CMC, the contents of total N, P and K in cherry orchard soil were 6.47 times, 2.92 times and 3.45 times of those in control group ($P < 0.01$), and the contents of total N, P and K in wheat field soil were 9.44 times, 1.94 times and 3.41 times of those in control group ($P < 0.01$). Compared with CK group, the activities of protease, urease and sucrase in cherry orchard soil increased by 13%, 57% and 57%, respectively. It increased by 54%, 62% and 52% in wheat field soil, respectively ($P < 0.01$). At the same time, the effect of compound bacterial fertilizer on protease activity in the wheat field group was more obvious than that in the cherry orchard group, and the effect on urease and sucrase activities in the cherry orchard group was more obvious than that in the wheat field group. ***Conclusion*** The compound bacterial solution can significantly improve the fertility of sandy loess soil.

Keywords Complex microbial community · Rhizobia WP1-3 · Pseudomonas RQ3-01 · Enzyme activity · Soil fertility

---

Y. Zhao (✉) · Y. Lu
School of Biological and Environment Technology, Xi'an University, Xi'an, China
e-mail: xazym1118@163.com

J. Zhang et al. (eds.), *Environmental Pollution Governance and Ecological Remediation Technology*, Environmental Science and Engineering,
https://doi.org/10.1007/978-3-031-25284-6_20

20.1 Introduction

Soil fertility is the ability of soil to provide various nutrients required for crop growth. The indicators to measure soil fertility include the content of available nitrogen (N), phosphorus (P) and potassium (K) in soil and the activity of soil enzymes. Soil enzymes mainly come from soil microorganisms, plant root exudates and enzymes released during the decomposition of animal and plant residues (Jun 2018). The main function of soil enzymes is to participate in various biochemical reactions in soil and play an important role in the transformation of soil and plant nutrients. Therefore, enzyme activity is an important index to evaluate soil fertility.

Compound microbial community (CMC) usually contains a variety of active enzymes, microbial flora and trace elements, which can provide or improve nutrients in plants, thus improving crop yield and quality. This experiment studied the effect of a CMC on soil fertility of sandy loess by breeding the coexisting compound bacteria of *Rhizobium* WP1-3and *Pseudomonas fluorescens* RQ3-01, to determine the application value of CMC in improving the quality of agricultural products.

20.2 Materials and Methods

20.2.1 Materials

Soil Sample and bacterial strain. Sandy loess soil, is taken from 10 to 20 cm below the surface of Cherry Orchard and wheat field in Fengjia village, Lantian County, dried and stored in the laboratory. Bacterial strain, *Rhizobium* WP1-3and *Pseudomonas fluorescens* RQ3-01 provided by the cell culture laboratory of Xi'an University.

Apparatus. Optical microscope, Ultra clean workbench, Automatic electrothermal pressure steam sterilizer (LDZX-40II-type), Water isolating electrothermal constant temperature incubator (PYX-OHS-40x-0), Digital constant temperature water bath oscillator (SHA-C), 722 s visible spectrophotometer and others.

20.2.2 Methods

Strain activation and expansion. Nutrient medium (3 g beef extract, 5 g peptone, 5 g NaCl, 18 g AGAR and distilled water in a total volume of 1L) was prepared and sterilized at 121 °C for 20 min for activation of *Rhizobium* WP1-3 and *Pseudomonas fluorescens* RQ3-01. After activation, the strains were inoculated into liquid nutrient medium without AGAR and placed on a 30 °C digital constant temperature distillation water bath oscillator with a speed of 180 r/min for 2 h to obtain the seed solution. The obtained seed solution was expanded and determined by violet light spectrophotometer (OD = 600 nm). The two seed solutions were diluted to the same concentration (effective viable count $\geq 3.5 \times 10^6$ CFU/g).

Antagonism experiment. The volume ratios of WP1-3 and RQ3-01 were 1:1, 1:2 and 1:3, respectively, and the inoculum volume of each plate was 0.1 mL. The symbiotic ratio of the two strains in the compound agent was screened after 20 h of mixed culture.

Determination of available N, P and K contents in sample soil. The soil samples were dried for 2 h, and 2.5 g of each sample soil was used for the measurement of active K content, and 5 g of each sample soil was used for the measurement of available N and P contents. 1 mL of CMC (effective viable count $\geq 3.8 \times 10^8$ CFU /g) was added, respectively, for the gradient culture for 3d, 6d and 9d. The content of available N in soil was determined by ammonium colorimetric method, the content of available P in soil was determined by molybdenum-antimony resistance method, and the content of available K in soil was determined by flame photometry (Liu et al. 2017).

Determination of protease activity in sample soil. The activity of protease in soil was measured by copper salt colorimetric method, and the activity of soil protease was expressed as the mass (μg) of NH_3-N released by enzymatic hydrolysis of protein from 1 g soil within 24 h at 37 °C.

Determination of urease activity in sample soil. The urease in soil was detected by phenol-sodium hypochlorite colorimetric method. The urease activity in soil was expressed as the concentration of NH_3-N (mg/L) measured within 24 h at 37 °C.

Determination of sucrase activity in sample soil. The sucrase activity was measured by 3.5 dinitrosalicylic acid colorimetry and expressed as the concentration of glucose (mg/mL) in 1 g of soil for 24 h at 37 °C.

Data analysis. SPSS25 software was used for data analysis. The test data are expressed as "mean ± standard deviation". The comparison between groups adopts analysis of variance (ANOVA), $P < 0.05$ indicates that the difference is statistically significant.

20.3 Results and Analysis

20.3.1 Antagonistic Screening Results of WP1-3 and RQ3-01 Bacteria

Antagonism phenomenon can reflect the mutual resistance and symbiosis of different strains in the process of culture on the same Petri dish. The results were shown in Table 20.1. Compared with the control group, the viable number of compound bacteria in treatment group 1 was 3.96×10^6 CFU/mL. The experimental indicated that the *Rhizobium* WP1-3 and *Pseudomonas fluorescens* RQ3-01 bacteria could coexist, and the ratio growth was significant at 10^{-6} gradient ($P < 0.01$). The results

Table 20.1 Antagonistic screening results of WP1-3 and RQ3-01 bacteria

Group	Number of compound bacteria (106 CFU/mL)
Group 1 (1:1)	3.96 ± 0.13 Bb
Group 2 (1:2)	2.31 ± 0.21 Bb
Group 3 (1:3)	2.87 ± 0.17 Bb
CK	28.44 ± 16.38a

showed that they showed similar nutritional requirements and resistance to antagonism. Therefore, the 1:1 volume ratio of bacteria WP1-3 and RQ3-01 bacteria were choose as the CMC to study the effect on soil fertility of sandy loess.

20.3.2 *Effect of CMC on N, P and K Contents in Sandy Loess Soil*

Among the four treatment groups, sample 1 group was from cherry orchard soil, sample 1 + CMC group was mixed with cherry orchard soil and CMC, sample 2 group was from wheat field soil, and sample 2 + CMC group was mixed with wheat field soil and CMC. Results As shown in Table 20.2. On the 9th day after application of CMC, the contents of total N, P and K in cherry orchard soil were 6.47 times, 2.92 times and 3.45 times of those in control group ($P < 0.01$), and the contents of total N, P and K in wheat field soil were 9.44 times, 1.94 times and 3.41 times of those in control group ($P < 0.01$). The results indicated that the combined application of compound bacterial WP1-3 and RQ3-01 fertilizer could dissolve the insoluble N, P and K in the soil and improving the soil fertility.

Table 20.2 Effect of CMC on N, P and K in sandy loess soil

Treat	Content	Sample 1	Sample 1 + CMC	Sample 2	Sample 2 + CMC
3d	N (mg/kg)	17.33 ± 0.10a	65.67 ± 0.56 Bb	15.44 ± 0.07a	76.76 ± 0.32Bb
6d		16.89 ± 0.31a	118.61 ± 0.41 Bb	15.23 ± 0.15a	111.22 ± 2.17Bb
9d		17.37 ± 0.14a	113.45 ± 0.12 Bb	13.35 ± 0.08a	127.55 ± 2.43Bb
3d	P (mg/kg)	21.33 ± 1.22a	42.27 ± 1.86 Bb	19.14 ± 0.11a	32.81 ± 0.82Bb
6d		22.10 ± 0.16a	46.73 ± 1.18 Bb	19.24 ± 0.06a	34.07 ± 0.93Bb
9d		21.27 ± 1.19a	62.13 ± 1.67 Bb	18.65 ± 0.12a	36.20 ± 1.58Bb
3d	K (mg/kg)	173.60 ± 1.16a	568.87 ± 4.50 Bb	181.44 ± 5.01a	557.76 ± 6.62Bb
6d		173.89 ± 2.35a	583.21 ± 3.51 Bb	188.11 ± 4.11a	576.82 ± 4.88Bb
9d		164.31 ± 3.02a	600.33 ± 3.19 Bb	173.25 ± 3.18a	591.25 ± 7.03Bb

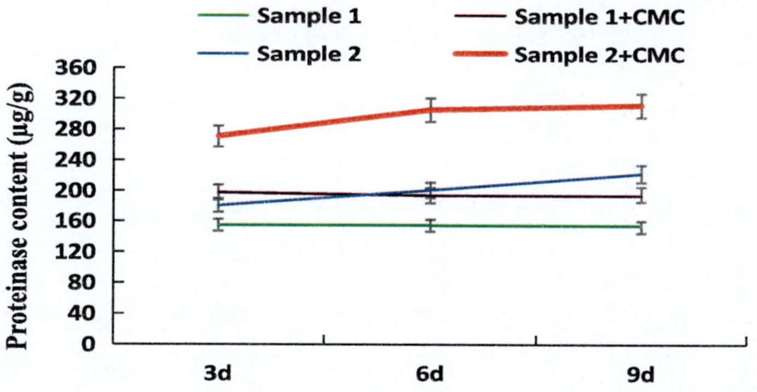

Fig. 20.1 Comparison of protease activity between cherry orchard and wheat field soil

20.3.3 Effects of CMC on Soil Protease Activities

The activity of protease in soil is very important to the transformation of nitrogen in soil. After adding CMC, the activities of protease in soil of two groups were significantly increased compared with CK group ($P < 0.01$). At 9 days, the activity of wheat field soil was 311.05 μg/g, which was increased by 54% compared with CK group. The protease activity of cherry orchard soil was 195.52 μg/g, which was 13% higher than that of CK group. At the same time, the effect of CMC on protease activity in wheat field soil was higher than that in cherry orchard soil (Fig. 20.1).

20.3.4 Effect of CMC on Soil Urease Activity

The urease activity in soil is measured by measuring the amount of ammonia produced after the enzymatic reaction with urea as the matrix. After adding CMC, the urease activity in soil of two groups was significantly increased compared with that of CK group during the experimental time ($P < 0.01$). At 9 days, the urease activity of wheat field soil was 2.70 mg/L, which was 62% higher than that of CK group. The urease activity in cherry orchard soil was 2.67 mg/L, which was 57% higher than that in CK group, and there was no difference in the urease activity between the two groups at 9 days (Fig. 20.2).

20.3.5 Influence of CMC on Soil Sucrase Activity

Sucrase plays an important role in increasing soluble nutrients in soil. The results showed (Fig. 20.3) that after adding CMC, the sucrase activity in the sample soil were significantly different from those in CK soil ($P < 0.01$), At 9 days, the sucrase

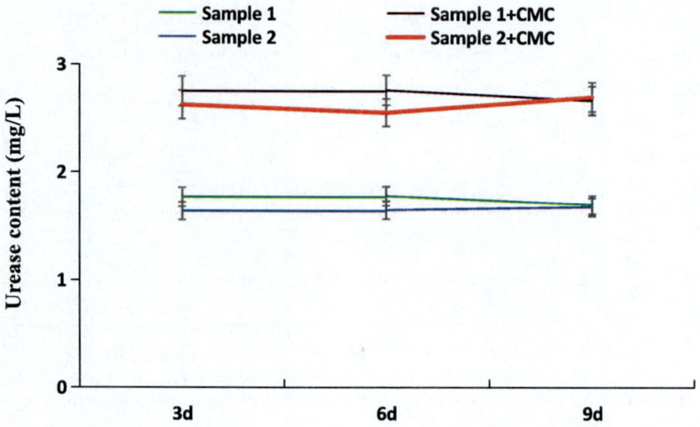

Fig. 20.2 Comparison of urease activity between cherry orchard and wheat field soil

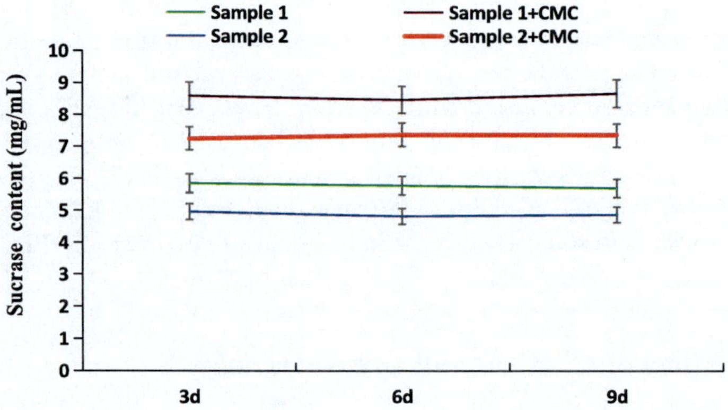

Fig. 20.3 Comparison of sucrase activity between cherry orchard and wheat field soil

activity in cherry orchard soil was 9.67 mg/mL, which was 57% higher than that in CK group, the sucrase activity of wheat field soil was 7.35 mgm/L, which was 52% higher than that of CK group. The effect of CMC on sucrase activity in cherry orchard soil was higher than that in wheat field soil.

20.4 Conclusion and Discussions

Many investigations have shown that the use of microbial fertilizers has a significant impact on soil enzyme activities. For example, protease is involved in the transformation of organic compounds such as amino acids and proteins in soil, and its hydrolysates are one of the nitrogen sources for higher plants. Urease is widely found in soil, which can hydrolyze urea in soil and produce carbon dioxide, ammonia and water. Sucrase is related to many soil factors, such as soil organic matter, nitrogen, phosphorus content, microbial quantity and soil respiration intensity. Soil enzymes can not only characterize the intensity of soil biological activities, but also be used as indicators to evaluate soil maturity and soil fertility level (Dacheng et al. 2018).

The results showed that soil microorganisms participate in the circulation of various nutrient elements in soil and the mineralization of mineral elements, and play an important role in the transformation and stability of soil enzymes. Previous studies have confirmed that *Rhizobia* WP1-3 and *Pseudomonas fluorescens* RQ3-01 have phosphorus and potassium solubilization, and that *Rhizobia* WP1-3 also has nitrogen fixation effect (Yongmei and Panpan 2021). CMC composed of WP1-3 and RQ3-01 bacteria could significantly promote soil fertility. The two bacteria could not only convert the insoluble N, P and K elements in soil into effective forms, so as to facilitate the absorption of crops. Meanwhile, CMC could increase the activities of protease, urease and sucrase in soil to varying degrees, and improve soil fertility. The reason was that the number and biomass of soil microorganisms were changed by fertilization, which was mainly caused by the change of enzyme content secreted by microorganisms (Xu 2019).

Further studies showed that the effect of CMC on protease activity in wheat field soil was more obvious than that in cherry orchard soil. The effect on sucrase activity of cherry orchard soil was more obvious than that of wheat field soil. It is speculated that the reason is related to different soil water content caused by different crop species (Liu et al. 2017).

In conclusion, this CMC composed of WP1-3 and RQ3-01bacteria is of great significance for the development and utilization of sandy loess soil. From the perspective of long-term development, the application of microbial compound fertilizer can improve soil physical and chemical properties and microbial living environment, so as to achieve the sustainable development of soil and obtain long-term good benefits of the dual effect.

Acknowledgements This work was supported by 2021 aerospace breeding project plan of Xi'an University (No.IITYZ2105).

References

Dacheng J, Shan G, Hailunl G (2018) Details in the determination of protein content by Coomassie brilliant blue method [J]. Exp Sci Technol 16(4):143–147

Jun H (2018) Rapid test of available nitrogen, phosphorus and potassium in soil [J]. Biochemistry 4(4):30–36

Liu WH, Zhang YJ, Shi SL (2017) Soil enzyme activities in alpine naked oat-artificial grassland in response to fertilizer and legume mix levels [J]. Acta Pratacul Sin 26(1):23–33

Liu G, Zhang X, Wang X (2017) Soil enzymes as indicators of saline soil fertility under various soil amendments [J]. Agricult Ecosyst Environ 237:274–279

Xu Y (2019) Research on microbial fertilizers [J]. Chem Des Commun 45(10):171–172

Yongmei Z, Panpan W (2021) Screening and preliminary identification of soil potassium and phosphorus-hydrolyzing bacteria WP1-3 [M]. E3S web of conferences, vol 293, p 02026. GCEECE 2021

Chapter 21
Correlation Ecological Impacts of Glacial Changes in the Hexi Corridor During the Historical Period

Guoquan Qian, Bin Yue, Kaiming Li, and Tianpeng Gao

Abstract The change of glaciers in Hexi Corridor has direct influence on river runoff and landscape in historical period. Glacier area and the runoff was positively related, melting ice river brings spring flood, Xia Hong, autumn and winter dry, the characteristics of a narrowing of the glacier area unified drainage divided, glacier area narrow lead to rump lake dried up. Glacier area and downstream desertification area was positively proportional relationship, melting glaciers periodically make desertification and oasis appear alternately, ice flake distribution determines the distribution of the oasis, rapid melting glaciers diluvium buried oasis, glacier area expands and freeze–thaw desertification caused by narrowing.

Keywords The hexi corridor · Glaciers · Ecological environment

21.1 Introduction

Due to the influence of climatic conditions and landscape structure, the average annual rainfall in Hexi Corridor is about 30–160 mm, but the evaporation is as high as 2600–3000 mm. The ecosystem is highly dependent on water resources, and the amount of water resources is closely related to glacier changes. Atmospheric precipitation, glaciers and seasonal snow in the southern Qilian Mountains are the main sources of recharge for the inland river basin, which serves as a water source for the river runoff formation area and the entire corridor region. The central plain corridor area is strongly evaporated, the topography turns, surface runoff and underground runoff are subject to climate, geological conditions and human activities, runoff replenishment is weak, and it is a runoff dissipation area, and it is also an important area where surface water and groundwater undergo multiple transformations and water resources are exploited and consumed on a large scale. The northern desert

G. Qian · B. Yue · K. Li · T. Gao (✉)
College of Urban Environment, The Engineering Research Center of Mining Pollution Treatment and Ecological Restoration of Gansu Province, Lanzhou City University, Gansu 730070, China
e-mail: zkgtp@163.com

J. Zhang et al. (eds.), *Environmental Pollution Governance and Ecological Remediation Technology*, Environmental Science and Engineering,
https://doi.org/10.1007/978-3-031-25284-6_21

area has a more arid climate, strong evaporation, very weak runoff recharge almost disappeared, and belongs to the downstream no-flow area.

21.2 Influence of Glaciers on River Runoff in Hexi Corridor

21.2.1 Glacier Area and River Runoff in the Hexi Corridor Are Positively Correlated

Hexi Corridor can provide meltwater glaciers a total of 2444, glacier area of 1657 km^2, accounting for 84% of the total area of the Qilian Mountains glaciers, water storage 801×10^8 m^3 (Li and Cheng 2002a). It is mainly distributed in the eastern section of Aerjinshan, Shule South Mountain, TurgenDaban, Corridor South Mountain, Danghenan Mountain, Daxue Mountain and Lenglongling Mountain Range. During the millennium time period, the glacier area is decreasing (Liu et al. 2010).

As a product of climate, glaciers' advance and retreat changes are closely related to regional climate, while changes in temperature and precipitation directly affect the amount of glacier accumulation and ablation. According to the results of the study, when the temperature increases by 1 K, a 25% increase in precipitation is needed to compensate for the material deficit caused by warming. Thus, climate change in the western part of the river affects the changes in the advance and retreat of the Qilian glaciers. And glacial changes affect the region's water resources first, and ultimately the region's ecological changes.

The glacier area of Shiyang River Basin is 64.8 km^2 with river runoff of 15.7×10^8 m^3, and the glacier area of Black River Basin is 410.6 km^2 with river runoff of 36.4×10^8 m^3. From the data analysis, the larger the glacier area, the larger the river runoff, and the two are positively related. Many glacial melt water in the Shule River basin does not converge into river runoff, so although this basin has the largest glacier area of 1171.8 km^2 among the three major inland rivers in the Hexi Corridor, the river runoff formed is less – 20.2×10^8 m^3 (Table 21.1). but within its tributaries, it still reflects the positive relationship between glacier area and river runoff.

Glacier melt is strongly influenced by temperature and precipitation factors. Under global warming, for upstream glacier-scale regions, glacier meltwater increases and thus contributes more to river runoff. For watersheds with smaller upstream glaciers, the contribution of glacial meltwater to river runoff will increase and then decrease, while the regulating effect on rivers is weakened. For example, Kangersi et al. combined the actual changes that have already occurred in the Qilian Mountains and the western part of the river, pointing out that precipitation may increase by 20% in the 2040s, with a corresponding increase in evaporation (Fig. 21.1a, b) (Kang et al. 1999).

Table 21.1 Discharge List of glacier conditions in Qilian Mountains (Li and Cheng 2002b)

Water system	River Name	Glacier area km^2	Glacial reserves 10^8 m^3	Glacial runoff 10^8 m^3	River runoff 10^8 m^3	Glacier recharge Specific gravity
Shiyang River	Zamu River	3.9	1.0	0.03	2.5	
	Jinta River	6.7	1.5	0.06	1.4	
	Xinyin River	19.8	7.1	0.17	3.2	
	Dongda River	34.4	11.8	0.32	3.0	
	Basin Subtotal	64.8	21.4	0.58	15.7	3.7%
Black River	Shandan River	9.6	2.3207	0.09		
	Dazhuma River	10.4	13.8	0.1	0.8	
	Black River	59.0	309	0.55	16.1	
	Liyuan River	16.2	3.9	0.12	2.4	
	Bailang River	15.1	4.8	0.15		
	Mayin River	19.5	5.4	0.18	1.2	
	fengle River	23.3	7.4	0.16	1.0	
	Hongshui River	130.8	53.3	0.83	2.6	
	Taolai River	136.7	43.1	0.81	6.6	
	Basin Subtotal	410.6	136.7	2.99	36.4	8.2%
Shule River	Baiyang River	10.2	2.4	0.08	0.5	
	Shiyou River	24.6	8.1	0.20	8.5	
	Shulei River	594.4	320.9	3.27	0.6	
	Tashi River	504	2.1	0.03	3.2	
	Dang River	232.7	111.2	1.23		
	Cuimutu River	27.1	12.7	0.13	4.3	
	Haerteng River	322.5	185.8	1.50		
	Basin Subtotal	1171.8	643.2	6.44	20.2	31.2

(continued)

Table 21.1 (continued)

Water system	River Name	Glacier area km^2	Glacial reserves 10^8 m^3	Glacial runoff 10^8 m^3	River runoff 10^8 m^3	Glacier recharge Specific gravity
Total		1647.2	801.3	10.01	74.3	13.8

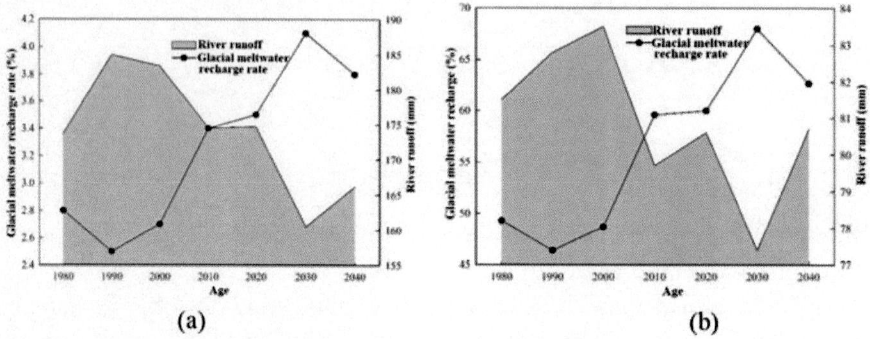

Fig. 21.1 Simulation results of the future 50-year chronological mean runoff composition and its changes in the control mountain area at Yingluoxia hydrological station **a** in the Heihe River basin and Changmabao hydrological station **b** in the Shule River basin

21.2.2 Glacial Ablation Makes the River Show the Characteristics of Spring Flood, Summer Flood, Autumn Flat, Winter Dry

The spatial and temporal distribution of surface water in the western part of the river mainly depends on the atmospheric precipitation in the Qilian Mountains and the Aljinshan Mountains, the spatial and temporal distribution of snow and ice melt water, and the hydro-meteorological vertical zonation subsurface conditions in the Qilian Mountains. In general, the intra-annual distribution of surface runoff in mountainous areas is basically consistent with precipitation processes and high temperature seasons, with runoff and precipitation concentrated in the warm season. Spring to ice melt water and groundwater recharge, summer and autumn to precipitation recharge, with "spring flood, summer flood, autumn flat, winter dry" characteristics, intra-annual changes by the temperature, the influence of vegetation, is a clear cycle pattern. River runoff formation, utilization, and disappearance zoning are obvious.

The upstream mountainous area is rich in precipitation and recharged by glacial meltwater, and the lower bedding surface is stony mountainous area with good vegetation, which is the surface runoff formation area, and the size of runoff is influenced by precipitation, ice melt and forest vegetation cover, etc. The runoff depth decreases from the ridge to the foot of the mountain and from east to west, and the intra-annual distribution is uneven and the inter-annual variation is large, but the variation is smaller than that of the single precipitation recharge type.

In the middle and lower reaches, precipitation is low and evaporation is strong, and the lower bedding surface is a deep quaternary sediment layer, which is a good underground water storage site. The average intensity of precipitation is dissipated by evaporation, and occasionally a larger intensity of precipitation, which also infiltrates to recharge groundwater, basically does not produce flow. The river water from upstream is quoted in large quantities, and the river runoff is reduced along the course, which belongs to the runoff utilization area.

In the downstream area, the underground runoff and the remaining river runoff evaporate in the form of soil diving layer and flow into the Coccyx Lakes, in the form of evaporation, which is consumed by the ecology of the coccyx area and belongs to the runoff disappearance area. River runoff is recharged in various forms, including precipitation, glacial meltwater and groundwater.

21.2.3 The Reduction of Glacier Area Leads to the Splitting of the Unified Water System

The three major water systems in the west of the river originated from the glaciers at the northern foot of the Qilian Mountains. Since the middle Holocene, the flow rate has decreased due to the aridity trend of the climate (Hu et al. 2013), and the glacier area has shrunk, resulting in the gradual shortening of the flow of each river, the loss of surface water connection between the tributaries and the mainstream, and the contraction of the river anadromous flow, gradually forming numerous independent water systems (Li and Cheng 2002b).

There is a migration and shrinkage process of the Heihe River water system from east to west. Before 3000 A.D., there was a unified water system in the Heihe Coccyx Lake, and two branches of the lower Heihe River still existed in the Han Dynasty, with the eastern branch flowing into the Juyanze and the western branch flowing into The Two Nur Lake Basins (Zhong 1995).Tang dynasty to the East River, the whistling horse camp—ancient day nai ancient river channel and the ancient Juyanze terminal lake; West Xia period weak water into the East and West branches, Sugo Nur (East Juyan Sea), Gashun Nur (West Juyan Sea) two lakes together as one, ancient Juyanze shrinkage; Yuan dynasty river water re-injected into the ancient Juyanze, Sugo Nur, Gashun Nur two lakes separated and shrinkage; late Yuan and early Ming dynasty, after the West River became the main over-water river, the East River downstream into the ancient Juyanze river dried up, to the north into the Sugo Nur river mainly, the ancient Juyanze gradually shrink dry, Sugo Nur, Gashun Nur become the main terminal lake; Qing after the water system has been in the process of shrinkage, disconnection, Coccyx Lake also in the 1960s after the gradual drying up (Zhang 1936).

The process of migration of the Shiyang River system from west to east and from downstream to upstream, Mr. Li Bingcheng believes that the massive desertification of the lower Shiyang River oasis began after 695 in Tang Dynasty at the latest. Since then the large downstream oasis has been empty, marking the desertification process of the whole downstream oasis has been obvious (Li 2003a). In the Ming Dynasty, the oasis was reclaimed downstream, and the scope of its reclamation spread throughout the dam oasis, and in the Qing Dynasty, it was expanded to the lake oasis. However, the oasis after the Ming Dynasty did not inherit the old oasis of the Han and Tang dynasties, but developed separately in its eastern part, that is, in today's Minqin Oasis. When the lower oasis was reopened during the Ming Dynasty, the main flow path of the river shifted eastward and began to move to the present-day location of the main stream.

Shule River that is the ancient end of the water, also known as meditation water, Blongier River, originated Jing Niwei (now Yumen Town) South Mountain, for the Changma River, north flow turn and west, by the old Liu Gouwei (now Anxi County four ditch) north, will be ten ditch water for Sule River. And west by the state north of 30 miles, and west by Dunhuang County, northwest of the party river from the south to inject it, and west of the flow of 30 miles, into the Haranor, its flow length of more than 700 miles, the pool square dozens of miles, forming the Han dynasty meditation, the Tang dynasty name Dazhe, the early Qing dynasty split for the Castle Peak Lake, Blu Lake and flower Haizi (Huang 1984a).

21.2.4 Drying Up of Tailrace Lakes Due to Shrinking Glacier Area

There used to be a large number of lakes in the western part of the river in history, the famous ones are Zhuye Ze, Guchangning Lake, Yabuli Salt Lake, Gaotai Salt Pond, Juyanhai, Huahai, Medze, Hala Nur, etc. Most of these lakes have gone through the process of shrinking, decomposing and drying up, and now they have basically disappeared.

ZhuyeZe, the terminal lake in the lower reaches of the Shiyang River, historically known as Zhuye Ze, had a maximum area of 1,260 km^2 in the last 20,000 to 30,000 years.During the pre-Qin period, the area was 525–575 km^2, and during the Han Dynasty and the North and South Dynasties, it was gradually divided into two lakes, the West Sea (Xiutu Ze) and the East Sea (Zhuye Ze in the narrow sense), which are not connected to each other. In the Sui and Tang dynasties, the East Sea was called Dadijiong sea (since the Qing dynasty, the name of the Baijian Lake), and the West Sea was called Baiting Sea (the Qing dynasty, the name of Liulin Lake, Qingtu Lake (Wang et al. 1999).The terminal lake of the lower Jinchuan River is Changning Lake (named Changning Lake in the Qing Dynasty) (Li 1999).

In the Kangxi thirty-ninth year (1700) there is still a record of "the town of Baiting seawater tide, the well flooded". However, by the twenty-sixth year of the

Qianlong reign (1761), the water area of Baitinghai Lake was less than 70 km^2. After 1840, the terminal lake downstream of the Shiyang River began to degrade and dry up completely, and before 1950, the remaining water from the Shiyang River was injected into Qingtu Lake and Baijian Lake. The vegetation of Qingtu Lake is quite lush, and the original landscape of the lake is full of reeds and wild ducks. Later, due to the development of irrigated agriculture in the middle reaches, the expansion of arable land, the lack of surplus water drainage, the disintegration of the aquatic ecosystem, and the emergence of a desert landscape, the lakeshore covered with quicksand (Zhang 2010).

GuyanSea, Historically, the maximum area of the Guyan Sea was as much as 2,000 km^2. During the Han Dynasty, the Juyan area was cultivated on a large scale, but because of the abundant water flowing into the downstream, three lakes, East and West Juyan Sea and Juyanze, existed in a large area. During the Yuan Dynasty and the middle of Ming Dynasty, as the economic activities were mainly concentrated in the west corridor, the large-scale construction of water conservancy in the middle reaches of the Heihe River reduced the surface runoff downstream accordingly, and the siltation of sediment made the bed of the East River silt higher, so the ancient Juyanze first tended to shrink and sandy. Since the twentieth century, the river mainly flows into the East and West Juyan Sea, and by 1958 the water surface of the West Juyan Sea still reached 262 km^2. Since it dried up in 1961, it has been covered by white alkali desert and barren sand, and has become one of the sources of flying sand and dust. The East Guyan Sea dried up six times after liberation and dried up completely in 1992.

The sea of flowers, the area of the lake used to reach 445 km^2 2000 years ago. According to Xu Song "Western Waterway Records" Volume 3 records: "Cha Hua (flower) Haizi, forced near the Chi, Jing and other places such as Yingxun, south to Qinghai, north to North Road Karen, convenient for animal husbandry, cover the seashore wide, so spare water plants, Haizi one hundred and sixty miles long, to the north, it links up with the Paidai Lake, its west for Bru Lake". After the natural diversion of the Shule River in the dam floodplain fan swing and artificially built a dam to open a canal, the amount of water into the lake is greatly reduced, so the lake is rapidly shrinking. Nowadays, there is basically no surface water distribution and a desertified landscape of dry lakes and sand dunes has been formed. The remnant of the Huahai Lakes, the downstream terminal lake of the Beishi River, Dryhaizi, has an original lake surface of about 300 hectares with a water depth of 0.1–1.5 m (Table 21.2). The plants in the lake are mainly broad-leaved reeds, and the plants around the lake are mainly red willow bushes. 26 kinds of tens of thousands of migratory birds and resident birds inhabit the lake, including white swans, great egrets, gray herons, gray cranes, ruddy ducks, etc. In April 1982, Ganghizi was established as a nature reserve for birds in Gansu Province. However, in recent years, the water source upstream of Ganhaizi was cut off and diverted to the south bank of the North Shihe River in the area of Sanduan Tan and Bijiatan, which dried up completely in June 1999, and the bird nature reserve also existed in name (Huang 1984b).

Table 21.2 Data Table of lake gain and loss (area unit: km^2)

	First Qin and Second Han	Early Qing Dynasty	The end of the Qing Dynasty	Current Status
Zhuye Ze	1200	210	70	Dry up
Hua Sea	445	49	10	Dry up
Juyan Sea	2000	695	352	Dry up

21.3 Impacts of Glacial Change on the Land Landscape

21.3.1 Wastewater Glacier Area and Downstream Desertification Area Are Positively Proportional

The reason is that the downstream area is located at the tail end of the oasis, with poor water resources and vulnerable to changes in water quantity, and the prevailing northerly winds, with strong effects of drought, high winds and other catastrophic weather.

With a runoff of 36.4×10^8 m^3, the Heihe River has the largest flow, so the area of desertification downstream of it is the largest, about 1200 km^2.In the southeast of Ejinna Banner in Inner Mongolia downstream of the Black River, there is the deserted ancient Juyan-Heicheng Oasis, which is an oasis from the Han Dynasty to the Yuan Dynasty, and the surface landscape today is a continuous wind erosion of ancient cultivation remains and blowing scrub sand piles, flowing sand dunes distributed between, the ancient channels that have long dried up are intermittent and vivid, and a number of ancient city sites remain. Ancient Guyanze is located in the Guyan depression in the northeastern part of the ancient oasis, and six ancient lakeshore dikes in the shape of parallel rings remain.

The runoff of Shiyang River is 15.7×10^8 m^3, the flow is smaller than that of Heihe River, so the desertification area is also smaller, there are about 1000 km, distributed in Minqin County, West Shawo, which was an ancient oasis in Han and Tang dynasties, but evolved into desert after Shengtang. The present-day surface landscape is that there are patches of semi-fixed white thorn scrub sand piles scattered on the ancient cultivated land that suffered from severe wind erosion, with a few exposed crescent-shaped sand monopolies, during which the remains of abandoned canal dikes and paths are intermittently recognizable, and scattered pottery, bricks, coins and other Han and Tang relics. In this vast sand dune, there are still several ruins of Shajing culture and several ancient cities such as Triangle City, Liancheng, Gucheng and Wenyi ancient city. The eastern part of the ancient oasis is the modern Minqin Oasis, and about 10 km northeast of it are the remains of the ancient terminal Zhuyeze, on the Shiyang River (Long et al. 2007).

The area of desertification formed by the 6.6×10^8 m^3 runoff from the river of Taolai River is smaller than the lower reaches of the above two rivers, with an area of only about 570 km^2, that is, the ancient oasis of Jinta East Tsavo, which is covered with white thorn scrub sand piles and tamarisk scrub sand piles of about 2–4 m in height, with crescent-shaped sand dunes and flowing sand beams between them, with patches of ancient cultivated land exposed between the sand piles and sand dunes, with clear remains. On the ancient oasis, there are several cultural sites of fire-burning ditch type and more than 10 ancient city sites of Han and Tang dynasties, such as West Triangle City, West Old City, Triangle City, Lower Great Wall, Broken City, Yellow Duck Dun City, Lower Broken City, North Triangle City, West Kiln Broken Village, etc., and more ancient pottery kilns are distributed.

In the case of the desertification of Dunhuang South Lake Oasis, the South Lake Oasis uses the outcrop spring as the irrigation water source, and its outcrop volume is 15×10^4 m^3, so the area of the ancient green oasis desertification is the smallest, only about 40 km^2.

According to Mr. Libingcheng statistics, the historical period of desertification in the Hexi Corridor is about 4600 km^2, of which about 1680 km^2 formed after the Han Dynasty, about 1760 km^2 formed after the Tang Dynasty extended to the Wei and Jin Dynasties, about 1160 km in the Ming and Qing Dynasties (Li 2003b).

21.3.2 Periodic Glacial Melt Alternates Desertification and Oasis

Oasis is the product of water resources in the arid desert area, and the size of the oasis and its change in size are mainly governed by the amount of water resources, which in turn depends directly on the melting of glaciers. The ablation of glaciers depends on the historical period of climate change, Mr. Zhu Kezhen "China in the past five thousand years, the preliminary study of climate change, located in the mid-latitude inland arid areas of the Hexi Corridor, the warm period will inevitably lead to a reduction in the area of glaciers, the ablation of snow water is reduced, the ecological environment of the Hexi Corridor deterioration.

First Warm Period: from 3000 to 1000 B.C. First Cold Period: from about 1000 to 850 B.C.

The Second Warm Period: began around 770 B.C. lasted until the beginning of A.D. The Second Cold Period: lasted from around the beginning of A.D. to A.D. 600, including the Eastern Han and Three Kingdoms to the Northern and Southern Dynasties period in China.

The third warm period: 600–1000 A.D., including the Sui and Tang dynasties (589–907 A.D.), and the third cold period: 1000–1200 A.D., corresponding to the Song dynasty.

The fourth warm period: 1200–1300 A.D. The early Yuan Dynasty (1268–1292), the Fourth Cold Period: from about the fourteenth century A.D. to the end of the nineteenth century, corresponding to the Little Ice Age in Europe (Zhu 1973).

The alternation of warm and cold periods leads to the alternation of desertification and oasis, but because the temperature does not change much on the millennial scale, the total area of the oases developed and formed in the Hexi Corridor does not change greatly, and the only thing that changes is the change in the location of the oases and desertification areas, in other words, there is only a shift of the oases.

For example, Minqin County, located in the lower reaches of Shiyang River, covers an area of 1,000 km West Shawo, which is the site of an ancient oasis in Han and Tang dynasties, but deserted after Han and Tang dynasties, but gradually formed a new oasis in its eastern part, which is the modern oasis that is still being used today after being gradually developed since Yuan, Ming and Qing dynasties. Another example is the ancient Han and Tang oasis in East Shawo, Jinta County, which is located in the lower reaches of the Beida River and covers an area of about 650 km. While the oasis was abandoned and desertified, a new Jinta oasis was formed on its western side which is still being used today. After the desertification of the ancient oasis of Camel City located at the border of Jiuquan and Gaotai in the downstream of Maying River and Pendulum River with an area of about 450 km, new oases of Tunsheng and Xinba appeared in its upper reaches, etc. After the desertification of the ancient oasis of HuahaiBijiatan in Yumen City, which is located in the lower reaches of the Petroleum and Beishi rivers and covers an area of about 480 km, new oases of Huahai have developed in the eastern part of the city.After the desertification of the ancient oasis of Lockyang City, located in the lower reaches of the ancient Changma River, covering an area of about 500 km, a new oasis of Ansi River East and the oasis of Yumen Town were formed in the northeastern part of the river at a large bend. The abandonment of these ancient oases has likewise not caused a large scale reduction in the total oasis area of the basin. It can be considered that the essence of desertification process is a kind of oasis transfer process triggered by human unreasonable development and management activities, due to the movement and redistribution of oasis water resources, and one of the basic directions of its transfer is the migration from downstream to middle and upper reaches, rather than the continuous shrinking or disappearance of oasis. This migration was accompanied by the abandonment of the original oases and the emergence of new ones.

This shows that the desertification process generally takes place in the downstream areas of the basin oasis, and that locally, the oasis is indeed greatly reduced or even destroyed. However, on a basin-wide basis, this does not imply a reduction in the total area of the oasis, but rather, to a large extent, a shift of the oasis. The essence of the desertification process is a kind of oasis shifting process mainly caused by human unreasonable development and management activities, due to the shifting and redistribution of oasis water resources, and one of the basic directions of shifting is the migration from downstream to middle and upstream, rather than the shrinking or disappearance of oasis.

21.3.3 The Distribution of the Oasis is Determined by the Lamellar Distribution of Glaciers

The glaciers of the Hexi Corridor are characterized by a patchy distribution, which also leads to a patchy distribution of various ecological landscapes, no matter which ecological environment is not connected into a complete large area. As for the distribution of oasis eco-regions, they are even more fragmented, mostly unconnected and far apart. The grassland eco-region is also distributed in pieces, with the eastern Pine Mountain, the central Rouge Mountain, and other places not very large in area and not connected.

21.3.4 Rapid Glacial Melt Buries the Oasis with Flood Deposits

It mainly occurs in the upper and middle reaches of the watershed where flood deposits are abundant. When the river flows out of the mountain, it carries a large amount of gravel and sediment, and once the river bank breaks in a flash flood, the flood deposits are dumped on the oasis farmland, thus causing it to be abandoned and barren. This kind of desertified land also has a small area. In Yongchang County, there is a thin layer of flooded gravel on the abandoned land in the area of messy dunzi beach, and the pebbles with axial length of several centimeters to ten centimeters are seen scattered everywhere, and the remains of the road are visible, and many Ming Dynasty celadon pieces and other things are seen. Here is the original Ming Yongning Fort reclamation area, located in the Shiyang River tributary Xiying River near the side, due to the Ming and Qing dynasties to destroy the Qilian Mountain forest, resulting in its ability to contain and store water is reduced, the river sand content increased, the water situation tends to be unstable, and finally this oasis was buried by flooding, and by wind erosion, fine-grained materials are more eroded away, leaving only larger gravel residue ground, the era of desertification occurred when the late Ming Dynasty.

21.3.5 Freeze–thaw Desertification Caused by Glacial Melt

The area of freeze–thaw desertification in Hexi Corridor is mainly distributed in the vicinity of Qiyi Glacier, mostly in patchy scattered distribution and piecewise scattered distribution, and the plateau permafrost area has the environmental matrix for the formation of freeze–thaw desertification. As the global climate continues to warm, the frequency and intensity of anthropogenic activities intensify, so that the freeze–thaw process of shallow multi-year permafrost intensifies, thus forming freeze–thaw desertified land. The formation process mainly includes the thickening of the permafrost seasonal thaw layer—the decline of the water table—the drying of the surface soil, the change of the surface cover or the increase of the underground melt water—the hot thaw of the permafrost thaw interface—the surface subsidence and fragmentation, the freeze–thaw process and the slope process are intensified, etc.

21.4 Conclusions

There is a positive correlation between glacier area and river runoff. The characteristics of spring flood, summer flood, autumn flat and winter dry are due to the melting of glaciers. The rump lakes dried up because glaciers shrank. There is a proportional relationship between the area of glaciers and the area of desertification downstream. The alternate occurrence of desertification and oasification is due to the periodic retreat of glaciers. The sheet-like distribution of glaciers determines the distribution of oases. The rapid retreat of glaciers has buried oases with diluvial deposits. Freeze–thaw desertification due to the expansion and shrinkage of glacier area.

Acknowledgements The research was funded by the National Natural Science Foundation of China (Grant No. 31860176), Key Research and Development Program of Gansu Province (No. 20YF3FA037)

References

Hu XF, Pan BT, Gao HS, Hu ZB, Cao B, Li QY, Geng HP (2013) The eastern section of qilian mountains was the holocene river terrace development and its relationship with climate change research. J Quat Res 33(4):723–736
Huang WH (1984a) Rebuild Xuzhou's history book. Gansu Province Library
Huang WH (1984b) Rebuild Xuzhou's history book. Gansu Province Library, p 283
Kang ES, Cheng GD, Lang YC, Jin HJ (1999) In continental river basin in arid zone of northwest eurvedresponse model of runoff change trend of climate change. China Sci (D) 29(1):47–54
Li PZ (1999) Wuwei's ancient history. Gansu People's Publishing House, Gansu
Li PZ (2003a) Studies of desertification in hexi corridor historical period. Science Press, Beijing
Li PZ (2003b) Studies of desertification in hexicorridor historical period. Science Press, p 298
Li SM, Cheng GD (2002a) The rational utilization of water resources in hexi corridor and ecological environment protection. TheYellow River Water Conservancy Press, p 71
Li SM, Cheng GD (2002b) The rational utilization of water resources in hexi corridor and ecological environment protection. TheYellow River Water Conservancy Press, p 9
Liu YS, Qin X, Du WT, Sun WJ, Hou DJ (2010) Qilian mountain tiger groove 12 glacier movement characteristic analysis. J Glacial Permafrost 32(3):475–479
Long H, Wang NA, Li Y, Ma HZ, Zhao Q, Cheng HY, Huang YZ (2007) Zhu Ye Ze record monsoon fringes of middle holocene climate environment evolution history. J Quat Res 27(3):372–381
Wang NA, Li GG, Cao JX, Cai WM (1999) Green lake soil study sedimentary climate records in recent 6000 years, and the theory of four or five centuries of climate warming. J Geogr Sci 19(2):119–124

Zhang W (1936) Recorded in GanSu province history book. Gansu Province Library Editions, Gansu, p 428
Zhang ZM (2010) Comprehensive prevention and control plan of heavy metal pollution in Gansu Province
Zhong GQ (1995) The history book of ancient Zhang Ye. Gansu Culture Press, Gansu
Zhu KZ (1973) China's nearly five thousand years, a preliminary study on climate change. Sci China S1:2–23

Chapter 22
Study on the Treatment of Cyanide Tailing Washing Wastewater by Acidification Sedimentation Process

Tengyue Gao, Guangsheng Li, Xingfu Zhu, and Qiang Ji

Abstract In this study, the solid–liquid separation washing wastewater from cyanide tailings was used as the research object, and the acidification iron salt precipitation process was used to purify the cyanide in the wastewater. The results show that the total cyanide content of the wastewater is greatly reduced after acidification stripping treatment. After acidification, ferrous sulfate is added to the filtrate for precipitation treatment, and the total cyanide content in the wastewater can be reduced to less than 50 mg/L. the treated wastewater can be used in the washing process of cyanide tailings. Therefore, acidification iron salt precipitation process can realize the recycling of wastewater after solid–liquid separation and washing.

Keywords Acidification · Iron salt precipitation · Cyanide containing wastewater · Washing and purification

22.1 Introduction

Due to the advantages of low production cost, stable process, high metal recovery rate and strong adaptability, cyanidation is the main method to extract gold from gold bearing ores at present (Bas et al. 2015) However, the leaching tailings produced by this method contain a small amount of cyanide leaching solution, which is a dangerous solid waste. When cyanidation process is used to treat flotation gold concentrate, because gold ore is usually associated with copper containing minerals, the dissolution of copper containing minerals in the cyanidation process leads to copper ions in the leaching solution, resulting in excessive heavy metal ions in the leaching toxicity test of cyanidation tailings (Oraby and Eksteen 2016; Zhihua 2015).

According to the requirements of relevant laws and specifications, cyanide residues need to be removed in advance before subsequent disposal (Yunliang et al.

T. Gao (✉) · G. Li · X. Zhu · Q. Ji
Metallurgical Laboratory Branch of Shandong Gold Mining Technology Co., Ltd., Yantai 261400, China
e-mail: gaotengyue@sd-gold.com

J. Zhang et al. (eds.), *Environmental Pollution Governance and Ecological Remediation Technology*, Environmental Science and Engineering,
https://doi.org/10.1007/978-3-031-25284-6_22

Table 22.1 Analysis of main components of washing wastewater

Analysis items	Free cyanide	Total cyanide	Cu	Zn	Fe
Value (mg/L)	967.12	1729.01	371.45	112.61	277.39

2018). In the existing cyanide removal process of tailings, oxidation method and precipitation method can effectively remove cyanide, but when treating cyanide tailings of flotation concentrate, copper ions in the leaching solution will remain in the tailings in the form of precipitation, which is easy to cause secondary pollution of heavy metals (Jing et al. 2012; Kitis et al. 2005; Kuyucak and Akcil 2013; Laifu et al. 2004; Shumei et al. 2016; Yong et al. 2013). The solid–liquid separation and washing method is to purify the cyanide tailings by filtration and washing, so it will not cause secondary pollution of heavy metals (Qiang et al. 2017). After the washing and decyanation treatment, only the washing wastewater needs to be purified to realize the recycling of washing water, so this process is more suitable for treating the tailings produced by the flotation cyanidation process. This study takes the wastewater after the solid–liquid separation and washing of the flotation cyanide tailings of a gold smelting company as the research object, adopts the acidification iron salt precipitation process to carry out the experimental research on the purification and treatment of wastewater, explores the low-cost treatment process of the washed wastewater, and uses the treated wastewater to carry out the washing test of the cyanide tailings to verify the feasibility of the recycling of washing water.

22.2 Experimental

22.2.1 Raw Materials and Reagents

The wastewater after solid–liquid separation and washing used in this test is from a gold smelter in Shandong, which uses copper bearing flotation gold concentrate as the main production raw material. The cyanide and heavy metals in the cyanide tailings washing wastewater are high. The main components of the wastewater are shown in Table 22.1. The pulp used in the tailings washing test is still from the plant, and the chemical reagents used in this test are analytical pure.

22.2.2 Test Process

The self-made acidification aeration device is used for experimental research at room temperature. Washed wastewater (1 L) is added with concentrated sulfuric acid, and the pH value of wastewater acidification is measured by Shanghai Instrument Electric PHS-2f pH meter; After aeration, filter, take the filtrate and put it into a sealed conical flask, add solid $FeSO_4 \cdot 7H_2O$, use magnetic stirring reaction for 20 min, add an appropriate amount of Cao to adjust the pH value, and continue stirring for 10 min.

Take the filter cake from the on-site cyanide tailings after pressure filtration, use the treated wastewater to adjust the pulp for filtration and washing, and take the washed filter cake for cyanide leaching toxicity detection.

22.2.3 Analysis Method

The content of free cyanide and total cyanide in the sample was determined by silver nitrate titration. The washed filter cake was extracted with distilled water, and the total cyanide content of the extract was still analyzed by silver nitrate titration. The contents of copper, iron and other metal ions in the samples were determined by atomic absorption spectrometry.

22.3 Results and Discussion

22.3.1 Effect of Acidification pH Value on Cyanide Removal Effect

Under the conditions of gas flow of 200 L/min and aeration time of 1 h, the author studied the influence of acidification pH value of wastewater after washing on the removal effect of total cyanide, as shown in Fig. 22.1.

It can be seen from Fig. 22.1 that with the reduction of acidification pH value, the total cyanide content in wastewater shows a significant downward trend; When the acidification pH value is lower than 2.0, the total cyanide content of the filtrate after treatment increases slightly, and the solution shows an obvious light blue. This is because the free cyanide and easily released cyanide in the wastewater will form HCN during acidification, which is very volatile and escapes with the help of air blowing, so reducing the acidification pH value is conducive to the removal of cyanide; When the pH value is lower than 2.0, the generated iron cyanide precipitate in the wastewater is dissociated, and the generated fine Prussian blue precipitate cannot be removed by filtration, resulting in the increase of total cyanide content and the blue color of the solution. Therefore, the pH value in the acidification process should not be too low and should be maintained at about 2.5.

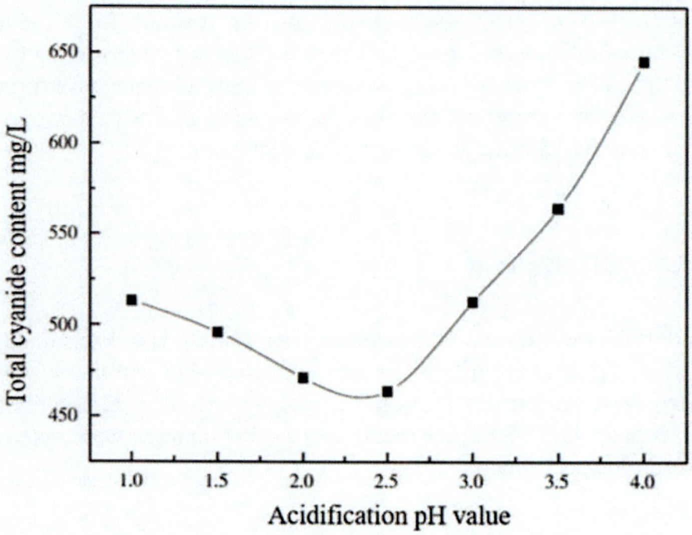

Fig. 22.1 Effect of acidification pH value on total cyanide content

22.3.2 *Effect of Aeration Time on Cyanide Removal Effect*

Under the conditions of gas flow of 200 L/min and acidification pH value of 2.5, the author studied the influence of aeration time on the removal effect of free cyanide and total cyanide, as shown in Fig. 22.2.

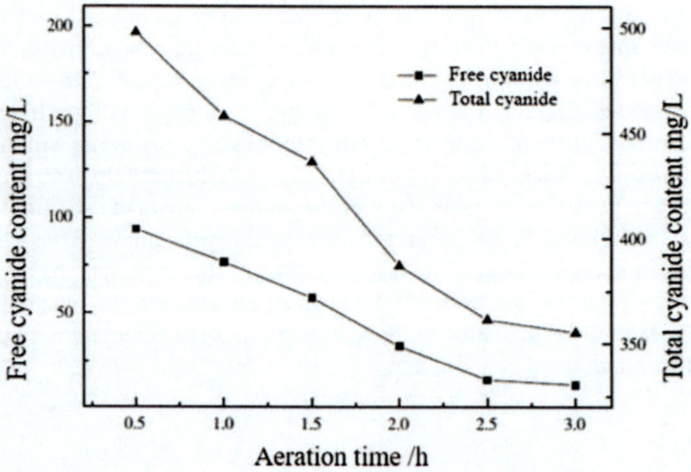

Fig. 22.2 Effect of aeration time on free cyanide and total cyanide content

It can be seen from Fig. 22.2 that with the extension of aeration time, the content of free cyanide and total cyanide in wastewater decreases significantly. When the aeration time reaches 2.5 h, the content of residual free cyanide and total cyanide in wastewater decreases to 15.14 and 346.18 mg/L respectively; The content of free cyanide and total cyanide continued to decrease with a small change after the aeration time was prolonged. After aeration stripping, the reduction of the residual cyanide content in the wastewater is conducive to the subsequent precipitation and purification treatment. However, the existing industrial equipment is difficult to achieve the above stripping effect while ensuring the treatment capacity. In order to be closer to the actual production status, this study selects the wastewater after aeration for 1.5 h for the subsequent test. At this time, the free cyanide and total cyanide content in the solution are close to the production data, which are 58.12 and 436.44 mg/L respectively.

22.3.3 *Effect of Iron Salt Dosage on Cyanide Removal Effect*

After acidification aeration treatment, $FeSO_4 \cdot 7H_2O$ was added to the wastewater for precipitation purification treatment. The author studied the influence of reagent dosage on the removal effect of total cyanide, as shown in Fig. 22.3.

It can be seen from Fig. 22.3 that using $FeSO_4 \cdot 7H_2O$ precipitation can effectively reduce the total cyanide content in the wastewater. When the amount of $FeSO_4 \cdot 7H_2O$ is 5 g/L, the total cyanide in the solution after stirring and filtration can be reduced to 23.54 mg/L; According to production experience, the total cyanide content in washing

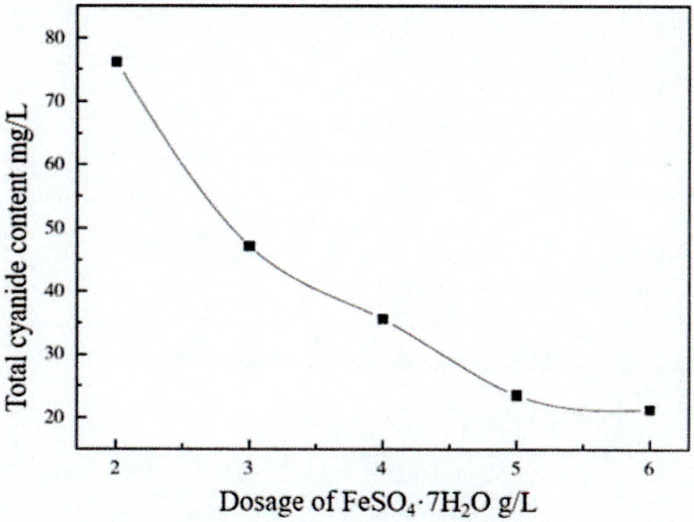

Fig. 22.3 Effect of $FeSO_4 \cdot 7H_2O$ dosage on total cyanide content

water is less than 50 mg/L, which can meet the requirements of cyanide residue washing water. In addition, the study found that when the amount of $FeSO_4{\cdot}7H_2O$ was 2 g/L, the generated iron cyanide sediment was not easy to settle and filter, and the resulting filtrate was light blue. This is because when the iron salt is added in excess, the residual Fe^{2+} is oxidized to form Fe^{3+} and hydrolyzed, which plays the role of flocculation and sedimentation, and improves the filterability of the sediment.

22.3.4 *Influence of Filtration pH Value on Cyanide Removal Effect*

When the addition amount of $FeSO_4{\cdot}7H_2O$ is 5 g/L, use NaOH to adjust different pH values and then filter. The relationship between total cyanide content and pH value in the filtrate is shown in Fig. 22.4. Since the pH of cyanide residue washing water is required to be kept at 6 ~ 8 in production, direct neutralization after iron salt precipitation can simplify the process and shorten the treatment time.

It can be seen from Fig. 22.4 that after iron salt precipitation, the total cyanide content in the filtrate gradually increases with the increase of filtration pH value. When the filtration pH value is lower than 6.5, the total cyanide content does not exceed 50 mg/L; However, when the pH value exceeds 7.0, the iron cyanide precipitate is dissolved back, resulting in a significant increase in the total cyanide content in the filtrate. In the actual production process, the use of lime to adjust the pH value has hysteresis, and the precise control of pH value cannot be achieved. Therefore, it is suggested to adopt a step-by-step neutralization method. First, adjust the pH to

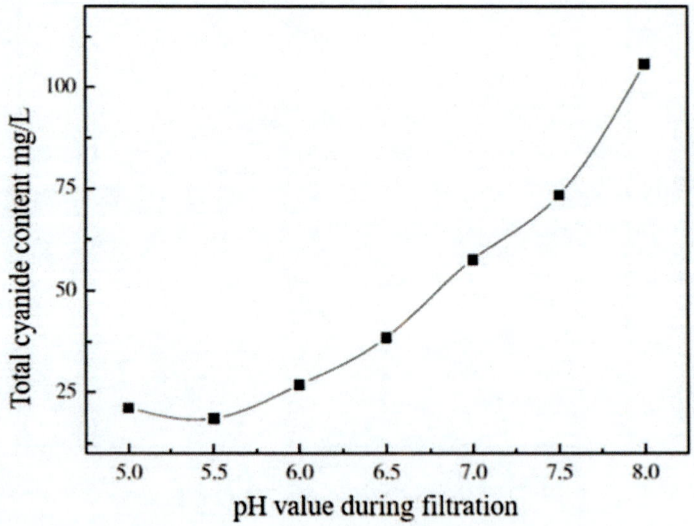

Fig. 22.4 Effect of pH value on total cyanide content during filtration

Table 22.2 Washing test results of cyanide tailings

Sample	Washing water consumption (L/kg)	Toxic leaching total cyanide content (mg/L)
1	Unwashed	127.42
2	1.0	5.19
3	1.5	2.71
4	2.0	3.12

about 5.0 to make Fe^{3+} fully hydrolyze and promote the filtration of iron cyanide sediment, and then adjust the pH value of the filtrate to keep it at 6 ~ 8 for the washing of cyanide tailings.

22.3.5 *Washing Test of Cyanide Tailings*

The cyanide tail residue washing test was carried out with the treated wastewater with a total cyanide content of 24.57 mg / L. under the conditions of different washing water dosage, the washing test results are shown in Table 22.2.

It can be seen from Table 22.2 that the use of washing water after treatment has no adverse effect on the washing process of cyanide tailings. When 1.5 times the amount of washing water is used, the cyanide content of the obtained filter cake is lower than the disposal requirements of 5 mg/L through leaching toxicity test. The test results show that the washing wastewater treated by acidification iron salt precipitation process can be returned to the cyanidation tailings washing process for recycling.

22.4 Conclusion

The acidification pH value of cyanide containing wastewater should be controlled at about 2.5. Too low pH value will lead to the formation of fine Prussian blue precipitation, resulting in the increase of total cyanide content in acidified filtrate. The neutralization process of wastewater after iron salt precipitation should be carried out step by step, and the pH value of filtration and neutralization should be maintained at about 5.0. Too high pH value will cause the iron cyanide sediment to reverse dissolve, resulting in the increase of total cyanide content. The treated washing wastewater can be used in the washing process of cyanide tailings, which has no adverse effect on the elution of cyanide. After leaching toxicity test, the cyanide content of filter cake after washing is less than 5 mg/L.

References

Bas AD, Safizadeh F, Zhang W et al (2015) Active and passive behaviors of gold in cyanide solutions [J]. Trans Nonferrous Met Soc China 25(10):3442–3453

Jing Y, Yali Z, Xianjin Yu et al (2012) Research status on treating methods of cyanide waste solution [J]. Hydrometallurgy China 31(5):278–280

Kitis M, Akcil A, Karakaya E et al (2005) Destruction of cyanide by hydrogen peroxide in tailings slurries from low bearing sulphidic gold ores [J]. Miner Eng 18(3):353–362

Kuyucak N, Akcil A (2013) Cyanide and removal options from effluents in gold mining and metallurgical processes [J]. Miner Eng 50–51:13–29

Laifu C, Xian L, Zhiqiang Q et al (2004) Treatment of highly concentrated cyanide-containing wastewater with ferrous sulfate and calcium hypochlorite [J]. Ind Water Treat 5(10):68–71

Oraby EA, Eksteen JJ (2016) Gold dissolution and copper suppression during leaching of copper-gold gravity concentrates in caustic soda-low free cyanide solutions [J]. Miner Eng 87:10–17

Qiang L, Xinhui L, Zhongkui C et al (2017) Experimental research on cyanide tailings rinsing [J]. Gold 38(5):69–72

Shumei Li, Fangxia S, Zifan C et al (2016) Treatment of high copper-bearing cyanide barren solution by chemical precipitation process [J]. Hydrometallurgy China 35(6):520–523

Yong Xu, Linan S, Xiaosong Y (2013) Treatment of gold cyanide wastewater by alkaline chlorination [J]. Nonferrous Met Eng 3(3):38–40

Yunliang F, Zhehao Li, Xinhui L et al (2018) Experimental research on non-hazardous treatment of cyanide tailings backfill technique [J]. Gold 39(6):65–68

Zhihua H (2015) Researches on copper behaviors in the cyanide leaching of gold for copper-containing gold ores [J]. Multipurpose Utilization Min Resour 1:35–38

Chapter 23
Experimental Study on the Removal of Xanthate Collectors from Mineral Processing Wastewater

Zhongbo Lu, Guangsheng Li, Xingfu Zhu, and Chao Xu

Abstract Magnetic activated carbon adsorbent was used to treat mineral processing wastewater to absorb xanthate collector in mineral processing wastewater. The effects of adsorption time, adsorbent dosage and pH value on adsorption performance were mainly investigated. The results showed that the removal rate of xanthate collector by magnetic activated carbon was 99.50% under the conditions of pH = 7, adsorption time of 2 h, concentration of Isoamyl xanthate of 100 mg/L and dosage of magnetic activated carbon of 0.7 g/L. At the same time, magnetic activated carbon can be recycled by magnetic separation, with a recovery rate of 97.60%, achieving the purpose of environmental protection and economic application, and has a good application prospect.

Keywords Mineral processing wastewater · Magnetic activated carbon · Adsorbent · Isoamyl xanthate

23.1 Introduction

The development of mineral resources not only creates great economic benefits, but also brings certain impacts on the environment. In the production process, due to the use of a large number of flotation agents, a large amount of mineral processing wastewater will be produced, which contains residual flotation agents, resulting in the exceeding of COD (Chengfang and Peng 2022; Chunyan 2020; Shangjun and Xian 2021; Xiaolian et al. 2018). As a large mining country, China produces 200 million tons of mineral processing wastewater every year, ranking in the forefront of the world (Bingyu et al. 2015; Hao et al. 2021; Jiyang and Yanli 2022; Ran et al. 2011; Qingfu et al. 2016). At present, the treatment methods of mineral processing wastewater mainly include hydrogen peroxide method, ozone method, photolysis method, Fenton

Z. Lu (✉) · G. Li · X. Zhu · C. Xu
Shandong Gold Mining Technology Co., Ltd. Smelting Laboratory Branch, Laizhou 261400, China
e-mail: 379195351@qq.com

J. Zhang et al. (eds.), *Environmental Pollution Governance and Ecological Remediation Technology*, Environmental Science and Engineering,
https://doi.org/10.1007/978-3-031-25284-6_23

Method, precipitation method, decomposition method, acid neutralization method, biodegradation method and adsorption method (Chao and Liping 2020; Hongying and Jin 2021; Xiaojing et al. 2020; Yanlan 2021; Zhaohui et al. 2019). In order to solve the above problems, the magnetic activated carbon with good adsorption effect and easy recovery was prepared from activated carbon and Fe_3O_4. By studying the adsorption performance of magnetic activated carbon adsorbent on xanthate collector and its recovery and reuse, it provides a theoretical basis for the application of magnetic activated carbon in the treatment of mineral processing wastewater.

23.2 Test Method

23.2.1 Test Agent

The test reagents include: ferric chloride hexahydrate, ferrous chloride tetrahydrate, sodium hydroxide, activated carbon powder and potassium butyl xanthate. The above reagents are all analytical pure.

23.2.2 Test Method

Preparation of Magnetic Activated Carbon

(1) Weigh 0.1 mol of ferric chloride hexahydrate and 0.05 mol of ferrous chloride tetrahydrate respectively, fully mix and dissolve in a beaker to obtain product 1.
(2) Weigh 47 g of activated carbon, add it to product 1, and stir it mechanically for 4H to make it fully mixed to obtain product 2.
(3) The constant temperature water bath is used to control the product 2 at 60 °C, add sodium hydroxide solution while mechanically stirring, control pH = 10, and continue stirring for 5 h under constant temperature to obtain the product 3.
(4) The product 3 is washed with distilled water to pH = 7 to obtain magnetic activated carbon, which is dried in a vacuum drying box at 100 °C for standby.

The reaction formula for preparing magnetic activated carbon is as follows:

$$Fe^{2+} + 2Fe^{3+} + 8OH^- \rightarrow Fe_3O_4 \downarrow + 4H_20 \tag{23.1}$$

Isoamyl Xanthate Adsorption Test

Prepare 0–10 mg/L isoamyl xanthic acid standard solution with different concentrations, and increase the concentration by 1 mg/L each time. Measure the absorbance of Isoamyl xanthate at the wavelength of 301 nm with ultraviolet spectrophotometer to obtain the relationship between absorbance and concentration, which is used to calculate the residual concentration of Isoamyl xanthate solution.

Take isoamyl xanthate solution of different concentrations into conical flask, add a certain amount of adsorbent, oscillate in constant temperature oscillation box, and filter with filter membrane after oscillation for a certain time. Measure the absorbance of the filtrate at the wavelength of 301 nm with ultraviolet spectrophotometer, and obtain the residual concentration of Isoamyl xanthate in the filtrate after the adsorption test according to the relationship between the absorbance and concentration of Isoamyl xanthate, and according to formula (23.2, 23.3). The adsorption capacity and removal rate of Isoamyl xanthate were calculated. The effects of adsorption time, amount of adsorbent and pH value of solution on the adsorption effect of adsorbent were investigated.

$$q = \frac{(C_0 - C_1)V}{M} \tag{23.2}$$

$$\eta = \frac{C_0 - C_1}{C_0} \times 100\% \tag{23.3}$$

where q is the amount of Isoamyl xanthate adsorbed by the adsorbent, mg/g; η Is the removal rate, %; C_0 and C_1 are the mass concentration of Isoamyl xanthate in the solution before and after adsorption, mg/L; V is the solution volume, l; M is the mass of adsorbent, g.

23.3 Results and Discussion

23.3.1 Relationship Between Adsorption Time and Removal Rate of Isoamyl Xanthate

Take a 100 ml conical flask, add 50 ml isoamyl xanthate solution with a concentration of 100 mg/L into the flask, add 0.05 g magnetic activated carbon under the conditions of 25 °C, pH = 7 and rotating speed of 200 R/min, and shake at constant temperature to study the influence of adsorption time on the adsorption effect. The results are shown in Fig. 23.1. It can be seen from Fig. 23.1 that the removal rate of Isoamyl Xanthate by magnetic activated carbon increases rapidly between 0 and 30 min; Between 30 and 60 min, the removal rate increased slowly with the continuous decrease of isopentyl xanthate content in the solution; After 60 min of adsorption, the adsorption basically reached a dynamic equilibrium state. When the adsorption time was 120 min, the removal rate of Isoamyl Xanthate by magnetic activated carbon reached more than 99.50%, and the removal effect was very obvious.

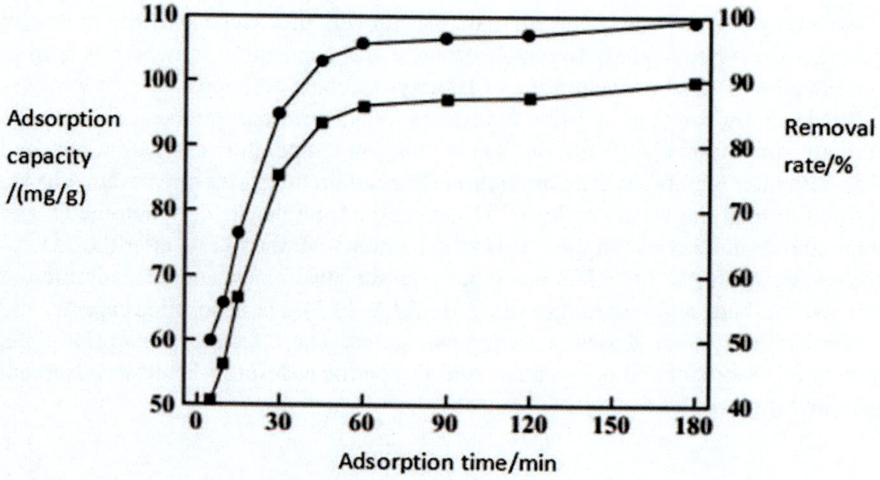

Fig. 23.1 Effect of adsorption time on adsorption effect

23.3.2 *Effect of the Amount of Magnetic Activated Carbon on the Removal Rate of Isoamyl Xanthate*

Add 50 ml of Isoamyl xanthate solution with a concentration of 100 mg/L into 10 100 ml conical flasks, and then add 0.1 to 1.0 g/L magnetic activated carbon respectively. The amount is increased by 0.1 g/L each time, and the adsorption is conducted for 120 min at pH = 7 and 25 °C. The Influence of the amount of magnetic activated carbon on the adsorption effect is studied. The results are shown in Fig. 23.2. It can be seen from Fig. 23.2 that with the increase of the amount of magnetic activated carbon, the removal rate increases from 29.05 to 99.73%. When the dosage of magnetic activated carbon is more than 0.6 g/L, the change of removal rate tends to be gentle. At this time, the adsorption reaches a balance. Considering economic, environmental protection and high efficiency, the dosage of magnetic activated carbon is 0.6 g/L in the subsequent test. At this time, the removal rate of Isoamyl xanthate by magnetic activated carbon is 99.60%.

23.3.3 *Effect of pH Value on the Removal Rate of Isoamyl Xanthate*

Add 0.030 g of magnetic activated carbon into a 100 ml conical flask, and then add 50 ml of Isoamyl xanthate solution with a concentration of 100 mg/L, adjust the pH value between 4 and 9, and adsorb at 25 °C for 150 min. The influence of pH value on the adsorption effect is studied. The results are shown in Fig. 23.3. It can be seen from Fig. 23.3 that with the increase of pH value, the removal rate of Isoamyl

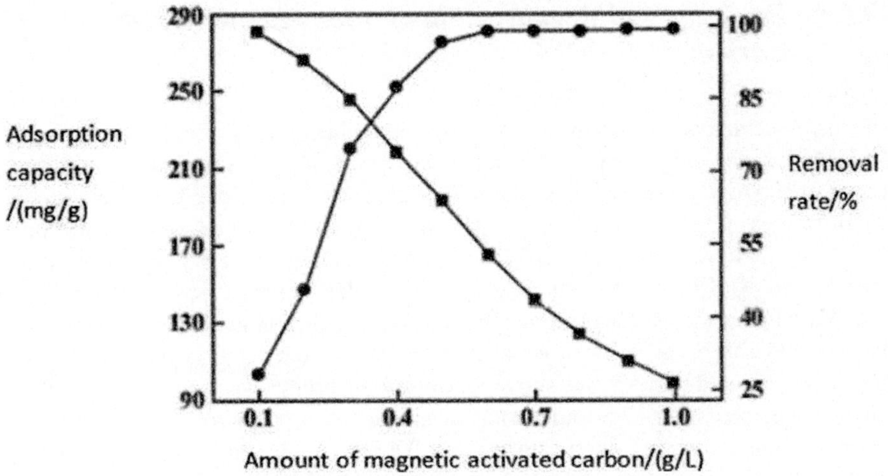

Fig. 23.2 Effect of adsorbent dosage on adsorption effect

xanthate decreases, and when the pH value is 6 to 7, the removal rate tends to be stable; When pH > 7, the removal rate decreased significantly. After comprehensive consideration, the subsequent tests are carried out in the solution with pH = 7.

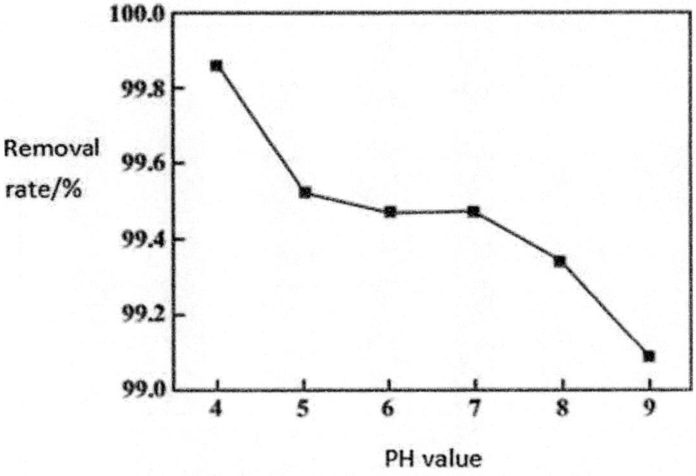

Fig. 23.3 Effect of pH value on adsorption effect

23.3.4 *Recovery and Regeneration of Magnetic Activated Carbon*

The magnetic activated carbon can be recovered by magnetic separation equipment. Add 0.6 g/L magnetic activated carbon to the isoamyl xanthate solution with concentration of 100 mg/L and pH = 7, shake at constant temperature for 3 h, take the supernatant to measure its absorbance, recover the magnetic activated carbon by strong magnetic separation (magnetic field strength of 600 Mt), and dry it in a vacuum drying oven after recovery. The recovery rate of the magnetic activated carbon is 97.60%, which indicates that the magnetic activated carbon is easy to recover, and the recovery rate is high. In order to investigate the regeneration performance of magnetic activated carbon, the above recovered adsorbent was soaked in absolute ethanol, separated after 12 h, and put into a vacuum drying oven for drying to prepare regenerated magnetic activated carbon. Take 0.6 g/L regenerated magnetic activated carbon and continue to add it to 100 mg/L isoamyl xanthate solution. Shake it for 3 h at constant temperature. Measure the absorbance of supernatant and continue to regenerate the magnetic activated carbon for 5 times. Compare the adsorption performance of the adsorbent. See Fig. 23.4 for the results. It can be seen from Fig. 23.4 that the removal rate of Isoamyl Xanthate by magnetic activated carbon decreases slightly with the increase of reuse times. After 5 times of reuse, the removal rate can still reach 77.30%, indicating that the magnetic activated carbon has good reusability.

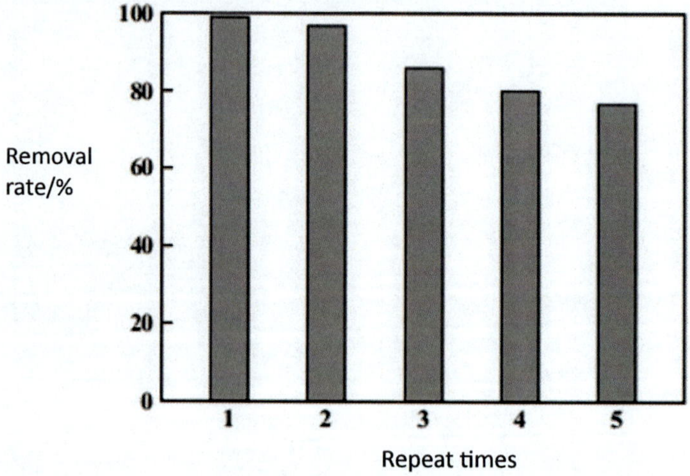

Fig. 23.4 Reuse effect of magnetic activated carbon

23.4 Conclusion

(1) Fe_3O_4 was attached to activated carbon by chemical precipitation method to prepare magnetic activated carbon adsorbent.
(2) Magnetic activated carbon has a strong adsorption capacity for isoamyl xanthate. Under the conditions of adsorption time of 120 min, initial concentration of 100 mg/L, adsorbent dosage of 0.7 g/L and reaction pH of 7, the removal rate of Isoamyl Xanthate by magnetic activated carbon is 99.73%.
(3) The magnetic activated carbon can be recovered by magnetic force, with a recovery rate of 97.60%. The removal rate of Isoamyl xanthate can reach 77.30% after 5 times of repeated use after desorption with absolute ethanol. It shows that magnetic activated carbon is an economical, efficient and environmental friendly adsorbent.

References

Bingyu M, Yubin T, Fangyan C et al (2015) Preparation of magnetic activated carbon and its adsorption on methyl orange in water [J]. J Environ Eng 9(4):1836–1868

Chao L, Liping W (2020) Research progress of mineral processing wastewater treatment technology [J]. Min Prot Utilization 40(1):72–78

Chengfang Y, Peng G et al (2022) Research progress on treatment technology and process of xanthate in mineral processing wastewater [J]. J Xuzhou Inst Eng Nat Sci Ed 6:62–70

Chunyan H (2020) Research progress of mineral processing wastewater treatment [J]. Chem Des Commun 46(6):226

Hao L, Zheng Y, Jianshe T (2021) Preparation of magnetic activated carbon and its adsorption mechanism for methylene blue [J]. Proceedings of the 18th Yangtze River Delta Science and Technology Forum Environmental Protection Sub Forum (2021 Academic Annual Meeting of Shanghai Environmental Science Society) and the 8th Member Representative Conference of Shanghai Environmental Science Society 10:144–155

Hongying Z, Jin L et al (2021) Study on preparation of magnetic activated carbon and removal of COD in river water [J]. Nonferrous Met Smelting Part 3:156–160

Jiyang Q, Yanli L et al (2022) Experimental study on the treatment of lead-zinc ore dressing wastewater by oxidation [J]. Value Eng 7:71–73

Qingfu Z, Zhimin W, Baoguo C et al (2016) XRD spectral analysis of the structural characteristics and carbonization mechanism of biochar prepared [J]. Spectro Spectral Anal 36(10):3355–3359

Ran W, Chunbao S, Huifeng Z et al (2011) Study on flotation reagent for treating tailings wastewater by ozone oxidation [J]. Water Treat Technol 9:41–43

Shangjun H, Xian X et al (2021) Research progress of xanthate treatment technology and process in mineral processing wastewater [J]. Miner Prot Utilization 8:43–48

Xiaojing L, Jun L, Haichen Z et al (2020) Influence factors and application of activated carbon adsorption for COD in high salt wastewater [J]. Appl Chem Ind 49(3):1–5

Xiaolian H, Linzhang Y, Shiying H et al (2018) Preparation of Fe_3O_4/BC composite and its adsorption and phosphorus removal performance [J]. Environ Sci Res 31(1):143–153

Yanlan Y (2021) Preparation of Fe_3O_4/halloysite composite and study on its degradation performance of organic matter in lead-zinc beneficiation wastewater [J]. Met Min 2:215–220

Zhaohui C, Jihong Z, Yongde L (2019) Research progress on preparation and application of magnetic activated carbon in water [J]. Green Sci Technol 16(8):144–146

Chapter 24
Research Status of Treatment of Cyanide Containing Wastewater and Comprehensive Recovery of Valuable Elements

Guanglin Qin, Guangsheng Li, Xingfu Zhu, and Qiang Ji

Abstract This paper briefly describes the source and harm of cyanide containing wastewater, introduces the main treatment processes of cyanide containing wastewater, such as sulfur oxide air method, dioxygen water method, alkaline chlorination method, adsorption method, membrane separation method, and analyzes the applicability of various processes. The comprehensive recovery methods of valuable elements in cyanide containing wastewater in recent years are summarized and prospected, and the viewpoint of comprehensive recovery of valuable elements in cyanide containing wastewater on the basis of purification is obtained. The analysis results show that the purification of cyanide containing wastewater is conducive to the leaching of gold, and the comprehensive recovery of valuable elements can improve the economic benefits of enterprises.

Keywords Wastewater · Cyanide · Purify · Recovery

24.1 Source and Harm of Cyanide Containing Wastewater

There are three main sources of cyanide containing wastewater, namely, cyanide production wastewater, cyanide utilization wastewater and other product wastewater. Cyanide production wastewater mainly refers to the wastewater from the production of inorganic cyanide and organic cyanide, such as sodium cyanide production wastewater from the gold leaching process. Because the cyanide ion has strong complexation, it can form metal complexes with a variety of metal ions. Cyanide is often used in electroplating, metallurgy, metal processing and other industries, thus forming cyanide utilization wastewater. Wastewater from other products mainly

G. Qin (✉) · G. Li · X. Zhu · Q. Ji
Beneficiation and Metallurgy Laboratory Branch Company, Shandong Gold Mining Technology Co., Ltd, Shandong, China
e-mail: 49260892@qq.com

J. Zhang et al. (eds.), *Environmental Pollution Governance and Ecological Remediation Technology*, Environmental Science and Engineering,
https://doi.org/10.1007/978-3-031-25284-6_24

comes from chemical and pharmaceutical industries, such as pesticide and pharmaceutical intermediate processing. Due to different sources of cyanide containing wastewater, its composition is different (Huijian 2016).

The toxic mechanism of cyanide to animals is mainly due to the strong complexation ability of cyanide ions. For example, cyanide ions enter human or animal bodies to generate HCN, and then quickly enter the blood, combine with iron ions and copper ions in the blood to form a stable complex, and further inhibit cytochrome oxidase, so that it cannot absorb dissolved oxygen in the blood, resulting in cell asphyxia and death. Low concentration of cyanide will cause headache, palpitation and other symptoms. When the concentration reaches above 0.1 mg/L, people will die in a very short time. The harm of cyanide to plants is also significant. When the concentration of cyanide in water is 50 mg/L, the growth of wheat and rice is significantly inhibited. Therefore, the state also imposes corresponding restrictions on the concentration of cyanide in farmland irrigation water (Chongbo et al. 2001).

24.2 Treatment Process of Cyanide Containing Wastewater

At present, the treatment methods of cyanide containing wastewater can be divided into sulfur dioxide air method, dioxygen water method, alkaline chlorination method, adsorption method, membrane separation method, etc.

24.2.1 Sulfur Dioxide Air Method

The principle of sulfur dioxide air method (Zhendong 1993) is to use SO_2–O_2 mixed gas as oxidant, add a small amount of copper sulfate into the solution as catalyst, and the reaction is as follows:

$$CN^- + SO_2 + O_2 + H_2O \rightarrow CNO^- + H_2SO_4 \quad (24.1)$$

$$CNO^- + 2H_2O \rightarrow CO_2 + NH_3 + OH^- \quad (24.2)$$

The technical operating conditions are: pH 8–10, normal temperature. Oxygen is provided in the form of air, in which the source of SO_2 can be liquid SO_2 or sodium sulfite, or baking furnace gas can be used. The process features: all cyanide is preferentially oxidized and decomposed, the cyanide is completely removed, and the reagent cost is low. The cyanide containing tail liquid treated by this method can meet the national discharge standard of industrial wastewater. SO_2 air method is one of the most economical and effective methods to treat cyanide containing wastewater at present.

24.2.2 Dioxygen Water Method

The principle of dioxygen water method is to use hydrogen peroxide as oxidant to oxidize cyanide and thiocyanide in solution into cyanate (Xikang 1997). The reaction is as follows:

$$CN^- + H_2O_2 \rightarrow CNO^- + H_2O \tag{24.3}$$

$$CNS^- + H_2O_2 \rightarrow S + CNO^- + H_2O \tag{24.4}$$

The operating condition is that under the ambient temperature, the pH value is controlled at 9.0–9.5, and a certain amount of H_2O_2 solution is added to the reaction, cyanide and thiocyanide are oxidized. The process is highly reliable and easy to operate. After treatment, the cyanide concentration can be reduced to 0.1 mg/L. The reaction products and intermediate products are non-toxic. The pH range applicable to the oxidation reaction is wide. The operation process does not release toxic gases harmful to health and polluting the environment. The cost is lower than that of SO_2 air method. H_2O_2 is used to decompose cyanide in the paradisepeak plant in the United States, so as to control the cyanide content in the tailings pond.

24.2.3 Alkaline Chlorination Process

Alkaline chlorination, also known as chlor alkali oxidation, is a chemical cyanide removal method with long-standing application and mature technology. Its principle is to utilize the oxidizability of chlorine containing oxides (such as sodium hypochlorite, chlorine dioxide, liquid chlorine, bleaching powder, etc.), under alkaline conditions (pH > 11), the cyanogen containing compounds are oxidized and hydrolyzed to cyanate (CNO^-) with low toxicity, and then further (pH value about 8.5) are oxidized to non-toxic bicarbonate and nitrogen (Shaowen et al. 2006), and the main chemical reactions are as follows:

$$CN^- + ClO^- + H_2O \rightarrow CNCl + 2OH^- \tag{24.5}$$

$$CNCl + 2OH^- \rightarrow CNO^- + Cl^- + H_2O \tag{24.6}$$

$$2CNO^- + 3ClO^- + H_2O \rightarrow 2CO_2 + N_2 + 3Cl^- + 2OH^- \tag{24.7}$$

The effect of alkaline chlorination on the removal of cyanide in 1250 m^3 of wastewater left over from a gold mine. When sodium dichloroisocyanurate (effective chlorine content 0%) is used as oxidant, the cyanide concentration in the wastewater can be reduced from 88.97 to 0.368 mg/L, and the removal rate is as high as

99.6%. Sodium hypochlorite is used as oxidant and alkaline chlorination is used as a two-stage series process to treat the effluent of cyanide containing wastewater from flocculation and complexation unit. The results show that the treatment effect of cyanide containing wastewater with initial concentration of 29.6–40.5 mg/L is remarkable, and the effluent cyanide concentration can be stably reduced to below 0.5 mg/L. Alkaline chlorination has simple equipment, various reaction forms and high treatment efficiency. However, because this method requires adding a large amount of chlorine containing oxides, a large amount of residual chlorine will be generated during the treatment process, which is easy to cause equipment corrosion, so the material requirements of the equipment are high.

24.2.4 Adsorption Method

Activated carbon or modified activated carbon is often used as adsorbent in the adsorption treatment of cyanide containing wastewater (Daomin et al. 2002). Activated carbon has developed microporous structure and high specific surface area, so it has a good effect on the removal of CN^- in low concentration cyanide containing wastewater. In the study of activated carbon adsorption treatment of gold mine cyanide containing wastewater, it is found that activated carbon has a significant effect on the treatment of cyanide containing wastewater with an initial total cyanide concentration of 389.9–420.0 mg/L, and the cyanide removal rate is 99.8–99.9%. However, in this study, the dosage of activated carbon is as high as 19.7–26.3 g water/g activated carbon, which is not technically economical. Through the modification of activated carbon with cationic flocculant, the CN^- removal rate is more than 99% at the dosage of 12 g/L activated carbon (the initial CN^- concentration is 28.7 mg/L), which is nearly 30% higher than that of unmodified activated carbon. It is found that the adsorption and removal capacity of activated carbon can also be further improved by adding additive $CaCl_2$ and blowing air. For example, under the conditions of 25 g/L of activated carbon, 0.3 g/L of $CaCl_2$ and 400 L/h of blowing air, the total cyanide concentration can be reduced from 318 to 0.4 mg/L, and the cyanide removal rate can be increased by 7.51% compared with that without additive and blowing air.

24.2.5 Membrane Separation Method (Semmens et al. 1987)

The commonly used physical transfer removal method of cyanide containing wastewater is membrane separation method. As one of the important separation methods, membrane separation uses its micron and nano pore size to realize the separation, purification or concentration of different components. The membrane absorption system of CN^- was constructed by using polypropylene hollow fiber membrane, that is, through the separation membrane with selective absorption of small molecules of CN^- and the absorption of alkaline solution on the other side of the membrane, the concentration difference of CN^- on both sides of the membrane was formed to promote the diffusion and removal of CN^-. When treating wastewater with initial CN^- concentration of 1227.5 mg/L, the CN^- concentration of effluent can be less than 0.5 mg/L in 80 min, and the removal rate can reach 99.9%. The chromium plating wastewater with an initial cyanide concentration of 23 mg/L was treated by ultrafiltration reverse osmosis series process. The cyanide concentration in the effluent was reduced to less than 10 mg/L, and the removal rate could also reach more than 95%. The process of treating cyanide containing wastewater by membrane separation is simple, fast and efficient, but cyanide containing wastewater often contains a variety of other organic and inorganic pollutants, so there are high requirements for the anti-pollution performance of the membrane.

24.2.6 Ferrous Salt Method (Ruizhong et al. 1994)

The treatment agent of ferrous salt method is ferrous salt, which has a wide range of sources, relatively low preparation cost, safe and convenient operation. Its principle is to add ferrous sulfate or ferrous chloride to cyanide containing wastewater to form insoluble ferrous cyanide, and then add alkaline substances to remove cyanide as Prussian blue sediment. According to the complexing principle of ferrous salt, researchers have also developed other heavy metal salt treatment agents, such as KAZM-R (SU462806AL) uses zinc carbonate to treat cyanide wastewater, and converts free cyanide ions into complex $Zn(CN)_4{}^{2-}$ for precipitation treatment.

24.2.7 Biological Method (Junliang et al. 1998)

In recent years, the degradation bacteria found in the research mainly include Nocardia, Pseudomonas fluorescens, colorless bacteria, xylose oxidation alkali producing bacteria, xylose oxidation subspecies PF3, Pseudomonas Schneider AK61, Fusarium oxysporum, Fusarium oxysporum N–10, Bacillus pumilus, Klebsiella acidogenes, etc., totaling more than 50 strains. At present, the microbial treatment method mainly studies the treatment of liquids with low cyanide content.

24.3 Comprehensive Recovery Process of Valuable Elements

24.3.1 Acidification Method (Chaohai and Jin 1991)

Acidification recovery method is first used in the treatment of cyanide containing wastewater in mines. Acidification recovery method was first adopted in 1930. After more than 80 years of continuous improvement and perfection, it has reached a relatively perfect level.

The principle of acidification recovery method is to adjust and control the pH of wastewater to be acidic by adding H_2SO_4 or SO_2 into wastewater, decompose metal complexes, and make hydrogen cyanide (HCN) generated by the combination of CN^- and H^+ in water escape from water by stripping or heating. This method is usually used to treat high concentration cyanide containing wastewater. The cyanogen containing wastewater from the cyanidation plant of Xinjiang gold chromium Mining Co., Ltd. is treated by acidification recovery method. The process is divided into four stages: acidification, blowing, absorption, sedimentation and filtration. The treated lean liquid can be reused for production. Acidification recovery method has the advantages of low cost, a wide range of reagent sources, can recover cyanide in wastewater, and can achieve good results in the treatment of high concentration cyanide containing wastewater. The disadvantage is that the total cyanide content in the purified wastewater cannot meet the requirements of the integrated wastewater discharge standard, which generally requires secondary treatment, and the investment and operation costs are high.

24.3.2 Step Precipitation Method

Stepwise precipitation method is a process of stepwise precipitation of copper and zinc in lean solution by adjusting the pH value of pulp in stages. Because the complex ions of copper and zinc in lean solution have different complex stability constants, the two complex ions have the possibility of step-by-step precipitation. Since the pH value of acidification method is directly adjusted to 2–3 in the production process, the copper and zinc in the lean solution precipitate into a product, so that the grade of copper and zinc in the product is low, and the pricing coefficient is not high at the time of sale. The distribution precipitation method can produce copper products and zinc products respectively, and the grade is higher than that of the acidification method. The pricing coefficient is higher, which is conducive to increasing the economic benefits of enterprises. In relevant studies, the precipitation pH value of copper cyanide complex ion in lean solution is about 3.5, and the precipitation pH value of zinc cyanide complex is about 6.0. After adding sodium sulfide (sodium hydrosulfide), copper products with a grade of about 55% and copper products with a grade of about 42% can be obtained.

24.3.3 Electrodeposition Method (Husheng et al. 2000)

Husheng et al. (2000) of Tsinghua University used electrowinning partial acidification method to electrodeposit copper and zinc from the high concentration copper cyanide solution produced by the treatment of gold ore lean solution by solvent extraction. They studied the influence of electrowinning process conditions, such as cathode current density, free CN concentration in electrobath, copper ion concentration, electrobath temperature, pH value and electrowinning time, on cathode current efficiency and copper content in alloy, and determined the best process conditions for electrowinning brass or crude copper, Brass (Cu Zn alloy) containing 71% Cu71 or crude copper containing 98.6% Cu 98.6% was obtained. Part of the electrolyte should be acidified to remove free cyanide every 5–6 h of electrowinning. The average cathode current efficiency can be maintained above 63% by alternating electrowinning and partial acidification.

24.4 Conclusion

(1) This paper briefly describes the source and harm of cyanide containing wastewater, introduces the main treatment processes of cyanide containing wastewater, such as sulfur oxide air method, dioxygen water method, alkaline chlorination method, adsorption method, membrane separation method, etc., and analyzes the applicability of various processes.
(2) The recovery methods of valuable elements such as free cyanide, copper and zinc in cyanide containing wastewater in recent years are introduced.
(3) Through the analysis of the treatment of cyanide containing wastewater and the current situation of comprehensive recovery of valuable elements in wastewater, it is concluded that the comprehensive recovery of valuable elements in cyanide containing wastewater on the basis of purification is conducive to gold leaching and can improve the economic benefits of enterprises.

References

Chaohai W, Jin X (1991) Development and review of cyanide containing wastewater treatment methods [J]. Ind Water Treat 11(2):3–7

Chongbo Z, Chenggong W, Bingchen C (2001) Review on the harm of cyanide and its treatment methods [J]. Met Min 36(5):44–47

Daomin Z, Shanlun F, Jia L (2002) Treatment method of cyanide containing wastewater [J]. Inorg Salt Ind 31(1):16–18

Huijian S (2016) Study on heterogeneous ozone photocatalytic oxidation of high salt cyanide containing wastewater [D]. Tianjin: Tianjin University

Husheng H, Mingde Y, Jie D et al (2000) Recovery of copper cyanide zinc from high copper cyanide solution by electrowinning acidification [J]. Nonferrous Met 52(3):61–64
Junliang L, Zhigang Z, Yuejiang H et al (1998) Treatment and recovery technology of cyanide containing wastewater [J]. Hebei Libr Sci Technol 2:29–31
Ruizhong L, Qing Z, Jianmin D (1994) New process for treatment of high concentration cyanide containing wastewater [J]. Hebei Chem Ind 4:57–58
Semmens MJ, Kenfield CF, Qin R (1987) A gas membrane-ion exchange process for cyanide recovery [J].J Met Finish 11:47–51
Shaowen Y, Yaohua C, Qi L et al (2006) Cyanide wastewater treatment test of a gold mine in Xiaoqinling [J]. Water Resour Prot 22(6):53–56
Xikang D (1997) Treatment of high concentration cyanide containing wastewater [J]. Shanxi Mach 1:47–48
Zhendong X (1993) Production practice of closed-circuit recycling of cyanide wastewater [J]. Gold 11:54–57

Chapter 25
Analysis of Heavy Metal Pollution Characteristics in Soil of a Sulfuric Acid Plant in Shandong

Shaoyong LI, Lina Wang, Yan Dong, and Chunrong Zhang

Abstract In order to analyze the characteristics of soil heavy metals (HMs) in soil of the sulfuric acid plant in Shandong, 6 HMs in these soil samples were analyzed. And the HMs in these soil samples were analyzed by the methods of statistical analysis and the potential ecological risk index (RI). The results showed that the mean concentration of As, Cd, Cu, Hg, Ni and Pb in the soil is 12.9, 0.30, 36.95, 0.088, 45.9, 38.5 $\mu g \cdot g^{-1}$, respectively. All of them exceeded their background value in Tai'an soil. The spatial distribution of HMs in the soil is uneven. On the whole, the contents of soil Cd, Cu, Pb are generally lower. On the other hand, the content of soil As, Hg, Ni are higher, especially in the production area. The mean Eri of HMs in the soil is Hg > Cd > As > Ni > Pb > Cu. Among them, the Cd and Hg have obvious ecological risks, especially Hg. There is an obvious risk of ecological pollution of comprehensive potential ecological risk of soil in some areas. These results of this study will provide technical support for the development, utilization and pollution control of the soil of the sulfuric acid plant.

Keywords Heavy metal · Soil · Sulfuric acid plant · Shandong

S. LI
The Second Prospecting Team of Shandong Coal Geology Bureau, Jining 272100, China

L. Wang
The Fifth Prospecting Team of Shandong Coal Geology Bureau, Jinan 250100, China

Y. Dong
Physical Exploration and Survey Team of Shandong Coal Geology Bureau, Jinan 250104, China

C. Zhang (✉)
College of Earth Science and Engineering, Shandong University of Science and Technology, Qingdao 266590, China
e-mail: zcrsdust@126.com

J. Zhang et al. (eds.), *Environmental Pollution Governance and Ecological Remediation Technology*, Environmental Science and Engineering,
https://doi.org/10.1007/978-3-031-25284-6_25

25.1 Introduction

With the implementation of the policies of "Suppress the second industry and develop the third industry" and "Exit city and enter industrial park", etc. in Chinese cities, a large number of polluting enterprises with relatively backward production technology, extensive management and lack of environmental protection equipment have moved out the area of city (Haoyu et al. 2021). These left over industrial sites may cause pollution to the soil environment because they have carried harmful substances through production, stacking and disposal (Dengkui et al. 2020; Haoyu et al. 2021). In particular, HMs have the characteristics of difficult degradation, easy enrichment, persistence and irreversibility, which will not only lead to soil ecological environment deterioration, and accumulate in the human body, which can cause cancer, teratogenesis and mutation, thus causing harm to human health (Chunrong et al. 2014; Mohammad et al. 2012).

Sulfuric acid is an important basic chemical raw material in the fields of metallurgy, petrochemical and chemical industry. Its production process will produce a large number of HMs pollutants, which has attracted the attention of experts and scholars at home and abroad, and some research has been carried out (Dengkui et al. 2020; Haiyan et al. 2014; Haoyu et al. 2021). Such as the pollution of HMs in the soil of sulfuric acid factories with raw materials of pyrite and Zn concentrate (Dengkui et al. 2020; Haiyan et al. 2014; Haoyu et al. 2021). Due to the different raw materials and industrial processes of sulfuric acid production, its pollution to soil also has obvious differences. Therefore, this paper intends to study the HMs in the soil of a sulfuric acid plant in Shandong, which uses sulfur as the production raw material, in order to provide technical support for the development, utilization and pollution control of the soil of the sulfuric acid plant.

25.2 Methods and Materials

25.2.1 Study Areas

The sulfuric acid plant in Shandong was founded in 2004 and closed down in 2010. And it produces 15,000 tons of concentrated sulfuric acid every year. The area of it is 17163.9 m^2. The production process of sulfuric acid is that sulfur is first ignited by diesel oil, and melted in the sulfur melting furnace. And then precipitated with trace solid impurities, including As. Finally, the purified liquid sulfur is transported to the sulfur incinerator for combustion to generate sulfur dioxide and sent to the sulfur dioxide conversion area. Under the action of V catalyst, it is converted and absorbed to obtain concentrated sulfuric acid. Among them, the unconverted trace sulfur dioxide is vented after reaching the standard through alkali absorption. The solid absorbed by alkali is the general solid waste produced by this process. Sulfur and lye, the raw and auxiliary materials used, are stored in the warehouse. The catalyst is regularly recovered and activated by the original manufacturer and replaced with the old one. The catalyst is not stored in the plant. Raw sulfur may produce a small amount of dust particles, but all production equipment are closed facilities, and sulfur is melted and burned. There is little dust generation, and there is a gas purification device.

25.2.2 Sample Collection

According to the distribution requirements of ≪ Technical guidelines for investigation on soil contamination of land for construction ≫ (HJ25.1–2019) and ≪ Technical guidelines for investigation and assessment of soil environment on construction land ≫, and according to the production process adopted in the sulfuric acid plant and the possible pollution, six soil points are set up in the remaining site, as shown in Fig. 25.1. Soil samples are collected in strict accordance with ≪ The technical specification for soil environmental monitoring ≫ (HJT166-2004), and the sample collection depth is 0–50 cm. All samples are collected and put into the low-temperature incubator, and sent to the laboratory for analysis in time.

25.2.3 Sample Analysis

The analysis of HMs was carried out in strict accordance with ≪ Soil environmental quality risk control standard for soil contamination of development land ≫ (GB36600-2018). Among them, Cd is detected by atomic absorption spectrophotometer (PerkinElmer PinAAcle900Z), and the detection limit is 0.01 $\mu g \cdot g^{-1}$; Hg is detected by mercury analyzer (LUMEX RA-915 M), and its detection limit is 0.0002 $\mu g \cdot g^{-1}$; Pb, Cu, Ni and As are detected by X-ray fluorescence spectrometer (S8 TIGER), and their detection limits are 2.0, 1.2, 1.5, 2.0 $\mu g \cdot g^{-1}$, respectively.

25.2.4 RI

RI method was proposed by Swedish scholar Hakanson, it reflects the impact of a single element on the environment and the combined impact of multiple elements on the environment (Dai et al. 2015; Hakanson 1980), with the formulaes (25.1), (25.2), (25.3):

$$P_i = \frac{C_i}{C_n^i} \tag{25.1}$$

$$E_r^i = T_r^i \times P_i \tag{25.2}$$

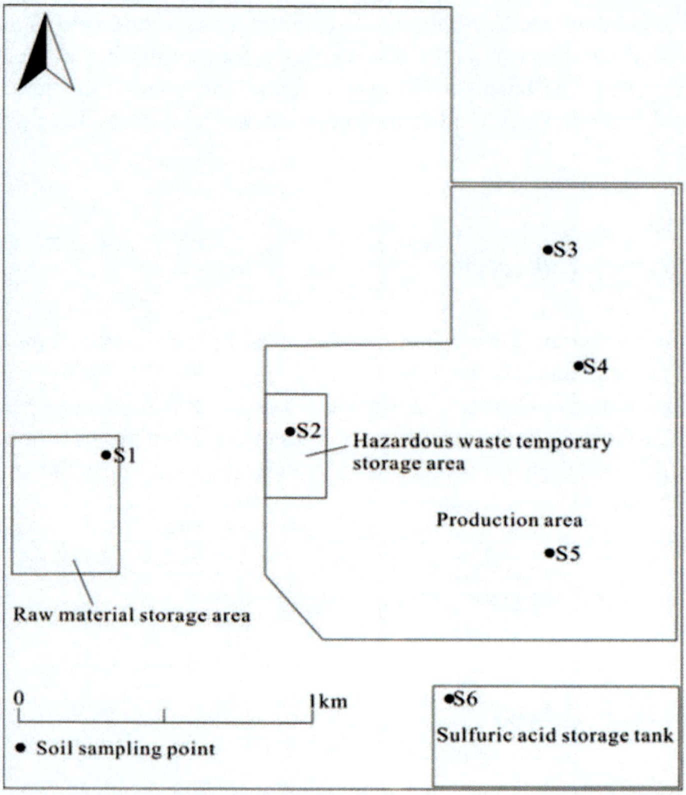

Fig. 25.1 Distribution of soil sampling points

$$RI = \sum_{i}^{n} E_r^i \tag{25.3}$$

The P_i is the i divided by the corresponding natural regional background value. The E_r^i is the pollution index of i multiplied by its toxicity coefficient, and it divided into 5 levels: such as low risk < 40, 40 ≤ moderate risk < 80, 80 ≤ higher risk < 160, 160 ≤ high risk < 320 and severe risk ≥ 320. The toxicity coefficient is $T_r^{As} = 10$, $T_r^{Cd} = 30$, $T_r^{Cu} = 5$, $T_r^{Hg} = 40$, $T_r^{Ni} = 5$, $T_r^{Pb} = 5$ (Dai et al. 2015). The RI is the sum of all pollution indexes and it divided into four levels: such as low risk < 150, 150 ≤ Moderate risk < 300, 300 ≤ high risk < 600 and severe risk ≥ 600.

25.2.5 *Data Analysis*

Statistical analysis was carried out with Excel 2007, and maps were drawn with surfer software and Coreldraw.

25.3 Results and Analysis

25.3.1 *Characteristics of HMs in Soil*

According to the analysis of HMs in soil, Table 25.1 is obtained.

According to Table 25.1, the mean concentration of soil As, Cd, Cu, Hg, Ni and Pb in the sulfuric acid plant in Shandong is 12.9, 0.30, 36.95, 0.088, 45.9, 38.5 $\mu g{\cdot}g^{-1}$, respectively. The max values of soil As, Cd, Cu, Hg, Ni and Pb are 1.9, 11.4, 2.3, 6.9, 1.4 and 2.8 times of the min values of them respectively. And all of them with obvious differences. The C.V. of soil As, Cd, Cu and Ni are less than 0.5. While Hg and Pb are greater than 0.5, and the spatial distribution of them is uneven, and may be affected by the sulfuric acid plant in Shandong. All the mean concentration of HMs in soil exceeded the background value of corresponding them in Tai'an soil (Xugui et al. 2019), which should be affected by the sulfuric acid plant in Shandong. All the HMs in the soil are lower than the screening value of industrial land in ≪ Soil environmental quality risk control standard for soil contamination of development land ≫ (Trial) (GB36600-2018), and meet the national standard for soil screening of construction land.

Table 25.1 HMs concentration in soil ($\mu g{\cdot}g^{-1}$)

Area	Eigenvalues	As	Cd	Cu	Hg	Ni	Pb
Study area	Mean	12.9	0.30	36.95	0.088	45.9	38.5
	Max	17.4	1.03	66.4	0.2	54	79
	Min	9.24	0.09	28.7	0.029	38	28.2
	Sd	2.80	0.37	14.51	0.07	6.38	19.89
	C.V	0.22	1.24	0.39	0.80	0.14	0.52
Taian	Background value	7.4	0.137	22.9	0.030	27.1	22.6
Industrial areas of GB36600-2018	Selected value	60	65	18,000	38	900	800

25.3.2 Spatial Distribution of HMs in Soil

According to the analysis of HMs in soil, its spatial distribution Fig. 25.2 was obtained.

According to Fig. 25.2, it can be seen that the spatial distribution of HMs in soil is uneven. On the whole, the content of soil Cd, Cu, Pb are generally lower, Only in the production area, the soil Cd, Cu, Pb at S5 sampling point and the soil Cd at S4 sampling point are higher. The sulfur melting furnace, the sulfur incinerator, etc. are made of steel, and Cd, Cu, Pb are the main elements of steel pollution (Xiao et al. 2015). Therefore, the soil Cd, Cu, Pb in the production area probably affected by the sulfuric acid plant. On the other hand, the content of soil As, Hg, Ni are higher. Among them, except for soil As in S5 sampling point, soil Hg in S3, S6 sampling point and soil Ni in S1, S5 sampling point, in the other sampling point are greater than 1.5 times of the possible change of them background value of soil in Tai'an City. In particular, in the production area, soil As in S4 sampling point and soil Hg in S5 sampling point. As sulfur is the main raw material of a sulfuric acid plant, diesel oil is its ignition supplies, the sulfur melting furnace, the sulfur incinerator, sulfuric acid storage tankis, etc. are made of steel, while sulfur contain As pollutant, diesel contain As, Hg pollutants (Dingchao et al. 2017) and steel contain Ni pollution (Xiao et al. 2015). Therefore, the soil As, Hg, Ni in the production area should be significantly affected by the sulfuric acid plant, and it is necessary to strengthen them pollution control and treatment to prevent damage to the soil ecological environment.

25.3.3 Potential Ecological Risks of HMs in Soil

According to the analysis of HMs in soil, the potential ecological risk of them are calculated by using the potential ecological hazard index formula (25.1), (25.2), (25.3), and Table 25.2 is obtained.

According to Table 25.2, the mean Eri of HMs in the soil is Hg > Cd > As > Ni > Pb > Cu. Among them, the potential risk of single HMs pollution of As, Cu, Ni and Pb are lower, which is low risk and will not cause ecological risk to the soil of the sulfuric acid plant in Shandong. Cd and Hg have obvious ecological risks, especially Hg, the low low risk only accounts for 16.7%, while others are moderate risk, higher risk and high risk. Attention should be paid to strengthening the treatment of soil Cd and Hg pollution to prevent further harm to soil ecology.

The RI of soil HMs is low risk accounting for 66.7% and high risk accounting for 33.3%. There is an obvious risk of ecological pollution in some areas, so the treatment of HMs pollution in the soil in the sulfuric acid plant in Shandong should be strengthened to prevent further harm to soil ecology.

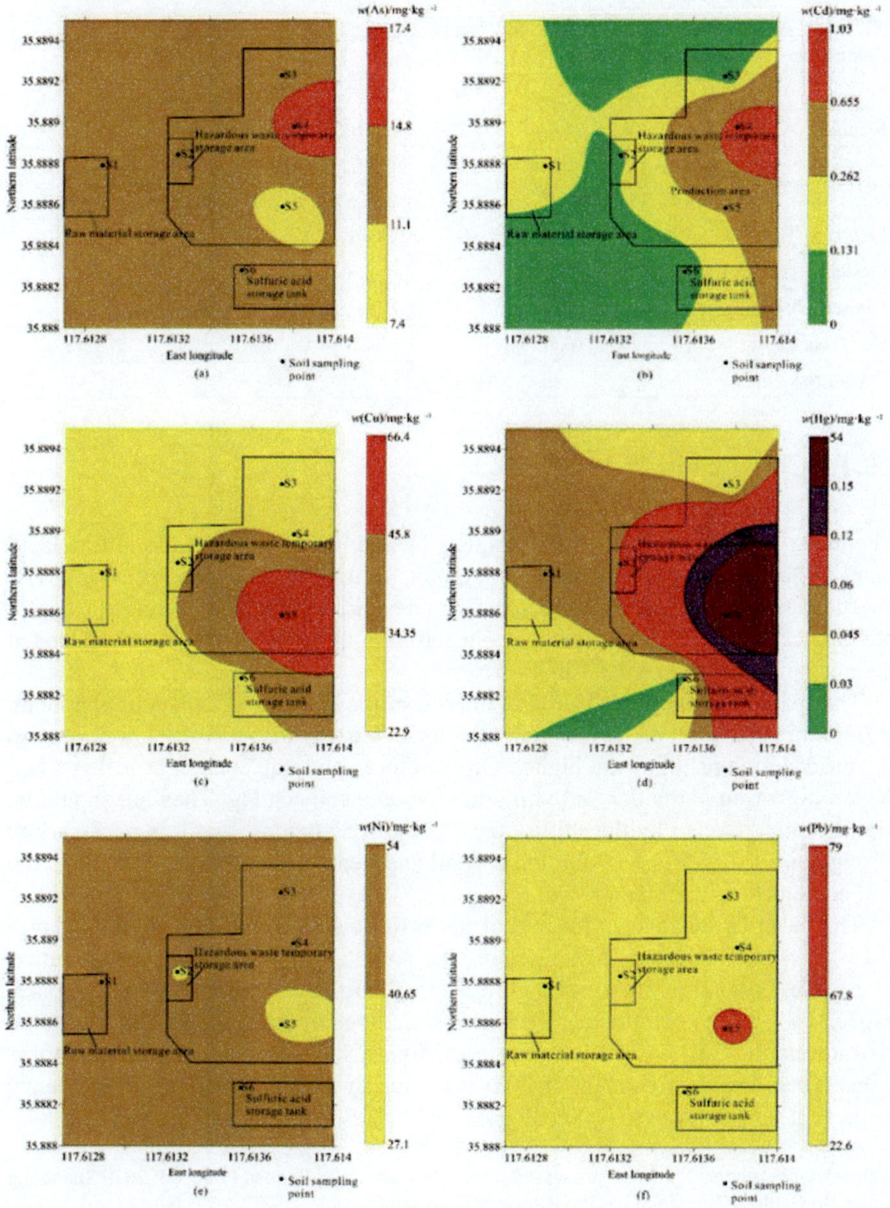

Fig. 25.2 Spatial distribution of HMs

Table 25.2 Potential ecological risks of HMs in soil

Eigenvalue	Eri						RI
	As	Cd	Cu	Hg	Ni	Pb	
Mean	17.40	64.96	8.07	116.67	8.47	8.53	224.09
Max	23.51	225.55	14.50	266.67	9.96	17.48	468.90
Min	12.49	19.71	6.27	38.67	7.01	6.24	100.67
Low risk (%)	100	66.6	100	16.7	100	100	66.7
Moderate risk (%)		16.7		33.3			
Higher risk (%)				16.7			
High risk (%)		16.7		33.3			33.3
Severe risk							

25.4 Conclusion

The mean concentration of As, Cd, Cu, Hg, Ni and Pb in the soil of the sulfuric acid plant in Shandong is 12.9, 0.30, 36.95, 0.088, 45.9, 38.5 $\mu g \cdot g^{-1}$, respectively. The coefficients of variation of soil As, Cd, Cu and Ni are less than 0.5. While Hg and Pb are greater than 0.5. All the average concentration of heavy metals in soil exceeded the background value of them in Tai'an soil.

The spatial distribution of HMs in the soil of the sulfuric acid plant in Shandong is uneven. On the whole, the content of soil Cd, Cu, Pb are generally lower, only in the production area, they are higher. On the other hand, the content of soil As, Hg, Ni are higher, in particular, in the production area soil As, Hg. They are should be significantly affected by the sulfuric acid plant in Shandong, and it is necessary to strengthen soil As, Hg, Ni pollution control and treatment to prevent damage to the soil ecological environment.

The mean Eri of HMs in the soil of the sulfuric acid plant in Shandong is Hg > Cd > As > Ni > Pb > Cu. Among them, the potential risk of single HMs pollution of As, Cu, Ni and Pb are lower. While, Cd and Hg have obvious ecological risks, especially Hg. The comprehensive potential ecological risk of soil HMs RI is high risk accounting for 33.3% and there is an obvious risk of ecological pollution in some areas. Therefore the treatment of HMs pollution in the soil should be strengthened to prevent further harm to soil ecology.

Acknowledgements This research was funded by the Specially Funded Project of 2019 Shandong Coalfield Geology Bureau Scientific Research (Lu Coal Geoscience Zi (2019) No.11)

References

Dai B, Lv J-S, Zhan J-C et al (2015) Assessment of sources, spatial distribution and ecological risk of heavy metals in soils in a typical industry-based city of Shandong province, eastern China [J]. Environ Sci 36(2):507–515

Dengkui G, Ganghui Z, Jian C et al (2020) Pollution characteristics and health risk assessment of heavy metal in soil of abandoned sulfuric acid plant site [J]. Environ Pollut Cont 42(9):1152–1157 (in chinese)

Dingchao L, Can S, Faliang T et al (2017) Study on the release characteristics of trace elements Hg, As and Cr during the combustion of petroleum coke [J]. Petrochemical Indus Technol 24(12):129–132 (in chinese)

Haiyan Z, Menghua C, Linling W et al (2014) Health risk assessment of a potential heavy metal contaminated site form a sulphuric acid plant [J]. Environ Eng 32(08):127–130 (in chinese)

Hakanson L (1980) An ecological risk index for aquatic pollution control.a sedimentological approach [J]. Water Res 14(8):975–1001

Haoyu Z, Zhong L, Haifeng C et al (2021) Investigation and research on soil pollution status of decommissioned sulfuric acid plant site [J]. China Resour Compr Utilization 39(5):125–128 (in chinese)

Mohammad O, Mohammad SK, Almas Z et al (2012) Soil contamination, nutritive value, and human health risk assessment of heavy metals: an overview [J]. Tox Heavy Metals Legumes Bioremediat, pp 1–27

Xiao Q, Zong Y, Lu S (2015) Assessment of heavy metal pollution and human health risk in urban soils of steel industrial city (Anshan), Liaoning, Northeast China [J]. Ecotoxicol Environ Saf 120:377–385

Xugui P, Jierui D, Lei C et al (2019) Soil geochemical background value of 17 cities in Shandong province [J]. Shandong Land Resour 35(1):46–56 (in chinese)

Zhang C-R, Wu Z-L, Yao C-H, Gao Z-J (2014) Health risk assessment of heavy metals in atmospheric dust of Qingdao city [J]. Environ Sci 35(7):2736–2741

Chapter 26
Study on Evaluation Method of Water Quality in Dianchi Lake

Xiaolin Kang, Peng Chen, and Hong Liang

Abstract In order to select a more suitable water quality evaluation method for Dianchi Lake, the water quality monitoring data of Dianchi Lake were studied. Four water quality detection indicators, namely dissolved oxygen, high acid salt index, ammonia nitrogen and total phosphorus, were selected as evaluation indicators. Three common water quality evaluation methods based on single factor analysis, principal component analysis and grey correlation analysis were studied in detail. The annual average detection data values of ten water quality monitoring points deployed in Dianchi Lake from 2019 to 2021 were selected, According to the national surface water evaluation standard, the water quality grade is evaluated and compared. The three methods have their own evaluation emphases, and the grey correlation analysis method is closer to the publicized water quality grade, which is more suitable for the water quality evaluation of Dianchi Lake, and can provide decision support for water environment treatment.

Keywords Water quality evaluation · Dianchi Lake · Single factor analysis · Principal component analysis · Grey correlation analysis

26.1 Introduction

Dianchi Lake Basin has a total area of 2920 square kilometers, a normal high water level of 1887.4 m, a lake surface water area of 309.5 square kilometers, and a lake capacity of 1.56 billion cubic meters. It is known as the "Pearl of Yunnan Guizhou Plateau". It is located at the lowest elevation of the catchment area in the lower reaches of Kunming City, and is the final water body in the basin (Pan and Gao

X. Kang (✉) · H. Liang
Yunnan University, Yunnan Province, Kunming City 650000, China
e-mail: sdlwkxl@163.com

P. Chen
Kunming Dianchi Lake Treatment and Development Co, Yunnan Province, Ltd, Kunming City 650000, China

J. Zhang et al. (eds.), *Environmental Pollution Governance and Ecological Remediation Technology*, Environmental Science and Engineering,
https://doi.org/10.1007/978-3-031-25284-6_26

2010). At the end of the 1990s, the water quality of the whole lake was worse than class v. Water pollution has caused a series of ecological problems, which not only restricted the economic development of Kunming, but also affected the normal life of the citizens. Therefore, only by selecting appropriate evaluation methods to evaluate the water quality of Dianchi Lake can we fully understand the water quality changes of the lake, formulate reasonable pollution prevention measures and strengthen the ecological restoration of the lake.

With the continuous in-depth study of water quality evaluation methods, a variety of water quality evaluation methods began to enter people's attention, such as single factor index method, Nemero method, comprehensive index method, analytic hierarchy process, fuzzy evaluation method and artificial neural network. The correctness and validity of lake water quality evaluation results are usually directly related to the evaluation methods selected. All kinds of evaluation methods have unique evaluation principles and calculation models, and their applicability and rationality of results have always been a hot topic discussed by scholars. Yingying et al. (2022) used the comprehensive water quality index method of the basin to carry out water quality evaluation and Research on key parameters. Taking Baiyangdian as the research water area, they evaluated the overall water quality of the basin; Yijing et al. (2022) explored the application of fuzzy comprehensive evaluation method, grey correlation analysis method, principal component analysis method and set pair analysis method in water quality evaluation of nine plateau lakes in Yunnan; Chao et al. (2017) selected the grey correlation analysis method to analyze the water quality of Dongping Lake. Due to the different characteristics of the lakes, the applicable evaluation methods will be different. Therefore, based on the research of scholars, the water quality evaluation methods suitable for Dianchi Lake are selected for statistical analysis and application verification, providing reference value for the water quality evaluation of Dianchi Lake and other similar lakes.

In this paper, single factor index method, principal component analysis method and grey correlation method are selected as water quality evaluation methods. Ten monitoring points are selected from the key control areas of cyanobacteria in Dianchi Lake. The annual average values of the ten monitoring points in 2019, 2020 and 2021 are calculated, and the water quality of Dianchi Lake in Yunnan Province is evaluated. Through comparison, the advantages, disadvantages and applicability of the four evaluation methods are further discussed. It provides reference for the exploration of water quality evaluation methods in Dianchi Lake in the future.

26.2 Data Sources

The grade is based on the annual average water quality monitoring data of ten monitoring points in Dianchi Lake from 2019 to 2021. According to the analysis of previous years' test reports in accordance with the environmental quality standard for surface water (GB3838-2002), it is found that the pollutants exceeding the standard are basically the same. Therefore, the water quality evaluation indicators of

Table 26.1 Evaluation results of single factor analysis

Year	Point	DO	PI	NH3-N	TP	Results	Comprehensive
2019	D4	1.2933	1.146	0.075	1.675	Inferior V	Inferior V
2020	D4	1.333	2.1	0.12	1.825	Inferior V	Inferior V
2021	D5	0.709	0.8	2.04	1.7	Inferior V	Inferior V

Dianchi Lake are selected as four indicators: dissolved oxygen (DO), high acid salt index (PI), ammonia nitrogen (NH3-N) and total phosphorus (TP).

26.3 Water Quality Assessment Results

26.3.1 Single Factor Analysis

Taking the data of 2019 as an example, the four evaluation indicators are calculated by the formula $P_i = C_i / S_i$, and rated according to the evaluation standard. Finally, the worst evaluation level among the four evaluation indicators is taken as the water quality level of the point in 2019. Finally, the comprehensive evaluation level of the whole year is obtained by synthesizing the evaluation results of the ten points. The water quality evaluation results obtained by single factor analysis are shown in Table 26.1. (S_i is the standard index of the evaluation index I at the sampling point; C_i is the measured value of the evaluation index at the sampling point, and C_{Si} is the standard value of the evaluation index (Rui 2009)).

It can be seen from Table 26.1 that the water quality monitoring indicators in 2019, 2020 and 2021 are all inferior to class v.

26.3.2 Principal Component Analysis

Taking the water quality data of each monitoring point in 2020 as an example, the principal component analysis calculation is carried out, and finally the water quality grade evaluation from 2019 to 2021 is given according to the principal component analysis method.

The initial factor load matrix is divided by the square root of the corresponding eigenvalue of each principal component to obtain the corresponding eigenvector, and the normalized data is multiplied by the eigenvector to obtain the following principal component expression (Hongmei and Xiao 2021)

$$F_K 1 = 0.211X_1 + 0.309X_2 + 0.284X_3 + 0.312X_4 \quad (26.1)$$

Table 26.2 Evaluation form of principal component analysis

Year	DO (%)	PI (%)	NH3-N (%)	TP (%)	Evaluation results
2019	18	44.34	10.09	27.58	V
2020	21.07	26.97	24.44	27.53	IV
2021	15.95	32.41	20.20	31.44	Inferior V

$$F_K 2 = 0.063X_1 - 0.082X_2 - 0.521X_3 - 0.164X_4 \tag{26.2}$$

$$F_K 3 = 0.741X_1 - 0.610X_2 + 0.992X_3 - 0.722X_4 \tag{26.3}$$

$$F_K 4 = 0.136X_1 + 0.488X_2 + 0.318X_3 - 0.336X_4 \tag{26.4}$$

Including: $F_K 1$, $F_K 2$, $F_K 3$ and $F_K 4$ is the score expression of principal component 1, principal component 2, principal component 3 and principal component 4, respectively; X_1 is the dissolved oxygen value (mg/L); X_2 is permanganate index (mg/L); X_3 is ammonia nitrogen value (mg/L); X_4 is the total phosphorus value (mg/L).

The comprehensive score f is obtained by accumulating the product of variance interpretation rate and component score. The calculation formula for the current data is:

$$F = \frac{(78.757 * F_K 1 + 17.434 * F_K 2 + 3.277 F_K 3 + 0.533 * F_K 4)}{100.000} \tag{26.5}$$

$$0.788 F_K 1 + 0.174 F_K 2 + 0.033 F_K 3 + 0.005 F_K 4 \tag{26.6}$$

Calculate the comprehensive evaluation value of the principal component analysis method of the ten monitoring points in 2020, and then summarize it for the whole year to obtain the comprehensive evaluation result of the monitoring points in 2020.

Finally, the evaluation results of the comprehensive monitoring points in 2019–2021 are obtained. As shown in Table 26.2, the water quality of some water areas of Dianchi Lake (the water around the ten monitoring points) in 2020 is improved compared with that in 2019, and the water quality in 2021 is deteriorated again; Among them, the weight of total phosphorus monitoring indicators is the largest, followed by high acid salt index and ammonia nitrogen index.

26.3.3 Grey Relational Evaluation

The basic principle is to determine the reference sequence and comparison sequence required for water quality assessment according to the monitoring value of water

Table 26.3 Evaluation results of grey management evaluation method

Year		D1	D2	D3	D4	D5	D6	D7	D8	D9	D10	Evaluation
2019	I	0.746	0.684	0.563	0.583	0.621	0.781	0.75	0.768	0.846	0.796	IV
	II	0.635	0.675	0.625	0.542	0.569	0.658	0.633	0.725	0.765	0.625	
	III	0.606	0.971	0.872	0.652	0.653	0.635	0.645	0.635	0.723	0.614	
	IV	0.605	0.896	0.756	0.685	0.543	0.611	0.598	0.619	0.638	0.546	
	V	0.564	0.652	0.631	0.482	0.598	0.526	0.577	0.544	0.462	0.498	
2020	I	0.855	0.976	0.863	0.782	0.756	0.895	0.866	0.632	0.862	0.896	IV
	II	0.826	0.856	0.821	0.765	0.721	0.822	0.756	0.689	0.812	0.812	
	III	0.725	0.746	0.763	0.852	0.633	0.752	0.721	0.785	0.752	0.762	
	IV	0.688	0.712	0.716	0.688	0.892	0.712	0.682	0.892	0.713	0.713	
	V	0.625	0.659	0.688	0.623	0.746	0.653	0.632	0.746	0.645	0.685	
2021	I	0.688	0.911	0.77	0.687	0.568	0.645	0.852	0.845	0.852	0.861	V
	II	0.657	0.854	0.759	0.754	0.635	0.891	0.743	0.746	0.698	0.765	
	III	0.612	0.816	0.561	0.892	0.756	0.712	0.712	0.721	0.625	0.637	
	IV	0.589	0.762	0.489	0.761	0.852	0.564	0.641	0.631	0.581	0.619	
	V	0.521	0.721	0.413	0.642	0.999	0.455	0.523	0.512	0.513	0.549	

quality assessment indicators and the water quality standards of various levels, and then determine the water quality subordination level according to the principle of maximum correlation degree (Cong 2019).

According to the principle and calculation steps of the grey correlation evaluation method, the grey correlation coefficient and weight of each evaluation index of the ten monitoring points are calculated. Finally, the correlation degree of the ten monitoring points in 2019–2021 is calculated by the weighting method, as shown in Table 26.3. It can be seen that the annual water quality assessment from 2019 to 2021 calculated by the grey correlation degree is the same as the calculation result of the principal component analysis method.

26.3.4 Comparison of Water Quality Assessment Results

Table 26.4 shows the comparison of the three water quality assessment results and the water quality grades from 2019 to 2021 published in the Dianchi Lake Water Environment Bulletin. It can be seen that the results obtained by the three evaluation methods are not completely the same, mainly because the principles and models of different evaluation methods have their own characteristics, and the evaluation indicators have their own emphasis; Among them, the water quality evaluation results of Dianchi Lake obtained by the grey correlation evaluation method are the closest to the public level.

Table 26.4 Comparison results of four methods

Year	Single factor analysis	Principal component analysis	Grey relational evaluation	Water quality publicity grade
2019	Inferior V	V	IV	IV
2020	Inferior V	IV	IV	IV
2021	Inferior V	Inferior V	V	IV

26.4 Conclusion

By introducing the basic situation of Dianchi Lake in Yunnan Province and the current water environment problems, combining the actual application of common water quality assessment methods in lake water quality assessment, taking the annual average data of Dianchi Lake from 2019 to 2021 as an example, this paper compares the water quality assessment results of the three assessment methods, and analyzes their advantages and disadvantages in the water quality assessment of Dianchi Lake. The following conclusions can be drawn:

(1) The calculation method of single factor analysis method is the simplest of the four water quality evaluation methods, and the results are intuitive. According to the detection data and evaluation standards, we can directly see whether a certain pollution factor exceeds the standard and whether the pollution is serious. However, during the whole evaluation process, it will amplify the dominance of the highest pollution index, it is impossible to objectively evaluate the overall water quality at the monitoring points of Dianchi Lake.

(2) The principal component analysis method is mainly based on the overall characteristics of the detection data, integrates the information overlap between all the detection indicator data at the monitoring point, and uses a few comprehensive data after dimension reduction to reflect the overall situation of the water quality at a monitoring point in Dianchi Lake, effectively avoiding the subjective factors in the selection of evaluation indicators, making the calculation results more objective and accurate. However, through the application of water quality assessment in Dianchi Lake, it can be seen that the selection of principal components still has errors that are not consistent with the actual situation, thus causing a false assessment of a monitoring point in Dianchi Lake.

(3) The grey correlation analysis method regards the water environment to be assessed as a grey system, and selects the water quality grade standard corresponding to the maximum correlation degree according to the multiple correlation degrees solved; It can effectively reduce the influence of uncertainty and fuzziness of water body, make the final calculation result fit the actual situation, have reference value, and meet the basic needs of water quality assessment of Dianchi Lake. At present, it is more suitable for water quality assessment of Dianchi Lake.

Based on the comprehensive data of the selected ten monitoring points in Dianchi Lake from 2019 to 2021, through the application and analysis of the three water quality assessment methods, it can be seen that the three methods have their own advantages and disadvantages; However, in the actual application, it is necessary to select a more suitable water quality assessment method in combination with the specific water environment characteristics to obtain a more reasonable and accurate assessment result. At present, among the four evaluation methods, the grey correlation analysis method is more suitable for the ten monitoring points of Dianchi Lake, but there are two problems: on the one hand, it is suitable for the ten monitoring points. Can the ten monitoring points be used to evaluate the water quality of the whole Dianchi Lake; On the other hand, with the deepening of research and the application of different evaluation methods, more suitable water quality evaluation methods for Dianchi Lake may be found.

References

Chao Y, Zhiguo Z, Huiqing Y (2017) Application of grey correlation analysis method in surface water quality evaluation [J]. Zhihuai 12:17–18

Cong Q (2019) Water quality evaluation of Fenhe River based on improved grey correlation analysis [J]. Water Resour Hydropower Express 40(5):35–38

Hongmei W, Xiao H (2021) Study on comprehensive evaluation and change trend of Dianchi Lake open sea based on principal component analysis [J]. Sichuan Environ 40(4):72–79

Pan M, Gao L (2010) Influence of social and economic development on water quality change in Dianchi Lake Basin [J]. Chin Eng Sci 12(6):117–122

Rui W (2009) Study on water environment quality assessment and water quality prediction model in Chaohu Lake Basin [D], Hefei University Technology

Yijing C, Daiqing Z, Mei CG, Hui Z, Sen M (2022) Application and comparison of four comprehensive methods in water quality evaluation of nine plateau lakes in Yunnan [J/OL]. Hydropower Gener, pp 1–7

Yingying H, Yicheng W, Yuyan Z, Yiran X, Fan L, Weihua X, Jianwei L, Wei X, Yuanhui Y (2022) Water quality evaluation of Baiyangdian basin based on comprehensive water quality index [J]. Water Conservancy Hydropower Technol 53(3):145–154 (Chinese and English)

Chapter 27
Adsorption Properties of Dyes on UiO–66 and UiO–66-NH_2

Mengqi Liu, Yongbo He, Tong Sun, and Linzhi Zhai

Abstract In this paper, UiO–66 and UiO–66–NH_2 were synthesized by a solvothermal method, and their adsorption properties for Methyl orange, Congo red, Methylene blue or Rhodamine B were evaluated. The results demonstrated that, when the pH value of the solution was 3–7, concentration of Methyl orange was 80 mg·L^{-1} and amount of UiO-66 was 200 mg·L^{-1}, the maximum adsorption capacity and removal rate of Methyl orange could separately reach up to 198.0 mg·g^{-1} and 99%. As for UiO–66–NH_2, the maximum adsorption capacity and removal rate of Congo red would exceed 153.0 mg·g^{-1} and 99% when the pH was 5–7, concentration of Congo red was 40 mg·L^{-1}, and dosage of UiO–66–NH_2 was 200 mg·L^{-1}.

Keywords Metal organic framework · UiO–66 · Dye · Wastewater · Adsorption

27.1 Introduction

A large amount of dye wastewater produced by printing and dyeing plants seriously affects human health (Wang et al. 2016). At present, the treatment methods for dye removal mainly include adsorption, coagulation precipitation, membrane separation, oxidation and electrochemical (Cai et al. 2020; Ma et al. 2021; Xie et al. 2020). Among them, the adsorption technology has become a promising avenue for the dye removal from the wastewater, because of its low cost, simple operation, and high efficiency. Researchers found that some adsorbents have no favorable long-term stability in the aqueous solution. Therefore, it is urgent to develop a novel porous material with ultra-high hydrothermal stability to remove dyes from wastewater (Chen et al. 2021; Wu et al. 2016). Based on the strong moisture stability records, UiO–66, a typical metal organic framework (MOFs), has been widely used in the high-efficiency dye removal from the wastewater (Cavka et al. 2008; Zhai et al. 2020). For example, Yao

M. Liu · Y. He · T. Sun · L. Zhai (✉)
School of Environmental and Chemical Engineering, Jiangsu University of Science and Technology, Zhenjiang 212100, China
e-mail: zhailinzhi@just.edu.cn

J. Zhang et al. (eds.), *Environmental Pollution Governance and Ecological Remediation Technology*, Environmental Science and Engineering,
https://doi.org/10.1007/978-3-031-25284-6_27

and co-workers reported that UiO–66 had an adsorption capacity of 84.8 mg·g^{-1} for Methyl orange (Qiu et al. 2017).

In order to explore systematically the adsorption performance of dyes on UiO–66-typed MOFs, herein, UiO–66 and UiO–66–NH_2 were synthesized by a solvothermal method. The effects of pH value, initial dye concentration and MOFs dosage on the adsorption capacities were discussed. The experimental results suggested that UiO-66-typed MOFs can remove 99% dyes from the wastewater.

27.2 Experiments

27.2.1 Materials

Terephthalic acid and 2-amino-terephthalic acid were obtained from Macklin Inc. Anhydrous zirconium chloride ($ZrCl_4$), Congo red (CR), and Rhodamine B (RhB) were purchased from Shanghai Aladdin Biochemical Technology Co., Ltd. Acetic acid, N,N–dimethylformamide (DMF), methanol, Methyl orange (MO) and Methylene blue (MB) were supported by Sinopharm Chemical Reagent Co., Ltd. The reagents were used without purification.

27.2.2 Preparation of UiO–66 MOFs

300 mg of $ZrCl_4$ and 270 mg of terephthalic acid were dissolved into 50 mL of DMF, followed by ultrasound treatment. 6.8 mL of acetic acid was subsequently added. Then the mixture was fallen into a sealed Teflon-lined autoclave, and cultivated for 72 h at 120 °C. After cooling, the fresh product was thoroughly washed with DMF and methanol, respectively. Finally, the sample was dried under vacuum for overnight.

For the preparation of UiO–66–NH_2, 300 mg of $ZrCl_4$, 297 mg of 2-aminoterephthalic acid and 6.8 mL of acetic acid were added into 50 mL DMF. After that, the mixture was heated at 120 °C for 72 h.

27.2.3 Characterization of UiO–66 MOFs

The microporous properties were measured by BELSORP-max adsorption instrument. The crystalline features of samples were tested by X-ray diffraction (XRD) on a Rigaku MiniFlex 600 diffractometer with a scan rate of 2 min^{-1} (Cu $K\alpha$ radiation, $\lambda = 1.54178$ Å). The morphology of the adsorbents were characterized by a JSM–7001F field-emission scanning electron microscope (FESEM, JEOL). Zeta potentials of the samples were determined by Zeta potential instrument.

27.2.4 Adsorption Experiments

The adsorption properties of the adsorbents under different pH value (3 ~ 11), different initial concentrations of the dye solution (20 ~ 80 mg·L^{-1}), and different dosage of the adsorbents (50 ~ 500 mg·L^{-1}) were discussed. A typical experiment, 100 mL of dye solution and 20 mg of MOFs powder were added into a flask, subsequently, the mixture was placed on a stirrer with the stirring speed of 150 r·min^{-1} at 25 °C. After the experiments, a certain amount of solution was taken, and the residual concentration were determined using a UV–Vis spectrophotometer (L5S, INESA). Each adsorption experiment was carried out for three times. The adsorption capacity and removal efficiency of the dyes were shown as Equs. (27.1) and (27.2):

$$Q_e = (C_0 - C_e)V/M \tag{27.1}$$

$$R/\% = (C_0 - C_e)/C_0 \times 100 \tag{27.2}$$

where Q_e and R are the adsorption capacity and removal efficiency of the dyes, respectively, mg·g^{-1} and %; C_0 and C_e are the initial, and equilibrium concentration of the dye solution, mg·g^{-1}; V is the volume of the solution, L; M is the amount of MOF adsorbent, g.

27.3 Results and Discussion

27.3.1 Characterization of Adsorbents

The XRD patterns of the adsorbents are presented in Fig. 27.1. The obvious diffraction characteristic peaks of UiO–66 and UiO–66–NH_2 appear at $2\theta = 7.34$ and 8.42°, which are attributed to the diffraction of (111) and (022) crystal planes of UiO–66–typed MOFs (Cavka et al. 2008). To further demonstrate the successful preparation of UiO–66 and UiO–66–NH_2, the SEM analysis are also performed, as shown in Fig. 27.2. The nanocrystals of UiO–66 and UiO–66–NH_2 exhibit smooth-surfaced octahedral structure with a diameter of about 600 and 400 nm, respectively.

The N_2 adsorption–desorption isotherms is employed to assess the micropore properties of UiO–66 and UiO–66–NH_2 at 77 K, as depicted in Fig. 27.3a. In the low-pressure region ($P/P_0 < 0.05$), N_2 adsorption rapidly increases, due to the strong interaction between N_2 molecules and micropores of MOFs. Therefore, N_2 adsorption–desorption curve can be classified as a type I isotherm, implying that UiO–66 and UiO–66–NH_2 samples have a plenty of micropores. The BET specific surface area of UiO–66 and UiO–66–NH_2 are separately 1569.2 and 1193.1 m^2·g^{-1}, as –NH_2 groups hinder the inner cavities of the framework to a certain extent. The pore

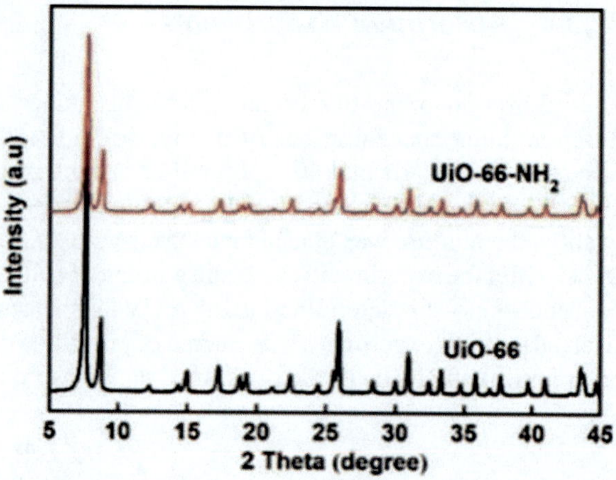

Fig. 27.1 XRD patterns of UiO–66 and UiO–66–NH_2

Fig. 27.2 SEM images of **a** UiO–66 and **b** UiO–66–NH_2

size distribution of UiO–66 and UiO–66–NH_2 was shown in Fig. 27.3b. The pore size of UiO–66 is about 6.0 Å, while the pore size of UiO–66–NH_2 reduces to 5.5 Å.

27.3.2 *Adsorption Experiments*

Adsorption Capacities of Dyes To estimate the adsorption performance of UiO–66 and UiO–66–NH_2 adsorbent, adsorption efficiencies of MB, MO, CR, and RhB are determined (Fig. 27.4). It can depict the phenomenon that UiO–66 has better adsorption property for MO, and the removal efficiency of MO gradually goes up with the increase of the adsorption time. After 15 min, the removal efficiency of MO remains steady at about 98.3%. For UiO–66–NH_2 adsorbent, it appears that the highest removal efficiency of CR attains 65.4%.

To consider the surface charge properties of the adsorbents, the Zeta potential of the adsorbent was tested (Fig. 27.5). The results shows that the isoelectric point

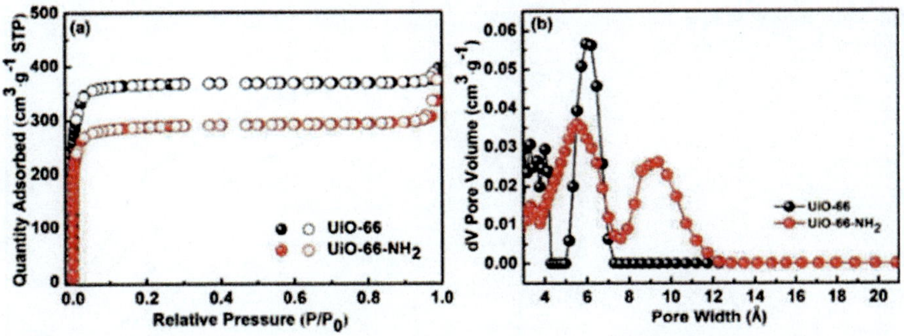

Fig. 27.3 **a** N_2 adsorption–desorption isotherms and **b** Pore size distribution of UiO–66 and UiO–66–NH$_2$

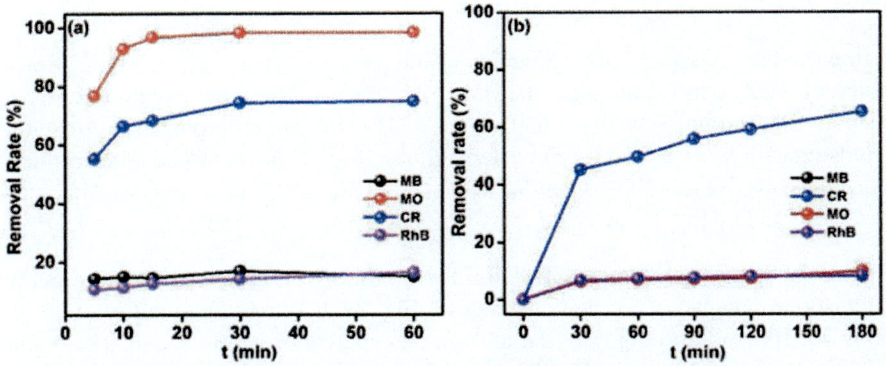

Fig. 27.4 Adsorption efficiencies of the dyes on **a** UiO–66 and **b** UiO–66–NH$_2$

(pH_{pzc}) of UiO–66 and UiO–66–NH$_2$ are approximately 7.2 and 7.8, respectively. When the pH value is below than pH_{pzc}, the surface of UiO–66 and UiO–66–NH$_2$ samples is positively charged, which helps to attract negatively charged MO^- and CR^-. When the pH value is higher than pH_{pzc}, the surface of the adsorbent is negatively charged, and anionic dye ions cannot be adsorbed by electrostatic attraction. Therefore, we performed the following adsorption experiments at pH≈6.8.

Effect of pH Value on Dye Removal The adsorption capacity of MO and CR are investigated with the pH value range of 3–11. It can be seen from Fig. 27.6a, the adsorption capacities of MO are 198.0, 198.4, 197.6 and 195.4 mg·g^{-1} under pH values of 3, 4, 5 and 7, respectively, and the removal efficiency of MO attains above 98%. Until the pH value was 11, the adsorption capacity of MO reduces to 133.8 mg·g^{-1}. This is because that MO molecule transfers into anionic MO^- when pH value is higher than pK_a ($pK_a = 3.5$). In an acid solution, H^+ ion adheres to the interface between the metal center and the organic ligands, a positive Zeta potential is

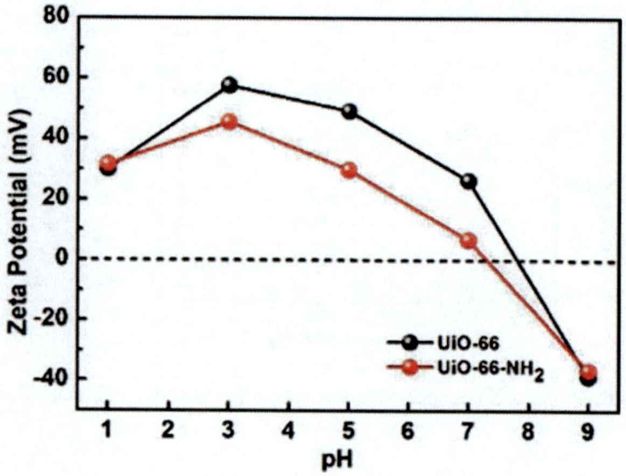

Fig. 27.5 Zeta potential of UiO–66 and UiO–66–NH_2

formed on the surface of UiO–66, which would produce a strong electrostatic attraction with MO^- ions (Zeng et al. 2018). In alkaline medium, more negative charges, such as OH^- group, deposit on the surface of UiO–66 which produce electrostatic repulsion with MO^- anion. It can be derived from Fig. 27.6b, the trend of adsorption capacities on UiO–66–NH_2 is similar with that of UiO–66. The adsorption capacities of CR are 153.0, 152.2 and 149.0 mg·g^{-1} under pH was 5, 6 and 7, respectively.

Effect of Adsorbent Dosage on Dye Removal The effects of the mass of adsorbents on adsorption capacity are demonstrated in Fig. 27.7. When the quantity of UiO–66 is 50, 100 and 200 mg·L^{-1}, the adsorption capacity of MO is 194.0, 196.0 and 190.8 mg·g^{-1}, respectively (Fig. 27.7a). After that, the adsorption capacity sharply decreases along with the increase of UiO–66 dosage. As you can observe from Fig. 27.7b, the adsorption capacity of CR exceeds more than 130 mg·g^{-1} under the dosage of 200 mg·L^{-1}. The results implicate that the adsorption sites increase as the quantity of the adsorbent increasing. The excess adsorbent, causing available

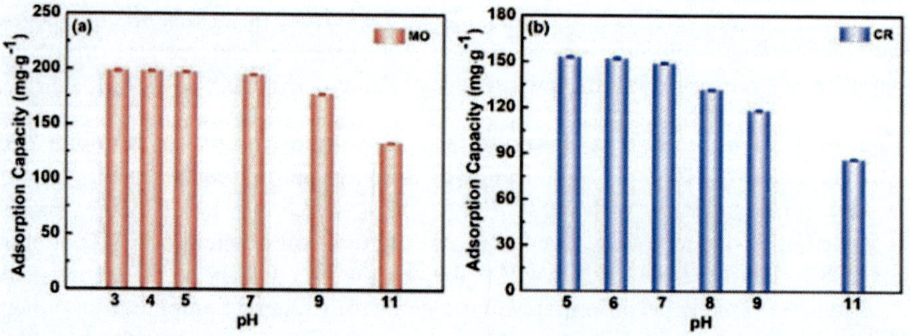

Fig. 27.6 The effect of pH value on the adsorption capacity of **a** UiO–66 and **b** UiO–66–NH_2

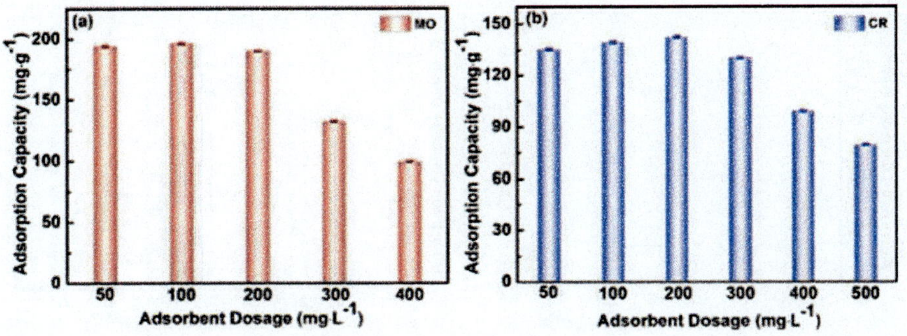

Fig. 27.7 The effect of adsorbent dosage on the adsorption capacity of **a** UiO–66 and **b** UiO–66–NH$_2$

adsorption site is not fully occupied, will reduce equilibrium adsorption capacity of dyes.

Effect of the Initial Concentration of Dye Solution on Dye Removal The effects of the initial concentration are represented in Fig. 27.8. The adsorption capacity gradually increases with the increasing of the dye concentration in the aqueous solution. The adsorption capacity of MO reaches up to 190.0 mg·g^{-1} from 97.7 mg·g^{-1}, with the range of initial concentration of 20 ~ 40 mg·L^{-1}. The maximum adsorption capacity of MO on UiO–66 is 198.0 mg·g^{-1}. For UiO–66–NH$_2$, it also shows a similar trend for the adsorption of CR, as emerged in Fig. 27.8b, and the maximum adsorption capacity is stable at about 153.0 mg·g^{-1}. With the lower initial concentration, the dye molecules will be easily anchored on the active sites of UiO–66 and UiO–66–NH$_2$. As the initial concentration of MO and CR increasing, more dye molecules occupy the adsorption sites of the adsorbents, causing a high adsorption capacity of MO and CR. At a higher concentration, the dye molecules are accelerated on the adsorbent, therefore, the adsorption sites are fully occupied, leading to an adsorption–desorption equilibrium.

27.4 Conclusion

In this study, UiO–66 and UiO–66–NH$_2$ are successfully prepared by a solvothermal method, and used for the dye removal from the wastewater. The maximum adsorption capacity of MO and CR on UiO–66 and UiO–66–NH$_2$ are 198.0 and 153.0 mg·g^{-1}, respectively, and the removal efficiencies are higher than 99%. Therefore, UiO–66-typed MOFs shed the light on the dye removal from the wastewater.

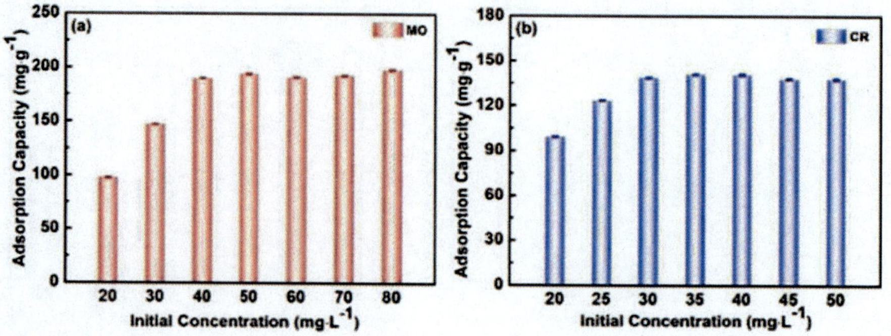

Fig. 27.8 The effect of initial dye concentration on the adsorption capacity of **a** UiO–66 and **b** UiO–66–NH_2

Acknowledgements The study is supported by the Industry-University-Institute cooperation program (2115082207).

References

Cai Y, Shi D, Liu G et al (2020) Polycrystalline zirconium metal-organic framework membranes supported on flexible carbon cloth for organic solvent nanofiltration. J Membr Sci 615:118551

Cavka JH, Jakobsen S, Olsbye U et al (2008) A new zirconium inorganic building brick forming metal organic frameworks with exceptional stability. J Am Chem Soc 130(42):13850–13851

Chen S, Zheng WL, Gu J et al (2021) Preparation and performance evaluation of methylene blue sludge based sorbents. J China Univ Pet Ed Nat Sci 45(2):173–180

Ma D, Yi H, Lai C et al (2021) Critical review of advanced oxidation processes in organic wastewater treatment. Chemosphere 275:130104

Qiu J, Feng Y, Zhang X et al (2017) Acid-promoted synthesis of UiO–66 for highly selective adsorption of anionic dyes: adsorption performance and mechanisms. J Colloid Interface Sci 499:151–158

Wang XS, Liang J, Li L et al (2016) An anion metal-organic framework with Lewis basic sites-rich toward charge-exclusive cationic dyes separation and size-selective catalytic reaction. Inorg Chem 55(5):2641–2649

Wu RL, Liu YF, Ren S et al (2016) Preparation of Fe_3O_4@carbon/graphene oxide composites for dye adsorption. China Environ Sci 36(10):2981–2987

Xie JW, Zhu GG, Xie JX et al (2020) Research progress in enhanced bioelectrochemical treatment of refractory wastewater. Ind Water Treat 40(10):1–7

Zeng L, Xiao L, Long Y et al (2018) Trichloroacetic acid-modulated synthesis of polyoxometalate@UiO–66 for selective adsorption of cationic dyes. J Colloid Interface Sci 516:274–283

Zhai L, Yu X, Wang Y et al (2020) Polycrystalline rare-earth metal-organic framework membranes with in-situ healing ability for efficient alcohol dehydration. J Membr Sci 610:118239

Chapter 28
The Long-term Diffusion of Dumped Wastewater from Japan in the Pacific Ocean

Anqi Wang

Abstract The wastewater dumped into the Pacific Ocean by Japan, which contains hard-removal tritium, in addition to other radioactive isotopes including carbon 14, cobalt 60 and strontium 90, polluted not only the sea-domain of Japan but also affected the whole Pacific Ocean or even further sea areas. This report aims to simulate the pollution trends caused by the wastewater produced in Fukushima Daiichi Nuclear Power Plant. In this context, it is assumed that the polluted concentration or residual pollutants will gradually decrease as it moves away from the source and diffuses in the Ocean with time. To test the hypothesis, after the complex traditional numerical and uncertain particle swarm optimization were compared, Lattice Boltzmann Method (LBM) was introduced into the model which benefits from the simple description of fluid interactions, easily sets complex boundaries and parallel calculations and etc., can implement the simulation program easily. The LBM simulation results showed a similar direction as the hypothesis, however, as the LBM model only simulates the pollution contains in the water, considered the transportation to further ocean areas by the large and highly migratory marine animals and native pollution locked by the marine flora and microorganisms, the real trends may worse. These results suggest that the further environmental pollution impact caused by Japanese Nuclear wastewater is unpredictable but profound and lasting.

Keywords Lattice Boltzmann method · Japanese nuclear wastewater · The Pacific Ocean · Radionuclide · Wastewater convection

28.1 Introduction

On 13th April 2021, Japan's government announced to dump more than a million tons of nuclear wastewater from Fukushima Daiichi Nuclear Power Plant into the Pacific Ocean (Chang and Zhao 2012). This wastewater with high activity, which

A. Wang (✉)
University of Nottingham, Nottinghamshire, UK NG7 2RD, England
e-mail: Wanganqi1771@outlook.com

J. Zhang et al. (eds.), *Environmental Pollution Governance and Ecological Remediation Technology*, Environmental Science and Engineering,
https://doi.org/10.1007/978-3-031-25284-6_28

is produced during the processes of nuclear fuels, is strongly acid with very long half-live isotopes, such as radium 226. Japan claimed that this action has no impact on Ocean health, the wastewater is safe enough to be discharged, though, the way to remove the radioactive properties of the components present in the wastewater is still unknown, all methods for the distillation of the water are to attenuate the concentration of the radioactive. Therefore, all substances remain unchanged and exist as a concentrated residue and locked permanently within the wastewater after treatment (Jørgensen 1979). Moreover, scientists have also discovered that there are some radioactive isotopes in the wastewater in addition to the hard-removal tritium, including carbon 14, cobalt 60, and strontium 90 (Liu 2021). The aim of this report is to develop a model to simulate the proliferation of radioactive caused by the nuclear wastewater situation in the Pacific Ocean and predict a long-run effect on the global ocean environment in order to circumvent or minimize the potential environmental risks. It first hypothesized an equation to calculate the convection and diffusion processes; then choose an effective method to simulate the pollution trends, in this part, in order to make the most vailed methods, numerical and particle swarm optimization methods, as well as LBM, were run and compared; finally, considered the diversified marine environments, more availabilities which may vary the results discuss, such as the marine animals carried and transported elements.

28.2 Methodology

28.2.1 Convection and Diffusion

The Convection–diffusion phenomenon is often applied to the distribution of pollutants in nuclear waste pollution. It is also used here for Japanese nuclear wastewater entering the sea. In order to study the physical process of material transport and molecular diffusion in the process of fluid flow, the author made a hypothesis. The author assumes that the concentration or residual pollutants of wastewater from the nuclear wastewater from the Fukushima Daiichi Nuclear Power Plant will gradually decrease as it moves away from the source.

Therefore, we simulate the transition process with the two-dimensional convection–diffusion equation in Cartesian coordinates:

$$\frac{\partial \emptyset}{\partial t} + u\frac{\partial \emptyset}{\partial x} + v\frac{\partial \emptyset}{\partial y} = \alpha\left(\frac{\partial^2 \emptyset}{\partial x^2} + \frac{\partial^2 \emptyset}{\partial y^2}\right) \tag{28.1}$$

28.2.2 Formulate a Scheme

In order to make an appropriate method to simulate the pollution trends, numerical solutions and particle swarm optimization were considered. The traditional purely numerical solution method, which expressing fluid variables as the value of each point, and lies in solving partial differential equations through approximating the domain. Besides, from a microscopic and mesoscopic perspective, particles represent atoms, molecules, molecular clusters or a part of a macroscopic fluid in a fluid. Three basic N-S equations and three particle swarm optimization methods were abandoned after test and comparison.

28.2.3 Simulate the Model

Lattice Boltzmann Method (LBM) method was introduced as a more suitable and simpler method to discuss the Convection and diffusion within the ocean by wastewater.

The LBM method consists of two steps, one is collision and the other is migration. For two-dimensional problems, the function of the collision process is:

$$f_k(x + \Delta x, y + \Delta y, t + \Delta t) = f_k(x, y, t + \Delta t) \tag{28.2}$$

The function of the migration process is

$$f_k(x, y, t + \Delta t) = f_k(x, y, t)[1 - \omega] + \omega f_k^{eq}(x, y, t) \tag{28.3}$$

Among them, f is the distribution function, and f^{eq} is the local equilibrium distribution function. The equilibrium is

$$f_k^{eq} = w_{k\emptyset}(x, t)\left[1 + \frac{c_k \cdot \vec{u}}{c_s^2}\right] \tag{28.4}$$

When $\Delta x = \Delta y = \Delta t = 1$,

$$\frac{c_k \cdot \vec{u}}{c_s^2} = \begin{cases} 2u & k = 1 \\ -2u & k = 2 \\ 2v & k = 3 \\ -2v & k = 4 \end{cases} \tag{28.5}$$

The basic quantity $f_i(x, t)$ in the equation is the discrete velocity distribution function, which represents the particle density with velocity c_i at position x and time t. The mass density and momentum density can be expressed as:

$$\rho(x, t) = i \sum fi(x, t), \rho u(x, t) = i \sum cif_i(x, t) \tag{28.6}$$

In the LBM method, D_nQ_m is a general term used for n to indicate the dimensionality of the problem, and m to indicate the number of speeds.

28.3 Results

28.3.1 The Test and Comparison of Numerical and Particle Swarm Optimization Methods

Three basic N-S equations are as follow

Finite Difference Method (FD)
The finite difference method (FD) solving the approximate progressively through the direction of increasing time with different formats. It starts with an initial value, for example, an initial time $t = t0$, the solution at the time $t > t0$ completely depends on the initial conditions and certain boundary conditions. It benefits from simplicity, flexibility, versatility and ease-implement with computers to describe and solve the physical phenomena that change with time.

Finite Volume Method (FVM)
FVM is based on an integral form of the conservation equation and focuses on constructing discrete equations to compute fluid dynamics with a numerical algorithm to define the computational grid with single control body. Every discrete equation expresses the conservation of a certain physical quantity on a finite volume. The discrete equation can be guaranteed due to the clear physical concept of coefficients from each process.

Finite Element Method (FEM)
Finite element analysis (FEM) solving the problem by structure an equilibrium condition by assuming a simpler but befitting approximate solution for each complex element. On account of the real problem is replaced, the final solution is not accurate but an approximate solution. It is still commonly used and has become an effective engineering analysis method profit from most practical problems are hard to obtain an accurate solution and the FEM calculation has high precision and adaption to various complex situations.

Three particle swarm optimization methods are as follow:

Molecular Dynamics (MD)
The principal rule of MD is Newtonian mechanics. It selects samples from the different statuses of the composed molecular system, calculates the configuration integral, further calculates the thermodynamic quantity and other macro properties.

$$xi(t + \Delta t) = 2xi(t) - xi(t - \Delta t) + mif_i(t)\Delta t_2 \tag{28.7}$$

Using the Verlet algorithm, the particle's past and present displacements are used to calculate the future displacement.

Lattice Mode (LGM)
The motion of the fluid is described by two criteria: collision and flow. The grid has only two states, which is good for describing collisions, so there is no rounding error inherent in floating-point operations in other CFD methods, thus, a large number of parallel operations can be performed. A Collision operator Ω_i is introduced into the calculation. However, because this LGM will gradually get out of control with the speed increases, when solving the N-S equation, the first successful model FHP in LGCA is only suitable for the flow with low Mach.

$$n_i(x + c_i \Delta t_a, t + \Delta t) = ni(x, t) + \Omega_i(x, t) \tag{28.8}$$

Direct Monte Carlo method (DSMC)
A particle method based on dynamics theory. The flow solution is obtained by simulating the collision of particles through a probability model. It is the main method to solve the problem of high Knudsen number flow. The algorithm can describe the physical effects well (even more than the LBM method), and its calculation results are determined by the sample size and averaging process, so there may be statistical errors, which are inversely proportional to the square root of the calculated particle number N, and the calculation cost It is proportional to the number of particles, so it needs to meet high computer resources while ensuring accuracy. At present, it has the highest accuracy in the calculation of rare gas flows. With the continuous development of computer technology, this algorithm has great potential.

Considering that the particle method does not directly solve the fluid mechanics equations, but uses particles to represent atoms, molecules, molecular clusters or a part of the macroscopic fluid, thereby reflecting the macroscopic laws of the fluid to a certain extent, moreover some particle methods also cannot reflect the macroscopic laws of fluid motion directly, and there are still difficult to quantify accuracy, noise issues, etc. Moreover, the numerical methods only tested the approximate values and the solution procedure is huge and complex. Both numerical and particle methods cannot be efficient and suitable models.

28.3.2 LBM and Its Applications

LBM is based on the principle of Lattice Gas Automaton (LGA), adopts different research objects and research systems (Lagrange method and Euler method), improves and breaks the theoretical framework of traditional calculation methods and classical mechanics, established a bright new pioneered probability and statistical mechanics based on the relationship between macro and micro, continuous and discrete. Conservation as the most fundamental law of the material world guides movement and development. The improvement and application of LBM just reflected it, in this case, means even if the movement of a single particle has no rules to follow, the irregular movement of several particles will affect the macroscopic parameters of fluid motion which is a certain internal connection between the two apparently opposing parties, which can be achieved in some way. Therefore, the macroscopic parameters of fluid motion can be obtained through statistical analysis of a large number of discrete particles.

The studies on the Chernobyl disaster in 1986 as well as the studies on nuclear weapon test during the 1950s and 1960s suggested that 137 Cs behaves in seawater approximately like a 'conservative' dissolvable tracer (Beck et al. 2009), with its distributed mainly caused by physical transportation processes on a decadal timescale (Nogrady 2021) which means the main mechanism of the noticed dispersion of 137 Cs concentrations is the ocean currents in the North Pacific surface water in the 1970s. From an oceanographic point of view, the spatial and temporal characteristics play an important role in distinguishing the differences between the weapon tests fallout in the 1960s and the Japanese discharge of artificial radionuclides off recently. Therefore, according to the models on studying the effect caused by Chernobyl disaster, could develop an ocean model to simulate the behaviour of the radionuclide caused by FDNPP wastewater diffusion with treat the wastewater as a 137 Cs 'conservative' or 'dye'. Trace a small range in a specific time scale use LBM based on Lagrange theory for a further analysis (Behrens et al. 2021).

The LBM model simulates the pollution trends in the Pacific Ocean. The wastewaters diffuse and dilute in the ocean, but unlike ordinary industrial wastewater, even the diluted wastewater, its half-lives are still decades or even hundreds of thousands of years. This also means that the clearing of nuclear pollution is hard-achieved, also the marine life will risk radiation. The discharge of wastewater into the ocean by Japan government leads to long-term environmental issues, not only risks the Japanese human's health but also risks the countries adjoin the Pacific Ocean even further nations.

28.4 Discussion

The traditional pure numerical solution method aims to solve fluid mechanics equations. These methods all express fluid variables, such as velocity and pressure, as the value of each point or node in the domain, and solve partial differential equations by approximating the domain, and the differences lie in the interpretation of the node value. The particle method does not directly solve the hydrodynamic equations but uses particles to represent atoms, molecules, molecular clusters or part of a macroscopic fluid, thereby reflecting the macroscopic laws of fluids to a certain extent. Thus, choosing the LBM method for modelling not only avoids the complicated solution and low precision of the traditional method to solve the fluid mechanic equation, but also avoids the noise problem of the particle method through the mesoscopic method, and has a strong physical foundation. LBM combines the dynamics of microscopic particles with the laws of macroscopic fluids and has good accuracy to simulate the convection and diffusion process of nuclear wastewater in the Pacific Ocean, and then infer the flow direction of nuclear wastewater dumped in the Pacific Ocean within a certain period of time and Pollution trend.

The LBM model could tract and simulate the wastewater Convection and diffusion in the ocean. Considered atmosphere, hydrosphere, biosphere and pedosphere both around Fukushima Daiichi Nuclear Power Plant and the entire northern hemisphere have been polluted by radionuclide at a large scale after the Fukushima Nuclear Accident (Nogrady 2021). The radionuclides will not only be delivered by atmospheric dry/wet deposition or precipitation (Okamura et al. 2016) but also through the zooplankton and mesopelagic fish (Nogrady 2021), it even transport to further ocean areas by the large and highly migratory marine animals (Nogrady 2021) such as Pacific Bluefin tuna took the radionuclides from original source in Japan to distant South and North pacific.

As Fig. 28.1 shows that the most of these radionuclides have higher biological concentration factors and seafloor sediment–water distribution coefficient (Lu et al. 2021) where the trophic level in the food web accumulated and could be absorbed easier by seabed sediments, in another word, these radionuclides entrench and accumulate in the seabed near Japanese seaside, and do not be transported by the ocean currents. The maximum values of 134 and 137 Cs on 67,000 and 68,000 Bq/L were reached in April 2011. The annual concentration means of 134 and 137 Cs declined 0.71 and 1.16 Bq/L in 2012 considering the seawater dilution. However, there is a relatively slow fall in the radioactive concentration of marine fish in Fukushima prefecture which means a longer substance half-life and a slower degradation. The radionuclides kept by the marine animals will finally affect human health through the food web.

28.5 Conclusion

The comparison among numerical methods, particle swarm optimization methods, and LBM have proved that LBM can simulate the convection and diffusion processes of wastewater more easily and effectively. The result shows that the numerical methods are comparatively complexed and not accurate, and the optimization methods are not a direct macroscopic calculation. Additionally, the hypothesis that the wastewater will attenuate and transport further ocean areas is relatively reliable, the wastewater was diluted but the radiant and toxic elements were not. Moreover, the environment also has an inner and outside loop, which means the radionuclides will not only be delivered by atmospheric dry/wet deposition or precipitation (Okamura et al. 2016) but also through the zooplankton and mesopelagic fish (Nogrady 2021), it even transport to further ocean areas by the large and highly migratory marine animals (Nogrady 2021). The LBM only simulate the polluted particle contains in the ocean water, ignoring other living or abiotic effect, therefore, the pollution trends may far lower than the reality. To simulate a more accurate model, it will need to

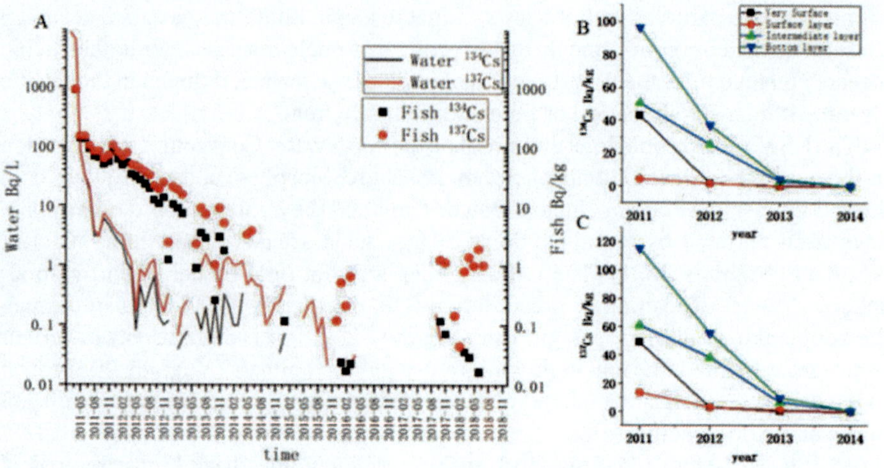

Fig. 28.1 Concentration of 134Cs and 137Cs in water and organisms near the Fukushima Daiichi Nuclear Power Plant (Based on results from the Radioactive Nuclide Analysis for Seawater performed near the Power Station (TEPCO) (TEPCO, 2011–2018), and the Levels of Radioactive Materials in Food Tested in Respective Prefectures (Ministry of Health, Labor and Welfare, Japan, 2011–2018). The classification of biological organisms into four layers was based on the report on monitoring of radionuclides in fishery products (Fisheries Agency of Japan, 2015))

include more variable factors, such as living habits, sea area characteristics, etc., and the data related to these factors will also become indispensable. The environmental impact caused by the dumping of wastewater is existed, profound and lasting.

Acknowledgements First and foremost, I would like to show my deepest gratitude to my supervisor, Professor David M Holland, a respectable, responsible, and resourceful scholar, who has provided me with valuable knowledge and guidance. Without his enlightening instruction, impressive kindness and patience, I could not have completed my thesis. His keen and vigorous academic observation enlightens me not only in this thesis but also in my future study.

I shall extend my thanks to Dr. Steven Hou and dear Hayley for all their kindness and help.

Last but not least, I'd like to thank all people who give me support and encouragement during this period.

References

Beck AJ, Cochran JK, Sañudo-Wilhelmy SA (2009) Temporal trends of dissolved trace metals in Jamaica Bay, NY: importance of wastewater input and submarine groundwater discharge in an urban estuary. Estuaries Coasts 32(3):535–550

Behrens E, Schwarzkopf FU, Lübbecke JF, Böning CW (2021) Model simulations on the long-term dispersal of ^{137}Cs released into the Pacific Ocean off Fukushima. Environ Res Lett 7(3):34004. https://doi.org/10.1088/1748-9326/7/3/034004. Accessed on 28 Jun 2021

Chang Y-C, Zhao Y (2012) The Fukushima nuclear power station incident and marine pollution. Mar Pollut Bull 64(5):897–901

Jørgensen S (1979) Industrial waste water management. Elsevier Scientific Pub. Co., Amsterdam, pp 373–376

Liu W (2021) Basic facts to know about Japan's nuclear wastewater dump. In: CGTN. Retrieved from https://news.cgtn.com/news/2021-04-12/Key-facts-to-know-about-Japan-s-nuclear-wastewater-dump--ZoVZBcuzDi/index.html. Accessed on 15 Jun 2021

Lu Y, Yuan J, Du D, Sun B, Yi X (2021) Monitoring long-term ecological impacts from release of Fukushima radiation water into ocean. Geogr Sustain 2(2):95–98

Nogrady B (2021) Scientists OK plan to release one million tonnes of waste water from Fukushima. In: Nature. Retrieved from https://www.nature.com/articles/d41586-021-01225-2. Accessed on 15 Jun 2021

Okamura H, Ikeda S, Morita T, Eguchi S (2016) Risk assessment of radioisotope contamination for aquatic living resources in and around Japan. Proc Natl Acad Sci 113(14):3838–3843

Chapter 29
Analysis of Soil Environmental Impact of a Waste Incineration Power Generation Project

Yanwen Liu, Yanjun Chang, Lin Wang, Guoxia Zheng, Jing Liu, and Zhiguo Liu

Abstract This research investigates and predicts the soil pollution problems caused by a waste incineration power generation project. By analyzing the types and pathways of the impact of the waste-to-energy project on the site soil environment, and according to the actual situation, mercury, lead, cadmium, dioxins, petroleum hydrocarbons, chromium, zinc, and nickel were selected as key predictors to analyze and predict the atmospheric deposition pollution and vertical infiltration pollution of the site soil. The results show that the concentration of the superimposed value is lower than the second-class land screening value in the Soil environmental quality risk control standard for soil contamination of development land (GB36600-2018).

Keywords Waste incineration · Heavy metals · Dioxin · Soil

Y. Liu · L. Wang (✉) · G. Zheng
School of Environmental and Chemical Engineering, Dalian University, Dalian 116622, China
e-mail: 309098628@qq.com

Y. Liu
e-mail: yanwenliu368@163.com

G. Zheng
e-mail: guoxiazheng2685@163.com

Y. Chang · J. Liu
Liaoning Environmental Planning Institute Co. Ltd, Shenyang 110037, China
e-mail: 709265538@qq.com

J. Liu
e-mail: 1176029163@qq.com

Z. Liu
School of Information, Dalian University, Dalian 116622, China
e-mail: liuzhiguo_dldx@163.com

J. Zhang et al. (eds.), *Environmental Pollution Governance and Ecological Remediation Technology*, Environmental Science and Engineering,
https://doi.org/10.1007/978-3-031-25284-6_29

29.1 Introduction

In recent years, China's domestic waste production has increased rapidly, and it is the largest waste-producing country in the world (Kaza et al. 2018). It is estimated that by 2030, the daily waste output will reach twice that of the United States (Fang and Li 2019), and the traditional landfill technology will lead to serious consumption of land resources (Cheng and Hu 2010; Gao and Yin 2019; Su et al. 2019). Therefore, waste incineration power generation has been widely promoted as a new resource utilization technology.

Soil pollution caused by heavy metal deposition in the incinerator flue gas is the focus of attention in the environmental impact assessment and environmental protection management of municipal waste incineration power generation projects (Liu et al. 2012; Tian et al. 2013; Zhang and Yin 2020).

29.2 Approach to Pollution of Soil by Waste Power Generation Project

The soil pollutants that may be released from waste disposal in waste power generation projects are mainly mercury, lead, cadmium, and other metal compounds, particulate matter, acid–base gases (HCl, SO_2, NO_2, NH_3, etc.), and organic highly toxic pollutants (dioxin, etc.) (Ferdan et al. 2019; Zhang 2020). These exhaust pollutants enter the surrounding soil in the form of atmospheric dry and wet deposition, which gradually affects the local soil environmental quality (Chen et al. 2021; Ferdan et al. 2019; Nie 2018).

The impact type and pathway of the project on the soil environment are shown in Table 29.1, and the identification of impact sources and impact factors on the soil environment is shown in Table 29.2.

Table 29.1 Soil environmental impact types and impact pathways of construction projects

Different periods	Pollution affected type			
	Atmospheric deposition	Overland full	Vertical infiltration	Other
Construction period	✓	–	✓	–
Operation period	✓	–	✓	–
After the service period	–	–	–	–

Table 29.2 Identification table of soil environmental impact sources and impact factors of polluting construction projects

Pollution sources	Processes	Pollution pathways	All pollutant indicators	Predictive factors	Remarks
Site	Waste incineration	Atmospheric deposition	Smoke, thallium, manganese, chromium, copper, nickel, antimony, cobalt, arsenic, ammonia, cadmium, mercury, lead, cadmium dioxins, mercury, lead, and dioxins	Cadmium, mercury, lead, and dioxins	Sensitive points are cultivated land and villages around
		Overland full	–	–	–
		Vertical infiltration	COD, BOD5, SS, NH3-N, total phosphorus, total nitrogen, cadmium, mercury, arsenic, lead, hexavalent chromium, petroleum, zinc, and nickel	Petroleum, chromium, zinc, and nickel	–
		Other	–	–	–

29.3 Prediction and Evaluation of Atmospheric Deposition Pollution

29.3.1 Predicted Area

Within the project area and within 1 km outside the project area, the predicted area is 3641907.377 m^2.

29.3.2 Predictor

According to the previous analysis, mercury, lead, cadmium, and dioxin are selected as predictors.

29.3.3 Prediction and Evaluation Methods

The project is pollution impact type, the pollutants in the exhaust gas enter the soil environment by way of atmospheric deposition, the prediction method adopts the surface source model recommended in Appendix E of Technical guidelines for environmental impact assessment-soil environment (trial) (HJ964-2018) for Prediction. The calculation formula for the increase of a substance in the soil through atmospheric deposition into the soil environment is as follows.

(1) Incremental formula of a substance in unit mass soil:

$$\Delta S = n(I_s - L_s - R_s) / (\rho_b \times A \times D)$$

In the formula

- ΔS Incremental formula of a substance in unit mass soil, g/kg
- I_s Predicted input amount of a substance in the surface soil per unit year within the evaluation range, g
- L_s The amount of a substance discharged by leaching from the surface soil per unit year within the predicted evaluation range, g
- R_s The amount of a substance discharged by runoff from the surface soil per unit year within the predicted evaluation range, g
- ρ_b Surface soil bulk density, kg/m^3
- A Scope of forecast evaluation, m^2
- D Surface soil depth, m
- N Duration Year, a。

This evaluation adopts the worst case to predict, that is, the maximum annual total settlement of heavy metals and dioxins is used to calculate the impact of the regional soil environment, regardless of the number of certain substances discharged by leaching and runoff in the soil. Therefore, the above formula can be simplified as follows:

$$\Delta S = \frac{nI_s}{(\rho_b \times A \times D)}$$

$$I_s = C \times V \times T \times A$$

In the formula

- C Maximum hourly landfall concentration of pollutants, mg/m^3
- V Pollutant sedimentation rate, m/s
- T Annual pollutant settling time
- S Annual project operation 8000 h
- t Taken as $8000 \times 60 \times \ \times 60 = 2.88 \times 10^7$ s
- A Scope of forecast evaluation, m^2

(2) Calculation formula of the predicted value of a substance in unit mass soil:

$$S = \Delta S + S_b$$

In the formula

S The predicted value of a substance in a unit mass of soil, g/kg
S_b Current value of a substance in a unit mass of soil, g/kg

29.3.4 *Selection of Prediction Parameters*

(1) The soil around the project area is mostly sandy soil with a soil capacity of 1.45–1.48 g/cm^3 and the average value of 1465 kg/m^3 is taken for this prediction, and the depth of the surface soil is taken as 0.5 m.
(2) The pollutants emitted from the project are mainly caused by atmospheric deposition to the surrounding soil, according to the Technical guidelines for environmental impact assessment-soil environment (trial) (HJ964-2018) technical Guidelines for Environmental Impact Assessment Soil Environment (Trial) (HJ964-2018), take the area 1000 m outside the plant boundary, i.e. 3,641,907.377 m^2.
(3) The annual settling time is taken as 8000 h, i.e. 2.88×10^7 s; the duration of the year is taken as 28 years for the predicted period.
(4) The current value of a substance in the soil per unit mass is taken as the maximum value in the monitoring value.
(5) Atmospheric deposition prediction results.

Based on the AERMOD model in the air pollution impact prediction model, the total deposition of heavy metals and dioxins is considered in the prediction process, and the total annual deposition of heavy metals and dioxins in this project is shown in Table 29.3.

Table 29.3 List of inputs of a substance in surface soil in the project unit year

Pollution factor	Average period	Maximum landing speed (μg/m$^{3)}$	Location	Input in soil (I_s) (g)
Lead	Average mean	0.0000382	Hanjialing1153, –659	5.712707916
Mercury	Average mean	0.00000001	Hanjialing1153, –659	0.006333752
Cadmium	Average mean	0.00000008	Hanjialing1153, –659	0.011265423
Dioxin	Average mean	0.00000012	Hanjialing1153, –659	0.030592022

Table 29.4 Increases in heavy metals and dioxins per unit mass of surface soil per year

Pollution factor	I_s (mg)	ρb (kg/m^3)	A (m^2)	D (m)	n (a)	ΔS (mg/kg)
Lead	5712.707916	1465	3,641,907.377	0.2	28	0.000149901
Mercury	6.33375196	1465	3,641,907.377	0.2	28	1.66197E-07
Cadmium	11.26542332	1465	3,641,907.377	0.2	28	2.95603E-07
Dioxin	30.59202197	1465	3,641,907.377	0.2	28	8.0273E-07

The evaluation calculates the impact on the regional soil environment based on the maximum annual total deposition of heavy metals and dioxins. The operation period of the project is calculated as 28 years. The selected list of soil deposition parameters is shown in Table 29.4.

29.3.5 *Forecast Result and Analysis*

The background value of the soil in this project is the background value of the soil at the maximum landing concentration. After the project runs for 28 years, the prediction results of heavy metals and dioxin pollution in the soil superimposed with the background value are shown in Table 29.5.

According to the above table, within 28 years after the project is put into operation, the superimposed predicted values of waste gas pollutants mercury, lead, cadmium, and dioxin discharged by the project in the soil within the total settlement maximum grid are all lower than the selected values of second-class land. The Soil environmental quality risk control standard for soil contamination of development land

Table 29.5 Prediction values of heavy metals and dioxins in soil

Project		Lead	Hydrargyrum	Cadmium	Dioxin
Contribution value	μg/m^3	0.0000382	0.00000001	0.00000008	0.00000012
	mg/kg (28 years)	0.000149901	1.66197E-07	2.95603E-07	8.0273E-07
Present value (mg/kg)		52.3	0.045	0.35	1.5 ng-TEQ/kg
28-year soil prediction value (mg/kg)		52.30015	0.04500017	0.3500003	1.5008 ng-TEQ/kg
Standard value (mg/kg)		170	3.4	0.6	40 ng-TEQ/kg

Note The standard values refer to the second-class land screening values in the Soil environmental quality risk control standard for soil contamination of development land (GB 36,600–2018)

(GB36600-2018). Therefore, the operation period of this project has little impact on the regional soil environment.

29.4 Analysis of the Influence of Vertical Infiltration on Soil Environment

29.4.1 Scenario Setting and Evaluation Factors

Combining the project type, pollution source and pollution path, the following two prediction scenarios are set.

(1) Under normal conditions: all areas of the project plant are hardened, all production equipment and structures take impermeable measures, and the plant takes rainwater and sewage diversion measures. The possibility of leakage of pollutants is very small. All kinds of raw materials and products are in the equipment and pipes, and the sewage is in the pipes and reinforced concrete ponds. Under normal conditions, there is no leakage of pollutants into the ground.
(2) Under abnormal conditions: According to the analysis of the actual situation of this project, if the production workshop or tank area impermeable ground and other visible places are damaged, it is easy to find and take corresponding repair measures in time, even if there is leakage of materials or sewage, the construction unit will take timely measures and will not let the materials or sewage roam and seep into the soil. In the cesspool, sewage pipeline, and other non-visible parts of these small area leakages, only a small amount of material or sewage may gradually seep into the soil through the leakage point. Therefore, this time, the breach of the impermeable layer of the regulating pond of the leachate sewage treatment station is used as the prediction scenario.

According to the above impact identification results, considering the correlation with the impact of the groundwater environment, and the reasons that heavy metals have more impact on the soil environment and are more difficult to dissipate, referring to the inlet concentration of wastewater from the leachate treatment station system of other projects, Hydrargyrum, Lead, and Cadmium are not detected among heavy metals, Chromium, Zinc and Nickel are detected, and petroleum hydrocarbon (C10–C40) is also considered. Therefore, petroleum hydrocarbons (C10–C40), Chromium, Zinc, and Nickel were selected as key predictors in this study (Table 29.6).

29.4.2 Prediction Model

According to the hydrogeological conditions of the project site and the results of soil sampling, the lithology of the vadose zone in this site is sandy soil, which is 70 ~

Table 29.6 Table of predicted source intensity of vertical soil infiltration

Leakage point	Pollution factor	Concentration (mg/L)	Leakage characteristics	Note
Conditioning tank	Petroleum	0.35	Continuous leakage for 100d, vertical infiltration	Abnormal working condition
	Chromium	3.89	Continuous leakage for 100d, vertical infiltration	Abnormal working condition
	Zinc	3.76	Continuous leakage for 100d, vertical infiltration	Abnormal working condition
	Nickel	0.72	Continuous leakage for 100d, vertical infiltration	Abnormal working condition

100 m according to the regional data and the well-drilling records in the plant area. The thickness of this predicted unsaturated zone is set to 1 m, the upper boundary of the model is set to variable flux head boundary, the lower boundary is set to free drainage boundary, and the upper boundary of the soil solute transport model is fixed concentration flux boundary, the lower boundary is free infiltration boundary.

Taking the surface as the zero datum, the direction of the coordinate axis is the same as the direction of the main permeability coefficient, and the coordinate axis is positive upward, the seepage area can be expressed as $Z \leq z \leq \ \leq 0$, where $Z =$ – 100 cm. The simulation time is 100d, that is $0 \leq t \leq T$, $T = 100$d. The control equations and boundaries are as follows.

One-dimensional Unsaturated Water Transport Control Equation

Based on soil contamination identification and construction engineering analysis, the one-dimensional unsaturated solute vertical transport control equation is determined as follows:

$$\frac{\partial(\theta c)}{\partial t} = \frac{\partial}{\partial z}\left(\theta D \frac{\partial c}{\partial z}\right) - \frac{\partial}{\partial z}(qc)$$

In the formula

c Concentration in pollutant medium, mg/L.
D Diffusion coefficient, m^2/d.
q Seepage velocity, m/d.
z Distance along the z-axis, m
T Time variable, d
Θ Soil moisture content, %

(1) Initial conditions:

$$c(z,t) = 0 \quad t = 0 \quad L \leq Z < 0$$

(2) Boundary conditions:

$$-k(h)\left(\frac{\partial h}{\partial z} + 1\right) = q_s$$

Dirichlet boundary condition:

Point Source Pollution Modelling:

$$c(z,t) = 0 \quad t > 0 \quad z = 0$$

Non-point Source Pollution Modelling:

$$c(z,t) = \begin{cases} c_0 \; 0 < t \leq t_0 \\ 0 \quad t > t_0 \end{cases}$$

Neumann boundary conditions:

$$-\theta D \frac{\partial c}{\partial z} = 0t > 0, \quad z = L$$

Model Parameter Setting

In this evaluation, HYDRUS software is used to solve the water and solute transport equation in the unsaturated zone.

The evaluation model selects ground to 1 m underground as the evaluation depth and sandy loam 0 to 1 m. the profile nodes are 101. Nine observation points are arranged in the predicted target layer, and the distances from the top of the model are 10, 20, 30, 40, 50, 60, 70, 80, and 90 cm respectively. five prediction time points are taken, in order of 20, 40, 60, 80, and 100d.

The upper boundary of the solute transport model is the concentration flux boundary, and the lower boundary is the zero concentration gradient boundary. The role of pollutants' own degradation and retention is not considered in this model. The values of soil hydraulic parameters for the sandy loam layer are shown in Table 29.7.

Table 29.7 Soil hydraulic parameters

Soil layers cm	Soil type	Residual water content θr/cm^3 cm^{-3}	Saturated water content θs/cm^3 cm^{-3}	Empirical parameter sα/cm^{-1}	Curve parameters	osmotic coefficients (cm/day)	Empirical parameters
0 ~ 100	Sandy loam	0.065	0.41	0.075	1.89	106.1	0.5

29.4.3 *Predicted Results*

According to the prediction results, the leachate vertically penetrates into the soil environment in the form of a point source during the operation period. The observation points are N1: 10 cm, N2: 20 cm, N3: 30 cm, N4: 40 cm, N5: 50 cm, N6: 60 cm, N7: 70 cm, N8: 80 cm and N9: 90 cm, and the prediction time is 100 days.

It can be seen from the prediction results that the concentration of pollutants at the N1 observation point (10 cm from the surface) and the N2 observation point (20 cm from the surface) will gradually increase with time. N3 ~ N9 did not continue to infiltrate after being adsorbed. At different times, the pollutants continuously migrate downward with time, and the concentration of pollutants decreases rapidly under the adsorption of soil, and the maximum impact depth of various pollutants is about 0.35 m. Therefore, after 100 days of continuous leakage from the regulating pond of the sewage treatment station, the main pollution zone is 0 to 0.4 m of soil, and the soil below 1 m has no obvious influence.

The volume concentrations of petroleum hydrocarbons (C10–C40), chromium, zinc, and nickel in each observation point gradually increased with time. The maximum volume concentrations of petroleum hydrocarbons (C10–C40), chromium, zinc and nickel were 0.08, 0.86, 0.82 and 0.16 mg/cm^3 respectively, and the soil bulk density was 1,465 kg/cm^3. The converted soil mass concentrations are 0.0546, 0.000587, 0.00056 and 0.00011 mg/kg, respectively. The present soil pollutants of petroleum hydrocarbon (C10–C40), chromium, zinc, and nickel in construction land are 238 mg/kg, undetected, 89, and 43 mg/kg. The content of petroleum hydrocarbon (C10–C40), chromium, zinc, and nickel in the soil after superposition of present values are 238.08, 0.86, 89.82, and 43.16 mg/kg respectively, were lower than the second-class land screening values in the Soil environmental quality risk control standard for soil contamination of development land (GB36600-2018) (Standard value: petroleum hydrocarbon (C10–C40) 4500 mg/kg, hexavalent chromium 5.7 mg/kg, zinc no standard, nickel 900 mg/kg).

29.5 Conclusion

According to the prediction results, within 28 years after the project is put into operation, the exhaust pollutants mercury, lead, and cadmium discharged by the project, after the cumulative predicted values in the soil in the total settlement maximum grid are superimposed with the background values, are all lower than the second-class land screening values in the Soil environmental quality risk control standard for soil contamination of development land (GB36600-2018).

The wastewater pollutants discharged from leachates, such as petroleum hydrocarbon (C10-C40), chromium, zinc, and nickel, were vertically infiltrated for 100d days. After superimposing the present soil values, the concentration of the superimposed values was lower than the second-class land screening values in the Soil environmental quality risk control standard for soil contamination of development land (GB36600-2018).

To sum up, the soil environmental impact of the project is acceptable.

References

Cheng HF, Hu Y (2010) Municipal solid waste (MSW) as a renewable source of energy: current and future practices in China [J/OL]. Bioresour Technology. 101(11):3816–3824

Chen L, Liao Y, Ma X (2021) Economic analysis on sewage sludge drying and its co-combustion in municipal solid waste power plant [J/OL]. Waste Manage 121:11–22

Fang ZD, Li KX (2019) Progress of domestic waste incineration power generation technology at home and abroad [J]. Power Gener Technol 40(4):367–376

Ferdan T, Pavlas M et al (2019) Greenhouse gas emissions from thermal treatment of non-recyclable municipal waste [J/OL]. Front Chem Sci Eng 12(4):815–831

Gao M, Yin Y (2019) Research on environmental benefits of waste incineration power generation project [J/OL]. IOP Conf Ser Earth Environ Sci 300(3):032031

Kaza S, Yao L et al (2018) What a waste 2.0: a global snapshot of solid waste management to 2050 [EB/OL]. In: Urban development series. Washington, DC: World Bank. https://doi.org/10.1596/978-1-4648-1329-0

Liu JW, Lei TZ, Yang SH et al (2012) Briefly discussed the present situation and development trend of waste incineration power generation in China [J]. Sino-Glob Energy 17(6):29–30

Nie YF (2018) Development and prospects of municipal solid waste (MSW) incineration in China [J/OL]. Front Environ Sci Eng China 2(1):1–7

Su Y, Zhang Z, Wu D et al (2019) Occurrence of microplastics in landfill systems and their fate with landfill age [J/OL]. Water Res 164:114968

Tian H, Gao J, Hao J, Lu L, Zhu C, Qiu P (2013) Atmospheric pollution problems and control proposals associated with solid waste management in china: a review [J/OL]. J Hazard Mater 252–253:142–154

Zhang F (2020) Study on domestic waste incineration power generation technology and waste gas pollution prevention and control countermeasures [J]. Environ Dev 32(9):41

Zhang ZW, Yin ZH (2020) Domestic and foreign domestic waste incineration power generation technology progress [J]. Mod Rural Sci Technol 3:95

Chapter 30
Spatial and Temporal Variations of Groundwater Quality in Guangrao City

Weiyan Pan, Xiuhong Qian, Shanshan Yang, Xiuping Tian, Riqing Zhu, Aipeng Fan, Hongfei Liu, and Zhen Jiao

Abstract This study takes Guangrao section of Xiaoqing River Basin as the research object, analyzes the annual and inter-annual variation of groundwater level, and carries out descriptive statistical analysis, hydrochemical analysis, and multivariate statistical analysis of groundwater quality according to the interpolation analysis of the spatial distribution map of groundwater level. The results show that the groundwater level shows an increasing trend between years and a certain periodicity within a year, with a low level in the south and a high level in the north. The main pollutants of groundwater quality are chloride, sulfate and nitrate, and the chemical properties of groundwater have a tendency to transform to weak acid water. The groundwater in Guangrao County is greatly affected by agricultural activities and human domestic sewage discharge. The government should strengthen the management and control of industry and agriculture, rationally discharge sewage and exploit groundwater, which is of great significance to the water environment protection of Xiaoqing River basin.

Keywords Guangrao country · Groundwater quality · Groundwater level · Spatial and temporal variations

30.1 Introduction

As an important water supply source, groundwater is of great significance to ensure people's drinking water safety. In recent years, the water quality of the main stream of

W. Pan (✉) · H. Liu · Z. Jiao
School of Water Conservancy and Environment, University of Jinan, Jinan 250022, Shandong, China
e-mail: weiyanpan@126.com

X. Qian · S. Yang · X. Tian · R. Zhu · A. Fan
Shandong Water Conservancy Management and Service Center of Haihe-Huaihe-Xiaoqing River Basin, Jinan 250014, China

J. Zhang et al. (eds.), *Environmental Pollution Governance and Ecological Remediation Technology*, Environmental Science and Engineering,
https://doi.org/10.1007/978-3-031-25284-6_30

Xiaoqing River has been significantly affected by the discharge of pollution industries such as Qilu Petrochemical and Guangrao oil refinery. Polluted river water can enter the groundwater through surface infiltration, agricultural irrigation and other ways, which may cause the risk of groundwater pollution. However, once the groundwater is polluted, it will be difficult to repair, and the ecological damage and human health will be more far-reaching. Groundwater environmental problems have attracted the attention of many scholars. Liu et al. (2017) analyzed the testing data of shallow groundwater quality in Guangrao County and found that due to the formation of shallow groundwater falling funnel, mineralization degree and chloride content generally have an increasing trend. Zhang and Wang (2022) analyzed the groundwater level in Huantai County, and found that the groundwater level was mainly affected by precipitation, mining volume and river recharge. Qin et al. (2022) carried out hydrochemical analysis on groundwater in the south of Zaozhuang City, and the results showed that the hydrochemical types of groundwater were mainly HCO_3–Ca and $HCO_3 \cdot SO_4$–Ca. Guang and Li (2019) analyzed the hydrochemical characteristics of karst spring in Jinan City and found that the large use of chemical fertilizers and pesticides in agricultural planting areas and the discharge of pollutants from enterprises were the main causes of water environmental problems of karst groundwater in Jinan City.

Groundwater level and water quality are affected by many factors, and the variation of groundwater characteristics is different in different time and space (Karimov et al. 2014; Yu et al. 2014). Taking the Guangrao section of Xiaoqing River basin as the research object, this paper analyzes and discusses the groundwater level and water quality characteristics, and the results are of great significance for improving the ecological environment along the coast, promoting the development of industry and agriculture, and using groundwater scientifically.

30.2 Analysis of Groundwater Dynamic Characteristics

Representative groundwater level observation Wells with continuous monitoring data in the research area of Guangrao County (see Fig. 30.1 for point distribution) were selected, and the collected historical data of groundwater level (2016–2020) were used to analyze the inter-annual and intra-annual variation of groundwater level. Figure 30.2 shows that the groundwater level in Guangrao County has an obvious rising trend after 2018. On the one hand, it is related to the heavy rainfall in that year, and on the other hand, the reduction of mining amount and irrigation by Yellow River diversion will also cause the groundwater level to rise. From the perspective of annual variation, the water level changes periodically, and the dynamic type belongs to the precipitation infiltration—runoff—mining type. The change of groundwater level is affected by atmospheric precipitation and artificial mining, which shows a downward trend and then an upward trend. The water level drops rapidly after spring irrigation in March every year, and reaches the lowest in May and June before the flood season. After that, the water level rose rapidly due to the increase of precipitation infiltration. After the autumn irrigation, the water level fell again weakly and then remained stable.

The groundwater depth of Guangrao County in 1980, 2001 and 2021 was analyzed by interpolation, as shown in Fig. 30.3. The results show that, in 1980, there was a great difference between the north and the south in the study area of Guangrao County, basically showing that the buried depth gradually increased from north to south, and the groundwater buried depth in the southwest was the largest, exceeding 20 m. Guangrao County in 2001, the embedded depth of groundwater spatial difference than twenty years ago, still show the meteorological north low spatial distribution, but can clearly see the blue areas in southwestern color deepened, that the

Fig. 30.1 Historical water level monitoring well distribution

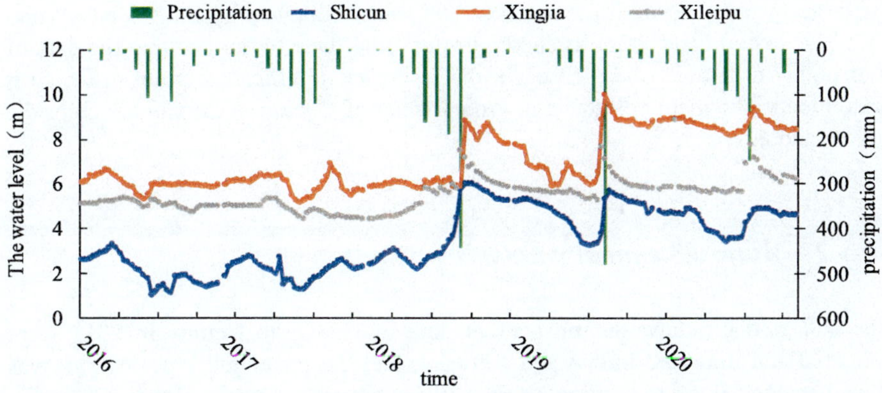

Fig. 30.2 Groundwater level change map of Guangrao County

further increase of groundwater depth, maximum buried depth more than 25 m, it has to do with the industrial farming activities on water demand is bigger. And the government reasonable exploitation of groundwater resources consciousness is not strong. In 2021, with the implementation of policies such as rational development and utilization of groundwater, the situation of groundwater overexploitation was improved. After the Yellow River diversion project was put into use, the demand for groundwater was reduced, the area of funnel area in the study area was significantly reduced, and the depth of groundwater was increased.

30.3 Groundwater Quality Analysis

30.3.1 *Describe the Statistical Analysis*

According to the test results of groundwater samples collected, the groundwater situation in Guangrao County was analyzed in terms of descriptive statistics and hydrochemical types. As shown in Table 30.1, the cation content in groundwater of Gugrao County is ranked as $Na^+ > Ca^{2+} > Mg^{2+} > K^+$, and the anion content is ranked as $HCO_3^- > Cl^- > SO_4^{2-} > NO_3^-$. The major groups in groundwater are HCO_3^-, Cl^-, SO_4^{2-} and Na^+. The coefficients of variation of K^+, TP, TN, NO_3^-, ammonia nitrogen, NO_2^-, SO_4^{2-} and other indicators are relatively large, indicating that their contents in groundwater are quite different. The coefficient of variation of pH, DO and HCO_3^- is small, the content changes little, and the distribution is uniform.

According to the indicators in Groundwater Environmental Quality Standard (GB/T14848-2017), the water quality factors that coincided with the detection indicators were selected, and the exceeding of the water quality indicators of all water samples was calculated according to the limit value of class III water standards. The results are shown in Table 30.2. In the groundwater of Guangrao County, Cl^- exceeded the standard by 60%, followed by F^- and SO_4^{2-}. Ammonia nitrogen with the largest exceeding multiple reaches 3.66 times, followed by SO_4^{2-}, NO_3^- and Cl^-. The spatial distribution is shown in Fig. 30.4. Due to the fact that the data of four points in F^- are one time each, the reference significance is not great, so it is excluded. The main pollutants in groundwater of Guangrao County are chloride, sulfate and nitrate.

30.3.2 *Water Chemical Analysis*

The collected groundwater ion content data of Guangrao County in 2017 were collected and imported into AqQa software, and the generated Piper tri-plot was shown in Fig. 30.5a.According to Fig. 30.5a, groundwater samples in Guangrao

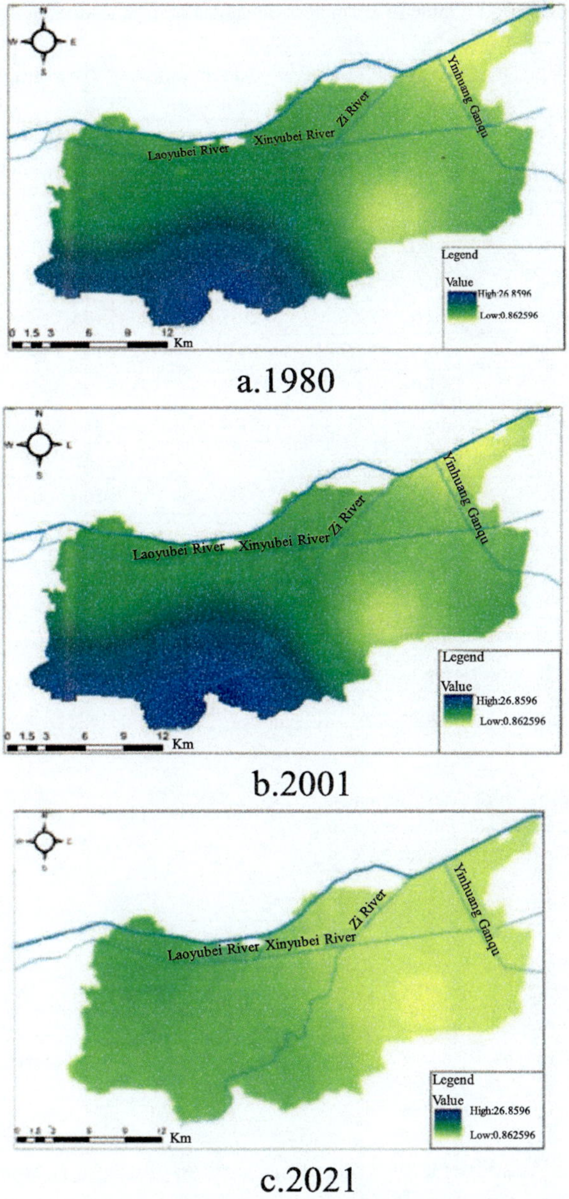

Fig. 30.3 Spatial distribution map of groundwater depth in Guangrao County

County are located above and to the right of the rhomboid area, and the hydrochemical types of groundwater belong to SO_4–Mg water and the combination of SO_4·Cl–Mg and SO_4·Cl–Na water, respectively. The dominant anions in the groundwater of Guangrao County are SO_4^{2-} and Cl^-, and the dominant cations are Na^+ and Mg^{2+}. The groundwater is greatly affected by human activities.

Table 30.1 Table of characteristic values of groundwater hydrochemical parameters in Guangrao County

Indicators	Unit	The maximum	The minimum	The average	The standard deviation	Coefficient of variation
pH	–	7.8	7.1	7.365	0.193	0.026
DO	mg/L	8	1.6	4.985	1.644	0.330
COD	mg/L	55	4	26.640	16.783	0.630
BOD_5	mg/L	5.8	0.5	1.670	1.283	0.768
Ca^{2+}	mg/L	203	35.3	102.250	63.525	0.621
Mg^{2+}	mg/L	97.6	29.1	75.125	27.671	0.368
Na^+	mg/L	316	89.8	194.450	95.944	0.493
K^+	mg/L	83.8	1	22.095	35.627	1.612
Cl^-	mg/L	721	21.4	345.840	232.152	0.671
$SO_4{}^{2-}$	mg/L	822	32	261.630	239.602	0.916
$HCO_3{}^-$	mg/L	835	338	506.500	195.395	0.386
$NO_3{}^-$	mg/L	59.6	0.45	17.123	20.250	1.183
$NO_2{}^-$	mg/L	0.449	0.005	0.109	0.119	1.095
$NH_4{}^+$	mg/L	2.33	0.026	0.614	0.699	1.139
TN	mg/L	86.3	1.03	22.850	26.390	1.155
TP	mg/L	1.94	0.03	0.477	0.568	1.190
F^-	mg/L	1.48	0.424	0.932	0.485	0.521

Table 30.2 Table of main groundwater index exceeding standard in Guangrao County

Indicators	Class III water limit (≤ mg/L)	Over standard rate (%)	Maximum superscalar multiple
Cl^-	250	60	1.88
$SO_4{}^{2-}$	250	35	2.29
$NO_3{}^-$	20.0	30	1.98
$NO_2{}^-$	1.00	0	–
$NH_4{}^+$	0.50	20	3.66
F^-	1.0	50	0.48

The Piper triplot of groundwater in Guangrao County in 2021 (Fig. 30.5b) shows that water samples are distributed in the upper and lower left regions of the rhombus. The main cations are Ca^{2+} and Mg^{2+}, and the main anions are $HCO_3{}^-$, $SO_4{}^{2-}$ and Cl^-.The chemical types of groundwater mainly belong to HCO_3–Ca·Mg and SO_4·Cl–Ca·Mg. Compared with 2017, the concentration of alkali metal ions (Na^+, K^+) in groundwater was gradually exceeded by the concentration of alkaline earth metal ions (Ca^{2+}, Mg^{2+}), indicating that the chemical properties of groundwater tended to transform to weak acid water.

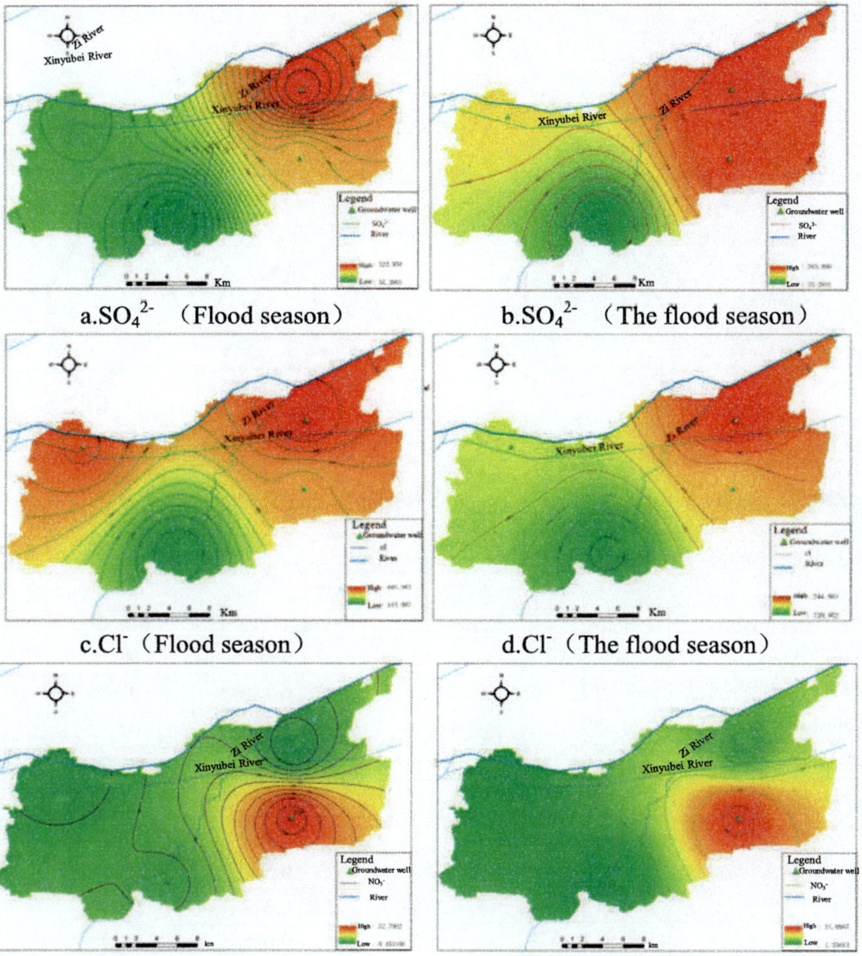

a.SO_4^{2-} (Flood season) b.SO_4^{2-} (The flood season)

c.Cl^- (Flood season) d.Cl^- (The flood season)

e.NO_3^- (Flood season) f.NO_3^- (The flood season)

Fig. 30.4 Spatial variation of typical groundwater pollutants in Guangrao County

30.3.3 Multivariate Statistical Analysis

Principal component analysis was conducted on the main pollutants in the groundwater of Gugrao County, and 12 water quality indicators (pH, salinity, DO, COD, BOD_5, ammonia nitrogen, NO_3^-, NO_2^-, TP, TN, Cl^- and SO_4^{2-}) were selected. According to the results, four principal components were extracted, as shown in Table 30.3.The cumulative variance contribution rate of principal component 1, principal component 2, principal component 3 and principal component 4 was 79.617%.

Principal component 1 has strong or moderate loading on NO_3^-, TN, NO_2^-, TP, ammonia nitrogen, Cl^-, salinity and pH. The principal component explained

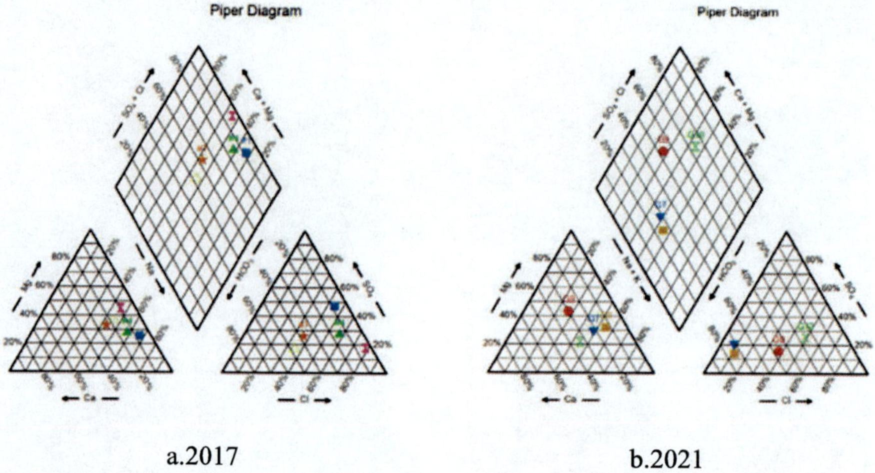

Fig. 30.5 Piper three-line map of groundwater in Guangrao County

Table 30.3 Principal component analysis table of groundwater quality parameters in Guangrao County

The water quality parameters	Principal component 1	Principal component 2	Principal component 3	Principal component 4
pH	− 0.772	− 0.112	− 0.130	0.340
Salinity	0.689	0.689	0.168	0.039
DO	− 0.300	0.119	0.583	0.644
COD	0.481	0.405	− 0.663	0.187
BOD_5	− 0.253	0.131	− 0.084	0.877
NH_4^+	0.634	− 0.454	− 0.521	0.072
NO_3^-	0.795	− 0.374	0.332	0.230
NO_2^-	0.713	− 0.340	− 0.058	0.297
TP	0.710	− 0.578	− 0.281	0.101
TN	0.764	− 0.353	0.362	0.220
Cl^-	0.712	0.617	0.185	0.022
SO_4^{2-}	0.592	0.734	− 0.187	0.059
Initial eigenvalue	5.164	2.537	1.815	1.678
The total variation (%)	39.724	19.518	13.960	12.905
Total variation (%)	39.724	59.242	73.202	86.107

39.724% of the total variance, and contributed significantly to groundwater ion concentration and nitrogen, indicating that groundwater in Guangrao County was greatly affected by agricultural activities and human domestic sewage discharge.

30.4 Conclusion

This paper discusses the dynamic characteristics of groundwater level and the analysis of water quality in the GuangRao section of Xiaoqing river basin. The results show that the groundwater level is increasing year by year, and the groundwater quality is seriously polluted, which is greatly affected by human activities. It is suggested to control the discharge of industrial wastewater, rationally control the application of nitrogen fertilizer in agriculture, and take a series of measures to protect the groundwater environment.

Acknowledgements This work was funded by the Scientific Research and Experimental Project of Xiaoqing River Comprehensive Flood control Project (XQHFHZL-KY202003) and the Science and Technology Program of University of Jinan (No. XKY1810).

References

Guang QH, Li FL (2019) Groundwater chemical characteristics and water environment evolution in karst spring region of Jinan City. China Karst 38(5):653–662

Karimov AK, Šimůnek J, Hanjra MA et al (2014) Effects of the shallow water table on water use of winter wheat and ecosystem health: implications for unlocking the potential of groundwater in the Fergana Valley (Central Asia). Agric Water Manag 131:57–69

Liu ZT, Sha FJ, Jiang SJ (2017) Study on hydrochemical evolution of shallow groundwater in Guangrao County. Shandong Land Resour 33(03):48–51

Qin Y, Tang XH, Li YL et al (2022) Hydrochemical characteristics and main controlling factors of groundwater in southern Zaozhuang City. Saf Environ Eng 1671:1–8

Yu Y, Song X, Zhang Y et al (2014) Identifying spatio-temporal variation and controlling factors of chemistry in groundwater and river water recharged by reclaimed water at Huai River, North China. Stoch Environ Res Risk Assess 28(5):1135–1145

Zhang YX, Wang XJ (2022) Analysis of dynamic variation characteristics and influencing factors of groundwater level in Huantai County. China Rural Water Hydropower. 7:110–115

Chapter 31
The Temporal and Spatial Variation Process of the Range Length of the pH Values in the Jiaozhou Bay

Dongfang Yang, Xiaoju Zhang, and Danfeng Yang

Abstract According to the research materials of the Jiaozhou Bay waters in May and October 1979, the range of pH variation, seasonal distribution, and length of range in the surface and bottom waters were studied. The surface waters and bottom waters of Jiaozhou Bay are composed of the waters of the inner side of bay mouth + bay mouth + outer side of bay mouth. The results showed that in May and October, the pH values in the surface and bottom waters with respect to seasonal changes from low to high are spring and autumn. When the pH in the surface waters was low, the pH in the bottom waters would also low and when the pH in the surface waters was high, the pH in the bottom waters would also high. In spring, there were only a few sea plants grown in the surface and bottom waters, while in autumn, there were many. Hence from spring to autumn, the growth of the sea plants changed from small amount to large amount in the surface and bottom waters of the whole waters from the inner side to the outer side of the bay mouth. According to the variation features of the pH in the surface and bottom waters, the author thought in the three areas of the surface waters, the bottom waters and the whole waters, if the pH value in one of the areas was known, the pH values in other two areas could be known; if the growth condition of sea plants in one of the areas was known, the growth condition in other two areas could be known. According to the range length of the temporal and spatial pH variation, the results could be obtained: (1) In May, the stability of the pH variation in the bottom waters was more conducive to the growth of marine organisms than in the surface waters. (2) In October, compared with the bottom waters, the pH variations in the surface waters were relatively stable, which were conducive to the growth of marine organisms. (3) In October, the stability of the pH variation in the surface waters was more conducive to the growth of marine organisms than in May.

D. Yang (✉) · X. Zhang
Xi'an Traffic Engineering Institute, Xi'an 710300, China
e-mail: dfyang_dfyang@126.com

D. Yang
North China Sea Environmental Monitoring Center, SOA, Qingdao 266033, China

D. Yang
The Fu Foundation School of Engineering and Applied Science, Columbia University, New York 10025, USA

J. Zhang et al. (eds.), *Environmental Pollution Governance and Ecological Remediation Technology*, Environmental Science and Engineering,
https://doi.org/10.1007/978-3-031-25284-6_31

(4) In October, compared with in May, the stability of the pH variation in the bottom waters was conducive to the growth of marine organisms. (5) In October, the stability of the pH variation in the whole waters was more conducive to the growth of marine organisms than in May. (6) In the bottom waters, the stabilities of the pH variation from May to October were more conducive to the growth of marine organisms than in surface waters. (7) The author validated the best time period and space area for the growth of sea plants in the Jiaozhou Bay waters: In October, the pH variation in the surface waters ranging from 8.28 to 8.34, and the range length of the sea pH variation was 0.06. The range and the range length of pH variation were the best condition in Jiaozhou Bay for the growth of sea plants. Hence with respect to the temporal and spatial variation, only divided time into different periods and divided space into different areas, can the best combination of time periods and spatial areas be confirmed in the combination of different time periods and spatial areas to let the pH and the range length of the pH be the most conducive to the growth of marine organisms.

Keywords pH value · The range length · The temporal and spatial variation process · The growth of sea plants · The Jiaozhou Bay

31.1 Introduction

pH is a sign for how acid or alkaline the seawater is, and the pH variation of seawater determines the ecological change of marine organisms. When the pH of seawater is higher than 8.1, the seawater is weakly alkaline, which is conducive to the growth of large amount of sea plants (Yang et al. 2014a, b, 2017, 2018a, b, c, d, e, 2019a; Yang and Gao 2009). Thus, the study of the pH variation, seasonal distribution and range length in the surface and bottom waters presented the effect of the pH variation on the growth of sea plants in different areas of waters and different time periods (Yang et al. 2014a, b, 2017, 2018a, b, c, d, e, 2019a, 2021; Yang and Gao 2009). According to the research materials in 1979, the range of the pH variation, the seasonal distribution and the range length in the surface and bottom waters of Jiaozhou Bay was confirmed; the temporal and spatial change process of the pH range length in the Jiaozhou Bay waters and its effect on the growth of sea plants were obtained. These results provide scientific theoretical basis for the studies of the effects of the temporal and spatial variation process of the pH intervel length in the Jiaozhou Bay waters on the growth of sea plants.

31.2 The Materials, Approaches and the Investigated Waters

31.2.1 The Natural Environment of Jiaozhou Bay

The location of Jiaozhou Bay is in the southern Shandong peninsula and its coordinates are between 120° 04′–120° 23′ E and 35° 58′–36° 18′ N. Its boundary connected the line between Tuandao Cape and Xuejia Island, and is near to the Yellow Sea. With an area of about 446 km^2 and an average water depth of about 7 m, it is a typical semi-closed bay. There are more than a dozen rivers entering the sea in Jiaozhou Bay, of which Dagu River and Yang River have large runoff and sediment content, and rivers such as Haibo River, Licun River and Loushan River around Jiaozhou bay presented all seasonal changes and hydrological characteristics (Yang et al. 2019b; Dongfang et al. 2019).

31.2.2 Materials and Approaches

The North Sea Monitoring Center of the State Oceanic Administration offered the dataset on Jiaozhou Bay in May and October 1979, resulting in the research results of the pH values in this paper. In May and October, eight stations were set in the Jiaozhou Bay waters for water samples: H34, H35, H36, H37, H38, H39, H40, H41 (Fig. 31.1).Three times of samples taking were respectively taken place in May and October 1979 according to the depth of the waters (when > 10 m, the surface and bottom waters were obtained; when < 10 m, only surface waters) to do the research. The research of the pH values of the Jiaozhou Bay waters was conducted through the national standards, and this way for Marine Monitoring was collected in the national Specification (State Oceanic Administration 1991).

31.3 Results

31.3.1 The Surface and Bottom Waters

The surface and bottom waters of Jiaozhou Bay are composed of the inner side of the bay mouth + the bottom water of the bay mouth + the outer side of the bay mouth.

In May, the pH values of the surface waters in Jiaozhou Bay waters from the inner side of the bay mouth including the bay mouth to the outer side of the bay mouth were 7.55–7.86, and the pH values of its bottom waters were 7.88–7.98. These results suggested that in the whole waters from the inner side to the outer side of the bay

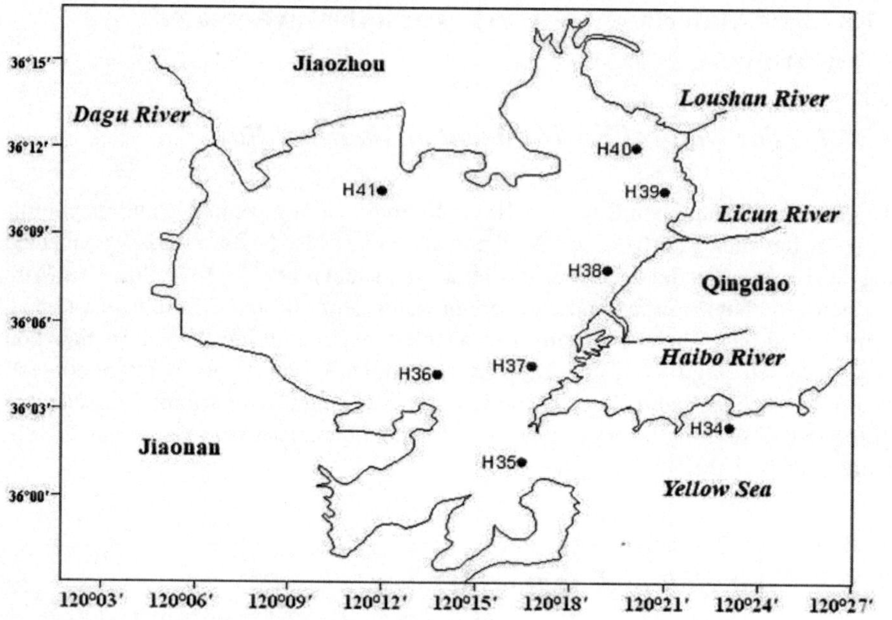

Fig. 31.1 The research stations in Jiaozhou Bay

mouth and from the surface waters to the bottom waters, the pH ranged from 7.55 to 7.98 and the range length of the pH variations of the seawater was 0.43.

In October, the pH values of the surface waters in Jiaozhou Bay waters from the inner side of the bay mouth including the bay mouth to the outer side of the bay mouth were 8.28–8.34, and the pH values of its bottom waters were 8.25–8.33. These results suggested that in the whole waters from the inner side to the outer side of the bay mouth and from the surface waters to the bottom waters, the pH changed from 8.25 to 8.34 and the range length of the pH variations of the seawater was 0.09.

31.3.2 The Seasonal Distribution of the pH in the Surface Waters

In May, the pH values of the surface waters in Jiaozhou Bay from the inner side to the outer side of the bay mouth ranged from 7.55 to 7.86; in October, changed from 8.28 to 8.34. These results suggested that in May and October, the pH values of the surface waters ranged from 7.55 to 8.34 which was a relatively small change, and the range length of the pH variation of the surface waters was 0.79. The pH values of the surface waters ranking from low to high were May and October. Hence the seasonal changes of the pH values in the surface waters ranking from low to high were: spring and autumn.

31.3.3 The Seasonal Distribution of the pH in the Bottom Waters

In May, the pH values of the bottom waters in Jiaozhou Bay from the inner side to the outer side of the bay mouth ranged from 7.88 to 7.98; in October, ranged from 8.25 to 8.33. These results suggested that in May and October, the pH values of the bottom waters ranged from 7.88 to 8.33 which was also a relatively small change, and the range length of the pH variation of the bottom waters was 0.45. The pH values of the bottom waters ranking from low to high were May and October. Hence the seasonal changes of the pH values in the bottom waters ranking from low to high were: spring and autumn.

31.3.4 The Variation Range in the Surface and the Bottom Waters

In May, the pH values of the surface waters in Jiaozhou Bay from the inner side to the outer side of the bay mouth were 7.55–7.86 which were relatively low, and the pH values in the bottom waters were 7.88–7.98 which were also low. In October, the pH values of the surface waters were 8.28–8.34 which were relatively high, and the pH values in the bottom waters were 8.25–8.33 which were also high. In this case, the variations were basically the same. Moreover, the pH values of the surface waters changed from 7.55 to 8.34 which were higher than that of the bottom waters which were 7.88–8.33. The range length of pH variations in the surface waters was 0.79, which was higher than that of pH variations in the bottom waters which was 0.45.

Therefore, if the pH values in the surface waters were relatively low, then the pH values in the bottom waters were also low; if the pH values in the surface waters were relatively high, then the pH values in the bottom waters were also high.

31.3.5 Spatial and Temporal Changes of the Range Length

In Jiaozhou Bay waters from the inner side of the bay mouth including the bay mouth to the outer side of the bay mouth.

In May, the pH values of the surface waters were 7.55–7.86, and the pH values of its bottom waters were 7.88–7.98. The range length of the pH variations of the surface waters was 0.31, and the range length of the pH variations of the bottom waters was 0.10. From the surface waters to the bottom waters and from the inner side to the outer side of the bay mouth, the pH in the whole waters changed from 7.55 to 7.98, and the range length of the pH variation was 0.43 (Table 31.1).

In October, the pH values of the surface waters were 8.28–8.34, and the pH values of its bottom waters were 8.25–8.33. The range length of the pH variations of the

Table 31.1 The range length of the pH variation in the waters

The range length of the pH variation in the waters	The surface waters	The bottom waters	The whole waters
May	0.31	0.10	0.43
October	0.06	0.08	0.09
May–October	0.79	0.45	0.79

surface waters was 0.06, and the range length of the pH variations of the bottom waters was 0.08. From the surface waters to the bottom waters and from the inner side to the outer side of the bay mouth, the pH in the whole waters changed from 8.25 to 8.34, and the range length of the pH variation was 0.09 (Table 31.1).

In the surface waters, the pH values from May to October were 7.55–8.34, and the range length of the pH variation was 0.79; in the bottom waters, the pH values from May to October were 7.88–8.33, and the range length of the pH variation was 0.45; in the whole waters, the pH values from May to October were 7.55–8.34, and the range length of the pH variation was 0.79 (Table 31.1).

31.4 Discussion

31.4.1 The Growth of the Sea Plants in the Whole Jiaozhou Bay

In Jiaozhou Bay waters from the inner side of the bay mouth including the bay mouth to the outer side of the bay mouth, and from the surface waters to the bottom waters, the pH values of the whole waters changed from 7.55 to 7.98 in May. A pH value of near 7.5 was considered a threshold in seawater, because generally, the content of free carbon dioxide in seawater cannot be increased any more. So in this case, the content of free carbon dioxide reached to a high value to the threshold, and the same as the acidity of the surface waters. Then in May, the content of free carbon dioxide in the whole waters was high, the acidity in the surface waters was also relatively high, and the surface waters extremely lacked oxygen. These results suggested that the pH of the whole waters was low, and then there had only a few sea plants grown in the waters.

From the inner side of the bay mouth including the bay mouth to the outer side of the bay mouth, and from the surface waters to the bottom waters, the pH values of the whole waters changed from 8.25 to 8.34 in October. When the pH value of was near 8.1, the seawater was conducive to the growth of sea plants, and sea plants grew better in the weak alkaline seawater. Therefor, in October, a large amount of sea plants flourished in the whole surface waters and the photosynthesis of marine plants produced a large amount of oxygen, which reduces the amount of free carbon dioxide in the surface waters. These results suggested that the pH was high in the whole waters, and a large amount of sea plants flourished in this condition.

31.4.2 Changes of the Growth of Sea Plants According to the Seasonal Distribution of pH Values

In Jiaozhou Bay, April, May and June were spring, and October, November and December were autumn.

In the surface waters from the inner side of the bay mouth to the outer side of the bay mouth, the pH ranged from 7.55 to 7.86 in spring. Thus only a few sea plants grew in the surface waters. In autumn, the pH ranged from 8.28 to 8.34, and thus a lot of sea plants grew in the surface waters. From spring to autumn, sea plants in the surface waters changed from little amount to large amount.

In the bottom waters from the inner side of the bay mouth to the outer side of the bay mouth, the pH ranged from 7.88 to 7.98 in spring. Thus only a few sea plants grew in the bottom waters. In autumn, the pH ranged from 8.25 to 8.33, and thus a lot of sea plants grew in the bottom waters. From spring to autumn, sea plants in the bottom waters changed from little amount to large amount.

Therefore, from spring to autumn, sea plants in the whole waters from the inner side to the outer side of the bay mouth changed from little amount to large amount.

31.4.3 Changes of the Growth of Sea Plants According to the Vertical Distribution of pH Values

From the inner side to the outer side of the Jiaozhou Bay waters.

In May, when the pH in the surface waters was relatively low, then the pH in bottom waters was low too. So when the surface waters with low pH was not conducive to the growth of sea plants, the same as the bottom waters because of the low pH. In this case, only a few sea plants grew in the whole waters.

In October, when the pH in the surface waters was relatively high, then the pH in bottom waters was high too. So when the surface waters with high pH was conducive to the growth of sea plants, the same as the bottom waters because of the high pH. In this case, a lot of sea plants grew in the whole waters.

Therefore, in the three areas of surface waters, bottom waters and the whole waters, if the pH of one of the areas was known, then the pH of the other two areas could be known; if the growth condition of sea plants in one of the areas was known, the growth condition in other two areas could be known.

31.4.4 The Stability of the Variation of the Range Length of pH Values

The smaller the range length of the pH variation was, the stabler the pH of the waters was, and then the waters were more conducive to the growth of sea plants. When the pH value of seawater was about 8.1, it was conducive to the growth of marine organisms and of course the growth of sea plants.

In May, from the surface waters to the bottom waters, the range length of the pH variation of the seawater largely decreased from 0.31 to 0.10. The smaller the range length of the pH variation was, the stabler the pH of the waters was. Hence in May, the stability of the pH variation in the bottom waters was more conducive to the growth of sea plants than in the surface waters.

In October, from the surface waters to the bottom waters, the range length of the pH variation of the seawater slightly increased from 0.06 to 0.08. The slightly larger the range length of the pH variation was, the basically stabler the pH of the waters was. Hence in October, the stability of the pH variation in the surface waters was more conducive to the growth of sea plants than in the bottom waters.

From May to October, the range length of the pH variation of the surface waters largely decreased from 0.31 to 0.06. The smaller the range length of the pH variation was, the stabler the pH of the waters was. Hence in October, the stability of the pH variation in the surface waters was more conducive to the growth of sea plants than in May.

From May to October, the range length of the pH variation of the bottom waters slightly decreased from 0.01 to 0.08. The smaller the range length of the pH variation was, the stabler the pH of the waters was. Hence in October, the stability of the pH variation in the bottom waters was more conducive to the growth of sea plants than in May.

From May to October, the range length of the pH variation of the whole waters largely decreased from 0.43 to 0.09. The smaller the range length of the pH variation was, the stabler the pH of the waters was. Hence in October, the stability of the pH variation in the whole waters was more conducive to the growth of sea plants than in May.

From the surface waters to the bottom waters, and from May to October, the range length of the pH variation of the seawater largely decreased from 0.79 to 0.45, and the range length became small. In this case, from May to October, the pH of the seawater was stabler in the bottom waters than in the surface waters. Therefore, from May to October, the stability of the pH variation in the bottom waters was more conducive to the growth of sea plants than in the surface waters.

From May to October, in the whole waters of Jiaozhou Bay from surface to bottom and from the inner side to the outer side of the bay mouth including the bay mouth, the range length of the pH variation was 0.79, which was relatively high. Therefore, the author validatedthe best time period and space range in Jiaozhou Bay for the growth of sea plants: In October, the variation range of the pH was 8.28–8.34, and the range length of the pH variation was 0.06. These were the best variation range and range length of the pH for the growth of sea plants. Hence with respect to the temporal and spatial variation, only divided time into different periods and divided space into different areas, can the best combination of time periods and spatial areas be confirmed in the combination of different time periods and spatial areas to let the pH and the range length of the pH be the most conducive to the growth of marine organisms.

31.5 Conclusions

The bottom waters of Jiaozhou Bay was composed of the bottom waters of inner side of the bay mouth + the bay mouth + the outer side of the bay mouth. In May, the pH values of the whole waters changed from 7.55 to 7.98, and the range length of the pH variation was 0.43; in October, the pH changed from 8.25 to 8.34, and the range length of the pH variation was 0.09. In May and October, the seasonal changes of the pH values in the surface and bottom waters ranking from low to high were: spring and autumn. When the pH in the surface waters was low, the pH in the bottom waters was low too; when the pH in the surface waters was high, the pH in the bottom waters was high too. From May to October, the temporal and spatial changes of the range length of the pH values in the whole waters from surface waters to bottom waters and from inner side to the outer side of the bay mouth were obtained by calculations.

In May, the pH values of the whole waters were relatively low, and only a few sea plants grew. In October, the pH values of the whole waters were relatively high, and a lot of sea plants grew.

In spring, there were only a few sea plants grown in the surface and bottom waters. In autumn, there were a lot of sea plants grown in the surface and bottom waters. Therefore, from spring to autumn, sea plants in the surface and bottom waters from the inner side to the outer side of the bay mouth changed from little amount to large amount.

In the three areas of surface waters, bottom waters and the whole waters, if the pH of one of the areas was known, then the pH of the other two areas could be known; if the growth condition of sea plants in one of the areas was known, the growth condition in other two areas could be known.

According to the range length of the temporal and spatial pH variation, the results could be obtained: (1) In May, the stability of the pH variation in the bottom waters was more conducive to the growth of marine organisms than in the surface waters. (2) In October, compared with the bottom waters, the pH variations in the surface waters were relatively stable, which were conducive to the growth of marine organisms. (3) In October, the stability of the pH variation in the surface waters was more conducive to the growth of marine organisms than in May. (4) In October, compared with in May, the stability of the pH variation in the bottom waters was conducive to the growth of marine organisms. (5) In October, the stability of the pH variation in the whole waters was more conducive to the growth of marine organisms than in May. (6) In the bottom waters, the stabilities of the pH variation from May to October were more conducive to the growth of marine organisms than in surface waters. (7) The author validated the best time period and space area for the growth of sea plants in the Jiaozhou Bay waters: In October, the pH variation in the surface waters ranging from 8.28 to 8.34, and the range length of the sea pH variation was 0.06. The range and the range length of pH variation were the best condition in Jiaozhou Bay for the growth of sea plants. Hence with respect to the temporal and spatial variation, only divided time into different periods and divided space into different areas, can the best combination of time periods and spatial areas be confirmed in the combination of different time periods and spatial areas to let the pH and the range length of the pH be the most conducive to the growth of marine organisms.

References

Dongfang Y, Lin D, Wang Q, Zhang L, Li H (2019) The rising and decreasing heat of waters. Earth Environ Sci 332(022034):1–6

State Oceanic Administration (1991) The specification for marine monitoring. China Ocean Press, Beijing

Yang D, Wang F, Zhu S, Wu Y, He H (2014a) The changing of atmospheric carbon determined by Silicon. Appl Mech Mater 687–691:4355–4357

Yang D, Wang F, He H, Wu Y, Zhu S (2014b) The cooling of the earth determined by orogenesis. Appl Mech Mater 675–677:1897–1900

Yang D, Yang D, Zhang Y, Qu X, Chen Y (2017) Seasonal variations of Pb sources in Jiaozhou Bay during 1979–1983. Adv Mater Sci Energy Technol Environ Eng 2017:295–298

Yang D, Miao Z, Zhu S, Fengyou W, Xiuqin Y (2018a) Examination of Yang's marine water filter phenomenon. Earth Environ Sci 153(062064):1–7

Yang D, Suo H, Zhu S, Wang M, Fan B (2018b) Source input and storage of petroleum hydrocarbon in Jiaozhou Bay. Earth Environ Sci 382(052043):1–6

Yang D, Zhu S, Fan B, Su C, Yang X (2018c) Horizontal loss rate modeling on matter content in marine bay with and without source input. IEEE 00091:410–412

Yang D, Zhu S, Yang D, Wu Y, Wang M (2018d) Yang's dynamic structural model for substance's content in Marine Bay. IEEE 00093:417–419

Yang D, Li H, Lin D, Zhang Y, Wang Q (2018e) Influence of source input on spatial-temporal variations of PHC in Jiaozhou Bay. Adv Comput Signals Syst 8:167–170

Yang D, Su C, Zhu S, Wang X, Wu Y (2019a) Vertical settlement results in homogeneous of Hg between surface and bottom waters in Jiaozhou Bay. Earth Environ Sci 218(012150):1–5

Yang D, Huang C, Wang X, Su C, Zhu S (2019b) Influence of water exchange on Pb contents in bottom waters in Jiaozhou Bay 1990. IEEE 00047:178–181

Yang D, Gao Z (2009) The application and study of the mathematical models in ecology (4). China Ocean Press, Beijing, pp 1–310

Yang D, Yang D, Li H, Lin D, Wang Q (2021) The stability of pH value in the waters of Jiaozhou Bay. Earth Environ Sci 829(012021):1–6

Part II
Environmental Pollution Control and Technology Research

Chapter 32
Research Progress of Partial Denitrification Coupled with Anammox: A Review

Zixuan Zhang, JunFeng Wan, Kaige Zhao, and Mingyuan Liu

Abstract Anaerobic ammonium oxidation (anammox) has been extensively investigated for cost-efficient nitrogen removal from wastewater. However, the full-scale application of the anammox process is limited because the anammox reaction produces an amount of nitrate, which cannot be removed further and deficient in nitrite supply. Therefore, it is urgent to develop a new process coupled with anammox. Partial denitrification (PD), first proposed by Chinese researchers, that is, the reduction of NO_3^--N to NO_2^--N is stopped, which is considered as a new method of substrate NO_2^--N for anammox bacteria. This review comprehensively reviews the new important achievements of partial denitrification and anammox and discusses the application of the coupled process of partial denitrification and anammox.

Keywords Anaerobic ammonium oxidation · Partial denitrification · Biological nitrogen removal

32.1 Introduction

Due to the rapid economic development and the increasing public awareness of environmental protection, water pollution has gradually become a hot topic in the academic circles. The transformation of nitrogen in wastewater treatment is almost entirely completed by microbial mediated redox reaction (Fig. 32.1), In the nitrogen cycle structure of sewage, nitrogen exists mainly in the form of NO_2^-, NO_3^- and NH_4^+ (Alloul et al. 2018). Traditional biological nitrogen removal technologies are based on microbial nitrification and denitrification, on which a series of new biological

Z. Zhang · J. Wan (✉) · K. Zhao · M. Liu
School of Ecology and Environment, Zhengzhou University, Zhengzhou 450001, PR China
e-mail: wanjunfeng@zzu.edu.cn

Henan International Joint Laboratory of Environment and Resources, Zhengzhou University, Zhengzhou 450001, PR China

J. Zhang et al. (eds.), *Environmental Pollution Governance and Ecological Remediation Technology*, Environmental Science and Engineering,
https://doi.org/10.1007/978-3-031-25284-6_32

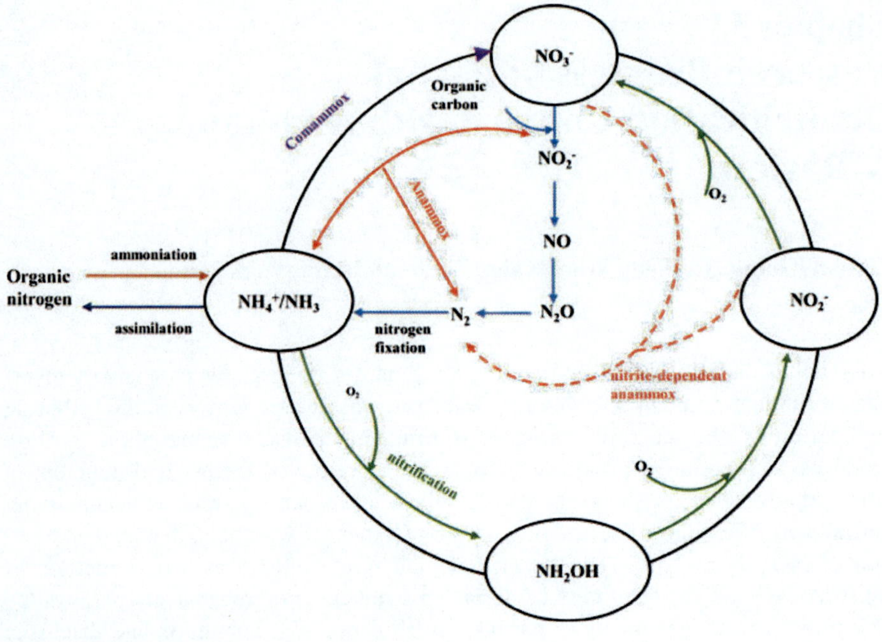

Fig. 32.1 Nitrogen transformation in wastewater

nitrogen removal processes have been developed, such as anaerobic ammonia oxidation (ANAMMOX), partial nitrification and denitrification (SNAD), simultaneous nitrification and denitrification (SND).

32.2 Anammox

In 1995, Mulder and Van de Graaf discovered in an anaerobic fluidized bed reactor (AFBR) in Netherlands, which was subsequently named anaerobic ammonia oxidation (Anammox). In this process, NH_4^+-N and NO_2^--N are used as electron acceptors for oxidation. Under anaerobic conditions, NH_4^+-N and NO_2^--N are converted into nitrogen gas by anammox bacteria when carbon dioxide is fixed, realized the completely autotrophic nitrogen removal (Eq. 32.1) (Kuenen 2008).

$$
\begin{aligned}
NH_4^+ + 1.32\,NO_2^- + 0.066\,HCO_3^- + 0.13\,H^+ &\rightarrow 1.02\,N_2 + 0.26\,NO_3^- \\
&+ 2.03\,H_2O + 0.066\,CH_2O\,0.5\,N_{0.15}
\end{aligned}
\quad (32.1)
$$

This process has significant advantages over the conventional process of nitrification/denitrification nitrogen removal, for example, no additional organic carbon is required, the aeration volume is reduced by about 60% and the sludge production is reduced by about 90% (Cao et al. 2017). It has great application prospects and is in line with the national "dual carbon" strategy. Several studies have reported that anammox processes can achieve high nitrogen removal rate (NRR). The highest NRR of high strength synthetic wastewater was 76.7 kg N/m^3/d (Yokota et al. 2018). However, the high nitrate content in effluent is an inherent problem because 11% of effluent TN is NO_3^--N. Therefore, the researchers try to overcome obstacles to anammox process in the challenges of the mainstream in the application. One of the keys to apply the advantages of this process to the denitrification treatment of municipal sewage is to solve the stable source of NO_2^--N (Yan et al. 2022). At present, two methods that can combine anammox to achieve nitrogen removal in municipal wastewater are partial nitrification coupled anaerobic ammonia oxidation (PN/A) and partial denitrification coupled anaerobic ammonia oxidation (PD/A) (Zhang et al. 2020). However, for most PN processes, the NO_2^--N accumulation rate (NAR) range < 5.2 kg N/m^3/d. Yongzhen Peng's team from Beijing University of Technology first proposed the process route of partial denitrification coupled with anammox, which aroused the academic circle's attention to the research and application of partial denitrification.

32.3 Partial Denitrification

Partial denitrification refers to the incomplete denitrification of NO_3^- under the action of denitrifying bacteria to generate NO_2^-, and the reduction product of NO_3^- is frozen in NO_2^- to form a large accumulation of NO_2^- (Eq. 32.2). Partial denitrification is considered to provide a more stable and efficient substrate for anammox reaction, it has received a lot of attention.

$$\begin{aligned} &1.32\,NO_3^- + 0.55\,CH_3COO^- + 0.088\,NH_4^+ \rightarrow 1.32\,NO_2^- + 0.088\,C_5H_7NO_2 \\ &\quad + 0.66\,HCO_3^- + 0.198\,H^+ + 0.246\,H_2O \end{aligned} \tag{32.2}$$

Partial denitrification can not only consume the NO_3^- produced by anammox reaction, but also provide the electron acceptor NO_2^- necessary for anammox reaction. In batch tests of activated sludge denitrification, researchers found that different degrees of NO_2^- accumulation occurred in the NO_3^- reduction process. Among them, activated sludge in the coupled denitrification system of sludge fermentation showed the highest NO_2^- accumulation efficiency, and the nitrate to nitrite transforming ratio (NTR) was up to 80% (Cao et al. 2013). These results open a new way for the popularization of anammox technology.

32.4 Application of Different PD and Anammox Coupling Processes in Wastewater Treatment

In recent studies, the combination of Partial denitrification and Anammox (PD/A) has received increasing attention. Some studies coupled SBR process of endogenous partial denitrification with UASB process of anammox, and the results showed that the TN removal rate reached 90%, and the contribution rate of anammox was 49.8% (Ji et al. 2018). All of the results indicate that the PD process is a feasible and development method to promote the stable and efficient implementation of anammox. There are two common PD/A process coupling modes, which are two-stage PD/A process (Fig. 32.2) and one-stage PD/A process respectively (Fig. 32.3).

The original PD/A process used a two-stage structure, which completed the reaction in two separate reactors. The influent is simulated wastewater with NO_3^--N and NH_4^+-N of 30 mg/L and 25 mg/L, respectively, the TN removal rate is up to 97% with acetate as electron donor (Cao et al. 2016). However, More and more one-stage PD/A reactors have been established by inoculating and acclimating PD and anammox sludge. The NRE of domestic wastewater and nitrate wastewater is 95.8% in a one-stage PD/A reactor (Du et al. 2017). In recent years, the segmented PD/A process has achieved remarkable results in phosphorus removal, which provides a new way for practical application (Wang et al. 2019).

In recent years, in the Blue Plains Advanced Sewage Treatment Plant (AWTP), the combination of PD and anammox for nitrogen removal in mainstream treatment has

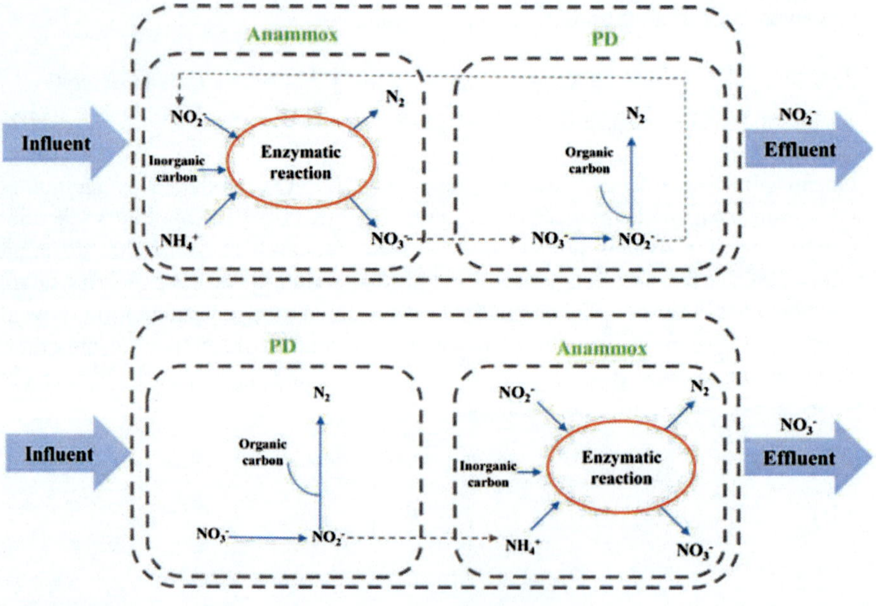

Fig. 32.2 Schematic diagram of two-stage PD coupled Anammox

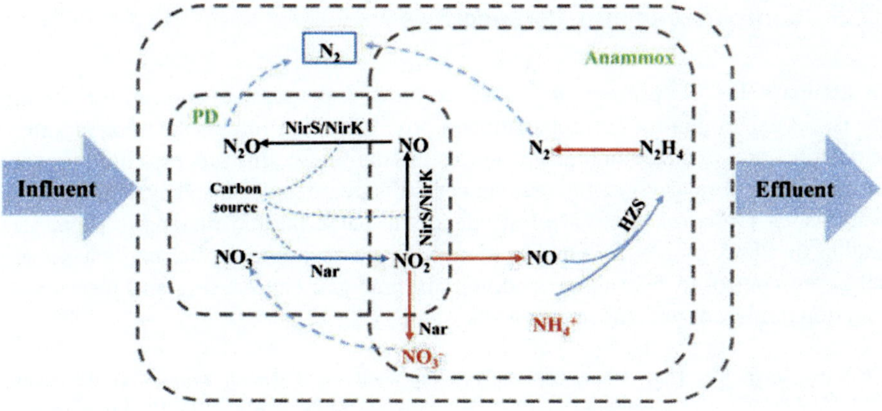

Fig. 32.3 Schematic diagram of integrated PD coupled Anammox

been successfully developed (Le et al. 2019). The combination of PD and anammox process provide a method for expending the application in practical engineering. As shown in Fig. 32.4, a flow chart of PD/A coupling process application in mainstream municipal sewage. Firstly, the municipal sewage is sent to the anaerobic/anoxic zone of the mainstream, secondary effluent and recirculating sludge digestion liquid are delivered together to the mainstream anaerobic/anoxic zone. In this process, NO_3^--N can be reduced to NO_2^--N, thus improving the removal rate of TN by anammox. This combination provides a method to the shortage of carbon sources in WWTPs. Studies have demonstrated that partial denitrification and anammox coupled pathway for nitrogen removal plays a pivotal role in anaerobic moving bed biofilm systems in full-scale mainstream treatments (Li et al. 2019).

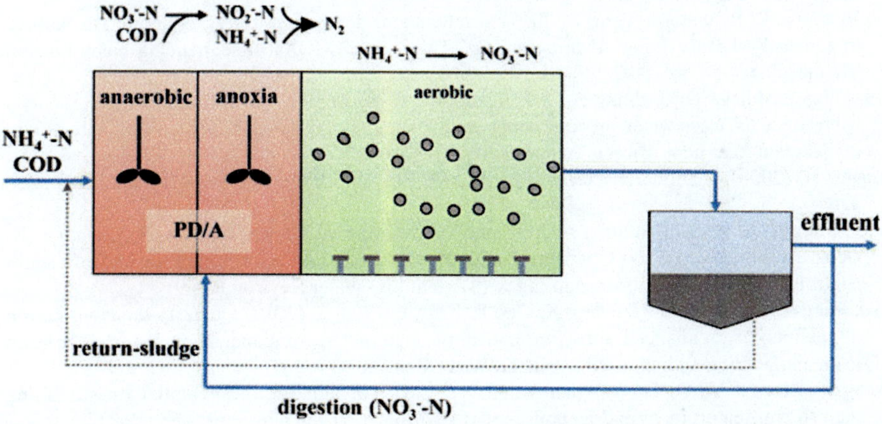

Fig. 32.4 Application of coupled partial denitrification and anammox in mainstream municipal sewage

32.5 Conclusions and Perspectives

In summary, the PD process of NO_3^--N to NO_2^--N provides a solution for solving the issues existing in the current anammox process, and promotes the popularization and application in different kinds of sewage. In the future, this process could provide economical and environmental treatment of mainstream sewage, and nitrate industrial wastewater. However, there are few functional bacterial and metabolic properties studies on PD/A processes. Therefore, it is also necessary to further understand and study the control of operating conditions in practical engineering and the various functional microorganisms in coupled systems.

Acknowledgements The authors gratefully appreciated the financial support of the Henan Provincial Science and Technology Development Program (Grant No. 212102110030).

References

Alloul A, Ganigué R, Spiller M, Meerburg F, Cagnetta C, Rabaey K, Vlaeminck SE (2018) Capture–ferment–upgrade: a three-step approach for the valorization of sewage organics as commodities. Environ Sci Technol 52(12):6729–6742

Cao S, Peng Y, Du R, Wang S (2016) Feasibility of enhancing the DEnitrifying AMmonium OXidation (DEAMOX) process for nitrogen removal by seeding partial denitrification sludge. Chemosphere 148:403–407

Cao S, Wang S, Peng Y, Wu C, Du R, Gong L, Ma B (2013) Achieving partial denitrification with sludge fermentation liquid as carbon source: the effect of seeding sludge. Biores Technol 149:570–574

Cao Y, van Loosdrecht MCM, Daigger GT (2017) Mainstream partial nitritation–anammox in municipal wastewater treatment: status, bottlenecks, and further studies. Appl Microbiol Biotechnol 101(4):1365–1383

Du R, Cao S, Li B, Wang S, Peng Y (2017) Simultaneous domestic wastewater and nitrate sewage treatment by DEnitrifying AMmonium OXidation (DEAMOX) in sequencing batch reactor. Chemosphere 174:399–407

Ji J, Peng Y, Mai W, He J, Wang B, Li X, Zhang Q (2018) Achieving advanced nitrogen removal from low C/N wastewater by combining endogenous partial denitrification with anammox in mainstream treatment. Biores Technol 270:570–579

Kuenen JG (2008) Anammox bacteria: from discovery to application. Nat Rev Microbiol 6(4):320–326

Le T, Peng B, Su C, Massoudieh A, Torrents A, Al-Omari A, Murthy S, Wett B, Chandran K, Debarbadillo C, Bott C, De Clippeleir H (2019) Nitrate residual as a key parameter to efficiently control partial denitrification coupling with anammox. Water Environ Res 91

Li J, Peng Y, Zhang L, Liu J, Wang X, Gao R, Pang L, Zhou Y (2019) Quantify the contribution of anammox for enhanced nitrogen removal through metagenomic analysis and mass balance in an anoxic moving bed biofilm reactor. Water Res 160:178–187

Wang X, Zhao J, Yu D, Du S, Yuan M, Zhen J (2019) Evaluating the potential for sustaining mainstream anammox by endogenous partial denitrification and phosphorus removal for energy-efficient wastewater treatment. Biores Technol 284:302–314

Yan G, Zibin L, Junhao S, Yu-You L (2022) The main anammox-based processes, the involved microbes and the novel process concept from the application perspective. Front Environ Sci Eng 16(7):84

Yokota N, Watanabe Y, Tokutomi T, Kiyokawa T, Hori T, Ikeda D, Song K, Hosomi M, Terada A (2018) High-rate nitrogen removal from waste brine by marine anammox bacteria in a pilot-scale UASB reactor. Appl Microbiol Biotechnol 102(3):1501–1512

Zhang Z, Zhang Y, Chen Y (2020) Recent advances in partial denitrification in biological nitrogen removal: from enrichment to application. Biores Technol 298:122444

Chapter 33
Analysis and Strategies on SCIP Database Notification Based on China Automotive Material Data System

Jing Qiu, Tian Li, Pan Wang, and Shenghai Lin

Abstract This paper systematically analyzes China's automobile export in recent years from three aspects: export region, export vehicle type and annual export volume, and explains in detail the SCIP database reporting obligations and reporting strategies that enterprises need to fulfill when exporting to the European Union.

Keywords Automobile · Export · SCIP database

33.1 Preface

Since China entries into WTO, the automobile industry has gradually become a strategic and pillar industry of our national economy and one of the most globalized industries (Foreign Trade Department 2019). In recent years, Chinese brands with independent innovation, independent intellectual property rights and ultra-high cost performance have been favored by overseas markets. The export momentum of Chinese automobile brands has gradually increased, especially new energy vehicles, with the strong support of the state, technology and quality have significant competitiveness in the global scale (Ministry of Commerce of the People's Republic of China Bureau of Industry Injury Investigation, China Automotive Technology Research Center Co., LTD 2004).

J. Qiu (✉) · T. Li · P. Wang · S. Lin
China Automotive Technology and Research Center Co., Ltd., Tianjin, China
e-mail: qiujing@catarc.ac.cn

Automotive Data of China (Tianjin) Co., Ltd., Tianjin, China

J. Zhang et al. (eds.), *Environmental Pollution Governance and Ecological Remediation Technology*, Environmental Science and Engineering,
https://doi.org/10.1007/978-3-031-25284-6_33

33.2 Current Situation of China's Industry

This paper collects the data of China's exported vehicle in recent four years, including three aspects: export region, export vehicle type and annual export volume (China Automobile Dealers Association 2021) (Figs. 33.1, 33.2 and 33.3).

It can be seen from the above data that, except for Asian countries, China is currently mainly exporting to Europe, and its main export types are passenger cars, and the overall export volume is keeping rising since 2016. However, behind the flashy data and growth rate, China's auto industry still faces some problems and challenges, one of which is the problem of environmental protection (China Association of Automobile Manufactures, China Automotive Technology Research Center Co., LTD, Toyota Motor Corporation, Annual Report on the Development of China Automotive Industry 2019). As the European Union and some other regions and countries have formulated and issued a series of relevant laws, regulations or standards, which have formed green trade barriers to Chinese enterprises when vehicle

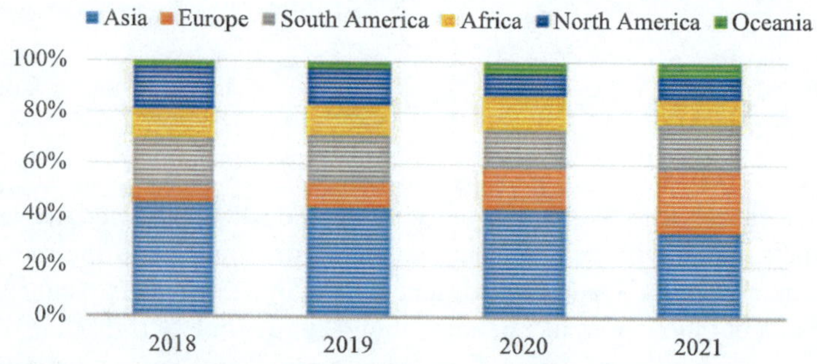

Fig. 33.1 China's automobile export volume by region, 2018–2021

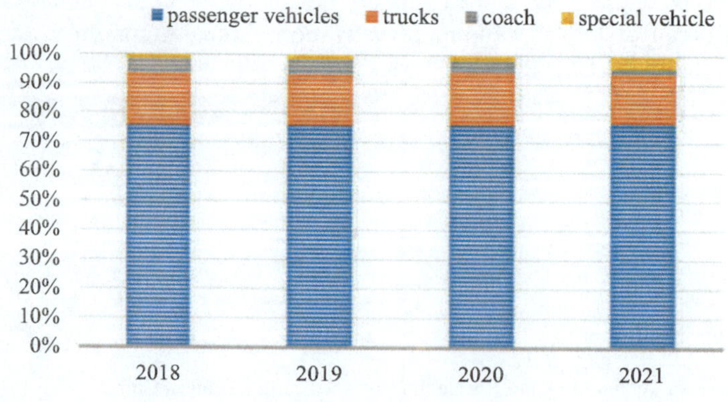

Fig. 33.2 Statistics of China's automobile export volume by vehicle type, 2018–2021

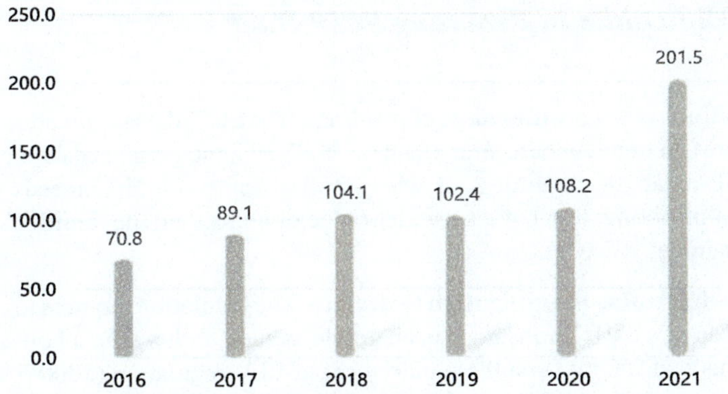

Fig. 33.3 China's annual export volume of passenger cars from 2016 to 2021 (unit: million)

exported to the European Union to a certain extent, how Chinese vehicle manufacturers should effectively comply with the EU environmental protection laws and regulations has become an important issue for enterprises at present.

This paper only makes a brief analysis of the Waste Framework Directive (WFD) issued by the European Union, and provides several coping strategies for automobile products planned to be exported to the European Union.

33.3 Introduction to SCIP Database

33.3.1 Background of SCIP Database

As part of the implementation of the EU's action plan for the circular economy adopted in 2015, the revised Waste Framework Directive entered into force in July 2018. It gave ECHA the task to develop a database with information on articles containing substances of very high concern (SVHCs) on the Candidate List. New substances are regularly added to the Candidate List under REACH.

Companies that produce, import or supply articles containing Candidate List substances have to submit information on these articles placed on EU market to the SCIP database, as from 5 January 2021. These articles can be produced in the EU or imported from non-EU countries.

The information in the database may help waste operators in sorting and recycling articles that contain Candidate List substances, and support consumers in making informed choices and considering how to best use and dispose of such articles. Overall the database should contribute to the progressive substitution of SVHCs in articles and to the development of safer alternatives (ECHA Homepage 2022).

33.3.2 Difficulties in Reporting SCIP Data

The regulations cover a wide range of products. The SCIP data notification requirement is that if an item contains more than 0.1% SVHC, the item needs to be notified to the SCIP database, regardless of whether the export volume exceeds one ton. Therefore, this notification covers a wider range of items than the traditional SVHC notification under REACH.

Difficult to determine the obligation to declare. The regulation requires to count the concentration of a SVHC substance in an article as such, rather than a homogeneous material, this is different from the requirement of ELV regulation (2000/53/EC) for regulating hazardous substances. Therefore, for the vehicle, it is not possible to directly check whether the hazardous substance needs to be declared in the way of coping with the ELV regulation, but to calculate based on the articles where the substance is located, which brings a lot of work to the SCIP notification of vehicle.

33.3.3 Data Analysis on Declared Vehicle Types

This paper selects the data of 50 vehicle types that have been declared in SCIP database for analysis. The analysis dimension includes two aspects: the average number of substances reported in vehicle and the most frequently reported substances, which can provide a reference for readers to report SCIP data later (Figs. 33.4 and 33.5).

It can be seen from the above two figures that, in the 50 types, each types reported average 19 kinds of substances. In the notified substance, Lead (CAS: 7439-92-1),

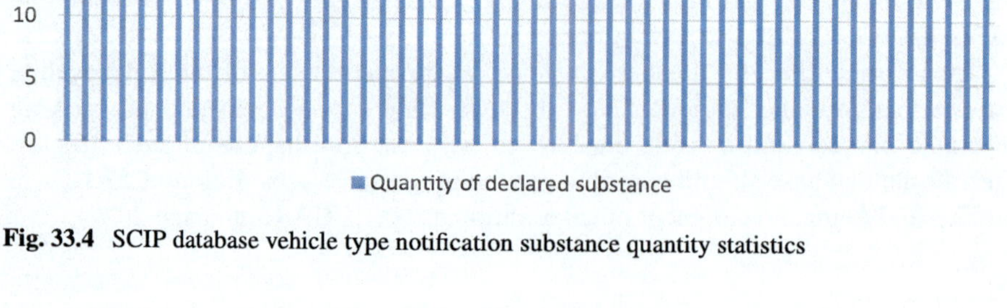

Fig. 33.4 SCIP database vehicle type notification substance quantity statistics

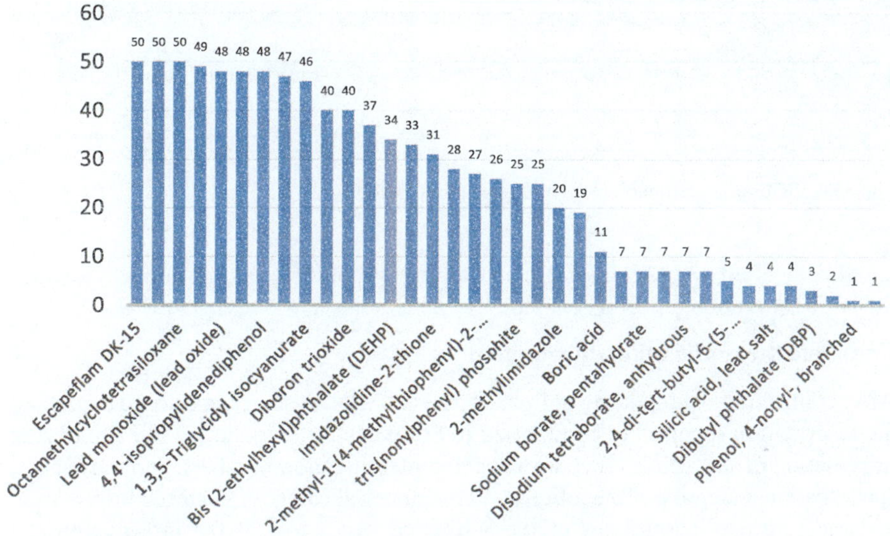

Fig. 33.5 SCIP database vehicle notification substance ranking

octamethylcyclotetrasiloxane (CAS: 556-67-2), and dechlorane A (CAS: 13560-89-9) are reported in each vehicle type, and other compounds such as lead compounds, organic compounds commonly used in ink, paint and some polymer materials preparations or industrial additives are also commonly declared substances.

33.3.4 Strategies Based on China Automotive Material Data System (CAMDS)

CAMDS can help vehicle manufacturers collect vehicle material data based on supply chain, and then realize the monitoring and management of vehicle hazardous substances using. The data collected based on the CAMDS system can be used for the analysis and notification of SCIP data in addition to responding to ELV regulations. The following is a brief analysis of the coping strategies.

Analysis of notification obligation

Since SCIP data reporting is based on articles, the content of SVHC should be the concentration in the article, which in the case of CAMDS is calculated as a percentage of the SVHC substance in the smallest component node where it is located. When this percentage is greater than or equal to 0.1%, the article should be reported to the SCIP database. The CAMDS system will calculate the percentage of SVHC substances in the minimum component node based on the information that the user

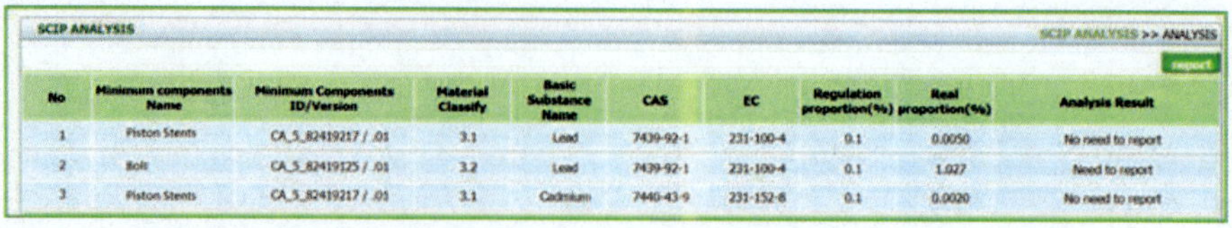
SCIP ANALYSIS

SCIP ANALYSIS >> ANALYSIS

report

No	Minimum components Name	Minimum Components ID/Version	Material Classify	Basic Substance Name	CAS	EC	Regulation proportion(%)	Real proportion(%)	Analysis Result
1	Piston Stents	CA_5_82419217 / .01	3.1	Lead	7439-92-1	231-100-4	0.1	0.0050	No need to report
2	Bolt	CA_5_82419125 / .01	3.2	Lead	7439-92-1	231-100-4	0.1	1.027	Need to report
3	Piston Stents	CA_5_82419217 / .01	3.1	Cadmium	7440-43-9	231-152-8	0.1	0.0020	No need to report

Fig. 33.6 SCIP notification obligation judgment in CAMDS

has already filled in CAMDS, and then determine whether the article needs to perform the notification obligation (Fig. 33.6).

Complete the notification information

After confirming the reporting obligation, the CAMDS system can be used to prepare the information required to be reported in SCIP database, including the identifying information of the article (such as name and identification number), and the source, specification, category of the article, and the material category, material information and the safe use instructions of the substance, etc. Most of the information can be directly read from the CAMDS data, such as article identification information, quantity, material category, SVHC information and its hierarchical position in the article, etc. Users only need to complete a few information, such as the safe use instructions to complete the SCIP dossier.

Dossier generation and submission

CAMDS can generate SCIP dossiers directly according to the reported information in CAMDS, and export the dossier to I6Z format for SCIP database notification. There are two ways to submit the dossier: directly submit the dossier in CAMDS through the S2S way provided by ECHA, or submit the dossier in the ECHA Submission Portal. After successful Submission, ECHA will return the Submission Number and SCIP Number corresponding to the dossier, thus the SCIP notification of the product is complete. In the future, if the product changes, the duty holders can use CAMDS to do some modification and update the dossier.

33.4 Conclusion

This article analyzes the Chinese automobile exports situation in recent years, which can be seen, with ever increasing competitiveness of Chinese independent brands, Chinese manufacturers are focusing the overseas markets, but it should be noticed that avoiding blind export, and thoroughly researching the importer market access laws regulations and standards as well as the conformity assessment procedures, to ensure correctly fulfill the obligations of the target countries, and make the product stand firm in foreign markets, avoid recalls, fines or other similar penalties, lead to the unnecessary expenses to the companies.

References

China Automobile Dealers Association official website. Analysis of China's automobile exports in 2021. Last accessed 2022/8/11

China Association of Automobile Manufactures, China Automotive Technology & Research Center Co., LTD., Toyota Motor Corporation, Annual Report on the Development of China Automotive Industry (2019) May 2019

ECHA Homepage (2022) https://echa.europa.eu/understanding-wfd. Last accessed 2022/8/11

Foreign Trade Department, Ministry of Commerce, China Automotive Technology & Research Center Co., LTD. (2019) Annual report on high-quality development of China's automobile trade, Dec 2019

Ministry of Commerce of the People's Republic of China Bureau of Industry Injury Investigation, China Automotive Technology & Research Center Co., LTD. (2004) Research report on international competitiveness evaluation of Chinese automobile industry, Dec 2004

Chapter 34
Response of Different Wetland Plants and Planting Strategies to Cadmium Pollution in Mines

Minjuan Zhao, Yuji Li, Lin Xiao, Wei Zhou, Shanshan Liu, Xiaofan Chang, Siyu Wang, Chi Wang, Jiatong Wang, and Tianpeng Gao

Abstract The mine's ecological environment is severely impacted due to the high concentration of heavy metals in the abandoned mining area. The motivation of this study is to determine the remediation of cadmium pollution in the mining area by the wetland super enrichment plants, so as to determine the enrichment capacity of the super enrichment plants to the heavy metals in the mining area. In this study, *Solanum nigrum* and *Pteris nervosa* were used as experimental plants. Soil from mines in Fugu County, Shaanxi province, which had been closed for mining, was used for pot culture in monoculture and intercropping, as well as examine the associated experimental data. The findings demonstrated that both monoculture and intercropping wetland hyperaccumulators dramatically raised plant diversity index and decreased cadmium concentrations in soil after 30 days of planting. Intercropping had the best ability to absorb heavy metals ($p < 0.05$), and the intercropping had the most obvious effect on the improvement of plant diversity index ($P < 0.05$). The discoveries exhibit that wetland hyperaccumulators beneficially affect the environmental reclamation of weighty metal mines, which can fundamentally bring down how much weighty metals in the dirt and increment species variety.

M. Zhao · Y. Li · L. Xiao · X. Chang · S. Wang · C. Wang · J. Wang · T. Gao (✉)
School of Biological and Environmental Engineering, Xi'an University, Xi'an 710065, China
e-mail: zkgtp@163.com

W. Zhou
Yangling Vocational and Technical College, Xianyang 712199, China

S. Liu
School of Biological and Pharmaceutical Engineering, Lanzhou Jiaotong University, Lanzhou 730070, China

T. Gao
Engineering Center for Pollution Control and Ecological Restoration in Mining of Gansu Province, Lanzhou City University, Lanzhou 730070, China

M. Zhao · T. Gao
Academy of Eco Xi'an, Xi'an University, Xi'an 710065, China

J. Zhang et al. (eds.), *Environmental Pollution Governance and Ecological Remediation Technology*, Environmental Science and Engineering,
https://doi.org/10.1007/978-3-031-25284-6_34

Keywords Hyperaccumulators · Cadmium pollution · Ecological restoration · *Solanum nigrum* · *Pteris nervosa*

34.1 Introduction

Heavy metal concentrations in abandoned mining areas are high, which is harmful to soil structure and fertility, biodiversity, and human health. As a result, cadmium pollution in mine soil has always been a hot topic in current environmental pollution research (Trace et al. 2020). Cadmium pollution is distinguished by significant harm, concealment, long-term nature, and difficulty in treatment, and it is the focus and challenge of soil pollution control (Fan et al. 2019). It is critical to treat and repair polluted land in order to reduce environmental and ecosystem pollution caused by cadmium pollution (Guo et al. 2019).

Superenriched plants generally refer to plants that absorb more than 10 times more heavy metals than ordinary plants without affecting normal life activities. They are more adaptable to heavy metals and can absorb an oversized quantity of heavy metals throughout their growth, so they have high research value for ecological restoration of heavy metal mines (Li et al. 2019). Generally speaking, the supernormal accumulation of heavy metal content in aboveground parts of plants is on top of that in roots. The accumulative quantity of specific heavy metals (Ni, Pb, Co, etc.) is generally 100 mg/kg (dry weight) (Hu 2020). In reality, adding soil amendments can considerably increase plants' capacity to absorb and enrich heavy metals in hyperaccumulator plants that have been domesticated artificially.

Both domestically and internationally, there are numerous techniques for reducing soil contamination, including chemical, physical, and biological techniques. Among these, phytoremediation technology offers a number of advantages for the economy and the environment in addition to cheap cost and minimal secondary pollution. It is a green and efficient remediation technology, and the screening of super-enriched plants is at the heart of phytoremediation technology (Wang et al. 2020). Previous studies have simulated wetland systems and set the concentrations of heavy metal cadmium in plant culture medium at 0.5, 10 and 20 mg/L. Mites (*Phragmites australis* (Cav.) trin.ex Steu) and *Pennisetum alopecuroides* (L.) Spreng.) were incubated in a medium, respectively. The results showed that the roots of *Phragmites communis* and Pennissima had a strong ability to enrich cadmium (Ning et al. 2019). Therefore, in this study, *Solanum nigrum* and *Pteris nervosa* were used as test plants, and the soil collected from the discontinued mine in Fugu County, Shaanxi Province was used for pot culture to observe the growth of different species of plants, so as to judge whether wetland super-enriched plants have a positive impact on the ecological restoration of heavy metal mines.

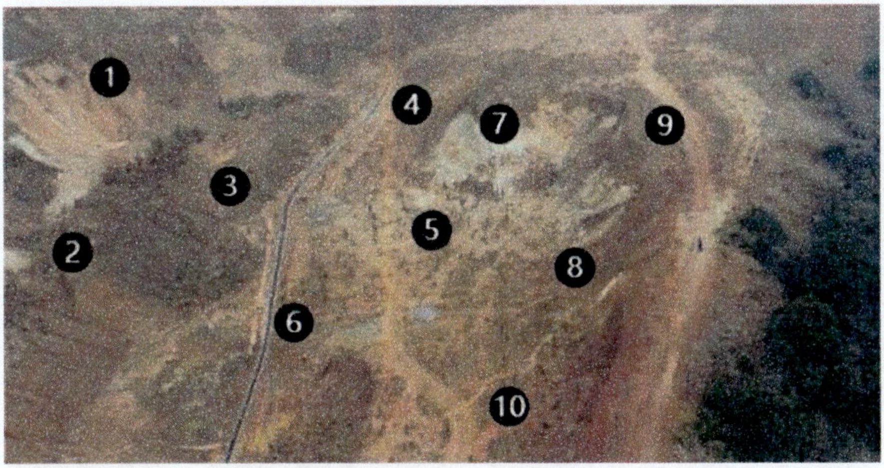

Fig. 34.1 Heavy metal mines in Fugu County, Shaanxi Province

34.2 Materials and Methods

34.2.1 The Experimental Area

In this study, a heavy metal mine (39° 13′ 17" N, 110° 38′ 46.30" E) in Fugu County, Shaanxi Province was selected for soil collection and used as the test area (Fig. 34.1). The rise of the mining region is 780–1426.5 m, and the surface layer of the mining region has weathering phenomenon. The dirt in the mining region is principally sandstone and sand shale, with low satisfied of natural matter and unfortunate water-holding limit. There are red and yellow soils at high altitudes. Mined-out area of the mine was selected as the specific transplant site, and there was soil leaching water in the area, with pH between 2 and 4. The test period was from April 10, 2019 to May 10, 2019. During the period, the average temperature was 17 °C and the precipitation was 435.5 mm.

34.2.2 Potted Culture of Superenriched Plants

In order to make the cultured superenriched plants grow in mine soil smoothly, the seedlings were first raised in uncontaminated humus soil. A PVC basin with a diameter of 13 cm and a height of 17 cm was used (Fig. 34.2 and 34.3). PVC basin added NH_4Cl, KH_2PO_4, KCl as the base fertilizer, the dosage is: N: P_2O_5: K_2O = 0.15: 0.10: 0.15 g/kg, and mixed with the medium liquid. *Solanum nigrum* grew to 2–3 small leaves and transplanted, while the *Pteris nervosa* grew to 5–7 cm and 3–5 small leaves and transplanted. Soil samples were selected from mines. Under monoculture,

Fig. 34.2 Pot culture of *Solanum nigrum*

Fig. 34.3 Potted culture of *Pteris nervosa*

6 plants of *Solanum nigrum* and 6 plants of *Pteris nervosa* were transplanted in each pot, and 3 plants of *Solanum nigrum* and *Pteris nervosa* were cultured in each pot under intercropping.

34.2.3 Analysis of Soil Properties in Mining Area

Through the determination of cadmium content in the soil of ten sample sites in Figure, it was found that the cadmium content in the soil exceeded the soil background value of Shaanxi Province to a large extent. The result showed that the mean substance of cadmium in the surface (0–10 cm) soil of Fugu County in Shaanxi Province was 0.857 ± 0.408 mg/kg, respectively, and the heavy metal cadmium content was high.

34.2.4 Experimental Methods and Evaluation Criteria

The examination endured 30 days. The plants were divided into two groups: Solanum and *Pteris nervosa*, and the control group was set up (no plants were planted). During the analysis, Soil tests were required every 5 days to decide Cd content in the area. And 30 days later, to recognize the non-trial plants filling in the test region, the ecological richness of the experiment (the total number of species in the quadrat) was taken as one of the assessment criteria. Finally, the plant samples were flushed with ordinary water first and a while later with deionized water for 2–3 times. The plant samples were disinfected by high tension (sterilization temperature was 121 °C) and dried to steady weight. The atomic absorption spectroscopy (ZCA-1000 SF) was used to determine the content of cadmium.

The sample plot area is 10 m^2, the data are recorded and summarized, and various indexes are calculated. The calculation formula is as Table 34.1.

34.3 Result and Analysis

34.3.1 Changes of Cadmium Content During Plant Growth

The change of soil cadmium content in the control group (without planting plants) was basically the same within 30 days of planting ($P < 0.05$). Be that as it may, monoculture of *Solanum nigrum*, monoculture of *Pteris nervosa*, intercropping of *Solanum nigrum* and *Pteris nervosa* all lessened the substance of cadmium in soil in a period subordinate way ($P < 0.05$). Compared with before planting (0 d), after 30 d of planting, soil cadmium content was reduced 66.67% by *Solanum nigrum* monoculture, 70.93% by *Pteris nervosa* monoculture, and 74.12% by *Solanum nigrum* and *Pteris nervosa* intercropping (Fig. 34.4). These results indicated that monocropping or intercropping of *Solanum nigrum* and *Pteris nervosa*could reduce soil cadmium content, while intercropping was more effective.

34.3.2 Comparison of Cadmium Uptake by Single and Interplanting Superenriched Plants

Three development methods of monoculture *Solanum nigrum*, monoculture *Pteris nervosa*, intercropping *Solanum nigrum* and *Pteris nervosa* were selected for the

Table 34.1 Calculation formula

Index	Formula	Remark
Importance degree **(IV)**	(Relative height + relative multiplicity + relative coverage)/300	Sparse standardization method was used to analyze the importance of regional vegetation IV
Diversity index **(H)**	$H = \frac{\sum P_i}{\ln P_i}$	P_i addresses the general significance worth of the primary species in Shannon–Wiener variety record, and $P_i = N_i/N$, where N_i addresses the quantity of individual species locally in the community, and N addresses the complete number of individual species in the exploratory region
Simpson Index of regional biology **(D)**	$D = 1 - \frac{N_i(N_i - 1)}{N(N-1)}$	
The evenness of biodiversity **(Jws)**	$Jws = \frac{D.H}{\ln S}$	S represents the first relative importance of species richness P_i
Breadth of niche **CSI**	$CSI = \frac{2C}{(A+B)}$	In light of the thorough use of various sorts of assets and different kinds of assets in the exploratory region, the biological specialty width of plant networks in the not set in stone, and the significant qualities were taken into Levins recipe as free factors to acquire the near connection between the natural specialty of predominant species locally
Overlap of ecological niches $\mathbf{Q_{ik}}$	$Q_{ik} = \sum_{j=1}^{r} P_{ij}P_{kj} / \sqrt{\left(\sum_{j=1}^{r} P_{ij}\right)^2 \left(\sum_{j=1}^{r} P_{kj}\right)}^{-2}$	P_{ij} and P_{kj} address the extent of individual number possessing individual number in populace I and populace K, separately, and R addresses the complete number of natural asset loci in the locale

study of mine ecological restoration. Under different planting conditions, the uptake of Cd by *Solanum nigrum* and macrophyllum were recorded in the plant culture pot (Fig. 34.5). The results showed that the uptake of Cd by *Solanum nigrum* under monoculture decreased by 2.37 ug·mg^{-1} kg^{-1} compared with intercropping. Under monoculture, the uptake of Cd by *Pteris nervosa* decreased by 0.23 ug·mg^{-1} kg^{-1}

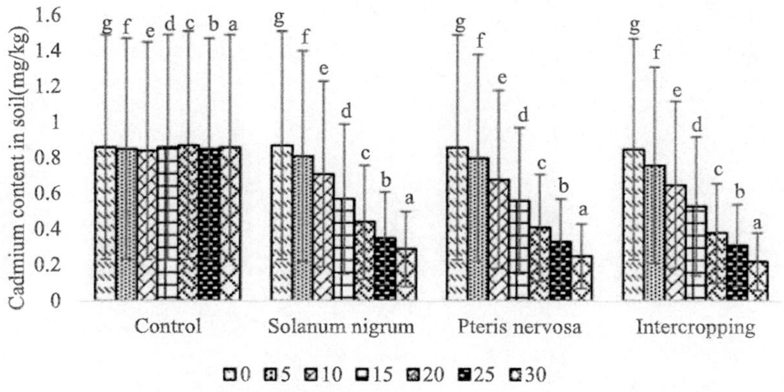

Fig. 34.4 Change trend of soil cadmium content

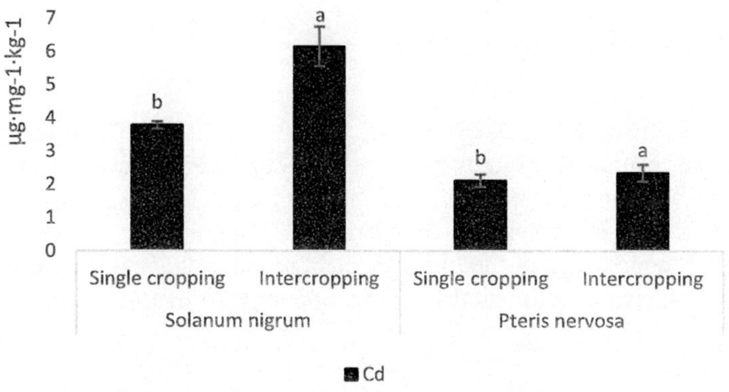

Fig. 34.5 Heavy metal uptake by plants under different planting patterns

compared with intercropping, indicating that *Solanum nigrum* and *Pteris nervosa* had better absorption effects of Cd under intercropping.

34.3.3 Plant Diversity Parameters in the Experimental Area

After the experiment was completed, the plant diversity parameters in the experimental area were checked (Fig. 34.6). The results showed that the 30th day was contrasted and the first day, the diversity index, Simpson index, biodiversity evenness, niche width, and inter group niche index of the control group increased by 4.34, 2.55, 2.20, 0.07 and 2.01; The diversity index, Simpson index, biodiversity evenness, niche width and inter group niche index of *Solanum nigrum* bunch expanded by 35.03, 21.96, 3.60, 13.69 and 29.30; The diversity index, Simpson index, evenness of biodiversity, niche width, and inter-group niche index of monoculture group of

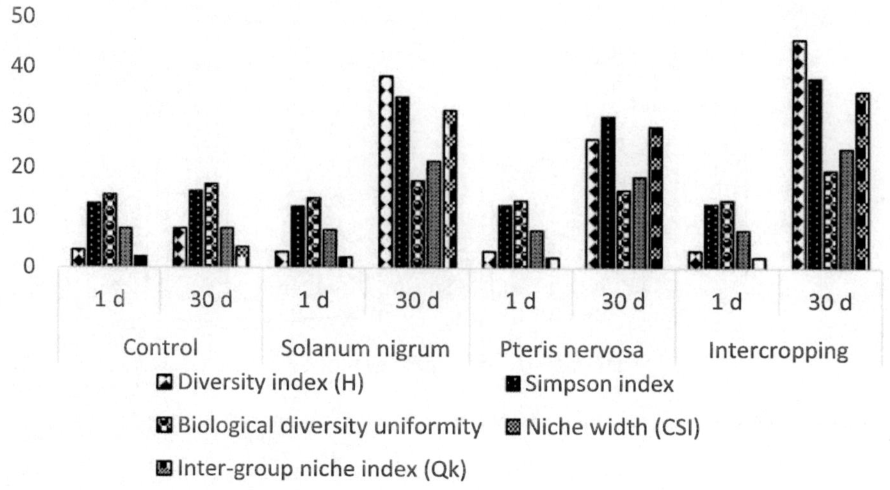

Fig. 34.6 Plant ecological index of experimental area

Pteris nervosa increased by 22.36, 17.80, 2.17, 10.69, 26.13. The diversity index, Simpson index, evenness of biodiversity, niche width, and inter-group niche index of intercropped *Solanum nigrum* and *Pteris nervosa* groups increased by 42.19, 25.18, 6.12, 16.18, and 33.10. Therefore, the planting of superenriched plants effectively improved the biodiversity of the experimental area, and the soil diversity index of the *Solanum nigrum* experimental group was better than that of the *Pteris nervosa* soil group.

34.4 Discussion

Through monoculture and intercropping, firstly, the content of heavy metals in the soil is reduced, the physical and chemical properties of the soil are improved, the growth of more plants is promoted, and the plant diversity index of the surrounding soil is improved. Secondly, the air pollution caused by tailings dust can be reduced by planting plants. Third, because open tailing deposits are prone to be oxidized, hydrolyzed and weathered, a large number of acidic wastewater containing metal ions will be generated, which will seriously pollute the surface and groundwater system. Plant growth can improve the soil and lessen water pollution. Therefore, intercropping *Solanum nigrum* and *Pteris nervosa* may be a successful method for reducing the soil contamination caused by heavy metals in mines.

Traditional ecological remediation methods are slow in ecological remediation of heavy metal mines and have low ecological benefits. Phytoremediation has been wide utilized in remedy of soils contaminated by heavy metals. However, planting alone requires a long time for soil remediation, and reasonable intercropping has advantages and complementary effects on soil heavy metal accumulation. *Solanum nigrum* and *Pteris nervosa* are cadmium and arsenic superenriched plants in wetland, respectively (Wei et al. 2019, 2002). Therefore, in this study, *Solanum nigrum* and *Pteris nervosa* were used as test materials to study the ecological restoration of heavy metal mines, and through the experiments of various planting strategies, we can judge the ecological restoration impact of wetland super enrichment plants.

At present, phytoremediation tests are still in the exploratory stage, and large-scale engineering applications are few. Experimental studies were mainly divided into nutrient liquid culture experiment and pot experiment (Park et al. 2020). Some studies adopted the method of indoor soil culture experiment. The soil collected from the field with Cd pollution was used as the cultivation soil, and the monoculture or intercropping of *Solanum nigrum* and *Pteris nervosa* were used as the cultivation soil. The results showed that intercropping was the better planting method to repair Cd polluted soil. Additionally, under the intercropping mode, the effects of exogenous addition of various amounts of NTA (aminotriacetic acid), EDDS (ethylenediamine disuccinate), and EDTA (ethylenediaminetetraacetic acid) were investigated on the uptake of Cd by plants were studied. The findings demonstrated that intercropping considerably facilitated the uptake of Cd by the above ground part of *Solanum nigrum* and *Pteris nervosa* (Xiong et al. 2011). This study supports the findings of Xiong Guohuan et al.'s investigation on the impact of intercropping circumstances on *Solanum nigrum* and *Pteris nervosa*'s absorption of heavy metals The findings demonstrated that *Solanum nigrum* and *Pteris nervosa* were both capable of efficiently regulating the amount of cadmium in soil, but that the intercropping technique was more efficient in lowering the amount of cadmium in the soil in the mining region.

In addition, it has been reported that intercropping or interplanting different plants can be used not only for soil remediation but also for crop production. For instance, research have demonstrated that intercropping wheat with Setophore can both lower the levels of heavy metals in wheat grains and in the soil, providing a protective double impact for both crop production and soil remediation (Ju et al. 2015). The interplanting of Sedum southeast and maize could promote the absorption of Zn and Cd by Sedum southeast without affecting the content of heavy metals in maize grains (Liang et al. 2007).In present study, *Solanum nigrum* and *Pteris nervosa* are wetlands, based on previous results, we speculated that these two kinds of plants and other crops and cash crops intercropping or intercropped planting, it will aid in the restoration of the ecological environment and the plant's expansion of economic value when soil restoration is protected and agricultural and economic crop output is encouraged.

To sum up, from present study we found that wetland superenriched plants have a positive effect on the ecological restoration of heavy metal mines, because in addition to effectively reducing the content of heavy metals in the soil and preventing cadmium pollution, they can also improve the local environment, contribute to the growth of more plants, and increase the biodiversity of the region. However, the heavy metals determined in this test are single, and it is difficult to consider the combined pollution of heavy metals in the mining area. Additionally, the joint effects of plants and microorganisms were not investigated, and the effects of the super enriched plants on other plants as well as the mechanism of their purification of cadmium pollution in the mining area were not able to be ascertained, so the study still needs to be further improved.

Acknowledgements The authors thank Key Laboratory for ecological restoration and high quality development of Qinling Mountains in the upper and middle reaches of the Yellow River, Xi'an Key Laboratory of plant stress physiology and ecological restoration technology.

Funding This work was funded by National Natural Science Foundation of China (31860176), Key Research and Development Program of Shaanxi (2020ZDLSF06-06) and Gansu (20YF3FA037), and XAWLKYTD012.

References

Boral S, Sen IS, Tripathi A, Sharma B, Dhar S (2020) Tracking dissolved trace and heavy metals in the ganga river from source to sink. A baseline to judge future changes. Geochem Geophys Geosyst 21(10)

Fan Y, Chen C, Ren Z, Wang Z, Zhang Z, Wang X (2019) Effects of heavy metals and mycotoxins on DNA methylation in animal cells. Genom Appl Biol 10:4483–4489

Guo D, Fan Z, Lu S, Ma Y, Nie X, Tong F, Peng X (2019) Changes in rhizosphere bacterial communities during remediation of heavy metal-accumulating plants around the Xikuangshan mine in southern China. Sci Rep 9(1)

Hu L, He Z (2020) Research progress of mine ecological restoration technology. Mineral Protection Util 04:40–45

Hei L, Wu Q, Long X, Hu Y (2007) Treatment effect of sedum and maize interplanting on Zn-contaminated sludge. Environ Sci 04:4852–4858

Ju S, Wang J, Mi Y, Li Z, Chen Y, Wu L, Luo Y (2015) Remediation of Sedum sedum/wheat-eggplant intercropping and crop rotation in heavy metal contaminated soil. J Ecol 08:2181–2186

Li H, Xu W, Dai M, Wang Z, Dong X, Fang T (2019) Assessing heavy metal pollution in paddy soil from coal mining area, Anhui, China. Environ Monitoring Assess 191(8)

Li X, Xia C, Liu Y (2020) Research progress on the status of Zn~(2+) pollution in soil and the effect of Zn~(2+) on cell apoptosis. Genom Appl Biol 07:3426–3430

Ning C, Li W, Xu Q, Li M, Guo S (2019) Arbuscular mycorrhizal fungi promote the absorption of cadmium in polluted water by wetland plants. J Appl Ecol 06:2063–2071

Park S, Choi M, Jang D, Joe D, Park K (2020) Distribution and sources of dissolved and particulate heavy metals (Mn, Co, Ni, Cu, Zn, Cd, Pb) in Masan Bay, Korea. Ocean Sci J 55(1)

Wang L, Dang QL, Tedla B (2020) Biochar and alternate partial root-zone irrigation greatly enhance the effectiveness of mulberry in remediating lead-contaminated soils. J Plant Ecol

Wei S, Xu L, Han R, Dou X, Yang W (2019) Research progress of electro-phytoremediation technology for heavy metal contaminated soil. J Nanjing For Univ (Nat Sci Edn) 01:154–160

Wei C, Chen T, Huang Z, Zhang X (2002) Dayejingkoubiancao: a newly discovered arsenic-rich plant. Acta Ecol 05:777–778

Xiong G, Gao J, Wang H, Pan, Jiao P (2011) Effects of chelating agents on the absorption of heavy metals by Solanum vulgaris and Daye wellhead grass under intercropping conditions. J Agric Environ Sci 04:666–676

Chapter 35
Research and Scheme Comparison Based on Zero Discharge Technology of Thermal Desulfurization Wastewater

Ruochen Sun

Abstract In 2017, the Ministry of environmental protection issued the technical policy on pollution prevention and control of thermal power plants (Guide to feasible technologies for pollution control in thermal power plants: HJ 2301, 2017), proposing to encourage the use of evaporation drying or evaporation crystallization and other treatment processes to ensure that desulfurization wastewater is not discharged. Desulfurization wastewater of thermal power plant, as the terminal wastewater of the whole plant, has the characteristics of high salt, high suspended solids, high heavy metal, and large fluctuations in water quality and quantity (Zhang et al. in J Shanghai Electr Power Univ 37(S1):64–66, 2021). It is one of the most difficult problems to achieve zero discharge of wastewater from power plants. This paper takes the zero discharge transformation of desulfurization wastewater in a thermal power plant as a case, compares and selects different schemes of pretreatment unit, thermal concentration unit and solidification unit, and puts forward the best scheme.

Keywords Desulfurization wastewater · Zero emission · Thermal concentration

35.1 Introduction

The power plant has two 125 MW generating units in phase I and two 300 MW subcritical generating units in phase II. The water taken in phase I comes from the reservoir water and reclaimed water. Part of the reservoir water is used as the water source of the boiler make-up water treatment system, and the other part is used as the water source of phase I. The other part is used as industrial water source for other purposes in the plant. Reclaimed water is mainly used to supply circulating water for phase II units (Table 35.1).

R. Sun (✉)
Northwest Electric Power Test and Research Institute of China Datang, Xi'An 7120000, Shaan'xi, China
e-mail: 1207011605@qq.com

J. Zhang et al. (eds.), *Environmental Pollution Governance and Ecological Remediation Technology*, Environmental Science and Engineering,
https://doi.org/10.1007/978-3-031-25284-6_35

Table 35.1 The quality analysis of desulfurization wastewater

Item	Unit	Desulfurization wastewater in phase I	Desulfurization wastewater in phase II
pH	/	6.62	6.85
Suspended solids	mg/L	121	118
Solution	mg/L	41,660	28,206
Sulfate	mg/L	8340	7058
Chloride ion	mg/L	15,516	17,659
Fluoride ion	mg/L	29.07	31.67
Chemical oxygen demand	mg/L	359	1236
Sulfide	mg/L	0.058	0.050
Calcium ion	mg/L	1070.94	1070.94
Magnesium ion	mg/L	4027.29	4663.86
Hardness	mmol/L	192.38	218.57
Ammonia nitrogen	mg/L	1062.568	938.243

35.2 Water Quantity Accounting of Desulfurization Wastewater

The calculation of water consumption and waste water of the desulfurization system is based on the water balance and salt balance of the desulfurization system, and is directly related to the unit load rate, coal quality and flue gas temperature at the outlet of the absorption tower.

The main sources of chloride ions in the slurry of the absorption tower are coal-fired flue gas and desulfurization process water. On the premise that the operating parameters of the desulfurization system remain unchanged, the discharge of desulfurization wastewater is inversely proportional to the equilibrium concentration of chloride ions in the slurry of the desulfurization absorption tower. In order to realize the cascade utilization and zero discharge of waste water in the whole plant, the waste water from the circulating water system and the chemical water production system, or the concentrated water after its advanced treatment, is used as the process water for the desulfurization system. See Table 35.2 for calculation of desulfurization wastewater.

35.3 Brief Introduction to Zero Discharge Process Scheme of Desulfurization Wastewater

The desulfurization wastewater treatment technology is arranged and combined according to the sequence of three units. To achieve zero discharge of wastewater, one-step, two-step and three-step processes can be selected (Jia 2021). At present,

Table 35.2 Accounting tables of desulfurization water consumption

Item	Units	Phase I	Phase II
Unit's load	MW	188	225
Gas flow (Standard state, dry basis)	m^3/h	690,000	782,190
Moisture content of raw flue gas	wt%	8.80	8.80
Moisture content of net flue gas	wt%	13.39	13.39
Total water volume of flue gas	m^3/h	31.67	35.9
External water content in gypsum	m^3/h	0.73	0.82
Internal water content in gypsum	m^3/h	1.27	1.44
Discharge amount of desulfurization wastewater	m^3/h	5.50	6.30
Consumption of desulfurization wastewater	m^3/h	39.2	44.5

there are many engineering application cases of desulfurization wastewater zero discharge with various process routes at home and abroad, most of which are thermal concentration and flue drying routes. The wastewater zero discharge process route shall be formulated and implemented according to the production characteristics of the power plant. Pretreatment technology and operation are the premise of zero discharge disposal of desulfurization wastewater. The concentration and reduction can effectively reduce the treatment load of the evaporation and solidification section and ensure the effective evaporation of the system in the later stage, which is the key to achieve zero discharge of desulfurization wastewater.

The heat source of the power plant is the flue gas extracted from the boiler or the steam from the auxiliary steam header, which is convenient to access (Fan 2021). It can be divided into high-temperature bypass flue evaporation and low-temperature bypass flue evaporation according to different flue gas extraction positions. In order to avoid the influence of high-temperature bypass flue evaporation on boiler efficiency, low-temperature bypass flue evaporation or low-temperature multi effect evaporation is usually used for pre-concentration, and then enters the high-temperature bypass flue evaporation for curing. The unit load drainage is limited by 1t/100 MW. When the drainage is greater than 1t/100 MW, it is recommended to add a concentration unit. When the drainage is less than 1t/100 MW, the concentration unit can be omitted.

The low-temperature bypass flue evaporation process has low requirements on the inlet water quality index and is attached to the desulfurization system. It has strong applicability to the inlet water quality. The concentration heat source is the tail gas after the boiler dust collector, which does not need to consume other heat sources. The water vapor evaporated from the concentration tower enters the desulfurization tower with the flue gas, reducing the flue gas temperature entering the desulfurization tower and reducing the process water consumption of the desulfurization system.

The high-temperature bypass flue evaporation process extracts the high-temperature flue gas in front of the air preheater, and the air temperature at the outlet of the air preheater decreases, thus affecting the thermal efficiency of the boiler, which is conducive to the increase of flue gas moisture content, reducing the evaporator volume and reducing investment. However, when the flue gas temperature is low or the amount of desulfurization wastewater is large, more flue gas needs to be extracted, which will affect the boiler efficiency (Xie et al. 2021).

35.4 Introduction to Treatment Scheme

35.4.1 Integrated Pretreatment + Low-Temperature Vacuum Multi Effect Evaporation (MED) + High-Temperature Bypass Rotary Atomization Evaporation Crystallization System (Three-Step Method)

The scheme is pretreated by integrated treatment device with integrated agent, and the suspended solids in the effluent are relatively low, and then it enters the low-temperature vacuum multi effect evaporation system for concentration and reduction. The advantages of the scheme are that the concentration unit does not produce solid waste, the operation adjustment is relatively simple, the concentration ratio adjustment range is large, and it is less related to the host system, which is relatively independent, and its fault handling, defect elimination and other work are not affected by the limit value of the host operating conditions.

35.4.2 Reuse Existing Wastewater Pretreatment Equipment (Primary Sedimentation Tank) + Seed Method Low Temperature Vacuum Multi Effect Evaporation (MED) + High Temperature Bypass Two Fluid Evaporation Crystallization System (Three-Step Process)

The pre-treatment unit of this scheme makes use of the original primary sedimentation tank, and the concentration unit selects the crystal seed method low-temperature vacuum multi effect evaporation. This process has low requirements on the inlet suspended solids, and SS can be controlled at 5000–20000 mg/L. The original pre-treatment unit can be bypassed, and the water intake position can be taken from the outflow water from the top of the desulfurization wastewater cyclone or the water with desulfurization gypsum. The disadvantage of this scheme is that the concentration ratio needs to be controlled within a certain range to avoid the problem of pipe blockage caused by sludge precipitation, which will adversely affect the solidification stage. It is required to configure a filter press that can press and filter high salt sludge, and there is no place to absorb the high salt sludge, which has subsequent disposal problems and environmental protection risks.

35.4.3 Reuse of Existing Wastewater Pretreatment Equipment + Low-Temperature Flue Gas Waste Heat Concentration + High-Temperature Bypass Rotary Atomization Evaporation Crystallization System (Three-Step Process)

In this scheme, low-temperature flue gas is extracted by the waste heat concentration and reduction process of low-temperature flue gas, and concentrated in the concentration tower through thc hcat exchange of flue gas and waste water. This process has low requirements for the water quality index. The pretreatment unit can use the original triple header pretreatment unit (or bypass). The concentrated heat source is the tail flue gas behind the boiler dust remover, without consuming other heat sources. The evaporated water from the top of the concentration tower enters the desulfurization tower with the flue gas, The flue gas temperature entering the desulfurization tower is reduced, and the process water consumption of the desulfurization system is reduced. This scheme has a certain impact on the water balance. Two 6 kV fans need to be added, so the power consumption is relatively high. The concentration tower and its supporting circulating pump, slurry tank, etc. for treating 20 m^3/h wastewater cover an area of about 240 m^2 and need to be placed near the desulfurization tower. It is not applicable to the shortage of land for retrofitting units.

35.4.4 The Optimum Option

On the basis of fully absorbing the investigation of other power plants, the objective evaluation of the desulfurization wastewater zero discharge reconstruction project faced by the plant is made. The design idea of the process technical scheme focuses on improving the stability of the process system, strengthening the buffer capacity of the system operation, and reducing the output of solid waste or crystal salt. The process route of integrated pretreatment + low-temperature vacuum multi effect evaporation + rotary high-temperature bypass flue evaporation (three-step method) is recommended.

35.5 Analysis of the Impact of the Recommended Scheme on the Main System

35.5.1 Accounting for Influence on Thermal Efficiency

According to the heat balance calculation, under the full load condition of desulfurization wastewater, when each unit consumes 2 m^3/h, it is necessary to extract about

Table 35.3 Calculation of influence of high-temperature bypass device on boiler efficiency

Item	Units	100% load	70% load	50% load	3% gas flow
Unit's load	MW	300	210	150	300
Gas flow (standard state, dry basis)	m^3/h	1,042,920	782,190	521,460	1,042,920
Flue gas temperature in front of air preheater (before)	°C	321.6	303.5	285.0	322.0
Flue gas flow bypass	m^3/h	27,910	22,300	18,600	31,288
outlet temperature of bypass flue gas	°C	175.0	175.0	175.0	191.2
Flue gas temperature in front of air preheater (before)	°C	316.9	298.5	278.8	316.3
Temperature changes of main-flue gas	°C	4.7	5.0	6.2	5.7
Proportion of bypass flue gas	%	2.7	2.9	3.6	3.0
Coal consumption of bypass flue gas	g/kWh	0.7061	0.7524	0.9412	0.8650

27910 Nm^3/h of high-temperature flue gas from the front end of the air preheater, accounting for 2.7% of the total flue gas volume, which affects the unit's coal consumption of about 0.7061 g/kw h, the cost of converting coal consumption into water per ton is 8.6 yuan/ton, and the flue gas temperature at the outlet of the air preheater decreases by 4.7 °C compared with that before operation. According to the calculation of 3% boiler flue gas, the flue gas temperature at the outlet of bypass flue is 191 °C. See Table 35.3 for accounting data under other load conditions.

35.5.2 Impact Accounting on Fly Ash Quality

The coal consumption of a single 300 MW unit is 132.5 t/h. The ash amount is calculated according to the proportion of pulverized coal ash in the total ash amount of 90%. The operating efficiency of the precipitator is 99.9%. The total amount of pulverized coal ash of the two units is 31.42 t/h. The chlorine ion concentration of desulfurization wastewater is calculated as 15000 mg/L, and the total amount of chlorine brought into fly ash is 30 kg. In fly ash Portland cement, fly ash is usually added in a proportion of about 20%. After mixing, the chloride ion concentration in the cement is 0.0097%, which meets the standard that the chloride ion concentration in general Portland cement (GB/t175-2007) is not higher than 0.06%. Therefore, after the high-temperature bypass flue evaporation process is implemented, the comprehensive utilization of fly ash will not be affected.

Table 35.4 Resistance calculation table in different load conditions

Item	Unit	97%BMCR	85%BMCR	60%BMCR	50%BMCR
Unit's load	MW	291	255	180	150
Flue gas Temperature at the inlet of air preheater	°C	334.45	321.03	296.36	287.41
Inlet resistance of air preheater	Pa	– 1640	– 1320	– 940	– 790
Flue gas Temperature at the outlet of air preheater	°C	116.51	112.26	106.33	105.35
Outlet resistance of air preheater	Pa	– 3040	– 2230	– 1570	– 1250
Resistance calculation of bypass flue	Pa	795	/	/	450

35.5.3 Commissioning Rate of Bypass Evaporation Device

Considering the on-site flue layout and other factors, the interface position of the bypass flue is selected from the flue from the outlet of the graded economizer to the front of the air preheater to extract high-temperature flue gas, and the flue gas returns to the flue in the mixing section in front of the dust collector after the air preheater after being atomized by rotation.

When the resistance of the air preheater is greater than that of the bypass system, the flue gas volume of the rotary atomizer can be adjusted through the regulating valve at the inlet of the bypass flue. With the reduction of the unit load, when the resistance of the air preheater is less than that of the bypass system, the rotary atomizer shall be out of operation. In order to ensure that the rotary atomizer can be put into operation stably for a long time, the first is to reduce the total resistance of the bypass evaporation device, which can adapt to the long-term low load operation condition of the unit; the second is to optimize the connection mode between the inlet bypass flue and the main flue of the unit, which can effectively reduce the fly ash content in the flue gas. Through calculation, the normal operation of high-temperature bypass evaporation device can be ensured under the condition that the load rate is more than 50% (see Table 35.4).

35.6 Conclusion

The steam or flue gas heat exchange is selected as the heat source, and the vacuum multi effect evaporator is used to realize the thermal concentration of wastewater. The flue gas in the high-temperature section is extracted to solidify the concentrated water after being atomized by the high-temperature bypass rotary atomizer through the high-speed centrifugal atomizer. The process of this scheme is stable and the operation is flexible. There have been many practical operation cases. To sum up, the zero discharge technical route of desulfurization wastewater adopts the process route of "integrated pretreatment + low temperature vacuum multi effect evaporation (MED) + high temperature bypass rotary atomization evaporation crystallization system", which is the first choice to achieve zero discharge of desulfurization wastewater in the plant.

References

Fan M (2021) Study on concentration and reduction process of desulfurization wastewater. Commun Chem Eng Design 47(12):172–173

Guide to feasible technologies for pollution control in thermal power plants: HJ 2301 (2017)

Jia X (2021) Analysis on the selection principle of zero discharge technology route for desulfurization wastewater. Water Treatment Technol 47(08):14–19+25. https://doi.org/10.16796/j.cnki.1000-3770.2021.08.003

Xie Z, Feng Y, Zhao N, Lin T, Chen T (2021) Comparative analysis of zero discharge technologies for desulfurization wastewater from power plants. Sci Technol Innov Appl 09:177–179

Zhang Y, Huang C, Yu L, Huang C, Chen J, Su M (2021) Research on desulfurization wastewater treatment technology of coal-fired power plants. J Shanghai Electr Power Univ 37(S1):64–66

Chapter 36
Study on Ecological Environment Control and Slope Stability Evaluation of Opencast Energy Mine in Plateau Area

Li Li, Jie Wang, XiaoYang Liu, CunHao Jiang, and Rong Xu

Abstract This paper systematically analyzes the eco-environmental problems existing in an energy mine on the plateau, such as topography and geomorphology destruction, vegetation and meadow degradation, permafrost destruction, pit stagnant water, and combined with the experiences and lessons of local ecological environment management, this paper puts forward the treatment scheme and suggestions for the current environmental problems in the study area, and puts forward some specific treatment measures, such as geomorphology reconstruction, soil reconstruction, vegetation reconstruction, fence closure. Simulate and analyze the overall strength reduction method for each slope that has been restored, and check the position and stability of the sliding surface, and then come to the conclusion that the stability of the treated slope is better. The purpose of this paper is to provide relevant reference for the treatment method of ecological environment and the evaluation method of slope stability after treatment in the same type of coal mining area.

Keywords Energy mine · Eco-environmental problems · Treatment measures · Ecological environment

36.1 Preface

The excessive mining and over-exploitation of coal resources and the random stacking of stripped materials will lead to the destruction of alpine meadows, permafrost and wetlands, the change of original water system, landslides and other geological disasters, and seriously damage the ecological environment of the mining area. In particular, the study area is located in high-altitude, high-cold areas, after the fragile ecological environment has been destroyed, its governance is difficult, the

L. Li (✉) · J. Wang (✉) · X. Liu · C. Jiang · R. Xu
China Aero Geophysical Survey and Remote Sensing Center for Nature Resources, Beijing, China
e-mail: dairywoman@126.com

J. Wang
e-mail: wangsnow2008@163.com

J. Zhang et al. (eds.), *Environmental Pollution Governance and Ecological Remediation Technology*, Environmental Science and Engineering,
https://doi.org/10.1007/978-3-031-25284-6_36

cost is high, and the cycle is long, and its eco-environmental problems have become a hot social issue to be solved urgently. In order to deeply adhere to the concept that "green water and green mountains are golden mountains and silver mountains", it is particularly necessary to carry out exploration and research on the measures for the restoration and control of the ecological environment in the mining area (Hai and Shaolin 2001).

There are experts at home and abroad to analyze and study the geological environmental problems caused by coal mining and the environmental protection measures after mining. Chen et al. (2020) took the Fengfeng mining area as an example to carry out the geological environment assessment of the closed coal mine, which provided the basis for the reuse of closed coal mine resources and the prevention and control of geological disasters. Liao et al. (2021) discussed the environmental remediation and treatment of acid pit drainage (AMD) from pyrite and high-sulfur coal mines in southwest China. Feng (2003) summarized the environmental problems existing in open-pit coal mining in China from the aspects of legal system, government management, mining discipline and technology, and discussed the prevention and control countermeasures. Britain, the United States and Australia began the research and practice of ecological reconstruction and land reclamation at the beginning of the twentieth century, focusing on the reclamation of mine solid waste and polluted land (Tianfeng et al. 2007). In 1965, Professor Zade put forward the law of maximum membership degree, and then combined with the principle of fuzzy transformation, established the fuzzy evaluation method (Zadeh 1965; Zhang and Qiang 2019; Chen et al. 2018). This method can make a more comprehensive and reasonable evaluation of things affected by many factors. The qualitative evaluation is transformed into quantitative evaluation, which is suitable for the problems that fuzzy is difficult to solve quantitatively. This method has been widely used in mine environmental assessment, and effectively solves the problems in mine geological environment assessment, such as many evaluation factors, fuzzy boundaries, difficult to quantify the evaluation index and so on. In the 1970s, Saaty, an American operations researcher, put forward Analytic hierarchy process (AHP) (Saaty 1988; Radbruch-Hall and Roger 1982), which combines qualitative analysis with quantitative analysis, which simplifies many complex fuzzy relations and is widely used in the comprehensive evaluation of mine geological environment problems. Previous studies on environmental control in energy mining areas have both contributions and limitations, and there are few studies on energy mine treatment measures, local slope stability and safety factor of open-pit mining in plateau area. This paper systematically analyzes the eco-environmental problems existing in a coal mining area in the plateau area, combined with the local ecological environment treatment experience, puts forward the treatment scheme suggestions, puts forward the specific treatment measures, and analyzes the slope stability after treatment.

36.2 General Situation of the Study Area

The study area is located in the northeast of Qinghai Province, located in the alpine zone, with a cold climate and a large temperature difference in the morning and evening, which is a typical continental climate of the plateau. This area belongs to the plateau meadow low swamp, the terrain is not undulating, alpine meadow grassland and alpine swampy meadow grassland are widely distributed. The coal mining pit in the study area was originally designed to cut and fill the whole slag mountain in the south, cut and fill the slag mountain in the north, renovate the topography, grade the platform, shape the slope, and the slope angle is 20°, with a total excavation volume of about 15 million m^3.

36.3 Past Governance Experiences and Lessons

The evaluation report of slag mountain degradation compiled by Chengdu University of Technology puts forward that the comprehensive environmental improvement work such as slag mountain brushing slope shaping and covering soil, recommended green restoration, illegal building demolition and environmental sanitation treatment should be carried out in different batches. According to the results of the survey, vegetation restoration has achieved certain results, but the vegetation coverage varies greatly, ranging from 0 to 40%. According to the investigation and comparison of 16 sample sites, the existing restoration experiences are summarized as follows: (1) the vegetation coverage and growth are significantly improved after proper soil cover; (2) the large slope and soil erosion are the important reasons for the low vegetation coverage. (3) crop gnawing caused by poor enclosure will seriously reduce the effect of ecological restoration. The phenomenon that the tender leaves of artificially planted grass are eaten by livestock and a large amount of cattle and sheep manure can be seen at the scene. According to the local people, cattle and sheep especially like to eat the young leaves of artificially planted grass. As a result, the early recovery effect is not sustainable.

The soil type of the slag mountain is a mixture of slag and soil, with high gravel content and serious soil erosion. The vegetation grows generally and tends to turn yellow, and it grows better after replanting and covering with plastic film. According to the comprehensive evaluation, the vegetation in the southern and northern dregs mountains is degraded obviously, the effect of vegetation restoration is poor, and there is a risk of soil erosion and collapse (Yin and Yunhai 2007).

36.4 Eco-Environmental Problems in the Study Area

Through the collection of relevant data, interpretation of remote sensing image data and field investigation, it is found that the ecological environment in the study area has been damaged to varying degrees due to the mining of coal resources. The main eco-environmental problems are topography and geomorphology destruction, vegetation and meadow degradation, soil and water loss, permafrost and rock damage, pit stagnant water and slope instability and so on (Jingping 2000).

36.4.1 Destruction of Topography and Geomorphology

One open pit and two slag hills have been formed in the study area due to coalfield mining. The total area of the open pit is 530000 m^2. The total area of the dregs mountain is 960000 m^2. The roads, temporary facilities and buildings in the mining area all cause varying degrees of damage to the topography and geomorphology. The land destroyed by mining is mainly natural pasture.

36.4.2 Vegetation and Meadow Degradation

The quality of grassland in the mining area is low, the ecological environment is fragile, and it is easy to degrade. The soil type of slag mountain is the mixture of slag and soil, high gravel content, large slope and soil erosion are the important reasons for the low vegetation coverage. At the same time, it can be seen that the tender leaves of artificially planted grass are eaten by livestock and a large amount of cattle and sheep droppings. According to local people, cattle and sheep especially like to eat the tender leaves of artificially planted grass. Poor fencing leads to serious crop eating, which reduces the effect of ecological restoration and leads to poor effect of vegetation restoration in the early stage (Fig. 36.1).

36.4.3 Frozen Soil Failure

The thickness of permafrost in the study area is 3086.65 m, which is distributed as an island, which results in slight erosion during melting, and the intensity of soil erosion is slight erosion. Although the destruction of permafrost in mining pit is beneficial to mining activities, it not only greatly affects the soil and eco-hydrological characteristics of the mining area, but also aggravates the risk of geological disasters such as thermal melting and collapse of open-pit slope. The slag mountain cover

Fig. 36.1 Vegetation degradation map of dregs mountain

changes the energy input of the original permafrost, while the slag mountain itself develops seasonal permafrost, which is easy to cause shallow landslides in this area.

36.4.4 Stagnant Water in the Pit

The study area is located in the Changjiang River basin, the mining coal seam in the area is below the groundwater level, and the stagnant water in the pit mainly comes from atmospheric precipitation. Due to the widespread distribution of permafrost in the mining area, the relationship between surface water and groundwater is isolated, and the surface drainage condition of the mining area is good, so it is less likely that the stagnant water in the mining pit comes from groundwater.

There are many irregular puddles distributed on the surface grassland in the study area, with a depth of about 0.4 m. There will be water in the pits in the rainy season, and all of them will dry up after the rainy season. The catchment area in the mining pit is large, the regional rainfall is abundant, and a large amount of atmospheric precipitation flows into the mining pit (Fig. 36.2).

36.4.5 Slope Instability

Due to open-pit mining, the permafrost layer is affected by frost heave, which will produce uneven settlement, and the hydrogeological conditions such as groundwater recharge and runoff are complex. The erosion of groundwater, the frost heave and thaw of permafrost and the erosion of surface water will aggravate the instability of mining pit slope.

The rock strata in the study area are mainly divided into two types: semi-hard rock group and soft rock group. Among them, the rock quality of the semi-hard rock

Fig. 36.2 Actual situation map of mining pit water accumulation area

formation is medium, the rock mass is moderately intact and the stability is good, while the compressive strength of the soft rock formation is small, and the rock quality, rock mass integrity and stability are very poor. Therefore, the weak rock formation is one of the main factors causing slope instability.

The structure of slag mountain waste is loose and the porosity is large. The surface water or short-term runoff formed by rainfall and snow melting will quickly flow into or infiltrate into the slag pile along the slag pile surface, resulting in erosion and infiltration erosion of the slag pile slope. Reduce the shear strength of the loose soil of the slag pile and the slope of the open-pit stope, resulting in slope instability, deformation and sliding.

The unstable slope is mainly located around the high and steep slope and slag pile formed by mining. Excavation produces a large number of waste rock, permafrost and gangue, which are stacked near the mining pit to form a slag pile up to 40–50 m high. Due to the lack of compaction treatment and untimely drainage, the slope is easy to produce tension cracks under the action of gravity to form an unstable slope. The slope in the pit is steep, the gravel in some areas accumulates and may collapse at any time, and there is a frozen soil thawing layer at the top of the slope, and there is a phenomenon of step collapse in summer (Li et al. 2005).

36.5 Eco-Environmental Protection Measures and Slope Stability Analysis

36.5.1 Eco-Environmental Protection Measures

According to geological survey and mapping, exploration and instability analysis, combined with the topography, rock mass structure and rock mass integrity of the slope, the protection measures can be divided into the following four aspects.

Geomorphological reconstruction: the geomorphological reconstruction is carried out in the exploration area, the top of the slag mountain and the current slope of the slag mountain are cut and fixed, and the slope shaping is carried out. A drainage trench is set at the foot of the slope to divert water from the top of the mountain and the slope to the surrounding wetlands. Ensure the overall stability of the dregs and provide conditions for greening at the same time. The present water body in the mining pit is retained, and the plateau lake is constructed through the natural restoration of water supply.

Soil reconstruction: fine-grained materials such as sandstone and mudstone obtained from slag slopes can be used as soil substitute materials for vegetation growth medium, and soil can be reconstructed by mixing and mixing with organic fertilizer.

Vegetation reconstruction: after the completion of mine backfilling, slag hill and road surface renovation, appropriate native grass species are selected, and vegetation restoration technology in alpine areas is adopted to restore vegetation. In the later stage, the purpose of vegetation restoration is achieved through fencing, animal husbandry prohibition, management and protection, ecological monitoring and other measures.

Fencing: after the completion of mine backfilling, slag hill and road surface renovation, the local grass species suitable for local natural conditions are selected, and the vegetation restoration technology in alpine area is adopted to restore vegetation. In the later stage, the purpose of vegetation restoration is achieved through fencing, animal husbandry prohibition, management and protection, ecological monitoring and other measures.

36.5.2 Slope Stability Analysis

According to the stability analysis of the slope after treatment, the construction slope (GB 50330-2013) and the technical code for slope engineering of non-coal open-pit mine (GB 51016-2014) are judged according to the safety registration load combination I of the second slope, and the safety factor is 1.20 (Table 36.1).

Engineering geological rock groups are mainly divided into two types: semi-hard rock groups and soft rock groups. The integrity of the rock layer is poor, it is a weak

Table 36.1 Design safety factor of overall slope under different load combinations of non-coal open-pit slope

Safety grade of slope engineering	Safety factor of slope engineering		
	Load combinationI	Load combinationII	Load combinationIII
I	1.25–1.20	1.23–1.18	1.20–1.15
II	1.20–1.15	1.18–1.13	1.15–1.10
III	1.15–1.10	1.13–1.08	1.10–1.05

layer, and its stability is poor. Therefore, according to the Engineering Rock Mass Classification Standard (GBT50218-2014), the parameters of V-grade rock mass with poor stability are reduced. At present, the relevant codes only give the range of suggested values of rock mass parameters according to the level of rock mass, and there are no specific provisions on rock mechanics parameters and reduction methods of rock mass mechanics parameters. This project is reduced according to the Hawke Brown rock mass strength reduction method which is widely used in the world.

Determine the reduction parameters s, a, D, GSI, MR according to the rock mass grade and excavation method.

$$s = \exp\left(\frac{GSI - 100}{9 - 3D}\right), a = \frac{1}{2} + \frac{1}{6}\left(e^{-GSI/15} - e^{-20/3}\right)$$

Compared with the BQ method in the Engineering Rock Mass Classification Standard (GBT50218-2014), according to the relevant data of V-grade rock mass, the GSI value is 40. The D value of large slope excavation is 1, and the value of MR is 300 according to the comprehensive consideration of rock layer distribution.

The uniaxial compressive strength of rock mass: $\sigma_c = \sigma_i \cdot s^a$.

Estimation of rock modulus: $E_i = MR \cdot \sigma_i$.

Rock mass modulus:

$$E_{rm} = E_i\left(0.02 + \frac{1 - D/2}{1 + e^{((60+15D-GSI)/11)}}\right)$$

After reduction, it is considered that the rock mass strength should correspond to the recommended value of rock mass strength in BQ method, and the five-grade rock mass parameters should be retained. According to the well field exploration report, the curves of rock mechanical parameters are calculated, and 215 rock stratification data from 5 boreholes are counted. After weighted average according to geological age, the parameters are calculated according to the above-mentioned reduction method. The slag mountain material is a mixture of sandstone, sandy shale, fine sandstone, mudstone, siltstone and Quaternary loose layer above the coal seam roof. According to the exploration report and engineering experience, the bulk density 2.05 t/m^3, internal friction angle 32° and cohesion 0 kPa are determined.

The slope stability analysis of the well pit treated according to the design is carried out, and the overall strength reduction method is used to simulate and analyze each slope to check the position and stability of the slip surface. The most dangerous slip surface of each section is on the north slope, and the most dangerous slip surface of each section is along the fill-undisturbed soil interface. The overall displacement trend of the slope is to move to the open surface, but the shear strain rate is small, and the through slip surface is not formed. The safety factors are all greater than 1.20.

36.6 Conclusion

The rational development and efficient utilization of coal mine resources are of great significance to the sustainable development of the national economy, and the predatory mining of coal mines brings great harm to the social and ecological environment. When restoring and harnessing the mine ecological environment in the mining area in different geographical positions, the main ecological environment problems existing in the mining area should be fully analyzed, combined with the key factors such as topography, vegetation cover, rainfall. Only through comprehensive analysis and research, the damaged ecosystem can be maximum restored.

This paper discusses the problems existing in the study area, such as the destruction of topography, vegetation degradation, permafrost destruction, stagnant water in the pit, slope instability. It is put forward that the restoration measures such as geomorphological reconstruction, soil reconstruction, vegetation reconstruction, fence closure and later management and protection in the study area which can restore the destroyed ecosystem in the mining area. Based on the stability analysis of the treated slope, it is concluded that the stability of the treated slope is better.

Acknowledgements I would like to thank my colleagues for their strong support and help in the preparation of the paper, which has brought me great inspiration. I would also like to thank the authors in the references for giving me a good starting point for choosing the topic of my thesis through their research articles. I would like to express my heartfelt thanks to my family and friends. I can finish this paper successfully with their encouragement and support.

Funding This research was funded by the China Geological Survey 'National mine development and ecological space remote sensing monitoring in key areas', grant number 20201200000210017.

References

Chen Z, Ming Z, Liu N et al (2020) Investigation and evaluation of geological environment problems of closed coal mine in Fengfeng mining area based on remote sensing technology. China Coal 46(11):112–117

Chen Z, Wu J, Guo Y et al (2018) Application of AHP and fuzzy mathematics in comprehensive assessment of mine environment. Resour Survey Environ 39(04):67–72

Duan Y, Duan Y (2007) Influence analysis of coal mining on ground water environment. Environ Sci Manage 32(8):176–178

Feng J (2003) Environmental problems caused by surface mining and preventive countermeasures in China. Safety Coal Mines 34(z1):122–123

Gu T, Wang J, Wang Y et al (2007) Comprehensive evaluation of urban geological environment in Xianyang City. Bull Soil Water Conserv 3:69–74

Liao L, Yang Z, Yang F et al (2021) Environmental problems and countermeasures of AMD in pyrite and high-sulfur coals in Southwest China. J Chengdu Univ Technol (Sci Technol Edn) 48(6):754–761

Li F, Hu K, Feng J et al (2005) Mining solid waste in China: present conditions and countermeasures. Sci Technol Manage Land Resour 3:66–70

Ren H, Peng S (2001) An introduction to restoration ecology. Science Press, Bei Jing

Radbruch-Hall DH, Colton RB (1982) Landslide overview map of the conterminous United Stated. United Stated Government Printing Office, Washington
Saaty TL (1988) Multi-criteria decision making: the analytic hierarchy process. Pittsburgh
Wang J (2000) The influence on the environment and its tientment policies of the development of coal resource—takes Xiazhuang coal mine of Shandong province as an example. J Hebei Normal Univ (Nat Sci) 24(4):544–547
Zadeh LA (1965) Fuzzy sets. Inf Control 8(1):340–352
Zhang F, Hu X, He Q (2019) Application of fuzzy analytic hierarchy process in water pollution treatment engineering performance evaluation. Ecol Environ Monitoring Three Gorges 004(002):1–8

Chapter 37
Research on Characteristics of Port Particulate Matter (PM_{10}) Fugitive Emissions Based on UAV Sniffing Technology

Liang Jing and Luhua Zhao

Abstract Under the severe environmental protection situation, port enterprises generally face the problem of failing to grasp current situation of environmental pollution comprehensively, resulting in ineffective environmental governance. In this study, the concentration of PM_{10} was monitored in the front and rear storage yard of a domestic dry bulk operation port through an unmanned aerial vehicle (UVA) equipped with a sniffer, and heat maps of PM_{10} concentration at 60 m above the storage yards were drawn. This study also analyzed the dust generation characteristics and the impact of dry bulk operations through a comparison between the measured and official released data. It is shown that the dry bulk operation has little impact on the atmospheric environment of ports and surrounding areas under conventional operating and meteorological conditions, and the distribution characteristics of PM_{10} in the specific space have a strong correlation with the instantaneous emission of dust generated by dry bulk cargo handling operations. The technical means and research results proposed in this study is proved to provide technical support for the optimization of port environmental management.

Keywords UAV · Sniffing · Particulate matter · Fugitive emissions

37.1 Background

As a key node in the waterway transportation of energy and raw materials, the port has greatly promoted the rapid development of the regional economy. At the same time, the subsequent environmental pollution problems have also attracted much attention. In particular, the fugitive emissions of particulate matter caused by dry bulk cargo

L. Jing (✉)
Environmental Technology Development of TIWTE (TIANJIN) CO., Ltd., Tianjin Research Institute of Water Transport Engineering, Ministry of Transport, Tianjin, China
e-mail: jingl@tiwte.ac.cn

L. Zhao
Shandong Port Group CO., Ltd., Qingdao, Shandong, China

J. Zhang et al. (eds.), *Environmental Pollution Governance and Ecological Remediation Technology*, Environmental Science and Engineering,
https://doi.org/10.1007/978-3-031-25284-6_37

such as ore and coal during its loading, unloading and transportation has made a great impact on the atmospheric environment of ports and surrounding areas, which is also considered to be one of the primary sources of atmospheric particulate matter pollution in coastal cities (Yunfei et al. 2014; Hong et al. 2013; Shan et al. 2006; Weiping 2006). With the proposal of a series of national environmental protection strategies in recent years, the environmental management of the dry bulk port has gradually transitioned from the factory boundary compliance to the whole operation process control. Therefore, port enterprises have also been increasing their efforts to control dust pollution, and tried to grasp the overall dust pollution situation in the port area with a more timely monitoring model, which contributes to the formation of a more targeted and systematic environmental management model (Chunyi 2020).

UAV sniffing technology is to use UAV equipped with remote sensing sensors, combined with telemetry and remote-control technology, edge computing technology, wireless communication technology, GIS geographic information technology and GPS positioning technology to obtain image data, and synchronously transmit the image to complete data processing, modeling and analysis (Yingjun 2021; Qing et al. 2020). It has the characteristics of wide field of view, flexible, controllable flight height, combination of fixed point and cruise, etc., and has been widely used in many fields (Meiling et al. 2008; Changhao et al. 2020; Long et al. 2005; Xuyu 2019; Qingzi et al. 2015). By applying the UVA sniffing technology to particle concentration monitoring, the concentration contour map within the local area of the port can be drawn in a short time, which can better reflect the characteristics of particle emission and provide technical support for the optimization of environmental management.

37.2 Monitoring Area

Dongjiakou Port is a comprehensive port with the largest planned land area and number of berths in Shandong Port. It is also a national bulk cargo distribution center as well as an important energy storage and transportation center. At present, it has realized the integrated operation of one 400,000-ton wharf, two 200,000-ton wharf and one 150,000-ton wharf. And the throughput of dry bulk cargoes including iron ore, bauxite, coal, coke, etc. will be about 85.8 million tons in the year of 2020.

The storage yard in Dongjiakou Port mainly includes the front and the rear storage yard, with a total area of more than 3.7 million square meters. In the process of storage and transshipment of dry bulk cargo, it will inevitably cause fugitive emission of dust. Generally, the operation process can be divided into specialized and general processes, of which the dust-generating nodes are roughly the same. And the dust production is mainly affected by its number. The operation process and dust-generating nodes are shown in Fig. 37.1.

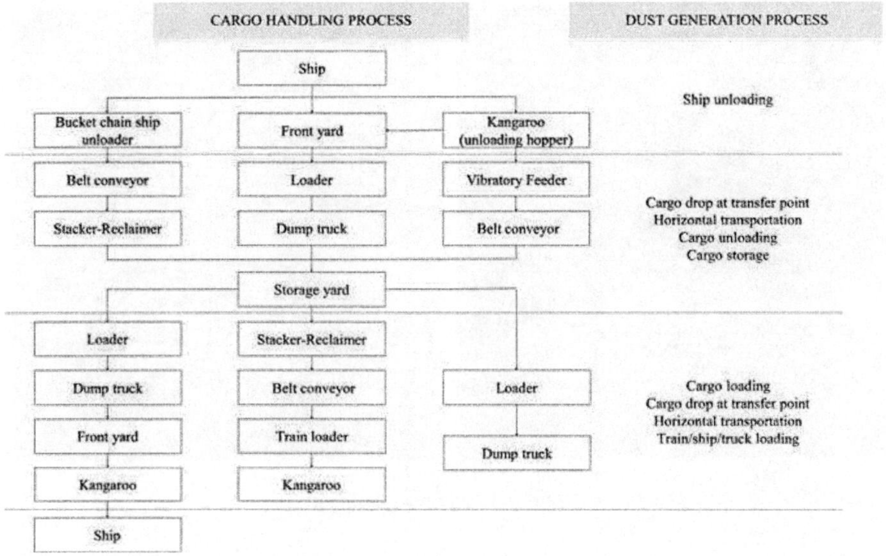

Fig. 37.1 Operation process and dust-generating nodes

37.3 Monitoring Equipment

This study used the DJI M210 quadrotor UAV flight platform, equipped with the host of Lingxiu ultra-fine grid atmospheric mobile monitoring system and Sniffer 4D Mapper data visualization and analysis software. It can display the concentration value and its real-time changes. And can visually display the concentration and distribution of $PM_{2.5}$, PM_{10}, O_3, NO_2, CO and SO_2 by visualization methods such as two-dimensional grid concentration distribution heat map, two-dimensional isoline concentration distribution heat map, and three-dimensional point cloud concentration distribution heat map. The system supports 4G long-distance data transmission. The monitoring time for a single flight is about 30 min.

37.4 Results and Analysis

37.4.1 Regional Environmental Conditions

This experiment was carried out on January 8, 2021. The meteorological conditions of Dongjiakou Port area were measured as northwest wind 4.7 m/s, relative humidity 51%. While the ambient air quality in Huangdao announced by Qingdao Ecology and Environment Bureau is moderate, and the average concentration of PM_{10} is 59 $\mu g/m^3$, which is used as background concentration.

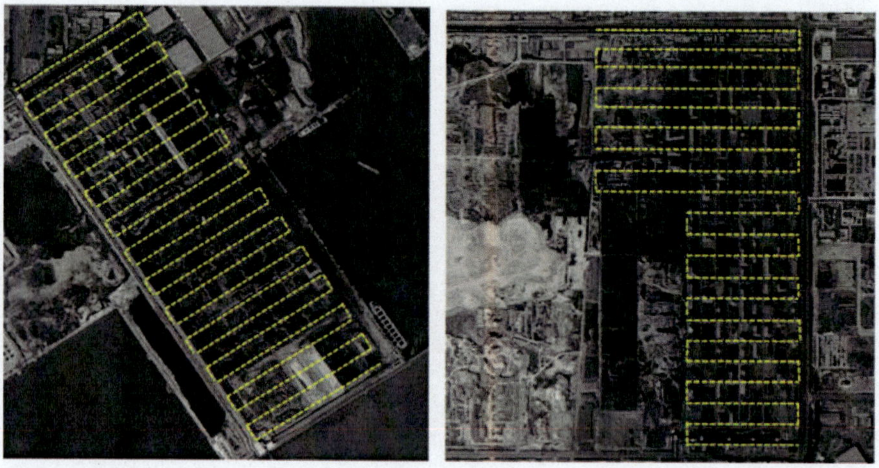

Fig. 37.2 Monitoring path of front yard (left) and rear yard (right)

37.4.2 Path Setting

The monitoring path covers the front and rear storage yards, and a 10-m extension of the factory boundary. The path settings are shown in Fig. 37.2. Considering that the operating height of large-scale machinery in the field is about 50 m, the scanning height was set at 60 m in order to avoid collision accidents.

37.4.3 Results Analysis

Front yard. The concentration ranges of PM_{10} within 10 m outside the front yard boundary and inside the yard are 8 $\mu g/m^3$–67 $\mu g/m^3$ and 10 $\mu g/m^3$–110 $\mu g/m^3$, respectively. The concentration inside the yard is slightly higher than that of the boundary. It can be seen that a certain amount of dust is generated by the dry bulk cargo operation inside the yard, but the 23-m-high windbreak net at the boundary has a good inhibitory effect on dust pollution.

The wind direction is northwesterly during the monitoring period. The northwest side of the front yard is bulk oil terminal, and there is no other particulate pollution source except for some bare ground. Based on the overall distribution of particulate matter concentration, it can be seen that the value in the area where the specialized and general processes are connected shows a significant increase. The operation of stacker-reclaimers and belt conveyors in the northernmost process line of the yard is considered to be the direct cause of fine particles escape. And the particulate matter cloud with a higher concentration gradually spreads to the southeast under northwesterly wind. The south side of the yard mainly adopts general technology operations.

The open-air stacking has been well covered and only small-scale operations were performed locally, therefore resulting in a relatively slight particle escape (Fig. 37.3).

Rear yard. The east and west sides of the rear yard are the specialized and general process operation areas, respectively. There were no large-scale dry bulk handling operations during the period of monitoring. And the concentration ranges of PM_{10} within the monitoring range is 10–66 $\mu g/m^3$. It can be seen from Fig. 37.4 that an instantaneous concentration increase appeared on the south side of the specialized operation area. This is mainly due to stacking bottom cleaning and ground sweeping, which caused local escape of particulate matter. Overall, the dust pollution of the rear yard has been well controlled, which is mainly attributed to the high coverage rate of static stacking and the continuous operation of sprinklers.

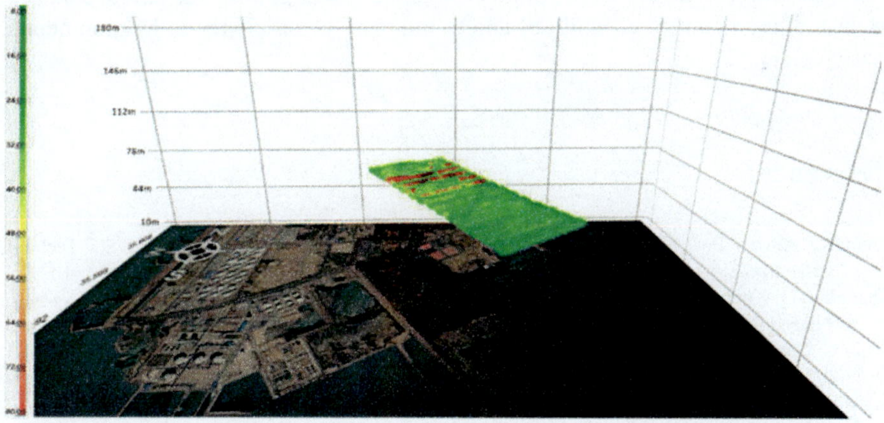

Fig. 37.3 Heat map of particle concentration in the front yard

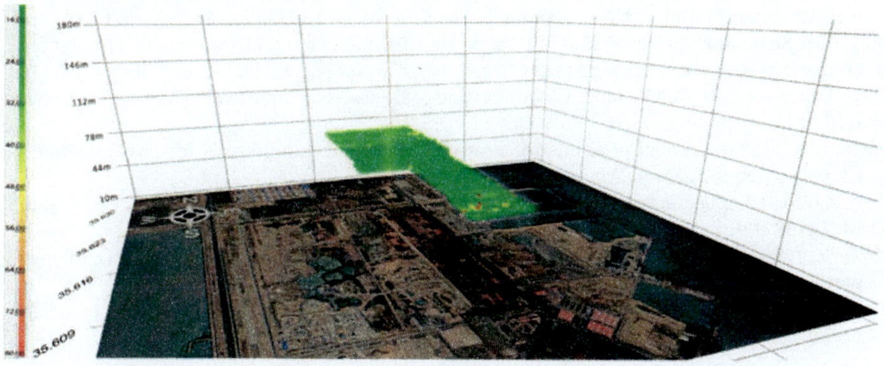

Fig. 37.4 Heat map of particle concentration in the rear yard

37.5 Conclusion

According to the results of UAV monitoring under conventional meteorological conditions, it is found that the application of environmental protection measures such as windbreak nets, stacking covering, and yard cleaning and watering has played a good role in suppressing the fugitive emission of dust from dry bulk cargo operations. Although some of the monitoring values of particulate matter concentration are higher than the average value in Qingdao at the same time period, they will not have a great impact on the ambient air quality in the port and surrounding areas. The thermal map of particle concentration in the front and rear yards based on monitoring data clearly reflects the characteristics of dust emissions from ground operations, which can provide an intuitive decision-making basis. The technical methods and research results proposed in this study are of great significance for guiding the port to enhance its air pollution control capabilities and to improve the efficiency of environmental management.

References

Chunyi Z (2020) Port ecological intelligent monitoring and control system. Port Sci Technol 176(10):20–22

Changhao D, Xuejun F, Yan Z, Sai M, Yunpeng Z (2020) Study on $PM_{2.5}$ distribution in dry bulk port based on unmanned aerial vehicle platform. Environ Pollut Control 337(12):63–67

Hong C, Guangru L, Yunhui D, Haizhen Z (2013) The analysis of impact factor of harbor dust and pollution controlling measures. Northern Environ 102(06):98–102

Long Y, Kebin H, Qiang Z, Qidong W (2005) Vertical distributive characters of $PM_{2.5}$ at the ground layer in autumn and winter in Beijing. Res Environ Sci 2:23–28

Meiling S, Huaibin M, Danzhu W (2008) Vertical distribution characteristics of $PM_{2.5}$ mass concentration in the atmosphere over Tianjin. Meteorol Monthly 10:60–66

Qing Y, Qiang Z, Yin L (2020) Ecological environment monitoring of Gannan Tibetan autonomous prefecture based on small UAV and remote sensing technology. J Green Sci Technol 10:143–147

Qingzi M, Shuxian F, Jiabao H (2015) Particle size distribution and characteristics of polycyclic aromatic hydrocarbons during a heavy haze episode in Nanjing, China. Particuology 1:127–134

Shan G, Shuifen Z, Binbin Z, Jun Z (2006) Review of Dedust technique in coal harbor. J Waterway Harbor 01:40–44

Weiping W (2006) Prevention & control strategies and it's existing status for dust pollution in Chinese harbour. Environ Prote Transp 4:1–6

Xuyu Z (2019) Application analysis of UAV in atmospheric environment monitoring. Environ Dev 31(7):138–139

Yunfei Z, Jiahai Z, Peng Z (2014) Research on dust pollution prevention and control technology in port bulk cargo handling process. China Water Transp 014(012):309–310

Yingjun X (2021) Research on the application of UAV remote sensing technology in the field of ecological environment. China Resour Comprehensive Utilization 39(2):3

Chapter 38
Novel Anionic Polymer-Modified Magnetic Chitosan-Based Composites for Efficient Adsorption of Ciprofloxacin Over a Wide pH Range

Moxi Wang and Xue-yi You

Abstract In this work, a novel magnetic chitosan-based adsorbent (MC-PS) was synthesized via anionic polymer grafting. MC-PS was proved to possess a large adsorption capacity of 416.67 mg g^{-1} for ciprofloxacin (CIP) and a wide pH adaptability. Chemisorption was the main rate-controlling step in the adsorption process, and CIP was monolayered on the adsorbent surface. Furthermore, MC-PS could be readily separated, regenerated, and reused. The ionic strength effect revealed that competitive adsorption existed between the salt ions and CIP molecules, confirming the role of electrostatic interaction. In addition, π-π stacking and hydrogen bonding were also involved in the removal mechanism. This work provides a new strategy for the preparation of adsorbents with excellent performance over a wide pH range and capable of being reusable.

Keywords Adsorbent · Ciprofloxacin · pH adaptability

38.1 Introduction

The extensive use of antibiotics has brought a series of negative impacts on the environment and human health. The removal of antibiotics from water bodies has long been a public concern due to their refractory biodegradability and bacterial resistance (Guo et al. 2021). Ciprofloxacin (CIP) is a broad-spectrum fluoroquinolone antibiotic widely used in humans or animals. Compared with other traditional removal processes, adsorption is increasingly used in the removal of antibiotics due to its high efficiency, low cost, and environmental friendliness (Wang and You 2021; Alsbaiee et al. 2016).

Chitosan is a natural organic polymer with wide sources, which contains abundant active amino and hydroxyl groups (Wang et al. 2021, 2018). Magnetic chitosan-based

M. Wang · X. You (✉)
Tianjin Engineering Center of Urban River Eco-Purification Technology, School of Environmental Science and Engineering, Tianjin University, Tianjin 300350, China
e-mail: xyyou@tju.edu.cn

J. Zhang et al. (eds.), *Environmental Pollution Governance and Ecological Remediation Technology*, Environmental Science and Engineering,
https://doi.org/10.1007/978-3-031-25284-6_38

composite is a new type of material that can obtain a large number of available active groups by functionalizing chitosan. It can effectively make up for the application limitations of chitosan on the premise of retaining its good physical and chemical properties. Moreover, compared with traditional water treatment materials, the introduction of magnetic materials greatly improves the separation efficiency of adsorbents (Zhao et al. 2021), and thus it is a potential adsorbent for removing antibiotics in water.

In this work, a novel magnetic cross-linked chitosan modified with poly(4-styrenesulfonic acid-co-maleic acid) sodium salt (PSSMA) with SiO_2 functionalized coating (MC-PS) was synthesized for the first time through a simple method and applied to adsorb ciprofloxacin in water. MC-PS was characterized by FTIR, N_2 adsorption–desorption isotherm, hysteresis loop, and zeta potential. Besides, pH effects, adsorption kinetics, isotherms, regenerability, and ionic strength effects of CIP on MC-PS were also investigated.

38.2 Materials and Methods

38.2.1 Synthesis of Adsorbents

45 mL of TEOS and 15 mL of ammonia solution were added to the Fe_3O_4 suspension at room temperature with rapid stirring. After continuous stirring for 10 h, SiO_2 functionalized coated Fe_3O_4 was obtained by separation with a permanent magnet and thorough washing with absolute ethanol and ultrapure water. Then 3.0 g CTS was dissolved in 40 mL of acetic acid solution. The above Fe_3O_4 was dispersed in the CTS solution through ultrasonic and mixed for 8 h. It was added to a mixture of 650 mL cyclohexane and 4 mL Span 80 in the form of a paste under stirring at 750 rpm, and then 3.0 mL of 50 wt% glutaraldehyde solution was added to cross-link for 2 h. The sample MC was obtained by magnetic separation, rinsing, drying, and grinding. Finally, a predetermined amount of PSSMA solution was added to the MC dispersed in 200 mL ultrapure water, and stirred under reflux for 8 h at 45 °C. The product was separated and collected by the magnetic field, and MC-PS was obtained after thorough washing and drying. Unmodified MC was used as a comparative adsorbent in the experiments.

38.2.2 Batch Adsorption Experiment

50 mg of the synthesized adsorbent was added to 100 mL of ciprofloxacin solution and shaken at 180 rpm for 12 h at room temperature. The adsorption kinetic (0–12 h) was carried out in 200 mg L^{-1} CIP solution; the isotherm experiments were performed at a constant optimal pH under 298.15, 308.15, and 318.15 K, respectively.

After adsorption, the adsorbent was separated from the water body by an external magnet. Subsequently, the residual CIP concentration was determined using a UV–Vis spectrophotometer (TU-1901, Beijing Purkinje General Instruments Co., Ltd., China) at the maximum absorption wavelength (λ_{max}) of 275 nm. Each test was performed in triplicate and the mean value was calculated.

The adsorption capacity q_e (mg g^{-1}) of the adsorbent at equilibrium and the adsorption capacity q_t (mg g^{-1}) at t were calculated by Eqs. (38.1–38.2), respectively.

$$q_e = \frac{C_0 - C_e}{W} V \tag{38.1}$$

$$q_t = \frac{C_0 - C_t}{W} V \tag{38.2}$$

where C_0 (mg L^{-1}), C_e (mg L^{-1}), and C_t (mg L^{-1}) are referring to the initial concentration of pollutants, the residual concentration of pollutants at equilibrium, and different time intervals, respectively; V(L) stands for the volume of the solution; W(g) represents the adsorbent mass.

38.3 Results and Discussion

38.3.1 Characterization

The FTIR spectra of MC, MC-PS, and CTS are shown in Fig. 38.1a. In the spectrum of MC, a characteristic peak representing imine at 1651 cm^{-1} was detected, which was attributed to the Schiff base reaction, indicating that the magnetic chitosan was successfully cross-linked (Zhao et al. 2021). All characteristic peaks representing CTS and PSSMA (marked with different colors in Fig. 38.1a) could be observed in the spectra of MC-PS, although their intensities and positions varied to varying degrees, confirming that the polymer branches were successfully grafted to the adsorbent surface.

The N_2 adsorption–desorption isotherms of MC and MC-PS are illustrated in Fig. 38.1b. According to the IUPAC classification, MC and MC-PS exhibited type IV isotherms with H3 hysteresis loops, confirming their mesoporous characters (Guo et al. 2021). MC-PS owned a high specific surface area of 173.81 m^2/g, which was attributed to its special magnetic core-silicon shell-polymer dendritic structure.

The zeta potential changes of MC and MC-PS are displayed in Fig. 38.2a. Among them, the pH_{pzc} of MC was confirmed to be 4.3, while the Zeta potential of MC-PS was negative at all pH values due to the introduction of plentiful negatively charged functional groups through polymer modification. Such properties of MC-PS also suggested its potential stronger electrostatic attraction to cationic contaminants.

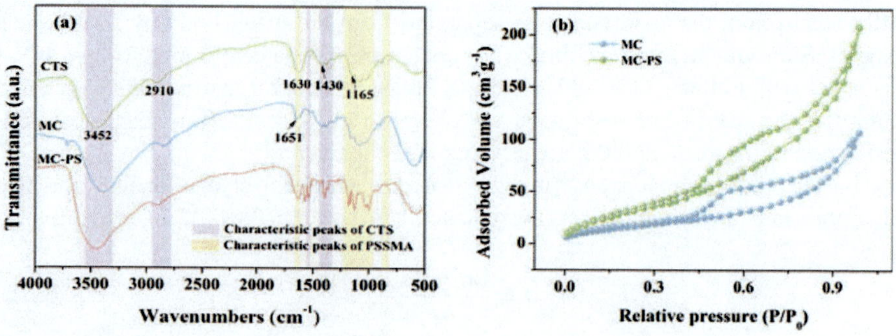

Fig. 38.1 Characterization of **a** FTIR spectra and **b** N_2 adsorption–desorption isotherms

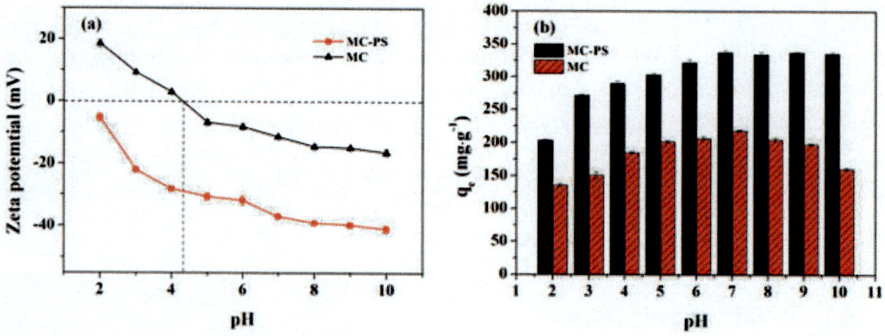

Fig. 38.2 **a** Zeta potentials and **b** pH effects of MC-PS and MC

38.3.2 pH Effect

In batch experiments, the pH effects during CIP removal by MC and MC-PS were first investigated. As shown in Fig. 38.2b, the adsorption capacity of MC-PS for CIP revealed an upward trend in the pH range of 2–7 and leveled off in the pH range of 7–10. At pH < 5.9, CIP presented a cationic state in aqueous solution. At the same time, the deprotonation degree of $-SO_3^-$ and $-COO^-$ on MC-PS increased with the increase of pH, thus the electrostatic interaction between MC-PS and positively charged CIP was enhanced. When pH > 5.9, the CIP^- species in the solution gradually increased, and the electrostatic attraction with MC-PS weakened. The adsorption capacity remained at a high level owing to the π-π interaction between the benzene ring as an electron donor in MC-PS and the aromatic ring in CIP, as well as the hydrogen bonding between the hydroxyl groups on the MC-PS surface and the nitrogen/oxygen-containing functional groups in the CIP structure.

38.3.3 Adsorption Kinetics and Isotherm Analysis

The adsorption capacity of the adsorbent for CIP as a function of time is illustrated in Fig. 38.3a. The contaminant was rapidly adsorbed at the initial stage of adsorption due to the abundant adsorption sites on the adsorbent surface and the large concentration gradient in the solution. Subsequently, the removal of CIP by MC-PS reached equilibrium after 110 min. In addition, in order to explore the adsorption mechanism, the obtained kinetic data were fitted by pseudo-first-order (PFO), pseudo-second-order (PSO), and Elovich models. The kinetic models are represented by Eqs. (38.3–38.5):

$$q_t = q_e - q_e e^{-k_1 t} \tag{38.3}$$

$$q_t = \frac{k_2 q_e^2 t}{1 + k_2 q_e t} \tag{38.4}$$

$$q_t = B \times \ln(XYt) \tag{38.5}$$

where q_t (mg g^{-1}) and q_e (mg g^{-1})are the pollutant adsorption capacities at equilibrium and t (min), respectively; k_1 (min^{-1}) and k_2 (g mg^{-1} min^{-1}) represent the adsorption rate constants of PFO and PSO, respectively; X and Y represent the desorption constants of the Elovich model.

The fitting results are shown in Fig. 38.3a. The PSO model had the highest correlation coefficient and fitted the experimental data better. Therefore, chemisorption was the main rate-controlling step.

The adsorption isotherms of MC-PS at three temperatures are displayed in Fig. 38.3b. With the increase of the initial CIP concentration, the adsorption capacity showed a trend of rapid increase at first and then tended to be saturated, and the high temperature was beneficial to the increase of adsorption capacity. In order to explore the adsorption process in depth, the experimental isotherm data were fitted by the Langmuir model and the Freundlich model with the following equations:

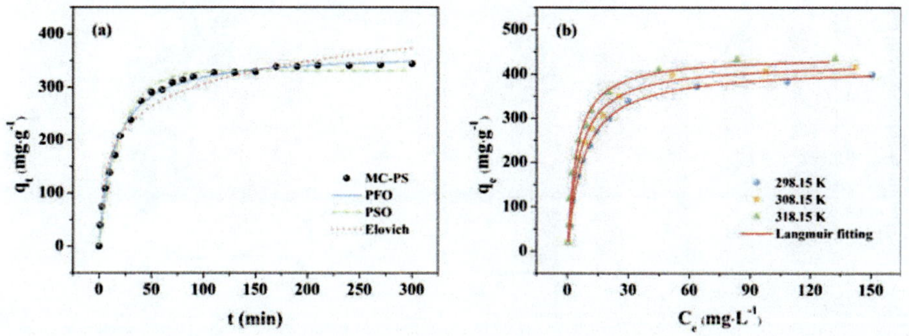

Fig. 38.3 **a** Adsorption kinetics and **b** isotherms of CIP adsorbed on MC-PS

$$q_e = \frac{q_{max} C_e K_L}{1 + C_e K_L} \tag{38.6}$$

$$q_e = K_F C_e^n \tag{38.7}$$

where C_e (mg L^{-1}) and q_e (mg g^{-1}) represent the residual concentration of pollutants in the supernatant after equilibrium and the equilibrium adsorption capacity of the adsorbent, respectively; q_{max} (mg g^{-1}) is the maximum adsorption capacity calculated according to Langmuir model; K_L (L mg^{-1}) and K_F (mg^{1-n} L^n g^{-1}) stand for the correlation constants of the Langmuir and Freundlich models, respectively; n is the heterogeneity factor.

Compared with Freundlich, the Langmuir model had the highest correlation coefficient $R^2 > 0.999$, indicating that the antibiotic molecules were covered by a monolayer on the adsorbent. The maximum adsorption capacity of MC-PS for CIP at 298.15 K was 416.67 mg g^{-1}, which was much higher than that of other types of adsorbents reported, thus it was a potential water treatment material.

38.3.4 *Reusability and Ionic Strength Effect of MC-PS*

The mixture of ethanol and HNO_3 in a volume ratio of 1:1 was used as the desorbent, and six adsorption–desorption cycle experiments were carried out continuously. As shown in Fig. 38.4a, the final adsorption capacity of MC-PS remained at 91% of the initial value, and its slight decrease was caused by the difficulty in desorbing the contaminants near the magnetic core of the adsorbent. The experimental results showed that MC-PS had good stability and excellent reusability.

The effects of NaCl and $NaNO_3$ under different ionic strengths on the adsorption performance of MC-PS were investigated. As shown in Fig. 38.4b, as the salt ion concentration gradually increased to 0.1 M, the adsorption capacity of MC-PS for CIP

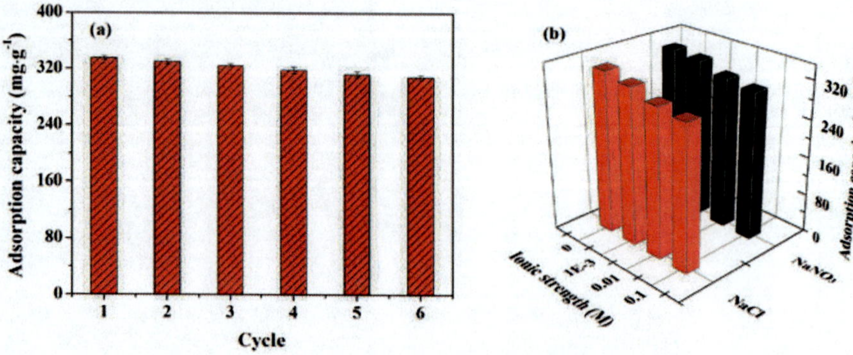

Fig. 38.4 **a** Reusability experiments and **b** ionic strength effects of MC-PS

gradually decreased, indicating that competitive adsorption existed between salt ions and CIP molecules, which further confirmed the vital role of electrostatic interaction during the adsorption process.

38.4 Conclusion

In this work, a PSSMA-functionalized magnetic chitosan adsorbent (MC-PS) with SiO_2-functionalized coating was prepared for the efficient removal of ciprofloxacin (CIP) from water. MC-PS revealed excellent adsorption performance for CIP over a wide pH range. The pseudo-second-order model and the Langmuir model could best describe the adsorption process, demonstrating that chemisorption was the dominant rate-controlling step and CIP was monolayered on the adsorbent. According to the Langmuir model, the maximum adsorption capacity of MC-PS for CIP was 416.67 mg g^{-1}, which was much higher than that of other reported adsorbents. Moreover, the adsorption capacity could be maintained above 91% after six cycles, confirming its excellent reusability. From the comprehensive experimental results, electrostatic interaction, π-π conjugation, and hydrogen bonding coexist in the adsorption mechanism of drug molecules by MC-PS.

Acknowledgements This work was supported by the key projects in the control and management of national polluted water bodies (2017ZX07106001).

References

Alsbaiee A, Smith BJ, Xiao L, Ling Y (2016) Rapid removal of organic micropollutants from water by a porous β-cyclodextrin polymer. Nature 529:190–194

Guo Z, Yang F, Yang R, Sun L, Li Y, Xu J (2021) Preparation of novel ZnO-NP@Zn-MOF-74 composites for simultaneous removal of copper and tetracycline from aqueous solution. Sep Purif Technol 274:118949

Wang M, You X (2021) Critical review of magnetic polysaccharide-based adsorbents for water treatment: synthesis, application and regeneration. J Clean Prod 323:129118

Wang M, Feng L, You X, Zheng H (2021) Effect of fine structure of chitosan-based flocculants on the flocculation of bentonite and humic acid: evaluation and modeling. Chemosphere 264:128525

Wang M, Feng L, Fan X, Li D, Qu W, Jiang S, Li S (2018) Fabrication of bifunctional chitosan-based flocculants: characterization, assessment of flocculation, and sterilization performance. Materials 11:2009

Zhao M, Cui Z, Pan D, Fan F, Tang J, Hu Y, Xu Y, Zhang P, Li P, Kong X, Wu W (2021) An efficient uranium adsorption magnetic platform based on amidoxime-functionalized flower-like Fe_3O_4@TiO_2 core-shell microspheres. ACS Appl Mater Interfaces 13:17931–17939

Chapter 39
Analysis of the Existing Problems in the Environmental Impact Assessment System of Industrial Park Planning and Suggestions for Countermeasures

Cuiyun Lu and Baoyuan Huang

Abstract Industrial park is the carrier of industrial agglomeration, and its planning environmental impact assessment assumes the function of pollution source prevention, which shows strong Chinese characteristics in practice. This paper systematically sorts out the problems and confusions faced in the current industrial park planning and environmental impact assessment work, combined with the current situation and requirements faced by China's industrial park planning and environmental impact assessment work, and puts forward targeted implementation of the park planning and planning environmental impact assessment work, and the construction of planning environmental impact assessment. The connection of working mechanisms, the realization of the synergy of pollution reduction and carbon reduction in the industrial park, and the implementation of the "three lines and one list" and the key links of the planning environmental assessment are the countermeasures and suggestions.

Keywords Industrial park · Planning environmental impact assessment · Environmental impact assessment linkage · Three lines and one list

39.1 Introduction

China has a vast territory, and industrial parks in different regions have diversity and differences in leading industries, socio-economic environment, resources and energy endowments (Yu 2018). As an important carrier of the real economy, industrial parks have the common characteristics of concentrated fixed pollution sources, high pollutant emission intensity, prominent environmental pollution problems, and

C. Lu · B. Huang (✉)
Ministry of Ecology and Environment, South China Institute of Environmental Science, Guangzhou, China
e-mail: huangbaoyuan@scies.org

C. Lu
e-mail: lucuiyun@scies.org

J. Zhang et al. (eds.), *Environmental Pollution Governance and Ecological Remediation Technology*, Environmental Science and Engineering,
https://doi.org/10.1007/978-3-031-25284-6_39

concentrated pollution risks (To give full play to the role of the EIA of park planning). Planning EIA assumes the function of preventing pollution sources in industrial parks, accurately grasps the evaluation focus of industrial park planning EIA, and uses the bottom line of pollutant discharge from parks, industries and key enterprises as the threshold for environmental access to improve the effectiveness of EIA (Huang 2021a), which is unbalanced in regional layout. The resulting environmental problems such as aggregated pollution and ecological damage can play an important role in source prevention and control.

In recent years, my country's planning system has undergone major adjustments, and the planning environmental impact assessment system has also made corresponding changes. The "Technical Guidelines for Regional Environmental Impact Assessment of Development Zones" (HJ/T131-2003) (hereinafter referred to as the 2003 version of the Guidelines) issued in 2003, as one of the supporting regulations of the EIA Law, is a follow-up to the "Technical Guidelines for Planning Environmental Impact Assessment (Trial Implementation).)" (HJ/T130-2003), the second industry standard for planning environmental impact assessment. Under the background and conditions at that time, the regional EIA of my country's industrial parks was standardized and guided. In order to adapt to the new requirements of ecological civilization construction and environmental protection under the new situation, the Ministry of Ecology and Environment initiated the revision of the 2003 version of the guidelines, which was renamed as "Technical Guidelines for Planning Environmental Impact Assessment Industrial Parks" (HJ/T131-2021), and finally published in 2003. It will be released in September 2021 to improve the pertinence and operability of the guidelines, and to provide technical support for further standardizing and strengthening the management of environmental impact assessment of industrial park planning. In actual work, there are still many problems with different types and levels of planning and planning EIA systems. The park planning EIA should play a role in preventing pollution at the source, and promote the improvement of the park's pollution control and protection level (Huang 2021b).

39.2 Main Existing Problems in the Environmental Impact Assessment of Industrial Park Planning

39.2.1 The Form of Planning EIA is not Clear

According to the "Regulations on Planning Environmental Impact Assessment": "Article 2 The relevant departments of the State Council, the local people's governments at or above the city level divided into districts, and their relevant departments shall organize and formulate relevant plans for land use and construction of regions, river basins and sea areas., development and utilization plans (hereinafter referred to as comprehensive plans), and relevant special plans for industry, agriculture, animal husbandry, forestry, energy, water conservancy, transportation, urban construction, tourism, and natural resource development (hereinafter referred to as special plans), environmental Impact assessment." "Article 10 When a comprehensive plan is prepared, an environmental impact chapter or description shall be prepared according to the possible impact on the environment after the implementation of the plan. When a special plan is prepared, an environmental impact report shall be prepared before the draft plan is submitted for approval. When formulating a guiding plan in a special plan, an environmental impact chapter or explanation shall be prepared in accordance with the provisions of the first paragraph of this article." According to the above provisions, an industrial park, economic and technological development zone or industrial park shall fall under the second category of the "Regulations on Planning Environmental Impact Assessment". Articles of "the relevant planning for land use and the planning for the construction, development and utilization of regions, river basins, and sea areas (hereinafter referred to as comprehensive planning)" shall be prepared in a chapter or statement on environmental impacts. In the actual implementation process, the planning EIAs carried out by industrial parks, economic and technological development zones or industrial parks are in the form of reports, especially the planning of industrial parks and industrial parks. Including the construction of agglomeration areas around statutory industrial parks (industrial parks), the preparation of planning environmental impact assessment reports is also generally carried out.

39.2.2 The Scope of Planning EIA is not Clear

According to the Regulations on Environmental Impact Assessment of Planning, "Article 2 Relevant departments of the State Council, local people's governments at or above the city level divided into districts, and their relevant departments shall organize and formulate relevant plans for land use and construction of regions, river basins, and sea areas, and relevant departments. Development and utilization planning (hereinafter referred to as comprehensive planning), as well as ……, shall be subject to environmental impact assessment." According to the above regulations, it is clearly stipulated that the objects that need to carry out planning environmental assessment are "local people at or above the city level with districts." government and its related departments." However, the main body responsible for compiling plans for industrial parks, economic and technological development zones or industrial parks is not only at the municipal level, but also at the district and county levels. EIA. In the developed areas of Guangdong Province, due to the initial stage of reform and opening up, they actively attracted investment and carried out industrial land development and construction. With the gradual deterioration of environmental quality, the increasing pressure on environmental protection, the increasingly stringent environmental protection policies, and the continuous improvement of the public's environmental protection demands, various regions are actively carrying out the renovation of village and town-level industrial parks. Carry out a planning EIA? The form and level requirements of planning EIA are not clearly defined. Different regions have different forms and different scale requirements in the implementation (Liu et al. 2021a).

39.2.3 The Planning Environmental Impact Assessment Review Level is Inconsistent with the Planning Approval Level

According to the "Regulations on Environmental Impact Assessment of Planning", "Article 17 For special plans approved by the people's governments at or above the city level with districts, their environmental protection departments shall convene representatives of relevant departments and experts to form a review team before approval to review the environmental impact report. Review. The review team shall submit written review opinions." The review unit of the planning environmental impact assessment is not the environmental protection department at or above the municipal level with districts, and as mentioned above, industrial parks, economic and technological development zones, industrial parks, and industrial agglomeration zones In addition to provincial and municipal industrial parks, economic and technological development zones, and industrial parks, the planning units are often prepared at the county and district level, and even include village and town-level industrial parks. Since the "Regulations on Planning Environmental Impact Assessment" does not give the county, district and environmental protection authorities the right to review planning, all these planning environmental assessments are reviewed by the environmental protection authorities above the prefecture-level city (except those assigned to the prefecture-level city and above). Approval level outside the districts and counties). Therefore, there is a common phenomenon that the review level of industrial parks, economic and technological development zones, and industrial park planning EIAs is inconsistent with the planning approval level.

39.3 Optimizing Management Strategies for Environmental Impact Assessment of Industrial Park Planning

39.3.1 Do Real Park Planning and Planning EIA Work

Establish the concept of "planning science is the greatest benefit" (Shi et al. 2020), and clarify that the main body of the park construction, the planning preparation department, the planning environmental impact assessment preparation department, the planning approval department, the planning environmental assessment review department, the planning environmental assessment review authority, and the construction subject are included in the planning environmental impact assessment. responsibilities and obligations, and give full play to the role of the planning EIA to realize the scientific benefits of planning. And establish a mechanism for investigating the illegal responsibility of the relevant responsible subjects in the process of planning preparation and implementation, and incorporate the government and departments' implementation of the planning EIA in accordance with the law into the performance assessment (Jia 2015).

Clarify the implementation of planning EIA results, and improve the legally binding force of planning EIA opinions. The opinions of the planning EIA are adopted by the planning and must be implemented in the whole process of planning implementation. It is suggested that the legal binding force of planning EIA opinions should be clarified and enhanced, and the linkage mechanism of "Three Lines and One List", planning EIA and construction projects should be clarified, and the environment of construction projects should be guided through the integration of "Three Lines and One List" and planning EIA. Access, EIA procedures are simplified and implementation is implemented (Li et al. 2020). Strengthen communication between planning departments and departments to jointly improve the scientific nature of planning and planning EIA. The management and control relationship between the industrial park and the three lines and one list is shown in Fig. 39.1.

Improve the regional total control and management system, and establish a refined total control and management framework of comprehensive control area-industrial park-fixed pollution sources based on the regional "Three Lines and One List" and the research results of the allowable emissions of the park planning environmental impact assessment, so as to realize the total amount control and management framework based on the capacity Quantitative indicators are implemented layer by layer from macroscopic regions to microscopic solid sources (Liu et al. 2021b). In order to improve the regional environmental quality, promote the efficient and differentiated

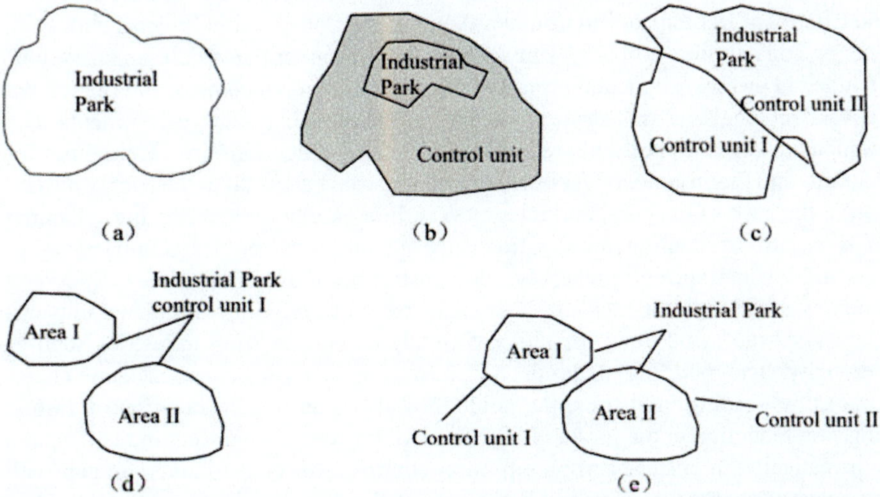

Fig. 39.1 The space of relevant key control units and industrial parks in the "three lines and one list"

Table 39.1 Park pollutant discharge and total control indicators (t/a)

Pollutants		Total environmental capacity	Environmental capacity to utilize	Remaining environmental capacity	Environmental capacity utilization %
Water environment	COD_{Cr}	579.26	36.65	542.61	6.33
	NH_3-N	14.17	1.83	12.34	12.91
	TP	1.87	0.37	1.5	19.79
Atmospheric environment	SO_2	1019	13.59	1005.41	1.33
	NO_X	498	63.34	434.66	12.72
	PM_{10}	260.8	55.49	205.31	21.28
	VOCs	487.6	86.2	401.4	17.68

management of the total amount control system, and dynamically adjust the total amount control indicators of the park and the total amount of pollutant discharge permits for fixed pollution sources. Take an industrial park as an example, The pollutant discharge and total control indicators of the park are shown in Table 39.1.

39.3.2 Realize the Synergy of Pollution Reduction and Carbon Reduction in the Industrial Park

In accordance with the requirements of the "Guiding Opinions on Overall Planning and Strengthening the Work Related to Addressing Climate Change and Ecological Environmental Protection" and "Guiding Opinions on Strengthening the Prevention and Control of Ecological Environment Sources for Construction Projects with High Energy Consumption and High Emissions", the carbon emission assessment will be included in the environmental impact assessment system, to connect and implement relevant regional and industry carbon peaking and other policy requirements, and promote synergies in pollution reduction and carbon reduction. The "Guidelines for Planning and Environmental Assessment of Industrial Parks" takes the energy utilization of the park as the core, and integrates carbon emission reduction into planning analysis, current situation investigation and evaluation, environmental impact prediction and evaluation, comprehensive demonstration and optimization of planning schemes, and mitigation measures and measures for adverse environmental impacts. chapter. At the same time, for parks with key carbon-emitting industries such as electricity, steel, building materials, non-ferrous metals, petrochemicals, and chemicals as the leading industries, the guidelines also require consideration of carbon emission reduction in the production process of key carbon-emitting industries, and to investigate the current carbon emission control level of the park. The gap with the industry's carbon peaking requirements and the carbon reduction potential, from the aspects of resource and energy utilization management and control constraints, compliance with regional and industry carbon peaking and carbon emission reduction requirements, resource and environment bearing status, etc., to demonstrate the park's industrial positioning, industry Structure, energy structure, and environmental rationality of the scale of key carbon-related industries. It provides technical support for controlling carbon source emission reduction and building the implementation path of carbon emission reduction in the park, which will effectively promote the low-carbon energy transformation of the park and the green development of the industry.

39.3.3 Implementing the "Three Lines and One List" and the Key Links of Planning EIA

Compared with the planning EIA, the "Three Lines and One List" determines the bottom line of environmental quality targets based on the positioning of regional ecological functions, and conducts comprehensive spatial management and control according to the spatial distribution of the ecological environment and the comprehensive evaluation results of environmental elements such as water environment, atmospheric environment, and soil environment, and proposes Differentiated management and access requirements of regional control units (Wang 2020). "Three lines and one list" has significant spatial characteristics, which improves the logic, science and systematicness of regional environmental access, and puts forward environmental access requirements from different perspectives from planning environmental assessment. Planning EIA is an access system that links the previous and the next (Liu et al. 2022). It not only needs to fully connect the results of the "three lines and one list" in the region, and demonstrate the environmental rationality of the industrial positioning, development scale and functional layout determined in the plan, but also to refine the ecological environment access. List management and control requirements, and clarify the content of the project EIA that can be simplified and should be strengthened. The "Three Lines and One List"—Planning EIA—Project EIA and other systems are implemented step by step, and the consistency and systematization of the access system requirements should be ensured. The project EIA should take the planning EIA conclusions and review opinions as an important basis. The site selection, scale, nature and process route of the construction project should be consistent with the planning EIA conclusions and review opinions. Realize the data dynamic management and intelligent application of the 'three lines and one list' data application platform.

39.4 Conclusion

My country's industrial park planning EIA work should follow the strategic deployment and problem-oriented management guidelines, and systematically promote the planning EIA work (Geng et al. 2019). At the park level, guided by the "Technical Guidelines for Planning Environmental Impact Assessment Industrial Parks", with the industrial park planning and environmental assessment as the starting point, the industrial positioning and access requirements, the coordinated development of pollution reduction and carbon reduction, and the connection of "three lines and one list" are all implemented. Go to the park planning EIA document as the criterion for the park's environmental management. At the local level, sort out the relationship between environmental quality and environmental stock within the jurisdiction, and connect the total control indicators of each industrial park according to regional conditions. From the industrial development of the park, industrial positioning, coordinated development of pollution reduction and carbon reduction, environmental pollution control priorities, etc., link the achievements of the "three lines and one list", consolidate the technical system of planning environmental assessment, and further give full play to the scientificity and effectiveness of planning environmental assessment.

References

Geng H, Li T, Xu H (2019) Research on the necessity and basic framework of EIA policy in China. China Environ Manage 2019(6):23–27

Huang L (2021a) Suggestions on the connection between the environmental impact assessment of industrial park planning and pollution discharge permit. China Environ News 06(003):1–2

Huang L (2021b) 14th five-year plan development new trend and industrial park planning EIA work suggestions. Sichuan Environ 40(4):234–237

Jia S (2015) Analysis of my country's planning environmental impact assessment and suggestions for improvement. Environ Impact Assess 2015(5):18–23

Li Y, Jiang Y, Han L (2020) Analysis on the connection between "three lines and one single" and environmental impact assessment. Environ Impact Assess 42(5):21–25

Liu L, Zhu X, Qiu X et al (2021a) Assumptions on reconstructing China's planning environmental impact assessment system. Environ Protection 49(12):17–21

Liu L, Han L, Li J et al (2021b) "14th five-year plan" environmental impact assessment and pollution permit reform situation analysis and prospect. Environ Impact Assess 43(1):1–6

Liu L, Li J, Cai Z et al (2022) Research on the application path of "three lines and one single" in industrial park planning EIA. China Environ Manage 2022(2):49–55

Shi J, Zhang M, Zhao Y et al (2020) The progress, challenges and countermeasures of China's planning environmental impact assessment system. China Environ Manage 06:43–46

To give full play to the role of the EIA of park planning

Wang Y (2020) The significance and key links of "three lines and one order" to the reconstruction of the environmental access system 2020(1):14–17

Yu X (2018) Research on the classification model of low-carbon development of national pilot industrial parks. China Population Resour Environ 8(09):32–39

Chapter 40
Research Trends of Carbon Sources in Denitrification: A Bibliometric Analysis During 2012–2022

Mingyuan Liu, Weijin Wang, Lin Guo, Wenfang Zhai, Junfeng Wan, Kaige Zhao, and Zixuan Zhang

Abstract The heterotrophic denitrification process requires the carbon sources to act as an electron donor. An external carbon source is required to increase denitrification efficiency when there aren't enough organic carbon sources already present in the system. We used the bibliometric method to conduct a statistical study of the papers on carbon sources in denitrification from 2012 to 2022 using the Web of Science Core Collection (WOSCC) database. An in-depth examination of the publication features of 2765 articles published between 2012 and 2022 found that the number of articles published on carbon sources in denitrification expanded fast, with China having the most publications (1419), followed by the USA (308). While the research on carbon sources in denitrification has primarily focused on nitrous oxide and the microbial community, the research trend has changed from ethanol, methane, membrane bioreactor to partial denitrification, solid-phase denitrification and aerobic granular sludge over the past ten years. The conclusions reached offer recommendations for further study into the use of carbon sources in denitrification and can help to avoid needless duplication of research in this area.

Keywords Carbon sources · Bibliometric analysis · Denitrification

40.1 Introduction

Excess nitrogen in sewage is a serious environmental problem across the world right now. High nitrate levels in drinking water can cause sickness and death in people

M. Liu · J. Wan (✉) · K. Zhao · Z. Zhang
School of Ecology and Environment, Zhengzhou University, Zhengzhou 450001, PR China
e-mail: wanjunfeng@zzu.edu.cn

Henan International Joint Laboratory of Environment and Resources, Zhengzhou University, Zhengzhou 450001, PR China

W. Wang · L. Guo · W. Zhai
Henan Institute of Geological Survey, Zhengzhou 450001, PR China

J. Zhang et al. (eds.), *Environmental Pollution Governance and Ecological Remediation Technology*, Environmental Science and Engineering,
https://doi.org/10.1007/978-3-031-25284-6_40

and animals. Nitrate binds to hemoglobin to restrict oxygen delivery to human cells, resulting in Methemoglobin, and nitrate can be converted to nitrite under specific conditions, causing sickness and death in people and animals. Biological denitrification is a proven process for removing nitrogen from wastewater, and it is often used in many wastewater treatment systems due to its low cost, high efficiency and ease of application (Fu et al. 2022).

Denitrification is an anoxic process in which the nitrate is reduced to nitrite and subsequently to nitrogen gas by means of the action of heterotrophic denitrifying bacteria in accordance with the following sequence:

$$NO_3^- \rightarrow NO_2^- \rightarrow NO(g) \rightarrow N_2O(g) \rightarrow N_2(g) \tag{40.1}$$

Heterotrophic denitrification requires the presence of an organic carbon source. Additional COD must be given to the system when there is insufficient COD in the wastewater being treated for denitrification to take place, such as in wastewaters with a low COD/N ratio or due to significant COD consumption in earlier processes like nitrification.

Important denitrification process parameters such as the denitrification rate, COD demand, biomass production, and biomass composition have been found to be significantly impacted by the characteristics of the additional carbon source. More than half of the entire cost of treated wastewater is attributed to waste management and carbon sources costs combined. Finding a reliable and economical sources of carbon is therefore essential (Kim et al. 2016).

Numerous scholars have examined and evaluated the impact of different carbon sources on denitrification. Articles already published have discussed the mechanisms of action of various carbon sources (Xu et al. 2018), their applications in various target water bodies (Zhang et al. 2021), and their effects on denitrifying microbial populations (Li et al. 2019). These publications were, however, condensed by the writers based on their personal experiences and typically centered on a particular and limited perspective. In the study, we used bibliometric and statistical approaches to conduct a thorough analysis of the literature on carbon sources and their use in the denitrification area from 2012 to 2022. In order to assess the state of research in a particular area and enable systematic analysis and evaluation. Bibliometric analysis uses the literature as the research object and applies mathematical, statistical, and other quantitative research methods to quantitatively analyze the distribution structure, quantitative relationship, and change law of the associated information (Xu et al. 2022). This method was employed in the current study to assess the hotspots and trends in the denitrification field's research in relation to carbon sources.

40.2 Data and Methods

40.2.1 Literature Search Strategy

Web of Science Core Collection (WoSCC) database was used to search the literature in fields associated with carbon sources research and its application in denitrification from 2012 to 2022. English was used as the basic language for the search keywords (including title, author keywords, abstract and keywords plus). The retrieval time is June 15, 2022, the Boolean operator was designed as follows: (TS = (carbon source)) AND TS = (denitrification*).

In the current analysis, we solely took into account "Article" kinds in order to more precisely analyze the developmental patterns and research hotspots in connected domains. From this strategy we obtained 2765 documents after filtering.

40.2.2 Bibliometric Analysis Methods

Bibliometrics considers the characteristics of the literature as the research object and uses mathematical and statistical methods to quantitatively analyze quantitative relationships, distribution structures, and information from the literature. We imported WoSCC classification data for the retrieved documents into Microsoft Excel 2019 and BiblioMetrix for manual screening and visual analysis.

40.2.3 The Social Network Analysis

We first perform a keyword filter, after which we manually remove irrelevant words, merge semantically repeated keywords, and finally the top 20 keywords with the highest frequency were selected for co-occurrence analysis. The approach utilized to determine keyword co-occurrence was social network analysis. In order to analyze the relationship between the co-occurrence of keywords and the intricate network of international cooperation, the processed keyword data were loaded into the Gephi visualization data network and the connections between nodes were employed to represent node importance.

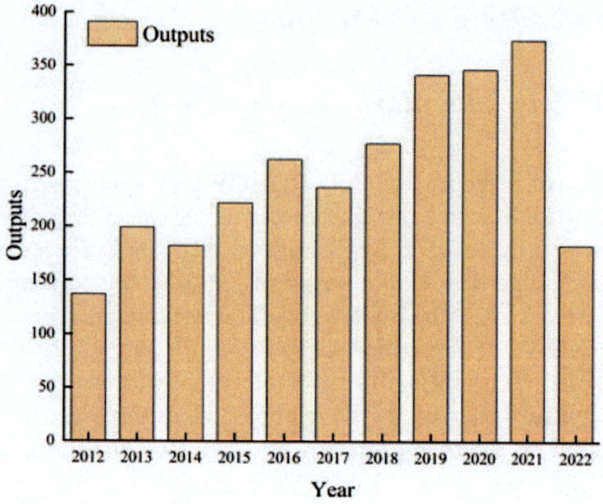

Fig. 40.1 Annual output of literature from 2012 to 2022

40.3 Results and Discussion

40.3.1 Temporal Evolution of Outputs

The number of paper outputs reflects, to some extent, the speed and extent of the field's development. The annual number of outputs related to carbon sources research during 2012–2022 is shown in Fig. 40.1.

The number climbed at a growth rate of 173.72% from 137 in 2012 to 375 in 2021. The number of outputs increased linearly over the research period. We further analyzed author number, number of references, and number of article source, revealing that from 2012 to 2022, the average number of authors and references increased along with increases in the number of published papers (Table 40.1). Between 2012 and 2021, there were an additional 1239 authors, and there were 1852 by 2021. The average number of references also grew, from 37.65 to 43.01. The consistent expansion of publications and related references demonstrates that cooperation and research on carbon sources have both steadily increased over the past 10 years.

40.3.2 Analysis of Literature Distribution in Different Countries

The contributions of different countries/regions to this field and the cooperative relationship between countries were evaluated by analyzing the characteristics of the articles according to country (Castro et al. 2022). When the co-authors were from different countries, the research was regarded as a publication demonstrating

Table 40.1 Characteristics of articles from 2012 to 2022

Year	NP	NAU	NAU/NP	NR	NR/NP	SO	NP/SO
2012	137	613	4.47	5158	37.65	78	1.76
2013	199	857	4.31	7123	40.10	91	2.19
2014	182	837	4.60	7388	40.59	105	1.73
2015	222	988	4.45	7695	34.66	103	2.16
2016	263	1162	4.42	9202	34.99	101	2.60
2017	237	1193	5.03	9026	38.08	110	2.15
2018	278	1385	4.98	10,606	38.16	127	2.19
2019	342	1603	4.69	13,607	39.79	136	2.51
2020	347	1653	4.76	14,465	41.69	133	2.61
2021	375	1852	4.94	16,128	43.01	125	3.00
2022	183	961	5.25	8647	9.00	74	2.47

NP number of published articles; *NAU* total number of authors in the year; *NR* number of references; *SO* source

international cooperation (Multiple Country Publications, MCP); otherwise, it is regarded as a publication of a single country (Single Country Publications, SCP). Among the 2765 articles, 710 articles involved international cooperation.

From 2012 to 2022, researchers and institutions from 65 countries published academic articles on carbon sources and denitrification according to the data we have compiled. China, USA, and Germany are the top three among the 10 nations with the most outputs (Fig. 2a).

The cooperative relationships among the top-10 countries with the largest number of publications in this field from 2012 to 2022 is shown in the cooperation network

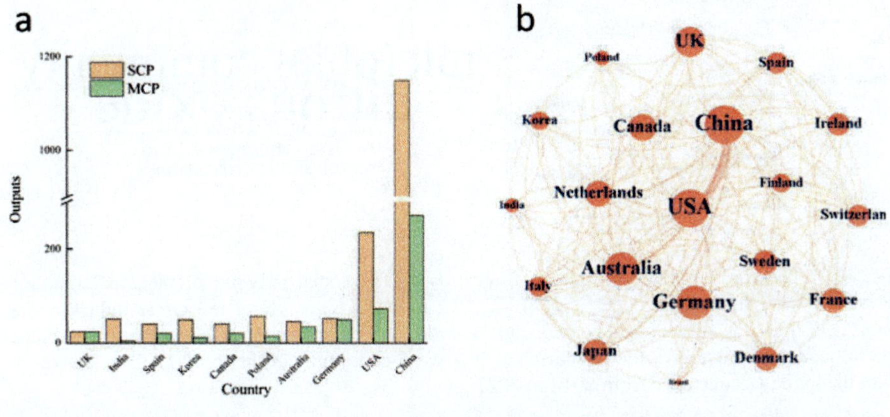

Fig. 40.2 **a** The 10 nations with the most outputs. **b** The cooperation network of the top-20 productive countries; the size of the circle indicates the frequency of outputs by a country and the width of the line between the two points represents the level of cooperation between countries

diagram (Fig. 2b). The two largest paper manufacturers among them are China and USA, with the level of cooperation between the two countries far exceeding that of the other countries. Additionally, China and USA collaborate internationally with other nations the most frequently. The results suggest that effective international cooperation plays an important role in promoting the application of carbon source in denitrification and the globalization of environmental governance.

40.3.3 Keyword Analysis Reveals Research Trends

According to keyword analysis of the 2765 articles retrieved in the last 10 years, the research of carbon sources in the field of denitrification today primarily focuses on nitrous oxide and microbial community. But there are also many articles related to its function in anammox, constructed wetland, aerobic denitrification, greenhouse gas, and groundwater. Methane, solid carbon source, and landfill leachate are the three carbon source keywords that have appeared the most during the previous 10 years (Fig. 40.3).

Use BiblioMetrix to examine keyword data after removing duplicate and irrelevant terms to create keywords emergence trend map (Fig. 40.4) (Aria and Cuccurullo 2017). A total of 22 words were extracted in this paper. An expanding area of study in recent years is partial denitrification, solid-phase denitrification and aerobic granular sludge. Additionally, studies on anammox, solid carbon sources, aerobic denitrification, and greenhouse gases are on the rise. Studies on ethanol, methane and membrane bioreactor, however, have decreased.

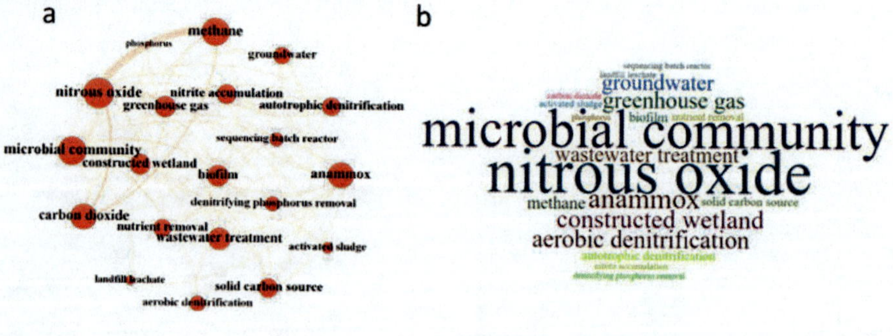

Fig. 40.3 The social network of author keywords. **a** The social network of the top-20 author keywords with the highest occurrence from 2012 to 2022. The size of the circle indicates the number of times the keywords appeared, and the width of the line between the two circles indicates the frequency of the keyword co-occurrence. **b** A word-cloud map of the top-20 author keywords with the highest occurrence from 2012 to 2022

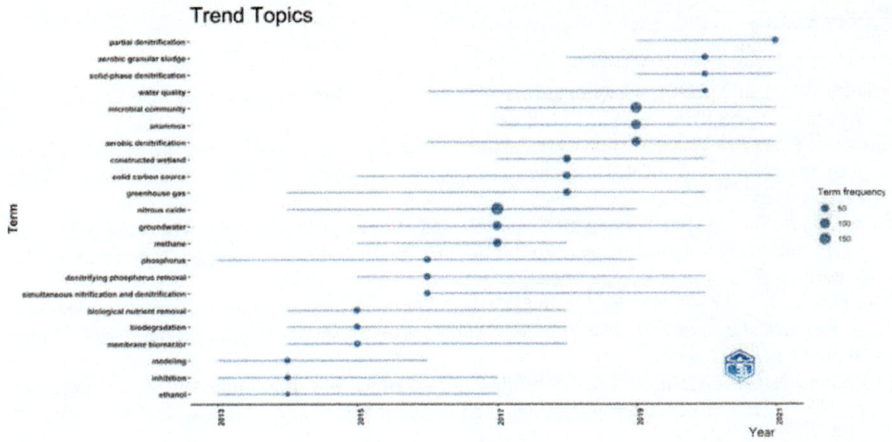

Fig. 40.4 Keywords emergence trend map from 2012 to 2022

40.4 Conclusion

(1) In the past 10 years, the total number of papers published in the field of carbon sources in denitrification has shown a steady growth trend, and the scientific community has paid more and more attention to the field.

(2) Research in this field has expanded and globalized in the past 10 years, with China among the most important contributors, by country, to the field due to having the largest number of published articles (1419). Moreover, China also played a vital role in the cooperation network of the top-20 producing countries or regions.

(3) The study of denitrification with carbon sources has been centered on nitrous oxide and the microbial community over the past 10 years. Over the last three years, the connection between external carbon sources with partial denitrification, solid-phase denitrification, and aerobic granular sludge has developed into a research hotspot. Since 2019, a number of keywords have emerged that are associated with this subject. Additionally, the research on such topics as ethanol, membrane bioreactor, biodegradation, methane, has become less popular, there are certain limitations in future research on these topics.

Acknowledgements The authors gratefully appreciated the financial support of the Development and application of key technologies for biogas slurry resource utilization in intensive farms (Grant No. 212102110030), Resource utilization and efficient utilization of Sino-French aquaculture manure (Grant No. NGD2021029).

References

Aria M, Cuccurullo C (2017) Bibliometrix : an R-tool for comprehensive science mapping analysis. J Inform 11:959–975

Castro L, Sganzerla WG, Ziero HDD, Mudhoo A, Martins G, Forster-Carneiro T (2022) Research progress, trends, and updates on anaerobic digestion technology: a bibliometric analysis. J Clean Prod 331:15

Fu X, Hou R, Yang P, Qian S, Feng Z, Chen Z, Wang F, Yuan R, Chen H, Zhou B (2022) Application of external carbon source in heterotrophic denitrification of domestic sewage: a review. Sci Total Environ 817

Kim H, Kim J, Shin SG, Hwang S, Lee C (2016) Continuous fermentation of food waste leachate for the production of volatile fatty acids and potential as a denitrification carbon source. Biores Technol 207:440–445

Li SY, Fei XN, Cao LY, Chi YZ (2019) Insights into the effects of carbon source on sequencing batch reactors: performance, quorum sensing and microbial community. Sci Total Environ 691:799–809

Xu P, Zhu X, Tian H, Zhao G, Chi Y, Jia B, Zhang J (2022) The broad application and mechanism of humic acids for treating environmental pollutants: Insights from bibliometric analysis. J Clean Prod 337

Xu Z, Dai X, Chai X (2018) Effect of different carbon sources on denitrification performance, microbial community structure and denitrification genes. Sci Total Environ 634:195–204

Zhang F, Ma C, Huang X, Liu J, Lu L, Peng K, Li S (2021) Research progress in solid carbon source-based denitrification technologies for different target water bodies. Sci Total Environ 782:146669

Chapter 41
Research on Decolorization Efficiency of Organic Wastewater with High-Concentration and High-Chromaticity

Xiao-yuan Zhang, Jia-yao Wang, Zhuo-xin Yin, Shu-yan Liu, Yuan-yuan Shen, and Tian-peng Gao

Abstract Discharge of organic wastewater with high-concentration and high-chroma is related to normal progress of many kinds of industries such as papermaking, food processing, printing and dyeing. The way to decolorize industrial organic wastewater is one of the focal points of water pollution treatment. Current treatment process of organic wastewater with high-concentration and high-chromaticity generally has some problems with its high contents of toxic and harmful substances, strong substantial contamination, and large emission capacity Direct discharge of such wastewater will cause serious pollution to the environment, pigments and toxic components in wastewater will also be adsorbed in soil, which has a great impact on discharge area. of wastewater in ecological environment. Therefore, according to the physical and chemical properties of wastewater, this manuscript briefly introduces the mechanism and research progress of current decolorization methods for wastewater, such as physical processes, chemical oxidation and biological methods. Meanwhile, reasonable and effective methods could be provided for decolorization treatment technology of organic wastewater with high-concentration and high-chromaticity from analysis in the future.

Keywords Organic wastewater with high-chromaticity · Physical process · Chemical oxidation method · Biological method

X. Zhang · J. Wang · Z. Yin · S. Liu · T. Gao (✉)
College of Urban Environment, Lanzhou City University, The Engineering Research Center of Mining Pollution Treatment and Ecological Restoration of Gansu Province, Gansu 730070, China
e-mail: zkgtp@163.com

Y. Shen · T. Gao
College of Biological and Environmental Engineering, Xi'an University, Xi'an 710065, China

Academy of Eco Xi'an, Xi'an University, Xi'an 710065, China

J. Zhang et al. (eds.), *Environmental Pollution Governance and Ecological Remediation Technology*, Environmental Science and Engineering,
https://doi.org/10.1007/978-3-031-25284-6_41

41.1 Overview of High-Chromaticity Organic Wastewater of High-Concentration

41.1.1 Resource of Wastewater with High-Chromaticity

Compliance discharge of wastewater in accordance with international and domestic standards has always been a significant topic in research area of environmental protection in terms of the rapid development of modern industries in China. Chromaticity is an important quality index in demonstrating contamination intensity of organic wastewater. Finding a way to decolorize organic wastewater with high-concentration and high-chromaticity reasonably is a central problem that must be valued, the requirement is not only the basic discharge requirements for industrial enterprises that need to achieve, but also an essential embodiment of ecological civilization thought.

Organic wastewater with high-chromaticity generally originated from a wide range of industrial wastewater of high-chromaticity with chemical oxygen demand (COD) of more than 2000 mg/L and pH value of strong acid/alkali discharged from papermaking, printing and dyeing, food processing and other kinds of industries (Zhou 2003). Chromaticity of wastewater is deepening with the increase concentration of colorable organic substances and metal ions which are the main reasons for high-chromaticity of wastewater (Hao et al. 2000). Indicators of wastewater quality such as CODcr, BOD_5, content of organic substances and so on., could all affect the chromaticity of water.

41.1.2 Characteristics of Wastewater Chromaticity

Chromaticity of organic wastewater mainly depends on the quantity of chromophoric groups and conjugated structures contained in organic components of wastewater, such as functional groups of vinyl (–HC=CH_2), azo (–N=N–), nitro (–NO_2), nitroso (–N=O–), azo-oxide (–N=NO–), carbonyl (>C=O), thiocarbonyl (–CO–S), thiocarbonyl (>C=S) and so on. They can absorb ultraviolet and visible light and make electronic transition. A functional group called a chromophore is mainly a group containing unsaturated bonds and unpaired electrons. Multiple chromophores conjugated due to the delocalization of n electrons in conjugated structure. The n electrons are easy to be excited, organic matters of this type can absorb light in the visible light region, so n electrons are revealing (Hu 2012).

Complex production process in various industries produced, different types of wastewater, many industries annual discharged large amount of sewage with high-chromaticity. Therefore, organic wastewater with high-chromaticity usually has high values of CODcr, BOD_5, and other water quality indicators which are exceeding the international and domestic standards, which can even cause problems of releasing excessive concentrations of certain heavy metals to the environment. Common metal ions can make high-chromaticity of wastewater, as divalent iron ions with light green, trivalent iron ions with brownish yellow, hexavalent chromium ions with orange-red, and so on. Most of heavy metal ions are biologically toxic, such as carcinogenic, teratogenic or mutagenic (Gong et al. 2017), wastewater with such poisonous and harmful heavy metal ions comes from the refractory industrial sewage.

41.1.3 Hazards of Wastewater with High-Chromaticity

Emissions of high-chromatic wastewater to the environment will increase the concentration of COD and affect the survival rate of microorganisms in wastewater. It is hard to make biochemically degrade, which can reduce the removal rate of chromaticity for wastewater (Xia 2003). Higher-chromaticity of dye wastewater could affect the sensory properties and reduce the light transmittance of the water body from the sunlight absorption of dye, which affects the photosynthesis of algae. In addition, it also can affect the diversity of the entire ecosystem (Jin 2014); and toxic organic substances in wastewater will continue accumulating and storing in groundwater and soil, which can cause pollution that influenced the growth of crops, it also has a profound impact on human health, and poses a severe threat to the ecological environment (Wang 2020). Therefore, decolorizing high-concentration and high-color organic wastewater is a significant problem in water pollution control (Forgacs et al. 2004).

41.2 Research Status for Decolorization Technology of Organic Wastewater with High-Chromaticity

Indicators for wastewater discharge is relatively strict in China and abroad in recent years. Advanced treatment processes are generally carried out for wastewater after traditional physical, chemical and biological treatment methods to achieve the purpose of water purification. Although physicochemical methods can reduce organic pollutants, new components can be generated at the same time and can cause secondary pollution. Requirements for biological treatment method are strictly on water quality and quantity, the results are also affected by the water condition. Biological treatment method is inefficient, costly and not recommended, so the actual operation effect is not ideal (Wang and Zhou 2020). Treatment methods usually used for decolorizing organic wastewater with high chroma mainly include the physical method (physical adsorption method, coagulation method), chemical oxidation method, biological method and so on (Zou and Chen 2021).

41.2.1 Physical Method

Physical Adsorption Method

Physical adsorption decolorization relies on adsorbents included common renewable adsorbents, such as activated carbon, zeolite, ion exchange fibers and so on) and non-renewable adsorbents such as various natural minerals (bentonite, diatomaceous earth) (Zakhama et al. 2011), industrial waste (cinder, fly ash), and biological waste (charcoal, sawdust) and so on. Particles of porous adsorbents can combine with components in wastewater, so colored substances in wastewater can adsorb on surface of porous sorbents and result to reduce chromaticity (Vandenbossche et al. 2014). One of the main advantages for adsorption decolorization is that it retains the original components of the water body and does not introduce other impurities.

Chen et al. (2017) used acidified attapulgite (which has significantly enhanced specific surface area and adsorption capacity for oil and pigment through acid activation) to degummed rice bran oil (mainly removes rice bran). Colloidal substances mainly composed of phospholipids in the crude oil can reduce the energy consumption of decolorization process, improve the decolorization effect, prevent color reversion, and carry out adsorption decolorization. Decolorization rate of rice bran oil reaches 96.6% in 30 min of decolorization time under the experimental condition. Guo et al. (2015) used activated carbon for decolorization treatment to reactive dye wastewater (with Drimaren Red X-2B, Drimaren Orange X-3LG, Drimaren Turquoise CL-B three reactive dyes simulated wastewater). The results showed that activated carbon had the best decolorization effect on Orange X-3LG wastewater and Red X-2B wastewater under the conditions of 40 g/L and 20 min, the decolorization efficiencies were respectively 79.48% and 91.91%; decolorization effect of activated carbon on Turquoise CL-B wastewater was perfect under the condition of 40 g/L, 30 min. the best decolorization result was more than 97%. Ma et al. (2013) used decolorization efficiency as an indicator for activated clay and attapulgite constituents (the proportion of mixture was 2:3) to decolorize oil. The experimental conditions were 1.9 g/100 mL of the used decolorant amount, 40 min of decolorization time, and 110 °C of decolorization temperature, The results showed that the decolorization efficiency of oil reached 85%. The results showed that the higher proportion of attapulgite in constituents, the more stable the oil was, and it is harder to return color in the subsequent constituents of different proportions for mixture of activated clay and attapulgite.

Coagulation method

Coagulation produced a certain amount of chemical coagulants to the wastewater after destabilization, bridging, agglutinate and flocculate and other reaction processes produced chromogenic organic substance in decolorization of wastewater after solid–liquid separation (Shang et al. 2020). Coagulants commonly used in decolorization of industrial wastewater are ferric chloride, ferrous salts, poly ferric sulfate, polyaluminum chloride, polyaluminum ferric sulfate, polyacrylamide and others (Wang et al. 2021). Coagulation decolorization technology has advantages of low investment cost, small equipment footprint, and large processing capacity. It is widely used because of its good removal effects on hydrophobic chromophoric organics in wastewater, while it has poor removal effect on water-soluble chromophores. Disadvantage of the coagulation method is it needs a large amount of sediment and dosage, sometimes as high as 1000 mg/L or more.

Xie et al. (2010) used polyaluminum chloride (PAC) as a kind of coagulant and polyacrylamide (PAM) as another kind of coagulant jointly to treat papermaking wastewater. Coagulation and precipitation are entirely carried out. As the dosage of PAC is large, it can change the charge of the colloidal particles to form a new stable system. Adding 10% aluminum content PAC 1.0 mL, 2 g/L PAM 0.5 mL to the 1000 mL papermaking wastewater, then stirring for 2 min and let it stand for 20 min; Chromaticity of wastewater reduced from 64 to 18 times, the removal efficiency was 71.9%, and COD was effectively reduced, and it reaching the national wastewater level emission standards. Hao et al. (2016) used Fe as the anode and C as the cathode to electrocoagulation the printing and dyeing wastewater from the printing factory, hydrolyzed and polymerized through the electrolytic metal anode (such as Fe, Al) to form a polymer polynuclear hydroxyl complex. These polynuclear hydroxyl complexes and hydroxides can coagulate and precipitate colloidal, suspended impurities in wastewater and separate them from wastewater with hydroxide. The study found that pH 5.5, the plates distance 3 cm, current density 11 mA/cm^2; under reaction time 30 min, the electrolysis result was the best. At this time, removal efficiency for COD and total suspended solids (TSS) was respectively 85.3% and 65.8% which provided significant function on the treatment of printing and dyeing wastewater. Zhang et al. (2006) used the homogeneous Fenton oxidation-coagulation method to process wastewater from printing and dyeing, with adding 3.6 mL H_2O_2, 1.8 mL $FeSO_4$, and 8 mL flocculant Polyaluminum silicate chloride (PASC) to wastewater, chromaticity of wastewater decreased to 35 after treatment, COD decreased to 103 mg/L, the chromaticity and COD removal efficiency from sewage reached 95% and 94.3% respectively, it showed that the decolorization effect of the method was remarkable.

41.2.2 Chemical Oxidation Method

Chemical oxidation decolorization uses oxidants such as sodium hypochlorite (NaClO), chlorine dioxide (ClO_2), hydrogen peroxide (H_2O_2), potassium permanganate ($KMnO_4$) and other reagents to achieve the purpose of wastewater decolorization (Zhang 2007), chromophore group and structure in the wastewater are broken or changed by the oxidizing properties of oxidants, the oxidative decolorization effect is powerful. As strong selectivity is closely related to substituents on organic compounds, the reason is the activation energy of reactions between oxidants and some organic compounds is too high, which make the response challenging. It is vital to choose a suitable catalyst. Chemical oxidation method is relatively mature, very likely to introduce new impurities during the treatment process, which can cause secondary pollution.

Liu et al. (2014) used Fenton as the oxidant in the chemical oxidation method, Fe^{2+} was used as a catalyst to make H_2O_2 generate hydroxyl radicals with more vital oxidizing ability under acidic conditions for oxidizing organic matters. COD reducing efficiency and chromaticity removal efficiency of the Fenton oxidation method can reach 89.7% and 91.1%, respectively for hair dyeing wastewater with COD of 3800 mg/L and chromaticity of 1210 times, when the dosing ratio of hydrogen peroxide is 0.00303. Chemical oxidation method has the characteristics of rapidity, non-toxicity, simple process, without secondary pollution, high removal efficiency of pollutants. Zhou et al. (2006) used ozone oxidation to disposal coking wastewater. The results showed that the sewage chromaticity continued decreasing with the increasing ozone reaction time. Color of wastewater changed from initial dark brown to light green when the reaction time reached 3 h. However, chemical oxidation method has high investment, large power consumption, and high processing cost. Zheng and Mei (2010) used chemical oxidation to decolorize turmeric saponin wastewater, using hydrogen peroxide solution to treat turmeric saponin wastewater under normal temperature conditions,, the removal efficiency of sewage chromaticity can reach 70%. COD removal efficiency in the wastewater can reach 80% at the same time, and the chromaticity removal efficiency for the sodium hypochlorite solution to disposal turmeric saponin wastewater can reach 80%. Meanwhile, the removal efficiency of COD in wastewater reaches 95%. Two oxidants were used in the effluent decolorization test of biochemical treatment of turmeric saponin wastewater, the chromaticity was about 50 times under the optimal conditions, which met the national industry discharge standard.

41.2.3 Biological Method

Biological decolorization method uses microbial enzymes to oxidize or to reduce colored substances Decolorization microbial degradation process is completed in two stages, the first stage is adsorption and enrichment of colored molecules, second stage is the biodegradation uses microbial enzymes to oxidize or reduce dye molecules by destroying its unsaturated bond chromophore group to remove chroma.

Biological methods are divided into aerobic physical oxidation method and anaerobic natural oxidation method. Aerobic biological oxidation method mainly includes ordered batch activated sludge method. Standard started sludge method, bodily contact oxidation method and so on. Anaerobic biological oxidation method mainly includes upflow anaerobic sludge bed method, compound anaerobic reactor and others (Liu et al. 2020). Physical form method is the most widely used treatment for wastewater decolorization and organic matter degradation. However, it is difficult to degrade for industrial wastewater biochemically, the natural process is not ideal for decolorization, and other methods are required to pretreat it first.

Huang (2004) used the cultured mixed strains isolated from the wastewater and sludge of printing and dyeing plants to decolorize the wastewater from printing and dyeing. Activated carbon was used as the carrier for the immobilization of decolorizing strains under anaerobic conditions of urea 0.45 g/L, yeast extract 0.25 g/L, temperature 30 °C, pH 7.0, inoculum size 1.25%, the average decolorization efficiency was over 70%. Zheng (2018) used the bacteria group enriched by activated sludge from a printing and dyeing factory that could efficiently degrade azo dyes (acid scarlet GR) and configured the mother liquor with the acid scarlet GR dye concentration of 5 g/L. Filter membrane of bacteria 0.22 μm was filtered and sterilized to treat the printing and dyeing wastewater. The results showed that the decolorization efficiency of the bacteria group reached 84.67% at pH 8.0 in 12 h; the decolorization efficiency reached the maximum 98.2% at 30 °C. Ye et al. (2010) treated pen-making wastewater by anaerobic biological oxidation sludge from the ABR reactor of a thickening sludge tank was used in a brewery to degrade after culturing the sludge with a high concentration of organic wastewater., COD removal and decolorization efficiency reached 75.54% and 78.53% respectively in the device with a water ratio of 1:2.5, the treatment effect was the best after 12 days of treatment. The amount of sludge input should suitable. Too large amount of sludge will also lead to the increase of suspended microorganisms and affect the result of effluent quality, too small amount would make the reaction incomplete or prolong the treatment time.

Summary for the various treatment methods of high-chromaticity wastewater mentioned above (see Table 41.1). Researchers focus on the modified adsorbent as the primary research direction and committee to making the adsorbent more efficient in the physical adsorption method. It can better absorb the color in wastewater, the average treatment efficiency reaches more than 90%. Development of the coagulation method is relatively perfect, the added coagulant will be affected by water environment indicators such as temperature and pH. Chemical oxidation method is relatively simple, all the listed methods use highly efficient chemical oxidants to oxidize certain types of components in wastewater, but other types of wastewater are not fully applicable or have lower decolorization efficiency. Breakthrough point in future should find two or more universality oxidants to maximize the decolorization efficiency by adding other oxidant components or changing several existing oxidant components for such method. Biological method is an emerging method for wastewater treatment in recent years which has been widely favored by scholars. The growth state and treatment efficiency of microorganisms or activated sludge will change in accoundance with temperature and pH, the treatment cycle is long, and the recovery of microorganisms takes a long time.

41.3 Conclusion and Outlook

Investigation of the paper obtained four decolorization methods of organic wastewater; they are physical adsorption, chemical oxidation, coagulation and emerging biological method, advantages and disadvantages of each method are listed below.

Table 41.1 Statistics of decolorization treatment methods for organic wastewater of high-concentration and high-chromaticity [author-drawing]

Approach	Sources	Decolorizer	Reaction conditions	Decolorization mechanism	Decolorization efficiency	Author of references
Physical adsorption method	Degummed rice bran oil	Acidified attapulgite	Decolorizing agent added: 5%. Decolorization at 110 °C, 30 min	Attapulgite adsorption	96.60%	13
	Reactive dye wastewater components: ① Drimaren Red X-2B; ② Drimaren Orange X-3LG; ③ Drimaren Turquoise CL-B	Activated carbon	① Wastewater of Red X-2B:40 g/L, 20 min;	Activated carbon adsorption	① Wastewater of Red X-2B: 79.48%;	14
			② Wastewater of Orange X-3LG: 40 g/L, 20 min;		② Wastewater of Orange X-3LG: 91.91%;	
			③ Wastewater of Turquoise CL-B: 40 g/L, 30 min		③ Wastewater of Turquoise CL-B: more than 97%	
	Grease	Composite of activated clay and attapulgite	Dosage of decolorizer: 1.9 g/100 mL, at 110 °C,40 min	Adsorption of compound materials	85%	15
Coagula-tion method	Papermaking wastewater	PAC: coagulant, PAM: coagulant aid combined treatment	PAC: 1 mL with 10% Al content, PAM: 0.5 mL at 2 g/L, fast/slow stirring: 2/10 min	① Add PAC to neutralize the charge.② Add PAM to form flocs	71.90%	18

(continued)

Table 41.1 (continued)

Approach	Sources	Decolorizer	Reaction conditions	Decolorization mechanism	Decolorization efficiency	Author of references
	Printing and dyeing wastewater	Fe as an anode, C as cathode, electrocoagula-tion treatment	pH: 5.5, electrode spacing: 3 cm, current density: 11 mA/cm^2, electrolysis: 30 min	Electrolytic metal anodes condense and separate colloids and suspended impurities in wastewater	–	19
	Dyeing Wastewater	Polyaluminum silicate chloride (PASC)	pH: around 4.0, H_2O_2: 3.6 mL, $FeSO_4 \cdot 7H_2O$: 1.8 mL, PASC: 8 mL	Homogeneous Fenton oxidation, coagulation, and Synergy of co-precipitation	95%	20
Chemical oxidation method	Hair dye wastewater	Fenton's reagent	pH: 8.2–9.5, dosing ratio of hydrogen peroxide: 0.00303	Fenton (Fe^{2+}, H_2O_2) acts as an oxidant, Fe^{2+} acts as a catalyst to make H_2O_2 generate hydroxyl radicals, thereby oxidatively degrading organic matter	91.10%	22
	Coking wastewater	Ozone	Reaction time 3 h, ozonated gas 0.11 m^3/h	Ozone oxidation	–	23
	Turmeric Saponin Wastewater	Hydrogen peroxide solution (30%), sodium hypochlorite solution	Normal temperature, wastewater pH: 7.5–8.0	Chemical oxidation, cleavage of the chromophore, change in chemical structure	70%; 80%	24

(continued)

Table 41.1 (continued)

Approach	Sources	Decolorizer	Reaction conditions	Decolorization mechanism	Decolorization efficiency	Author of references
Biological method	Dyeing Wastewater	Decolorizing bacteria were screened and fixed with activated carbon as carrier	Urea: 0.45 g/L, yeast paste: 0.25 g/L, temperature: 30 °C, pH: 7, inoculum volume:1.25%, immobilized and cross-linked time: 10 min	Microbial decolorization	More than 70%	26
	Dyeing Wastewater	Activated sludge enriched flora of a printing and dyeing plant	pH: 7.5, temperature: 30 °C	Activated sludge degradation	98.20%	27
	Pen wastewater	ABR reactor of the thickening sludge tank of a brewery	Ratio of sludge to water: 1:2.5, reaction time: 12 d	Activated sludge degradation	78.53%	28

Physical adsorption method uses adsorbent to efficiently adsorb water-soluble organic substances of high-chromaticity. However, the regeneration of adsorbents is relatively complicated and high cost, which makes the relative high running cost, so it is still not widely used.

Chemical oxidation method has the advantages of short reaction time, high removal efficiency of organic matter with high-chromaticity, and efficient controllable reaction. While chemical reagents added in wastewater for chemical reaction occurring can cause secondary pollution, and post-processing of wastewater is complicated.

Coagulation method has simple technological process, easy operation and management, small occupied area, excellent decolorization effect on wastewater of hydrophobic dyes. Convenience of adding coagulants has also received extensive attention from academia. However, the operating costs of coagulation method are high like physical adsorption method. Coagulation method could produce a large amount of sludge which is difficult to dewater, and the decolorization efficiency of hydrophilic dye wastewater is not ideal. Developing composite coagulants with high-efficiency is still the current research focus.

Emerging biological method matured after proposed and improved by various researchers. As microorganisms have a wide range of resources, the operation cost of various reactors is low, adsorption is effective. But microorganisms have strict requirements on the living condition, the processing ability of biological method for some organic substances with high toxicity, strong acidity and alkalinity is low. Processes of biological method could affect the growth of microorganisms, which could prolong the recovery time of microorganisms.

Now the researches for decolorizing organic wastewater of high-concentration and high-chromaticity are relatively simple in China, the following conditions should be considered for decolorization research in accordance with the advantages and disadvantages of the decolorizing methods for organic wastewater above.

(1) Modified adsorbent has high adsorption efficiency and can adsorb diversified adsorbate compared with common adsorbents under the same condition of unit adsorbent mass and adsorption time. However, the modified adsorbents are still not widely used because of their high cost.
(2) Currently, many researchers adopt physical–chemical or physical-biological methods as new methods coupled with multiple process to decolorize or further degrade organic wastewater with high-chromaticity. It should be considered that the new methods should try not to introduce new components harmful in order to avoid secondary pollution to environmental water, and to prevent the increasing difficulty of wastewater treatment.
(3) Wastewater treatment methods should consider the chromaticity reduction of organic wastewater, separation of other toxic and harmful substances, operations of the wastewater treatment process in order to achieve high-efficiency on harmful organic matter removing in wastewater treatment.

Researchers in fields of wastewater treatment can adopt combined methods in future to improve efficiency and capacity for decolorizing organic wastewater of high-concentration and high-chromaticity. Methods can not only reduce the chromaticity, but also can decrease the quantities of other contaminants in the wastewater, and gradually improve the water quality. Methods can also provide preliminary treatment of wastewater with high-concentration and high-chromaticity. Technologies of wastewater decolorization treatment with high efficiency and low cost can be used to promote environmental protection, and make contribution to the construction of ecological civilization in our country.

Acknowledgements This research was funded by the Science and Technology Project of Gansu Province (20YF3FA037), Doctoral Research Fund of Lanzhou City University (LZCU-BS2019-26), and 2021 National Undergraduate Innovation and Entrepreneurship training Program (No.202110737003).

References

Chen XF, Wang C, Zhang HR, Guo HJ, Chen XD (2017) Degumming and bleaching with acid-modified attapulgite of rice bran oil. China Oils Fats 42(02):15–17+37 (in Chinese)

Forgacs E, Cserháti T, Oros G (2004) Removal of synthetic dyes from wastewaters: a review. Environ Int 30(7):953–971

Gong YM, Jiang WL, Li AM, Fan YM, Chang WJ (2017) Present research situation in the treatment of highly concentrated organic wastewater by wet oxidation. Ind Water Treatment 37(5): 20–25+49 (2017) (in Chinese)

Guo DY, Sun TZ, Wu DD, Tian Y, Li XQ (2015) Treatment for dyeing wastewater decolorizing of reactive dye. Wool Technol 43(04):41–44 (in Chinese)

Hao OJ, Kim H, Chiang PC (2000) Decolorization of wastewater. Crit Rev Environ Sci Technol 30(4):449–505

Hao YP, Tong SH, Wang B (2016) Electric coagulation method and treatment of the experimental study of printing wastewater. Shanxi Architect 42(31):123–124 (in Chinese)

Hu XB (2012) Study on characterics analysis of coloured organics in ferment pharmic production wastewater and decolorization mechanism. Nanjing University (in Chinese)

Huang JM (2004) Microbial decolorization treatment on printing and dyeing wastewater. Yunnan Environ Sci 02:51–53 (in Chinese)

Jin ZY (2014) Study on the treatment efficacy of different dye wastewater by bioelectrochemical technology. Shenyang Jianzhu University (in Chinese)

Liu W, Yin J, Kai Y (2014) A preliminary study on chemical treatment of hair dyeing wastewater. Compr Utilization Resour China 32(03):24–26 (in Chinese)

Liu ZL, Hu DY, Sun Y (2020) Research progress of pollution control technology for high concentration organic wastewater. Jiangxi Chem Ind 36(06):111–114 (in Chinese)

Ma LN, Zhang H, Liu F, Sun LB, Yu DY (2013) Applications of Freundlich equation in oil bleaching process and its color reversion. Food Sci 34(20):27–31 (in Chinese)

Shang M, Jiang C, Wu PF, Wu Y (2020) Advances in high concentration organic wastewater treatment technology. Rural Pract Technol (04):177 (in Chinese)

Vandenbossche M, Vezin H, Touati N, Jimenez M, Casetta M, Traisnel M (2014) Cysteine-grafted nonwoven geotextile: a new and efficient material for heavy metals sorption–Part B. J Environ Manage 143:99–105

Wang W (2020) Research on chemical synthesis high concentration organic wastewater treatment technology. China's Petrol Chem Stand Qual 40(18):225–226 (in Chinese)

Wang C, Zhou HD (2020) Advances in high-concentration organic wastewater treatment technology. Pract Technol Rural Areas 03:161–162 (in Chinese)
Wang H, Wang M, Qiao RP, Ma Y, Duan DY (2021) Study on enhanced coagulation treatment of phosphating cleaning wastewater. Chem Manage 22:54–57 (in Chinese)
Xia P (2003) Treating the waste water of high chromatic dyestuff through chemical coagulation. J Huzhou Vocat Tech College 04:81–84 (in Chinese)
Xie L, Liu WJ, Tu Y, Guo FZ, Li DS, Lu X (2010) Experimental study on advanced treatment of wastewater from waste paper papermaking by coagulation sedimentation. J Environ Eng 4(10):2190–2194 (in Chinese)
Ye YG, Chi Y, Zhang YL, Xu W (2010) A study on the optimal process conditions of anaerobic treatment of the pen making wastewater. Guangdong Chem Ind 37(04):252–253 (in Chinese)
Zakhama S, Dhaouadi H, M'Henni F (2011) Nonlinear modelisation of heavy metal removal from aqueous solution using *Ulva lactuca* algae. Biores Technol 102(2):786–796
Zhang LH (2007) Research progress on the treatment of dye wastewater by chemical oxidation. Hebei Chem Ind 10:70–71 (in Chinese)
Zhang GL, Xu XJ, Guo JM, Jia J (2006) Enhanced treatment of dyeing wastewater by homogeneous phase fenton oxidation-coagulation process. Chem Environ Prot 01:38–40 (in Chinese)
Zheng J (2018) A study of wastewater decolorization of azo dyes by biological method. Sci Technol Innov Appl 32:64–65 (in Chinese)
Zheng R, Mei M (2010) Decolorization of biochemical treatment effluent of the sapogenin wastewater by chemical oxidation. Water Purif Technol 29(02):42–44+48 (in Chinese)
Zhou YE (2003) Reviews on treatment of high concentration organic wastewater. Energy Eng 03:40–42 (in Chinese)
Zhou T, Wei SB, Wu XH, Wang J, Wu GM, Lu XH (2006) Study of coagulation assisted by ozonation on bio-treatment outflow of coking wastewater. Appl Chem Ind 02:110–112 (in Chinese)
Zou BB, Chen Y (2021) Research progress of high concentration organic wastewater treatment technology. Chem Manage 09:40–41 (in Chinese)

Chapter 42
Methods and Research Advances in the Preparation of Covalent Organic Frameworks (COFs) Membranes

Yiyun Shu and Liqing Zhang

Abstract Covalent organic framework materials (COFs) are a new class of organic materials linked by covalent bonds and having a high specific surface area and regular pore structure. Channel membranes based on these materials can have great potential in areas such as desalination, dye removal and gas separation. In this paper, the properties of COFs membranes are investigated and membrane preparations based on them, such as co-blending and layer stacking methods, as well as their applications in gas separation and dye separation, are reviewed, and the future prospects of COFs membranes are also presented.

Keywords Covalent organic frame film · Film preparation · Dye wastewater · Fuel cell · Gas separation

42.1 Introduction

With the development of today's world, the lack of fresh water resources and how to manage environmental pollution has become a major issue for all countries. Since the 1980s, the problem of environmental pollution has become more and more serious, and it is important to develop stable, efficient and high-throughput membrane separation technologies. Covalent organic frameworks, designed in the twenty-first century as a new class of porous crystalline materials, have highly ordered and adjustable pore structures, which have important applications in environmental management.

This material was first designed by Yaghi's group in 2005(Cote et al. 2005) as a six-membered macrocyclic ring linked by boron ester rings, distributed and extended periodically in the plane, with π interactions between the macrocyclic skeletons resulting in a tightly stacked structure of regular pores resembling a honeycomb. In 2009, Yaghi's group again synthesised C=N linked COFs by condensing aniline with benzaldehyde to form imine COFs, which further increased the stability of COFs and

---

Y. Shu · L. Zhang (✉)
School of Civil Engineering and Architecture, University of Jinan, Jinan 250022, China
e-mail: cea_zhanglq@ujn.edu.cn

J. Zhang et al. (eds.), *Environmental Pollution Governance and Ecological Remediation Technology*, Environmental Science and Engineering,
https://doi.org/10.1007/978-3-031-25284-6_42

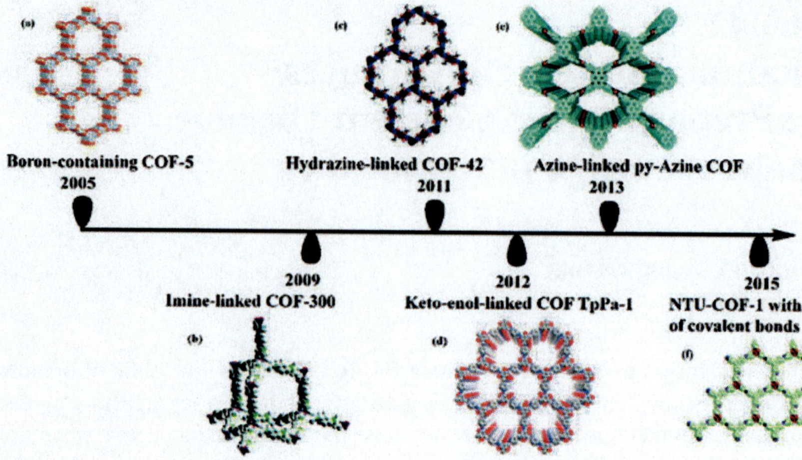

Fig. 42.1 Development history of COFs membrane

broadened their applications. The history of the application of COFs in separation membranes is shown in Fig. 42.1.

42.2 Preparation of Covalent Organic Frame Membranes

42.2.1 Mixed Method

Conventional polymer membranes are not resistant to high temperatures, are prone to ageing and swelling, and are subject to the trade-off effect (Wu et al. 2019). However, hybrid membranes prepared by the co-blending method can accept the diversity of materials while overcoming the contradiction between permeate flux and ion retention rate. This is achieved by blending COFs with polymers in the form of nanoparticles or nanosheets so that the COFs are incorporated into the polymer matrix membrane as porous fillers, thereby improving permeate flux and selectivity. Yang et al. prepared SNW-1 based membranes by incorporating 50–70 nm diameter COF SNW-1 particles into a sodium alginate (SA) matrix for ethanol dehydration (Yang et al. 2016). However, the discontinuous COFs-based membranes were unable to exploit the advantages of their pore structure, making it difficult to synthesise membranes with high permeation selectivity using only co-blending.

42.2.2 *In Situ Growth Method*

In general, discontinuous COFs-based membranes do not fully exploit the potential of COFs pore structure for membrane separation, and therefore it is difficult to prepare advanced membranes with high permeation selectivity by co-blending. Therefore, the establishment of synthetic methods for continuous COFs-based membranes is essential to obtain higher separation performance. With the development of COFs synthesis techniques, it has become possible to grow COFs in situ on modified or unmodified porous support membranes by improving existing COFs synthesis methods. using this strategy, Caro and coworkers prepared continuous and high quality COF-LZU1 membranes on commercial ceramic tube membranes for the removal of dyes from water (Fan et al. 2018).

42.2.3 *Layer Upon Layer Stacking Method*

The method of preparing membranes by layering sheets or monolayers originally came from the synthesis of graphene and graphene oxide monolayers (Abraham et al. 2017). In this method, the nanosheets are first obtained by exfoliating large pieces of material. The nanosheets are then stacked on a porous substrate by pressure or vacuum assisted filtration or dip coating to form a continuous film. This method breaks down COF layer interactions, does not require specific substrates or interfaces and harsh film formation conditions, has fewer COF film formation defects than methods such as in-situ polymerisation (Li et al. 2017), is thinner, can recycle insoluble and difficult to process COF microcrystals, and is a simple and economical preparation process. The main technology of the layer stacking method involves the exfoliation and assembly of COF nanosheets (CONs), and the technical distinction mostly focuses on the different exfoliation methods, such as direct synthesis of nanosheet CONs, solvent ultrasonic exfoliation of CONs, chemical exfoliation, mechanical exfoliation of CONs and self-exfoliation, etc. Zhang et al. (2019a) used hot solvent in situ polymerisation of p-phenylene diboronic acid with the aid of graphene oxide sheet dispersion to form 2D GO/2D COFs. The composite functional membrane GO/COF-1/PAN with a 100 nm thick COF layer was prepared by ultrasonic dispersion in PVA/PVAm solution and filtered through surface carbonation in PAN-based membrane.

42.2.4 *Interfacial Polymerization*

COFs can also be prepared by interfacial polymerisation of traditional PA and polyester films, which is one of the main methods for the synthesis of COF crystals. The interfacial polymerisation method has mild synthesis conditions and a variety of substrates, which can be modified with polymer substrates or self-supported into pure COFs functional films of controlled thickness. Dey et al. (2017) used direct polymerisation at the liquid/liquid interface by dissolving COF aldehyde monomers with chloroform, forming an aqueous buffered interlayer on the liquid surface, followed by the dropwise addition of two COF amine monomers, catalysed by toluene sulfonic acid (PTSA) acid, and controlled liquid/liquid interfacial polymerisation, respectively. The Tp-Bpy membrane had a water flux of 211 L/(m^2·h) and an acidic magenta retention of 99%.

42.3 Application of a Covalent Organic Frame Film

42.3.1 Dye Separation

Approximately 7 × 105 tonnes of dyes are currently used annually in textiles, printing, leather, packaging and cosmetics. 10–15% of these dyes are discharged into water bodies during the colouring process and these dye effluents can cause serious health and ecological problems due to their carcinogenic and mutagenic potential (Mohan et al. 2004). Most COFs materials have their own pore size of about 1–3 nm, which makes them particularly suitable for dye separation, and in recent years, there have been many reports of COFs being used as membrane materials to prepare COFs membranes for dye removal processes. Zhang's team implanted COF-1 crystals onto the surface of graphene oxide (GO) membranes by in situ growth, and the size sieving of COF-1 and the proper interlayer spacing between the graphene oxide nanosheet layers acted synergistically to retain up to 99% of Congo Red, Methylene Blue, Reactive Black 5 and Direct Red (Zhang et al. 2019b).

42.3.2 Gas Separating

Hydrogen is a clean energy alternative to conventional fuels for automobiles, but current methods of hydrogen production inevitably require the separation of hazardous gases. Both discontinuous and continuous COFs-based membranes can be used to separate H_2 from other gases. Based on the multi-interface engineering strategy, the preparation of multilayer COF membrane was applied to the isolated of the gas molecule H_2/CO_2 (Ying et al. 2020). The researchers used in situ interface polymerization to grow COF films with similar pore structure but different pore sizes on the alumina buffer layer, which improved the separation performance of small molecule gas. In addition to this, CO_2/N_2 separation has attracted significant interest in capturing CO_2 to address global warming. A 2016 study used COF-based membranes to separate CO_2/N_2 for the first time (Wu et al. 2017). In this study, the interlayer channels formed between the stacked sheets had a 'gate closing' effect on the selective transport performance of the ultra-thin COF-based membrane. By varying the stacking of COF sheets, high CO_2 permeability and good CO_2/N_2 selectivity could be achieved.

42.3.3 *Fuel Cell*

With growing concerns about fossil fuel depletion and environmental degradation, hydrogen fuel cells are an important alternative to renewable and clean energy sources (Peng et al. 2016). In fuel cells, proton exchange membranes act as electron insulators and anti-butters to block oxygen and hydrogen while conducting protons. The performance of proton exchange membranes depends heavily on the proton conductivity. COFs have well-defined nanochannels that provide space to accommodate proton carriers, conferring them with tunable proton conductivity. Furthermore, the organic linkers used for the synthesis of COFs are chemically versatile and thus strong acid groups can be easily immobilised on COFs to enhance the proton conductivity of COFs. As a result, COFs-based membranes exhibit improved performance when used as proton exchange membranes in fuel cells.

42.4 Conclusion and Further Works

Environmental pollution has become a serious environmental problem today. Dye wastewater contains high concentrations of salts in addition to large amounts of organic dyes, so the treatment of composite dye wastewater and its desalination is essential and new membrane materials and composite membrane designs allow for more accurate screening than other technologies. Similarly, the preparation of hydrogen using COF materials can lead to more clean energy and reduced environmental pollution. The preparation of fuel cells can also further enhance electrical conductivity and prevent resource depletion.

In the future, covalent organic framework membranes may be developed in such a way that: (1) surface charge properties can be achieved through molecular structure design to enable efficient and low cost preparation of membranes at room temperature; (2) the application areas of COFs membranes can be broadened to investigate how more clean energy can be extracted; (3) a systematic theory of membrane preparation can be developed and COF membrane materials can be industrialised and no longer limited to small-scale processes.

References

Abraham J, Vasu KS, Williams CD et al (2017) Tunable sieving of ions using graphene oxide membranes. Nat Nanotechnol 12: 546–550

Cote AP, Benin AI, Ockwig NW et al (2005) Porous, crystalline, covalent organic frameworks. Science 310(5751):1166–1170

Dey K, Pal M, Rout KC et al (2017) Selective molecular separation by interfacially crystallized covalent organic framework thin films. J Am Chem Soc 139(37):13083–13091

Fan H, Gu J, Meng H et al (2018) High-flux membranes based on the covalent organic framewofk COF-LZU1 for selective dye separation by nanofiltration. Angew Chen Int ED 57(15):4083–4087

Li G, Zhang K, Tsuru T (2017) Two-dimensional covalent organic framework (COF) membranes fabricated via the assembly of exfoliated COF nanosheets. ACS Appl Mater Interfaces 9(10):8433–8436

Mohan SV, Bhaskar YV, Karthikeyan JJIJOE et al (2004) Biological decolourisation of simulated azo dye in aqueous phase by algae Spirogyra species. Asian J Microbiol Biotechnol Environ Exp Sci 21(3):211–222

Peng Y, Xu G, Hu Z et al (2016) Mechanoassisted synthesis of sulfonated covalent organic frameworks with high intrinssic proton conductivity. ACS Appl Mater Interfaces 8(28):18505–18512

Vyas VS, Vishwakarma M, Moudrakovski I et al (2016) Exploiting noncovalent interactions in an imine-based covalent organic framework for quercetin delivery. Adv Mater 28(39): 8749–8754

Wu J, Zhan Z, Wang L et al (2019) Analysis of anti-fouling performance of wider spacer RO membrane module. J Chem Ind Eng (China) 70(4): 1446–1454

Wu XS, Tian Z, Wang S et al (2017) Mixed matrix membranes comprising polymers of intrinsic microporosity and covalent organic framework for gas separation. J Membr Sci 528:273–283

Yang H, Wu H, Pan F et al (2016) Highly water-permeable and stable hybrid membrane with asymmetric covalent organic framework distribution. J Membr Sci 520:583–595

Ying Y, Tong M, Ning S et al (2020) Ultrathin two-dimensional membranes assembled by ionic covalent organic nanosheets with reduced apertures for gas separation. J Am Chem Soc 142(9):4472–4480

Zhang XK, Li H, Wang J et al (2019a) In-situ grown covalent organic framework nanosheets on graphene for membrane-based dye/salt separation. J Membr Sci 581(7):321–330

Zhang X, Li H, Wang J et al (2019b) In-situ grown covalent organic framework nanosheets on graphene for membrane-based dye/salt separation. J Membr Sci 581:321–330

Chapter 43
Study on Regeneration of Waste Powdered Activated Carbon by CO_2 Supercritical Extraction

Zhifeng Zhang, Ming Wang, Kan Hong, Ying Wu, Peng Zeng, Xuanxu Zhang, Zhongqi Li, Jianbing Xu, and Xin Liang

Abstract In this study, the activated carbon regeneration of saturated organics in wastewater treatment of a cobalt production enterprise was studied by using CO_2 supercritical extraction method. The effects of supercritical regeneration pressure, duration, temperature, cycle times and other factors on the regeneration and adsorption performance of activated carbon were investigated. The results show that the CO_2 supercritical extraction method can effectively regenerate the waste powdered activated carbon, with a single regeneration rate of 84.8%. After four cycles of adsorption regeneration, the adsorption performance of the regenerated powdered activated carbon is only reduced by 3.52%, Carbon loss rate is 4.86%. The experimental results can provide an effective new idea for the regeneration of powdered activated carbon.

Keywords CO_2 · Supercritical extraction · Powdered activated carbon · Regeneration

43.1 Introduction

Activated carbon adsorption technology is a widely used sewage treatment technology and one of the most commonly used adsorption methods to remove toxic, refractory and non biodegradable compounds from aqueous solutions (Jennifer et al. 2015; Margot et al. 2013). After the activated carbon is used as adsorbent, its pores are finally saturated, so it needs to be regenerated. How to develop a simple and effective method to regenerate activated carbon is one of the most concerned problems in recent years. These methods should significantly increase the cost, increase the service life of the adsorbent, and reduce the amount of waste. The regeneration methods of activated carbon mainly include thermal regeneration (Clifford et al. 1983), chemical regeneration (Li et al. 2015), biological regeneration (Dewalle and Chian 1977), microwave regeneration (Dąbek 2007), photocatalytic regeneration

Z. Zhang (✉) · M. Wang · K. Hong · Y. Wu · P. Zeng · X. Zhang · Z. Li · J. Xu · X. Liang
Ganzhou Nonferrous Metallurgy Research Institute Co., Ltd., Ganzhou 341000, P.R. China
e-mail: zhangzf78@163.com

J. Zhang et al. (eds.), *Environmental Pollution Governance and Ecological Remediation Technology*, Environmental Science and Engineering,
https://doi.org/10.1007/978-3-031-25284-6_43

(Gulyas 1990). Supercritical CO_2 extraction technology is a new separation technology developed in recent years. It is used for the separation and purification of physiologically active substances and natural products, and has been industrialized in medicine, food, environmental protection, essence and other fields (Raventós et al. 2002). Because supercritical CO_2 has the characteristics of high density, high solubility for organic matter, low viscosity and high mass transfer rate, it can penetrate into the microporous system of activated carbon. Some scholars have carried out research on regeneration of waste activated carbon by supercritical CO_2 extraction technology. Tan and Liou (1989) used supercritical CO_2 extraction technology to regenerate waste activated carbon loaded with benzene and toluene. Heidari et al. (2013) studied the regeneration of cyclohexane loaded granular activated carbon by supercritical carbon dioxide at different temperatures and pressures. However, these studies focused on the regeneration of activated carbon adsorbed with a single organic substance, while there were few studies on the regeneration of waste activated carbon adsorbed with complex pollutants.

43.2 Experimental

43.2.1 Reagents and Instruments

CO_2 is industrial grade (steel cylinder), H_2SO_4, NaOH, ethanol, etc. are all analytically pure, detergent is commercially available, and all drugs are directly used after purchase without further purification.

CO_2 supercritical extractor: SCFE120-50-1L, Nantong Kexin Supercritical Equipment Co., Ltd. (Fig. 43.1); COD testing instrument: LH-T725 multi parameter water quality analyzer, Hangzhou Luheng Biotechnology Co., Ltd.

43.2.2 Test Methods

Regeneration method of powdered activated carbon by supercritical CO_2 extraction

Take 400 g of waste powder (with a moisture content of 50%, agglomerated, not dried, directly used), crush it through a 30 mesh sieve, and then load it into a supercritical hopper to investigate the influence of CO_2 flow, reaction pressure, regeneration temperature, regeneration duration, etc. on regeneration. After the reaction, take out the regenerated activated carbon, weigh it, measure the moisture content, and obtain the regenerated powdered activated carbon.

Fig. 43.1 CO_2 supercritical regeneration device

Adsorption performance test of regenerated powdered activated carbon

Take 0.3 g of regenerated powdered activated carbon, add 100 ml of the actual wastewater of the enterprise, mix it on the magnetic stirrer at a rate of 250 r/min for 2 h, and filter the solution after the adsorption is completed; Determine COD in the adsorbed tail solution. Take the same amount of unused activated carbon, carry out reference adsorption experiment, and compare and calculate the regeneration rate of activated carbon. Calculation formula of regeneration rate:

$$\text{Regeneration}\% = \text{COD}_{(N)} * 100/\text{COD}_{(R)} \tag{43.1}$$

$COD_{(N)}$: COD value in wastewater by adsorption with new activated carbon;

$COD_{(R)}$: COD value in wastewater by adsorption with regenerated activated carbon.

43.3 Results and Discussion

43.3.1 Comparison of Regeneration Methods of Activated Carbon

The effects of pure water, ethanol, sulfuric acid (1 mol/L), NaOH (1 mol/L), detergent (1%), ultrasound (353 K for 2 h) and supercritical CO_2 extraction (333 K, 20 MPa, 30 L/h) on the regeneration performance of activated carbon were compared at room temperature and 1:5 volume ratio. As COD removal by activated carbon adsorption is physical adsorption, the organic substances used in the cobalt extraction process

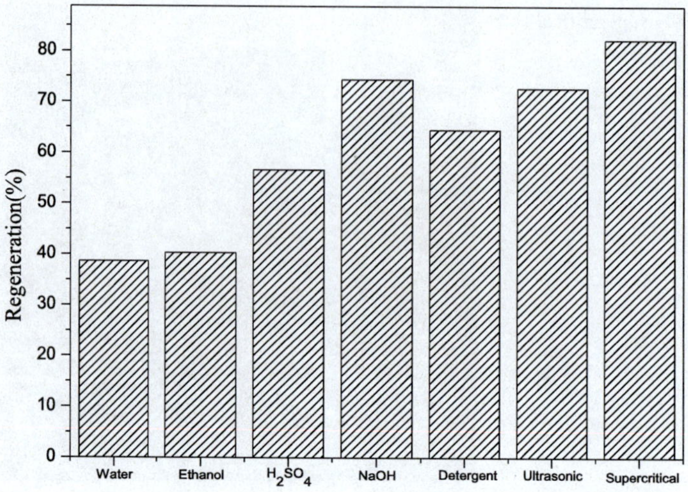

Fig. 43.2 Comparison of regeneration rate of activated carbon treated with different solvents

are extractant P507, kerosene, secondary octanol, etc. According to the principle of similar phase dissolution, ethanol, detergent, CO_2 supercritical fluid are all organic solvents or surfactants, which have good solubility to organic substances, so the purpose of activated carbon regeneration can be achieved; Sulfuric acid is selected because activated carbon has the same adsorption capacity for metal ions in wastewater, and acid can be used to dissociate metal ions from carbon powder. The reason for choosing NaOH is that organic matter has a large solubility in alkali, so as to realize the regeneration of carbon powder. The results are as shown in the figure. It can be seen from Fig. 43.2 that the regeneration effect is pure water < ethanol < H_2SO_4 < detergent < ultrasound < NaOH < CO_2 supercritical extraction. After CO_2 supercritical extraction, the active regeneration rate can reach 82.2%, and the regeneration rate of activated carbon washed with pure water is only 38.6%. Therefore, CO_2 supercritical extraction method is used in this experiment to regenerate saturated activated carbon.

43.3.2 Effect of Temperature on Activated Carbon Regeneration by Supercritical CO_2

Temperature is a major factor affecting the regeneration performance of activated carbon. The influence of supercritical CO_2 temperature on the regeneration of activated carbon is related to pressure. The moisture content and regeneration rate of activated carbon were measured under the conditions of 30 meshes of activated carbon at 318–343 K, supercritical CO_2 pressure of 20 MPa, flow of 30 L/h, and time of 2 h, as shown in Fig. 43.3. It can be seen from the figure that with the

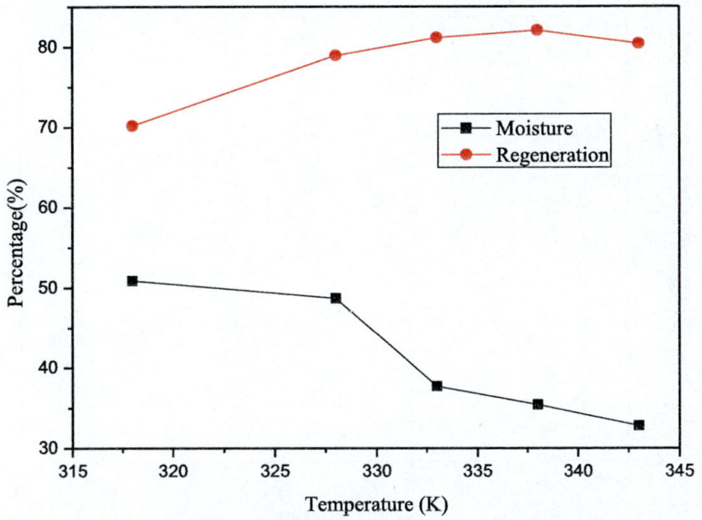

Fig. 43.3 Effect of supercritical CO_2 extraction temperature on activated carbon regeneration rate

increase of temperature, the water content decreases, which is 50.9% at 318 K and 32.8% at 343 K, indicating that the higher the temperature is, the more favorable the water is to be removed from the activated carbon. The regeneration rate of activated carbon increases with the increase of temperature. At 318 K, the regeneration rate is 70.2%. At 333–343 K, the regeneration rate is between 80 and 82%. This shows that with the increase of temperature, the volatility of organic matter adsorbed by activated carbon increases, that is, the driving force of organic matter leaving the surface of activated carbon increases. Therefore, the supercritical CO_2 extraction temperature used in subsequent experiments is 333 K.

43.3.3 *Effect of Pressure on Activated Carbon Regeneration by Supercritical CO_2*

Under the conditions of setting supercritical CO_2 extraction temperature of 333 K, flow rate of 30 L/h and time of 2 h, the regeneration rate of activated carbon within the pressure range of 15–40 MPa was investigated. It can be seen from Fig. 43.4 that the regeneration rate of activated carbon increases with the increase of pressure. This is because with the increase of supercritical CO_2 pressure, CO_2 supercritical fluid is more likely to penetrate into the pores of activated carbon, leaching the organics adsorbed therein, and increasing the ability to dissolve volatile organics. However, when the pressure exceeds 35 MPa, the regeneration rate decreases again, which may be due to the increase of the pressure and the extrusion pressure on the activated carbon, which increases the resistance of the fluid entering the pores, thus reducing

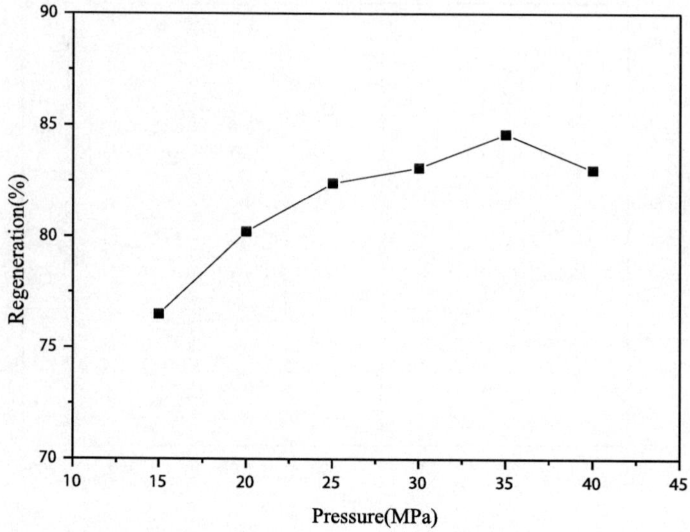

Fig. 43.4 Effect of supercritical CO_2 extraction pressure on activated carbon regeneration rate

the leaching of organic matters. In addition, the higher the pressure is, the higher the requirements for equipment are, and the improvement of 20–35 MPa regeneration rate is not particularly obvious, so the supercritical CO_2 extraction pressure used in this experiment is 20 MPa.

43.3.4 Effect of Time on CO_2 Supercritical Regeneration of Activated Carbon

The extraction temperature is set at 333 K, the pressure is 20 MPa, and the flow rate is 30 L/h. 400 g of 30 mesh activated carbon is separated. The influence of 1 h, 2 h, 3 h, and 4 h regeneration time on the regeneration rate is investigated. It can be seen from Fig. 43.5 that the regeneration rate of activated carbon increases with the increase of reaction time. The regeneration rate increases from 79.3 to 84.8% when the reaction time is 1 h to 4 h. Considering that the effect of 2–4 h regeneration rate increase is not great, the longer the time is, the greater the energy consumption is, so 2 h is appropriate for the reaction time in this experiment.

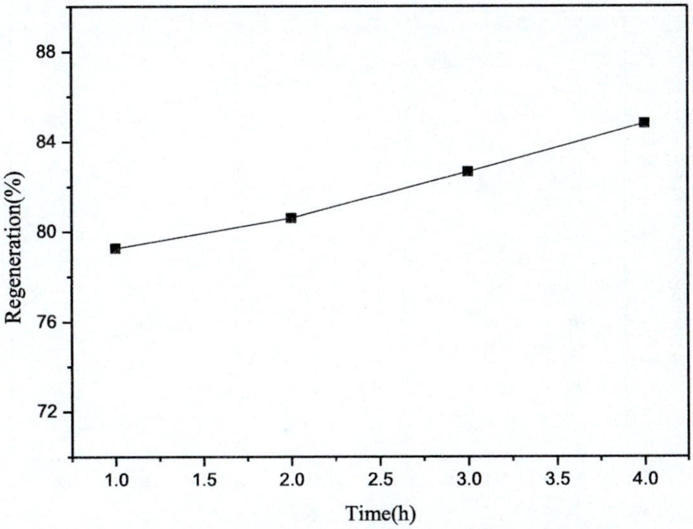

Fig. 43.5 Effect of supercritical CO_2 extraction time on activated carbon regeneration rate

43.3.5 Effect of Particle Size on Activated Carbon Regeneration by Supercritical CO_2

The extraction temperature is set at 333 K, the pressure is 20 MPa, the flow rate is 30 L/h, and the reaction time is 2 h. The regeneration rates of activated carbon (400 g) of 5 mesh, 10 mesh, 30 mesh, 80 mesh and 300 mesh are investigated respectively. It can be seen from Fig. 43.6 that the regeneration rates of activated carbon (400 g) of 5 mesh to 80 mesh increase significantly with the increase of the particle size of activated carbon. The regeneration rates of 5 mesh are 66.4%, 80 mesh are 84.6%, and 300 mesh are slightly reduced. As the water content of waste powdered activated carbon is 50%, it is easy to agglomerate and cannot be sieved under 300 meshes. It can only be sieved after drying. During the experiment, it was found that the dry powder was very easy to lift ash. Under the washing of CO_2 fluid, the dried powdered activated carbon was carried to the instrument pipeline with the fluid, and powdered activated carbon was found at both the pressure relief port and the liquid outlet.

43.3.6 Effect of Regeneration Times on Regeneration Performance of Activated Carbon

The conditions for each regeneration of activated carbon are the same: 333 K, 20 MPa, 2 h, 30 meshes, 30 L/h. The regenerated carbon obtained will be adsorbed with wastewater, and the activated carbon after adsorption will be regenerated under the

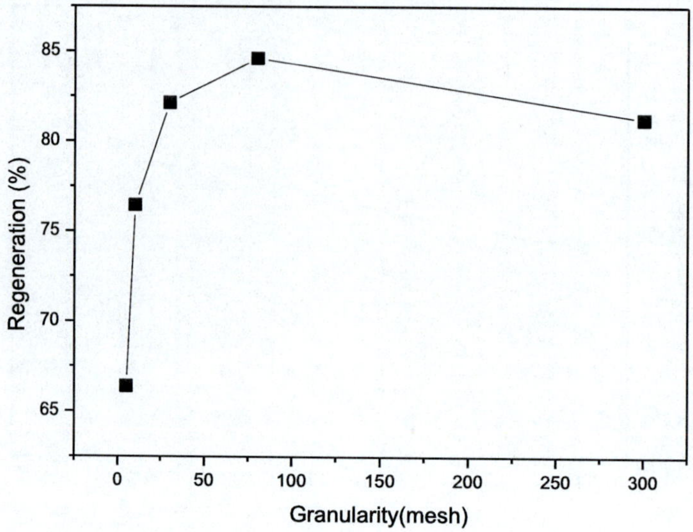

Fig. 43.6 Effect of particle size of activated carbon on its regeneration rate

above conditions for four cycles to investigate the regeneration performance of the activated carbon. It can be seen from Fig. 43.7 that after the first two regenerations, the regeneration rate has little change. After the third and fourth regenerations, the regeneration rate decreases slightly. After four cycles, the adsorption performance decreases from 82.13 to 78.61%, a decrease of 3.52%. This may be partly because some organics cannot be completely eluted through multiple regenerated, and some of them remain in the pores. As the supercritical regeneration is a static regeneration, and solid materials do not turn and stir, the loss rate is low. After four cycles of regeneration, the total loss rate in the regeneration process is 4.86% (Fig. 43.8).

43.3.7 *Effect of Supercritical Alkali Method on Regeneration Performance of Activated Carbon*

Conditions of regenerated activated carbon: 333 K, 20 MPa, 2 h, 30 meshes, 30 L/h. Taking 25 g of regenerated carbon, 2.5 g of NaOH, 50 ml of water, 333 K, stirring for 2 h, suction filtration, washing, drying at 373 K, and then carrying out wastewater adsorption experiments. The regeneration rate can reach 94.28%, which is significantly higher than 84.8% of the non alkali treated.

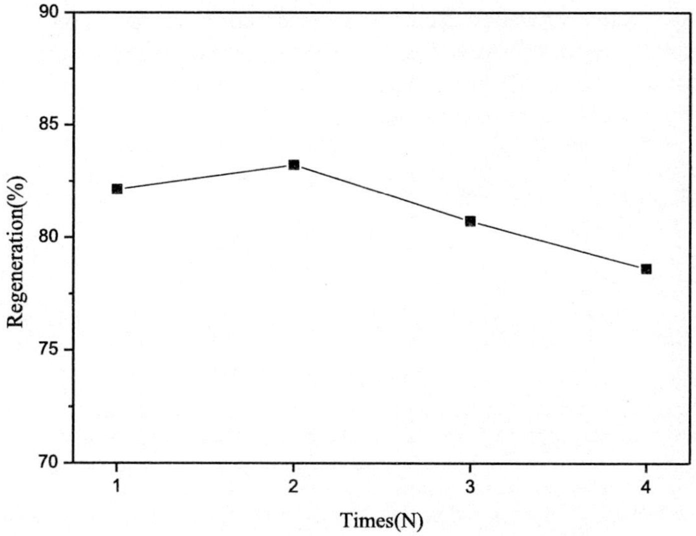

Fig. 43.7 Effect of activated carbon recycling times on its regeneration rate

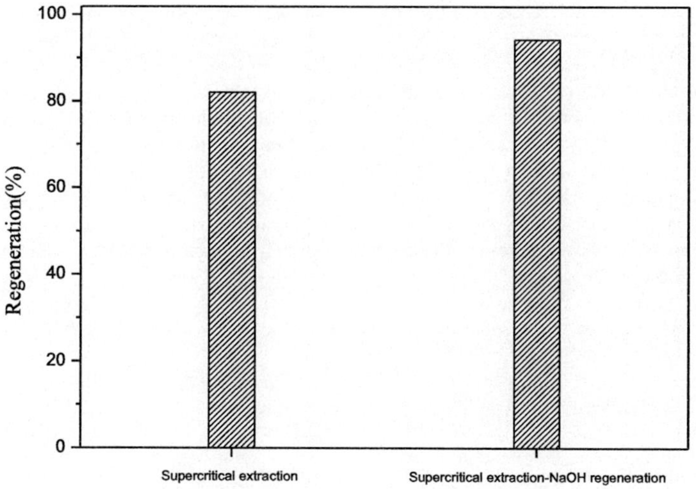

Fig. 43.8 Effect of supercritical alkali combined method on regeneration performance of activated carbon

43.4 Conclusions

In this paper, supercritical CO_2 extraction technology is used to regenerate the waste powdered activated carbon produced by the adsorption extraction separation of actual wastewater from enterprises. The results show that the CO_2 supercritical extraction method can effectively regenerate the waste powdered activated carbon, with a single regeneration rate of 84.8%. After four cycles of adsorption regeneration, the adsorption performance of the regenerated powdered activated carbon is only reduced by 3.52%, and the carbon loss rate is 4.86%.

Acknowledgements The project was supported by the Science and Technology Project of Jiangxi Tungsten Holding Group (JXTC-22B-11).

References

Banuelos JA, García-Rodríguez O, Rodríguez-Valadez FJ et al (2015) Cathodic polarization effect on the electro-Fenton regeneration of activated carbon. J Appl Electrochem 45:523–531

Clifford D, Chu P, Lau A (1983) Thermal regeneration of powdered activated carbon (pac) and pac-biological sludge mixtures. Water Res 17(9):1125–1138

Dąbek L (2007) Microwave regeneration of activated carbon. Environ Prot Eng 33(2):107–115

Dewalle FB, Chian E (1977) Biological regeneration of powdered activated carbon added to activated sludge units. Water Res 11(5):439–446

Gulyas H (1990) Photocatalytic oxidation of biologically treated greywater in the presence of powdered activated carbon

Heidari A, Lotfollahi MN, Baseri H (2013) Regeneration of activated carbon loaded with cyclohexane using supercritical carbon dioxide: experimental results and modeling. Chem Eng Technol 36(2):315–322

Li Q, Qi Y, Gao C (2015) Chemical regeneration of spent powdered activated carbon used in decolorization of sodium salicylate for the pharmaceutical industry. J Clean Prod 86:424–431

Margot J, Kienle C, Magnet A et al (2013) Treatment of micropollutants in municipal wastewater: ozone or powdered activated carbon? Sci Total Environ 461:480–498

Raventós M, Duarte et al (2002) Application and possibilities of supercritical CO_2: extraction in food processing industry: an overview. Food Sci Technol Int

Tan CS, Liou DC (1989) Supercritical regeneration of activated carbon loaded with benzene and toluene. Ind Eng Chem Res 28(8)

Chapter 44
Spatial–Temporal Evolution and Spatial Differentiation of Ecosystem Service Value in Ningxia at the Grid Scale

Wei Chen

Abstract To provide scientific reference for ecological protection and high-quality development of the Yellow River basin, the spatial and temporal evolution pattern and spatial differentiation characteristics of ESV in the upper Yellow River Reaches were investigated. On a 3 km × 3 km grid scale, the ecosystem service value (ESV) of different land use types in Ningxia was calculated using 4 periods remote sensing data from 2000 to 2018. The spatial–temporal evolution characteristics, spatial correlation, and heterogeneity of ESV were discussed in conjunction with spatial exploratory analysis. (1) The land use in Ningxia increased significantly during the study period, with increases of 143.37%, 16.14%, and 3.71% for construction land, forest land, and water area, respectively. The unused land and cultivated land decreased by 8.89% and 5.56%, respectively, whereas the grassland decreased only slightly. (2) From 2000 to 2018, the ESV of Ningxia increased from 39.436 billion yuan to 39.776 billion yuan. Grassland had the greatest ESV, followed by arable land, forest land, and water area. The percentage of unused land was less than 1%, and the change in each ESV across time periods was obscure. (3) The results of the spatial analysis indicate that at the grid scale, the ESV in Ningxia displays a clear "high in the south and low in the north" distribution pattern, with significant positive spatial correlation and dependence, and a significant agglomeration areas effect of low-low and high-high types. The changing of ESV value can provide a scientific basis and decision support for rational layout and precise regulation of land use pattern and ecological barrier function in Ningxia.

Keywords Ecosystem service value · Grid scale · Space exploratory analysis · Ningxia

W. Chen (✉)
Collage of Architecture and Design, New York Institute of Technology, New York, USA
e-mail: cw19960601@gmail.com

J. Zhang et al. (eds.), *Environmental Pollution Governance and Ecological Remediation Technology*, Environmental Science and Engineering,
https://doi.org/10.1007/978-3-031-25284-6_44

44.1 General Instructions

Ecosystem services are products and services derived from ecosystem structure, processes, and functions (Hua et al. 2003). Ecosystem Service Value is part of the UN Millennium Ecosystem Assessment, which promotes sustainable development and human well-being. Ecosystem Service Value (ESV) accounting is essential for assessing the efficacy of environmental protection (Huang et al. 2018) and for building regional ecological safety networks, ecological compensation, and ecological civilization (Peng et al. 2018). Human activities are among the primary causes of the degradation and loss of 60% ecological functions, according to the UN Millennium Ecosystem Assessment Plan (Lin et al. 2010). Land use changes indicate human impact on the ecosystem. By affecting the ecosystem's structure and process, it modifies the supply of ecosystem products and services, influencing ESV (Fu and Zhang 2014). The interaction between urban and rural factors is changing China's land use types and patterns (Liu and Long 2017). Global change research focuses on evaluating ecosystem service function changes based on land use change.

Different fields have studied land use and ecosystem services for years. Global, national, regional, urban agglomeration, river basin, provincial, and county scales are studied. The research focuses on the theoretical framework of ecosystem services assessment (Ouyang et al. 1999), evaluation methods (Xie et al. 2008), and ESV's dynamic evolution and influencing factors (Hou et al. 2019), among others. The most common research methods are ecosystem service value quantity, matter quality, and energy (Li et al. 2018). The value equivalent factor method based on unit area is the most widely used ESV evaluation tool. ESV has become a mature component of the ecosystem based on land use change value assessment and time comparative analysis of different land types (Hou et al. 2019). Most ESV studies evaluate large administrative regions. Due to space dependence and improper evaluation scale selection, land use and ESV spatial information expression is often inadequate, and ESV spatial heterogeneity cannot be revealed. ESV can conduct micro research with grid technology. Improved evaluation precision and reliability. Domestic scholars have also conducted useful research (Hou et al. 2019; Jun et al. 2014), which promotes the widespread use of GIS grid technology and methods and enriches the ESV research system. Most relevant research focuses on southeast China's rapidly developing coastal areas and key ecological protection areas, while northwest China's arid regions, ecologically fragile areas, and key ecological function areas receive less attention. In Ningxia, where "Saishang Jiangnan" is a strategic source site and ecological barrier area, the study on regional land use change and ESV reveals the space–time evolution characteristics and spatial distribution of ESV, which is essential to maintaining regional ecological security and achieving Yellow River basin ecological protection and development.

This paper used Ningxia as the research object, gathered four periods of remote sensing image data from 2000 to 2018, adopted grid technology to analyze ESV in the study area, and combined spatial exploratory analysis to provide a scientific basis for ecological protection, rational land resource use, and ecological civilization.

44.2 Overview of Study Area

Ningxia Hui Autonomous Region lies between 35°15 and 39° 25 north and 104° 15 and 107° 40 east. It's 250 km east-to-west and 456 km north-to-south. It's an ecological barrier in the Yellow River Basin and northwestern China. Ningxia's landform is complex and varied, primarily consisting of hills and plains. The climate is temperate continental at 1100 m. The annual frost-free period is 180 days, and the annual average temperature is 7.8 °C. Annual precipitation averages 230 mm, but north–south distribution is uneven. Ningxia's ecosystem is delicate. The southern mountainous region is prone to soil erosion, the central arid zone to desertification, and the northern Yellow River region to rapid economic growth. Ningxia is a major hub for east–west transport and commerce along the Silk Road. It's a key strategic resource site and ecological barrier region for China's western, eastern, and central regions.

44.3 Research Methods

44.3.1 Data Sources and Processing

ENVI 5.2 was used to process Landsat images from 2000 to 2018 in Ningxia. Manual interactive interpretation was used to classify and extract land use data using ArcGIS10.2. Agricultural land, forest land, grassland, water area, construction land, and unused land were categorized based on the study area. Using field survey data and historical image images, sample areas were created. Phase 4 image data accuracy was 86.4%, 87.2%, 87.8%, and 89.7%; Kappa coefficients were 0.84, 0.86, 0.85, and 0.88; and interpretation accuracy met research requirements. National Meteorological Information Center of China provides NPP data, while Ningxia Statistical Yearbook provides socioeconomic data.

44.3.2 Dynamic Attitude of Land Use

The dynamic attitude of land use is a classic paradigm for studying land use change that can reflect the area change of a particular land use type over a specific time period in the study area (Cheng and Huang 2021). Calculation formula:

$$K = \frac{U_a - U_b}{U_a} \times \frac{1}{T} \times 100\% \tag{44.1}$$

K represents a land use type's dynamic disposition; U_a and U_b represent its area of a particular land use at the beginning and end of the study. T is the research period.

44.3.3 Valuation of Ecosystem Services

Xie et al. (2008) proposed the most widely used value equivalent conversion method in China. This method quantifies ecosystem service value based on the research results of Costanza et al. (1997), which has been unanimously recognized by the academic community. This paper refers to Xie et al. (2008)'s revision of China's Ecosystem Service Value Scale per unit area (Xie et al. 2008, 2015). Since the scale uses the national average ecosystem level, it cannot reflect spatial heterogeneity or socioeconomic imbalance. This paper modifies the ESV coefficient using the coefficients of regional spatial heterogeneity and social and economic development.

$$ESV = \sum (A_i \times VC_i \times S_i \times PI) \tag{44.2}$$

ESV is the value of ecosystem services; A_i is area of land use class i; VC_i is ecosystem service value coefficient. S_i is the spatial heterogeneity coefficient and PI is socioeconomic factors coefficient. Modification of spatial heterogeneity coefficient.

Ecosystem services' value is directly related to their quality and biomass. As biomass increases, so do ecosystem services and ESV (Zuo et al. 2021). This paper used NPP as a spatial heterogeneity coefficient to reflect ecosystem biomass. In this paper, the Thornthwaite Memorial model is used to calculate data due to its accuracy and ease of use.

$$S_i = \frac{NPP'}{NPP''} \tag{44.3}$$

$$NPP = 3000 \times \left[1 - \mathrm{e}^{-0.0009695 \times (V-20)}\right] \tag{44.4}$$

$$V = \frac{1.05r}{\sqrt{1 + \left(\frac{1.05r}{L}\right)^2}} \tag{44.5}$$

$$L = 300 + 25 \times T + 0.05 \times T^3 \tag{44.6}$$

N is the spatial heterogeneity coefficient; NPP is net primary productivity of vegetation [kg/($hm^2 \cdot$ year)]; NPP′ is the study area's net primary vegetation productivity, and NPP″ is the whole country's; V is the actual evapotranspiration (mm); R is annual rainfall (mm); L is the average evaporation (mm); T is the annual average temperature (°C). Social and economic development coefficient revision.

Due to differences in natural environment, social, and economic conditions, there are spatial differences in socio-economic development, ecosystem service functions, and people's perception of ecosystem service value (Zhang et al. 1988). This paper calculates the regional socioeconomic development coefficient using willingness to pay and ability to pay based on existing research. Calculation formula:

$$PI = W \times A \tag{44.7}$$

PI represents the coefficient of social and economic development, W represents residents' willingness to pay for ESV, and A represents their ability to pay for ESV.

Residents' willingness to pay for ESV varies with the social economy. The Peel growth curve model depicts social and economic development level change:

$$W = \frac{Z_s}{Z_g} \tag{44.8}$$

$$Z = \frac{1}{\left(1 + e^{3 - \frac{1}{E_n'}}\right)} \times \frac{h}{H} \tag{44.9}$$

$$E_n' = E_a \times P_a + E_b \times P_b \tag{44.10}$$

W is the social and economic development coefficient; Z_s and Z_g are the study area's and country's socioeconomic development levels; E_n' is Engel coefficient; E_a and E_b are Engel's coefficient of town and country respectively. P_a and P_b are the proportions of urban and rural population in the total population respectively. H and h are the urbanization level of the study area and the whole country respectively.

Ability to pay is the concrete manifestation of economic capacity at the individual, regional, and national levels. Consequently, the capacity to pay can be represented and revised using the per capita GDP level of the study area and the entire nation, with the following calculation formula:

$$A = \frac{GDP_s}{GDP_g} \tag{44.11}$$

In this formula, GDP_s is the per capita GDP of the study area; GDP_g stands for national per capita GDP.

44.3.4 Spatial Exploratory Analysis Method

The spatial exploratory analysis method is a compilation of analyses of spatial data. The degree of spatial correlation is central to the model (Liu et al. 2014). Through the description and visual representation of spatial distribution patterns, spatial agglomeration and differentiation characteristics (Li and Yuan 2017) are investigated, such as global spatial autocorrelation and local spatial correlation.

Global spatial autocorrelation.

The global spatial autocorrelation can reflect ESV values' spatial correlation at the study grid scale. In this paper, Moran's I was used to measure and test ESV's global autocorrelation. Calculation method:

$$I = \frac{n\sum_{i=1}^{n}\sum_{j=1}^{n} w_{ij}(x_i - \overline{x})(x_j - \overline{x})}{\sum_{i=1}^{n}(x_i - \overline{x})\sum_{i=1}^{n}\sum_{j=1}^{n} w_{ij}} \tag{44.12}$$

I is Moran's index; n is the total number of the study area; X_i and X_j are the attribute values of a research area i and j, respectively. $\overline{x}$ is the average value of attributes; W_{ij} is the weight of spatial matrix.

Local spatial correlation

Local spatial correlation can reflect the same attribute value under a certain degree of agglomeration of the region and the surrounding area, identify the space of local imbalance and the presence of spatial correlation and heterogeneity, and measure a similar or dissimilar degree of the region and the surrounding area. To evaluate the significance of local spatial agglomeration (Chen 2015), its computation formula is as follows:

$$I_i = Z_i \sum_j w_{ij} z_j, \tag{44.13}$$

$$z_i \frac{n(x_i - \overline{x})^2}{\sum_i (x_i - \overline{x})} \tag{44.14}$$

$$z_j = (x_j - \overline{x}) \tag{44.15}$$

Z_i and Z_j are the standardization of observed values on region i and j respectively.

44.4 Results and Analysis

44.4.1 Analysis of Land Use Pattern Change in Ningxia

Table 44.1 shows how land use in Ningxia changed from 2000 to 2018. Grassland and arable land make up more than 78% of Ningxia's total area, followed by unused land, forest land, and construction land. Water makes up only 1.9%. During the study period, cultivated land decreased 2.11% and unused land decreased 0.91%. Forest, water, and construction land all increased by 0.75%, 0.07%, and 2.80%. The grassland area decreased by 0.60% during the study.

In the study area, construction land, forest land, unused land, and water area increased by 7.965% annually. The national policy of converting farmland to forest and grassland had a significant impact on forestland and cultivated land from 2000 to 2005, accounting for 2.075% and -1.046%, respectively. Construction land, grassland, and undeveloped land changed the most from 2005 to 2010. From 2010 to

Table 44.1 Area changes of different land use types in Ningxia from 2000 to 2018

Year		Arable land	Forest land	Grassland	Water area	Construction land	Unused land
2000		35.45	4.63	4005.91	1.87	1.95	10.18
2005		33.59	5.11	46.45	1.91	2.28	10.65
2010		34.04	5.34	45.51	1.91	3.28	9.91
2018		33.33	5.37	45.32	1.94	4.75	9.27
2000–2005	Variation rate/%	−1.85	0.48	0.53	0.04	0.33	0.47
	Dynamic rate/%	−1.05	2.08	0.23	0.46	3.34	0.92
2005–2010	Variation rate/%	0.45	0.23	−0.93	−0.00	1.00	−0.74
	Dynamic rate/%	0.27	0.92	−0.40	−0.02	8.75	−1.39
2010–2018	Variation rate/%	−0.70	0.03	−0.20	0.03	1.48	−0.63
	Dynamic rate/%	−0.26	0.08	−0.06	0.19	5.63	−0.80
2000–2018	Variation rate/%	−2.11	0.75	−0.60	0.07	2.80	−0.91
	Dynamic rate/%	−0.33	0.90	−0.07	0.21	7.97	−0.49

2018, construction land area changed the most (5.632%), followed by unused land (−0.801%) and cultivated land (−0.259%). Forest land and grassland changed little.

Several ecological projects and urban development drove the significant land use change in Ningxia during the study period. Due to ecological engineering, Ningxia's ecological consciousness and land protection concept will be improved and changed, and sustainable implementation of returning farmland to forest, grazing prohibition, and sand prevention engineering will ensure its effective and sustainable growth. As urbanization and urban–rural integration construction grew rapidly, Ningxia construction land also rose.

44.4.2 Analysis on the Change of ESV in Ningxia

Ningxia's ESV increased by 0.86% from 39.436 billion yuan in 2000 to 39.942 billion yuan in 2005 and then decreased to 39.776 billion yuan in 2018 (Table 44.2). The ESV's change rate was 1.28%, −0.02%, and −3.39% over different time periods. Different land uses had different ESVs during the study period. Grassland ESV was 48.22% of the total, followed by cultivated land, forest land, and water area. Less than 1% of ESV was unused land, and construction land reduced ESV. The ESV

Table 44.2 Changes in ecosystem service value of different land use types from 2000 to 2018

Project		Arable land	Forest land	Grassland	Water area	Construction land	Unused land	Total
Ecosystem service value	2000	88.09	59.15	194.37	50.25	0.00	2.50	394.36
	2005	83.48	65.29	196.63	51.40	0.00	2.61	399.42
	2010	84.59	68.28	192.68	51.35	0.00	2.43	399.33
	2018	82.84	68.70	191.84	52.11	0.00	2.28	397.76
2000–2005	Variation volume/10^8 CNY	−4.61	6.14	2.26	1.14	0.00	0.12	5.06
	Variation rate/%	−5.23	10.38	1.16	2.28	0.00	4.61	1.28
2005–2010	Variation volume/10^8 CNY	1.11	2.99	−3.95	−0.05	0.00	−0.18	−0.09
	Variation rate/%	1.32	4.58	−2.01	−0.09	0.00	−6.96	−0.02
2010–2018	Variation volume/10^8 CNY	−1.75	0.42	−0.84	0.76	0.00	−0.16	−1.57
	Variation rate/%	−2.07	0.62	−0.44	1.49	0.00	−6.40	−0.39
2000–2018	Variation volume/10^8 CNY	−5.25	9.55	−2.53	1.86	0.00	−0.22	3.40
	Variation rate/%	−5.96	16.14	−1.30	3.70	0.00	−8.90	0.86

of construction land, cultivated land, grassland, and unused land decreased over the study period. Agricultural land, grassland, and unused land have decreased due to urban and rural construction. The ESV of forestland and water area increased each year, with forestland increasing by 955 million yuan due to ecological projects and policies such as the returning farmland project, the ecological migration project, and the mountain grazing policy.

From the standpoint of service value of various service functions in the study area, regulation service ESV is the greatest, accounting for 55.95% of the total ESV, followed by support service and supply service, accounting for 33.68 and 7.79%, and cultural service accounting for just 2.58%. Adjustment service climbed by 215 million Yuan, followed by cultural service by 70 million Yuan, and supply service by just 23 million Yuan, showing a consistent ESV trend in the studied region.

Individual ESV in the research region changed somewhat over time (Fig. 44.1). During the study period, food production, waste treatment, and soil conservation decreased at rates of −3.86%, −0.49%, and −0.12%, respectively. Other individual ESVs showed a slight upward trend, with raw material production and aesthetic landscape showing the highest growth rates, at 10.84% and 7.36%, respectively. Soil conservation had the largest ESV in Ningxia, contributing for 20.51% of the total, followed by waste treatment, water conservation, biodiversity, and climate control. Aesthetic landscape only contributed for 2.58%.

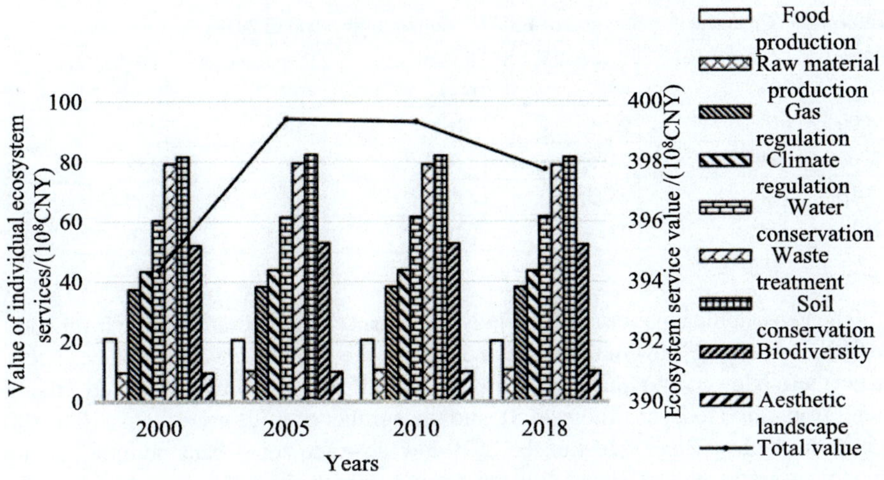

Fig. 44.1 Changes of individual ecosystem service values in Ningxia from 2000 to 2018

44.4.3 Spatial and Temporal Patterns of ESV in Ningxia

To explore the geographical and temporal differentiation features of ESV at the micro scale in Ningxia, a grid was used as the research unit for spatial visualization and analysis. Based on relevant research (Jun et al. 2014), the spatial and temporal pattern properties of ESV at different grid sizes were compared. 3 km × 3 km grid scale best represents ESV geographical distribution in studied region. The ESV was classed using the natural breakpoint approach and grouped into five grades: highest, higher, medium, lower, and lowest. Finally, ESV's geographical distribution in Ningxia was determined (Fig. 44.2).

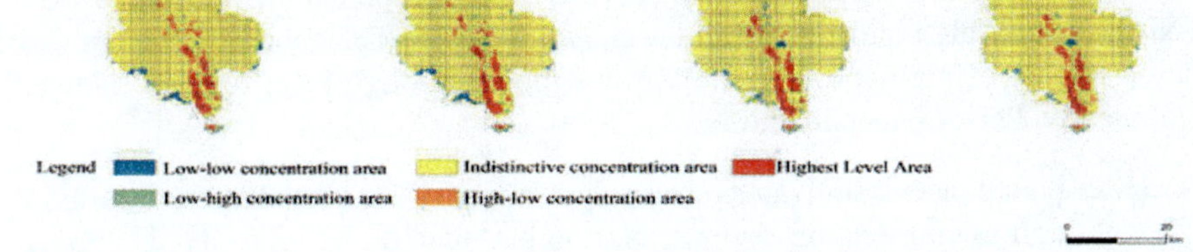

Fig. 44.2 Spatial distribution pattern of ESV in Ningxia at grid scale from 2000 to 2018

Table 44.3 Changes of spatial grid of ESV in Ningxia from 2000 to 2018

Year	Lowest-level area	Lower-level area	Intermediate area	Higher-level area	Highest-level area
2000	610	2579	2161	510	125
2005	624	2495	2197	537	132
2010	702	2393	2175	576	139
2018	778	2234	2219	608	146

The geographical pattern of ESV in Ningxia has changed dramatically from 2000 to 2018, becoming "low in the north and high in the south." Low-grade zones occur in northern Ningxia and Shapotou. Low-grade ESV zones in Ningxia showed a rising trend from 2000 to 2018 (Table 44.3), and the number of grids grew from 610 to 780 along the Yellow River. Most of the 2579 lowest-grade zones became intermediate and lower-grade regions. Expanding regions were mostly in wooded areas with ample resources.

44.4.4 Spatial Autocorrelation Analysis of ESV

To explore the geographical variation and differentiation of ESV in Ningxia, the global Moran's I index and LISA cluster analysis were employed to assess ESV at the grid size in ArcGIS. The global Moran's I index is computed using contiguous border and ROW. Ningxia ESV's Moran's I index is 0.4836, 0.4868, 0.4857, and 0.4987 in 2000, 2005, 2010, and 2018. $Z = 52.26$, 52.60, 52.47, and 53.87. Z values were 52.26, 52.60, 52.47 and 53.87, respectively. All P values are 0.000, showing that ESV exhibits high spatial autocorrelation at the 3 km $\times$ 3 km grid size, that is, positive agglomeration and dependence characteristics.

Based on the considerable spatial connection of ESV at the research area's grid, in the 95% confidence interval, Moran's I index and LISA clustering map classify ESV into four types: (1) Low-low agglomeration areas, low ESV of grid itself and surrounding grid; (2) High-high agglomeration areas; (3) Low–high agglomeration areas, low ESV of grid itself, but high surrounding grid; (4)High-low agglomeration areas.

Ningxia ESV has considerable geographical agglomeration features, according LISA (Fig. 44.3). During the research period, the geographical distribution of ESV in Ningxia was "high in the south and high in the north." The geographical agglomeration impact of high and low kinds was strong, although local regions changed significantly. Performance includes:

(1) High-high concentration areas. Jingyuan County in the south has the most ESV high-high agglomeration regions. Northern Dawukou, Pingluo, Helan, Xixia, Xingqing, and Zhongning. High-high agglomeration zones around the confluence of Yongning County, Lingwu City, and Qingtongxia City are steady, while

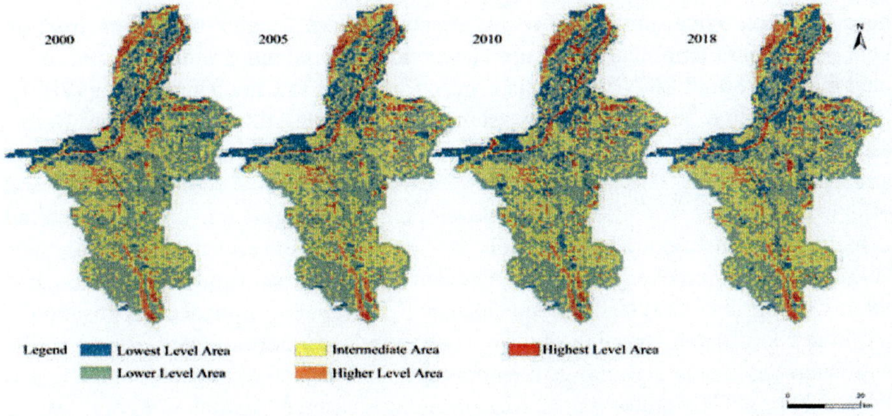

Fig. 44.3 LISA spatial agglomeration of Ningxia ESV from 2000 to 2018

dispersion is erratic elsewhere. From the geomorphic features in the study area, high concentrated areas are mainly distributed in Liupan mountain, Helan mountain area, Roshan and Xiangshan areas with relatively abundant ecological resources, forest land and high coverage grass distribution dense, Jingyuan county forest coverage rate of more than 86%, is the important ecological headwater, is also an important source of Ningxia ESV.

(2) Low-low concentration areas. Low concentrations of ESV are found west of Shapotou, north of Jinfeng, Yongning, and Litong. Jinfeng's low-low concentration regions grow while Shapotou's shrink. Rest of low concentration area is county seat, villages, towns, and industrial land. Low-concentrated regions in Ningxia are mostly unused land, urban and rural development land, and cultivated land.

(3) Low–high and high-low agglomeration areas. In the research region, ESV agglomeration areas are tiny and point-like. It's mostly in high-high and low-low agglomeration boundaries. From the research region's geomorphic features, these areas are generally the interleaved zone of forest land, grassland, and cultivated land and the border area of water, cultivated land, and grassland. In Huangxian County, agricultural land and building land are densely dispersed around water regions, and high-low agglomeration areas surround high-high agglomeration areas. In Jinfeng District and Lingwu, low–high agglomeration regions surround low-low ones.

44.5 Discussion

Natural, social, and economic influences cause land use changes, which cause regional ecological changes. Social and economic considerations determine its volume and quality (Verburg et al. 2004). Due to social and economic development needs and ecological protection requirements, Ningxia's quick urban construction land growth has complicated land use conversions. Grassland, arable area, and forest land make up about 86% of Ningxia ESV, comparable to Cheng and Huang (2021). More people live on farms and grassland due to urban–rural convergence. Reforesting crops will impair ecological engineering, soil conservation, food production, and waste treatment. This will assist Ningxia maintain national ecological safety and offer ecological commodities. Future development in Ningxia must control the blind expansion of construction land, consolidate land, strictly observe the red line of cultivated land, delimit the red line of ecological protection, protect and restore ecological land, and improve the system and mechanism of regional ecological compensation to promote coordinated and high-quality whaling industry development. Based on the equivalent factor method and grid technology, this paper investigated land use and ESV in Ningxia Hui Autonomous Region and discussed its spatial–temporal pattern and evolution characteristics. This will help reveal the internal correlation characteristics between the two and provide a new technology and way to investigate the spatial change of ESV. The spatial–temporal evolution of ESV is affected by natural elements like topography and climate, as well as social and economic factors like urbanization, industrial development, and people density (Jun et al. 2014). Future research should focus on the spatial–temporal development pattern and properties of ESV under various scenario scenarios, as well as critical elements impacting this process. These principles give new avenues for ecosystem services research and will be investigated in this work.

References

Chen Y (2015) Spatial-temporal variation and driving factors of land ecological service value in Hubei Province. China Univ Geosci

Cheng J, Huang Y (2021) Res Soil Water Conserv 28(02):382–389

Costanza R, d'Arge R, De Groot R et al. (1997) The value of the world's ecosystem services and natural capital. Nature 387(6630):253–260

Fu BJ, Zhang LW (2014) Land-use change and ecosystem services: Concepts, methods and progress. Prog Geogr 33(4):441–446

Hou M, Yao S, Deng Y et al (2019) Spatial-temporal evolution pattern and differentiation characteristics of ecological service value in Yan 'an city under grid scale: based on the implementation background of farmland for forest project. J Nat Resour 34(03):539–552

Hua Z, Zhiyun O, Tongqian Z, Zhenxin L, Weihua X (2003) The impact of human activities on ecosystem services. Journal of Nat Resour 18(1):118–126

Huang L, Cao W, Xu XL et al (2018) Linking the benefits of ecosystem services to sustainable spatial planning of eco-logical conservation strategies. J Environ Manage 222:385–395

Jun MA, Peng MA, Changxiao LI et al (2014) Spatial-temporal changes of ecosystem service value based on land use in the Three Gorges Reservoir Area (Chongqing). Sci Silvae Sin 50(05):17–26

Li J, Yuan W (2017) Evaluation of land ecological security based on GIS grid model in Yinchuan City. J Nat Resour 32(6): 988–1001

Li L Wang X, Luo L et al (2018) A review of valuation methods for ecosystem services. Chin J Ecol 37(04):1233–1245

Liu JY, Kuang WH, Zhang ZX et al (2014) Basic characteristics and spatial pattern of land use change in China since the late 1980s. Acta Geogr Sin 69(1):3–14

Liu Y, Long H (2017) Study on the spatio-temporal patterns of land use transition and its impact on ecological service function of the middle of Yangtze River economic belt. Econ Geogr 37(11):161–170

Ouyang Z, Wang X, Hong XK (1999) A primary study on Chinese terrestrial ecosystem services and their ecological-economic values. Acta Ecol Sin 5(2):19–25

Peng J, Yang Y, Liu Y, Du Y, Meersmans J, Qiu S (2018) Linking ecosystem services and circuit theory to identify ecological security patterns. Sci Total Environ 644:781–790

Verburg PH, Schot PP, Dijst MJ et al (2004) Land use change modelling: current practice and research priorities. GeoJournal 61(4):309–324

Xie GD, Zhen L, Lu CX, Xiao Y, Chen C (2008) A value-based approach to ecosystem services based on expert knowledge. J Nat Resour 23(5):911–919

Xie GD, Zhang CX, Zhang CS et al (2015) The value of ecosystem services in China. Resour Sci 37(9):1740–1746

Zhang Q, Gao M, Yang L et al (2017) Spatial structure and ecosystem service value change of ecological land use in nine districts of Chongqing city from 1988 to 2013. Acta Ecol Sin 37(02):566–575

Zhen L, Liu XL, Li F, Wei YJ, Koenig H (2010) Consumption of ecosystem services and eco-compensation mechanism in ecological sensitive regions: progress and challenges. Resour Sci 32(5):797–803

Zuo LL, Peng WF, Tao S et al (2021) Dynamic changes of land use and ecosystem service value in the upper reaches of the Minjiang River. J Ecol 41:6384–6397

Chapter 45
Research on Calibration of Air Quality Data Based on Principal Component Analysis and Multiple Linear Regression Model

Xiaoming Jiang

Abstract Based on the accurate monitoring data of national control points, this paper analyzes the difference factors of monitoring data of self-built points through data analysis, and establishes a mathematical model to calibrate the data of self-built points. More accurate and real-time monitoring of self-built monitoring data are collected and processed in this paper. First, the principal component analysis (PCA) method is used to analyze the data of self-built points, and identified six main factors (pressure, PM10, O_3, NO_2, SO_2 and humidity) that affect air quality among the 11 factors. Then, the daily mean value of overseas control points and self-built points from November 14, 2018 to June 11, 2019, is calculated and compared with the data of national control points and self-built points to draw a broken line trend chart, this data makes it more intuitive to show China's progress in controlling smog, and the air quality continues to improve. Then, the data difference between the self-built point and the state-controlled point is analyzed in days. The data difference between the self-built point and the state-controlled point is taken as the dependent variable, and the temperature, humidity, wind speed, air pressure and precipitation are taken as the independent variables for multiple function linear regression analysis. It is concluded that the factors affecting PM2.5 are pressure and humidity, the factors affecting PM2.5 are pressure and temperature, the factors affecting CO are pressure and temperature, and the factors affecting NO_2 are pressure and temperature, The factors affecting SO_2 are pressure and temperature, and the factors affecting O_3 are pressure and temperature.

Keywords Air quality · Calibration · Principal component analysis · Multiple linear regression

X. Jiang (✉)
Liaoning Institute of Mechatronic Technology, Dandong 118009, Liaoning, China
e-mail: mingbeibei@yeah.net

J. Zhang et al. (eds.), *Environmental Pollution Governance and Ecological Remediation Technology*, Environmental Science and Engineering,
https://doi.org/10.1007/978-3-031-25284-6_45

45.1 Introduction

Air pollution does great harm to the ecological environment and human health (Wu et al. 2022; Tong et al. 2021; Li et al. 2022; Ryu et al. 2021; Kim et al. 2021). Through the real-time monitoring of the concentration of "two dusts and four gases" ($PM_{2.5}$, PM_{10}, CO, O_3, NO_2 and SO_2), we can timely grasp the air quality and take corresponding measures for the pollution source (Iribagiza et al. 2021; Schneider et al. 2021; Richards and Brozell 2021). The monitoring data of "two dusts and four gases" from national monitoring and control stations (national control points) are real, but the distribution and control are few, the release time lag is long, and the cost is large (Ketu and Mishra , 2021; Betancourt et al. 2021; Solanki et al. 2021; Majewski et al. 2021; Zeng and Jin 2021). The micro air quality detector independently developed by the company can carry out real-time grid monitoring of air quality in a certain area, and simultaneously monitor meteorological parameters such as temperature, humidity, wind speed, air pressure and precipitation. However, the monitoring data of self-built sites are different due to weather and other reasons (Diaz et al. 2021; Turiel and Kaufmann 2021; Kuhr 2015; Environmental Building; Reports from Kyung Hee University Provide New Insights into Environmental Building 2020; Maji et al. 2020).

Since the electrochemical gas sensor used will produce certain zero drift and range drift after long-term use, the concentration change of unconventional gaseous pollutants (gas) will have cross interference on the sensor, and the influence of weather factors on the sensor (Environmental Health - Air Quality; New Air Quality Data Have Been Reported by Investigators at North-West University 2020). It is necessary to use the hourly data of national control points to calibrate the data of self-built points close to national control points (Morris et al. 2020). Air pollution, also known as air pollution, usually refers to the phenomenon that some substances enter the atmosphere due to human activities or machine learning approaches for urban natural processes, showing sufficient concentration for sufficient time, and thus endangering human comfort, health and welfare or the environment (Information Technology; Investigators at Northeast Normal University Discuss Findings in Information Technology 2020). In other words, as long as the quantity, nature and time of the existence of a certain substance are sufficient to affect human beings or other organisms and properties, we can call it an air pollutant; The phenomenon caused by its existence is air pollution (Rahimpour et al. 2020). The harm of air pollutants to human body is multifaceted, mainly manifested in respiratory diseases and physiological dysfunction, as well as diseases caused by irritation of mucous membrane tissues such as eyes and nose. It is a chronic factor causing asthma in the elderly. Lack of lung qi leads to physical decline (Teles et al. 2020). When the concentration of pollutants in the air is very high, it will cause acute pollution poisoning, or worsen the disease, and even kill thousands of people in a few days. In fact, the human body breathing this polluted air for many years will also cause chronic bronchitis, bronchial asthma, emphysema, lung cancer and other diseases (Teles et al. 2020). This paper has collected the monitoring data of "two dusts and four gases" from the state control points, and the air quality, temperature, humidity, wind speed, pressure, precipitation and other data from the self-built points. It is intended to conduct exploratory analysis on the data from the state control points and the self-built points, and establish a mathematical model to calibrate the data from the self-built points.

45.2 Data Analysis

45.2.1 Principal Component Analysis

First of all, by analyzing the data given by the national control points and self-built points, we use the principal component analysis method to reduce the dimension of 11 factors in the data of self-built points as independent variables. According to the analysis results of MATLAB, we select 6 main factors (pressure, PM_{10}, CO, O_3, NO_2, SO_2 and humidity) that affect the air quality.

We use MATLAB to conduct principal component analysis on the data of self-built points. Principal component analysis (PCA) is a data statistical method. It uses a mathematical statistical dimensionality reduction method to find out several comprehensive variables to replace the original many variables.

$$\begin{aligned} y_1 = {} & 0.2314x_1 + 0.4873x_2 + 0.007x_3 + 0.1491x_4 - 0.0019x_5 + 0.0748x_6 \\ & + 0.0002x_7 + 0.0280x_8 + 0.8218x_9 - 0.0283x_{10} + 0.0651x_{11} \end{aligned} \tag{45.1}$$

$$\begin{aligned} y_2 = {} & 0.3973x_1 + 0.203x_2 - 0.0002x_3 + 0.0482x_4 + 0.0235x_5 - 0.00863x_6 \\ & - 0.000009\,x_7 + 0.0282x_8 - 0.0549x_9 - 0.0431x_{10} + 0.0912x_{11} \end{aligned} \tag{45.2}$$

$$\begin{aligned} y_3 = {} & -0.0022x_1 + 0.1243x_2 + 0.0025x_3 - 0.2498x_4 - 0.0373x_5 + 0.8716x_6 \\ & + 0.0024x_7 - 0.0364x_8 - 0.0722x_9 + 0.1122x_{10} - 0.3767x_{11} \end{aligned} \tag{45.3}$$

$$\begin{aligned} y_4 = {} & -0.0425x_1 - 0.0866x_2 + 0.0026x_3 + 0.9282x_4 + 0.2155x_5 + 0.2273x_6 \\ & - 0.0043x_7 - 0.0709x_8 - 0.1134x_9 + 0.0711x_{10} - 0.0895x_{11} \end{aligned} \tag{45.4}$$

$$\begin{aligned} y_5 = {} & 0.0140x_1 + 0.0046x_2 + 0.0016x_3 - 0.2130x_4 + 0.9738x_5 - 0.0336x_6 \\ & + 0.0009x_7 + 0.0280x_8 + 0.0395x_9 - 0.0186x_{10} - 0.0473x_{11} \end{aligned} \tag{45.5}$$

$$\begin{aligned} y_6 = {} & -0.1784x_1 - 0.0050x_2 + 0.0022x_3 - 0.0346x_4 + 0.0546x_5 + 0.3816x_6 \\ & 0.0035x_7 - 0.0608x_8 - 0.0460x_9 - 0.0482x_{10} + 0.9001x_{11} \end{aligned} \tag{45.6}$$

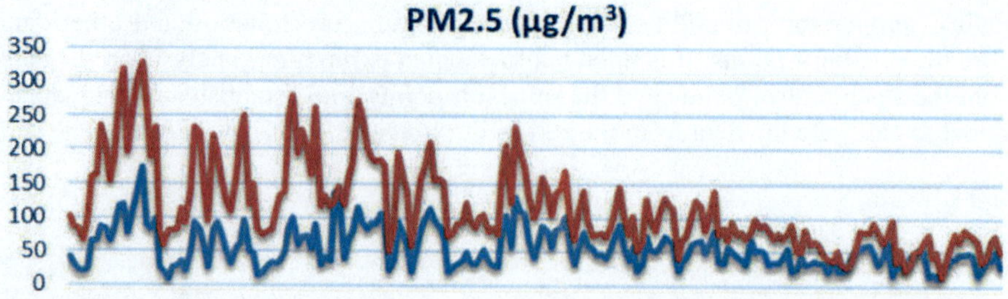

Fig. 45.1 The tendency of concentration of PM2.5 at overseas control points and self-built points

It can be seen that the main factors affecting air quality are pressure, PM_{10}, O_3, NO_2, SO_2 and humidity. Let's consider the pressure at self-built points, PM_{10}, O_3, NO_2, SO_2 and other factors as the main factors.

45.2.2 Trend Analysis of Air Quality

Then we calculate the mean values of pressure, PM_{10}, O_3, NO_2, SO_2, humidity and air quality at overseas control points and self-built points through Excel, and analyze the indexes of various pollutants to obtain the corresponding trend, and obtain the broken line trend chart and analyze the characteristics. This paper analyzes the daily average concentration changes of $PM_{2.5}$, PM_{10}, CO, NO_2, SO_2 and O_3 from November 4, 2018 to June 11, 2019. The changes are shown in the following Fig. 45.1.

It can be seen that the concentration of most pollutants detected at the national control points at the self-built points has decreased since February 2019. Therefore, the self-built points may differ greatly from the data detected at the national control points due to winter, but the concentration of CO monitored at the national control points and the self-built points has always been very different. We need to further strengthen it. At the same time, we can see China's progress in controlling smog from the broken line diagram of $PM_{2.5}$ and PM_{10}.

45.3 Results and Discussion

45.3.1 Multiple Linear Regression Model

By processing the data of national control points and self-built points, we get the data difference between national control points and self-control points, and use MATLAB to conduct multiple function linear regression analysis with the data difference between self-built points and national control points as the dependent variable and

Table 45.1 The coefficients, coefficient confidence intervals and statistics of the output regression model

Regression coefficient	Estimated value of regression coefficient	Confidence interval of regression coefficient
β0	−496.7872	[−918.6952, −74.8793]
β1	−3.0455	[−9.3405, 3.2495]
β2	0.4856	[0.0806, 0.8906]
β3	0.0121	[−0.0027, 0.0270]
β4	0.1249	[−0.3513, 0.6011]
β5	0.2888	[0.1916, 0.3860]

$R^2 = 0.3601$, $F = 22.6270$, $P < 0.0001$, $S^2 = 80.6811$

temperature, humidity, wind speed, air pressure and precipitation as the independent variables to obtain the coefficient of regression equation.

We use Excel to calculate the average value of daily air quality of overseas control points and self-built points. In order to determine the relationship between the difference between the data of self-built points and the data of national control points and wind speed, pressure, precipitation, temperature and humidity, we make the scatter diagrams of Y_1, Y_2, Y_3, Y_4, Y_5 and X_1, X_2, X_3, X_4, X_5 respectively. If the scatter diagram shows an approximate linear relationship between them, we can set Y_1, Y_2, Y_3, Y_4, Y_5, Y_6 and X_1, X_2, X_3, X_4, X_5 is a five variable linear regression model. The difference of $PM_{2.5}$ is taken as an example for calculation.

$$y_1 = \beta_0 + \beta_1 x_1 + \beta_2 x_2 + \beta_3 x_3 + \beta_4 x_4 + \beta_5 x_5 + \varepsilon \tag{45.7}$$

As shown in Table 45.1 for the coefficients, coefficient confidence intervals and statistics of the output regression model.

The model thus obtained is:

$$y = -496.7872 - 3.0455x_1 + 0.4856x_2 + 0.0121x_3 + 0.1249x_4 + 0.2888x_5 \tag{45.8}$$

The model is improved. The figure is as shown in Fig. 45.2. Eight abnormal points are found. The following is the procedure for re modeling after deleting the abnormal points.

Delete outliers in the file and re model. After deleting the outliers, there are no outliers at this time. See Table 45.2 for the coefficients, coefficient confidence intervals and statistics of the improved regression model.

Comparing Table 45.1 with Table 45.2, we can find that the determinable coefficient is increased from 0.3601 to 0.4613, and the F statistic is increased from 22.6270 to 33.0493, which shows that the improved model is significantly improved.

The optimal model is:

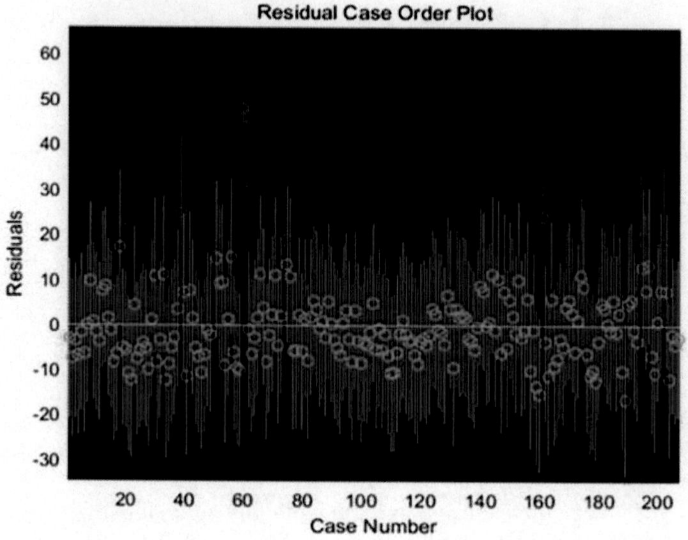

Fig. 45.2 The residual case order plot

Table 45.2 The coefficients, coefficient confidence intervals and statistics of the improved regression model

Regression coefficient	Estimated value of regression coefficient	Confidence interval of regression coefficient
β0	−427.0848	[−750.0846, −104.0850]
β1	−3.3666	[−8.3512, 1.6180]
β2	0.4181	[0.1081, 0.7280]
β3	0.0121	[0.0024, 0.0258]
β4	0.0141	[−0.2733, 0.4560]
β5	0.2653	[0.1905, 0.3401]

$R^2 = 0.4613$, $F = 33.0493$, $P < 0.0001.0001$, $S^2 = 46.2305$

$$y_1 = -427.0848 - 3.3666x_1 + 0.4181x_2 + 0.0121x_3 + 0.0121x_4 + 0.2653x_5 \tag{45.9}$$

It can be seen from the model that pressure and temperature are the biggest factors that affect the difference of PM2.5 between self-built points and national control points. Similarly, we can get another five models.

$$y_2 = -427.0848 - 0.0426x_1 + 0.0026x_2 + 0.0001x_3 + 0.0011x_4 + 0.0009x_5 \tag{45.10}$$

$$y_3 = 38.1496 - 0.3741x_1 - 0.0357x_2 + 0.0002x_3 - 0.0492x_4 - 0.0055x_5 \tag{45.11}$$

$$y_4 = -2.1829 - 0.0059x_1 + 0.0021x_2 + 0.0001x_3 + 0.0022x_4 + 0.0006x_5 \tag{45.12}$$

$$y_5 = -1.0504 - 0.0235x_1 + 0.0010x_2 + 0.0001x_3 + 0.0008x_4 + 0.0002x_5 \tag{45.13}$$

$$y_6 = -1.5745 - 0.0435x_1 + 0.0016x_2 + 0.0001x_3 + 0.0010x_4 + 0.0003x_5 \tag{45.14}$$

It can be seen from the model that pressure and temperature are the biggest factors affecting the difference of PM_{10} difference between self-built points and national control points, precipitation is the biggest factor affecting the difference of CO difference between self-built points and national control points, pressure and temperature are the biggest factors affecting the difference of NO_2 difference between self-built points and national control points, and pressure and temperature are the biggest factors affecting the difference of SO_2 difference between self-built points and national control points Temperature is the biggest factor that affects the difference of O_3 between self-built points and national control points.

45.3.2 Pollutant Concentration Calibration Model

We use the data of national control points to calibrate the data of self-built points. Firstly, due to the complexity of data, one hour is selected as the calibration time period (10:00–11:00, November 24, 2018). To sum up, the data cleaning and analysis by Excel in this paper can save a lot of time and improve efficiency, but it can only analyze the existing data There are many factors that affect air quality, and we have not considered them comprehensively.

We mainly use MATLAB algorithms such as data interpolation and data fitting to obtain the data. There is still a lot of room for optimization. For further improvement of the model, we should consider reducing the mean value to better calibrate the data.

45.4 Conclusion

In conclusion, we selected PM2.5 data from 10:00 to 11:00 on November 24, 2018 as an example for data analysis to calibrate the self-built point data. Then we use MATLAB to plot the data image of 24 data of self-built points by using the interpolation method, plot the line of 2 data points of foreign control points, calculate the average value of the data of national control points and self-built points at the same time to calibrate 24 scattered point data, and use the data fitting method to give the data calibration function of self-built points. The data of the other five indicators shall be processed in the same way. We used Excel to calculate the daily mean value of overseas control points and self-built points from November 14, 2018 to June 11, 2019, and compared the data of national control points and self-built points to draw a broken line trend chart, this data makes it more intuitive to show China's progress in controlling smog, and the air quality continues to improve. Then, the data difference between the self-built point and the state-controlled point is analyzed in days. The data difference between the self-built point and the state-controlled point is taken as the dependent variable, and the temperature, humidity, wind speed, air pressure and precipitation are taken as the independent variables for multiple function linear regression analysis.

Acknowledgements The authors express thanks to the support of the first batch of national vocational education teachers' teaching innovation team research project named "research and Practice on building a new mode of cultivating professional and compound technical talents in Higher Vocational Education Based on new formats" (SJ202010201).

References

Betancourt C, Hagemeier B, Schröder S, Schultz MG (2021) Context aware benchmarking and tuning of a TByte-scale air quality database and web service. Earth Sci Inform (prepublish)

Díaz JJ, Mura I, Franco JF, Akhavan-Tabatabaei R (2021) aiRe—A web-based R application for simple, accessible and repeatable analysis of urban air quality data. Environ Model Softw (prepublish)

Environmental Building; Reports from Kyung Hee University Provide New Insights into Environmental Building (2020) Imputing missing indoor air quality data via variational convolutional autoencoders: implications for ventilation management of subway metro systems. In: Global warming focus

Environmental Health - Air Quality; New Air Quality Data Have Been Reported by Investigators at North-West University (2020) (Characterizing light-absorbing aerosols in a low-income settlement in South Africa). Ecol Environ Conserv

Information Technology; Investigators at Northeast Normal University Discuss Findings in Information Technology (2020) (Airexplorer: visual exploration of air quality data based on time-series querying). Comput Technol J

Iribagiza C, Sharpe T, Coyle J, Nkubito P, Piedrahita R, Johnson M, Thomas EA (2021) Evaluating the effects of access to air quality data on household air pollution and exposure—an interrupted time series experimental study in Rwanda. Sustainability 13(20):11523

Kuhr K (2015) Advances in the visualization of urban air quality data and environmental monitoring using TIBCO Spotfire (R) and the Elm sensor network. In: Abstracts of papers of the American chemical society, vol 250

Ketu S, Mishra PK (2021) Scalable kernel-based SVM classification algorithm on imbalance air quality data for proficient healthcare. Complex Intell Syst 7(5):2597–2615

Kim T, Kim J, Yang W, Lee H, Choo J (2021) Missing value imputation of time-series air-quality data via deep neural networks. Int J Environ Res Public Health 18(22):12213

Li K, Bai K, Li Z, Guo J, Chang NB (2022) Synergistic data fusion of multimodal AOD and air quality data for near real-time full coverage air pollution assessment. J Environ Manage 302:114121

Majewski G, Szeląg B, Mach T, Rogula-Kozłowska W, Anioł E, Bihałowicz J, Dmochowska A, Bihałowicz JS (2021) Predicting the number of days with visibility in a specific range in Warsaw (Poland) based on meteorological and air quality data. Front Environ Sci 9:623094

Maji S, Ahmed S, Ghosh S, Garg SK (2020) Evaluation of air quality index for air quality data interpretation in Delhi, India. Curr Sci 119(6)

Morris E, Liu X, Manwar A, Zang DY, Evans G, Brook J, Rousseau B, Clark C, MacIsaac J (2020) Application of distributed urban sensor networks for actionable air quality data. ISPRS Ann Photogrammetry Remote Sens Spat Inf Sci VI-4/W2-2020

Rahimpour A, Amanollahi J, Tzanis CG (2020) Air quality data series estimation based on machine learning approaches for urban environments. Air Qual Atmos Health (prepublish)

Richards J, Brozell T (2021) Compilation and evaluation of ambient respirable crystalline silica air quality data near sand quarries and processing facilities. Atmosphere 12(7):903

Ryu B, Yoo S, Kim S, Choi J (2021) Thirty-day hospital readmission prediction model based on common data model with weather and air quality data. Sci Rep 11(1):23313

Schneider SR, Lee K, Santos G, Abbatt JP (2021) Air quality data approach for defining wildfire influence: impacts on $PM_{2.5}$, NO_2, CO, and O_3 in Western Canadian Cities. Environ Sci Technol

Solanki N, Bruckman D, Tang AS, Khatri SB (2021) Air quality data for nitrogen dioxide and pediatric hospital admission rates in selected cities. Am J Respir Crit Care Med 203(9)

Teles B, Mariano P, Santana P (2020) Game-like 3d visualisation of air quality data. Multimodal Technol Interact 4(3):54

Tong PF, Chen SX, Tang CY (2022) Detecting and evaluating dust-events in north china with ground air quality data. Earth Space Sci 9(1):e2021EA001849

Turiel JS, Kaufmann RK (2021) Evidence of air quality data misreporting in China: an impulse indicator saturation model comparison of local government-reported and U.S. embassy-reported PM2.5 concentrations (2015–2017). PloS one 16(4)

Wu Z, Ma C, Shi X, Wu L, Dong Y, Stojmenovic M (2022) Imputing missing indoor air quality data with inverse mapping generative adversarial network. Build Environ (prepublish)

Zeng G, Jin R (2021) Predicting Beijing air quality data based on LSTM method. J Trend Sci Res Dev 5(3)

Chapter 46
Study on Zero Discharge Process and Utilization of Coking Wastewater

Ying Tian, Chong Zhao, Xia Zhu, Zhengkun Li, Xiaoyu Liu, Shuai Du, Zheng Chen, Can Wang, Jilin Teng, and Ruozheng Li

Abstract In this study, the composition analysis of coking wastewater organic was conducted through a serious of experiments. And then, salt separation process of "evaporative concentration—freezing crystallization—evaporative crystallization" was investigated. The experiment results show that the main components are sodium chloride as well as sulfate sodium. It's feasible to carry out the salt separation process, which can be applied to prepare the crystal salt of sodium chloride and sodium sulphate. The obtained salt products of sodium sulphate satisfies the standard of Coal Chemical Industry—By-product industrial sodium sulfate (T/CCT 001–2019) with first in Class A, meanwhile, the obtained salt products of sodium chloride satisfies the standard of Coal Chemical Industry—By-product industrial sodium chloride (T/CCT 002–2019) with first class.

Keywords Coking wastewater · Evaporation · Crystallization · Organic compounds · Sodium chloride · Sodium sulphate

46.1 Preface

Coking wastewater is produced by recovering tar, benzene and other by-products in the process of obtaining coke and gas from coal pyrolysis at high temperature (Shan et al. 2007). The main sources include the period of coal dry distillation, coal washing, blue carbon coke quenching, raw gas recovery and purification, and chemical product recovery. The complex upstream process and various sources lead to high organic content, large fluctuation of water quality and extremely complex composition (Zhang et al. 2021; Chen 2011; Mi and Sun 2021; Gao and Zhang 2021; He et al. 2020).

Y. Tian (✉) · C. Zhao · X. Zhu · Z. Li · X. Liu · S. Du · Z. Chen · C. Wang · J. Teng · R. Li
Beijing Guodian Futong Science and Technology Development Co., LTD., Beijing 100070, China
e-mail: ttsherry@126.com

J. Zhang et al. (eds.), *Environmental Pollution Governance and Ecological Remediation Technology*, Environmental Science and Engineering,
https://doi.org/10.1007/978-3-031-25284-6_46

At present, the commonly used reduction treatment method is membrane concentration process, including RO, STRO and others. During the membrane concentration process, membrane concentrated water with high salt content, high pollutant concentration and poor biochemical performance will be produced (Li et al. 2018; Wu 2017; Shen et al. 2021). In order to meet the requirements of energy conservation and discharge standards for water pollutants, the membrane concentrated water needs to be treated urgently (Wang and Wang 2020; Wang 2022).

In this study, based on the component analysis and salt separation theory analysis of coking wastewater, the process of "evaporative concentration—freezing crystallization—evaporative crystallization" is investigated. And then, the process of salt separation from coking wastewater is obtained, which conforms to standard sodium sulfate and sodium chloride products so as to achieve "zero discharge" of coking wastewater membrane concentrated water.

46.2 Experiments

46.2.1 Sample

Coking wastewater membrane concentrated water in a factory.

46.2.2 Experimental Setup

See Fig. 46.1.

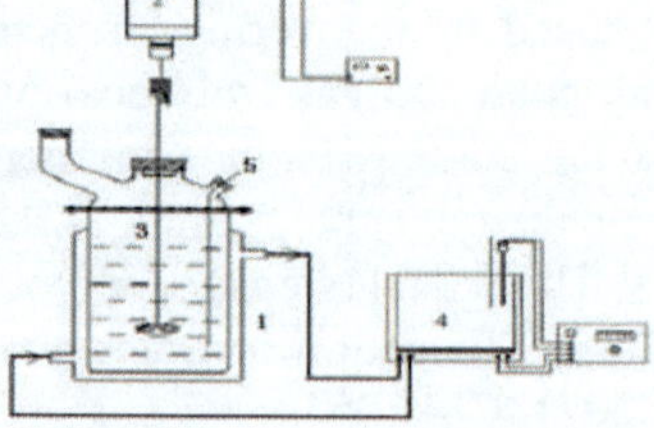

1—Crystallizer; 2—Stirrer; 3—Agitator; 4—Thermostatic oil bath; 5—thermocouple

Fig. 46.1 Schematic diagram of evaporation/crystallization device

46.2.3 *Experimental Methods*

Evaporative concentration

Add materials into the evaporator and start the stirrer. And then turn on the thermostatic water bath to heat for evaporation and concentration. The evaporation process occurs at atmospheric pressure.

Evaporative crystallization

Add materials into the evaporative crystallizer. And then turn on the thermostatic water bath to heat for evaporative crystallization. After evaporation, the crystal slurry was dewatered with a centrifuge to obtain crystal salt.

Freezing crystallization

Add materials into the freezing crystallizer for freezing crystallization. Subsequently, the crystal slurry was dewatered with a centrifuge to obtain sodium sulfate decahydrate crystal salt.

46.3 Results and Discussion

46.3.1 *Wastewater Component Analysis*

Physical–chemical properties

The coking wastewater with high salt and organic matter. The wastewater has a high chroma with 128–200 times. The TDS is 20,000–70,000 mg/L. The composition of inorganic salt in the wastewater is complex, mainly including sodium chloride and sodium sulfate. And other inorganic ions are potassium ion, calcium ion, magnesium ion, nitrate, fluoride ion and bromine ion. As shown in Fig. 46.2.

Fig. 46.2 Coking wastewater

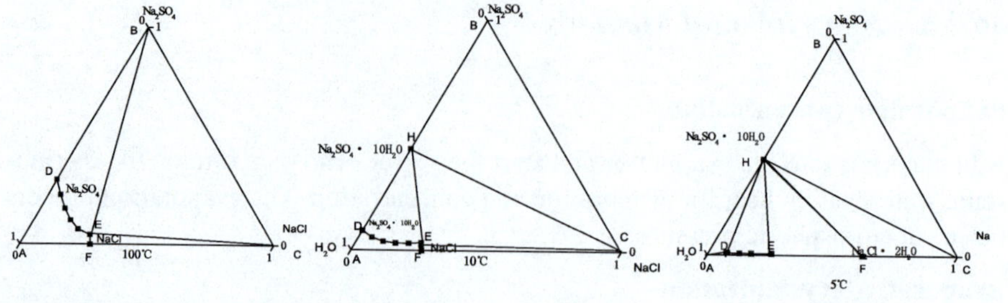

Fig. 46.3 Phase diagram of NaCl-Na_2SO_4-H_2O ternary water-salt system at 100 °C,10 °C, −5 °C

46.3.2 Process Route Analysis

The system can be simplified into a ternary system, namely, the NaCl-Na_2SO_4-H_2O ternary water-salt system (Fig. 46.3).

From Fig. 46.3a, b and c, it can be found that the composition point of coking wastewater is located in the liquid phase area. If the wastewater is evaporated and concentrated to some extent, then cooled to the crystallization area of sodium sulfate decahydrate. Therefore, sodium sulfate decahydrate can be prepared by the process of "evaporative concentration + freezing crystallization", meanwhile, anhydrous sodium sulfate crystal salt can be prepared by thermal crystallization of sodium sulfate decahydrate.

The mother solution of frozen crystallization can be evaporated sequentially at high temperature which falls within the crystallization zone of sodium chloride. Therefore, it can be used to prepare the sodium chloride crystal salt.

46.3.3 Experimental Study

Evaporative concentration

The condensed water of coking wastewater is good in quality (with 38 mg/L COD and 130μS/cm Conductivity) by evaporative concentration at atmospheric pressure.

Freezing crystallization

Sodium sulfate decahydrate crystal salt can be prepared by freezing crystallization. Sodium sulfate decahydrate crystal salt are shown in Fig. 46.4.

The filtration performance of sodium sulfate decahydrate is good in quality by freezing crystallization.

Evaporative crystallization of sodium sulfate

Sodium sulfate crystal salt (Fig. 46.5) can be prepared by evaporative crystallization at atmospheric pressure.

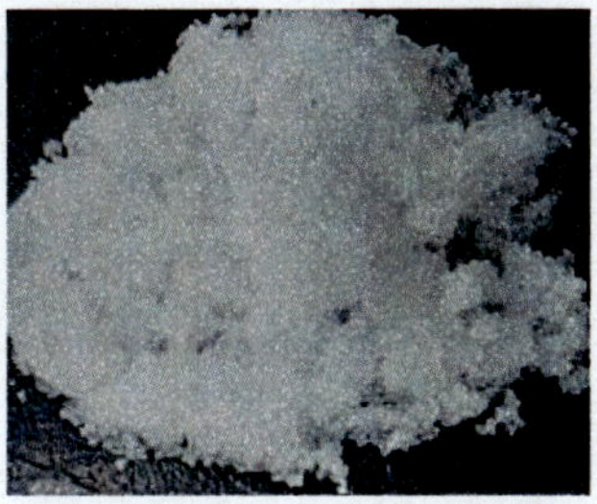

Fig. 46.4 Left: sodium sulfate decahydrate crystal salt

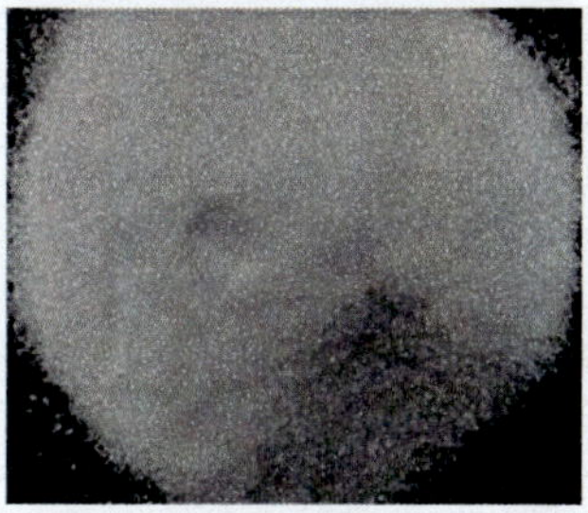

Fig. 46.5 Sodium sulfate crystal salt

According to Fig. 46.5 and Table 46.2, the quality of the prepared sodium sulfate crystalline salt has satisfied the standard of Coal Chemical Industry—By-product industrial sodium sulfate (T/CCT 001-2019) with first in class A.

The crystal habits of the prepared sodium sulfate crystal salt with different magnifications are shown in Fig. 46.6. It can be seen that the sodium sulfate crystals are single crystals, mostly in irregular shape.

Evaporative crystallization of sodium chloride

The sodium chloride crystal salt (Fig. 46.7) can be prepared from frozen mother solution by evaporative crystallization at atmospheric pressure. The composition analysis of sodium chloride crystal salt is shown in Table 46.3.

Table 46.2 Composition analysis of sodium sulfate crystal salt

Items	Detection results
Na_2SO_4 (%)	98.9
NaCl (%)	0.08
Water content (%)	0.22
Water insoluble substance (%)	0.03
Calcium and magnesium (%)	0.07
Whiteness	83.8
TOC mg/kg	22.7

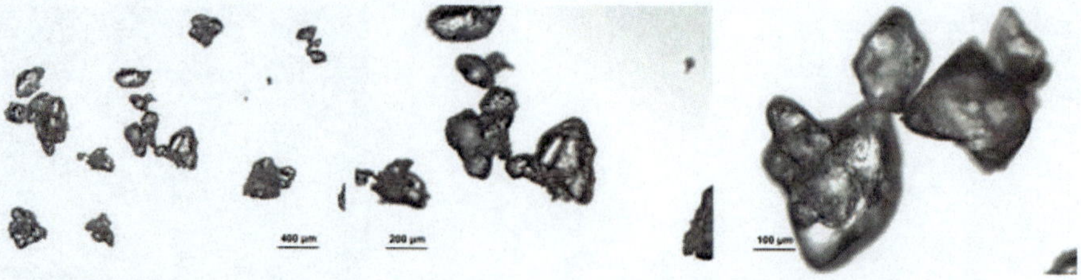

（a）2.5 times （b）5 times （c）10 times

Fig. 46.6 Crystal habit of sodium chloride crystal salt product

Fig. 46.7 Sodium chloride crystal salt

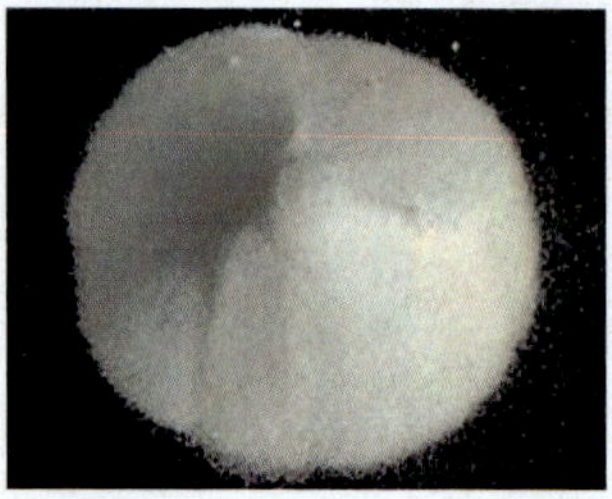

Table 46.3 Composition analysis of sodium chloride crystal salt

Items	Detection results
NaCl (%)	98.6
Na_2SO_4 (%)	0.12
Water content (%)	0.21
Water insoluble substance (%)	0.05
Calcium and magnesium (%)	0.14
Whiteness	76.3
TOC mg/kg	29.5

According to Fig. 46.7 and Table 46.3, the quality of the prepared sodium chloride crystal salt has satisfied the standard of Coal Chemical Industry—By-product industrial sodium chloride T/CCT 002-2019) with first class.

The crystal habits of the prepared sodium chloride crystal salt with different magnifications are shown in Fig. 46.8. It can be seen that the sodium chloride crystals are single crystals, mostly in square shape.

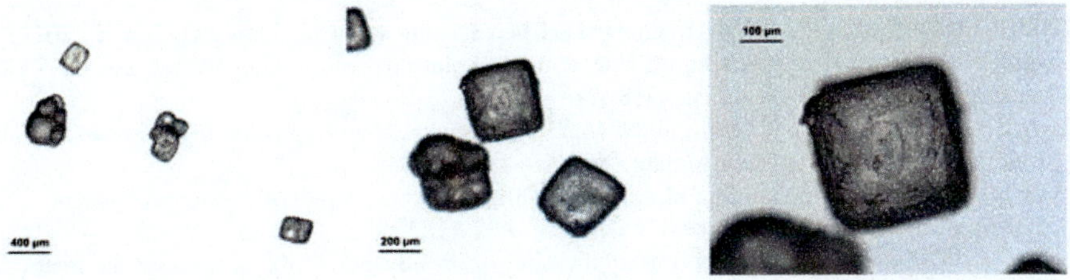

(a) 2.5 times (b) 5 times (c) 10 times

Fig. 46.8 Crystal habit of sodium chloride crystal salt

46.4 Conclusion

Based on the component analysis of coking wastewater, the processes of evaporative concentration, freezing crystallization and evaporative crystallization were investigated in this study, the main results are as follows:

(1) Coking wastewater appears yellow with high chromaticity and salinity. The composition of the wastewater is complex, sodium chloride and sodium sulfate are in majority.
(2) Through the process of "evaporative concentration—freezing crystallization—evaporative crystallization", sodium sulfate and sodium chloride crystal salt products can be prepared.

The prepared sodium sulfate crystal salt meets the standard of Coal Chemical Industry—By-product industrial sodium sulfate (T/CCT 001–2019) with first in Class A. The prepared sodium chloride crystal salt meets the standard of Coal Chemical Industry—By-product industrial Sodium Chloride (T/CCT 002–2019) with first class.

References

Chen C (2011) Research on coking wastewater treatment by Fenton oxidation combined with coagulation process. Fuel Chem Process 52(6):56–58

Gao D, Zhang C (2021) The concentration reduction and near-zero discharge process of concentrated salt water of coking wastewater. Water Wastewater Eng 57(11):92–97

He Z, Fang Z, Xu Z, Mao T, Tan J, Han K (2020) Effect of an advanced treatment of coking wastewater by an improved membrane-process. Min Metall Eng 40(3):111–115

Li H, Tao R, Sun B, Xu Y, Yao H (2018) Case study on design and operation of a coking wastewater treatment project. China Water Wastewater 34(4):97–101

Mi Y, Sun H (2021) Development and technology trend of coking wastewater technology. Shanxi Chem Ind 41(1):215–217

Shan M, Lu Y, Cong L (2007) Coking wastewater treatment technology. Chemical Industry Press, Beijing

Shen L, Liang Y, Zhang Z, Liu D, Zhang H, Chen L, Wu S, Fu K, Zhang Q, Liu X (2021) A pilot-scale treatment on membrane concentrated solution from coking wastewater by STRO membrane. Ind Water Treatment 41(8):118–121

Wang B (2022) Study on ozone combined with heterogeneous fenton process in coking wastewater treatment. Water Resour Dev Manage 8(1):40–45

Wang S, Wang J (2020) Treatment of membrane filtration concentrate of coking wastewater using PMS/chloridion oxidation process. Chem Eng J 379:122361

Wu Y (2017) Design and engineering application of an advanced coking wastewater treatment process. Water Wastewater Eng 53(12):62–66

Zhang C, Zhang T, Wang F (2021) Deep treatment and reuse of coking wastewater. In: 2021 Science and technology annual meeting of the Chinese society of environmental sciences. Environmental Engineering, Tianjin, , pp 311–313

Chapter 47 Advances in Solid Carbon Sources for the Treatment of Low C/N Wastewater

Shuyuan Luo and Ying Fu

Abstract The current phenomenon of low C/N in domestic wastewater in China has serious implications for the extensive use of biological methods to treat domestic wastewater, making it difficult to meet treatment water quality standards. Researchers have considered the use of additional carbon sources such as sodium acetate, glucose and ethanol. However, these sources are costly and do not provide a sustainable source of carbon, so solid carbon sources have come into focus. Solid carbon sources are generally crop straw, fruit peel and some artificial polymers. This type of carbon source has the advantages of sustainable carbon release, environmental protection, economic safety and practicality, while using plant straw and peel as a solid carbon source will release a certain amount of nitrogen, but compared to the amount of carbon released is very small.

Keywords Solid carbon sources · External carbon source · Nitrogen removal

47.1 Introduction

Urban domestic wastewater presents the characteristics of low C/N. It is generally believed that influent C/N > 4 is required to meet the growth requirements of denitrifying bacteria, while at present most urban wastewater in China has an organic matter concentration (COD) of less than 200 mg/L and a C/N of around 3.5 or even lower (Liu 2018; Wang et al. 2015). In China, 70% of urban wastewater treatment plants have an influent BOD_5/TN of less than 4 (Wei and Zhang 2013). According to 127 urban wastewater treatment plants surveyed, 114 plants had BOD_5/TN < 4, accounting for 90% of the total, and more than 50% of them had a serious shortage of carbon sources (Guo et al. 2018).

For the problem of low C/N, mainly affecting the effect of biological methods of denitrification, China's urban wastewater treatment plants generally use SBR,

S. Luo · Y. Fu (✉)
School of Civil Engineering and Architecture, University of Jinan, Jinan 250022, China
e-mail: cea_fuy@ujn.edu.cn

J. Zhang et al. (eds.), *Environmental Pollution Governance and Ecological Remediation Technology*, Environmental Science and Engineering,
https://doi.org/10.1007/978-3-031-25284-6_47

oxidation ditch and other processes, the lack of carbon source is the main bottleneck of denitrification technology of wastewater, to add an additional carbon source to improve the C/N ratio to achieve rapid denitrification of wastewater. Traditional carbon sources are generally sodium acetate, ethanol and glucose. The advantage is high denitrification efficiency and good denitrification effect, the disadvantage is high cost and easy to cause nitrite accumulation (Yu et al. 2022). The advantages of solid carbon sources are non-toxic and harmless, good denitrification effect, and the disadvantages are large individual differences and high pretreatment costs.

47.2 Solid Carbon Sources in Water Treatment Applications

Commonly used solid carbon sources can be divided into three categories: first, natural carbon-containing materials; second, synthetic polymer materials that can be degraded by microorganisms; third, natural materials mixed with synthetic materials.

47.2.1 Natural Carbonaceous Materials

Natural carbonaceous materials generally include crop straw, fruit hulls, fruit peels, starch, bamboo flour, etc. Boussaid et al. (1988) used rice straw as an additional solid carbon source in the late 1980s to treat groundwater containing nitrate, which resulted in rapid removal of 10–20 mg/L of nitrate. In 1996, Volokita et al. (1996) used newspaper as a carbon source and filler to treat wastewater containing 100 mg/L nitrate, removing nitrate without nitrite accumulation, and the COD concentration in the treated water was only 4–10 mg/L. Jing et al. (2019) used lignite as a solid carbon source to remove groundwater. The simulated groundwater had a nitrate concentration of 30 mg/L. The nitrate removal effect was superior, with the highest nitrate removal (34%) characterised as follows: effluent nitrate concentration below 20 mg/L and effluent nitrogen dioxide concentration below 0.01 mg/L. Zhang et al. (2017) used corn cobs as an external solid carbon source, and the potential for release of organic matter from corn cobs, the amount released and the use safety were evaluated. The effect of different amounts of solid carbon sources on simultaneous nitrification and denitrification was investigated in a sequencing batch biofilm sub-reactor (SBBR). The results showed that the concentration of raw water NH_4^+-N varied in the range of 19.8–27.5 mg/L. Dissolved oxygen levels were 4.0 ± 0.2 mg/L when the temperature was in the range of 28–31 °C and pH conditions were in the range of 7.5–8.0. The optimum dosage of corn cob was 5 g per 1.5L of low carbon wastewater. After treatment, the average effluent concentrations of NH_4^+-N and TN were 2.85 mg/L and 4.51 mg/L respectively. The effluent concentrations of NH_4^+-N and TN have met the Class A national standard for the discharge of pollutants from wastewater treatment plants. Fan et al. (2012) studied the biological denitrification process using wheat straw and PLA as carbon sources, and showed through their research that: the biofilm formed on the surface of wheat straw earlier than PLA, and the denitrification rate was also significantly faster, reaching 0.067 mg-N/(g-wheat straw·h), while PLA only had 0.0026 mg-N/(g-PLA·h), and the effect of straw as a carbon source to remove nitrate was better than PLA.

47.2.2 Synthetic Polymeric Solid Carbon Sources

In addition to using natural carbon containing materials as a solid carbon source, the most commonly used synthetic carbon sources studied so far are poly-β-hydroxybutyrate valerate (PHBV), polybutylene succinate (PBS), poly-β-hydroxybutyrate (PHB), polylactic acid (PLA), polycaprolactone (PCL), etc. Peng et al. (2021) used PBS and PVA-SA as carriers to prepare a modular composite solid The prepared 3P was an ideal carbon source for the denitrification secondary effluent, releasing 92.11% of acetic acid in short-chain fatty acids with a semisaturation constant of 21.28 mg/L. The effluent DOC concentration below 20 mg/L not only improved denitrification performance, but also increased the relative abundance of the gamma-anamorphic phylum from 39.32 to 43.58%. Yang et al. (2022) prepared a PCL composite carbon source using polycaprolactone (PCL) as the backbone combined with different low-cost carbon sources. The static release test screened the composite carbon source suitable for deep denitrification of wastewater, and the results showed that the carbon release performance of PCL/corn starch was stable for 30d, and the release amounts of nitrogen and phosphorus were lower than 1.0 mg/L, which had less impact on water quality and could be used for deep denitrification of domestic wastewater. Chu and Wang (2013) used PCL as a filler in a fixed bed bioreactor to treat nitrate in groundwater and found that the removal rate of total nitrogen could reach more than 95% at a hydraulic retention time of 3–6 h during long-term operation. Zhang et al. (2016) used several biodegradable polymers (PHBV, PCL, PBS) as carbon sources in denitrifying biofilters and found that the removal rates of both nitrate and total nitrogen increased with increasing HRT when different polymers were used as carbon sources, while the increase gradually decreased.

47.2.3 Mixed Solid Carbon Sources

Due to some of the limitations of natural carbon sources and the high cost of synthetic carbon sources, researchers have looked at hybrid solid carbon sources, i.e. a mixture of natural and synthetic sources. Shen and Wang (2011) added starch to polycaprolactone (PCL) as a co-blended carbon source and achieved 90% nitrate removal at an influent nitrate concentration of 50 mg/L and an HRT = 2 h. The nitrate removal rate reached more than 90%at HRT = 2 h. Zhang et al. (2015) used thermoplastic starch/PCL blended plate material as a solid carbon source and compared the effect of the solid carbon source filling method on the denitrification effect and found that filling in a folded and tilted manner could achieve better denitrification than vertical suspension filling method. The average denitrification rate of the folded reactor was 4.9 mg/(L·h) compared to 2.5 mg/(L·h) for the vertically suspended reactor, and the average effluent concentrations of total nitrogen and COD were lower.

47.3 Limitations of Solid Carbon Sources and Potential Research

Solid carbon sources are less conducive to direct use by denitrifying bacteria than traditional carbon sources (methanol, ethanol, sodium acetate). While glucose and sucrose are effective in denitrification, they tend to cause nitrite accumulation in water. Solid carbon sources are slow releasing and can address the spatial and temporal unevenness in the distribution of carbon sources, but this also suggests that solid carbon sources cannot be used as emergency carbon sources and cannot address large amounts of carbon demand over short periods of time. It can be seen that the combination of solid carbon sources and traditional carbon sources can solve the problems of high cost and excessive carbon input in a short period of time, as well as solving the problem of insufficient carbon sources in special circumstances.

47.4 Conclusion and Further Work

The main purpose of solid carbon sources is to serve as an efficient, environmentally friendly carbon source that can meet the low C/N conditions in wastewater treatment. For low C/N effluents, solid carbon sources can be effective, but there is a large variation between different solid carbon sources and some solid carbon sources need to be further optimised. Natural carbon sources, which release a small amount of nitrogen along with carbon, and synthetic sources, which contain a single substance, are not conducive to increased microbial population abundance. In the future, solid carbon sources will be developed in the following directions.

(1) Development of solid carbon sources that can release carbon stably over a long period of time.
(2) Mixing multiple solid carbon sources to complement a single source.

References

Boussaid F, Martin G, Morvan J et al (1988) In situ denitrification of groundwater by solid carbon matter. Environ Technol Lett 9(8):803–816

Chu L, Wang J (2013) Denitrification performance and biofilm characterization using degradable polymer PCL as a carrier and carbon source. Chemosphere 91(9):1310–1316

Fan Z, Hu J, Wang J (2012) Biological nitrate removal using wheat straw and PLA as substrate. Environ Technol 33(19–21):2369–2374

Guo HL, Li XW, Ren QY et al (2018) Characteristic analysis of influent water quality of typical urban wastewater treatment plants in China. Water Supply Drainage 54(06):12–15

Jing L, Sun Y, Wang H et al (2019) Denitrification in simulated groundwater using lignite as a solid-phase organic carbon source. Tcnologia Y Ciencias Del Agua 10(4):238–255

Liu WH (2018) Pilot study on MPSR treatment of low temperature low carbon source municipal wastewater. Changchun College of Engineering

Peng ZX, Jiang K, Lou TY, Niu NQ, Wang J (2021) Enhanced denitrification of secondary effluent using composite solid carbon source based on agricultural wastes and synthetic polymers. Water Sci Technol 83(4):886–893

Shen Z, Wang J (2011) Biological denitication using cross-linked starch/PCL blends as solid carbon source and biofilm carrier. Biores Technol 102(19):8835–8838

Volokita M, Belkin S, Abeliovich A et al (1996) Biological denitrification of drinking water using newspapers. Water Res 30(4):965–971

Wang X, Wang S, Xue T et al (2015) Treating low carbon/nitrogen (C/N) wastewater in simultaneous nitrification-endogenous denitrification and phosphorous removal (SNDPR)systems by strengthening anaerobic intracellular carbon storage. Water Res 77:191–200

Wei QX, Zhang XC (2013) Study on the key water quality parameters and spatial distribution characteristics affecting biological denitrification capacity of wastewater. Water Supply Drainage 49(09):127–131

Yang HL, Zhang D, Lan SH et al (2022) Preparation of polycaprolactone composite solid carbon source and its deep denitrification performance. J Environ Sci 42(05):263–273

Yu P, Yu FT, Zhang XM, Chen XX (2022) Prospects for the application of wastewater denitrification plus carbon source. Chem Des Lett 48(06):197–199

Zhang Q, Fan HX, Zhang YZ et al (2015) Study on the denitrification performance and mechanism of solid carbon source biofilm reactor. China Water Supply Drainage 31(19):27–32

Zhang HP, Li SB, Zhang CJ et al (2016) Study on denitrification of urban wastewater plant tailwater denitrification using biodegradable solids as a carbon source. Environ Eng 34(07):11–15

Zhang LQ, He PF, Wu D, Li SG, Huang YL, Huang YW, Wang DM (2017) Study of nitrogen removal performance when treating low carbon sewage using external solid carbon sources in SBBR systems. Biomed Res Int 2017:3691819

Chapter 48
How Does FDI Affect Carbon Emissions?—Heterogeneity Analysis Based on Spatial Econometric Models

Yuehan Dong

Abstract Using a two-way panel fixed effects model and a spatial econometric model, an empirical study examines the effect of FDI on carbon emissions based on inter-provincial spatial panel data in China from 1998 to 2016. The study shows that: Overall, there seems to be a significant positive spatial spillover effect on carbon emissions, and there is a high-high aggregation of carbon emissions across provinces. Among them, the direct effect and spatial spillover effect of FDI on carbon emissions have significant temporal and regional differences. From the analysis of regional heterogeneity, the direct impact of FDI on carbon emissions in the province exists only in the central region; the spatial spillover effect of carbon emissions in other provinces on carbon emissions in the province exists in the eastern region and the central region; the spatial spillover effect of FDI in other provinces on carbon emissions in the province exists in the eastern region and the western region, but the eastern region has a positive promoting effect and the western region has a negative inhibiting effect. Therefore, the quality of FDI should be further optimized and governed according to local conditions to achieve the goal of energy conservation and emission reduction, so as to promote the coordinated development of China's economy and environment.

Keywords FDI · Carbon emissions · Spatial econometric models · Heterogeneity

48.1 Introduction

In the face of the growing carbon emissions crisis, China has responded positively to the international community's demand for China to reduce carbon emissions and has pledged to strive to reach peak carbon by 2030 and carbon neutrality by 2060 (Intensifying action on climate change - China submits its national autonomous contribution to address climate change 2015). As the largest developing country, China's rapid

---

Y. Dong (✉)
China University of Petroleum (East China), Qingdao, China
e-mail: dyhupc@163.com

J. Zhang et al. (eds.), *Environmental Pollution Governance and Ecological Remediation Technology*, Environmental Science and Engineering,
https://doi.org/10.1007/978-3-031-25284-6_48

growth is mainly driven by export-oriented development, and the expansion of FDI and foreign trade is a strong engine of economic growth (Yi et al. 2015). However, along with the introduction of advanced technologies and economic development, the massive inflow of FDI has been accompanied by increasing energy consumption and environmental pollution problems in China (Xiong et al. 2012). Therefore, it is important to study in depth the relationship between provinces and regions.

48.2 Study Design

To analyze the impact of FDI on carbon emissions, the following panel fixed-effects base econometric model is developed.

$$C_{it} = c_0 + \beta \mathrm{FDI}_{it} + \gamma X_{it} + \alpha_i + \delta_t + \varepsilon_{it} \tag{48.1}$$

(1) Where i denotes province, t denotes year, C_{it} is carbon emissions, FDI_{it} is FDI, X_{it} is other control variables affecting carbon emissions, α_i is province-specific effects, δ_t is time effects, and ε_{it} is a random disturbance term.

Since carbon emissions have spatial spillover effects, this paper constructs this paper's spatial lag model (SAR) and spatial Durbin model (SDM) to examine the spatial spillover effects of carbon emissions from other provinces on carbon emissions in this province and the spatial spillover effects of FDI from other provinces on carbon emissions in this province, respectively, and the expressions are.

$$C_{it} = c_0 + \beta \mathrm{FDI}_{it} + \gamma X_{it} + \rho \sum_j \omega_{ij} \mathrm{CO}_{2jt} + \alpha_i + \delta_t + \varepsilon_{it} \tag{48.2}$$

$$C_{it} = c_0 + \theta Z_{it} + \rho \sum_j \omega_{ij} \mathrm{CO}_{2jt} + \sigma \sum_j \omega_{ij} Z_{jt} + \alpha_i + \delta_t + \varepsilon_{it} \tag{48.3}$$

where, ω_{ij} in Eq. (48.2) is the element of the spatial weight matrix W of in $\times$ n, and ρ is the spatial autoregressive coefficient, which reflects the spatial dependence of carbon emissions among provinces, implying the degree and direction of influence of carbon emissions of a province by carbon emissions of neighboring provinces. (3) The equation Z is all independent variables including FDI and other control variables X, and σ is the influence of carbon emissions of a province by the independent variables of neighboring provinces. Similarly, X is the other control variables that affect carbon emissions. The variables in the model are specified below.

The sample selected in this paper is 29 provinces of China (excluding Hainan, Tibet and Hong Kong, Macao and Taiwan) from 1998 to 2016, and the data sources

Table 48.1 Description of data and statistical description of variables

Variable	Symbolic	Average value	(Statistics) Standard deviation	Minimum value	Maximum value	Unit
Carbon footprint	$\ln CO_2$	5.1832	0.8839	0.5933	7.348	mt
FDI	FDI	0.4095	0.4153	0.0477	2.3182	%
Output level	lnGDP	9.6441	0.7614	7.7681	11.454	Yuan/person
Economic structure	INDUS	47.1175	6.7027	19.3	61.5	%
Technological progress	TECH	1.2071	1.1101	0.0926	6.7772	%
Environmental regulation	ENVR	0.1822	0.1413	0.0068	0.992	%
Energy mix	ENE	70.2594	24.9418	8.6966	164.7331	%
Urbanization rate	URBAN	48.481	16.0250	21.89	89.6	%
Population density	InPOP	431.4465	609.2075	7.0292	3847.1297	People/km^2

are China Statistical Yearbook (1998–2016), China Environmental Yearbook (1998–2016), China Energy Statistical Yearbook (1998–2016), and China Science and Technology Statistical Yearbook (1998–2016).The data descriptions and variable statistics descriptions of the above explanatory variables, core explanatory variables and a series of control variables are shown in Table 48.1.

48.3 Exploratory Spatial Analysis of FDI and Carbon Emissions

In order to examine the spatial relationship more intuitively, this paper draws a Moran scatter plot of the distribution of CO_2 emissions by province for three representative years based on the spatial weight matrix, as shown in Fig. 48.1. Its horizontal axis represents carbon emissions and the vertical axis is the spatial lagged value of carbon emissions.

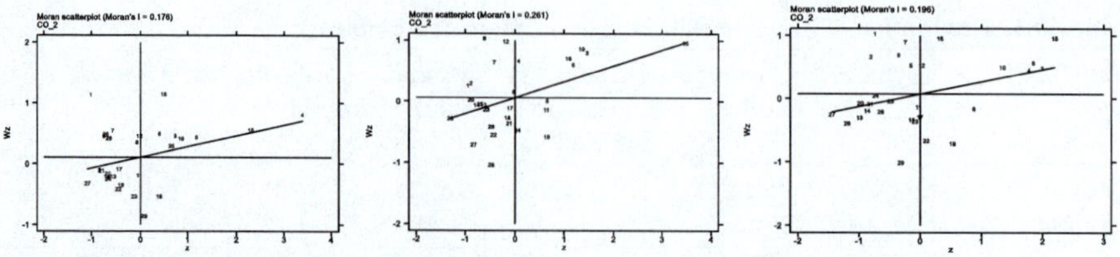

2016 2007 1998

Fig. 48.1 Scatterplot of carbon emissions by province in China for selected years under the spatial geographical weight matrix

48.4 Analysis of Empirical Results

48.4.1 Analysis of Baseline Regression Results

Table 48.2 shows the results of the benchmark regressions of the impact of FDI on carbon emissions in 29 Chinese provinces (excluding Hong Kong, Macao, Taiwan, Tibet and Hainan).

First, the estimation results in column (1) in Table 48.2 do not pass the significance test, i.e., FDI in the province has no direct effect on the level of carbon emissions in the province. The estimated coefficient rho of the spatial weight matrix in column (2) is significantly positive, indicating that there is a significant positive spatial spillover effect of carbon emissions from other provinces on carbon emissions in the province. The estimated results in column (3) show that there is no spatial spillover effect of FDI levels in other provinces on carbon emissions in this province.

48.4.2 East, Central and West Comparison

In order to examine the impact of FDI on carbon emissions in different provinces more comprehensively, this paper further estimates and tests the relationship between FDI and carbon emissions in three regions: east, central and west (Table 48.3).

Table 48.2 Baseline regression results: direct and spatial spillover effects of FDI on carbon emissions

Variables	(1)	(2)	(5)
	$lnCO_2$	$lnCO_2$	$lnCO_2$
	FE	SAR	SDM
FDI	0.060	−0.035	−0.046
	(0.078)	(0.078)	(0.083)
lnGDPp	−0.200	0.261***	0.264***
	(0.255)	(0.086)	(0.086)
INDUS	0.011***	0.012***	0.013***
	(0.004)	(0.003)	(0.003)
TECH	−0.048	0.019	0.019
	(0.034)	(0.038)	(0.038)
ENVR	−0.273	−0.274**	−0.275**
	(0.279)	(0.114)	(0.114)
ENE	0.014***	0.013***	0.013***
	(0.004)	(0.001)	(0.001)
URBAN	0.037***	0.025***	0.025***
	(0.007)	(0.006)	(0.006)
POP	0.000*	0.000	0.000
	(0.000)	(0.000)	(0.000)
rho		0.319***	0.320***
		(0.053)	(0.053)
W. FDI			0.070
			(0.166)
Year FE	Yes	Yes	Yes
Province FE	Yes	Yes	Yes
Observations	532	532	532
R-squared	0.924	0.311	0.307

Note ***, *, and * denote significant at the 1%, 5%, and 10% levels of significance, respectively, and numbers in parentheses are standard errors, all using Robust robust standard errors. rho is the estimated coefficient of the SAR model spatial weight matrix, and W. FDI denotes the FDI spatial weight matrix. "YES" means that the fixed effect is controlled for

Table 48.3 Spatial heterogeneity of the direct impact of FDI on carbon emissions and spatial spillover effects

Variables	(1)			(2)			(3)		
	$\ln CO_2$	$\ln CO_2$	$\ln CO_2$	$\ln CO_2$	$\ln CO_2$	$\ln CO_2$	$\ln CO_2$	$\ln CO_2$	$\ln CO_2$
	FE			SAR			SDM		
Area	Eastern part	Middle part	Western part	Eastern part	Middle part	Western part	Eastern part	Middle part	Western part
FDI	0.041	0.653**	−0.126	−0.020	0.777***	−0.464	−0.049	0.777***	−0.198
	(0.039)	(0.258)	(0.585)	(0.028)	(0.267)	(0.380)	(0.030)	(0.268)	(0.376)
lnGDPp	0.343**	0.594**	−1.548**	0.127***	0.297***	0.464	0.131***	0.298***	0.623**
	(0.144)	(0.263)	(0.749)	(0.046)	(0.100)	(0.297)	(0.046)	(0.104)	(0.289)
INDUS	−0.005*	−0.010**	0.026***	0.003	0.006*	0.019**	0.002	0.006*	0.011
	(0.003)	(0.004)	(0.008)	(0.002)	(0.003)	(0.008)	(0.002)	(0.003)	(0.008)
TECH	−0.051*	−0.502***	0.094	−0.029*	−0.280***	0.257	−0.031*	−0.281***	0.305*
	(0.027)	(0.127)	(0.182)	(0.016)	(0.086)	(0.171)	(0.016)	(0.087)	(0.167)
ENVR	−0.120	−0.153	−0.475	−0.041	−0.278*	−0.369	−0.017	−0.278*	−0.393*
	(0.085)	(0.277)	(0.463)	(0.065)	(0.160)	(0.240)	(0.065)	(0.160)	(0.232)
ENE	−0.001	0.006**	0.013***	0.001	0.008***	0.016***	0.001	0.008***	0.015***
	(0.001)	(0.002)	(0.005)	(0.001)	(0.002)	(0.002)	(0.001)	(0.002)	(0.002)
URBAN	−0.007*	0.049***	0.065**	−0.003	0.036***	0.015	−0.001	0.036***	−0.009
	(0.004)	(0.009)	(0.032)	(0.003)	(0.007)	(0.023)	(0.003)	(0.007)	(0.023)
POP	−0.000	0.006**	0.013**	0.000	0.002***	0.003	0.000	0.002***	0.003
	(0.000)	(0.003)	(0.005)	(0.000)	(0.001)	(0.002)	(0.000)	(0.001)	(0.002)
rho				0.739***	0.396***	0.128	0.741***	0.396***	0.051
				(0.033)	(0.069)	(0.101)	(0.033)	(0.069)	(0.104)

(continued)

Table 48.3 (continued)

Variables	(1)			(2)			(3)		
	$lnCO_2$	$lnCO_2$	$lnCO_2$	$lnCO_2$	$lnCO_2$	$lnCO_2$	$lnCO_2$	$lnCO_2$	$lnCO_2$
W. FDI							0.102**	0.013	−3.732***
							(0.045)	(0.383)	(0.992)
Year FE	Yes	Yes	Yes	Yes	Yes	Yes	Yes	Yes	Yes
Province FE	Yes	Yes	Yes	Yes	Yes	Yes	Yes	Yes	Yes
Observations	190	171	190	190	171	190	190	171	190
R-squared	0.973	0.937	0.880	0.103	0.765	0.387	0.075	0.765	0.461

48.5 Conclusions and Recommendations

48.5.1 Main Findings

Spatial autocorrelation of carbon emissions and FDI between provinces in China

China's provincial carbon emission intensity has a significant spatial autocorrelation characteristic and an increasing trend over time. The distribution in the Moran scatter plot shows that there is a positive spatial spillover effect of China's carbon emissions among the provinces. Among them, Beijing-Tianjin-Hebei, Yangtze River Delta and the central region connecting them have been the "hardest hit" in terms of carbon emissions in China, while the neighboring provinces centered on Qinghai are less.

There is spatial and temporal heterogeneity in the impact of FDI on carbon emissions in China

Based on the national sample, There is significant heterogeneity between regions, the direct effect of FDI in this province on carbon emissions in this province exists only in the central region; the spatial spillover effect of carbon emissions from other provinces on carbon emissions in this province exists only in the eastern region and the central region; the spatial spillover effect of FDI in other provinces on carbon emissions in this province exists in the eastern region and the western region, but the effects embodied in the two are different, with positive The spatial spillover effect of FDI on carbon emissions in other provinces exists in the eastern and western regions, but the effect of the two is different, with the positive effect in the east and the negative inhibiting effect in the west.

References

Intensifying action on climate change—China submits its national autonomous contribution to address climate change (2015). Chsina Energy 37(07):1

Xiong L, Permit, Wang J (2012) Has FDI brought low carbon to China-an empirical analysis based on China's time series data from 1985–2007. Macroecon Res (05):68–75. https://doi.org/10.16304/j.cnki.11-3952/f.2012.05.003

Yi Y, Guan W, Gao Y (2015) The relationship between FDI and carbon emissions in China—an empirical study based on ARDL. J Guizhou Univ Fin Econ (03):58–65

Chapter 49
A Comparative Study and Visualization of Water Quality in Calumet Beach and Montrose Beach

Elaine Gu

Abstract Water quality is something that is often ignored or considered trivial, but it is a primary indicator of human health. With a continuing decline of water supplies on our planet, it is vital that we research the root of the issue and how it is affecting our safety. Calumet Beach and Montrose Beach, two public beaches in Chicago, are especially vulnerable to environmental consequences due to human activity, both indirect and direct. While they are largely susceptible to water quality deterioration, the results showed that with the help of human preservation and care for environment protection, Calumet Beach displayed a much cleaner water state as opposed to Montrose Beach, suggesting that humans play a significant role in the health of water sources.

Keywords Water quality · Data visualization · Comparative study

49.1 Introduction

Out of the many factors essential to human survival, water is one of the most crucial substances. It hydrates the body and delivers vital nutrients to cells, along with other purposes critical for the body to function. While over 70% of the Earth's surface is covered in water, with around 95% of that water held in oceans, only a small portion of that water is available for human consumption–and its ever-decreasing sources from an increased number of human activities. Water pollution is the primary cause of disruptions and variants of water quality. Stemming from waste disposal (industrial, chemical, radioactive, etc.), fertilizer run-off, etc. (Denchak 2022), these resulting pollutants contaminate fresh bodies of water with bacteria and parasites, making the sources unsafe for drinking.

This is where scientific research on water quality comes into play. The numerical and qualitative data can help assess whether the body of water is safe for human

E. Gu (✉)
Concordia International School Shanghai, Shanghai 201203, China
e-mail: elaine2023163@concordiashanghai.org

J. Zhang et al. (eds.), *Environmental Pollution Governance and Ecological Remediation Technology*, Environmental Science and Engineering,
https://doi.org/10.1007/978-3-031-25284-6_49

consumption. All over the world, studies are being conducted, with a combination of on-site data collection and laboratory data analysis, for the most accurate and current results. Several studies, including the recent U.S. Geological Survey (USGS n.d.), indicate that the water quality of rivers has been drastically dropping in the past decades, mainly due to unnatural interference with the environment. Currently, organizations such as the Clean Water Act (CWA) (Environmental Protection Agency n.d.) have been contributing long-term to control and reduce water pollution, and similar research funded by the Environmental Protection Agency (EPA) directly supports the sustainability of clean water sources and human health in general.

Focusing on the water quality of Calumet Beach and Montrose Beach in Chicago, Illinois, the primary goal is to analyze past trends, predict future outcomes, and discover their significance in other prominent environmental issues such as climate change and pollution. With Montrose Beach being Chicago's largest public beach, and Calumet Beach located nearby the steel industry, they are most vulnerable to the effects of industrial pollution and human activities, hence making it a great research area as a representative of the entire city's water quality.

49.2 Data

The dataset used in this research is taken from Kaggle (Eschlbeck n.d.), an online community platform that connects data scientists and machine learning enthusiasts and primarily allows users to publish datasets. According to Sven Eschlbeck, a visiting scientist at Institut de Ciències del Mar Barcelona, certified sensors are lined up in the water along the Lake Michigan lakefront. The data were collected from six locations: 63rd Street Beach, Calumet Beach, Montrose Beach, Ohio Street Beach, Osterman Beach, and Rainbow Beach, with a total of 34,923 unfiltered samples. With every entry, the water temperature, turbidity (opaqueness of matter), transducer depth (depth of sensor), wave height (difference between the elevations of a crest), wave period (time for two successive wave crests to pass at a single point), are recorded and automatically sorted within each column. The data were then grouped by each respective location and the order was chronologically rearranged through machine learning. From this, it can be determined that the granularity of the measurements is rather low, with an hourly collection rate throughout the day from June to September (although not always accurate and varying over each location) over a three-year period. However, it should be noted that sensors are only fully operational in the summer. At other times, sensors are unavailable, thus unable to provide consistent data.

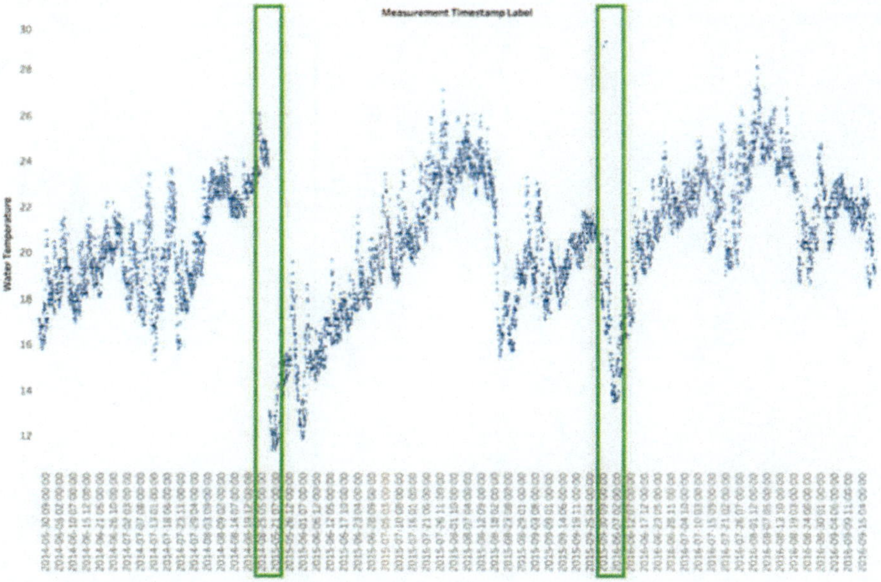

Fig. 49.1 Water temperature and time in Calumet Beach

49.3 Method

To analyze the dataset, I used Tableau (n.d.)–an interactive visual analytics platform–as my primary tool. Allowing any combination of variants to be graphically represented, I experimented with different independent and dependent variables and observed the general relationship on the surface level. Wanting a direct and reliable comparison between Calumet Beach and Montrose Beach, I picked a few essential combinations with the most apparent relationships and set them side to side. Figures 49.1 and 49.2 show the relationship of water temperature and time in Calumet Beach and Montrose Beach. Figures 49.3 and 49.4 show the relationship between turbidity and time. Figures 49.5 and 49.6 show the relationship between wave height and time. Figures 49.7 and 49.8 show the relationship between wave period and time. Figures 49.9 and 49.10 show the relationship between turbidity and temperature. Figures 49.11 and 49.12 show the relationship between turbidity and transducer depth.

49.4 Result

Looking at the graphs from both Calumet Beach and Montrose Beach, general trends and patterns can be directly seen. For both locations, water temperature is impacted mainly by seasonal changes, thus increasing as the surrounding climate becomes

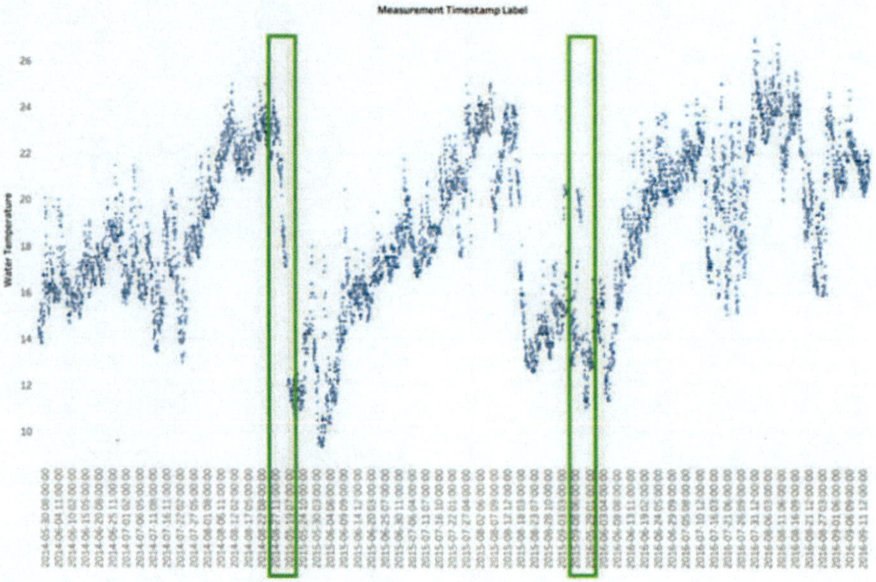

Fig. 49.2 Water temperature and time in Montrose Beach

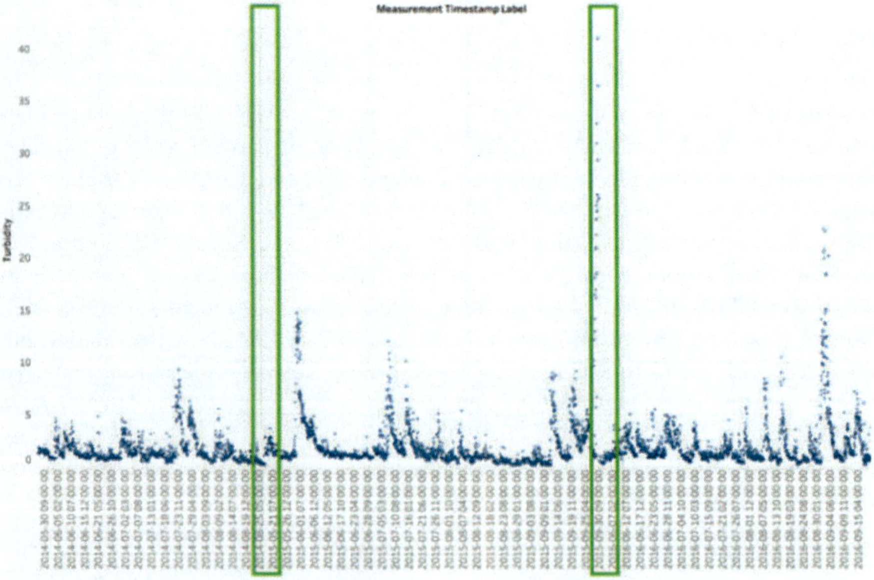

Fig. 49.3 Turbidity and time in Calumet Beach

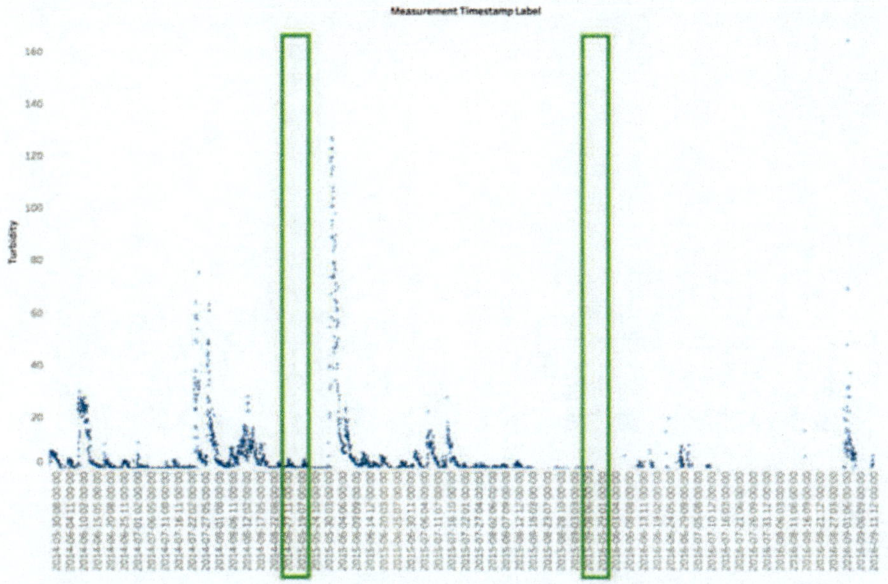

Fig. 49.4 Turbidity and time in Montrose Beach

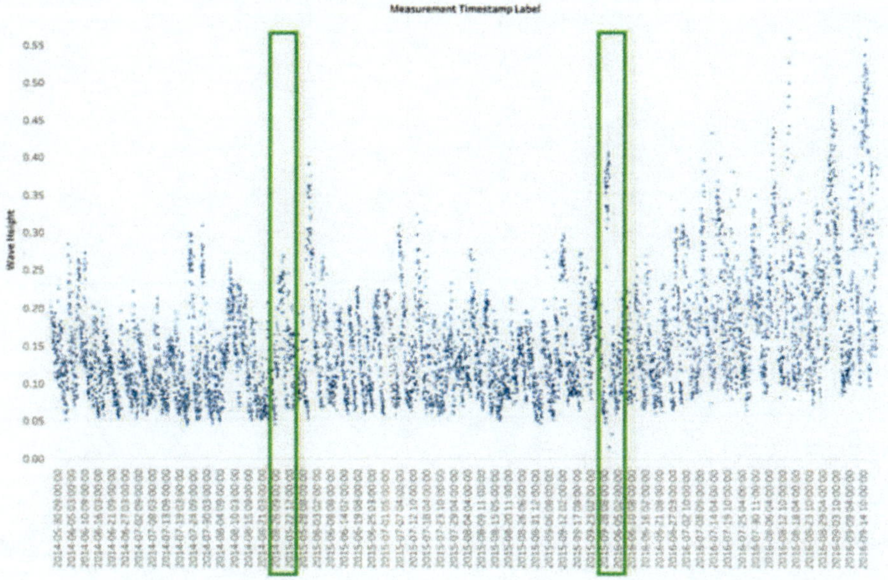

Fig. 49.5 Wave height and time in Calumet Beach

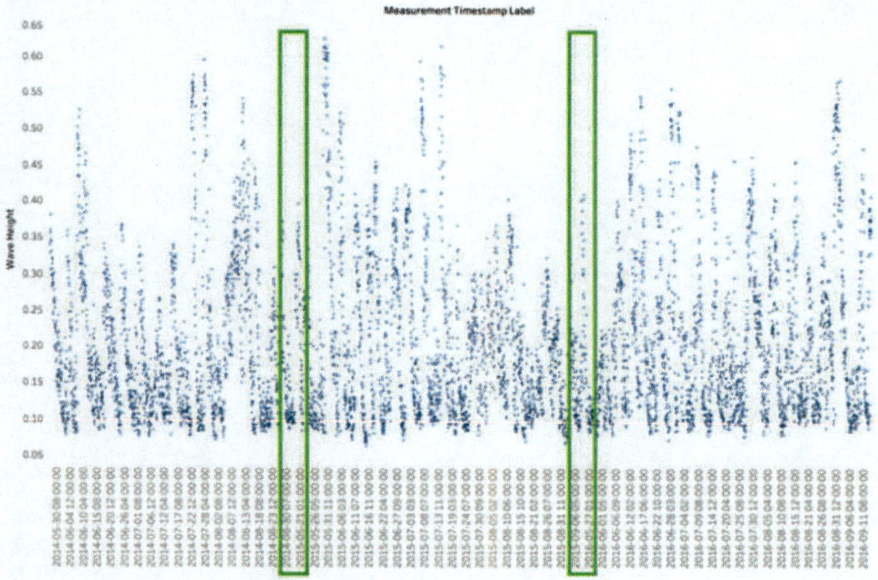

Fig. 49.6 Wave height and time in Montrose Beach

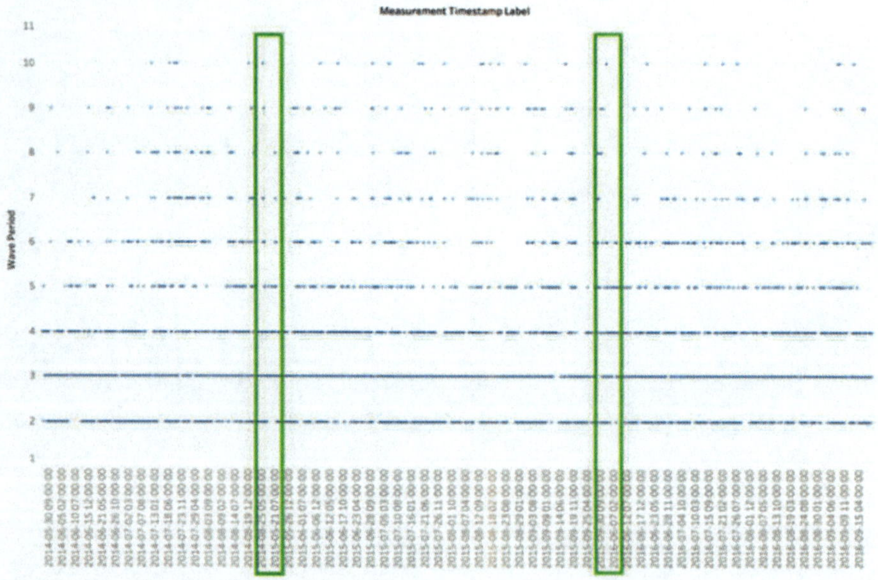

Fig. 49.7 Wave period and time in Calumet Beach

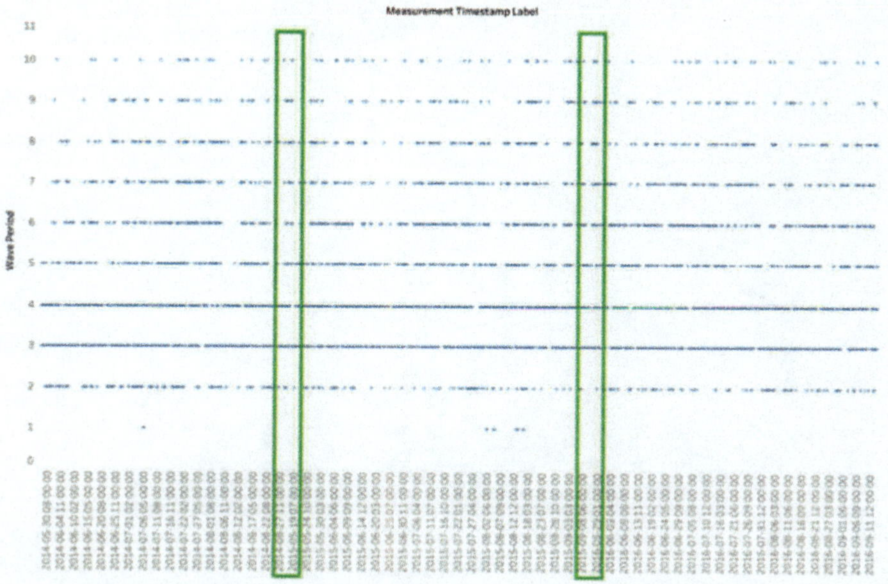

Fig. 49.8 Wave period and time in Montrose Beach

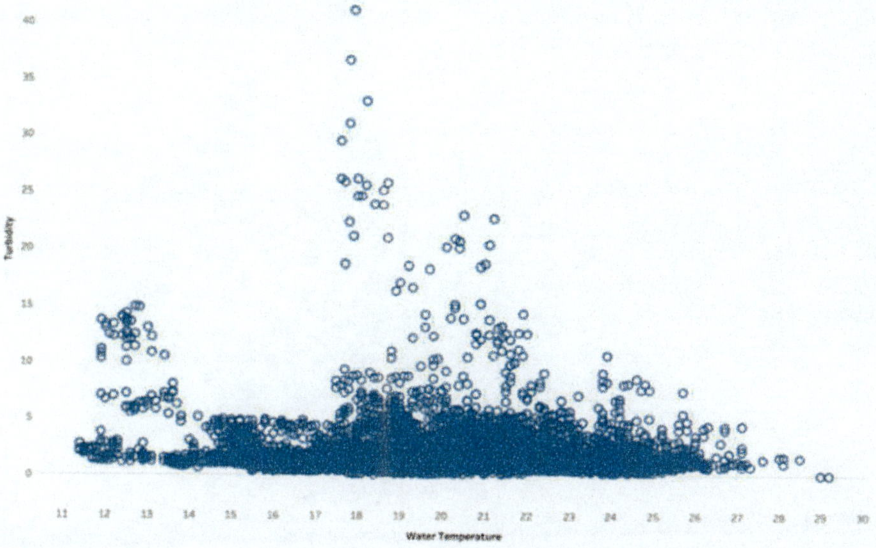

Fig. 49.9 Turbidity and water temperature in Calumet Beach

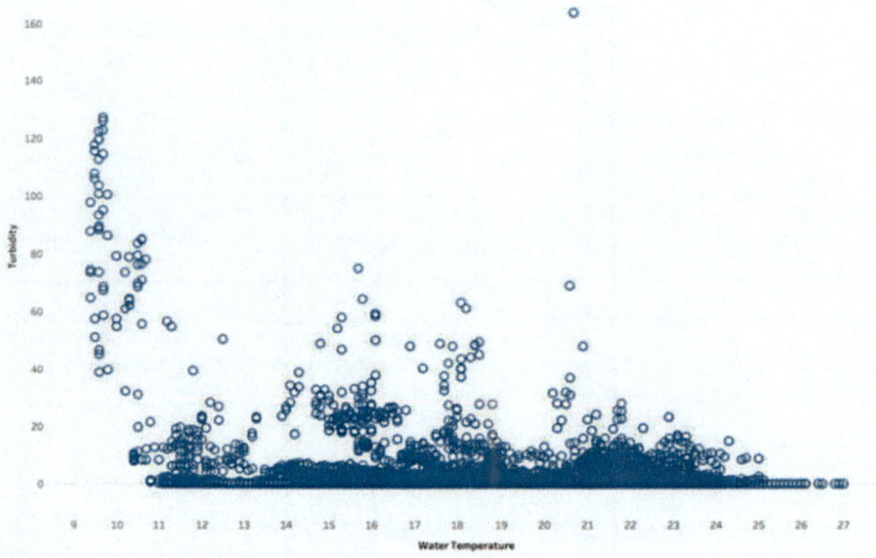

Fig. 49.10 Turbidity and water temperature in Montrose Beach

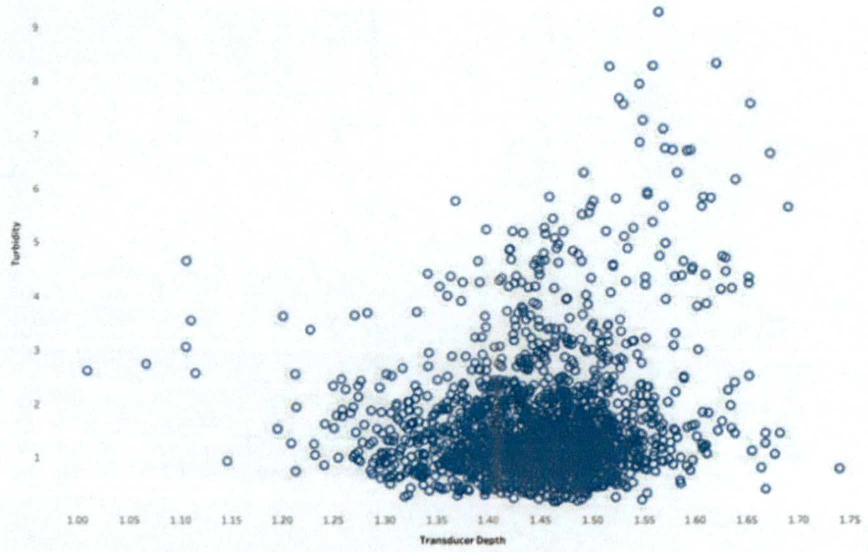

Fig. 49.11 Turbidity and transducer depth in Calumet Beach

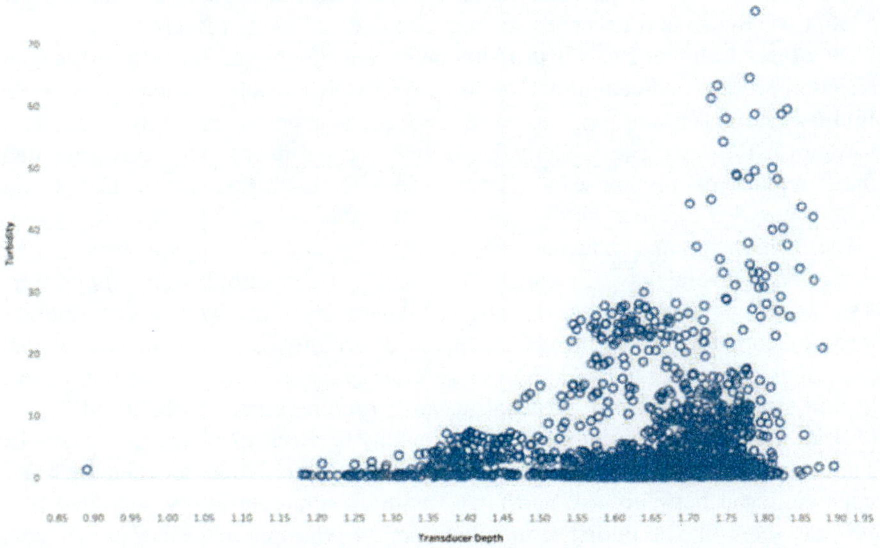

Fig. 49.12 Turbidity and transducer depth in Montrose Beach

warmer (the lowest water temperatures appear in May, reaching their maximum around September). While this is a predictable pattern of change, and its range each successive year stays relatively similar, the temperature peak is getting higher and higher. With climate change becoming an increasing threat to our planet as time goes on and a minimum amount of effort to address the issue, it is not surprising that bodies of water are beginning to heat up. It is even predicted by the World Bank that, due to the influence of climate change, our world will start running out of water by 2050, only three decades away. However, because the sensor deactivates near the end of September, it is unfair to fully assume the trends and their implicated meanings solely based on the provided observations.

Even though there are no clear trends regarding the relationship between wave height and time for Montrose Beach, the data gathered in Calumet Beach suggests that the average range and variation of wave height recorded substantially increases over time. In September 2016, the average range of wave height appears to be more than double the average range of its initial recordings in May of 2014. It can also be observed that the highest readings of wave height in Calumet Beach occurred on August 12th, 2016, and September 14th, 2016, whereas the lowest reading of wave height occurred near the beginning and end of the "off-period" around September 28th, 2015 and June 5th, 2016. As for Montrose Beach, the range and variation of wave height seems to stay relatively the same throughout the two-and-a-half-year period, but is equivalent to the maximum ranges of wave height in Calumet Beach. This suggests that the body of water in Calumet Beach was calmer and flatter at the beginning of the data collection process, getting progressively more unstable, likely due to human activity or other natural influences. It also means that the wave height of Montrose Beach is consistently greater than that of Calumet Beach.

Over time, Calumet Beach's turbidity peaks are slowly increasing over the data collection process, indicating that the water quality is lessening as time goes on. In Montrose Beach, a major gap can be seen between data points from July 16th, 2016, to August 27th, 2016. This is likely due to unforeseen sensor failures that generated a large number of missing values. Thus, a definite conclusion on the relationship between turbidity and time for Montrose Beach cannot yet be accurately determined.

The relationship and influence between turbidity and water temperature can be interpreted in several ways. Scientifically speaking, water turbidity directly impacts its surface temperature because the suspended particles in the water absorb sunlight more efficiently than water. It can be seen that as turbidity decreases–in other words, water clarity increases–the variations of water temperature also increase (Paaijmans n.d.). In Calumet Beach, the peaks of turbidity (and maximum turbidity of 41) all occurred when the water temperature was roughly between 17 °C and 19 °C. As for Montrose Beach, its highest occurrences of turbidity were recorded when the water temperature was between 9 °C and 11 °C., with maximum turbidity of around 165. This may seem like an outlier at first, but upon viewing the majority of datapoints that lie across the upper ranges of turbidity, unseen by those of Calumet Beach, it can most reasonably be assumed that the water quality of Montrose Beach is unclean and likely representative of other sources of water near the area unsafe for human consumption. Since the range of water temperature of the two locations is around 11–27 °C, the possibility that the high variants of turbidity of Montrose Beach are determinate of abnormal events in water temperature is completely eliminated.

Observing the graph of turbidity and transducer depth, it is clear that as transducer depth increases, turbidity also increases. Similar to the observations previously stated, the average and range of turbidity of Montrose Beach's water are significantly higher than that of Calumet Beach. While the maximum turbidity of 9.2 was seen when the transducer depth was around 1.56 m below Calumet Beach's surface water, a maximum turbidity of 76 was recorded when the transducer depth was around 1.79 below Montrose Beach's surface water.

49.5 Conclusion

From all the data gathered and carefully analyzed, it is reasonable to conclude that although the water quality of both Calumet Beach and Montrose Beach has slowly decreased over the two-and-a-half-year period, Calumet Beach still displays a substantially clearer and cleaner state of water, presumably from greater care for maintenance and effort for beach sustainability. This suggests that human conservation of water plays a significant role in preserving their health. Regardless of whether the body of water is safe for human consumption, it is no doubt that water sources as a whole are rapidly declining. In a matter of decades–or even less–there might not be enough water to sustain the 7.7 billion people on Earth, so it is necessary that the general public is made aware of this issue in order to urge them to act to preserve our precious water sources.

References

Denchak M (2022) Water pollution: everything you need to know. NRDC. Retrieved 6 July 2022, from https://www.nrdc.org/stories/water-pollution-everything-you-need-know

Environmental Protection Agency (n.d.) Summary of the clean water act. EPA. Retrieved 10 July 2022, from https://www.epa.gov/laws-regulations/summary-clean-water-act#:~:text=(1972),quality%20standards%20for%20surface%20waters

Eschlbeck S (n.d.) Chicago beach water quality | Kaggle. Retrieved 21 Dec 2021, from https://www.kaggle.com/sveneschlbeck/chicago-beach-water-quality

Paaijmans KP, Takken W, Githeko AK, Jacobs AF. The effect of water turbidity on the near-surface water temperature of larval habitats of the malaria mosquito Anopheles gambiae. Int J Biometeorol. Retrieved 3 Sept 2022, from https://pubmed.ncbi.nlm.nih.gov/18633650/#:~:text=Water%20turbidity%20affects%20water%20temperature,the%20extinction%20of%20solar%20radiation

Tableau (n.d.) Business intelligence and analytics software. Retrieved 26 Dec 2022, from https://www.tableau.com/

USGS (n.d.) USGS water-quality data for the nation. National water information system: web interface. Retrieved 26 June 2022, from https://waterdata.usgs.gov/nwis/qw

Chapter 50
Study on the Influence of Particulate Matter Concentration and Baffle on Clarifier

Shuo Zhang, Jiajiong Xu, Yan Wang, and Ruhua Wang

Abstract In order to adapt the high-efficiency clarification tank to the switch of the raw water from high turbidity to low turbidity, CFD was used to simulate the overall tank body, and optimization measures were proposed. The three-dimensional simulation results showed that if the optimized partition clarifier was used to treat raw water with low turbidity, the velocity distribution and the sediment distribution were similar to the raw water with high turbidity in the transition section between the flocculation tank and the sedimentation tank. But the distributions were obviously more uniform. The floc flow ratio and water flow ratio of hole 1 to hole 2 were both higher than 90%, which could achieve uniform distribution of water and flocs.

Keywords Clarifier · Turbidity · Baffle · Optimization · CFD

50.1 Introduction

In the past fifteen years, various types of clarification tanks have been used in many Chinese water plants (Xanthos et al. 2013). High-efficiency clarification tanks such as Densadeg, Multiflo and Actiglo have been introduced such as Nanshi Water Plant, Yangshupu Water Plant 7# Production Line, Jinhai Water Plant and Linjiang Water Plant. The study of tests and CFD to clarifiers in water treatment has been advancing (Xanthos et al. 2013; Fouad and Hassan 2018; Romphophak et al. 2016).

The high-efficiency clarification tanks in a water treatment plant were designed and operated according to the raw characteristics of high turbidity and high organic matter. After the water source had been switched from Huangpu River to Qingcaosha Reservoir, the turbidity, chromaticity, oxygen consumption and ammonia nitrogen of the raw water have been significantly reduced. The flocs of the clarifier are greatly affected by the concentration of influent particles, and it is necessary to verify the

S. Zhang (✉) · J. Xu · Y. Wang · R. Wang
Shanghai Municipal Engineering Design Institute (Group) Co.,Ltd., 200092 Shanghai, P.R. China
e-mail: zhangshuo@smedi.com

J. Zhang et al. (eds.), *Environmental Pollution Governance and Ecological Remediation Technology*, Environmental Science and Engineering,
https://doi.org/10.1007/978-3-031-25284-6_50

treatment effect and adjust the parameters according to the changes in the raw water quality characteristics.

In recent years, the full-scale flow field and suspended particle distribution of the clarifiers had been studied based on CFD to guide the unit operations (He et al. 2008). Therefore, CFD was used to optimize the hydraulic conditions and the operations of the high-efficiency clarifier to adapt the low turbidity water in the water plant.

50.2 Parameters

50.2.1 Mixing Bank

Fast mixing time was 94 s. The effective volume of the mixing tank was 102.5 m^3. The plane size was 5.0 m × 5.0 m, and the effective water depth was 4.1 m. There was a mechanical mixing stirrer with a diameter of 2.0 m in the tank.

50.2.2 Clarifier

The clarification tank included a flocculation zone and an inclined pipe sedimentation zone. The plane size of the flocculation zone was 7.0 m × 7.0 m, and the effective water depth was 6.4 m. There was a mechanically stirred flocculator with a diameter of 3.5 m. The plane size of the sedimentation zone was 17.0 m × 17.0 m, and the area of the inclined pipe area was 169 m^2. The maximum rising velocity in the sedimentation zone was 23.1 m/h.

50.2.3 CFD

The geometric structure and calculation domain included the inlet pipe at the bottom of the flocculation tank, agitating paddles, and a sedimentation tank, Fig. 50.1. The total number of computational domain grids was 2,815,499. The static domain included the inlet section and the sedimentation tank, and the rotating domain and the static domain were connected by the interface interface, Figs. 50.2, 50.3 and 50.4.

The inlet was a velocity boundary of 0.1 m/s. The blade speed was 74 r/min. The inlet particle concentration was given a two-phase volume concentration of 50 mg/L or 20 mg/L.

In the transition section between the flocculation tank and the sedimentation tank, the plane X = 4.5 m along the flow direction was intercepted, Fig. 50.5. Then the x-axis velocity distribution and floc concentration distribution on the cross section of

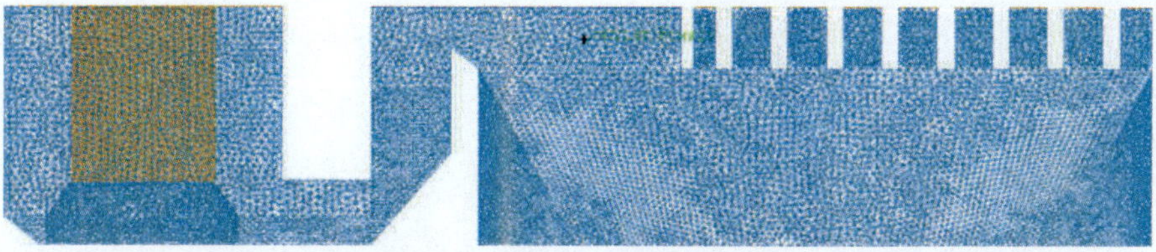

Fig. 50.1 Grids of flocculation tank and sedimentation tank

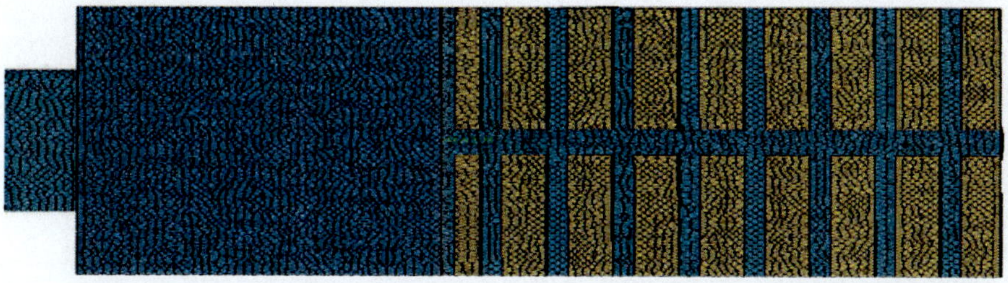

Fig. 50.2 Grids on top of sedimentation tank

Fig. 50.3 Grids along the inlet pipe

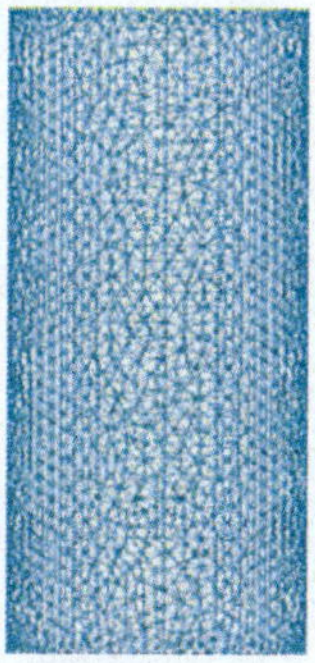

Fig. 50.4 Grids of the inlet pipe section

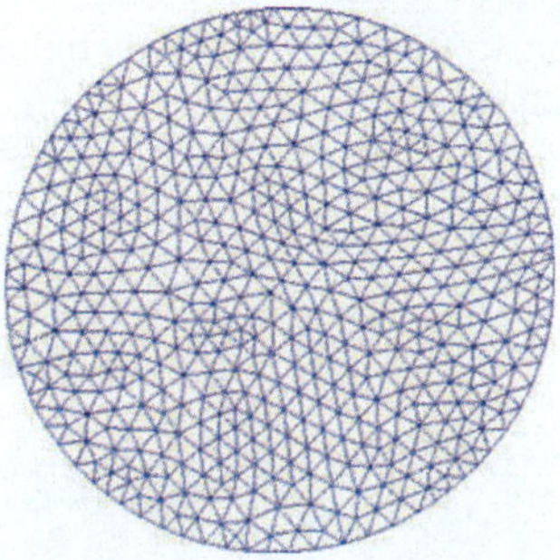

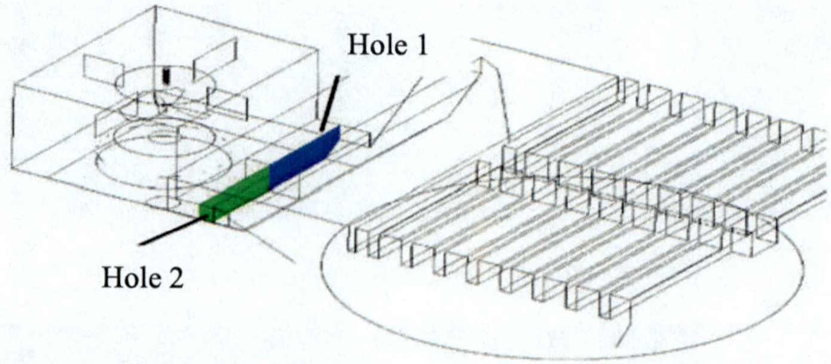

Fig. 50.5 The position of the hole section

the original scheme, the partition scheme and the optimized partition scheme were compared and analysed.

50.3 Results and Discussions

50.3.1 Partitions Effect

Without a partition, it was shown that the speed distribution on the section was extremely uneven. The reason was that the speed loop of the agitator rotating around the axis could not be eliminated. There was a large backflow in the speed on the section. There were both positive and negative flow along the x-axis in the cross section. When the partition was added and optimized, the velocity distribution on the section was significantly improved. The sediment concentration on the cross section was relatively high, and it was obviously biased to one side. However, when the partition and other measures were adopted, the sediment concentration on the cross section was significantly reduced, and the sedimentation to one side was significantly alleviated, Figs. 50.6 and 50.7.

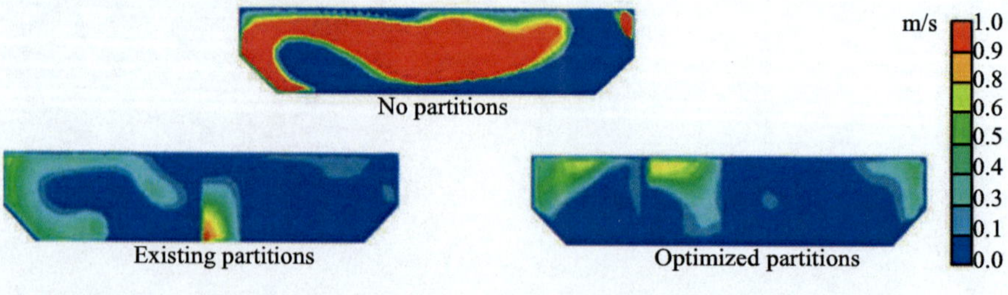

Fig. 50.6 Water velocity on X = 4.5 m cross-section under 50 mg/L of influent particles

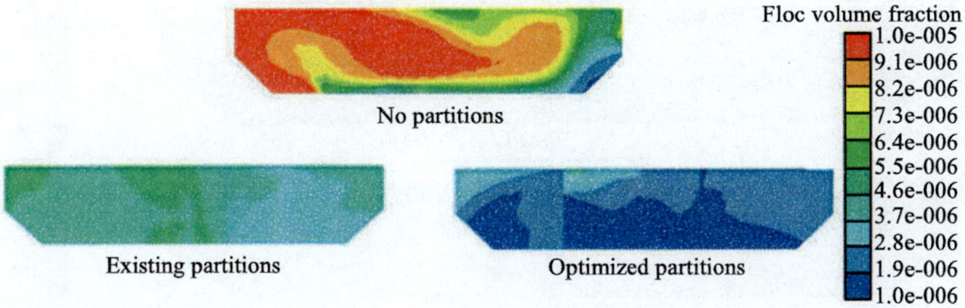

Fig. 50.7 Floc concentration on X = 4.5 m cross-section under 50 mg/L of influent particles

Table 50.1 Flow through two holes

Partitions	Influent particles (mg/L)	Substance	Hole 1 flow (kg/s)	Hole 2 flow (kg/s)	Hole 1/Hole 2 (%)
No	50	Floc	0.00022	0.00538	4.09
		Water	57.5	54.5	105.50
Existing	50	Floc	0.00199	0.00361	55.12
		Water	54	58	93.10
Optimized	50	Floc	0.00275	0.00285	96.49
		Water	54.9	57.1	96.15
Optimized	20	Floc	0.0011	0.0012	91.67
		Water	55.1	56.9	96.84

The difference of water through the holes was very small, but the flow of flocs was very different. Without a partition, the floc flow via holes was 24 times different. After the partition was optimized, the floc flow of the two holes was basically the same, Table 50.1.

50.3.2 *Influent Particle Concentration*

The concentration of particles in the raw water of the Qingcaosha Reservoir was relatively low, about 20 mg/L. In order to study the influence of particle concentration on floc deposition, the two cases of influent particle concentration of 20 and 50 mg/L were compared based on the optimized partition model.

When the concentration of influent particles was reduced, the water flow velocity distribution on the cross section was obviously more uniform and the sediment drift was further improved, Figs. 50.8 and 50.9. The floc flow ratio and water flow ratios of hole 1/hole 2 were both higher than 90%, Table 50.1.

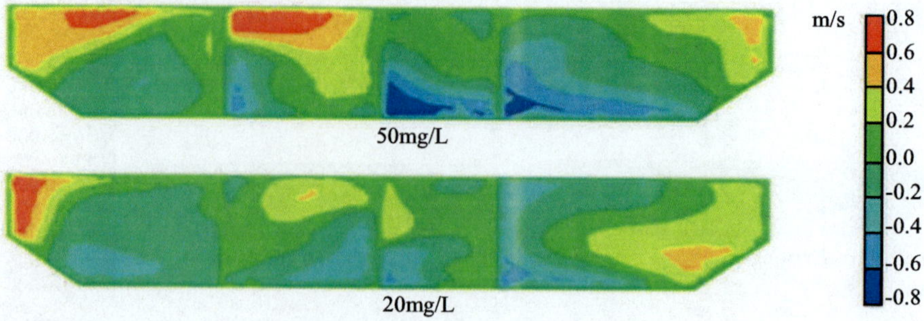

Fig. 50.8 Water velocity on X = 4.5 m cross-section under optimized partitions and 50 mg/L of influent particles

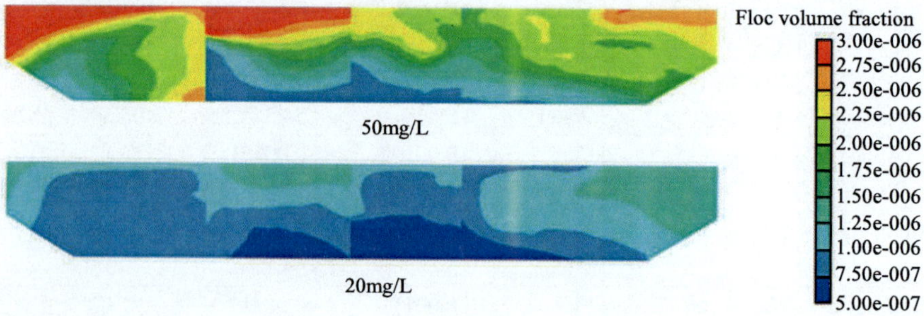

Fig. 50.9 Floc concentration on X = 4.5 m cross-section under optimized partitions and 50 mg/L of influent particles

50.4 Conclusion

CFD was adopted to study hydraulic numerical simulation of the clarifier to reveal the uneven flow through the two holes. The main conclusions could be summarized as follows:

(1) It was shown that when the optimized partition clarifier treated raw water with a low particle concentration, the velocity distribution and the sediment distribution were similar to the raw water with a higher particle concentration in the transition section between the flocculation tank and the sedimentation tank.
(2) The distributions were obviously more uniform that could accommodate the switch of the raw water from high turbidity to low turbidity.
(3) Under optimizing the partition, the floc flow ratio and water flow ratio of hole 1/hole 2 were both higher than 90%, which could achieve uniform distribution of water and flocs.

Acknowledgements This work was financially supported by the key Special Program on the S&T for the Pollution Control and Treatment of Water Bodies (2017ZX07501001-05).

References

Fouad M, Hassan S (2018) The performance of sludge blanket clarifier against conventional settler under high water turbidity conditions. Water Pract Technol 13:642–653

He C, Wood J, Marsalek J, Rochfort Q (2008) Using CFD modeling to improve the inlet hydraulics and performance of a storm-water clarifier. J Environ Eng 134:722–730

Romphophak P, Wongwailikhit K, Chawaloesphonsiya N, Samornkraisorakit P, Painmanakul P (2016) Study of flow pattern in jet clarifier for removal of turbidity by residence time distribution approach. Eng J 20:17–27

Xanthos S, Ramalingam K, Lipke S, McKenna B, Fillos J (2013) Implementation of CFD modeling in the performance assessment and optimization of secondary clarifiers: the PVSC case study. Water Sci Technol 68:1901–1913

Chapter 51
Evaluation Method of Urban Air Quality Based on AERMOD Model—A Case of Guangdong Province

Shihao Zhang

Abstract Due to the rapid development of economy and technology, the consumption of fossil energy in China has increased significantly, resulting in the emission of various large amounts of pollutants into the atmosphere. Apart from the six common pollutants, the secondary pollutants in the atmosphere are generated by the interaction of various primary pollutants in the atmosphere or the chemical reaction with the normal components of the atmosphere, and the photochemical reaction caused by the participation of solar energy, which is completely different from the physical and chemical properties of primary pollution. In this paper, AERMOD model is used to monitor the concentration of six conventional pollutants in Guangdong Province, and GIS technology is used for reasonable analysis.

Keywords GIS · AERMOD model · Air pollution · Monitoring

51.1 First Section

In recent years, with the continuous advancement of urbanization and the emergence of various respiratory diseases, people have noticed the rapid increase in the types and concentrations of pollutants in the atmosphere. In the field of urban planning and construction, people have also continuously proposed the need to coordinate environmental, energy, finance, laws, regulations and social factors (Wan et al. 2022). Therefore, how to detect and ensure a safe atmospheric environment has become a topic of concern to the public in the contemporary era. Atmospheric environment changes very fast, and the derived atmospheric chemistry research also focuses on the migration, diffusion and consumption processes of various pollutants (Ponomarev et al. 2020). On April 1, 2022, the China Meteorological Administration issued the 'Atmospheric Environmental Weather Bulletin (2021)', which pointed out that

S. Zhang (✉)
College of Resources and Environment, Shanxi University of Finance and Economics, Taiyuan, China
e-mail: 1156187672@qq.com

J. Zhang et al. (eds.), *Environmental Pollution Governance and Ecological Remediation Technology*, Environmental Science and Engineering,
https://doi.org/10.1007/978-3-031-25284-6_51

although the meteorological conditions were more favorable to dust weather than 2020.

51.1.1 Introduction

For the detection of atmospheric pollutants, Li (2019), Cheng (2018) and Hao (2017) studied various methods for determining the health protection distance of construction projects and the comprehensive analysis of pollution. Based on the AERMOD model, this paper predicted the concentrations of six common pollutants and detected the data. The observation points involved in the study refer to the division method of regional encrypted grid points, and draw the maps of each province and city according to the grid of 2 km * 2 km. Finally, 133 positions that can relatively stably represent the overall air quality of the region are selected, and meet the requirements of "automatic monitoring technology specification for ambient air quality".

51.1.2 GIS and AERMOD Model

Geographic information system can simulate the relative surface height of pollutants in various gaseous environments, and analyze the air quality in some regions or the whole city (Shareef et al. 2016). The evaluation of air quality at a certain height on the ground based on GIS is an important method for the detection of atmospheric environmental quality. It ensures a atmospheric environment in a certain buffer zone (He and Duan 2015), and then points-to-point plays a certain warning (Li 2014) and visual analysis (Sun et al. 2011) for air pollution, and then carries out effective governance.

AERMOD monitoring and forecasting system has been a practical case (Zhang 2010), abroad by combining other models have been more accurate guess results. Its application fields are not only limited to traditional pollution sources such as refineries, steel plants, nuclear power plants and transportation hubs, but also used for speculation of natural sources such as volcanoes. This model can not only guess the concentration range of atmospheric gaseous pollutants and particulate pollutants, but also measure the concentration of heavy metals such as Pb and Hg. It can be seen that the guess is more extensive.

51.1.3 Sources of Data

The data of this paper comes from the '2021 China Ecological Environment Bulletin', 'China Statistical Bureau', 'Guangdong Provincial Atlas', Guangdong Provincial

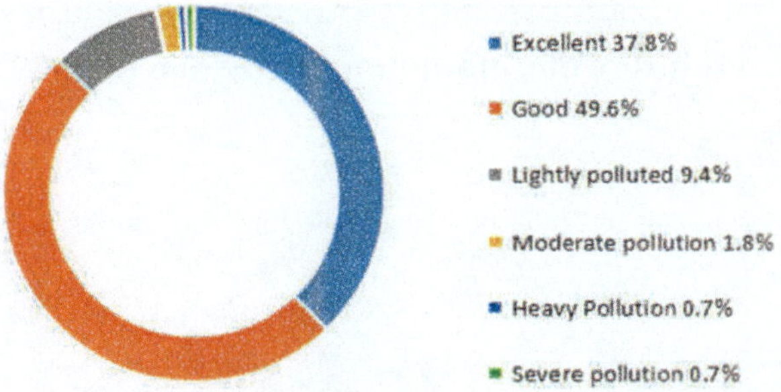

Fig. 51.1 Comparison of days at all levels of ambient air quality in 339 cities in 2021

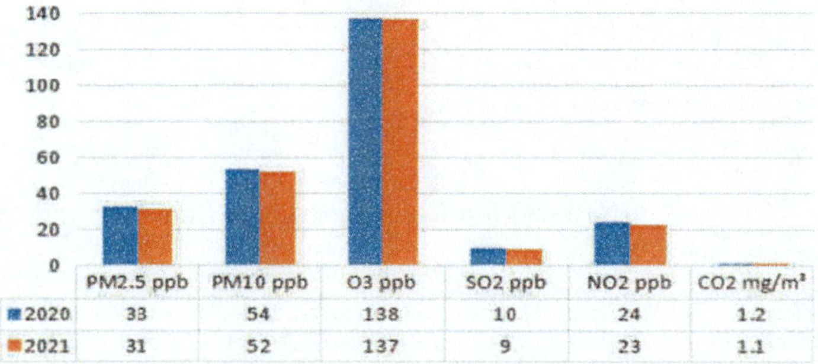

Fig. 51.2 Inter-annual comparison of concentrations of six pollutants in 339 cities in 2021

Department of Health website part of the statistical results report and the relevant information on the Internet control site website (Figs. 51.1, 51.2 and 51.3).

51.2 AERMOD Model

AERMOD was developed by the Regulatory Model Improvement Committee established by the National Environmental Protection Agency and the American Meteorological Society, which is a steady-state plume model. In the stable boundary layer, the concentration distributions in the vertical and horizontal directions are regarded as Gaussian distributions. In the convective boundary layer, the horizontal distribution is also regarded as Gaussian distribution (Wu 2010), but the vertical distribution is described by the probability density function. In addition, AERMOD considers the "plume lift" phenomenon: some plume materials from the buoyancy source first rise to the top of the boundary layer and stay there for a period of time, and then mix into the convective boundary layer. AERMOD calculates the partial plume penetrating into the stable layer, allowing it to return to the boundary layer in some cases. Both in stable boundary layer and convective boundary layer, AERMOD takes into account

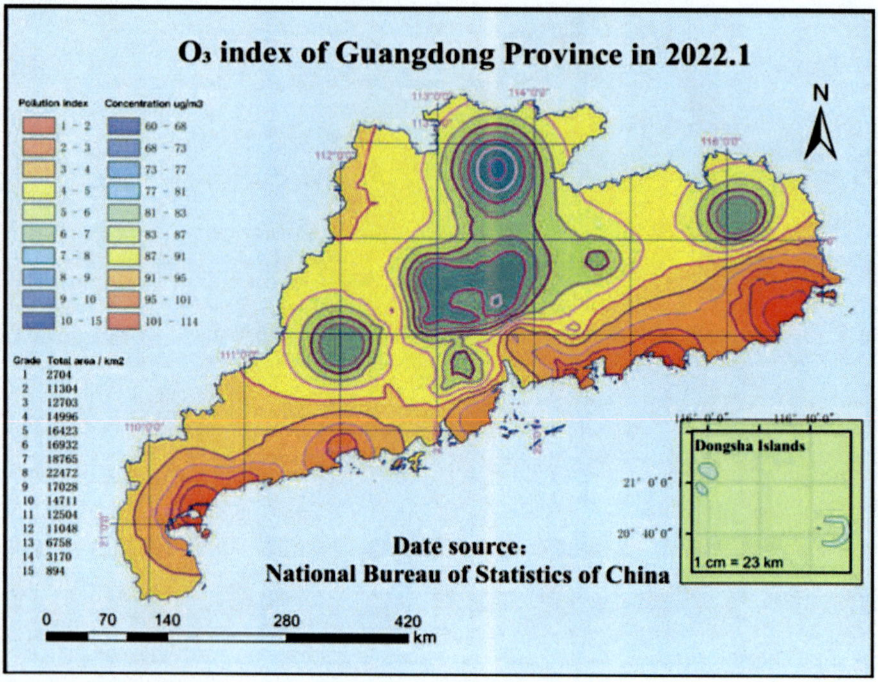

Fig. 51.3 O_3 analysis for Guangdong Province, January 2021

the horizontal diffusion enhancement caused by curved plume. That is, the pollutant concentration of any grid point in the AERMOD model is the sum of plume concentration weighted in the upper and lower layers of the diffusion flow field. Both in stable boundary layer and convective boundary layer, AERMOD takes into account the horizontal diffusion enhancement caused by curved plume. That is, the pollutant concentration of any grid point in the AERMOD model is the sum of plume concentration weighted in the upper and lower layers of the diffusion flow field.

Assuming that the pollutant mass concentration of grid coordinate point (x, y, z) is C (x, y, z), the basic calculation formula of the total pollutant mass concentration C (x, y, z), λ is the plume state weight function of the upper and lower layers of diffusion:

$$\xi = \frac{\int_0^H C(x, y, z)\mathrm{d}z}{\int_0^\infty C(x, y, z)\mathrm{d}z}$$

The basic data needed for the model operation include terrain data, emissions, receiving points, etc. At present, due to the imperfect collection of meteorological data, the model can not accurately represent the turbulent characteristics of the stable

Classification	Type of pollutant					
	$PM_{2.5}$	PM_{10}	SO_2	NO_2	O_3	CO
Industrial Sources	√	√	√	√	√	√
Flow source	√	√	√	√	√	
Surface source	√	√	√	√	√	

Fig. 51.4 Identification of Pollutants

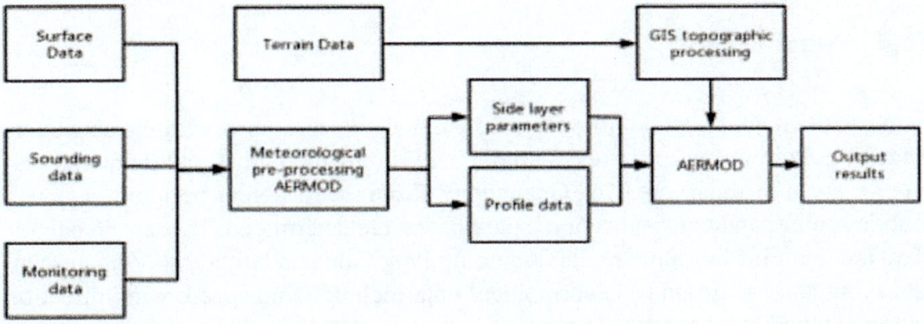

Fig. 51.5 Flow chart of AERMOD model system

boundary layer. In the process of atmospheric guess, due to the characteristics of turbulent instability, the model guess deviation (Zhang et al. 2009) (Figs. 51.4 and 51.5).

51.3 AERMOD Defect

The correlation coefficient between AERMOD model prediction results and measured values is high, but there are also defects, mainly in the following aspects:

AERMOD application conditions are limited, which can only be used for air quality model prediction within 50 km. Not applicable under complex terrain and special climatic conditions (Ding et al. 2007).

Imperfect development of meteorological collection systems or inaccurate meteorological data will increase the deviation of forecast results. The meteorological data at home and abroad are mainly wrapped by mesoscale meteorological models. Based on the monitoring data of weather stations where they are predicted, the accuracy is tested by comparing the prediction and monitoring data of major pollutants.

Through the output analysis of the collected meteorological data, the above difficulties and limitations can be avoided, and the accuracy of prediction can be improved. WRF output high altitude meteorological data, coupled with WRF and AERMOD to improve the performance of the model, so as to improve the prediction accuracy of the model. This method can solve the problem that meteorological data is insufficient and difficult to obtain in some areas of China. In addition, AERMOD model can also establish a coupling model with CALPUFF, so as to analyze the scope and flexibility of prediction, and solve the problem of inaccurate model prediction under wind speed.

In addition, AERMOD is relatively weak in spatial diffusion and comprehensive prediction analysis. GIS software can be used for overlay analysis of AERMOD prediction results to enhance the predictability of the model.

51.4 Verification

In the AERMOD model pollution source parameters, combined with the characteristics of pollution sources in residential areas, commercial areas and industrial and mining areas in urban areas of Guangdong Province, different pollution sources, mobile sources and non-point source parameters are determined. The terrain parameters are input into aermod model, including longitude and latitude coordinates and elevation, and the ground meteorological data include wind speed, wind direction and total cloud amount (Fig. 51.6).

The AERMOD model has a high requirement for the setting of surface parameters. In addition to the surface parameters that have an important impact on the diffusion of atmospheric pollutants such as surface reflectivity, and because the geographical location of Guangdong Province is related to the water system, and the population setting will also have a certain impact on the simulation results (Figs. 51.7 and 51.8).

In order to verify the rationality and reliability of the model and parameter selection, the direct comparison method is used to compare and analyze the predicted data and monitoring data. It can be seen from table 4 that the predicted concentration of PM2.5 at Nanhaizi monitoring station is high, which may be due to the serious impact of typhoons and rains in flood season and the inability of closed citizens to go out. The ratio of different monitoring stations is 0.9–1.3. The main reason is not only the system error of the model itself, but also the pollution source parameters and the background concentration of the actual monitoring points and local small pollution source emissions (GIS-based Analysis of Urban Air Pollution Emissions 2007). The ratio of the average predicted value and the average monitored value of

Classification	Type of pollution			Total
	Industrial Sources	Flow source	Surface source	
$PM_{2.5}$	2458.9936	136.0616	1398.9401	3993.9953
PM_{10}	2427.4745	1165.4433	2519.0900	6112.0078
SO_2	831.2328	23.1000	94.1674	948.5002
NO_2	2442.6700	1527.9103	41.4243	4012.0046
O_3	4167.3123	3719.7777	3006.6157	10893.7057
CO	110.8400			110.8400

Fig. 51.6 January 2021 Pollution Emission Inventory of Guangdong Province

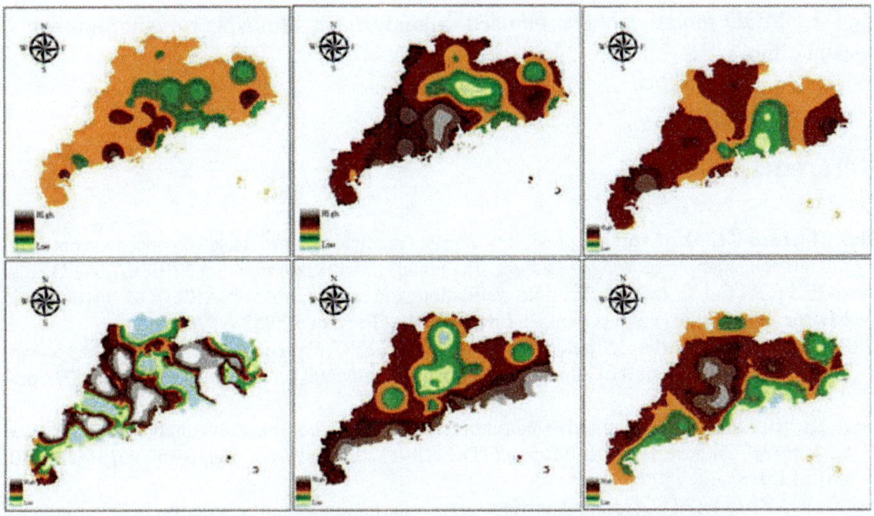

Fig. 51.7 PM2.5, PM10, SO_2, NO_2, O_3 and CO concentration in Guangdong Province in January

Monitoring station locations	PM2.5 (μg/m3)		
	Projections	Monitoring	Proportion
New City Hall of Yunfu City, Guangdong	28.7400	30.0000	0.9580
Guangdong Zhanjiang City Environmental Monitoring Station	41.4720	31.9550	1.2978
Nanhaizi Station, Shenzhen, Guangdong	45.9900	25.3000	1.8178
Guangdong Shanwei Environmental Protection Bureau	39.5320	40.6350	0.9729
Guangdong Qingyuan Environmental Protection Building	46.1500	49.1600	0.9388

Fig. 51.8 PM2.5 prediction results of five random monitoring points

the monitoring point is close to 1, which has high consistency. The verification results show that the prediction evaluation based on AERMOD model is representative for pollutant diffusion simulation and air quality prediction in the study area.

The comprehensive and objective evaluation of urban air quality by GIS technology will provide the necessary scientific basis for the effective control and treatment of urban air pollution. Through the comprehensive use of GIS and GCTM model, the advection, diffusion, chemistry, subgrid scale cumulus convection and dry–wet deposition process can be analyzed in depth, and the causes of the formation of complex vertical ozone structure observed on Java Island can be successfully explained by the sources of anthropogenic and biomass combustion (Nishizawa and Chino 2004).

Conclusion: In this paper, by establishing the AERMOD model, we analyzed the concentrations of six pollutants in the atmosphere of Guangdong Province, and gives a reasonable explanation for the anomalies of PM2.5 sampling data, verifying that the AERMOD model can also be used to analyze the atmospheric environment of coastal cities.

References

Cheng M, Lian Y (2018) Spatial accessibility of urban medical facilities based on improved potential model—an example of Yangpu District, Shanghai[J]. Adv Geograph Sci 37(02):266–275

Ding F, Li S, Cai F (2007) Example validation and application of AERMOD in domestic environmental impact assessment[J]. Environ Pollut Prevent 12:953–957

GIS-based analysis of urban air pollution emissions[C] (2007) Excellent proceedings of the annual academic conference of the Chinese Society of environmental science (next volume) 2007:766–773

He X, Duan C (2015) GIS quantitative measurement of built environment for suitable buffer distance analysis[J]. Chinese J Sports Med 34(11):1115–1120. https://doi.org/10.16038/j.1000-6710.2015.11.018.

Huchang H, Zhao Y (2017) Comprehensive analysis of urban air pollution based on GIS[J]. Mapping Spatial Geograph Inf 40(10):43–47

Li Z (2014) Research on urban air pollution early warning and display based on GIS [D]. Xi'an University of Science and Technology

Li B (2019) Application of AERMOD model in the accounting of atmospheric environmental health protection distance[D]. Xi'an University of Architecture and Technology. https://doi.org/10.27393/d.cnki.gxazu.2019.001464

Nishizawa M, Chino M (2004) Development of a chemical transport model coupled with a mesoscale meteorological model and application to atmospheric diffusion of volcanic gases from Miyake Island[J]. J Japan Soc Atmos Environ/Taiki Kankyo Gakkaishi 39(1)

Ponomarev NA, Elansky NF, Kirsanov AA, Postylyakov OV, Borovski AN, Verevkin YM (2020) Application of atmospheric chemical transport models to validation of pollutant emissions in Moscow[J]. Atmos Ocean Optics 33(4)

Shareef MM, Husain T, Alharbi B (2016) Optimization of air quality monitoring network using GIS based interpolation techniques[J]. J Environ Protect 7(6)

Sun QZ, Dong XM, Ren ZB (2011) GIS visualization study of atmospheric point source dispersion model[J]. Surv Mapp Sci 36(01):24–25+142. https://doi.org/10.16251/j.cnki.1009-2307.2011.01.073

Wan S, Xu L, Qi QiYang H, Zhou Y (2022) Building a multi-objective optimization model for Sponge City projects[J]. Urban Clim 43

Wu YM (2010) GIS-based simulation of Gaussian plume dispersion of atmospheric point source pollution[D]. East China Normal University

Zhang X (2010) Normative use of AERMOD model in atmospheric environmental impact assessment[D]. Lanzhou University

Zhang C, Chen J, Li H (2009) Simulation of industrial point source air pollution dispersion based on GIS and Surfer[J]. Safety Environ Eng 16(05):48–52

Chapter 52
Start-Up Partial Nitrification Based on Simultaneous Fill/Draw SBR

Kaige Zhao, Weijin Wang, Ruiyang Li, Junfeng Wan, Yixing Tian, Zixuan Zhang, Mingyuan Liu, and Lin Guo

Abstract In this study, a simultaneous fill/draw SBR (SFD-SBR) was operated for 59 days to started and operated PN. The results showed that the start-up period of PN lasted 27 days, and operated stably on day 28–59. The ammonium removal efficiency (ARE) and nitrite accumulation rates (NAR) were 98.83 and 89.60% when the effluent NO_2^--N concentration was 468.9 mg N/L. Due to the inhibition of nitrite-oxidizing bacteria (NOB) by free nitrous acid (FNA), free ammonia (FA) and intermittent aeration, the stable performance of PN was achieved. Therefore, the SFD-SBR could be used to operate PN.

Keywords Partial nitrification · Simultaneous fill/draw SBR · Ammonium-oxidizing bacteria

52.1 Introduction

The traditional nitrification and denitrification process is one of the most practical and efficient means to removal nitrogen, while is facing about the large aeration for nitrification, insufficient COD for denitrification and huge disposal pressure for excess sludge. Compared to nitrification, partial nitrification (PN) is only carried out until the stage of ammonia (NH_4^+-N) oxidation to nitrite (NO_2^--N), as a result, it is theoretically possible to reduce 25% of aeration energy consumption. PN can

K. Zhao · J. Wan (✉) · Z. Zhang · M. Liu
School of Ecology and Environment, Zhengzhou University, Zhengzhou 450001, PR China
e-mail: wanjunfeng@zzu.edu.cn

Henan International Joint Laboratory of Environment and Resources, Zhengzhou University, Zhengzhou 450001, PR China

W. Wang · R. Li · L. Guo
Henan Institute of Geological Survey, Zhengzhou 450001, PR China

Y. Tian
School Environment and Safety Engineering, Jiangsu University, Zhenjiang 212013, PR China

J. Zhang et al. (eds.), *Environmental Pollution Governance and Ecological Remediation Technology*, Environmental Science and Engineering,
https://doi.org/10.1007/978-3-031-25284-6_52

provide NO_2^--N for partial nitrification and denitrification, partial nitrification and anaerobic ammonia oxidation (anammox), and saves 40% COD and 100% COD, respectively.

As the functional bacteria, the enrichment of ammonia oxidizing bacteria (AOB) is the key step to achieve PN. However, AOB usually coexists with nitrite oxidizing bacteria (NOB), NOB will lead to oxidate NO_2^--N to NO_3^--N, which make the PN performance decrease or even disappear. So far, most studies published on initiated and maintained PN used conventional sequencing batch reactors (C-SBR) (Wei et al. 2014). Compared with the conventional SBR, the simultaneous fill/draw SBR (SFD-SBR) means that as the original wastewater enters the reactor, the treated wastewater is discharged in the form of overflow, simultaneously, the volume is constant, avoiding the use of pumps and saving operation time (Pronk et al. 2015).

As far as we know, there were few researches on the application of SFD-SBR in PN. Under this background, the primary aim of this study was to start-up the PN in SFD-SBR with high NH_4^+-N concentration rapidly. In this 59-day study, (1) monitor water quality of effluent; (2) figure out variation trend of pollutant concentration in the system and analyze the mechanism of PN realization. The research results would be a useful reference value for practical engineering application of PN in SFD-SBR.

52.2 Materials and Methods

The working volume of the SFD-SBR was 2.4 L, and height / diameter (H/D) ratio was 20. The temperature was controlled at 25–28 °C through the temperature controller. The 300 ml sludge were inoculated into the system. The wastewater used in the experiment was artificially prepared, and the target pollutant in the water was 500 mg N/L NH_4^+-N, provided by $(NH_4)_2SO_4$, the alkalinity was provided by $NaHCO_3$, and other components and concentrations according to Xu et al. (2021). The operation mode of SFD-SBR followed this: Firstly, simultaneous fill / draw stage: the 600 mL original wastewater (volumetric exchange ratio was 25%) was pumped into the bottom of the reactor, and the effluent flowed out from the water outlet above the reactor, this stage lasted 55 min; Secondly, aeration stage: the 600 mL/min of aerobic aeration for 60 min; Finally, settling stage: five minutes for the sludge to settle to the bottom of the reactor (Fig. 52.1). The three stages took 2 h for a cycle, and run continuously for 12 cycles every day. Ammonium removal efficiency (ARE), Nitrite accumulation rate (NAR), Nitrogen load rate (NLR), FA and FNA were calculated as from Eqs. (52.1–52.5), respectively.

$$\text{ARE}(\%) = \frac{([\text{NH}_{4\text{inf}}^+] - [\text{NH}_{4\text{eff}}^+])}{\text{NH}_{4\text{inf}}^+]} \cdot 100\% \tag{52.1}$$

$$\text{NAR}(\%) = \frac{[\text{NO}_{2\text{eff}}^-]}{[\text{NO}_{2\text{eff}}^-] + [\text{NO}_{3\text{eff}}^-]} \cdot 100\% \tag{52.2}$$

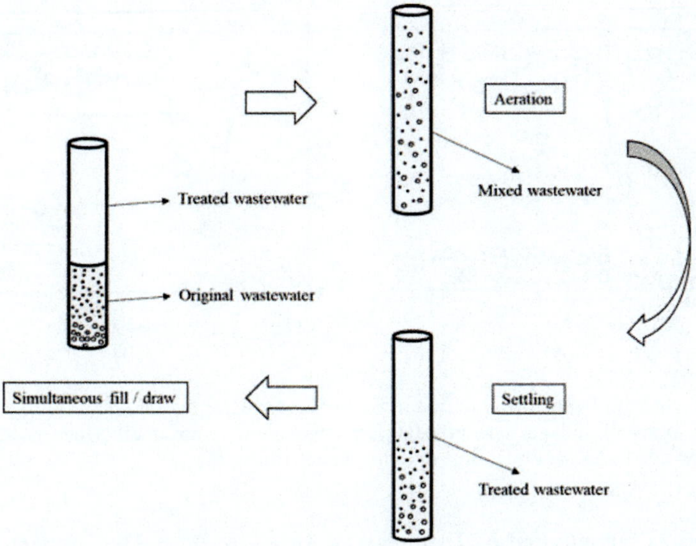

Fig. 52.1 Schematic diagram of the simultaneous fill/draw SBR mode

$$\mathrm{NLR}\left(\mathrm{kg\,N/m^3\cdot d^{-1}}\right)=\frac{24\cdot[\mathrm{TIN_{inf}}]}{1000\cdot\mathrm{HRT}} \tag{52.3}$$

$$\mathrm{FA}=\frac{17}{14}\cdot\frac{[\mathrm{NH_4^+}]\cdot 10^{\mathrm{pH}}}{\exp^{\left(\frac{-2300}{273+T}\right)}+10^{\mathrm{pH}}} \tag{52.4}$$

$$\mathrm{FNA}=\frac{64}{14}\cdot\frac{[\mathrm{NO_2^-}]}{\exp^{\left(\frac{-2300}{273+T}\right)}\cdot 10^{\mathrm{pH}}} \tag{52.5}$$

where, $[NH_{4inf}{}^+]$ was $NH_4{}^+$-N concentration in the influent; $[NH_{4eff}{}^+]$, $[NO_{2eff}{}^-]$, and $[NO_{3eff}{}^-]$ was $NH_4{}^+$-N, $NO_2{}^-$-N and $NO_3{}^-$-N concentrations in the effluent, respectively; and the FA and FNA concentrations were calculated as functions of temperature, pH, and $NH_4{}^+$–N or $NO_2{}^-$-N for FA or FNA, respectively.

52.3 Results and Discussion

52.3.1 Partial Nitrification Performance of the Reactor

As shown in Fig. 52.2, the whole 59-day experiment was divided into two phases. In phase I (1-27d), the influent $NH_4{}^+$-N concentration was 506.23 ± 16.65 mg N/L, the influent $NaHCO_3$ concentration (5900 mg/L) was added according to the theoretical alkalinity on day 1–13, but the removal value of $NH_4{}^+$-N during this period was

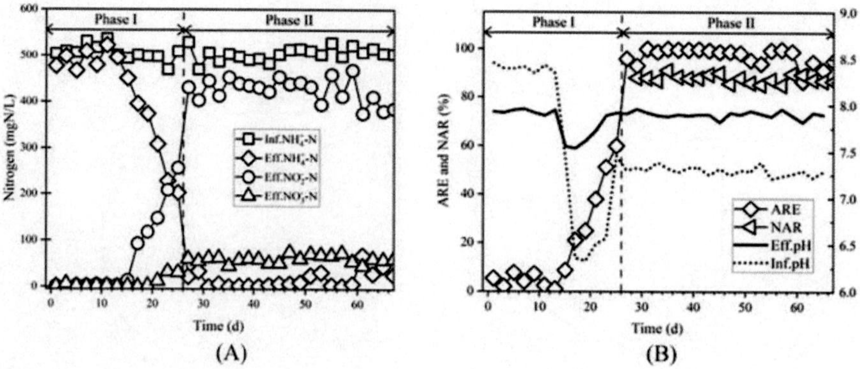

Fig. 52.2 Partial nitrification overall performance of the reactor over 60 days: nitrogen concentration and forms of influent and effluent (**a**); ammonium removal efficiency (ARE), nitrite accumulation rate (NAR) and pH value of influent and effluent (**b**)

22.18 ± 13.27 mg N/L and ARE only was 4.31 ± 2.61%, and the effluent NO_2^--N and NO_3^--N concentration were 3.03 ± 1.23 and 1.93 ± 2.15 mg N/L. On day 14, the influent $NaHCO_3$ concentration decreased to 500 mg/L, the content of NO_2^--N was 13.65 mg N/L on day 15 and on day 17 increased rapidly to 93.33 mg N/L. The effect of ARE gradually improved with the increase of in the influent. On day 24, the $NaHCO_3$ in the influent increased to the theoretical amount. Surprisingly, the concentration of NH_4^+-N, NO_2^--N and NO_3^--N in effluent were respectively 21.82, 431.37 and 60.40 mg N/L. ARE and NAR reached 95.88 and 87.72%. That suggested it is feasible to realize PN by increasing alkalinity step by step. PN was successfully started up in SFD-SBR.

In phase II (28–59), influent NH_4^+-N and $NaHCO_3$ remained at the theoretical level. In the whole phase, PN performance was excellent, and the average concentration of effluent NO_2^--N was 424.4 ± 41.11 mg N/L. The nitrite was the main form of nitrogen in the effluent. The ARE and NAR were 98.83 and 89.60% with a concentration of 468.9 mg N/L on day 59. The results fully proved the feasibility of SFD-SBR to achieve PN rapidly and efficiently.

52.3.2 *One Cycle Performance of the in the Reactor*

The one cycle experiment (120 min) in the reactor revealed the variation trend of nitrogen forms and concentration in settling (S), simultaneous fill/draw (SFAD) and aeration (A) stage. At interface of sludge and water, pH, DO and water samples were taken and measured during the S and SFD period, and these were also done at the same position during the A stage.

Figure 3a showed the trend of nitrogen concentration and forms in minute 0–120. In minute 0–60, the NO_2^--N, NO_3^--N, and FNA concentrations decreased, but

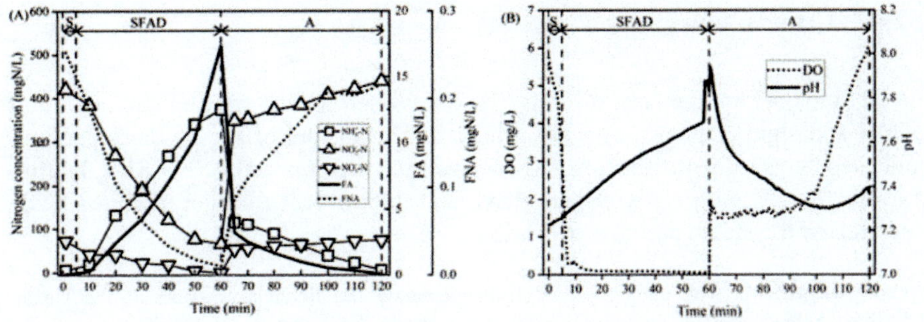

Fig. 52.3 The variation trend of nitrogen forms, FA and FNA (**a**), pH and DO (**b**) in the reactor within 120 min

NH_4^+-N, FA, and pH increased as original wastewater (high NH_4^+-N concentration) was introduced into the reactor. However, the DO decreased below 0.1 mg/L and stabilized until the end of SFAD stage (Fig. 3b). During the A period, after mixing of original wastewater and treated water (low NH_4^+-N concentration) in the reactor completely, the NH_4^+-N concentration decreased to 114.58 mg N/L, while NO_2^--N and NO_3^--N increased to 347.54 mg N/L and 53.67 mg N/L from minute 60–65. NH_4^+-N was first oxidized into nitrite under the action of AOB, resulting in a large accumulation of NO_2^--N. When the aerobic phase was over, the NH_4^+-N concentration had decreased to 6.2 mg N/L, and NO_2^--N and NO_3^--N had increased to 438.73 and 74.21 mg N/L. The DO concentration in the system was still maintained at a low value, which was only 1.71 ± 0.20 mg/L in minute 60–100, the minimum concentration was 1.46 mg/L and the maximum was only 2.42 mg/L. The pH decreased from the initial 7.906 to 7.346 in the aeration stage as ammonia oxidation proceeded.

Intermittent aeration has been observed to suppress NOB, because the AOB is more sensitive to DO and preferentially uses it than NOB (Yang et al. 2011). In this study, DO continued to rise in the late aeration stage (Fig. 3b), but during the simultaneous fill/draw stage, the DO decreased rapidly to below 0.1 mg/L, it was one of the factors for the realization of PN. In addition, FA and FNA were also important factors that contributed to the inhibition of NOB and achievement of PN (Pourbavarsad et al. 2022), during A stage, the FA concentration decreased from 3.21 to 0.09 mg N/L (minute 65–120), FNA increased from 0.09 to 0.22 mg N/L in the same period, indicating FNA also played a role in suppressing NOB. Furthermore, FA also can produce inhibitory effect on NOB under anaerobic environment (Qian et al. 2017). FA increased from 0.04 to 19.89 mg N/L from minute 1 to 60, and anaerobic conditions for most of this stage, so FA could have also contributed to the inhibition of NOB under anaerobic conditions.

Therefore, due to the special operation mode of the process, FA, FNA and intermittent aeration contributed to the realization and stable operation of PN.

52.4 Conclusions

In our study, a simultaneous fill/draw SBR was used for the start-up and operation of PN and significant results were achieved. The results of the 59-day operation indicated that partial nitrification was achieved and maintained by SFD-SBR. During a cycle, the FA and FNA concentrations, and intermittent aeration inhibited NOB and created the stable partial nitrification.

Acknowledgements The authors gratefully appreciated the financial support of the Henan Provincial Science and Technology Development Program (Grant No. 212102110030).

References

Pronk M, de Kreuk MK, de Bruin B, Kamminga P, Kleerebezem R, van Loosdrecht MC (2015) Full scale performance of the aerobic granular sludge process for sewage treatment. Water Res 84:207–217

Pourbavarsad MS, Jalalieh BJ, Landes N, Jackson WA (2022) Impact of free ammonia and free nitrous acid on nitritation in membrane aerated bioreactors fed with high strength nitrogen urine dominated wastewater. J Environ Chem Eng 10(1)

Qian W, Peng Y, Li X, Zhang Q, Ma B (2017) The inhibitory effects of free ammonia on ammonia oxidizing bacteria and nitrite oxidizing bacteria under anaerobic condition. Biores Technol 243:1247–1250

Wei D, Xue X, Yan L, Sun M, Zhang G, Shi L, Du B (2014) Effect of influent ammonium concentration on the shift of full nitritation to partial nitrification in a sequencing batch reactor at ambient temperature. Chem Eng J 235:19–26

Xu W, Zhang T, Wan J, Li H, Chen Y, Wang Y (2021) Phosphorus recovery via the formation of hydroxyapatite crystals at various nitrogen loading rate in an anammox-based UAFB. Biores Technol 326:124628

Yang S, Yang F (2011) Nitrogen removal via short-cut simultaneous nitrification and denitrification in an intermittently aerated moving bed membrane bioreactor. J Hazard Mater 195:318–323

Chapter 53
A Review of Research on Uranium Aerosol Formation Under Fire Conditions

Yanjun Wang and Ming Guo

Abstract Uranium materials are widely used in the nuclear industry. Uranium-containing materials will release uranium aerosols under fire conditions. Uranium aerosol seriously harms the health of workers. This article summarizes the related research of uranium aerosol under fire conditions from the aspects of experimental device design, sample processing and analysis, parameter characterization and results, and generation mechanism.

Keywords Uranium aerosol · Fire condition · Generation mechanism

53.1 Introduction

Uranium materials are widely used in the nuclear industry due to their unique physical and chemical properties. The application of uranium materials and related technologies has also caused certain safety hazards to colleagues who have brought progress to the society. For example, uranium aerosol may be generated due to fire in the production, transportation, storage and application of uranium materials. Uranium aerosol will seriously pollute the environment and cause internal exposure to workers. To study the hazards of uranium materials to the environment and workers in a fire, we must first obtain the amount of uranium aerosol produced and the particle size distribution. Studying the law of aerosol generation of uranium materials under fire conditions is of great significance to the health of workers.

Y. Wang · M. Guo (✉)
College of Nuclear Science and Technology, Naval Engineering University, Wuhan 430033, Hubei, China
e-mail: morpheusgwo@163.com

Y. Wang
Detachment of PLA 91515, Sanya 572000, Hainan, China

J. Zhang et al. (eds.), *Environmental Pollution Governance and Ecological Remediation Technology*, Environmental Science and Engineering,
https://doi.org/10.1007/978-3-031-25284-6_53

The current research on the generation law of uranium aerosol under fire conditions can be roughly divided into several processes such as experimental device design, sample processing and analysis, parameter characterization and results, and generation mechanism. The following focuses on these aspects to explain the research progress.

53.2 Research Status

53.2.1 Design of Experimental Device for Uranium Aerosol Under Fire Conditions

The uranium aerosol experimental device under fire conditions mainly includes combustion device and sampling device.

(1) Combustion device. The current research on uranium aerosol under fire conditions, the literature that can be referred to is the experimental report of aerosol generation under fire and combustion conditions of depleted uranium bombs carried out by Elder and Tinker in 1980 during storage and transportation (Elder and Tinkle 1980; Hanson et al. 1974), Mishima in 1985 Reports obtained from experiments conducted by others under similar conditions (Mishima et al. 1985) and in 2015, Clark et al. used three metallographic phases of depleted uranium, uranium-titanium alloy and uranium-niobium alloy to simulate the formation of aerosols under fire conditions (Clark 2015).

In 1980, Elder and Tinkle et al. used the combustibles in the storage and transportation of depleted uranium bombs such as depleted uranium metal shavings, artillery propellant, pine and paper materials as fuel to burn XM774 depleted uranium bombs to generate aerosols. Studied the particle size distribution characteristics and source data of uranium aerosols generated under different combustion conditions of materials (Elder and Tinkle 1980). In 1985, Mishima et al. placed M829 depleted uranium bombs in sleepers stacked on outdoor metal supports, poured fuel oil for combustion experiments, and simulated the scenes of fires during the transportation and storage of depleted uranium bombs. The formation characteristics of uranium aerosol at 800–1200 °C (Mishima et al. 1985). Since the experiment was carried out in an open-air environment, and only a metal bracket was used as the experimental site, the simulation of the actual fire scene was the most suitable. The deficiencies of the experimental methods are therefore concentrated in the lack of control over the experimental conditions. In 2015, Clark et al. used charcoal as fuel to study the formation of aerosols from three different alloy phases of depleted uranium, uranium-titanium alloy, and uranium-niobium alloy at a temperature of about 600 °C (Clark 2015). Compared with the foregoing two sets of experimental devices, this device better solves the problem of airtightness, and at the same time, the ability to control the experimental

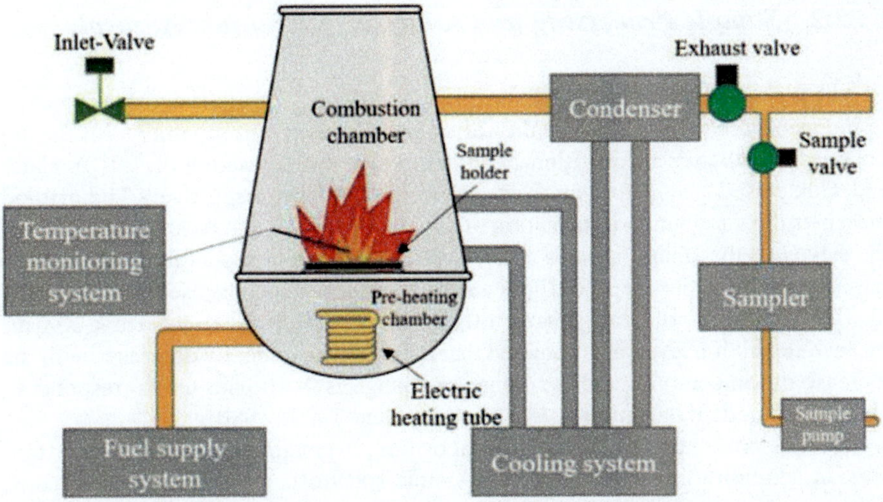

Fig. 53.1 The aerosol fire experiment device designed by Mao Hanyuan

conditions has been greatly improved. The main problem is the integrated design, which makes it difficult to take samples and fuels during the experiment.

In 2020, Mao Hanyuan analyzed the disadvantages of precious fire experimental devices problems such as difficult variable control and great environmental hazards (Hanyuan 2021). Mao Hanyuan designed a closed aerosol generation experimental device using kerosene as fuel to solve the problem of poor sealing and maneuverability of previous experimental devices. The experimental device is shown in Fig. 53.1. The experimental device adopts an integrated design according to the actual requirements of the fire accident scene, which can effectively monitor and control the experimental conditions such as temperature and gas flow rate. The fire experiment verified that the device has good reliability and versatility.

(2) Sampling device. The most commonly used sampling techniques are filtration and inertial separation. Filtration technology uses filter paper to trap particles larger than a certain size, and is mainly used for air purification and total sampling in the experimental system. Inertial separation technology uses the large inertia of large particles at a constant flow rate. When the direction of the airflow changes, the large particles will rush out of the airflow and be collected on a suitable substrate (Baron and Willeke 2006).

In the study of uranium aerosols, most experiments use cascade impactors to separate and sample colloidal particles, and a small number of them use cyclones. In the study of the particle size distribution of depleted uranium aerosols, Cheng et al. used an eight-stage cascade impactor to classify the generated aerosol particles, and compared with the results obtained by Guilmette et al. using a cyclone separator, they found that the impactor produced the particle size distribution of particles is narrower (Cheng et al. 2009; Guilmette and Cheng 2009).

53.2.2 *Sample Processing and Analysis of Uranium Aerosol*

Prasad et al. used ultraviolet-fluorescence spectroscopy and alpha activity detectors to independently analyze the quality of uranium at different separation levels and obtained distribution information of uranium aerosols (Prasad et al. 2010). Elder and Tinkle used gravimetric analysis and γ spectroscopy, respectively The method measures the distribution information of the large particles that have passed through the sieve and the small particles collected by the different separation stages of the impactor. In the experiments of Elder and Tinkle, the researchers used the lognormal distribution to perform least squares fitting on the particles (particle size $< 10\ \mu m$) on each separation level, and obtained that the MMAD gradually decreases with the increase of temperature, and the range of change is 4–10 μm, the corresponding geometric standard deviation is 1.7–3.6 (Elder and Tinkle 1980).

In the determination of chemical composition, the methods used are more extensive, in addition to general ultraviolet–visible spectroscopy, Raman spectroscopy, mass spectrometry, and XRD. Yang reported a method of using the color reaction between uranium and arsenazo to determine trace uranium in the atmosphere by simple ultraviolet–visible spectroscopy (Yang et al. 2013); Di Lemma et al. reported the use of mass spectrometry and Raman spectroscopy to measure gas Methods of chemical information such as elemental composition and phase composition of sol particles (Lemma et al. 2014a, b); Surya Narayana et al. reported the use of XRD and a special Davis-Gray method to analyze the phase composition and content of uranium aerosol (Surya Narayana et al. 1994). Elder et al. conducted XRD analysis on the samples obtained in the high-temperature combustion experiment, and found that the composition was mainly U_3O_8, and there was also a small amount of UO_2. The composition of uranium aerosols obtained in the simulation experiment of Water et al. is mainly U_3O_8, UO_2, and UO_3 (Shui et al. 2003a, b), which is the same as that in the Capstone Depleted Uranium Program (Capstone Depleted Uranium Program) depleted uranium aerosol research by Krupka et al. through XRD The analysis results are consistent (Krupka et al. 2009).

In terms of particle morphology observation, SEM is mainly used to observe the collected particles, and EDX is used to qualitatively analyze the chemical composition of the particles (Krupka et al. 2009; Hansson et al. 2017). In the report of the CDUT project written by Krupka et al., the SEM photos show that the morphology of the particles is very diverse. There are particles with regular shapes similar to crystal planes and particles with irregular surfaces similar to amorphous surfaces; some spherical particles show Shallow cracks are called "brains and grains", and some show the plume structure of dendrites and grains, and are called "flakes and flowers", as shown in Fig. 53.2 (Krupka et al. 2009). Surya Narayana et al. used SEM to characterize the aerosol generated by the UO_2 block under the action of a strong current pulse, and found that within 0.7–2.1 μm, the particles showed a cluster-like aggregate morphology; within 2.1–4.7 μm, the aerosol It shows obvious spherical particles, as shown in Fig. 53.3 (Surya Narayana et al. 1994).

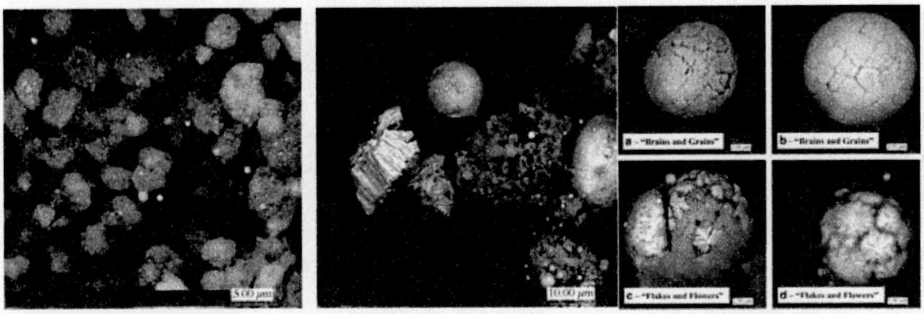

Amorphous particles Regular particles Spherical particles

Fig. 53.2 The morphology of aerosol particles generated in the CDUT project

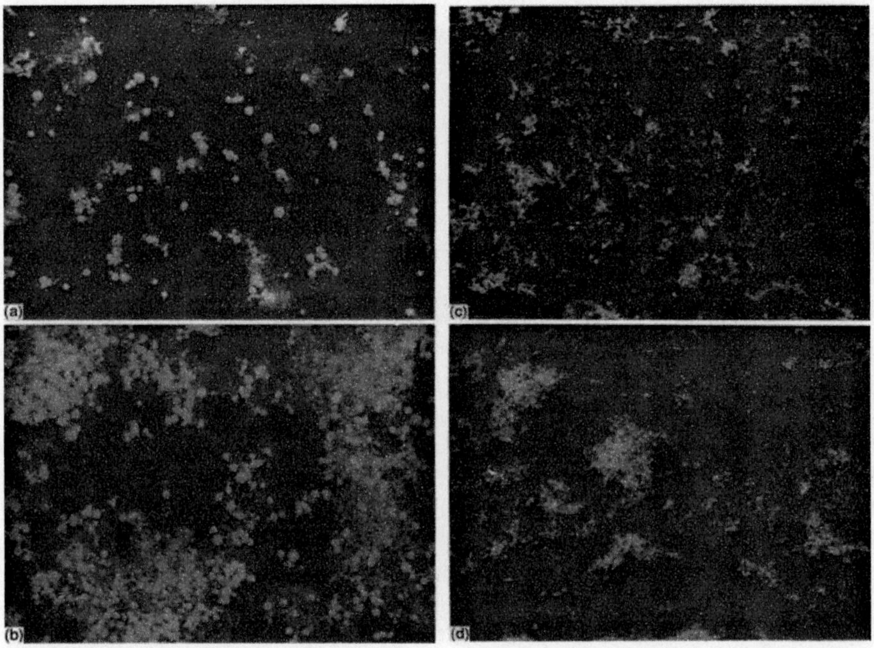

Fig. 53.3 The morphology of aerosol colloidal particles produced by Surya Narayana et al. heating UO2 block (**a**, 3.3–4.7 μm; **b**, 2.1–3.3 μm; **c**, 1.1–2.1 μm; **d**, 0.7–1.1 μm)

53.2.3 Characterization and Results of Uranium Aerosol Parameters

In the study of uranium aerosols, the most widely used characterization parameters mainly include aerosol rate (ARF), aerosol release rate (ARR), respirable rate (RF), particle size, etc. These parameters better describe the source term information of the generated aerosol. In addition, in most studies, the morphology information of the generated aerosol will be given. Based on the morphology information and the corresponding experimental conditions, the approximate reaction mechanism can be inferred to provide help for follow-up research. According to the document "Airborne Release Fractions/Rates and Respirable Fractions for Nonreactor Nuclear Facilities" (DOE 1994) issued by the U.S. Department of Energy, the current general aerosol source term calculation formula includes MAR (Material at risk) DR (Damage ratio) ARF/ARR (Airborne release fraction/rate) RF (Respirable fraction) LPF (Leak path factor).

The ARF is the ratio of the mass of material released in the form of aerosol to the total mass of the material; the aerosol release rate ARR is the mass of the material released into the air per unit time (Lemma et al. 2014a). The two describe the ability and rate of materials to release aerosols from macro-thermodynamics and kinetics respectively, and they are the most important parameters in the study of aerosol generation. RF generally refers to the cumulative share of particles with an aerodynamic diameter of less than 10 μm. This part of aerosol can enter the human body through breathing and be deposited inside the respiratory system, which is the most harmful to the human body (DOE 1994; Roszell et al. 2009; Laul et al. 2006). RF is mainly related to the type, shape and quality of materials, and is also affected by environmental factors.

In terms of ARF and RF research, Carter and Stewart carried out an experiment in 1970 in which metal uranium chips (9 cm^2/g) fell freely and burned in a column with countercurrent air flow. The experimental results are shown in Table 53.1. The column height ranges from 0.2 m to 3.5 m. According to the results of Carter and Stewart tests, the recommended value of ARF $\times$ RF for the combustible form of metallic uranium (specific surface area greater than 1.0 cm^2/g) is 1.1×10^{-4} (Carter and Stewart 1971). Elder and Tinkle conducted a depleted uranium bomb combustion experiment in 1980 and provided the best matching data that can be used for the expected combustion of massive metallic uranium. Table 53.2 shows the ARF and RF values obtained by Elder and Tinkle experiments (Elder and Tinkle 1980). Clark used α-phase pure depleted uranium (100% by weight U), β-phase U-Ti alloy, and γ-phase U-Nb alloy for combustion tests in 2015, and compared the difference between ARF and RF of aerosol produced by the combustion of different phase uranium materials. The results obtained are shown in Table 53.3 (Clark 2015). The results of the depleted uranium combustion experiment carried out by Mao Hanyuan and others in China (Table 53.4) show that the ARF range is 0.53–0.68, and the RF result is close to 1, which is quite different from the previous experimental research results.

In terms of particle size research, the commonly used standard is the aerodynamic equivalent diameter (AED), which is defined It is the diameter of a spherical particle with a unit density (1 g/cm^3) that has the same settling velocity as the particle (Baron and Willeke 2006; Zhengyong 2000; Hinds 1999). The commonly used radioactive aerosol particle size distributions are mass particle size distribution and activity particle size distribution, both of which are obtained by measuring the total mass or total activity of particles on each separation stage after particle size classification. After the distribution form of a certain physical quantity of the aerosol is determined,

Table 53.1 Experimental results of Carter and Stewart

	Uranium < 10 μm part of the release in the form of aerosol
Ignite and burn in the air	1.1×10^{-4} [a]
	3.6×10^{-4} [b]
Partially melted liquid turns into droplets (1)	1.9×10^{-3} [a]
	6.0×10^{-3} [b]

Note [a] geometric mean value; [b] 95% confidence limit; (1) drop height is about 0. 75 m

Table 53.2 Experimental results of Elder and Tinkle et al

	ARF	RF	ARF × RF
500 °C air	3.2E-05	0. 045 45	1.5E-06
600 °C air	2.6E-05	0. 106 25	2.7E-06
700 °C air	8.6E-05	0. 030 00	2.6E-06
800 °C air	5.1E-05	0. 041 11	2.1E-06
900 °C air	2.1E-04	0. 115 15	2.4E-05
600 °C air + CO_2	7.1E-05	0. 055 56	4.0E-06
700 °C air + CO_2	2.7E-04	0. 146 55	4.0E-05
800 °C air + CO_2	5.0E-05	0. 015 33	7.6E-07
	Arithmetic mean		9.7E-06
	Geometric mean		4.0E-06

Table 53.3 Experimental results of Clark et al

Parameter	Geometric mean			Arithmetic mean		
	U	U-Ti	U-Nb	U	U-Ti	U-Nb
ARF	1×10^{-6}	2×10^{-5}	5×10^{-8}	4×10^{-5}	1×10^{-6}	6×10^{-8}
RF	33%	47%	77%	49%	33%	78%

Table 53.4 Experimental results of Mao Hanyuan et al

Serial number	Initial quality (g)	Product quality (g)	ARF			RF
			O/U = 2	O/U = 2.67	O/U = 3	
U-1	2.41	1.30	0.53	0.54	0.55	1.00
U-2	2.40	0.93	0.66	0.67	0.68	0.99

the corresponding characteristic parameters can be given by statistical methods to indicate the distribution state of the colloidal particles with the physical quantity. For mass particle size distribution and activity particle size distribution, the mass median aerodynamic diameter (MMAD) or activity median aerodynamic diameter (AMAD) and the corresponding geometric standard deviation are commonly used to express (Zhengyong 2000).

The common particle size distribution types of uranium aerosol are lognormal distribution and bimodal distribution, which can be roughly characterized by corresponding parameters. Experiments conducted by Megaw in 1961 showed that the mass median diameter of uranium aerosol particles varies with different experimental conditions. The particle size range in air is 42–350 μm, and the particle size range in CO_2 is 10 μm-500 μm (Megaw et al. 1961). This result is quite different from the subsequent experimental results. In the experiment of heating UO_2 released by Surya Narayana et al. in 1994, the obtained colloidal particles showed a bimodal distribution, with large particles accounting for 15–36% of the total, with a particle size range of 2.2–11 μm, and MMAD of 4.8 μm, the geometric standard deviation is 2.3, small particles account for 64–85%, the particle size range is 0.43–2.1 μm, the MMAD is 0.78 μm, and the geometric standard deviation is 1.8(Surya Narayana et al. 1994). In the experiments of Elder and Tinkle, they used the lognormal distribution to perform least squares fitting on the particles (particle size < 10 μm) on each separation level, and found that the MMAD gradually decreases with the increase of temperature, and the range is 10–4 μm, the corresponding geometric standard deviation is 1.7–3.6 (Elder and Tinkle 1980).

The particle size distribution of aerosol can directly display the influence law of different experimental conditions, which is closely related to the type of material, the existence state, the generation mechanism of aerosol and environmental factors. The distribution form widely used in current research is the log-normal distribution, because it can fit particles in a larger particle size range (Distributions from active particle sampling at the Los Alamos national laboratory uranium foundry 2010). Plionis et al., pointed out that the lognormal distribution has the most probable particle size, and the number or mass of particles larger or smaller than this particle size decreases exponentially with the particle size, which can better describe the aerosol distribution formed by a single physical and chemical process (Distributions from active particle sampling at the Los Alamos National Laboratory uranium foundry 2010). For aerosols generated through multiple physical and chemical processes, multiple log-normal distributions can be used to respectively fit the aerosols, and finally show the characteristics of multimodal distribution.

In terms of morphology research, except for a small part of which are standard spherical, most of the particles show irregular morphology. Some particles show obvious crystal planes, while some particles show an amorphous state; some particles are dispersed, and some particles form clusters or aggregates (Lemma et al. 2014a).

53.2.4 The Formation Mechanism of Uranium Aerosol

Wang Yan et al., reported on the research status of uranium metal combustion mechanism and aerosol rate. Uranium metal will be oxidized in the presence of oxygen. The temperature is higher than the ignition point and a continuous direct oxidation reaction can occur. The oxides produced are mainly atomized particles U_3O_8. If the temperature is lower than the ignition point, the oxidation reaction requires a continuous external heat source. In order to proceed, the rate and progress of oxidation are proportional to temperature, and $UO_{1.X}$ and $UO_{2.X}$ particles are formed under the mechanical stress/tension of uranium (Yan et al. 2019). Liu Zhiyong and others have studied the formation mechanism of uranium aerosol under the action of explosive load. The study shows that under the action of explosive load, uranium material is in extreme conditions, and when subjected to high temperature and high pressure during the extreme time, aerosol will be produced quickly. When the specific internal energy reaches 1.9 MJ/kg, it can be considered as a complete aerosol (Zhiyong et al. 2021).

The US DOE report pointed out that for uranium exposed to the air, the first thing that occurs is the oxidation of the surface to produce UO_2 (DOE 1994). Because the density of uranium metal is 19.1 g/cm^3 and the density of UO_2 is 10.97 g/cm^3, the oxide layer will expand (Lide and Haynes 2010). On the one hand, the generation of the oxide layer limits the diffusion of oxygen into the material and reduces the reaction rate; on the other hand, the internal stress causes part of the oxide layer to crack, and oxygen can contact the metal to continue to react. The two sides work together to make the oxidation rate at a certain level. Maintaining a constant value at temperature, this characteristic always exists in the oxidation process of uranium metal (Antill and Peakall 1959).

As the temperature increases, the UO_2 in the oxide layer continues to be oxidized to produce higher proportions of oxides (U_3O_7, U_3O_8, etc.). The experimental results of Rakitskaya and Panov show that it is calculated by the cumulative mass method within 22–300 °C (Rakitskaya and Panov 2000). When the temperature reaches 300°C or higher, the U_3O_8 produced will increase the internal stress of the oxide layer due to the density difference with UO_2, produce dust, small particles, and observable products such as gaps in the oxide layer.

When the temperature continues to rise to 600 °C, UO_2 in the oxide layer will gradually be converted into U_3O_8, forming a black fine powder and falling off the oxide layer (DOE 1994). Antill and Peakall observed in experiments that when the temperature reaches about 725 °C, the oxidation rate and temperature will suddenly increase. It is speculated that uranium metal has undergone a β-γ phase transition, which causes the oxidation rate and exothermic rate to accelerate (Antill and Peakall 1959). A similar phenomenon was also observed in the experiments of Megaw et al. (Megaw et al. 1961). In 800–1000°C, the oxidation rate decreases with the increase of temperature. It is speculated that the protective layer formed by UO_2 attached to the metal surface under high temperature conditions prevents further oxidation of the internal metal (DOE 1994).

53.3 Summary

From the point of view of the experimental equipment, the existing equipment may be poorly airtight, and there is a problem of aerosol leakage; or the control of the experimental conditions is weak and the experiment is relatively rough; or the experimental equipment is troublesome to operate and is not convenient for large-scale development Experimental research. Although the current experimental devices have made significant progress in terms of airtightness and experimental condition control, they have insufficient consideration of aerosol deposition and interference between adjacent experiments, and the collection and treatment of sampling exhaust gas is not perfect.

From the perspective of the experimental process, the methods used are different in the selection of experimental materials, characterization instruments and methods, aerosol generation and collection, and sample post-processing and analysis. The accuracy of the experimental procedures established by Maohanyuan in sampling, chemical analysis and measurement needs to be improved. In addition, the data processing and error analysis of the previous experiments are relatively simple, resulting in the current experimental research results magnitude difference.

The current research on the law of uranium aerosol generation under fire conditions focuses on obtaining the effects of different types of experimental materials, different combustion materials, temperature and other conditions on the generation parameters and particle size distribution of uranium aerosols. Research on burning time, gas flow rate and other aspects is slightly inadequate. In actual scenarios, the influence of burning time on the formation of uranium aerosol is of great significance for the emergency rescue of accidents, and the gas flow rate has a direct impact on the diffusion rate of uranium aerosol, which is of great significance for pollution prevention and control.

Acknowledgements We are very grateful for the funding of the National Defense Advance Research Fund project. We are also very grateful to the Institute of Nuclear Physics and Chemistry, Chinese Academy of Engineering Physics for their support.

References

Antill JE, Peakall KA (1959) Kinetics of the oxidation of uranium by carbon dioxide. J Less-Common Metals 1(3):227–231

Baron PA, Willeke K (2006) Aerosol measurement: principles, techniques, and applications. Chemical Industry Press

Carter RF, Stewart K (1971) On the oxide fume formed by the combustion of plutonium and uranium". Inhaled Part 2(2):819

Cheng YS, Kenoyer JL, Guilmette RA et al (2009) Physicochemical characterization of capstone depleted uranium aerosols II: particle size distributions as a function of time. Health Phys 96(3):266–275

Clark DK (2015) Characterization of respirable uranium aerosols from various uranium alloys in fire events. Aerosol Sci Technol 49(3):188–195

Distributions from Active Particle Sampling at the Los Alamos National Laboratory Uranium Foundry (2010) IOP Conference Series: Materials Science and Engineering

DOE (1994) Airborne release fractions/rates and respirable fractions for nonreactor nuclear facilities

Elder JC, Tinkle MC (1980) Oxidation of depleted uranium penetrators and aerosol dispersal at high temperatures

Guilmette RA, Cheng YS (2009) Physicochemmical characterization of capstone depleted uranium aerosols IV: in vitro solubility analysis. Health Phys 96(3):292–305

Hanson WC, Elder JC, Ettinger HJ et al (1974) Particle size distribution of fragments from depleted uranium penetrators fired against armor plate targets

Hansson E, Pettersson HBL, Fortin C et al (2017) Uranium aerosols at a nuclear fuel fabrication plant: Characterization using scanning electron microscopy and energy dispersive X-ray spectroscopy. Spectrochim Acta, Part B 131(2017):130–137

Hanyuan M (2021) Review of studies on generation of uranium aerosols under high temperature. J Ordnance Equipment Eng 06(2021):1–5

Hinds W C. (1999) "Behavior and Measurement of Airbone Particles." *Aerosol Technology: Properties*, (1999)

Krupka KM, Parkhurst MA, Gold K et al (2009) Physicochemical characterization of capstone depleted uranium aerosols III: morphologic and chemical oxide analysis. Health Phys 96(3):276–291

Laul JC, Foppe TL, Mishima J (2006) Applicability of airborne release fraction and respirable fraction values to particulate toxic chemical material releases at DOE sites. J Chem Health Safety

Lemma Di FG, Colle JY, Ernstberger M et al (2014a) Rades an experimental set-up for the characterization of aerosol release from nuclear and radioactive materials. J Aerosol Sci 70(2014):36–49

Lemma Di FG, Colle JY, Benes O et al (2014b) Chemistry studies for radioactive aerosol release in simulated severe accidents

Lide DR, Haynes WMM (2010) CRC handbook of chemistry and physics

Lu Z (2000) Introduction to aerosol science. Beijing Atomic Energy Press

Megaw WJ, Chadwick RC, Wells AC et al (1961) The oxidation and release of iodine-131 from uranium slugs oxidizing in air and carbon dioxide. J Nuclear Energy Parts a/b Reactor Sci Technol 15(4):176–184

Mishima J, Parkhurst MA, Scherpelz RI (1985) Potential behavior of depleted uranium penetrators under shipping and bulk storage accident conditions

Prasad KV, Balbudhe AY, Srivastava GK et al (2010) Aerosol size distribution in a uranium processing and fuel fabrication facility. Radiat Prot Dosimetry 140(4):357–361

Rakitskaya EM, Panov AS (2000) The behavior of uranium dioxide in various gases. At Energ 89(5):890–894

Roszell LE, Hahn FF, Lee RB et al (2009) Accessing the renal toxicity of capstone depleted uranium oxides and other uranium compounds. Health Phys 96(3):343–351

Shui Y, Xiaoying Z, Fa Z et al (2003a) The research of physical and chemical properties of DU aerosol. Chin J Radiol Med Prot 05:317–319

Shui Y, Xiaoying Z, Fa Z et al (2003b) Chemical composition analysis of aerosol oxides from depleted uranium. Chin J Radiol Med Prot 23(3):212–213

Surya Narayana DS, Sundararajan AR, Harvey J (1994) Characterization of uranium oxide aerosols. J Aerosol Sci 25(5):909–922

Yan W, Bing L, Yangjun Z et al (2019) Research status and prospect of combustion mechanism of metal uranium and airborne release fraction. Environ Sci Manage 44(5):75–79

Yang Y, Xiao S, Zhang Y et al (2013) In situ detection of trace aerosol uranium using a handheld photometer and solid reagent kit". Anal Methods 5(18):5

Zhiyong L, Jintao W, Bin H et al (2021) Study on the formation mechanism of uranium aerosol under explosion load. Explosion and Shock Waves

Chapter 54
Micro-polluted Water Treatment Technology and Research Progress

Jiewen Zheng and Ying Fu

Abstract Micro-polluted water is difficult to degrade the composition of complex substances, characterized by low concentration and high toxicity. These pollutants directly threaten the safety of people's drinking water. A variety of micro-polluted water treatment technologies such as pretreatment technology, enhanced conventional process technology and depth treatment technology are analyzed, and the current research progress of micro-polluted water treatment technology is discussed, among which the most applied is the membrane combination technology. It is hoped that the membrane combination technology can be further developed in the field of micro-polluted water treatment to improve the removal of organic pollutants while reducing membrane pollution.

Keywords Natural coagulants · Wastewater treatment · Coagulant

54.1 Introduction

With the improvement of people's living standard, people's requirements for drinking water quality are constantly increasing, and micro-polluted water treatment technology has become a hot spot in current research. Surveys show that more than 50 percent of China's water sources are polluted. Micropollutants in water sources are composed of a variety of refractory chemicals with low concentrations and high toxicity. These chemicals directly threaten the safety of people's drinking water and pose serious health risks to our residents (Li et al. 2020). Traditional water treatment processes, including coagulation, precipitation, filtration and disinfection, mainly remove bacteria and turbidity in the water, but the removal effect of dissolved micro-pollutants is not obvious (Han 2018), therefore, it is necessary to add some pretreatment process on the basis of conventional treatment process and strengthen conventional process technology or advanced treatment technology.

J. Zheng · Y. Fu (✉)
School of Civil Engineering and Architecture, University of Jinan, Jinan 250022, China
e-mail: cea_fuy@ujn.edu.cn

J. Zhang et al. (eds.), *Environmental Pollution Governance and Ecological Remediation Technology*, Environmental Science and Engineering,
https://doi.org/10.1007/978-3-031-25284-6_54

54.2 Pre-treatment Technology

At present, the pretreatment technology of slightly polluted water mainly includes biological pretreatment and chemical pretreatment. Biological pretreatment technology is to add some biological treatment technology before the conventional water treatment process, using microbial metabolism, initially remove the soluble organic matter, ammonia, nitrite, iron, manganese and other pollutants in polluted water. Biological pretreatment is difficult to realize in practice. First of all, biological pretreatment technology requires high concentration of organic matter, while the concentration of organic matter in micro-polluted water is very low, which cannot meet the survival needs of microorganisms, resulting in low pretreatment efficiency. Secondly, microorganisms and their metabolites will have a certain impact on human health after entering water. Therefore, biological oxidation pretreatment has not been widely used in practical engineering. Chemical oxidation pretreatment technology is through the addition of chlorine, potassium permanganate, potassium permanganate, ozone and other strong oxidation substances, so that the structure of organic pollutants in water decomposition or destruction, and then coagulation precipitation and subsequent filtration and disinfection and other conventional treatment. Adding a large amount of chlorine to water will produce trihalomethane and other toxic and harmful substances in water, which have carcinogenic effects and is potentially dangerous to human health. Potassium permanganate oxidation of water will produce base replacement mutagens precursors and other intermediate products, which will be converted into mutagens in the subsequent chlorination process, so as to improve the mutagenic activity of effluent(Longyu and Hongyuan 2010).Ozone has strong oxidation ability, can be rapidly decomposed in water, produce hydroxyl radical with strong oxidation, can oxidize and degrade most organic matter in water, without leaving other by-products in the water. At present, ozone pretreatment technology has been widely studied and applied.

54.3 Enhanced Conventional Treatment Technology

The enhancement technology of micro-polluted water is mainly to improve the removal of organic pollutants by adjusting coagulation conditions and flocculation conditions. According to the quality of sewage, change the type and concentration of coagulant, use appropriate flocculant, increase the adsorption and bridging of organic particles, improve the removal rate of flocculant. Slightly polluted water commonly used coagulants are polyferric sulfate, aluminum sulfate, aluminum sulfate, aluminum chloride, powder activated carbon, potassium permanganate, potassium permanganate and so on. These substances can remove low molecular weight organic matter, NH_3-N, nitrite and other pollutants dissolved in water. Enhanced coagulation technology is usually to add new polymer flocculants such as polyacrylamide, through a series of processes such as net capture, cleaning, adsorption, co-precipitation, to increase the coagulation effect of flocculants, so as to improve the removal effect of pollutants.

54.4 Deep Treatment Technology

54.4.1 Photocatalytic Oxidation Technology

The photocatalytic oxidation technology uses H_2O_2 and O_3 as oxidants, and the organic matter is oxidized and degraded by the free radicals generated in the system under UV radiation. UV/H_2O_2system is a homogeneous advanced oxidation system, and the mechanism of this system is that H_2 O_2 generates hydroxyl radicals under UV irradiation, and the hydroxyl radicals can remove most of the hard-to-degrade organic pollutants without selectivity, and the decomposition products of H_2O_2 are water and oxygen. The decomposition products of H_2O_2 are water and oxygen, and the process will not produce secondary pollution, which has great application value. Du Zhenqi et al. (2019) have comprehensively discussed the progress of UV/H_2O_2 process for degrading organic micropollutants in drinking water, by outlining the effects of UV, natural organic concentration, solution pH, initial concentration of H_2O_2, initial concentration of target and inorganic anions in aqueous solution on the degradation of organic micropollutants by UV/H_2O_2 process, and analyzed the current situation of conversion products and operation cost of UV/H_2O_2 process. The feasibility of the UV/H_2 O_2 process for treating micropolluted water was demonstrated.

54.4.2 Membrane Filtration Technology

Membrane filtration technology plays an important role in micro-polluted water treatment. The membrane can effectively remove the disinfection byproducts, chroma, bacteria and odor in sewage, sewage depth treatment. A large number of studies have shown that the turbidity of slightly polluted water after membrane filtration technology treatment can reach the relevant national water standards, and less bacteria, after subsequent disinfection treatment, the water can completely meet the drinking water standards. The commonly used membrane filtration technologies include ultrafiltration, microfiltration, infiltration and so on. Membrane are easy to block and polluted during operation, and membrane filtration technology has a high investment cost. However, with the development of technology and the improvement of membrane cleaning methods, the problems of membrane blockage and membrane pollution have been effectively solved, and the market price of membrane has gradually decreased, so membrane filtration technology has been more and more widely used.

54.4.3 Activated Carbon Depth Treatment Technology

Activated carbon advanced treatment technology comprises activated carbon adsorption technology, biological activated carbon technology, etc. Among them, biological activated carbon technology is more commonly used, which uses the adsorption of activated carbon and the oxidation and degradation of microorganisms adsorbed on activated carbon to achieve the degradation of organic pollutants in water. Biological activated carbon technology can enhance the removal of dissolved organic pollutants in water, so as to enhance the quality of wastewater. At present, the technology is comparatively sophisticated in engineering application.

54.4.4 Micro-polluted Water Treatment Technology Research Progress

Membrane separation technology is commonly used in combination with other water treatment technologies because of its tendency to clog and contaminate during operation. Numerous studies have indicated that incorporating membrane technology with other water treatment technologies can decrease membrane contamination and enhance membrane life cycle. On the other hand, it can strengthen the treatment of organic pollutants and improve the removal rate of pollutants. Li, Bowen et al.(Li et al. 2018) studied micro-polluted raw water by combined ozone-ceramic membrane process, and the results showed that the removal rates of COD by ceramic membrane reached 292%, 247% and 195%, and the removal rates of UV_{254} were 400%, 367% and 413%, respectively, when the ozone dosing was 2 mg/L. Compared with the membrane filtration alone, the membrane fluxes of ceramic membranes (15,000, 50,000, 150,000 molecular weight) were increased by 12–26%, 60–105%, 42–84% after pretreatment with 1–4 mg/L ozone, which effectively reduced the membrane contamination and improved the removal of micro-pollutants. Li et al. (Weiying et al. 2018) treated micro-polluted water by a combined micro-flocculation-metal microfiltration membrane process, and the experimental results showed that the removal efficiency of turbidity, UV_{254} and COD_{Mn} of micro-polluted water by the combined micro-flocculation-metal microfiltration membrane process were 976%, 800% and 631%, respectively. When 03 μm metal membrane cartridge was selected, the specific flux of membrane gradually increased from 44.44L/m^2-h-kPa to 5833L/m^2-h-kPa. Ou Siying (2017) studied the effect of the combined process of powdered activated carbon and ceramic membrane loaded with catalyst on the treatment of various pollutants in micro-polluted raw water. turbidity, COD_{Mn}, and TOC were increased by 21.3%, 3.2%, 27.5%, and 37.0%, respectively, compared with the single ceramic membrane process. This indicates that the combined process of activated carbon and membrane can improve the removal of micro-polluted water and has a great mitigation effect on the membrane contamination problem after the activated carbon treatment.

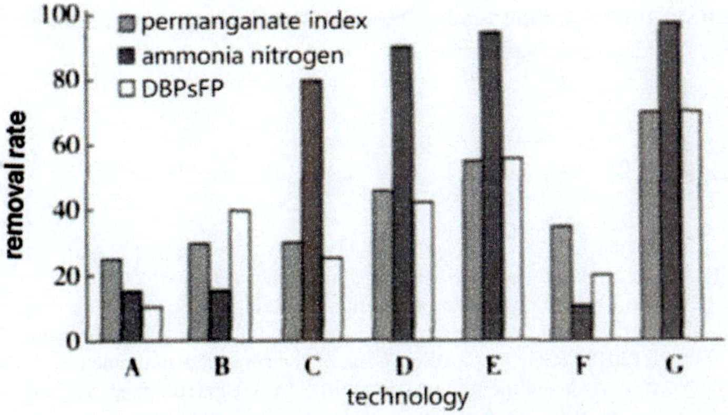

Fig. 54.1 Removal efficiency of pollutants by different treatment processes

54.4.5 Removal Effect of Each Proces

According to the literature reviewed, the removal effect of each process on pollutants (including the removal effect of DBPsFP, a precursor of disinfection byproduct) was investigated. The specific results are shown in Fig. 54.1.

As can be seen from Fig. 54.1, the removal effects of various water treatment processes on permanganate index, ammonia nitrogen and DBPsFP are quite different. The removal rates of permanganate index and DBPsFP are generally not high, and only some processes (C, D, E and G) have high removal efficiency on ammonia nitrogen. Therefore, ACCORDING TO THE pollution degree of raw water and the economic capacity of each region, some of the processes can be selected as the recommended treatment process for the slightly polluted water mainly with PERMANganate index and ammonia nitrogen pollution.

54.5 Conclusion

With the rising complexity of pollutant components in micro-polluted water, micro-polluted water treatment technology has become a research hotspot of widespread concern among specialists and scholars in the water treatment industry. Conventional water purification technology can no longer meet the relevant water standards. Membrane filtration technology can effectively remove the preaccumulation of disinfection by-products, bacteria, odor and chromaticity in polluted water, which cannot be achieved by chemical pretreatment methods. The experiment indicates that the combination of membrane technology and other water treatment technologies can decrease membrane pollution, prolong membrane life, and enhance the removal impact of organic pollutants. In the subsequent research, we should try to combine other pretreatment technologies with membrane filtration, and continue to enhance the research on micro-polluted water treatment technology to guarantee the water quality safety of drinking water sources.

References

Du ZQ, Wang YL, Tian LP (2019) Progress of UV/H2O2 process for degradation of organic micropollutants in drinking water. J Shandong Jianzhu Univ 34:50–57

Han M, Zhao ZW, Cui FY et al (2018) Pretreatment of contaminated raw water by a novel double-layer biological aerated filter for drinking water treatment. Desalination Treatment 37:308–314

Li BW, Li L, Zhou L (2018) Study on the effect of ozonated ceramic membrane process for treating micro-polluted raw water and membrane contamination. Water Treatment Technol 1:114–117

Li X, Wang L, Ma W et al (2020) Evaluation of water quality purification efficiency and water quality stability of micro-polluted water source pretreatment project of different river types 40:2771–2784

Longyu C, Hongyuan L (2010) Research status and progress of chemical biological pretreatment technology for micro-polluted water. Guangdong Chem Indus 37:142–145

Ou S (2017) Treatment of slightly polluted raw water by powdered activated carbon and ceramic membrane (supported with different catalysts) 36:38–43

Li W, Wu X, Qi W (2018) Analysis of membrane contamination mechanism in the combined microflocculation-metal membrane water purification process. J Tongji Univ (Natl Sci Edition) 47:79–84

Chapter 55
Study on Influence of Thermal Insulation Coatings on Small Room Temperature

Yunuo Chen, Peng Chen, and Tao Zhang

Abstract Under the background of today's rapid urbanization development, the application of thermal insulation coatings on roofs has become one of the methods of energy saving and environmental protection. This study mainly discusses the influence of thermal insulation coatings on the small room temperature of the carton, and records the relevant meteorological data and surface temperature. Based on the measured data, the small room temperature performance of the reflective coating chamber under different conditions is analyzed, and the relationship with the thermal radiation force is studied. Also the influence of thermal insulation coating on the small room temperature and the feasibility of applying thermal insulation coating on roof are analyzed.

Keywords Urbanization · Thermal insulation materials · Energy saving · Environmental protection

55.1 Introduction

The rapid urbanization development process has brought a series of environmental problems such as the intensification of the urban heat island effect and the deterioration of the thermal environment (Qing 2012). Due to the increase in the intensity of the heat island, the cooling energy consumption of buildings and vehicles in cities in summer increases significantly, and the energy consumption intensifies (Chen Jitao and Wulong 2011; Hui and Hongwei 2009). The increase in energy consumption

Y. Chen (✉) · P. Chen · T. Zhang
College of Civil Engineering and Architecture, Zhejiang University of Water Resource and Electric Power, Hangzhou 310081, China
e-mail: 1511129484@qq.com

P. Chen
e-mail: chenp@zjweu.edu.cn

T. Zhang
e-mail: 1051550285@qq.com

J. Zhang et al. (eds.), *Environmental Pollution Governance and Ecological Remediation Technology*, Environmental Science and Engineering,
https://doi.org/10.1007/978-3-031-25284-6_55

leads to an increase in greenhouse gas emissions, which further aggravates the deterioration of the outdoor thermal environment, forming a vicious circle. In addition, the urban heat island effect also promotes the return of polluted air and stagnates over the city, degrading the urban air quality and endangering human health (Suming et al. 2012).

In summer, reducing the surface temperature of the urban underlying surface can effectively alleviate the urban heat island effect and improve the urban microclimate. The thermal insulation coating roof refers to the roof with high solar reflectivity, that is, by coating the surface of the ordinary roof with a light-colored, high-reflectivity elastic coating to improve the solar reflectance of the roof and reduce the absorption of solar radiation by the roof, to achieve the purpose of reducing the roof temperature (Xiaofeng and Yanyan 2011). Thermal insulation paint roofs can reduce the roof surface temperature in summer, and its large-area application can also increase the reflectivity of urban areas. In China, the relevant research has just started. And there are relatively few studies on the practical application effect of thermal insulation coating roofs. Based on this, this paper uses ordinary cardboard boxes to simulate the environment of a small room, conducts research on the effect of thermal insulation coating on the small room temperature, and evaluates the effect of thermal insulation coating on the temperature in the small room, in order to provide a reference for urban roof energy-saving research.

55.2 Materials and Methods

55.2.1 Materials and Equipment

In this study, three 45 * 35 * 35 cm carton boxes were used to simulate a small room under the sun, two of which were coated with 1.5 and 3 mm thermal insulation materials respectively, and the other carton was used as a blank reference. The specific photos are shown in Fig. 55.1. The thermal insulation material is a commercial waterproof and thermal insulation paint with a solar reflectance of 86%, a near-infrared reflectance of 88%, and a hemispherical emissivity of 88%, in line with GB/T 25,261–2018 "Reflective Thermal Insulation Coatings for Buildings" the specified requirements.

The above-mentioned small room was placed in the sunken square of the school, and the air temperature, surface temperature and indoor temperature above it were measured. The measuring instrument is DL333380 infrared thermometer with a range of −30–380 °C and an accuracy of ± 2 °C. Hangzhou's real-time temperature and other data come from China Weather Network.

Fig. 55.1 Experimental materials—simulation chamber

55.2.2 Experimental method

This research adopts the method of combining experimental test and theoretical calculation. Firstly, the roof temperature and indoor temperature of the chamber simulated by ordinary carton and the chamber coated with different thicknesses of thermal insulation coatings are measured, and the temperature changes in different weather and different time periods are analyzed. Then, according to the temperature data during the experiment, the change of the thermal radiation force of the chamber at different temperatures is calculated, the influence of the thermal insulation coating on the regional thermal radiation force is analyzed, and the effect of applying the thermal insulation coating to the roof is studied.

55.3 Experimental Results and Discussion

55.3.1 Analysis of Sunny Daytime Temperature Performance

It can be seen from Table 55.1 and Fig. 55.2 that on sunny days, the surface temperature of the carton with thermal insulation coating is lower than that of the reference group, and the average is 16.4 °C lower. The temperature inside the carton coated with thermal insulation coating was also lower than the temperature inside the reference HDB, with an average lowering of 8.7 °C. At the same time, both the surface temperature difference and the indoor temperature difference decrease with the decrease in temperature, the difference is the highest at 12:30 and the lowest at 17:30. It can be seen from the research results that the thermal insulation coating can have lower surface and indoor temperature in sunny weather, which proves the application feasibility of thermal insulation coating.

Table 55.1 Temperature test values in sunny weather (°C)

Weather	Time	Weather temperature	Reference surface temperature	1.5 mm coating surface temperature	Refer to the room temperature	1.5 mm paint indoor temperature
Clear	10:30	31	58.1	39.3	50.3	41.4
Clear	11:30	32	66.4	42.6	54.6	44.2
Clear	12:30	33	68.6	45.3	57.4	45.2
Clear	13:30	34	63.2	42.6	56.1	44.6
Clear	14:30	34	55.8	38.4	53.2	42.3
Clear	15:30	34	52.4	37.6	50.6	40.8
Clear	16:30	33	49.3	37.2	43.4	38.1
Clear	17:30	32	33.8	33.1	33.9	33.5

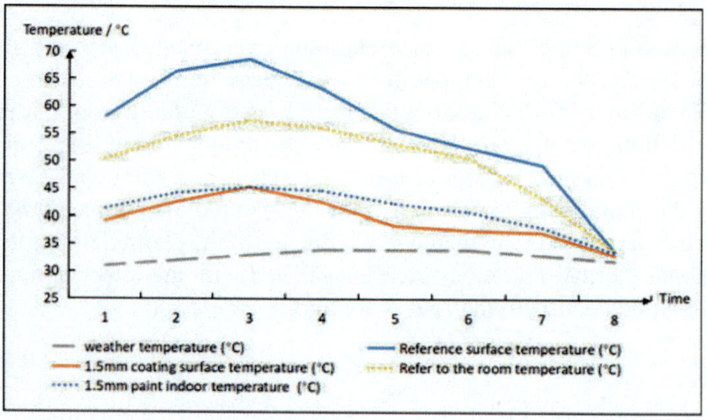

Fig. 55.2 Temperature chart in sunny weather

55.3.2 *Analysis of Temperature Performance under Cloudy Weather*

It can be seen from Table 55.2 and Fig. 55.3 that the experimental data changes on cloudy days are the same as those on sunny days. In light rain and cloudy weather, the surface temperature of the carton with thermal insulation coating is lower than the reference group, and the average is 2.75 °C lower. The temperature inside the carton coated with thermal insulation coating was also lower than that of the reference group, 2.63 °C lower on average. At the same time, both the surface temperature difference and the indoor temperature difference decrease with the decrease of air temperature, and the difference is the highest at 13:00 and the lowest at 17:00.

Table 55.2 Temperature test value in cloudy weather (°C)

Weather	Time	Weather temperature	Reference surface temperature	1.5 mm coating surface temperature	Refer to the room temperature	1.5 mm paint indoor temperature
Cloudy	10:00	26	33.4	31.2	35.8	32.6
Cloudy	11:00	26.6	36.1	33.8	37.5	35.4
Cloudy	12:00	27	36.8	33.6	38.7	36.2
Cloudy	13:00	28	37.2	34.1	39.6	36.7
Cloudy	14:00	29	36.4	33.8	38.6	35.4
Cloudy	15:00	29	35.6	32.5	37.8	34.8
Cloudy	16:00	28.7	34.2	31	36.5	33.4
Cloudy	17:00	28	30.6	28.3	32.2	31.1

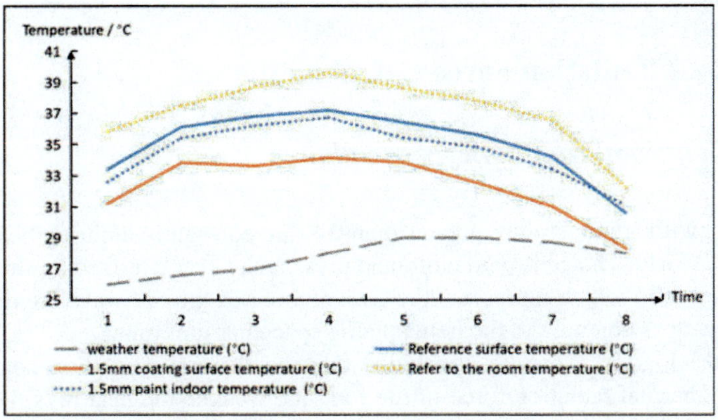

Fig. 55.3 Temperature chart in cloudy weather

55.3.3 *Performance Analysis of Coatings with Different thicknesses*

It can be concluded from Fig. 55.4 that the temperature and indoor temperature of the carton with thermal insulation coating are lower than the reference, and through the comparison of the coating thickness of 1.5 and 3 mm, it can be concluded that the thicker the coating, the lower the temperature and the temperature in the small room. Effect is better. However, as the thickness increases, the added effect will gradually weaken.

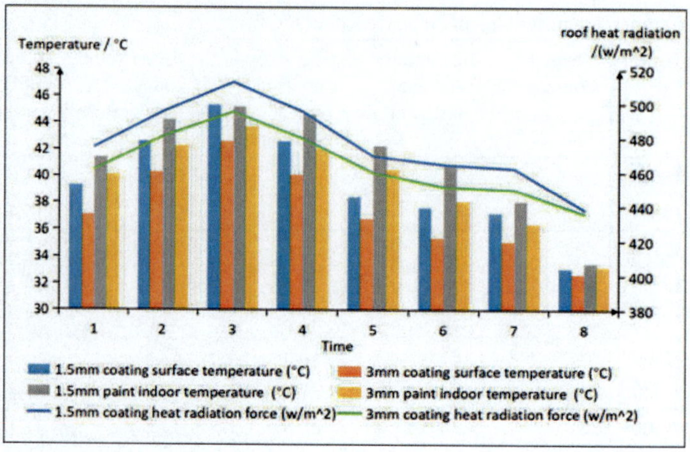

Fig. 55.4 Analysis chart of paint with different thickness

55.4 Roof Radiation Force Calculation

55.4.1 Thermal radiation force value

All objects with a temperature greater than 0 K are constantly emitting heat rays to the outside world. This ability to emit heat rays can be characterized by the thermal radiation force E, which represents the total amount of the full wavelength emitted by the object per unit area to the hemispheric space per unit time.

From the Fig. 55.5 can be concluded that the thermal insulation coating can reduce the thermal radiation force of the roof and weaken the heat rays radiated to the surrounding, thereby alleviating the problems of urban heat island effect and harsh environment, and making an effective contribution to the energy conservation of urban buildings.

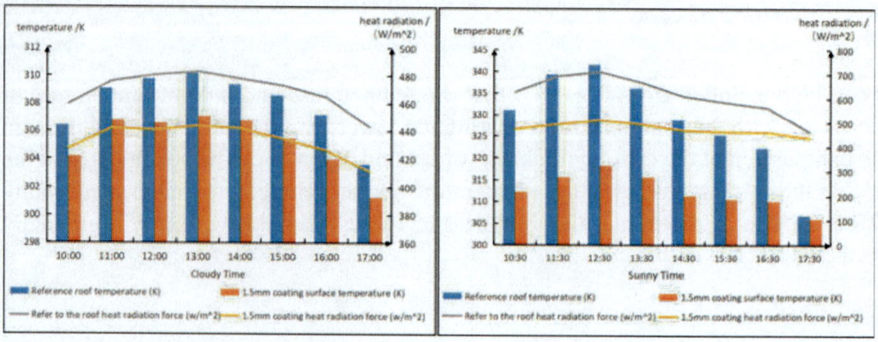

Fig. 55.5 Thermal radiation force value

55.5 Conclusion

In this study, the effects of thermal insulation coatings on small room temperature under different conditions were discussed through actual measurements and theoretical calculations. The conclusions are as follows:

- Under sunny conditions, the surface temperature and indoor temperature of the thermal insulation coating are lower than the reference roof. The temperature difference is more pronounced in sunny weather than in cloudy weather, and the variation is larger.
- Study the cartons coated with 1.5 mm and 3 mm coatings respectively. It can be concluded that the thicker the coating, the lower the surface temperature and the temperature in the small room. It is confirmed that the thermal insulation coating roof does have the effect of reducing the temperature, and the thicker the coating is, the more cooling effect will be gradually weakened.
- The research data proves that the radiation force of the thermal insulation coating roof is weaker than that of the ordinary roof, which proves that the thermal insulation coating roof really emits less heat to the outside world, which is helpful for urban energy saving.

References

Luo Q, Xie M, Tang X et al (2021) Influence of thermal insulation roof coatings on building energy consumption[J]. Mater Rev z1:295–297

Chen J, Xu W (2011) Research on properties of heat reflective coatings for roofing [J]. China Build Waterproof 22:26–30

Hui S, Hongwei T (2009) Study on the effect of solar heat reflective paint on the roof of factory buildings in areas with hot summer and warm winter [J]. Build Sci 3:49–53

Suming C, Wenhong H, Tingyi L et al (2012) Establishment and analysis of temperature field model of heat reflective coating pavement [J]. J Chongqing Jiaotong Univ (Natl Sci Edition) 1:44–47

Xiaofeng W, Yanyan W (2011) Acrylic roof reflective energy-saving coating technology [J]. New Build Mater 8:54–57

Chapter 56
Application and Research Progress of Membrane Technology Based on Lithium Extraction from Salt Lake

Chao Zhang and Liqing Zhang

Abstract As the extraction process of lithium resources from Salt Lake brine is simpler than lithium extraction from ore, lower cost, less environmental pollution and can meet the current demand of lithium products market, countries around the world have turned to the development of lithium extraction from Salt Lake brine as a major direction. At present, the main membrane related technologies for lithium extraction from Salt Lake brine are: nanofiltration membrane filtration, electrodialysis, ion exchange adsorption and covalent organic skeleton membrane separation. We will review the research progress of membrane technology for lithium extraction from Salt Lake starting from the above membrane related technologies.

Keywords Salt Lake lithium extraction · Membrane technology · Lithium resources

56.1 Introduction

Lithium and compounds have high electrochemical activity and the highest specific heat capacity of all solid elements, making them particularly useful in rechargeable lithium-ion batteries (LIBs) for electronics, electric vehicles (EVs), and energy storage. There is also a huge demand for chemical and pharmaceutical industries, and global lithium consumption is expected to increase significantly to 95,000 tons by 2025 (Xu et al. 2021a; Paredes and San 2020). Therefore, lithium extraction is imminent to meet the strong future demand of the lithium element market.

Lithium resources in nature are mainly found in mineral sands, seawater and Salt Lake brines. Because the proportion of total lithium in water is about 60% (Jaskula

C. Zhang · L. Zhang (✉)
School of Civil Engineering and Architecture, University of Jinan, Jinan 250022, China
e-mail: cea_zhanglq@ujn.edu.cn

L. Zhang
State Key Laboratory of Green Building in Western China, Xian University of Architecture and Technology, 13 Central Yanta Road, Beilin District, Xi'an, Shanxi Province, China

J. Zhang et al. (eds.), *Environmental Pollution Governance and Ecological Remediation Technology*, Environmental Science and Engineering,
https://doi.org/10.1007/978-3-031-25284-6_56

2014) and the extraction cost is lower than that of solids Seawater contains the most lithium resources, but it is not economical to extract lithium from seawater because the concentration of lithium ions is too low (below the ppm level) and is subject to excessive interference from other metal ions (Yang et al. 2018; Wu et al. 2017). In contrast, lithium extraction from Salt Lake brine has been progressing due to its large percentage and ease of extraction as a Salt Lake water source for production. This review summarizes the research of related domestic and international groups in recent years with the classification of membrane technology as the main line, and provides an outlook on the future development of membrane technology for lithium and magnesium ion separation.

56.2 Membrane Separation Technology

The main challenge of lithium extraction from Salt Lake brine is the interference of high concentration of magnesium ions, so the development of new technologies for lithium extraction mainly focuses on the separation of lithium and magnesium (Xu et al. 2021b). Compared with the traditional evaporation, precipitation, solvent extraction and other methods, membrane separation technology has the advantages of low energy consumption, environmentally friendly, simple operation, etc., and is expected to become a promising alternative method for lithium extraction (Table 56.1).

For the current state of research, it is widely believed that an ideal membrane module for Li^+/Mg^{2+} separation should not only possess a high Li^+/Mg^{2+} separation factor, but also exhibit a high Li^+ permeability (Andreeva et al. 2021; Bing et al. 2021a). In order to investigate membrane modules with better performance, spatial potential resistance, the Donnan effect and special separation mechanisms are among the keys, and many researchers at home and abroad have devoted themselves to them (Bing et al. 2021b; Binder and Zschörnig 2002).

56.2.1 Nanofiltration Membrane Separation

The magnesium-lithium separation performance of different nanofiltration membranes varies greatly, and there is no nanofiltration membrane on the market specifically for magnesium-lithium separation in high magnesium-lithium ratio brine. Therefore, to apply the nanofiltration membrane in the brine magnesium-lithium separation, still need to carry out a lot of research work in membrane selectivity, membrane process strengthening, different ions on the magnesium-lithium separation. In addition, the description and simulation of the nanofiltration membrane separation process can not only explain the nanofiltration membrane separation mechanism, but also make prediction judgment on the nanofiltration membrane separation performance. The development of nanofiltration membrane separation models

Table 56.1 Comparison of Li^+/Mg^{2+} separation performance of existing studied membranes

Method	Membrane	Experimental Conditions	Permeance mol m^{-2} h^{-1} (Li^+)	Li^+/Mg^{2+}	References
Nanofiltration	PAN hollow fiber	Li^+/Mg^{2+} = 0.17 (0.0033 mol L^{-1} for Li^+ ion)	0.03766	2.6	Li et al. (2015)
Nanofiltration	PA-B2-E3	Li^+/Mg^{2+} = 0.093 (0.0024 mol L^{-1} for Li^+ ion)	0.0677	12.7	Li et al. (2017)
Electrodialysis	PET films	Li^+/Mg^{2+} = 1/1 (1.0 mol L^{-1})	0.0362	634	Wen et al. (2016)
Electrodialysis	PSS@HKUST-1-6.7	Li^+/Mg^{2+} = 1/1 (0.5 mol L^{-1})	6.75	1815	Xu et al. (2021b), Guo et al. (2016)
Diffusion dialysis	DMBP-TB	Li^+/Mg^{2+} = 1/1 (1.0 mol L-1)	0.06	14.6	Tan et al. (2020)
Diffusion dialysis	COF-4eo-PAN	Li^+/Mg^{2+} = 1/1 (0.1 mol L^{-1})	0.034	65	Andreeva et al. (2021), Bing et al. (2021a)

is the cornerstone of the continuous progress of nanofiltration membrane separation technology (Li et al. 2017; Wen et al. 2016; Wang et al. 2021).

The low-pressure and high flux of nanofiltration membrane, as well as the selective permeability property of intercepting divalent ions by monovalent ions, can help to solve the key technology for the separation of magnesium and lithium from Salt Lake brine. The key of nanofiltration separation lies in the special pore size range of nanofiltration membrane, and then after compounding and charging treatment, the nanofiltration membrane has special separation performance, which can effectively separate monovalent ions and multivalent ions. Some researchers have used nanofiltration membranes for ion separation in seawater and brine systems and found that DL and DK nanofiltration membranes have better separation effects between monovalent and divalent ions and are more suitable for lithium separation in Salt Lake brines. The Salt Lake Institute of the Chinese Academy of Sciences has conducted a systematic analysis of the effects of membrane elements, temperature, pressure and other operating conditions on lithium separation for brines with high calcium-to-lithium ratios (Xu et al. 2021c).

For the composite modification of nanofiltration membranes, the influence of various optimization conditions on the performance of magnesium-lithium separation from composite nanofiltration membranes was explored. Hou et al. have explored the effect of oil phase solvent on the magnesium-lithium separation performance of composite nanofiltration membranes under optimal preparation conditions. The composite nanofiltration membranes prepared by mixing n-dodecane in the oil-phase solvent improved the retention of lithium ions.

At the same time, nanofiltration membranes with positive charge have more excellent separation performance, based on the influence of membrane surface zeta potential and Dornan repulsion effect. The analysis shows that the permeability of nanofiltration membranes to total salt shows an increasing trend as the concentration of salt solution containing magnesium ions increases, and this phenomenon is caused by the transformation of the surface charge properties of nanofiltration membranes after the adsorption of magnesium ions on the membrane surface. Due to the high valence of magnesium ion, it will have a shielding effect on the membrane surface charge after being adsorbed by the nanofiltration membrane, therefore, it will neutralize or change the membrane effective charge. Also, due to the presence of magnesium ions, the dielectric repulsion effect can cause the phenomenon and is a non-negligible factor. So far, research scholars have been able to make positive charge modification on commercially available nanofiltration membranes as well as directions that can provide additional lithium ion-promoting transport capacity as hot spots of research (Gang et al. 2011).

The experimental results of Xu et al. showed that the structure and chargeability of NF membranes have a great influence on the separation performance of magnesium and lithium, and the ultrathin composite NF membranes with high charge positivity have higher selective separation performance for magnesium and lithium. During the preparation of NF membranes, the interfacial polymerization conditions parameters such as monomer concentration and reaction time will directly affect the structure and performance of the final prepared NF membranes. The aim of this experiment is to prepare highly charged positively charged ultrathin composite NF membranes containing a large number of protonated amine groups ($-NH^{3+}$ and $-NH^{2+}$) on the surface, so as to improve the separation efficiency of Mg^{2+} and Li^+ in brine of high Mg-lithium ratio brine. The influence of the interfacial polymerization process on the structure of NF membranes and the separation performance of Mg^{2+} and Li^+ was investigated by modulating the parameters in the interfacial polymerization process, so as to optimize the structure and performance of the functional layer of charged NF membranes. Ma et al. also designed a mild secondary interfacial grafting method and EDC/NHS amidation method based on the phenomenon that the nanofiltration separation process depends on Donnan equilibrium theory, and constructed a charged amine substrate on the surface of polyamide nanofiltration membrane to prepare surface charged nanofiltration membrane. The rejection of multivalent cations by the charged nanofiltration membrane was used to improve the separation selectivity of the membrane for the binary mixed solution of $MgCl_2/LiCl$.

56.2.2 *Electrodialysis*

Electrodialysis (ED), as one of the electrically driven ion membrane processes, can use the electric field to carry out the targeted migration of target ions in the liquid phase, thus achieving the concentration and purification of target ions. When applied to lithium extraction technology, under the action of the applied DC electric field, the potential difference is used as the driving force to make the Li^+ in the brine in the desalination chamber migrate through the ion exchange membrane to the concentration chamber, while the high-valent ions such as Mg^{2+} are retained in the desalination chamber, thus achieving the purpose of separating magnesium and concentrating lithium, i.e., realizing the extraction of lithium from brine (Guo et al. 2016; Nie et al. 2017). Many conductive membranes with the function of lithium and magnesium ion separation have been utilized in the exploration of lithium extraction from salt lakes because of their function and high permeability.

Donald et al., used electrodialysis to enrich lithium from $Mg^{2+}/Li^+ = 60:1$ high-Mg brine with a membrane stack consisting of acidic sulfate-based cation exchange membranes and trimethylamine derivative anion exchange membranes in alternating arrangements. Mg^{2+} in the brine was first precipitated by $CaCO_3$ to bring Mg^{2+}/Li^+ down to 5:1, and then after a multi-stage ED process, the Li^+ content of the obtained lithium-enriched solution could reach 15.5 g/L and the Li^+ recovery could reach 78.9%. Chang et al. combined the ED process with an ion-exchange adsorption process to extract Li^+ from the brine by first using a manganese-based. The brine was first adsorbed with a manganese-based lithium-ion sieve adsorbent to achieve a Li^+ content of 1.2–1.5 g/L, and then the Li^+ content was increased by another 1.5% through a two-stage ED concentration process, resulting in a Li^+ recovery of 85%. This method is a relatively economical and environmentally friendly process technology, and provides a relatively efficient way to extract lithium from Salt Lake brines (Xu et al. 2021b; Zhao et al. 2020).

Ma et al., transported the concentrated brine containing Li^+ concentration of 0.2–20 g/L and Mg^{2+}/Li^+ of 1:1–300:1 after solarization and evaporation in the salt field to the desalination chamber of E. After passing through the ED membrane stack equipped with monovalent selective cation exchange membrane and monovalent selective anion exchange membrane, the lithium concentration operation was carried out under continuous circulation process conditions, and the concentrated lithium-rich solution was obtained in the concentration chamber. The concentrated lithium-rich solution is obtained in the concentration chamber, while Mg^{2+}, ${BO_3}^{3-}$ and ${SO_4}^{2-}$ are retained in the desalination chamber to separate Li^+ from Mg^{2+}, ${BO_3}^{3-}$ and ${SO_4}^{2-}$, and finally Na_2CO_3 is added to precipitate and produce Li_2CO_3. At present, electrodialysis is still not mature enough for lithium extraction from the Salt Lake, and further technological enhancement is needed to truly achieve scientific, standardized, rational, efficient and comprehensive utilization of lithium resources development in the Salt Lake.

56.2.3 *Diffusion Dialysis*

With the continuous research on bionic membranes, the preparation of covalent organic framework (COF) membranes with ion sieving effect has been proposed to be applied to lithium extraction from salt lakes. Compared with current nano-filtration membranes, COF has the advantage of high modularity and can form active layers with designable pore structures and adjustable functions, while nano-filtration membranes are empirically optimized. In addition, the discogens of 2D COF are arranged in a columnar fashion due to the strong p-p interactions between aromatic nuclei and the proximity (Tan et al. 2020). Distance oriented arrangement of the pro-lithium function provides a unidirectional path for fast ion diffusion and enhances the queue of phase communication between ions.

Sun et al. prepared membranes with ion selectivity by introducing COF with oligoether branched chains, which possessed high Li^+/Mg^{2+} separation coefficients along with relatively good permeation effects. On this basis, the designable and tunable backbone structure of COF will be integrated into the base membrane with good stable point performance to achieve the compensation effect of Li^+/Mg^{2+} separation and the sieving ability of nanopore size (Fig. 56.1).

Because of its transmission principle of using the concentration difference as the driving force of separation and saving energy, it exists as an extremely negative prospect in the application of lithium extraction from brines possessing high concentrations in salt lakes (Andreeva et al. 2021).

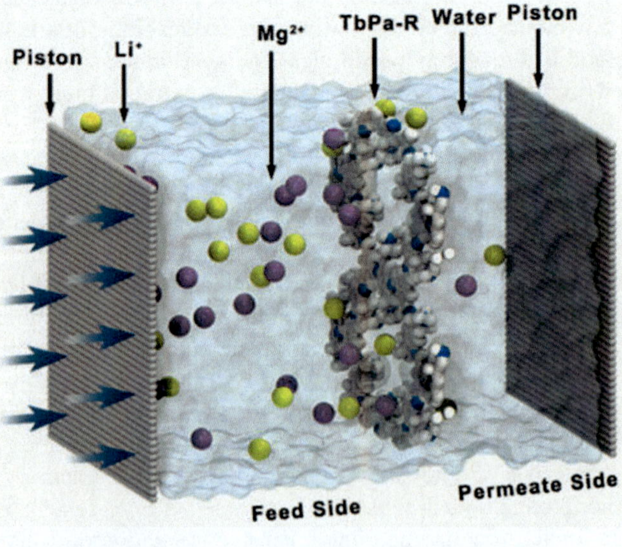

Fig. 56.1 Schematic diagram of Li^+/Mg^{2+} separation simulation system

56.3 Summary

So far, the application of membrane for lithium extraction from Salt Lake is mainly using nanofiltration membrane with good salt separation ability, for ion exchange adsorption, electrodialysis and other technologies, the cost and difficulty of use are much higher than nanofiltration. With the continuous development of new membrane species and research of better membrane lithium technology, there will be a better prospect for the development of the technology related to lithium extraction in the Salt Lake.

References

Andreeva DV, Trushin M, Nikitina A et al (2021) Two-dimensional adaptive membranes with programmable water and ionic channels[J]. Nat Nanotechnol 16(2):174–180

Binder H, Zschörnig O (2002) The effect of metal cations on the phase behavior and hydration characteristics of phospholipid membranes[J]. Chem Phys Lipid 115(1–2):39–61

Bing S, Xian W, Chen S et al (2021b) Bio-inspired construction of ion conductive pathway in covalent organic framework membranes for efficient lithium extraction[J]. Matter 4(6):2027–2038

Bing S, Xian W, Chen S, Song Y, Hou LX, Liu XL, Ma SQ, Sun Q, Zhang L (2021a) Bio-inspired construction of ion conductive pathway in covalent organic framework membranes for efficient lithium extraction

Gang Y, Hong S, Wenqiang L et al (2011) Investigation of Mg2+/Li+ separation by nanofiltration[J]. Chin J Chem Eng 19(4):586–591

Guo Y, Ying Y, Mao Y et al (2016) Polystyrene sulfonate threaded through a metal–organic framework membrane for fast and selective lithium-ion separation[J]. Angew Chem 128(48):15344–15348

Jaskula BW (2014) Mineral commodity summaries: lithium [R], U.S. Geological Survey

Li X, Zhang C, Zhang S et al (2015) Preparation and characterization of positively charged polyamide composite nanofiltration hollow fiber membrane for lithium and magnesium separation[J]. Desalination 369:26–36

Li W, Shi C, Zhou A et al (2017) A positively charged composite nanofiltration membrane modified by EDTA for LiCl/MgCl2 separation[J]. Sep Purif Technol 186:233–242

Nie XY, Sun SY, Song X et al (2017) Further investigation into lithium recovery from Salt Lake brines with different feed characteristics by electrodialysis[J]. J Membr Sci 530:185–191

Paredes C, de San MER (2020) Selective lithium extraction and concentration from diluted alkaline aqueous media by a polymer inclusion membrane and application to seawater[J]. Desalination 487:114500

Tan R, Wang A, Malpass-Evans R et al (2020) Hydrophilic microporous membranes for selective ion separation and flow-battery energy storage[J]. Nat Mater 19(2):195–202

Wang C, Zheng M, Zhang X et al (2021) Geothermal-type lithium resources in Southern Xizang, China[J]. Acta Geologica Sinica-English Edition 95(3):860–872

Wen Q, Yan D, Liu F et al (2016) Highly selective ionic transport through subnanometer pores in polymer films[J]. Adv Func Mater 26(32):5796–5803

Wu Q, Liu XF, Zheng MP et al (2017) Present situation, existing problems and countermeasures of development of Salt Lake lithium resources in China[J]. Modern Chem Indus 37(5):1–5

Xu F, Dai L, Wu Y et al (2021b) Li+/Mg2+ separation by membrane separation: the role of the compensatory effect[J]. J Membr Sci 636:119542

Xu S, Song J, Bi Q et al (2021c) Extraction of lithium from Chinese salt-lake brines by membranes: design and practice[J]. J Membr Sci 635:119441

Xu S, Song J, Bi Q et al (2021a) Extraction of lithium from Chinese salt-lake brines by membranes: design and practice[J]. J Membrane Sci 119441

Yang S, Zhang F, Ding H et al (2018) Lithium metal extraction from seawater[J]. Joule 2(9):1648–1651

Zhao Z, Liu G, Jia H et al (2020) Sandwiched liquid-membrane electrodialysis: Lithium selective recovery from Salt Lake brines with high Mg/Li ratio[J]. J Membr Sci 596:117685

Chapter 57
Study on the Adsorption of Cr(VI) by Carbonated Fe3O4@UiO-66

Xue Li, SiTan Wang, and Wenlu Che

Abstract The adsorption process is consistent with Langmuir and the pseudo-first order kinetics, indicating that the adsorption is a monolayer adsorption process. The pH effect study shows that at $pH = 3$, the adsorption capacity of Cr(VI) is 356.8 mg/g as the highest, and electrostatic interaction is the main mechanism of action. The carbonized magnetic UiO-66 adsorbent showed good adsorption capacity as well as the advantage of the easy recovery, showing great potential for application in the field of wastewater chromium ion treatment.

Keywords Magnetic UiO-66 · Carbonization · Cr removal

57.1 Introduction

Chromium (Cr) is widely used in human industrial production such as electroplating, steel production, and wood preservation. Cr(VI) is one of the most common metal contaminants in the environment and has a high risk of environmental contamination (Altundogan 2005). It causes allergic contact dermatitis, liver damage, and other diseases in humans as well as the presence of carcinogenic risks (Dhal 2013; Song 2017). Therefore, it is necessary to remove Cr(VI) pollutants from the water environment. The methods for removal of Cr(VI) from water include electrochemical methods, biological treatment, ion exchange, etc., (Cheng 2012; Hao 2021; Yuan 2022) among which adsorption is one of the most cost-effective methods due to its simplicity of operation, lower cost, high removal rate, no toxic substances produced

X. Li · S. Wang (✉) · W. Che
Liaoning Technical University, Fuxin, Liaoning, China
e-mail: wangsitan@lntu.edu.cn

X. Li
e-mail: lx1831617818@163.com

W. Che
e-mail: chewenlu@163.com

J. Zhang et al. (eds.), *Environmental Pollution Governance and Ecological Remediation Technology*, Environmental Science and Engineering,
https://doi.org/10.1007/978-3-031-25284-6_57

and recyclability. Conventional adsorbents such as zeolites, activated carbons, and resins have small and irregular pore sizes leading to poor adsorption efficiency for Cr(VI) (Rowsell 2004). Therefore, metal–organic skeletons (MOFs) with large surface area, tunable and good pore structure stand out among many adsorbents (Li 2016; Wang 2015). However, most MOFs are limited in their use as adsorbents in the liquid phase due to their poor stability in water (Arabkhani 2019). Therefore, the calcination of MOFs under specific conditions to obtain porous carbon composites has received wide attention (Liu 2017; Wen 2017). The carbonized materials have good adsorption properties and chemical stability and show promising applications in adsorption because they combine the properties of MOFs and the properties of carbon materials such as dispersion and hydrophobic properties (Hao 2014; Zhang 2015; Zhou 2016).

In this research, UiO-66, a typical Zr-MOFs material, was chosen as a precursor to obtain mesoporous porous carbon by calcination (Zr-C), while the composite Fe_3O_4 imparted magnetic properties to the adsorbent to prepare a magnetic and efficient porous carbon adsorbent for easy separation and recovery (Hao 2021).

57.2 Eaperimental Instruments and Methods

57.2.1 Experimental Reagents and Apparatus

Main reagents: Ferrous chloride tetrahydrate ($FeCl_2$-$4H_2O$), Ferric chloride hexahydrate ($FeCl_3$-$6H_2O$), sodium hydroxide, hydrogen chloride, N–N-dimethylformamide (DMF), terephthalic acid (H_2BDC), zirconium chloride, anhydrous ethanol. The experimental water was pure.

Main instruments: Tube furnace, UV–Vis spectrophotometer, HJ-3 constant temperature magnetic stirrer, water bath thermostat, shaking shaker, oven, magnet.

57.2.2 Preparation of Zr-C

Firstly, Fe_3O_4 magnetic nanoparticles were prepared by co-precipitation method: 2.4 g $FeCl_2$-$4H_2O$ and 6.5 g $FeCl_3$-$6H_2O$ were dissolved in 50 ml of ultrapure water, and the pH was adjusted by slowly dropping 1.5 mol/L NaOH solution to make the solution pH = 11 ~ 12, and the mixture turned from orange-red to black rapidly. After the reaction, the reaction was separated by the magnet, washed repeatedly with distilled water to pH = 7.0, removed the supernatant, dried under vacuum at 60 °C for 24 h, and ground to obtain magnetic Fe_3O_4 nanoparticles. Using the layer self-assembly method, 1 g of Fe_3O_4 sample was dispersed in 80 ml of DMF solution of $ZrCl_4$ (10 mmol/L), stirred at 120 °C for 1 h, and then magnetically separated. The solid was then dispersed into 80 ml of DMF solution of terephthalic acid (10 mmol/L) and reacted at 120 °C for 1 h. Each of the above steps was magnetically separated and washed 3 times with DMF solution. After repeating the above operation 10 times, the solid was dried at 150 °C to prepare Fe_3O_4@UiO-66; 500 mg of Zr-60 was weighed and placed in a porcelain boat, flat in a tube furnace, under argon atmosphere, 120 min to 600 °C, constant temperature for 30 min and then naturally cooled to room temperature to obtain a black powder, which was recorded as Zr-C sample.

57.3 Adsorption Experiments of Cr(VI)

The chromium standard solutions of 1000 ug/ml were taken to configure 1 mg/L, 2 mg/L, 5 mg/L, 8 mg/L, 10 mg/L solutions to make the Cr(VI) standard curve. The determination of hexavalent chromium ion concentration was carried out by atomic absorption spectrophotometry. To obtain the adsorption isotherm, 25 ml of solution with Cr(VI) concentrations of 50 mg/L, 75 mg/L, 100 mg/L, 125 mg/L, 150 mg/L, 175 mg/L was added to each of the conical flasks, and 10 mg of adsorbent was weighed in each of the above six conical flasks, which were shaken for 12 h, and the upper clear layer was filtered and measured. The absorbance of the solution in each conical flask was measured. To study the effect of adsorption time on the adsorption performance, 10 mg of Zr-C was accurately weighed and placed in a 50 ml conical flask, 25 ml of 175 mg/L Cr(VI) solution was poured into the conical flask and then placed on a shaker for 12 h (25 °C). At 2 min, 5 min, 10 min, 30 min, 60 min, 120 min, 240 min, and 720 min, the solution was removed from the shaker, filtered and the supernatant was aspirated, and the absorbance was measured and recorded. Cr(VI) solution of 175 mg/L was taken for the pH effect experiment. The pH was adjusted to 3, 4, 5, 6, 7, 8, 9, and 10, with 0.1 mol/L sodium hydroxide solution and hydrochloric acid, respectively. They were shaken at room temperature for 12 h and then filtered to obtain the supernatant. The concentration of Cr(VI) in the filtrate was measured to obtain the adsorption characteristics at different pHs. The adsorption characteristics were obtained for different pH values.

57.4 Results and Discussion

57.4.1 Adsorbent Characterization

The scanning electron microscopy (SEM) image (Fig. 57.1) shows that the carbonized Zr-C loses its UiO66 cubic shape and shows an amorphous form, which is caused by the collapse of the MOF framework during calcination. Figure 57.2 shows the nitrogen-adsorption desorption isotherm and pore size distribution of Zr-C. The specific surface area of Zr-C is 57.3 m^2/g, the average pore size is 18.78 nm, the

Fig. 57.1 Scanning electron microscope image of Zr-C

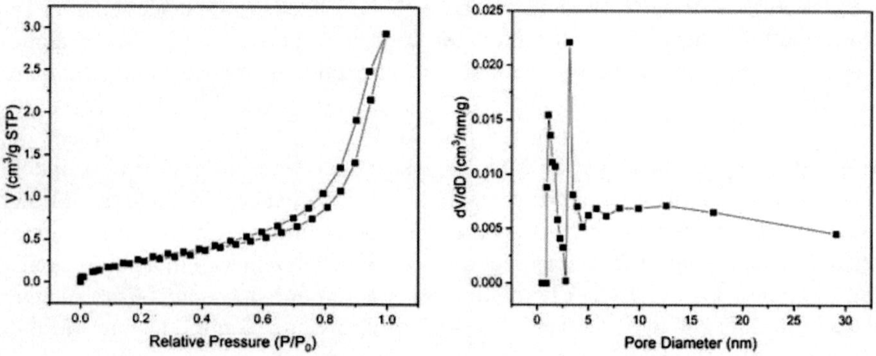

Fig. 57.2 Nitrogen-adsorption desorption isotherm and pore size distribution of Zr-C

total pore volume is 0.27 cm^3/g, and the mesoporous pore volume is 0.203 cm^3/g, which shows that the surface of Zr-C is dominated by mesopores.

57.4.2 Adsorption Isotherm Study

To investigate the adsorption separation capability of the adsorbent Zr-C the adsorption thermodynamic model was used to study the partitioning of the adsorbed mass between the liquid and solid phases during the adsorption process. The adsorption thermodynamic experiments were carried out according to the adsorption experimental method introduced above, and the adsorption isotherms are shown in Fig. 57.3. The obtained experimental data were fitted using Langmuir and Freundlich adsorption models, respectively, and simulated using Langmuir and Freundlich models, and the two isotherms are as follows.

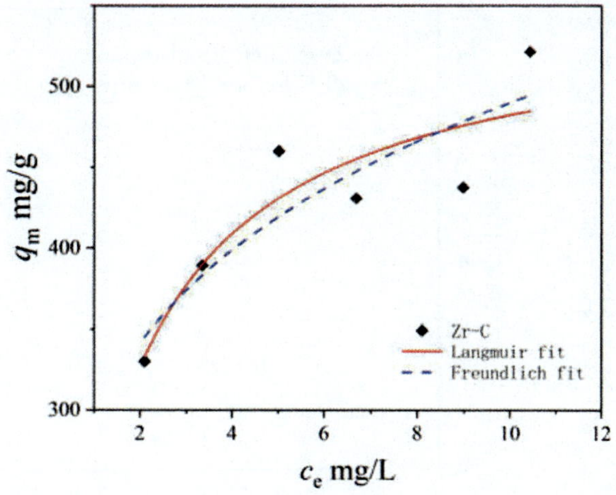

Fig. 57.3 Isotherm of Zr-C adsorption on Cr(VI)

$$\frac{c_e}{q_e} = \frac{c_e}{q_{\max}} + \frac{1}{k_L q_{\max}}$$

$$\ln q_e = \ln K_F + \ln \frac{c_e}{n}$$

The above equation is the adsorption capacity of the adsorbent at adsorption equilibrium, (mg/g); *Ce* is the equilibrium adsorption concentration of Cr(VI) solution, (mg/g); k_L is the Langmuir adsorption constant (1/min); K_F is the Freundlich adsorption constant, (1/min); q_{max} is the maximum adsorption capacity of the adsorbent, (mg/g); 1/n is the deviation of Freundlich adsorption isotherm.

As shown in Table 57.1, the normalized coefficient R^2 of the Langmuir model is higher than that of the Freundlich model after fitting with the Langmuir and Freundlich isotherms, respectively. The Langmuir model can better describe the adsorption process of the Zr-C adsorbent for Cr(VI), indicating that the adsorption process is monolayer adsorption with homogeneous active adsorption sites on the surface of the adsorbent.

The IR spectra of Zr-C before and after adsorption of Cr(VI) are shown in Fig. 57.4, from which it can be seen that the absorption peak at 1535.58 1/cm is relatively weaker after adsorption, and the peak here is the amide II band, which is caused by the bending vibration of N–H and the stretching vibration of C-N (Lu 1989). The absorption peak at 1727.53 1/cm is relatively weaker after adsorption, and this peak is the C = O stretching vibration peak of the ketone. infrared spectra before and after

Table 57.1 Langmuir and Freundlich model parameters of adsorption of Cr(VI)

Samples	Langmuir model		Freundlich model		
	q_m (mg/g)	R^2	K_F ($mg^{1-n} \cdot L^n/g$)	n	R^2
Zr-C	547.3	0.798	292.2	0.22	0.784

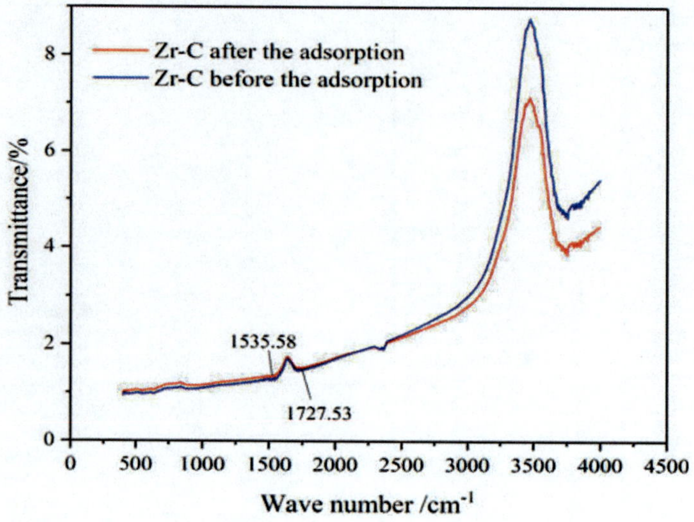

Fig. 57.4 Infrared spectra of Zr-C before and after adsorption of Cr(VI)

the adsorption of Cr(VI) by Zr-C indicate that the amide II band, C = O double bond, etc. may have important contributions in the Cr(VI) adsorption process.

57.4.3 Adsorption Kinetics Study

In this section, the kinetic experiments of Cr(VI) adsorption on Zr-C was carried out, and the adsorption capacity of Cr(VI) on Zr-C versus time is shown in Fig. 57.5. The data of the kinetic experiments were fitted with pseudo-first order kinetic and pseudo-second order kinetic modes and the kinetic equations of the two models are shown below.

pseudo-first order kinetic mode

$$\ln(q_e - q_t) = \ln q_e - k_1 t$$

pseudo-second order kinetic mode

$$\frac{t}{q_t} = \frac{1}{k_2 q_e^2} + \frac{t}{q_e}$$

q_e is the adsorption capacity of the adsorbent at adsorption equilibrium, (mg/g); q_t is the adsorption capacity of the adsorbent at moment, (mg/g); k_1 is the pseudo-first order kinetic rate constant (1/cm); k_2 is pseudo-second order kinetic rate constant, (1/cm).

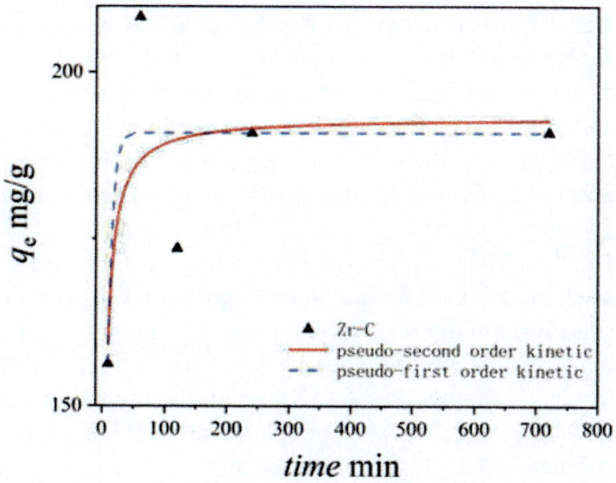

Fig. 57.5 Adsorption capacity of Cr(VI) on Zr-C versus time

Table 57.2 pseudo first-order kinetic and pseudo second-order kinetic model parameters of adsorption kinetics of Cr(VI) on Zr-C

Samples	Pseudo-first-order			Pseudo-second-order		
	q_e (mg/g)	k_1 (1/min)	R_1^2	q_e (mg/g)	k_2(1/min)	R_2^2
Zr-C	190.9	0.15	0.61	193.3	90.2	0.52

As shown in Fig. 57.5, the adsorption rate increased rapidly at the beginning of adsorption, the adsorption rate slowed down at 25 min and the adsorption equilibrium is reached within 240 min, at the same time, the adsorbent reached saturation. The rapid adsorption occurred within the initial 1 h. The kinetic study of the adsorption process is mainly used to describe the rate of solute adsorption by the adsorbent, and the data are now fitted using pseudo-first order kinetic and pseudo-second order kinetic modes, as can be seen in Table 57.2. For Zr-C adsorbent, the correlation coefficient R_1^2 of pseudo-first order kinetic mode is higher. so pseudo-first order kinetic mode can better describe the adsorption process of Cr(VI) on Zr-C, indicating that the adsorption is monolayer adsorption accomplished by boundary diffusion (Wang 2018).

57.4.4 Effect of pH

For Cr(VI)-solid phase interactions under aqueous solution conditions, the pH of a given system is a key parameter affecting not only the conversion of Cr(VI) species on the adsorbent surface but also the charge density on the adsorbent surface.

Hexavalent chromium generally exists in water as H_2CrO_4, $HCrO^{4-}$, CrO_4^{2-}, $Cr_2O_7^{2-}$, etc. When the pH is less than 1, the solution is dominated by H_2CrO_4, when the pH is between 1 and 6, $HCrO^{4-}$ occupies the main position, and when the pH is greater than 6, the solution generally contains CrO_4^{2-}, $Cr_2O_7^{2-}$ (Wu 2017). Due to the special form of Cr(VI) in the environment, there is a unique adsorption mechanism, which can be summarized as follows: surface reduction, Cr(VI) is converted to Cr(III) by reduction on the adsorbent surface, and then Cr(III) is adsorbed by the adsorbent; Cr(VI) is adsorbed through a combination of electrostatic gravity, ion exchange, and surface complexation; (Plummer 2018; Pan 2016; Li 2019) Cr(VI) and the Cr(III) produced by reduction are deposited on the activated carbon surface as chemical precipitation under appropriate pH conditions, and $Cr(OH)_3$ is the only solid form in the form of precipitation under alkaline conditions (Yu 2017). The pH range of this experiment was 3–10. The effect of pH on the adsorption of Cr(VI) by Zr-C is shown in Fig. 57.6. With the increase in pH, the adsorption capacity of Cr(VI) on the Zr-C showed an overall trend of decreasing and then increasing, but still reached the highest point of adsorption capacity at pH = 3. Under the condition that the initial concentration of Cr(VI) was 175 mg/L, the adsorption capacity of Zr-C reached 356.8 mg/g at this time. Combined with the zeta potential of Zr-C for further analysis, when Zr-C was used for the adsorption of Cr(VI) in solution, the possible adsorption mechanism under initially acidic conditions included a decreasing trend of the adsorbent for Cr(VI) as the pH increased when the pH was between 2 and 6. The reason is that the adsorbent surface is positively charged and the number of positive charges on the adsorbent surface increases with increasing pH (Zi 2013). The main form of Cr(VI) present is $HCrO^{4-}$, and electrostatic attraction occurred between Zr-C and $HCrO^{4-}$, when electrostatic gravity is the main mechanism of action. Part of Cr(VI) is reduced to Cr(III) on the adsorbent surface and the electrostatic repulsion between positively charged Zr-C and Cr(III) weakens the adsorption effect. The main adsorption mechanism is through the ion exchange between the oxygenated acidic functional groups on the Zr-C surface and Cr(III). With the gradual increase of pH > 5.87, the adsorbent surface is negatively charged, meanwhile the main form of Cr(VI) exists as CrO_4^{2-}, so the electrostatic repulsion occurred and adsorption is weakened. However, the adsorption effect of the reduced Cr(III) is enhanced due to electrostatic attraction. At the same time, chemical precipitation in the form of $Cr(OH)_3$ may occur under alkaline conditions. Under the combined effect of the above mechanisms, although the adsorption capacity of Zr-C under alkaline conditions is enhanced, it is still lower than the adsorption capacity at pH = 3, thus proving the existence of electrostatic repulsion between the adsorbent and the adsorbent mass (Fan 2017).

57.5 Conclusion

In this research, metal–organic framework material UiO-66 was selected as the base material and modified by controlled synthesis conditions and post-treatment methods in combination with magnetic separation technology, the core–shell structured Fe_3O_4@UiO-66-based porous carbon adsorbent was successfully synthesized.

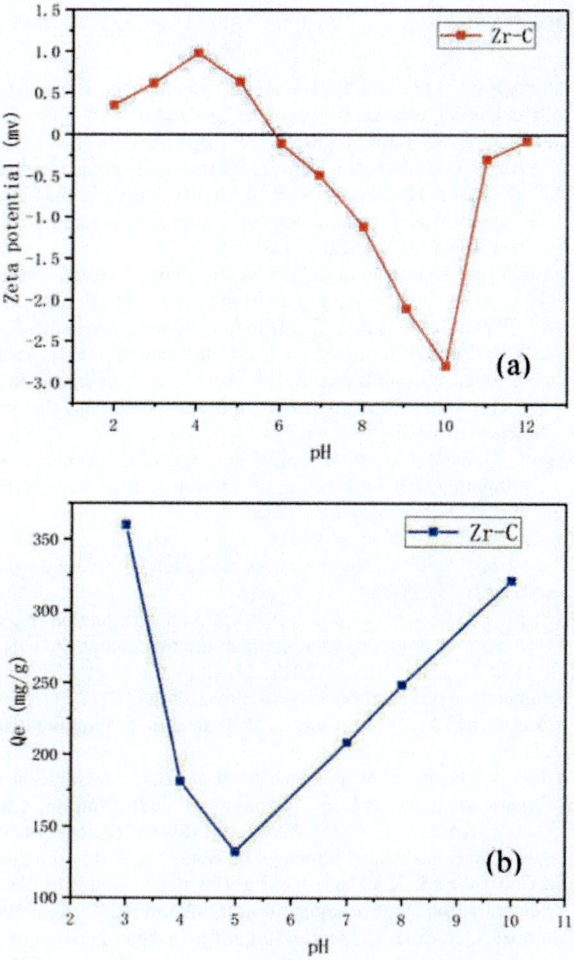

Fig. 57.6 Zata potential of Zr-C (**a**) and Effect of pH (**b**). (Ion Concentration of Cr(VI) = 175 mg/L; T = 298 k; adsorption time:20 h)

A series of experiments demonstrated that the high-temperature carbonized core–shell structure Fe_3O_4@UiO-66 can effectively remove Cr(VI) from water. The adsorption capacity can reach 494.6 mg/g at the initial concentration of Cr(VI) of 175 mg/L. The adsorption process of Zr-C is by the quasi-level kinetic model and Langmuir isothermal model, indicating homogeneous monolayer adsorption. the main adsorption mechanism of Zr-C on Cr(VI) is electrostatic interaction. This study shows that Fe_3O_4@UiO-66-based porous carbon adsorbent has obvious advantages and application prospects for the adsorption of Cr(VI) in water.

Acknowledgements This work was supported by Science and Technology Research Project of Liaoning Provincial Education Department (LJ2019QL009).

References

Altundogan HS (2005) Cr(VI) removal from aqueous solution by iron (III) hydroxide-loaded sugar beet pulp[J]. Process Biochem 40(3/4):1443–1452

Arabkhani P, Asfaram A (2019) Development of a novel three-dimensional magnetic polymer aerogel as an efficient adsorbent for malachite green removal[J]. J Hazard Mater 384:121394

Dhal B, Thatoi HN, Das NN et al (2013) Chemical and microbial remediation of hexavalent chromium from contaminated soil and mining/metallurgical solid waste: a review[J]. J Hazard Mater 250–251(30):272–291

Fan S, Yi W, Yang L et al (2017) Facile synthesis of tea waste/Fe3O4nanoparticle composite for hexavalent chromium removal from aqueous solution[J]. RSC Adv 7(13):7576–7590

Hao L, Wang C, Wu Q et al (2014) Metal–organic framework derived magnetic nanoporous carbon: novel adsorbent for magnetic solid-phase extraction[J]. Anal Chem 86(24)

Hao W, Shuai W, Swa B et al (2021) Adenosine-functionalized UiO-66-NH2 to efficiently remove Pb(II) and Cr(VI) from aqueous solution: thermodynamics, kinetics and isothermal adsorption—ScienceDirect[J]

Hao J, Guo C, Gao X, Li M, Long H (2022) Synergistic mechanism of Cr(VI) removal by thiourea/sodium alginate adsorption and photocatalytic reduction[J/OL]. J Composites 1–12[2022–03–06].20210622.001

Li B, Zhu X, Hu K et al (2016) Defect creation in metal-organic frameworks for rapid and controllable decontamination of roxarsone from aqueous solution[J]. J Hazardous Mater 302(JAN.25):57–64

Li P, Fu T, Gao X et al (2019) Adsorption and reduction transformation behaviors of Cr(VI) on mesoporous polydopamine/titanium dioxide composite nanospheres[J]. J Chem Eng Data

Liu Y, Li G, Guo Y et al (2017) Flexible and binder-free hierarchical porous carbon film for supercapacitor electrodes derived from MOFs/CNT[J]. ACS Appl Mater Interfaces

Lu YQ, Deng ZH (1989) Analysis of IR spectra. Beijing, Publishing House of Electronics Industry 21

Maleki A, Hayati B, Naghizadeh M et al (2015) Adsorption of hexavalent chromium by metal organic frameworks from aqueous solution[J]. J Ind Eng Chem 28:211–216

Owlad M, Aroua MK, Daud WAW et al (2009) Removal of hexavalent chromium-contaminated water and wastewater: a review[J]. Water Air Soil Pollut 200(1–4):59–77

Pan C, Troyer LD, Catalano JG et al (2016) Dynamics of chromium(VI) removal from drinking water by iron electrocoagulation[J]. Environ Sci Technol 50(24):13502

Plummer S, Gorman C, Henrie T et al (2018) Optimization of strong-base anion exchange O&M costs for hexavalent chromium treatment[J]. Water Res 139(AUG.1):420–433

Rowsell J, Yaghi OM (2004) Metal–organic frameworks: a new class of porous materials[J]. Microporous Mesoporous Mater 73(1–2):3–14

Shiyang C, Zhou S, Xuerui L, Lin D, Yong S (2012) Study on the removal of Cr(VI) from drinking water by ion exchange membrane chemical reactor[J]. J Safety Environ 12(04):72–75

Song JY, Jhung SH (2017) Adsorption of pharmaceuticals and personal care products over metal-organic frameworks functionalized with hydroxyl groups: quantitative analyses of H-bonding in adsorption. Chem Eng J 322:366–374

Wang TT, Cui QL, Wang LL et al (2018) Adsorption characteristics of Al-modified lemon biochar on P and its mechanism[J]. China Environ Sci 38(6):13

Wang, Yuan XZ et al (2015) Synthesis and applications of novel graphitic carbon nitride/metal-organic frameworks mesoporous photocatalyst for dyes removal[J]. APPL CATAL B-ENVIRON 174:445–454

Wen C, Xiao M et al (2017) MOFs-derived carbon covered alumina (CCA) supported Pt nanoparticles as catalyst for enantioselective hydrogenation[J]. Curr Appl Phys: Official J Korean Phys Soc 17(10):1347–1352

Wu S, Ge Y, Wang Y et al (2017) Adsorption of Cr(VI) on nano Uio-66-NH2 MOFs in water[J]. Environ Technol 39(15):1–40

Ye Y, Zhao et al (2017) Efficient removal of Cr(III)-organic complexes from water using UV/Fe(III) system: negligible Cr(VI) accumulation and mechanism.[J]. Water Res

Yuan F, Zhang Y, Xu H, Gong G (2022) Exploration and prospect of the mechanism of Cr(VI) removal from the environment by humic acid[J/OL]. Environ Chem 1–10[2022–03–06].20220214.1402.006

Yu Z, Xiaodan et al (2013) Magnetic chitosan–iron(III) hydrogel as a fast and reusable adsorbent for chromium(VI) removal[J]. Indus Eng Chem Res

Zhang J, Zhao Z, Xia Z et al (2015) A metal-free bifunctional electrocatalyst for oxygen reduction and oxygen evolution reactions[J]. Nat Nanotechnol 10(5):444–452

Zhou W, Lu J, Zhou K et al (2016) CoSe2 nanoparticles embedded defective carbon nanotubes derived from MOFs as efficient electrocatalyst for hydrogen evolution reaction[J]. Nano Energy 28:143–150

Chapter 58
Research Progress of Natural Polymer Coagulants in Water Treatment

Yuanyuan Zhang and Ying Fu

Abstract Water turbidity can be decreased through coagulation, effectively removing suspended particles and some dissolved solids from water. The coagulant employed has a significant impact on the outcome of the coagulation process. Inorganic and organic coagulants are the two primary divisions of commonly used coagulants. Due to their practical application and favour able economics, these flocculants are frequently used. However, these flocculants also have drawbacks, including organic synthetic monomers that can present health and safety issues and residual metal ions in the water. Natural polymer coagulants are economical, safe, non-toxic, practical and environmentally friendly. Developing green coagulants based on natural polymers is crucial since they have received widespread acclaim and have a wide range of potential applications.

Keywords Natural coagulants · Wastewater treatment · Coagulant

58.1 Introduction

The use of natural polymer coagulants to treat water and wastewater has been documented since ancient times and remains popular today (Vijayaraghavan et al. 2011). Natural polymers have been utilized for a long time as efficient flocculants and coagulants to cure excessive water turbidity (Anastasakis et al. 2009; Yin 2010). Natural polymer flocculants are derived from various sources (Hussain and Hydar 2019), primarily from plants, seeds, marine crustacean and shellfish (shrimp and crab) biomass, and microorganisms. Most research has concentrated on natural flocculants from plant and marine biomass, such as chitosan, cellulose, starch, and alginates. Natural coagulants are new, renewable, ecologically friendly, affordable, non-toxic, and easily degradable coagulants with good application potential (Choy et al. 2014). Natural POlymer flocculants are being researched at an unprecedented level under

Y. Zhang · Y. Fu (✉)
School of Civil Engineering and Architecture, University of Jinan, Jinan 250022, China
e-mail: cea_fuy@ujn.edu.cn

J. Zhang et al. (eds.), *Environmental Pollution Governance and Ecological Remediation Technology*, Environmental Science and Engineering,
https://doi.org/10.1007/978-3-031-25284-6_58

the environmental protection principle of "turning waste into treasure," which creates new opportunities for the recycling of natural waste but also addresses public demand for environmental protection.

Chemically powerful reactive groups, such as carboxyl and hydroxyl, are present in natural polymers. These groups form the nucleus of coagulation-inducing natural polymer coagulants. Currently, research on chemically altering the active groups of natural polymers to create coagulants is quite adequate, and the research findings have excellent outcomes for actual water treatment applications (Lee et al. 2014). However, research on producing coagulants from unaltered, pure natural raw materials is somewhat lacking. Most studies are still in the laboratory and have not been translated into actual technological applications (Ang and Mohammad 2020).

58.2 Application of Natural Polymer Coagulants in Water Treatment

Using natural polymer coagulants in treating different wastewaters is currently gaining much attention. Agricultural, food processing, aquaculture, beverage manufacturing, paint and varnish, adhesives, cement, soap and detergents, cosmetics, plastics, steel, leather, textile and dyeing, pulp and paper industries have all made extensive use of natural coagulants. In recent years, natural polymer coagulants have been studied to treat various wastewaters, as shown in Table 58.1.

Table 58.1 Application of natural coagulants in wastewater

Natural coagulant	Type of wastewater	Concentration	Removal performance	References
G. ulmifolia	Dairy wastewater	775 mg/L pH:5	Turbidity:95.8% BOD:81.2%	Muniz et al. (2020)
Guar gum	Landfill leachate	44.39 mL/L pH:8.56	COD:22.57%	Cheng et al. (2020)
Moringa seeds	Pharmaceutical wastewater	4 g/L	Turbidity:64% COD:38%	Omna et al. (2010)
Chitosan	Aquaculture wastewater	30 mg/L	Turbidity:96.3% Algae:98%	Fareza, et al. (2017)
Banana skin	Municipal wastewater	0.4 g/L	Turbidity:59.6%	Maurya and Daverey (2018)
Broad bean	Algal wastewater	0.5 g/L pH:5	Cyanobacterial cell removal rate: > 85%	Bouaidi et al. (2020)

58.2.1 Dairy Wastewater

The worldwide dairy business has grown due to the rising demand for milk and dairy products, making it one of the agricultural sector's most economically significant leaders (Ekka et al. 2022). Cheng et al. (2020) employed a brand-new all-natural coagulant from G. ulmifolia's stem bark to filter contaminants from dairy wastewater. It was discovered that the coagulant of G. ulmifolia removed 95.8, 76.0, 81.2, and 85.6% of turbidity, COD, BOD, and UV 254 compounds from dairy wastewater, respectively, at the optimum dose rate of 775.8 mg/L and pH 5.00. Additionally, the sludge created during the treatment process is safe and can be used on the soil to improve it and serve as fertilizer.

58.2.2 Landfill Leachate

Waste leachate post-membrane concentrate is a highly concentrated organic waste with a complex composition that is difficult to biodegrade due to its high chemical oxygen demand, high ammonia nitrogen, and high salt content produced during the treatment of waste leachate by membrane technology (Zouboulis et al. 2004). Landfill leachate can contaminate groundwater and streams by leaching into the soil and subsoil layers. Because of the complexity of waste leachate and the high variability of its content, it is critical to treat waste leachate because it can pose serious environmental and health risks if not adequately treated (Gupta and Singh 2007). Wang's (2020) treatment of waste leachate with guar gum revealed that at a guar gum dosage of 44.39 mg/L, pH 8.56 (the natural pH of leachate), and a mixing rate of 79.27 rpm, the COD removal rate was maximized at 22.57%. Guar gum coagulant could be a future development path for waste leachate treatment because structural characterization using Fourier transform infrared analysis revealed that hydrogen bonds between guar gum and pollutant particles were involved in the coagulation and flocculation.

58.2.3 Pharmaceutical Wastewater

Pharmaceutical manufacturing generates large volumes of highly concentrated organic waste with a complex composition, and its release into the environment endangers human and environmental systems (Alazaiza et al. 2022). The effectiveness of a natural coagulant made from Moringa seeds in treating hospital wastewater was examined by Nonfodji et al. (2010). According to the findings, the coagulant was 64 and 38% effective at removing turbidity and COD, respectively. In a subsequent study, Thirugnanasambandham and Karri (Riaz et al. 2018) compared the coagulant between a natural coagulant derived from Moringa seeds and a typical chemical coagulant, aluminum sulfide. According to the findings, natural-based coagulants may not only be efficient for removing COD, turbidity, and color but may also be economically viable. For hospital wastewater, moringa seed polymers are promising bio-coagulants.

58.2.4 Aquaculture Wastewater

Aquaculture is expanding quickly, and at the same time, more and more farm effluent is being discharged. Aquatic excrement, leftover bait, and other solid particle waste, as well as other contaminants such as pesticides, anti-microbial agents, and antibiotics used in aquaculture, are the main constituents of farm wastewater. In addition to seriously degrading the environment and disrupting the ecology, these pollutants contaminate aquatic products, spread viruses and epidemics, cause fish and shrimp to grow slowly or even die, and eventually destroy the farming business (Riaz et al. 2018). Fareza et al. (2017) investigated the use of chitosan papaya as a natural coagulant in the treatment of aquaculture wastewater. They found that at an optimum coagulant dose of 30 mg/L, the algae removal efficiency was 98%, the turbidity of the treated wastewater was reduced from 60 NTU to 2.2 NTU, and the further discharged supernatant water could be used as recycled water or removed into the water body.

58.2.5 Municipal Wastewater

Banana peel powder, banana stem juice, papaya seed powder, and neem leaf powder were the four natural plant coagulants that Maurya (2018) investigated for their effects on the treatment of municipal wastewater. The results revealed that the best turbidity removal rate was achieved at 0.4 g/L dosages of banana peel powder coagulant; the best COD removal rate was 66.7% for banana stem juice coagulant; and the best COD removal rate for papaya seed powder mix. The coagulant achieved the best removal of total suspended particles, which had a removal rate of 66.66%. This result shows that low-cost natural plant coagulants have a lot of potential for treating municipal wastewater in this study.

58.2.6 Algae Wastewater

With the expansion of agriculture and industry, a significant amount of organic or inorganic nitrogen and phosphorus enter lakes and reservoirs, resulting in eutrophication of water bodies and algal blooms (Teixeira et al. 2022). Of the complicated species, vast number, low specific gravity, and significant negative charge of the water's algae cells. However, the prevalent treatment methods used today rely on pricey membrane filtering or chemicals. According to ElBouaidi et al. (2020), broad bean seed and cactus coagulants were employed to treat algal wastewater, and both could eliminate 85% of Pseudomonas aeruginosa at 0.5 g/L and 1 g/L dosage and pH 5, respectively.

58.3 Limitations and Potential Study in the Use of Natural Coagulants

There haven't been detailed studies on the specific coagulation mechanisms each plant-based procoagulant uses (Kannankai and Devipriya 2022). One or more coagulation mechanisms may bring colloidal particle accumulation during water clarification. Because of the complexity of the process and the potential for synergistic effects between the components present, it is challenging to pinpoint the precise chemical composition that causes particle aggregation (Oladoja 2015). Additional research is required to characterize the active substances that would be useful in identifying the essential substances that are important in the coagulation-flocculation process.

Due to barriers to commercialization, natural polymer coagulants have not been commonly utilized in studies on industrial-scale wastewater treatment. Funding, R&D, market awareness, and regulatory approval requirements hinder commercialization. Market reaction and demand will also impact the outlook and demand for natural coagulants as a potential replacement for chemical coagulants. Target market estimates may be incorrect due to consumer and supplier misconceptions about the viability of natural coagulants (Hewawasam 2021). With the aid of pilot plant studies, it is possible to evaluate elements that may act as sustainability indicators in the economic, social, and environmental spheres. Any new product's successful commercialization will also need the support of local regulatory bodies (Btia et al. 2021).

58.4 Conclusion and Further Works

The main goal is to increase the coagulation effect of coagulants and treatment efficiency, so water treatment coagulants need to be more efficient and environmentally friendly. The creation of a non-toxic, non-hazardous water treatment coagulant should be done as part of the process of developing the future application of natural coagulants. Based on the concept of green and sustainable development, natural polymer coagulants will likely develop in one of two main ways: on the one hand, research and development of coagulants will gradually favor secure, non-toxic, non-polluting natural types and composite types (Jiang et al. 2021); on the other hand, this will be combined with in-depth research and analysis of the coagulation mechanism to prepare a new kind of effective coagulant. In comparison, more research on natural polymer coagulants has been conducted abroad, while the development in China has started a little later and is slower. In the future, the development of natural polymer coagulants will develop in the following directions.

(1) Enhancing and stabilizing the coagulation performance of current coagulants by combining the characteristics of natural polymer coagulants to expand the scope of application;
(2) Inoculating bacterial strains and cultivating them into microbial flocculants that can treat specific wastewaters;
(3) Continuing research and development of cationic or amphoteric coagulants.

References

Alazaiza MYD, Albahnasawi A, Ali GAM et al (2022) Application of natural coagulants for pharmaceutical removal from water and wastewater: a review. Water 14(2)

Anastasakis K, Kalderis D, Diamadopoulos E (2009) Flocculation behavior of mallow and okra mucilage in treating wastewater. Desalination 249(2):786–791

Ang WL, Mohammad AW (2020) State of the art and sustainability of natural coagulants in water and wastewater treatment. J Clean Prod 262:1–18

Bouaidi WE, Essalhi S, Douma M et al (2020) Evaluation of the potentiality of Vicia faba and Opuntia ficus indica as eco-friendly coagulants to mitigate Microcystis aeruginosa blooms 195:198–213

Btia B, Vtoa B, Sar C (2021) Eco-friendly approaches to aquaculture wastewater treatment: assessment of natural coagulants vis-a-vis chitosan. Bioresource Technol Rep 15:1–9

Cheng SY, Show PL, Juan JC et al (2020) Sustainable landfill leachate treatment: Optimize use of guar gum as natural coagulant and floc characterization. Environ Res 188:109737–109737

Choy SY, Prasad KMN, Wu TY et al (2014) Utilization of plant-based natural coagulants as future alternatives towards sustainable water clarification. J Environ Sci 26(11):2178–2189

Ekka B, Mieriņa I, Juhna T et al (2022) Synergistic effect of activated charcoal and chitosan on treatment of dairy wastewaters. Mater Today Commun 31:103477

Mohd FH et al (2017) Harvesting of microalgae (Chlorella sp.) from aquaculture bioflocs using an environmental-friendly chitosan-based bio-coagulant. Int Biodeterioration Biodegradation 124:243–249

Gupta SK, Singh G (2007) Assessment of the efficiency and economic viability of various methods of treatment of sanitary landfill leachate. Environ Monit Assess 135(1–3):107–117

Hewawasam C (2021) Effectiveness of natural coagulants in water and wastewater treatment. Global J Environ Sci Manage 1–16

Hussain G, Hydar S (2019) Exploring potential of pearl millet (Pennisetum glaucum) and black-eyed pea (Vigna unguiculata subsp. unguiculata) as bio-coagulants for water treatment. Desal Water Treat 143:184–191

Jiang XC, Li YS, Tang XH (2021) Biopolymer-based flocculants: a review of recent technologies. Environ Sci Pollut Res 28(34):46934–46963

Kannankai MP, Devipriya SP (2022) An introduction to cost-effective technologies for solid waste and wastewater treatment 1–8

Lee CS, Robinson J, Chong MF (2014) A review on application of flocculants in wastewater treatment. Process Saf Environ Prot 92(6):489–508

Maurya S, Daverey A (2018) Evaluation of plant-based natural coagulants for municipal wastewater treatment 8(1):1–4

Muniz GL, Silva T, Borges AC (2020) Assessment and optimization of the use of a novel natural coagulant (Guazuma ulmifolia) for dairy wastewater treatment. Sci Total Environ 744:1–11

Oladoja NA (2015) Headway on natural polymeric coagulants in water and wastewater treatment operations. J Water Process Eng 6:174–192

Omna B, Jkfa C, Taa B et al (2010) Performance of Moringa oleifera seeds protein and Moringa oleifera seeds protein-polyaluminum chloride composite coagulant in removing organic matter and antibiotic resistant bacteria from hospital wastewater—ScienceDirect. J Water Process Eng 3:1–11

Riaz M, Ijaz B, Riaz A, Amjad M (2018) Improvement of waste water quality by application of mixed algal inocula. Bangladesh J Sci Indus Res 53:77–82

Teixeira M, Speranza L, Silva ID (2022) Tannin-based coagulant for harvesting microalgae cultivated in wastewater: efficiency, floc morphology and products characterization. Sci Total Environ 807:1–10

Vijayaraghavan G, Sivakumar T, Kumar AV (2011) Application of plant-based coagulants for waste water treatment. Int J Adv Eng Res Stud 1(1):88–92

Yin CY (2010) Emerging usage of plant-based coagulants for water and wastewater treatment. Process Biochem 45(9):1437–1444

Zouboulis AI, Xiao-Li C, Katsoyiannis IA (2004) The application of bioflocculant for the removal of humic acids from stabilized landfill leachates. J Environ Manage 70(1):35–41

Chapter 59
Study on the Treatment of Domestic Sewage of Home Stay in Moganshan Town

Tao Zhang, Peng Chen, and Xiao Chai

Abstract In recent years, with the rapid development of the home stay industry in China, domestic sewage which can only be discharged after efficient treatment are produced more and more by a large number of tourists. This paper takes Moganshan Town, Deqing, where the home stay industry is relatively developed, as an example, to carry out field research, describe the general situation of the "zero direct sewage discharge" domestic sewage treatment project in Moganshan Town, and select some typical home stays to understand their domestic sewage treatment measures. The research results show that Moganshan Town has basically realized the decentralized treatment of domestic sewage. The home stay with more than seven rooms and the home stay in the secondary water source protection area have all been installed with sewage treatment facilities as required, effectively reducing the environmental pollution caused by the indiscriminate discharge of sewage.

Keywords Home stay · Domestic sewage · Treatment process · Small unit

59.1 Introduction

With the vigorous development of personalized tourism, home stay experience has attracted more and more attention. The concept of home stay originated from abroad. Residents provide redundant rooms for people who need short-term accommodation. In China, B&B has developed from roadside inns and farmhouses in scenic spots in the early days. In recent years, it has gradually grown and formed a certain scale

T. Zhang (✉) · T. Zhang (✉) · P. Chen · X. Chai
College of Civil Engineering and Architecture, Zhejiang University of Water Resource and Electric Power, Hangzhou 310081, China

P. Chen
e-mail: chenp@zjweu.edu.cn

X. Chai
e-mail: chenp@zjweu.edu.cn

J. Zhang et al. (eds.), *Environmental Pollution Governance and Ecological Remediation Technology*, Environmental Science and Engineering,
https://doi.org/10.1007/978-3-031-25284-6_59

(Zhang Haizhou and Lin 2019). Compared with traditional standard hotels, B&B prefer immersive experience with local characteristics and customs, so most of them are located in villages with high air quality and good landscape ecology (Jie 2020).

One of the attractions of rural home stay is to appreciate the natural scenery and local environment of different villages. Therefore, from design, construction to operation, the home stay must integrate the local natural environment and terrain conditions to excavate and inherit the regional culture (Liamlaem et al. 2019). The rapid development of the home stay market has not only greatly boosted the development of the countryside, but also has a certain impact on the local ecological environment, such as the discharge of domestic waste and domestic sewage (Feng et al. 2021). Due to the relatively concentrated flow of people in the home stay, it is inevitable that more kitchen wastewater, domestic sewage and other sewage will be generated during its operation. Therefore, how to scientifically and efficiently treat these sewage has attracted much attention. Taking Moganshan Town, Deqing City, where the home stay industry is relatively developed, as an example, this paper expounds the local domestic sewage treatment projects and the domestic sewage treatment methods of typical home stay, and puts forward suggestions for the future development trend.

59.2 Domestic Sewage Management Objectives of Moganshan Home Stay

Moganshan Town, Huzhou, Zhejiang Province, which has a profound history and culture, is close to Hangzhou, Nanjing, Shanghai and other economically developed areas. Also it has unique ecological environment, cultural literacy and regional advantages. This area is the origin of "foreign home entertainment". In 2011, Naked Heart Valley Resort was opened and achieved great success. Subsequently, the "foreign home entertainment" model was rapidly copied, and the home stay industry has become the leading industry of the town.

The development of the home stay industry has not only brought a large number of tourists, but also affected the ecological environment of Moganshan Town. The sewage and noise generated during the operation of the home stay have troubled the surrounding residents, and the local villagers have complained a lot. The discharge characteristics of rural domestic sewage are small water volume, large change, relatively dispersed distribution and low pollutant concentration. Therefore, Moganshan Town has actively built a rural domestic sewage treatment project, and since June 2016, it has started the centralized treatment of home stay sewage.

59.3 Moganshan Town "Zero Direct Sewage Discharge" Rural Domestic Sewage Treatment Project

59.3.1 Project Overview

The sewage treatment project in Deqing County includes the treatment of domestic sewage and industrial wastewater. The sewage treatment project in Moganshan Town focuses on the treatment of domestic sewage in administrative villages and home stays.

Administrative villages are divided into 10–15 natural groups with one processing terminal for each group. Sewage interceptor and receiver, and the terminal shall be treated uniformly. The sewage treatment system project in Moganshan Town starts from Houwu Village, and the pipe network goes through Miaoqian Village and Fatou Village to Lanshukeng Village along Provincial Highway 304, then turns to the Lan Line, passes through He Village to Laoling Village, and finally turns to Gaonan Line to connect with the existing sewage main pipe network of Sanmo Line. The tail water of the villages along the line is collected through the pipe network and finally discharged to Shishan for sewage treatment. There are more than 100 sewage treatment terminals in Moganshan Town.

Home stay is more than that of local farmers and has a large passenger flow. Each household uses micro power treatment equipment. All home stay with a scale of more than seven rooms and in the secondary water source protection zone are required to install sewage treatment facilities, and all home stay with catering is also required to install oil separator. Effluent shall meet Class I A discharge standard as far as possible, and artificial wetlands shall be added when micro power treatment equipment cannot meet the requirements.

59.3.2 Treatment Scale

There are about 18 villages in Moganshan Town, covering an area of 185.77 km^2, with more than 8000 households and 31,000 registered population. The sewage pipe network system construction project in Moganshan Town mainly collects the sewage from 13 administrative villages and discharges it to the urban sewage treatment plant for treatment. The No. 4 pump station is a Phase I project, with a scale of 160 tons/day, involving 6 administrative villages, namely Houmu, Miaoqian, FaTou, Lanshukeng, Laoling and He villages. The designed length of the sewage pipe network is about 31 km, including 5 booster pump stations and 1 accident pool. Since 2018, the pipeline network has been constructed, and all 18 villages in Moganshan Town have been connected to the largest terminal in Deqing. Moganshan Town has more than 700 B&Bs with complete licenses, of which more than 500 have independent sewage treatment facilities.

59.3.3 Effluent Quality

The effluent quality shall meet the Class I A standard in the Discharge Standard of Pollutants for Municipal Wastewater Treatment Plants (GB18918-2002). If the project involves a reservoir, the effluent quality shall meet the requirements of water source protection.

59.3.4 Sewage Treatment Process and Technical Principle

Xiantan Village There are 574 farmers in Xiantan Village, about 2000 people. There are more than 130 home stays, and the number of tourists reaches 100,000 to 120,000 person times per year. The sewage treatment unit involves 4 buildings, including 9, 7, 6 and 4 rooms. The total sewage treatment capacity of the sewage treatment unit is 13 tons/day.

Laoling Village Shiyi New Village Use A^2O Treatment process: "grid regulation—anaerobic tank—facultative tank—aerobic tank—sedimentation tank—constructed etland—water outlet well". The photo of the project facilities is shown in Fig. 59.1.

Fig. 59.1 Photo of sewage treatment facilities in Xiantan Village and Laoling New Village

Fig. 59.2 Photo of oil separator and drain pipe

59.4 Typical Domestic Sewage Treatment Cases of Home Stay

59.4.1 《Qingqi》

The home stay Qingqi has 17 rooms. It is equipped with oil separator and three compartment septic tank, and the cleaning frequency is about 7–10 days and once a month respectively.

The sewage treatment equipment is made of concrete. On the surface of the filter in the facility, a large number of microorganisms form a biofilm, which can decompose organic pollutants in the sewage to achieve the purpose of purification. The treated water quality can meet the Class I B discharge standard, and the effluent is directly discharged into the stream through the pipeline (Fig. 59.2).

59.4.2 《Remote Mountains》

The home stay "Remote Mountain" has 17 rooms and a total sewage treatment capacity of 13 tons/day. The micro power sewage treatment equipment installed in the home stay adopts A^2O Treatment process, the specific technology is biological purification tank, and the equipment material is glass fiber reinforced plastic, which has excellent anti-corrosion performance.

There is an oil separator, and the oil separator shall be cleaned by special maintenance personnel. The septic tank is treated by a septic truck. Two vehicles are treated each time. The treatment capacity of each vehicle is 5.6 tons, and the treatment cost

Fig. 59.3 Aeration pipe and micro power device of B&B Remote Mountain

is 1000 yuan/time. The treatment equipment shall be cleaned every half month at a cost of 800 yuan/time. There is no place to deal with the cleaned materials, and occasionally tea gardens recycle them (Fig. 59.3).

59.4.3 Environmental Supervision

The construction of sewage treatment project in Moganshan Town has greatly improved the water environment quality of Moganshan, built a "zero direct sewage discharge", and promoted the friendly and harmonious development of rural environment. The environmental protection department of the government regularly checks the sewage treatment equipment independently installed by the home stay, and supervises whether the equipment operates normally and the water quality of the incoming and outgoing water. At the same time, Moganshan Town has established a long-term monitoring mechanism for home stay wastewater. The supervisor will regularly carry out spot checks and tests on the effluent quality of the unit, and has strict requirements on the setting of the oil separator.

59.4.4 Suggestions

If a large amount of domestic sewage is directly discharged into the natural water body without any treatment, it will cause the organic matter, N, P and other pollutants to exceed the environmental capacity and ecological carrying capacity of the water body, and seriously damage the ecosystem (Song et al. 2020). The characteristics of domestic sewage of home stay are as follows: ① The distribution is scattered and random; ② The sewage flow is small and the water quality fluctuates greatly; ③ The pollutants have complex components, low concentration and strong biodegradability.

Therefore, the research and development of small domestic sewage treatment devices with low energy consumption, easy installation and operation, and the matching of appropriate sewage biochemical treatment process are the key technical issues of domestic sewage treatment of home stay. At present, the mature market mainly includes purification tank technology and SBR technology. In addition, micro power devices are also widely used.

59.5 Conclusion

The development of home stay industry is an effective way to coordinate urban and rural integration and promote farmers' income. However, the industrial development may also cause damage to the ecological environment, so it is necessary to actively carry out home stay management and strengthen comprehensive environmental governance. Moganshan Town, as a relatively developed area of home stay industry, has basically realized the decentralized treatment of domestic sewage, effectively reducing the environmental pollution caused by indiscriminate sewage discharge. In the future, with the development of the home stay industry, each can realize the combination of human settlement environment, human environment and natural environment.

Acknowledgements This study is supported by Scientific and Technological Innovation Activity Plan for College Students in Zhejiang Province (No. 2022R423A002) and National College Student's Innovation and Entrepreneurship Training Program (No. 202111841011).

References

Feng Y, Peng Y, Wang B et al (2021) A continuous plug-flow anaerobic/aerobic /anoxic/aerobic(AOAO) process treating low COD/TIN domestic sewage: Realization of partial nitrification and extremely advanced nitrogen removal[J]. Sci Total Environ 771:145387

Wang Jie, Guoliang Study on the dilemma and breakthrough path of sustainable development of Moganshan home stay [J] China Business Theory, 2020 (7): 3.

Liamlaem W, Benjawan L, Polprasert C . Sustainable wastewater management technology for tourism in Thailand: case and experimental studies[J]. Water Science & Technology, 2019, 79(10).

Song P, Huang G, Hong Y et al. A biophysiological perspective on enhanced nitrate removal from decentralized domestic sewage using gravitational-flow multi-soil-layering systems[J]. Chemosphere, 2020, 240(Feb.): 124868.1–10

Zhang Haizhou L, Lin ZD et al (2019) Spatial and temporal distribution characteristics and causes of home stay around Mogan Mountain [J]. Geographic Research 38(11):21

Part III
Environmental Restoration and Green Ecological Governance

Chapter 60
Research on Forest Carbon Sequestration and Management Strategies

Bo Wang, Xintong Yao, Ziyi Liu, and Hongzhou Wang

Abstract Living trees in forests and forest products make an important contribution to slowing climate warming. Meanwhile, deforestation can also bring economic benefits. Our purpose is to develop a predictive model of forest carbon sequestration. We hope to make recommendations for forest managers to make appropriate decisions under the constraints of different benefit measures. Two models were developed in this paper: Model 1: Carbon sequestration model; Model 2: Forest management decision model. Besides, we derived the relationship between the average economic value and ecological value with the age of the trees to determine the optimal period for cutting trees. However, the vertical structure of the forest, the competitive relationship between populations and the influence of insect pests were not considered. Therefore, improvements on these aspects need to be made.

Keywords Forest management · Carbon sequestration · Logging strategy · Sustainable development

---

B. Wang · X. Yao (✉) · Z. Liu · H. Wang
Beijing Institute of Technology, Beijing, China
e-mail: yxintong@yeah.net

B. Wang
e-mail: wb18810589907@163.com

Z. Liu
e-mail: Lzy2001813@163.com

H. Wang
e-mail: wanghongzhou@bit.edu.cn

J. Zhang et al. (eds.), *Environmental Pollution Governance and Ecological Remediation Technology*, Environmental Science and Engineering,
https://doi.org/10.1007/978-3-031-25284-6_60

60.1 Introduction

60.1.1 Research Background

Forest ecosystem is the main body of terrestrial ecosystem and plays an important role in maintaining ecological balance. In order to cope with global climate change, the carbon sequestration ability of forest ecosystem has gradually become the focus of attention. In October 2021, The State Council issued the Carbon Peak 2030 Action Plan, which has become a national strategy to actively promote carbon neutralization. It is of great practical significance to establish forest carbon sequestration model and formulate forest management strategy to achieve maximum ecological and economic benefits.

60.1.2 Review of Research Status

At present, the main research methods of forest carbon storage include biomass method, stock method, biomass inventory method, eddy correlation method and vorticity covariance method, all of which have certain applicability but also have their limitations. At present, researches on forest management strategies mainly focus on forest resource protection, but few of them comprehensively consider the carbon sequestration benefits and economic value of forests.

60.1.3 Main Content

In this paper, we calculate the carbon sequestration of the forest and discuss the best forest management strategies under different scenarios based on the changes in the carbon sequestration of the forest over time. Our work mainly includes the following.

- Predictive models of forest carbon sequestration over time for each temperature zone based on biomass data collected for major tree species in four different temperature zones.
- Recommendations on forest management were given for optimal CO2 sequestration, optimal economic benefits, and optimal environmental benefits, respectively.

60.1.4 Assumptions and Justifications

Assumption 1 The survival rate of tree seedlings accepted after one year is 100%. **Justification**: We assume that after transplanting the saplings will be regularly observed and the seedlings found dead will be replanted immediately.

Table 60.1 Data source collation

Data we use	Data source
Biomass of cold-temperate tree species	Reference (Ignacio 2001)
Biomass of temperate tree species	Reference (Jokela 1981)
Biomass of subtropical tree species	Reference (Yang 2004)

Assumption 2 The economic, social and ecological benefits of the forest can be averaged over each tree. **Justification**: The benefits of the forest are a whole and do not satisfy the relationship $1 + 1 = 2$. We do averaging only in a mathematical sense and assume reasonable.

Assumption 3 The forest is a plantation with a capacity of 1000 trees. **Justification**: According to the distribution density of cold-temperate and temperate trees is about 5 m $\times$ 5 m, it is estimated that the cold-temperate and temperate plantation forest occupies about 2.5 hm^2. Similarly, according to the distribution density of subtropical and tropical trees, it is estimated that the plantation forest occupies about 6 hm^2.

60.2 Model Preparation

The data we used mainly include data on tree biomass in different temperature zones, variation of diameter at breast height with age in Horsetail pine. The data sources are summarized in Table 1.

60.3 Model I: Carbon Sequestration Estimation Model

60.3.1 Linear Regression Analysis of Diameter at Breast Height and Biomass

A scatter plot matrix of the data for these four latitudes was plotted, as shown in Fig. 60.1.

According to the relationship between the diameter at breast height and biomass of live trees in different temperature zones in 4, it can be observed that there is a clear linear relationship between diameter at breast height and biomass after taking the logarithm.

$$\ln M = b \ln D + \ln a \tag{60.1}$$

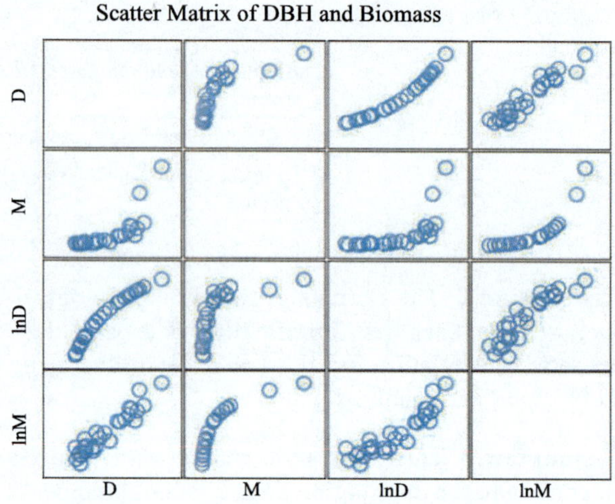

Fig. 60.1 Scatter matrix of DBH and biomass

where M is the biomass, D is the diameter at breast height, and a, b are the coefficients to be fitted by least squares method. If the logarithmic form is not used, then Eq. (60.1) can be expressed as

$$M = aD^b \tag{60.2}$$

Based on the characteristics of biomass data of living trees in different temperature zones, different coefficients a, b can be obtained according to the anisotropic growth equation shown in Eq. (60.2), and then the variation curves of total above-ground dry biomass with diameter at breast height of the main tree species in each temperature zone can be fitted.

60.3.2 Growth Model of the Diameter at Breast Height

The Von Bertalanffy model was used to simulate the assimilation and alienation processes of living trees. Let y(t) be the diameter at breast height in year *t*. According to the rate characteristics of organic matter generation or decomposition by assimilation and dissimilation, the rate of organic matter accumulation by assimilation is assumed to be ay^{m}, where α is the assimilation coefficient and m is the assimilation power index; the rate of organic matter decomposition by dissimilation is assumed to be cy, where c is the dissimilation coefficient.

The growth rate equation can be expressed as Eq. (60.3). After the variable substitution and other steps, the general solution of Eq. (60.3) is obtained as Eq. (60.4).

$$\frac{dy}{dt} = \alpha y^m - cy \tag{60.3}$$

$$y = \frac{\alpha}{c}(1 - Ce^{-c(1-m)t})^{\frac{1}{1-m}} \tag{60.4}$$

$$y = k(1 - e^{-\gamma t})\beta \tag{60.5}$$

where, $k = \frac{\alpha^{\frac{1}{1-m}}}{c}$, $\gamma = c(1 - m)$, $\beta = \frac{1}{1-m}$.

60.3.3 Calculation of Carbon Sequestration Amount

There are various methods to estimate the carbon sequestration of living trees, and here we chose the biomass method to calculate the carbon sequestration. The biomass method is a carbon estimation method based on forest biomass data. Biomass is the total amount of organic matter that includes all plants, animals and microorganisms existing on a unit area. Since the proportion of microorganisms is very small and animal biomass is less than 10% of plant biomass, plant biomass is usually used as a proxy (He 2005). The principle is based on large-scale field surveys of forests, obtaining actual measurement data, establishing a standard set of measurement parameters and biomass databases (Nie 1997), using sample plot data to obtain the average carbon density of vegetation, and then multiplying the carbon density of each vegetation with the area to estimate the carbon volume of the ecosystem (Foley 1997). Different forest vegetation has slightly different conversion rates due to differences in their community composition and age structure, and the internationally commonly used conversion rate of 0.5 is used in this paper based on Richard A. Houghton et al. (Levine 1995). The carbon curing volume was over time as follows

$$C = 0.5M\rho \tag{60.6}$$

We only consider the carbon stocks of in-use wood products, and the decomposition cycles and decomposition rates of the two types of in-use wood products differ significantly. Since it is not possible to track the flow and rate of products obtained from harvesting, we classify end-use wood products into solid wood products and paper products based on their use. The life span of wood products is related to many factors, and in order to simplify the calculation, it is assumed that there is no loss in the processing of harvested wood products from primary products to final products, and the decomposition rate is constant, and the information is reviewed to determine that the decomposition rate of wood products basically follows the first-order decay law, so this paper uses the First-Order Decay (FOD) formula to calculate the carbon stock of the existing pool (Houghton 1996).

$$C_{Ht} = C_0 e^{-k_1 t} \tag{60.7}$$

Table 60.2 Parameter values for first-order decay formula

	k1	C0
Wood products	0.023	0.85Ct
Paper products	0.347	0.15Ct

where C_{Ht} is the HWP pool carbon stock (kg-hm-2) at the end of year t, C_0 is the initial carbon stock (kg-hm-2), t is time (a), k_1 is the annual loss rate. For parameter k_1, the definition of the half-life is given by

$$k_1 = \frac{\ln 2}{T} \tag{60.8}$$

where T is the half-life of harvested wood products. Since the half-lives in paper products and solid wood products are very different, according to the excellent 2003 IPCC guidelines, the half-lives of solid wood products and paper products are specified as 30 a and 2 a, respectively, so their annual loss rates k_1 are 0.023 and 0.347, respectively. According to the study of Yujing Ye et al. the ratio of solid wood products and paper products was determined to be 17:3, which was converted to percentages corresponding to 85 and 15%. Therefore, C_0 was obtained by multiplying Eq. (60.7) by the respective ratios of 85 and 15% for solid wood products and paper products. The values of k_1 and C_0 are shown in Table 60.2.

60.3.4 Results

Anisotropic Growth Equation
The data of the major tree species in four different temperature zones were used to fit the parameters a, b in the anisotropic growth equation by least squares in the form of Eq. (60.1), respectively. Based on the parameters obtained by fitting, we draw the corresponding fitting curve. The obtained fitting results are shown in Fig. 60.2.

Von Bertalanffy Growth Model
The parameters in Eq. (60.6) were fitted iteratively using the collected growth of the diameter at breast height of the Marengo pine during 80 years using a multivariate parameter fitting method, and the obtained parameter fitting results are shown in Table 60.3. The model parameters of acacia cited from reference are also shown in Table 60.3.

Since horsetail pine is a subtropical tree species, the temperatures in the subtropics and tropics are similar. Therefore, the model parameters obtained by fitting the data of Masson pine can be approximated to be applicable to both subtropical and tropical species, and the growth curve trends of major tree species in these two temperature zones can be considered to be approximately the same as those of Masson pine. Similarly, since Acacia sylvestris is a temperate species and the temperatures in the cold and temperate zones are closer, the growth curve trends of the major tree species

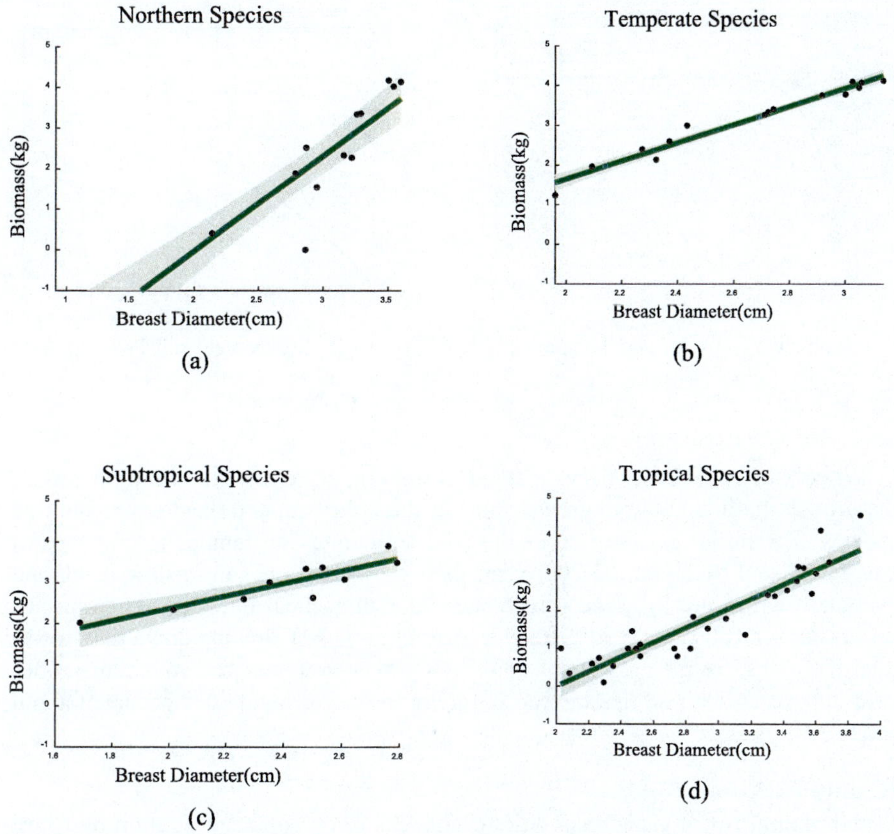

Fig. 60.2 Fitting of anisotropic growth equations for different temperature zones

Table 60.3 Model parameters of Masson pine and Robinia Pseudoacacia

Parameter name	Parameter value	
	Masson pine	Robinia Pseudoacacia
k	65.3181	15.2063
γ	0.0198	0.1786
β	1.0108	2.7087

in these two temperature zones can be approximated to be approximately the same as those of Acacia sylvestris. However, since there are differences in the maximum diameter at breast height that can be reached by different tree species during growth, the model parameters for several major tree species in each temperature zone need to be corrected.

The growth curves of live trees at breast height over 100 years, obtained using the Von Bertalanffy Growth model predictions for each temperature zone, are shown in Fig. 60.3.

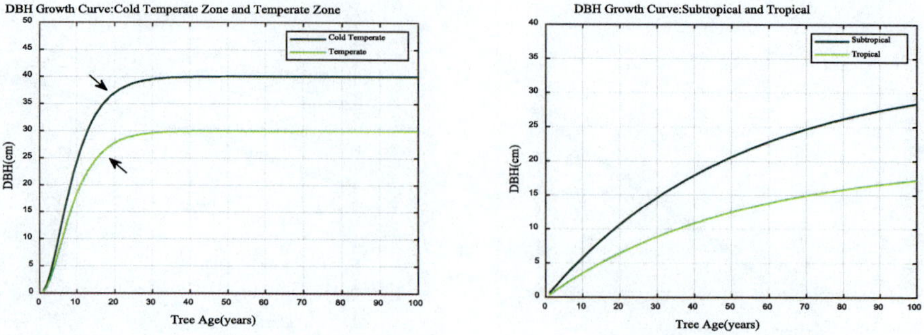

(a) Cold Temperate and Temperate (b) Subtropical and Tropical

Fig. 60.3 Predicted diameter at breast height growth curve of live trees

Where, the two black arrows marked in the Fig. 60.3a are the turning points of the growth prediction curve growth rates of the cold-temperate and temperate tree species. The initial growth rate of the cold-temperate and temperate tree species was faster, and by about the 20th year, their growth rate began to slow down and eventually converged to 0. The growth rate of the subtropical and tropical tree species was faster for 100 years, and there was a certain trend of slowing down the growth rate. If we observe the growth curves of living trees in these two temperature zones over 200 years, we find that the turning point in their growth rates occurs at about year 120.

Result Prediction

Under natural forest growth conditions, changes in carbon sequestration by forest populations are observed so that the optimal period of carbon sequestration can be determined as a reference for the best cutting policy. Firstly, we assessed the trends of carbon sequestration of forests in temperate and subtropical zones, and their unharvested carbon sequestration over 100 years is plotted over time, as shown in Fig. 60.4.

Observing the curves in Fig. 60.4a, it can be found that for temperate forests, the growth rate of total carbon solidification increases rapidly and then decreases gradually with the increase of tree age, and finally tends to level off. Looking at the curve of the average annual carbon solidification, we find that there is a great value at the 15th year. We took the age of the tree corresponding to this extreme value as the optimal carbon sequestration period, and thus selected 15 years as the optimal cutting cycle; while for subtropical forests, observing the average annual carbon solidification curves in Fig. 60.4b, we found a great value at the 61st year, so we selected it as the optimal cutting cycle. For temperate and subtropical forests, we simulated the changes of total carbon sequestration for 100 years at 10–100% cutting rates, under the premise that the optimal cutting period was determined to be 15 years and 61 years, respectively. We chose a deforestation rate simulation step of 10%, and the results are shown in Fig. 60.5 for deforestation rates of 40, 50, and 60%. For temperate forests, comparing the total carbon sequestration values in

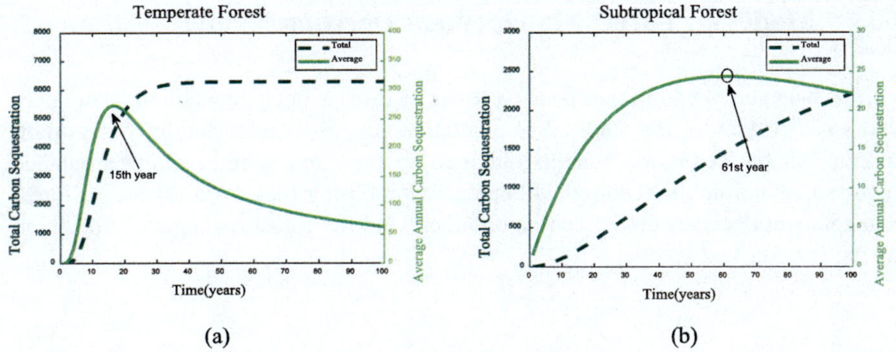

Fig. 60.4 Predicted change curve of forest carbon sequestration

Fig. 60.5a, a 60% cutting rate was obtained to have the best carbon sequestration effect. Therefore, the 15-year cutting cycle and 60% cutting rate were chosen as the best cutting strategy for temperate forests. For subtropical forests, comparing the values of total carbon sequestration in Fig. 60.5c, we found that the values of the three curves were very similar within 100 years. We choose the middle value of 50% as the optimal cutting rate. Therefore, the 61-year cutting cycle with 50% cutting rate was chosen as the optimal cutting strategy for subtropical forests.

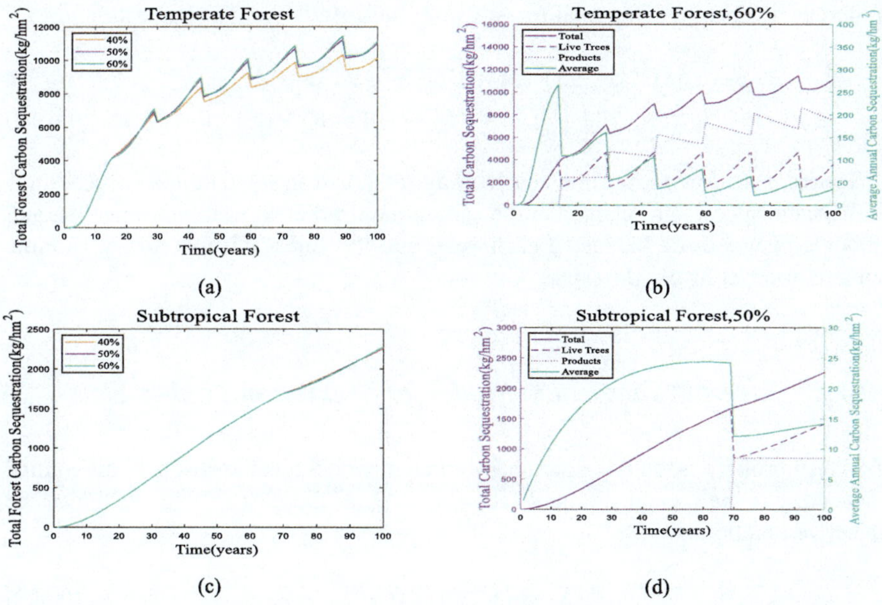

Fig. 60.5 Simulation of different deforestation rates

60.4 Model II: Forest Management Decision Model

Forest management strategies need to consider three aspects: economic, ecological and social values of the forest. For plantation forests, it is particularly important to consider the economic benefits obtained by investors, while making a balance between economic and ecological benefits. Combining these two aspects, a forest management decision model can be obtained, and the forest management decision model is described below.

60.4.1 Management Model Under the Principle of Intergenerational Equity

Assuming that the ecological and social effectiveness of a forest is positively correlated with its wood storage, the above principle is simplified to ensure that the amount of wood that can be harvested each year does not decrease and that the amount of wood stored in the forest can be maintained so that the various effectiveness of the forest does not decrease. Under the principle of intergenerational equity, forest managers need to grasp the speed of harvesting and afforestation. (Jiang 2013). Assume that the optimal harvesting time for a certain tree is T years, the trees are planted at x hm2 per year, and the total standing tree stock in the forest at the end of T years is H(t). If there is no change in the total forest wood stock after 1 year of harvesting, the

$$
\begin{aligned}
&nx\{V(1) + [V(2) - V(1)] + [V(3) - V(2)] \\
&\quad + \ldots + [V(t) - V(T-1)] = nxV(t)\}
\end{aligned}
\tag{60.10}
$$

Trees are cut down when they reach the age of T, and an equal number of seedlings of the same species are planted at the same time. In this way, approximately equal amounts of wood are harvested each year, and the forest's wood storage volume remains approximately the same.

60.4.2 Forest Management Model for Maximum Timber Yield

If S is net profit, p is the price per unit volume of wood, q is the cost per unit volume of wood, r is the capital discount factor, and Q_0 is the cost of tree seedlings, the objective function is

$$
\max(S) = (p - q)V(t)e^{-rt} - Q_0 \tag{60.11}
$$

$$
dS/dt = (p - q)e^{-rt}[dV(t)/d(t) - V(t) \cdot r] = 0 \tag{60.12}
$$

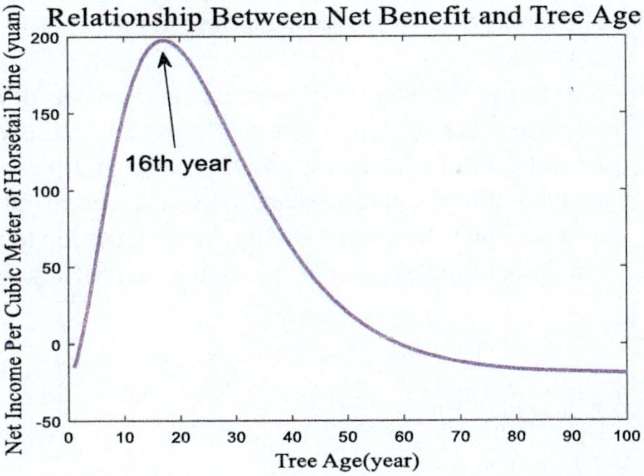

Fig. 60.6 Average yield per horsetail pine as a function

$$dV(t_1)/dt = r \cdot V(t_1) \tag{60.13}$$

We assumed that the tree growth pattern is first accelerated, then decelerated, after which it enters a saturated state and eventually dies. For a forest where the dominant species is horsetail pine, the initial timber price is p = $1400 per cubic meter, the initial value of logging costs and planting is $20 per cubic meter, the capital discount factor is 0.1. The price of horsetail fir seedlings is $110 per tree. Observing the curve in Fig. 60.6, it can be seen that the greatest economic benefit from harvesting is obtained when the age of horsetail pine reaches 16 years.

60.5 Conclusion

We calculate the amount of forest carbon sequestration and discuss the optimal forest management strategy under different scenarios based on the change of forest carbon sequestration over time. Prediction models for forest carbon sequestration over time in each temperature region based on biomass data of major tree species in four different temperature regions suggest optimal carbon dioxide uptake, optimal economic benefit and optimal environmental benefit for forest management. Sensitivity and stability tests were carried out for each model and applied to a real forest to give our recommendations for forest management. Through model establishment and analysis, the following conclusions are drawn: Different tree species have different optimal harvesting periods, and the optimal harvesting periods of tree species with similar living habits are similar. In addition, this report also explores the differences in carbon uptake and growth conditions between forests in cold regions and forests in warm regions. The model proposed in this paper still has many deficiencies, and the hypothesis of the research problem is too ideal. In the future work, we will make improvements in the following aspects:

- The model was improved according to the difference of survival rate of different saplings.
- To improve the balance of benefits in all aspects of forest, adopt more reasonable balance of benefits and improve the model. In this paper, reasonable suggestions are put forward for forest growth model and management model, and effective analysis of forest carbon sequestration is carried out with a view to global climate change. The future work will pay more attention to the optimization of the model and the solution of practical problems, and get a more perfect model and planning.

References

Daoping' N et al (1997) Research on the relationship between the global carbon cycle and forests–problems and progress. World Forestry Res 5:33–40

Foley JA (1997) An equilibrium model of the terrestrial carbon budget. Tellus 47(B):310–319

He Y (2005) A review of forest carbon sequestration estimation methods[J]. World Forestry Res 01:22–27

Houghton RA (1996) Converting terrestrial ecosystems from sources to sinks of carbon[J]. Ambio 25(4):267–272

Jiang X, Wang XJ (2013) Optimal harvesting decision model for forests-a new framework for forestry economic policy analysis[J]. Forestry Sci 9

Jokela EJ, Shannon CA, White EH (1981) Biomass and nutrient equations for mature Betula papyrifera arsh. Canadian J Forest Res 11(2):299–305

Levine JS, Cofer Ill WR, Jr CDR, Winstead EL (1995) Biomass burning: a driver for global change. Environ Sci Technol 29(3):120A – 125

Peng C, Michael JA (1997) Contribution of China to the global cycle since the last glacial maximum reconstruction from Palaeovegetation. *Maps and an Empirical Biosphere Model.* Tellus 49(B):393~408

Santa Regina I, Tarazona T (2001) Nutrient pools to the soil through organic matter and throughfall under a Scots pine plantation in the Sierra de la Demanda, Spain 37(2):0–133

Wang D, Gu J, Holmes TP (1992) A stochastic dynamic programming model for optimizing forest harvesting strategies [J]. Inf Control 3

Yang H, Zan Y, Zeng Z, Long W, Luo Y, Zhou X, Tang L (2004) Study on the biomass and growth of three silvicultural models in Jishui[J]. J Jiangxi Agricult Univ (natl Sci) 02:164–168

Chapter 61
Study on Biomass Gasification Char Purify Domestic Sewage

Peijun Ma, Ru Yang, Sheng Kang, Keyu Chen, and Jun Teng

Abstract Domestic sewage was one of the important reasons for groundwater pollution. Now it's expensive to treat groundwater pollution, so it's significant to find cheap and renewable adsorbents to purify domestic sewage. This paper would use the industrial solid waste—biomass gasification char (BGC) of biomass gasification power plants to treat groundwater pollution, it was a by-product of biomass gasification power generation. It had strong load capacity and developed pore structure, and was a potential adsorbent for water purification. This paper provided a theoretical basis for BGC to purify domestic sewage.

Keywords Biomass gasification char · Purify · Domestic sewage · Influencing factors

61.1 Introduction

China was a country with frequent water disasters and insufficient water resources. While the total amount of water resources was decreasing, the amount of sewage discharged increases sharply. More than 50% of the urban groundwater had been polluted and 90% of the rivers flowing through the city didn't meet the drinking water standard. Water pollution was caused by industrial wastewater and domestic wastewater discharge. Domestic sewage usually referred to the large amount of sewage produced and discharged in the daily production and living activities of residents. The sources of domestic sewage mainly included urban and rural domestic water, agricultural wastewater, medical wastewater. The composition of domestic sewage was complex, and its organic matter accounts for a relatively high proportion. Most of the organic substances were easily biodegradable, such as carbohydrates, proteins, lipids, etc. Along with water soluble inorganic salts such as chloride, phosphate, sulfate, carbonate, sodium, potassium, calcium and magnesium were also found in

P. Ma · R. Yang (✉) · S. Kang · K. Chen · J. Teng
Zhejiang University of Water Resources and Electric Power, Hangzhou, China
e-mail: yangru@zjweu.edu.cn

J. Zhang et al. (eds.), *Environmental Pollution Governance and Ecological Remediation Technology*, Environmental Science and Engineering,
https://doi.org/10.1007/978-3-031-25284-6_61

domestic sewage. The general characteristics were high organic matter, nitrogen, sulfur and phosphorus (Li 2015). In addition, there were many pathogenic microorganisms in domestic sewage which take organic matter in domestic sludge as feed for a large number of proliferation and growth. If domestic sewage was not treated in time, it would pose a great threat to the natural environment and human health. With the call of lucid waters and lush mountains were golden mountains and silver mountains getting louder and louder, the promotion of urbanization had brought about 5% annual growth of domestic sewage discharge, which had brought serious environmental problems to human survival. How to deal with domestic sewage efficiently and cleanly was one of the main subjects of dealing with environmental problems caused by urbanization in China.

The adsorption method could obtain better water quality. At the same time, its process had a larger adjustment space in the design and operation process, some adsorption processes were reversible, which could be regenerated through appropriate desorption conditions, it is environmentally friendly to energy and environment. Activated carbon with stable physical and chemical properties, large specific surface area, developed pore structure and abundant surface chemical functional groups, it had become the mainstream adsorbent material for wastewater treatment. In water treatment, biomass activated carbon mainly taked activated carbon as the main carrier to play its adsorption function. Using the strong adsorption effect of main activated carbon on organic matter and dissolved oxygen, the microorganism was enriched on the surface of biomass activated carbon by natural enrichment or artificial culture fixed method to form biofilm, at the same time the adsorption function and biodegradation of biomass activated carbon could be carried out in coordination to realize the goal of physical and chemical treatment and biological treatment, but its cost was high and so to find efficient and cheap adsorbents was important (Chu et al. 2018).

China was rich in biomass resources. With the improvement of biomass gasification technology and the increasing scale of power generation, the industrial solid waste–BGC had drawn people's attention. BGC had strong load capacity, high carbon content and developed pore structure, which was a potential adsorbent. Using by-products of biomass gasification power generation to treat domestic sewage not only solved the problem of resource utilization of waste, saved energy and reduced environmental pollution, but also reduced the treatment cost to achieve the purpose of making waste by waste. It was significant to improve the utilization rate of resources and improved the ecological environment. Using BGC as adsorbent could realize many benefits.

61.2 Materials and Methods

This paper used the rice husk as material of BGC, which was prepared in the biomass gasifier. BGC was taken from Jiangsu Gaoyou Gasification Power Plan, which was prepared in a gasification furnace. By controlling the excess air coefficient, RHGC

Fig. 61.1 Biomass gasification char (BGC)

was incompletely burned to produce CH_4, CO and H_2, and simultaneously produce BGC, as shown in Fig. 61.1. Putting RHGC in a drying box, dried, grinded and crushed it with secondary deionized water at room temperature and stirred with a magnetic stirrer for 30 min, then let the stirring solution stand at room temperature for 24 h, filtered and dried at 90 °C for 30 h, bottling spare.

Scanning electron microscopy (SEM) was an important tool for the determination of surface morphology and surface elements of adsorbent samples, which could provide high resolution observation requirements, as shown in Fig. 61.2. It could be seen that the surface structure of BGC was relatively loose and the pore structure was relatively developed, which provided an excellent adsorption basis for the treatment of sewage.

Elemental analysis could quantitatively analyze the specific elements contained in the sample. The elemental analysis was determined by the Vario EL cube element analyzer produced by Elementar, Germany (accuracy C, H, N, S $\leq$ 0.1 abs, O $\leqq$ 0.2% abs). Table 61.1 listed the element composition and the mass percentage of BGC. RHGC also contained elements such as H, O, N, S, Cl, Si, K, Ca, O element content was 21.82%. The C was 43.58%, which could be seen that C proportion was very large. The high carbon content provided important adsorption basis for treating sewage. So RHGC could be used to purify sewage.

BGC adsorbed and accumulated organic matter through its own carbon-based adsorption material characteristics to improve the content of dissolved oxygen in

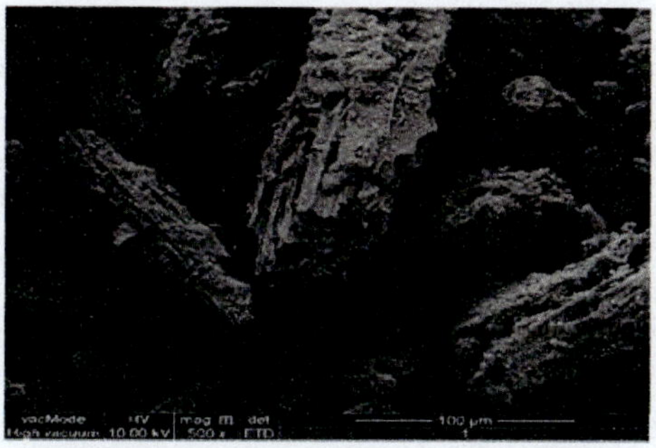

Fig. 61.2 Scanning SEM image of BGC

Table 61.1 Elemental analysis of RHGC

Ingredient	C	H	S	N	O	Cl	Si	Ca	P	K
Content (%)	43.58	1.84	0.30	1.59	21.82	0.90	3.92	1.36	0.29	6.52

the surface layer, which provided rich organic matter for microorganisms that play degradation functions in the surface layer. The rough surface could also effectively reduce the impact of water flow shear force on microorganisms, the small depression provided a good adhesion position for microorganisms, which created favorable conditions for microbial survival and growth and it was beneficent to microbial enrichment (Chu and Chi 1999). Because BGC itself belonged to carbon-based adsorption material, it could adsorb heavy metals, which reduced the inhibition of some inorganic toxic substances on microorganisms. In addition, the organic substances in BGC and obtained the adsorbed biomass could provide breeding nutrients for the microorganisms regeneration on the surface of BGC, which also increased the total amount of adsorbing pollutants and extended service cycle and life (He 2011).

The process of surface microbial treatment of pollutants was divided into three stages, the first stage was the adsorption and absorption of organic matter, this stage was slow, it was because microorganisms synthesized some organic compounds into new cellular substances and carried out metabolic activities. The second stage was to consume and utilize the organic matter in the first stage to obtain the energy required for the synthesis of new cells by oxidative decomposition, which would supply the decomposition and metabolism for dissolved organic matter. When the microbial adsorption reached saturation state, it would not have adsorption capacity, but it could consume a large amount of organic matter absorbed by itself through the oxidation stage and the external factors did not have a great influence on the living environment of the surface microorganism. Under these conditions the microorganism would show activity again and restore the adsorption oxidation decomposition ability until the end of its life cycle (Wei et al. 2005).

Like most carbon-based adsorption materials, BGC had large specific surface area and developed pore structure, which could expand its contact with sewage and improve the physical adsorption capacity. Developed pore structure was closely related to its outstanding adsorption performance. Organic substances contained in domestic sewage not only had small molecular substances, but also had various macromolecular substances. Larger pores provided diffusion channels, so that the adsorption material flowed into the pores to complete the initial adsorption process and then the adsorption of solute molecules was completed by the pores to ended the adsorption process (Jiang et al. 2008). BGC also had rich chemical functional groups, which could react with heavy metal components in domestic sewage to form chemical adsorption to remove its inorganic toxic substances.

In the adsorption process of BGC, a mass transfer phenomenon occurs on the surface layer of the BGC, which was caused by the solute passing through the solvent. The driving force for this mass transfer could be caused by the hydrophilic and hydrophobic characteristics of the solute or a characteristic of the affinity of the solute to the adsorbent surface, it could also be caused by a combination of the two. The pollutants removed by BGC adsorption in water treatment were usually organic compounds with hydrophobic and hydrophilic groups. The affinity of solute to adsorbent surface could be divided into two types: solubility of solute in solvent and van der Waals force between solute and adsorbent, chemical bond force and electrostatic attraction (Xu and Long 2000). The main mechanism of BGC adsorption of metal ions is: electrostatic mechanism and ion exchange mechanism. The adsorption of trace organic matter was mainly explained by the following two mechanisms: one wan the role of adsorbate and surface oxygenated group electrons, the other was electrostatic attraction and repulsion between ions. The adsorption carried out by BGC was a complex process that utilized its physical adsorption, chemical adsorption, exchange adsorption, oxidation and reduction properties, as well as the concurrency of various effects (Li et al. 2009).

61.3 Application of Biomass Coke in Domestic Sewage Treatment

After treating domestic sewage with activated carbon technology, the pollutant removal rate in water was high, which was far superior to the removal effect of polar compounds such as alkyl benzene compounds and it was non-toxic and harmless, which was mainly applied in the following aspects (Zuo 2016).

61.3.1 Remove Organic Matter

Most of the organic matter in domestic sewage was easily biodegradable, such as carbohydrates, fat, protein, etc., which was easy to decay and produce stench under the action of microorganisms. At the same time, domestic sewage also contained a large number of pathogenic microorganisms, which fed on the organic matter in sewage and multiply in large numbers. Cong et al. (2009) used modified biomass coke to treat bath wastewater and the average removal rates of turbidity and COD reached 95.2% and 84.8%.

61.3.2 Deodorize, Deodorize and Remove Residual Chlorine

A large number of algae reproduction because domestic sewage eutrophication, so the water with odor, it was more intense in the summer high temperature season, powdered biomass char could be adsorbed to improve the smell of water. In the process of water purification, chlorine disinfection was often used. When chlorine dosage was high, biomass char could be used to remove trace residual chlorine in water (Shen and Lu 2007).

61.3.3 Remove Synthetic Detergent (ABS)

With the extensive use of synthetic detergent in daily life, the pollution of ABS to water quality was increasingly serious. When the content was higher that the foam would appear in the water, which affected the self-purification of water and water treatment. Biomass char adsorption was one of the most effective treatment methods.

61.3.4 Treat Heavy Metal Salts

The heavy metal salts contained in water WERE difficult to degrade, which resulted heavy metal ions existence in the environment for a long time, it affected the physiological health of animals and plants and when they transmitted in the food chain. All of these materials could be accumulated in suspended solids, sediments, and aquatic organisms. It might cause cancer if you consume food containing these substances for a long time.

61.4 Factors Affecting the Treatment of Domestic Sewage by BGC

61.4.1 Solubility

BGC was a carbon-based adsorption material, its excellent adsorption performance could adsorb most organic pollutants, heavy metal elements and large suspended particles in domestic sewage. Because BGC was a porous structure material, if the solubility of organic pollutants in domestic sewage to be treated was low, flocculent or the concentration of suspended particles was large, the pores of BGC are easily blocked. The adsorption capacity of BGC decreased, and the difficulty of subsequent surface microbial treatment of pollution was improved. At the same time, too much particle blockage would affect the contact between domestic sewage and BGC to a certain extent, which would reduce the reaction area of microorganisms and organic pollutants on the surface of BGC, resulting in the weakening of the treatment effect of biological treatment (Zhong et al. 2014). Therefore, the greater the solubility of organic pollutants, the more easily biological adsorption; the smaller concentration of suspended particles, the easier physical adsorption (Ma et al. 2002). The solubility of most pollutants was large, which could have a positive impact on the adsorption process.

61.4.2 PH Value

Enriched microorganisms attached to the surface of BGC play its biodegradation function, so the PH value would affect the adsorption and degradation of organic pollutants from water. The PH value suitable for survival was generally between 6.5 and 8.5 for aerobic microorganisms, it decreased with the increase of PH value of the solution. When the PH value was higher than 9.0, it was not conducive to the adsorption and degradation of microorganisms and the effect was better when the PH value was low. At the same time the PH value not only affected the existence of heavy metal ions in sewage but also affected the chemical properties of active groups on BGC. In the study of pomelo peel-based activated carbon preparation and adsorption mechanism, Hua (2013) concluded that the increase of PH would lead to the decrease of Cr (VI) removal rate. The existence of Cr (VI) and the surface electrical properties of the activated carbon formed different states because of different PH value. When the PH value was in the range of (1, 5.67), the surface of the activated carbon was positively charged, and electrostatic attraction was generated between the activated carbon and Cr (VI) anion. The adsorption amount was increased due to the electrical adsorption of the two. In this range, the positive charge on the surface of activated carbon would be neutralized due to the increase of PH value, resulting in the decrease of electrostatic attraction and the decrease of adsorption rate. When the PH value was greater than 5.67, the surface of activated carbon would be negatively charged, and the Cr (VI) anion would be rejected. The same electrical repulsion between the two led to the decrease of adsorption capacity and the rejection would increase with the increase of PH value, which led to the decrease of Cr (VI) removal rate of the activated carbon.

61.4.3 Temperature

Temperature was one of the key factors affecting the adsorption rate of most adsorption materials. In liquid phase adsorption, the higher the liquid temperature, the more intense the liquid molecular movement, the more difficult the adsorption process. Moreover, when the temperature was higher the greater the solubility in water for general pollutants. At the same time the movement of pollutant molecules dissolved in liquid would intensify, so the adsorption effect would be weakened. Adsorption heat referred to the thermal effect of pollutants on the surface of the adsorption material. When the adsorption heat greater, the impact was greater on the adsorption effect (Zang and Bi 2015). However, when the heat exceeded the temperature limit of BGC, it would lead to irreversible changes in its internal structure and physical and chemical properties, which would destroy the internal structure stability of the adsorption material and led to the decrease of adsorption rate.

61.5 Conclusions

Water pollution had brought many unsafe factors to people's production and life. Sewage treatment was imminent. It was significant to find an efficient, inexpensive and renewable adsorbent for domestic sewage treatment instead of high-cost activated carbon. This paper used the industrial solid waste of biomass gasification power plant–BGC to treat domestic sewage, which was renewable and environmental protection, it could achieve the purpose of making waste by waste.

A high carbon content of 43.58% provided important adsorption basis for treating sewage. At the same time, the loose surface structure and large pore structure provided an excellent adsorption condition for the treatment of sewage. The BGC treatment of domestic sewage was in its infancy, and the following issues need to be further explored:

(1) The pore size distribution and arrangement of BGC and the specific influence of surface modification on the formation of biofilm and the removal rate of organic matter need to be further studied.
(2) The influence of the thickness of biofilm formed by BGC on the adsorption effect in the wastewater treatment process. If the biofilm formed is too thin, the adsorption effect cannot reach the target; if the biofilm is too thick, it will lead to filter slag blockage, small effluent and other adverse effects. It will increase the head loss in the reactor and increase the scouring frequency of BGC. Because the scouring process must be completed together with hydraulic backwash and gas scouring, the corresponding operation and management complexity will also increase.
(3) Biofilm formed by BGC's own biomass supplement and adsorption, the specific influencing factors of biofilm formation rate and how to improve the biofilm formation rate need further research.
(4) Surface modification is a key link to expand its application environment, but the research on surface modification of BGC was not much, and its application potential needs to be further explored.

Acknowledgements This work is supported by 2021 College Student Innovation and Entrepreneurship Training Program of Zhejiang University of Water Resources and Electric Power in China References.

References

Chu MK, Chi TK (1999) The effects of some inhibitor component son biologically activated carbon processes. Water Res 33:2609–2616

Chu XS, Chen ML, Su CY, Huang Z, Nong ZW (2018) Research and application progress of biological activated carbon technology in water treatment. Water Treat Technol 44:5–10

Cong Q, Xiao ZL, Qu J (2009) Study on the treatment of bath wastewater by immobilized biological activated carbon fiber. J Saf Environ 9:44–46

He NT (2011) Microbial regeneration of spent activated carbon dispersed with organic contaminants: mechanism, efficiency, and kinetic models. Environ Sci Pollut Res 18:534–546

Hua Z (2013) Preparation and adsorption mechanism of pomelo peel-based activated carbon. College of Light Industry and Food Engineering, Guangxi University, NanNing

Jiang WN, Wang YE, Li J (2008) Application and development of BAC technology in wastewater treatment. Shanxi Archit 34:8–10

Li WX (2015) Introduction to leather environmental engineering. China Light Industry Publishing, Beijing

Li ZL, Ma SF, Wang D, Wang CY, Wei MY (2009) Study on adsorption modes and mechanisms of metal ions and organic compounds in water by activated carbon. Environ Sci Manage 3:89–92

Ma QL, Wang ZZ, Li MM, She BS (2002) Study on activated carbon purification wastewater treatment. New Carbon Mater 1:59–61

Shen WY, Lu SZ (2007) Application of carbon in water treatment. Ind Water Treat 27:13–16

Wei HT, Liu XJ, Li T (2005) A review of activated sludge treatment of domestic sewage and wastewater. Hebei Electric Power Technol 24:40–42

Xu BJ, Long TR (2000) Contemporary water supply and wastewater treatment principle. Higher Education Publishing, Beijing

Zang P, Bi DS (2015) Application status and prospect of powdered activated carbon in water treatment. J Shanghai Inst Technol (nat Sci Ed) 15:344–348

Zhong MG, Wang Z, Cheng XX, Sun Q, Jing J, Zhai YJ (2014) Research progress on the application of biological activated carbon filter in water treatment. Coal Chem Ind 37:156–160

Zuo YJ (2016) Research and application of biological activated carbon technology in water treatment. Qual Inspection 5:36–37

Chapter 62
Saihanba Forest Ecological Environment Evaluation Model Based on AHP

Xinyang Zhang, Jiawen Li, and Zhirui Wang

Abstract The ecological environment evaluation model aims at evaluating the impact of Saihanba's forest restoration on the ecological environment. The previous ecological environment evaluation models did not use forest ecological environment indicators, so the accuracy of forest environmental assessment was poor. To address this issue, this study proposes a new model for assessing forest ecological impacts. We divided the influencing factors into three indicators: environmental status, environmental function and environmental impact, and constructed a four-level evaluation model of Saihanba's impact on the ecological environment based on AHP. From this, it can be concluded that ecological environment condition of the Saihanba has been significantly improved.

Keywords AHP · Evaluation model · Ecological indicators · Saihanba · Ecological environment

62.1 Introduction

After the founding of the People's Republic of China, the Saihanba Forest Area has carried out ecological restoration. Saihanba ecological environment evaluation model is a model for evaluating the ecological environment quality of the restored Saihanba forest area.

Although Saihanba Forest Farm has been restored, it has not yet been determined whether its restoration experience will have a positive ecological impact on the surrounding areas, and how to refer to its restoration experience based on different geographical locations and vegetation factors, the construction of ecological protection areas still faces many challenges. The first is that the existing analytic hierarchy process (AHP) (Lin 2019) ecological environment evaluation models (Deng et al.

X. Zhang (✉) · J. Li · Z. Wang
Mathematics and Applied Mathematics, Shandong Normal University, Jinan 250014, Shandong, China
e-mail: youqiongqiong@mathor.com

J. Zhang et al. (eds.), *Environmental Pollution Governance and Ecological Remediation Technology*, Environmental Science and Engineering,
https://doi.org/10.1007/978-3-031-25284-6_62

2012) are not suitable for the ecological environment evaluation of forest areas and lack evaluation indicators for evaluating human factors such as forest structure and comprehensive forestry benefits. The second is that due to the construction of ecological protection areas in real life, it is faced with the influence of vegetation factors, climatic factors, and social factors.

To solve the first challenge, we summarized three goal tiers which are the current status of the ecological environment (Shen 2018), the function of the ecological environment and the impact of the ecological environment, and quantify them, established Saihanba ecological environment evaluation model based on AHP. Substituted the corresponding data into the model to calculate the index of the ecological environment status of Saihanba in the past 20 years, and verified the rationality of the ecological protection model of Saihanba. To solve the second challenge, we selected the factors that can affect the resistance to sandstorms and fitted them with the index obtained from the model (Ning and She 2021), and concluded that the restoration of Saihanba improves the condition of environment. In summary, the main contributions of this paper are:

(1) We are the first to propose a four-level forest ecological assessment model based on AHP.
(2) The evaluation indicators of our model are more comprehensive, which can comprehensively evaluate the quality of forest ecosystems, analyze environmental factors in different regions.
(3) Our model calculation results are more accurate because we define the weights (Pan 2021) as the entropy weight, the arithmetic mean, the geometric mean and the mean of the eigenvalues, which can minimize the individual equation.

62.2 Related Work

This section describes the relevant work of environmental assessment, the environmental evaluation model is roughly divided into the TOPSIS and entropy weight method evaluation model (Liang 2021), and the RSEI (Zhao and Peng 2021) evaluation model.

The TOPSIS and entropy weight method evaluation model uses the weighted model of entropy weight method to calculate the weight vector of the indicators affecting the environment. However, in the process of establishing this model, the entropy weight method is cumbersome to determine the weight, the amount of calculation is relatively large.

The evaluation model based on RSEI is to monitor environmental changes. In the process of calculating the RSEI index, due to the different dimensions of different indicators, it needs to be normalized, also, values obtained lack accurate quantification of the environment and is not convincing.

All in all, there are various differences in the above three environmental assessment models, but they all have certain limitations. Therefore, we propose a new model, that is, the environmental evaluation model based on AHP (Yan et al. 2022).

62.3 Saihanba's Ecological Evaluation Model Based on AHP

In this section, we introduce our models Saihanba's ecological evaluation model based on AHP. The ecological evaluation model mianly contains the following key modules: index system, judgement matrix (Dong 2022) and consistency test (Zhao 2020), Weights of indicators. The establishment is mainly divided three steps: First summarize the indicators that can comprehensively evaluate the ecological environment, propose three target levels (Ma et al. 2013), determine the feature layers and quantitative indicators in each target layer and determine the standard values of the indicators. Next, use the entropy weight method to calculate the weight of the evaluation indicators in each target layer, and perform a consistency test. Fianlly assign the model values of relevant data and calculate to obtain the quantitative index of the ecological environment (Chen et al. 2018).

62.3.1 Index System

In this process, we first refer to the domestic ecological environment evaluation indicators and consider the indicators that are relative to Saihanba's environment to build up the evaluation index system. The index system of this model is made up of four-level system, which are the target layer (N), the feature layer (D), the indicator layer (S) and the influencer for the indicator (Z) (Guo et al. 2008). The factors influencing the analytic hierarchy process are (Fig. 62.1).

62.3.2 Judgement Matrix and Consistency Test

Using the 1–9 scale method proposed by Saaty, we compare each pairs of indicators in the same layer, and the comparison result is expressed in the importance scale value 1–9 (and its reciprocal), that is, the judgment matrix is obtained. Then use the root square method to solve the matrix's eigenvalue and eigenvector. The resulting eigenvector is the weight of each factor.

To determine whether the weight obtained in the matrix is reasonable, it is necessary to perform a consistency test on the judgment matrix and calculate the consistency ratio CR. The equation of it is:

$$CR = \frac{CI}{RI} \tag{62.1}$$

CI is the average stochastic consistency indicator; RI is the deviation from the full consistency indicator. If CR < 0.1, it means the judgement matrix passes consistency

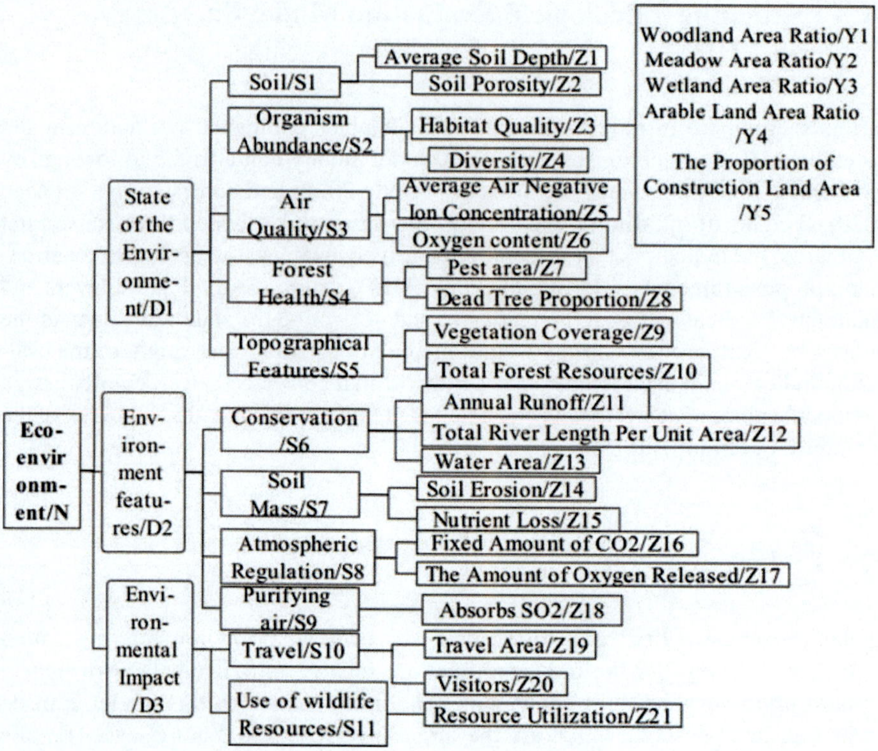

Fig. 62.1 The factors influencing the analytic hierarchy process

test, if CR $\geq$ 0.1, we need to modify our judgement matrices. Because Second-order matrices are always consistent, so we don't need to test them (Lu et al. 2021) (Fig. 62.2).

First calculate the eigenvalue of the N-D matrix, the result is $\lambda_{max} = 3$, and the eigenvector is $\omega_i = (0.4545, 0.4545, 0.0909)^T$, so weights of D1, D2 and D3 in this layer are 0.4545, 0.4545 and 0.0909. By the formula

$$CI = \frac{\lambda_{max} - n}{n - 1} \tag{62.2}$$

Fig. 62.2 The judgment matrix N-D

N	D1	D2	D3
D1	1	1	5
D2	1	1	5
D3	0.2	0.2	1

Fig. 62.3 The judgment matrix D1-S

D1	S1	S2	S3	S4	S5
S1	1	0.2	0.25	0.2	0.25
S2	5	1	2	1	2
S3	4	0.5	1	2	1
S4	5	1	0.5	1	2
S5	4	0.5	1	0.5	1

we obtain that $= 0 < 0.1$, the consistency test is passed (Figs. 62.3, 62.4, 62.5 and 62.6).

Then calculate the eigenvalue of the D1-S, the D2-S, the S-Z, and the Z-Y, all of them pass the consistency test.

Fig. 62.4 The judgment matrix D2-S

D2	S6	S7	S8	S9
S6	1	2	0.25	0.5
S7	0.5	1	0.67	0.67
S8	4	6	1	1
S9	2	6	1	1

Fig. 62.5 The judgment matrix S-Z

S6	Z11	Z12	Z13
Z11	1	3	1
Z12	0.33	1	0.33
Z13	1	3	1

Fig. 62.6 The judgment matrix Z-Y

Z3	Y1	Y2	Y3	Y4	Y5
Y1	1	6	5	7	9
Y2	0.67	1	0.5	4	5
Y3	0.2	2	1	5	7
Y4	0.14	0.25	0.2	1	3
Y5	0.11	0.2	0.14	0.33	1

Indicators	Total Score	Rank	Indicators	Total Score	Rank
Absorbs SO2/Z18	0.1651	1	Annual Runoff/Z11	0.0270	11
Fixed Amount of CO2/Z16	0.1308	2	Soil Erosion/Z14	0.0201	12
Diversity/Z4	0.1115	3	Travel Area/Z19	0.0189	13
Pest area/Z7	0.0910	4	Dead Tree Proportion/Z8	0.0182	14
Oxygen content/Z6	0.0712	5	Soil Porosity/Z2	0.0170	15
The Amount of Oxygen Released/Z17	0.0654	6	Woodland Area Ratio/Y1	0.0160	16
Visitors/Z20	0.0568	7	Resource Utilization/Z21	0.0151	17
Woodland Coverage/Y6	0.0530	8	Wetland Coverage/Y7	0.0106	18
Average Air Negative Ion Concentration/Z5	0.0356	9	Nutrient Loss/Z15	0.0100	19
Water Area/Z13	0.0270	10	Total Forest Growing Stock/Y8	0.0096	20

Fig. 62.7 Top 20 factors that directly influence the environment

62.3.3 Weights of Indicators

We calculate the weight coefficients of each layer element and the combined weight system of the same level from the upper level to the bottom layer and summarize top 20 factors that directly influence the environment in the Fig. 62.7.

So first we determine the factors in the index system, then we establish judgment matrices for each layer and verify that if they meet the consistency test, finally, we calculate the weight of each factor for N. And now, we have established Saihanba's ecological evaluation model based on AHP.

62.4 Experiments

According to the total weight ranking of ecological environment influence factors, we choose the top 20 factors that account for the significant weight ratio as branches of the indicator system (Jin et al. 2004).

By consulting the data, we multiply the data by the weight obtained, and obtain the Saihanba 2008–2019 ecological environment indexes as follows (Table 62.1):

According to the table, the trend of ecological environment level change in the Saihanba in the past two decades is: the ecological environment evaluation index increased year by year, there is the near-equal increase in 2008–2015, and the growth rate after 2015 significantly accelerated. It can be seen that with the increase of the year, the ecological environment condition of the Saihanba has been significantly improved.

Table 62.1 Ecological environment evaluation index of each year

Years	Ecological environment evaluation index
2008	80.9371
2009	100.7467
2010	114.2293
2011	139.0487
2012	167.6355
2013	185.6204
2014	202.6797
2015	230.1896
2016	313.5463
2017	369.9212
2018	442.5172
2019	507.4534

62.5 Conclusion

We propose the Saihanba's ecological environment evaluation model based on AHP to do quantitative evaluation of the ecological status before and after environmental restoration in Saihanba area. We use the forest ecological environment indicators to establish the evaluation model and obtain the ecological indexes from 2008 to 2019 which indicate the restoration of the Saihanba area has effectively improved the level of local forest ecological environment. We show that the evaluation model can comprehensively evaluate one area's forest ecological environment level.

References

Chen ZF, Wu J, Guo YB (2018) Application of AHP and fuzzy mathematics in comprehensive assessment of mine environment. East China Geol 39:305–310

Deng X, Li JM, Zeng HJ (2012) Research on computation methods of AHP wight vector and its applications. Math Practice Theory 42:93–100

Dong YY (2022) Research on the legalization of ecological environment health evaluation system based on DPEHA model. China Environ Manage 14:130–136

Guo JY, Zhang ZB, Sun QY (2008) Study and applications of analytic hierarchy process. China Saf Sci J 18:148–153

Jin JL, Wei YM, Ding J (2004) Fuzzy comprehensive evaluation model based on improved analytic hierarchy process. J Hydraul Eng 35:65–70

Liang MY (2021) A study on ASEAN investment environment assessment based on entropy right TOPSIS method. Guangxi Qual Supervision Herald 03:171–172

Lin L (2019) Teaching evaluation and excellence management evaluation based on analytic hierarchy method. J High Teach 19:150–152

Lu YQ, Wang J, Cao MQ (2021) Evaluation method of shale gas development selection based on improved analytic hierarchy method. Reservoir Eval Dev 11:70–77

Ma LL, Tian SF, Wang N (2013) Ecological environment evaluation of the mining area based on AHP and fuzzy mathematics. Remote Sens Land Resour 25:165–170
Ning J, She ZG (2021) Quantitative evaluation model of railway safety risk based on protective layer analysis method. Railway Comput Appl 30:6–10
Pan CJ (2021) Investigation on the causes of fire on the comprehensive surface based on analytic hierarchy method. Shaanxi Coal 40:71–74+117
Shen J (2018) A study on the evaluation model of teaching quality in colleges and universities based on analytic hierarchy method. Technol vis 28:133–135
Yan K, Liu R, Yang YT (2022) Ecological environment assessment of giant panda national park based on RSEI. Geospatial Inf 20:26–30
Zhao YL (2020) Study and application of analytic hierarchy process of mine geological environment: a case study in Hainan Island. Remote Sens Land Resour 32:148–153
Zhao GL, Peng PH (2021) An analysis of ecological environment change in Panzhihua City, Sichuan Province, based on RSEI. J Mt Sci 39:842–854

Chapter 63
Study on the Factors Affecting the Adsorption/Desorption Performance of Activated Carbon in the Application of Industrial VOCs Treatment Technology

Zhipeng Zhuang, Xingui Yi, and Mingxin Xu

Abstract The effect of different adsorption temperature and different space velocity on the adsorption performance of activated carbon was investigated by experiments. After the adsorption of activated carbon was saturated, the effects of different desorption temperature, different desorption air volume and desorption times on the desorption performance of activated carbon were studied. The effect of water vapor on the adsorption and desorption performance of activated carbon was studied. The experimental results show that the higher the adsorption temperature is, the lower the saturated adsorption capacity of the activated carbon adsorbent is, the adsorption capacity of the adsorbent can still reach more than 60%, and the desorption amount can reach more than 80%.

Keywords Activated carbon · VOCs · Adsorption · Desorption · Waste gas treatment

63.1 Introduction

Activated carbon is the most widely used adsorbent for VOCs adsorption in our country because of its large specific surface area, adsorption capacity, wide pore size distribution and strong selectivity (Jin et al. 2006; Song 2020; Zhu et al. 2020; Zhu 2021). However, at present, there is still unclear research on the performance of activated carbon adsorbent itself, such as adsorption/desorption performance, working conditions, stability, resistance, which leads to the problems of low utilization rate of adsorbent, small adsorption capacity, easy penetration and so on in the actual

---

Z. Zhuang · X. Yi · M. Xu (✉)
Guangzhou Huake Environmental Protection Engineering Co. Ltd., Guangzhou 510655, China
e-mail: 383689450@qq.com

South China Institute of Environmental Sciences, Guangzhou 510655, China

J. Zhang et al. (eds.), *Environmental Pollution Governance and Ecological Remediation Technology*, Environmental Science and Engineering,
https://doi.org/10.1007/978-3-031-25284-6_63

industrial application of activated carbon adsorption technology (Zhou 2018), which seriously restricts the further application and development of this technology.

This experimental study, from the perspective of practical application, takes the characteristic adsorbent activated carbon as the research object, systematically explores the influence of adsorption/desorption conditions (space velocity, adsorption/desorption temperature, flow rate, etc.) on the adsorption performance, adsorption/desorption process, adsorption capacity, etc. of the adsorbent, and obtains the best working conditions of the adsorbent through relevant research, it is of great significance to the industrial application of VOCs treatment technology and the academic improvement in the field of VOCs adsorption.

63.2 Materials and Methods

63.2.1 Experimental Materials

With industrial activated carbon as adsorbent, the honeycomb activated carbon is crushed and ground into 40–60 mesh, packaged and stored for use. The adsorbent is high-purity nitrogen and toluene (500 ppm standard gas, Dalian date Gas Co., Ltd.). The adsorption bed is a fixed bed catalytic reaction experimental device (Tianda Beiyang Equipment Co., Ltd.).

63.2.2 Evaluation Standard of Adsorption Performance

The adsorption capacity of activated carbon adsorbent is generally calculated by the integration method of adsorption breakthrough curve. The increase of adsorption capacity in the adsorbent should be equal to the decrease of the amount of adsorbate in the inlet and outlet adsorbate gas. The adsorption capacity of adsorbate under this condition can be calculated by calculating and counting the decrease of inlet and outlet adsorbate. The calculation formulas (Song 2014) (63.1) and (63.2) are as follows:

$$Q_{\mathrm{e}} = \frac{K \times Q_{\mathrm{g}} \times \int_{\mathrm{t_0}}^{\mathrm{t}} (C_{\mathrm{in}} - C_{\mathrm{out}})\mathrm{dt}}{m} \tag{63.1}$$

$$q_{\mathrm{m}}\left(\mathrm{mg/m^3}\right) = \frac{q_v(\mathrm{ppm}) \times M}{24.5} \tag{63.2}$$

wherein, Qe refers to the saturated adsorption capacity of adsorbent per unit mass, unit mg/g, K refers to the conversion coefficient between volume concentration and mass concentration (under standard conditions, toluene coefficient is 4.11, under

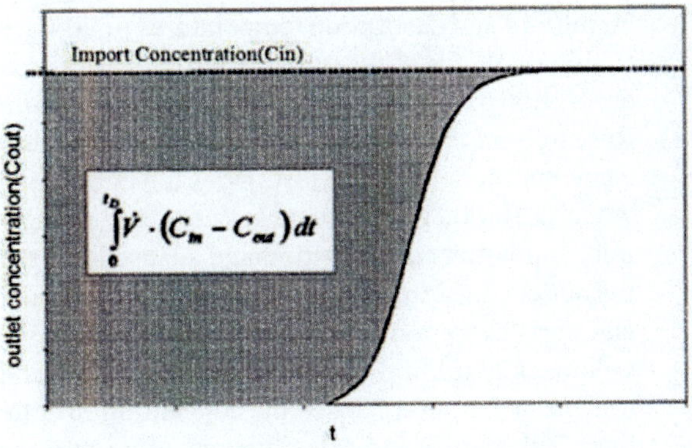

Fig. 63.1 Adsorption penetration curve integral method

normal temperature and pressure, toluene coefficient is 3.76), Qg refers to the gas flow, unit m^3/min, Cin refers to the inlet concentration. Cout refers to the outlet concentration, M refers to the adsorbent mass, and M refers to the molecular weight of toluene.

As shown in Fig. 63.1, the shaded area is the integral area s calculated by the adsorption penetration curve.

63.2.3 Experimental and Analytical Methods

In the application of industrial VOCs waste gas treatment, activated carbon has better adsorption capacity and relatively low cost. This study takes activated carbon as the research object, and systematically explores the relationship between the adsorption performance of activated carbon under different adsorption conditions. The main research contents are as follows:

(1) The influence of adsorption temperature on the adsorption effect of adsorbent (adsorption process): under the condition of a certain VOCs concentration (taking the common toluene gas as VOCs gas, the concentration is 50 ppm) and space velocity of 10,000 h^{-1}, the toluene concentration at the inlet and outlet of the adsorption device is measured respectively at the adsorption temperature of 30 °C and 40 °C, the adsorption process of adsorbent is explored, and the adsorption penetration curve is obtained. Then, the adsorption capacity of the adsorbent at different temperatures is explored comparatively, and the adsorption rate of the adsorbent at a certain inlet concentration is evaluated through the penetration curve, and the optimal adsorption rate site of the adsorbent is obtained.

(2) Research on the influence of desorption temperature on desorption effect: when the adsorbent is saturated, under the conditions of desorption air volume of

20 ml/min and desorption temperatures of 70 °C, 90 °C and 100 °C, the concentration of VOCs at the outlet of different desorption times is detected, and the desorption performance of activated carbon at different temperatures is obtained.

(3) Research on the adsorption/desorption cycle performance of activated carbon: carry out the adsorption test under the conditions of temperature 30 °C, space velocity 10,000 h^{-1}, toluene gas concentration 50 ppm, measure and record the inlet and outlet concentration and adsorption time of the adsorption bed; When the outlet concentration exceeds 30% of the inlet concentration, stop adsorption and start desorption. The desorption temperature is 100 °C, and the desorption air volume is 20 ml/min. test and record the desorption time and the concentration of toluene at the outlet. When the concentration at the outlet of desorption is lower than 100 ppm, it indicates the end of an adsorption/desorption cycle. Repeat the above experiment for 10 times, and compare the changes of adsorption curve and adsorption capacity of activated carbon to obtain the cyclic adsorption performance of activated carbon.

63.3 Results and Discussion

63.3.1 *Influence of Adsorption Temperature on Adsorption Effect*

It can be seen from Fig. 63.2 that the adsorption temperature has a great influence on the adsorption penetration time of activated carbon and the overall adsorption capacity of adsorbent. When the adsorption temperature is 30 °C, the adsorption penetration time of activated carbon is close to 732 min, while when the adsorption temperature rises to 40 °C, the adsorption penetration time is about 482 min, which is about 250 min less than that of the two. It can be explained that the adsorption temperature can greatly affect the penetration time of adsorbent within a certain range. With the increase of temperature, the adsorption penetration time of activated carbon decreases. At the same time, the adsorption capacity at different temperatures is calculated by formula (63.1), and the results are shown in Table 63.1. At 30 °C, the saturated adsorption capacity of activated carbon is 30.88 mg/g, while at 40 °C, the saturated adsorption capacity of activated carbon is 25.18 mg/g. the adsorption process temperature of adsorbent is increased by 10 °C, and the adsorption capacity of adsorbent is reduced by 18.5%. The adsorption temperature has a great impact on the adsorption capacity of adsorbent. In the industrial adsorption process, the adsorption temperature of the adsorption process should be reduced as much as possible to improve the adsorption capacity of the adsorbent.

The adsorption rate of adsorbent is a key parameter to characterize the speed of adsorbent adsorption process. In the process of adsorption, the adsorption penetration curve is differentiated, and the adsorption rate of activated carbon when cout/cin = 0.5 is calculated. When the temperature is 30 °C, the slope of the curve at this position is 0.16. When the temperature is 40 °C, the slope of the curve at this position is 0.86.

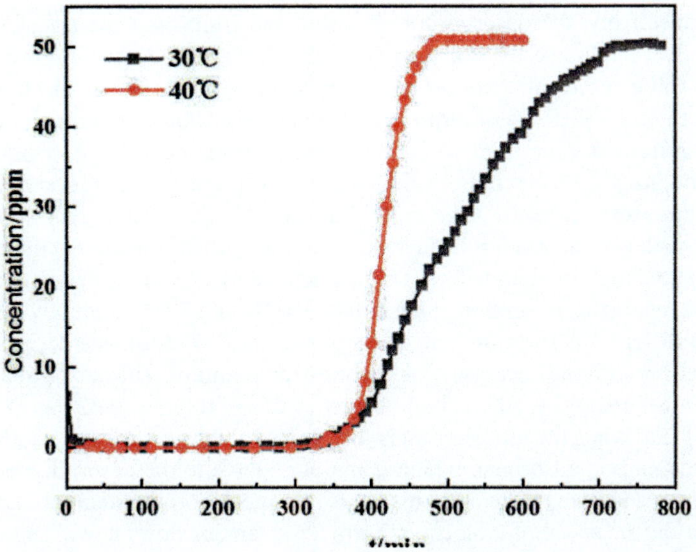

Fig. 63.2 Adsorption breakthrough curves of activated carbon at different temperatures

Table 63.1 Calculation results of saturated adsorption capacity of activated carbon at different temperatures

Adsorption temperature	Integral area	Qg	m (g)	Qe (mg/g)
30	25391.37	120×10^{-6}	0.3710	30.88
40	20708.00	120×10^{-6}	0.3710	25.18

It can be seen that the adsorption capacity decreases with the increase of adsorption temperature, while the adsorption rate increases with the increase of temperature. Therefore, in industrial applications, considering the cost of tail gas treatment and energy consumption, the best condition for activated carbon adsorption is adsorption at room temperature, and the treatment efficiency is the best.

63.3.2 *Influence of Desorption Temperature on Desorption Effect*

The desorption performance of adsorbent is mainly affected by desorption temperature and desorption wind speed. The desorption rate can reflect the strength of the force between toluene and activated carbon to a certain extent (Luo 2019; Yue 2020). When the adsorbent is saturated at 30 °C, the space velocity is 5000 h^{-1}, the toluene concentration is 50 ppm, and the total flow is 120 ml/min, the desorption air volume is set at 20 ml/min, and the desorption temperature is adjusted to 70, 90, and

100 °C respectively, the desorption curve obtained from the experiment is shown in Fig. 63.3. It can be seen from the figure that during the desorption process, the maximum value of the outlet concentration increases with the increase of the desorption temperature. When the desorption temperature is 100 °C, the maximum value of the desorption concentration is 3700 ppm, while when the desorption temperature is 70 °C and 90 °C, the maximum value of the desorption concentration is 850 ppm and 2100 ppm respectively. At the same time, at 100 °C, the emergence rate of the maximum outlet concentration is the fastest, and the rate of the outlet concentration falling from the maximum to below 100 ppm is the fastest, and the desorption time is the shortest, while the desorption curve obtained at 90 and 70 °C gradually tends to be flat. Under different desorption temperatures, the amount of toluene desorbed gradually decreases with the decrease of desorption temperature. The desorption amount of toluene (18.29 mg/g) at 100 °C is almost twice that of toluene (9.52 mg/g) at 70 °C. This shows that when the temperature is low, some toluene cannot be desorbed from the activated carbon adsorbent, which is not conducive to the desorption activation process of the adsorbent. The main reason is that the desorption temperature gradually decreases, the speed of molecular thermal movement slows down, and the force between toluene and activated carbon increases, so the desorption time increases. Therefore, in the industrial VOCs treatment technology, the desorption temperature of activated carbon can be appropriately increased within a certain range, thereby increasing the activation rate of adsorbent.

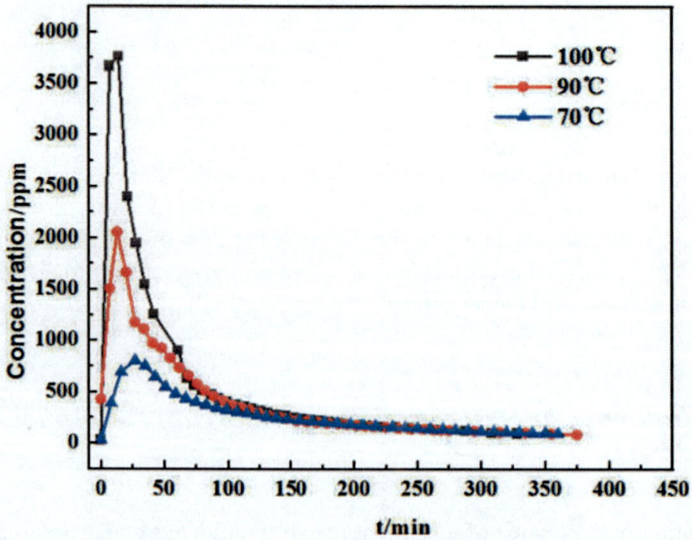

Fig. 63.3 Desorption curves of activated carbon at different temperatures

63.3.3 Influence of Adsorption/Desorption Times on VOCs Treatment Effect

Adsorption saturated activated carbon is generally highly enriched with VOCs harmful substances, which must be recycled as hazardous waste (Wang 2020; Wu 2020). In practical industrial applications, the adsorption/desorption cycle performance of activated carbon is a key factor in the recycling of activated carbon (Lv 2020). In this study, combined with the activated carbon adsorption/desorption process in practical application, the adsorption process temperature is 30 °C, the toluene concentration is 50 ppm, the adsorption flow rate is 120 ml/min, and the space velocity is 10000 h^{-1}. When the toluene concentration at the outlet of the adsorbent is 30% of the imported toluene concentration, the adsorption stops and the desorption starts. The desorption temperature is 100 °C, and the flow rate is 20 ml/min. Cycle the adsorption/desorption for 10 times, and measure the adsorption/desorption process of the adsorbent respectively. The experimental results are shown in Figs. 63.4 and 63.5. It can be seen from Fig. 63.4 that the maximum first adsorption capacity of activated carbon adsorbent is 7.87 mg/g, and the adsorption penetration curve of activated carbon for 2–10 times is basically the same, and the adsorption capacity remains at 5.2 mg/g, which indicates that after the first adsorption/desorption, the adsorption capacity of activated carbon is reduced to 67.09% because some adsorbates cannot be desorbed, while the adsorption capacity of activated carbon remains basically unchanged during the 2–10 cycle adsorption desorption process, This shows that in the process of cyclic adsorption/desorption, the adsorption efficiency of activated carbon can be maintained at nearly 70%, with good cyclic adsorption performance. At the same time, it can be seen from Fig. 63.5 that after ten times of desorption, the activated carbon can still be desorbed in 150 min, and the desorption efficiency reaches 80%, indicating that after ten times of desorption, the activated carbon can still maintain a good desorption rate.

63.4 Conclusion

In this paper, the adsorption/desorption performance of activated carbon (AC) under different working conditions (adsorption bed space velocity, desorption temperature) was studied systematically, and the adsorption characteristics of AC were summarized, the effects of adsorption stability, resistance and cycle number of adsorption/desorption on the adsorption efficiency of activated carbon were investigated. The results are as follows:

(1) By controlling different adsorption temperatures, the adsorption breakthrough curves of activated carbon at different adsorption temperatures were measured, and the corresponding saturated adsorption capacity and adsorption rate were calculated. The results showed that the higher the adsorption temperature was,

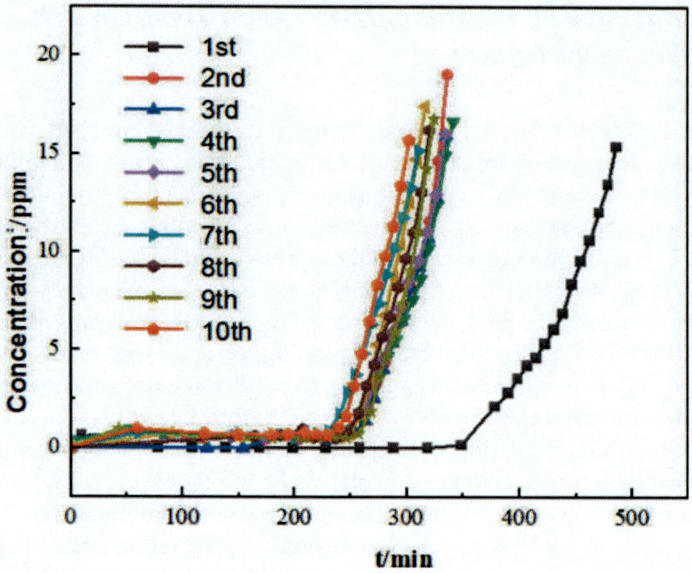

Fig. 63.4 Adsorption breakthrough curve of activated carbon with different times of adsorption and desorption

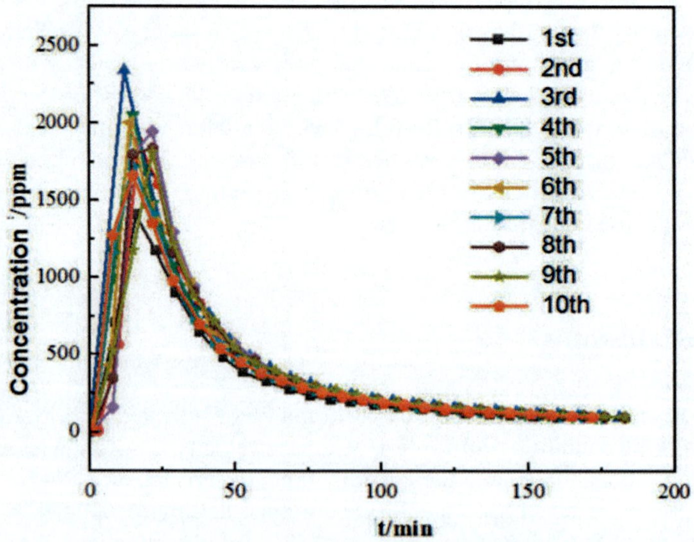

Fig. 63.5 Desorption curves of activated carbon with different adsorption times

the lower the saturated adsorption capacity of activated carbon adsorbent was, while the complete adsorption time of adsorbent was basically not affected.

(2) When the desorption temperature increased from 70 to 100 °C, the desorption amount of toluene on activated carbon doubled, while the increase of desorption air volume had little effect on the desorption amount of activated carbon adsorbent.

(3) After ten times of adsorption/desorption, the adsorption capacity of the adsorbent can still reach more than 60%, and the desorption capacity can reach more than 80% (except the first desorption), indicating that the activated carbon has good adsorption/desorption cycle performancc.

References

Jin X, Ma LM, Wang LW (2006) Research progress of organic matter adsorption on chemically modified activated carbon [J]. Jiangsu Environ Sci Technol (S2):43–45

Luo XH (2019) Study on the adsorption and desorption properties of VOCs on honeycomb ceramics loaded with molecular sieves [D]. Zhejiang Normal University

Lv SL (2020) Preparation and modification of rice hull based activated carbon and its adsorption mechanism for phenol [D]. Beijing University of Chemical Technology

Song JF (2014) Adsorption of VOCs on activated carbon and its structure-activity relationship [D]. Central South University

Song ZM (2020) Terminal treatment technology and selection method of volatile organic compounds [J]. Environ Dev 32(09):84–86

Wang JY (2020) Study on vacuum desorption behavior and flame retardant modification of activated carbon for VOCs adsorption [D]. South China University of Technology

Wu SC (2020) Study on VOCs law and regeneration characteristics of activated carbon thermal nitrogen cyclic desorption coating [D]. South China University of Technology

Yue X (2020) Study on dynamic adsorption/desorption performance of VOCs adsorbent and its modification [D]. Taiyuan University of Technology

Zhou EL (2018) Study on regeneration and adsorption properties of waste powdered activated carbon [D]. Beijing University of Chemical Technology

Zhu YR (2021) Application of volatile organic compounds treatment technology [J]. Resour Conserv Environ Protect (02):109–110

Zhu L, Shen D, Luo KH (2020) A critical review on VOCs adsorption by different porous materials: species, mechanisms and modification methods. J Hazard Mater 389:122102

Chapter 64
Correlation Analysis of China's Renewable Energy and Carbon Emissions

Jian Huang, Jiansheng Hou, Yingcong Wang, and Chong Ji

Abstract CO_2 emissions are of global concern because of climate change. China has become the largest CO_2 emitter in the world and presently accounts for 30% of global emissions. This paper selects the energy consumption from 1990 to 2018, combined with the carbon emission coefficients of various energy sources to calculate the carbon emissions of each year to obtain the carbon emission time series, and then selects the renewable energy consumption from 1990 to 2018 to obtain the time series. The two series use Eviews to perform unit root test, cointegration test and Granger causality test to conclude that there is a long-term cointegration relationship between renewable energy consumption and total carbon emissions; there is only between renewable energy and carbon emissions for a single cause and effect relationship, an increase in renewable energy consumption will reduce carbon emissions, and carbon emissions will not have a significant impact on renewable energy.

Keywords Renewable energy · Carbon emissions · Unit root test · Cointegration test · Granger causality test

64.1 Introduction

Since the Industrial Revolution, the economy has developed rapidly and scientific and technological progress has been extremely rapid. At the same time, the industrial revolution has brought many negative impacts on society, such as excessive

J. Huang · J. Hou · Y. Wang
Jinhua Power Supply Company, State Grid Zhejiang Electric Power Company, Jinhua 321000, China
e-mail: wyc20084752@163.com

C. Ji (✉)
College of Economics and Management, Shanghai University of Electric Power, Shanghai 200090, China
e-mail: zhao_wenhui@shiep.edu.cn

J. Zhang et al. (eds.), *Environmental Pollution Governance and Ecological Remediation Technology*, Environmental Science and Engineering,
https://doi.org/10.1007/978-3-031-25284-6_64

consumption of energy and excessive carbon emissions, this is a great threat to the long-term survival and development of mankind (Bikse et al. 2022).

Constrained by the increasingly serious problems of climate warming and energy shortage, countries all over the world are focusing on promoting carbon emission reduction and developing renewable energy for the sustainable development of mankind. China has always been a big energy consuming country, especially in recent years, with the rapid economic development of our country and the drawbacks brought about by the industrial revolution gradually exposed to the public, makes the international community to China, the rising power of the east to give high attention. In order to adapt to the international situation, our government has been analyzing the factors affecting carbon emission, so as to achieve carbon emission reduction at the root, and has formulated a series of carbon emission reduction policies and regulations (An and Zhou 2021).

With China's increasing emphasis on carbon emission reduction and renewable energy development, a group of scholars focused on this research. Based on the panel data model of the eight asia–pacific countries, NIU Shuwen and others draw the following conclusions: in developed countries, gross national product (GNP) is the cause of energy consumption, while there is no correlation in developing countries. Energy consumption is the cause of carbon dioxide emissions, but carbon dioxide emissions have no effect on energy consumption (Niu et al. 2010). Wang Liang et al. used variance decomposition and impulse response function to analyze data from 1980 to 2009 for renewable energy consumption, carbon dioxide emissions and economic growth, it is found that long-term use of energy-saving and emission-reducing policies can improve economic efficiency, and the impact of renewable energy on carbon emissions is relatively weak (Wang and Zhao 2013). Zhao Jinwen and fan jitao used the STP model and Granger causality test found that China's economic growth and energy consumption growth from 1956 to 2005 were nonlinear, and the economic growth rate was lower than the growth rate of energy consumption, but the study of 1977–2005 found a linear relationship between economic growth and energy consumption (Zhao and Fan 2007). Chen et al. analyzed the relationship between renewable energy consumption and carbon emission intensity in order to control carbon emission intensity, and concluded that there is a negative correlation between renewable energy consumption and carbon emission intensity in China (Chen and Huang 2016). Zhao et al. calculated and analyzed the energy consumption and carbon emissions of Guangzhou from 1992 to 2007, and found that the decrease of energy intensity can restrain the growth of carbon dioxide emissions (Zhao et al. 2010). Peng Jiawen et al. demonstrated the weak decoupling of economic and energy carbon emissions for most of 1980–2008 by constructing a decoupling analysis model of economic and energy carbon emissions (Peng et al. 2011). Ang studied the dynamic relationship among France's per capita carbon dioxide emissions, trade exports, and energy consumption (Ang 2007); Halicioglu used the ARDL boundary cointegration test method to study Turkey's per capita carbon dioxide emissions, energy consumption, and national income. The relationship between the four parties in international trade (Halicioglu 2009).

Therefore, it is necessary for China's current carbon emissions and renewable energy consumption before the existence of the relationship between further research. This paper selects the energy consumption from 1990 to 2018, combined with the carbon emission coefficients of various energy sources to calculate the carbon emissions of each year to obtain the carbon emission time series, and then selects the renewable energy consumption from 1990 to 2018 to obtain the time series. Through unit root test, co-integration test and Granger causality test, it is concluded that there is a long-term co-integration relationship between renewable energy consumption and total carbon emissions.

64.2 Research Methodology

64.2.1 Data Source

The data for this article is from the National Bureau of Statistics of the People's Republic of China's China Statistical Yearbook, which covers the period from 1990 to 2018. China's statistical agencies have not carried out specific statistics on total carbon emissions. Considering the reliability and availability of the data, this paper uses the energy emission data published on the official website of National Bureau of Statistics of the People's Republic of China. The specific method is as follows: first, the total amount of energy consumption from 1990 to 2018 and the total amount of various energy categories constituting the total energy consumption are collected from the China Statistical Yearbook, calculate the respective share of coal, oil and natural gas in the energy category, and take these three sources because they are the total energy component, and then multiply the total and the share, multiply the corresponding carbon emission coefficient, add the three together, the estimated value is the total carbon emissions.

$$TPL = \sum_i (TPL_i/S_i)(S_i/S)S = \sum_i F_i H_i S \tag{64.1}$$

where TPL represents the total carbon dioxide emissions resulting from various energy consumption; i indicate the type of energy chosen, selected as three in the total amount of energy consumption accounted for a larger type of energy, respectively: coal, oil, natural gas; S represents the total energy consumption from 1990 to 2018 collected from the China Statistical Yearbook; F_i represents coal, oil, natural gas and other corresponding energy consumption type I carbon emission coefficient; H_i represents the share of Category I Energy in total energy.

The FI value comes from Table 64.1.

Table 64.1 Carbon emission factor

	Coal	Oil	Gas
DOE	0.702	0.478	0.389
Japan institute of energy economics	0.756	0.586	0.449
Chinese academy of engineering	0.680	0.540	0.410
The greenhouse gas control project of the state environmental protection bureau	0.748	0.583	0.444
Ministry of science and technology of the People's Republic of China	0.726	0.583	0.409
National development and reform commission	0.748	0.583	0.444
Mean value	0.727	0.559	0.424

64.2.2 Maintaining the Integrity of the Specifications

Unit root test

Classical economic theory is mostly based on the stability of time series and assumes that the correlation coefficient between variables is normal distribution. But research from the econometrics suggests that most economic variables are not as smooth as they should be. We use the Monte Carlo method method to analyze the correlation coefficients of nonstationary economic variables, and conclude that when the time series is nonstationary, the correlation coefficients actually follow an inverted U-shape and a U-shape distribution, and when we increase the sample size and the single integer order of the time series, the probability of rejecting the hypothesis that the coefficient of explanatory variable is 0 will also increase. In layman's terms, when the sample is large and the single integer order is high, we can test any two independent and unrelated variables at will and get very significant results for their correlation coefficient, this leads to false conclusions when we analyze the data later. Therefore, if we use the non-stationary time series to carry on the regression analysis under the condition that the sample is very big and the single integral order is very high, we can get the result that they have the relativity. But in reality, it's just a false regression, a false regression. We must consider the data can not be analyzed in this case, so, in the non-stationary variable regression analysis, we have to do unit root test to test the time series stability.

Cointegration test

Cointegration is that if a sequence of two or more non-fixed variables has a linear combination and the linear combination sequence is stationary, we say it has a cointegration relationship. The regression of single integral variable without cointegration relation is still pseudo-regression. So before we analyze the Granger causality, we do cointegration validation. The function of co-integration theory is to correctly explain economic phenomena and forecast phenomena. The economic significance of cointegration is that although each of the two variables has its own long-term fluctuation law, it may also be the long-term stable relationship between them, that is to say, cointegration reveals this long-term stable equilibrium relationship, regression without a unified integration variable is still pseudo-regression, so a cointegration test is needed before regression analysis. AEG two-step method is commonly used to do co-integration test, this paper also used this method to do co-integration test for LNTPL and LNNHL time series to verify whether there is a co-integration relationship between carbon emissions and renewable energy consumption.

Granger causality test

The Granger causality method is used to analyze the causal relationship between variables from a statistical point of view. The method works like this: if *XT* and *YT* are stable time series variables:

$$x_t = c_1 + \alpha_1 x_{t-1} + \alpha_2 s_{t-2} + \ldots + \alpha_p x_{t-p} + \beta_1 y_{t-1} + \beta_2 y_{t-2} + \ldots + \beta_p y_{t-p} + u_t \tag{64.2}$$

OLS can then be used to estimate the sum of the square residual RSS of the above formula and compare this result with the square and RSS of the unit autoregressive residuals of the unit, as

$$F_1 = \left[(RSS_0 - RSS_1)/p\right]/\left[RSS_1/(T - 2p - 1)\right] \tag{64.3}$$

If *F1* is greater than the 5% critical value of the F (p, T-2p-1) distribution, then Y is accepted as the Percy Grainger cause of X, whereas the Percy Grainger cause of X is rejected as Y. If *XT* and *YT* are non-stationary time series variables, the error correction model is needed to determine the causal relationship between the variables.

In the Granger causality test, the most important thing is to determine the lag time length. If a random decision is made, an error will occur in the test result. In this study, the length of the lag time was determined according to the Akaike information criterion and Schwaz's criterion.

$$AIC = \log\left[\frac{\sum_{t=1}^{T} ut}{T}\right] + \frac{2k}{T} \tag{64.4}$$

$$SC = \log\left[\frac{\sum_{t=1}^{T} ut}{T}\right] + \frac{k \log T}{T} \tag{64.5}$$

where *UT* is the residuals and *k* is the lag. The first term on the right side of the two equations decreases as *K*, and as the second term in *k* increases. Therefore, we determine the optimum value by continuously increasing the value of K until *AIC* and *SC* get the minimum value.

64.3 Result

64.3.1 Time Series Analysis

The total amount of carbon emissions is calculated from Formula (64.1), which is shown in Figs. 64.1 and 64.2.

Logarithm for total carbon emissions and renewable energy consumption is shown in Fig. 64.3. Both *LNTPL* and *LNNHL* are on the rise. This shows that there is a certain correlation between the two variables, and the time series of the two variables can be preliminarily judged to be non-stationary. As can be seen from the graph, the dependence between the two variables is strong, but if we directly analyze them, it may lead to some problems such as Pseudo-regression, and the statistic distribution of many parameters is no longer standard, so we need to do the unit root test on the sequence of two variables.

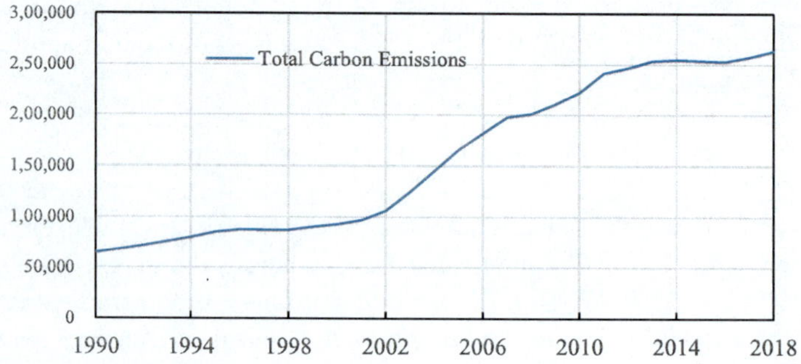

Fig. 64.1 Total carbon emissions

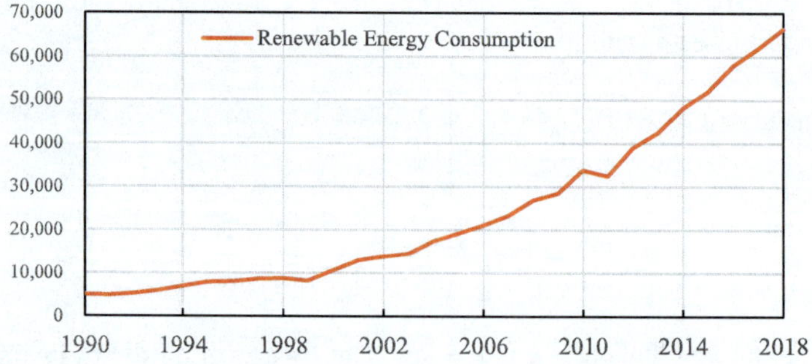

Fig. 64.2 Renewable energy consumption

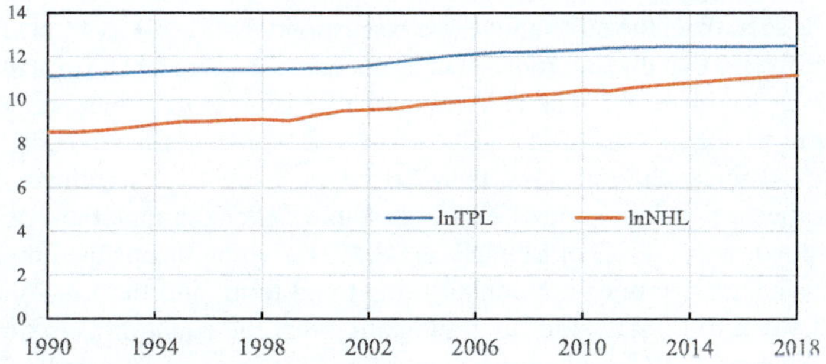

Fig. 64.3 Variable sequence diagrams for LNTPL and LNNHL

64.3.2 Unit Root Test

Because causality is based on the premise that the time series is stable, it is necessary to verify whether the time series is stable before analyzing the causality. The stationary time series is a stochastic process in which the random features of the data do not change with time. On the contrary, if the random features change with time, the sequence is proved to be non-stationary. If the time series is stable, we can make a model of the time series, and then regression analysis of the model, otherwise there may be pseudo-regression.

In the analysis of the stationarity, we can use three conditions to prove whether the time series is stable or not: First of all, the covariance of two periods is only related to the interval; Second, the covariance of two periods is not related to the selected period; Finally, mean and variance remain constant. When the time series satisfies the above three conditions, then it is stationary. In this paper, the unit root test is used to verify the stationarity.

First, the total carbon emission of the original sequence TPL was tested by ADF unit root test method. The results are as Table 64.2.

Table 64.2 Results of TPL unit root test

Variable	Ln TPL	Ln NHL
Difference number	2	1
(C, T, K)	(n, n, 1)	(C, n, 1)
DW	1.93	2.09
ADF	−2.9	−4.68
5% threshold	−1.96	−2.98
1% threshold	−2.66	−3.71
Result	I(2)***	I(1)***

*Represents $P \leq 0.05$; ** Represents $P \leq 0.01$; *** Represents $P \leq 0.001$

As can be seen from the table above, the corresponding (C, T, K) of LNTPL is (N, N, 1), which means that the unit root test does not contain a constant term, a time trend term, and a lag number of 1. The DW value of 1.93 is close to 2, indicating that the equation is in the correct form. The value of ADF-2.90 is less than the critical value of "5% level"-1.96, and it is also less than-2.66 at the level of 1% significance, which shows that the tested time series LNTPL is stable under the quadratic difference. The corresponding (C, T, K) of LNNHL is (C, N, 1), which means the unit root test contains constant term, does not contain time trend term, and the LAG number is 1. The DW value of 2.09 is close to 2, indicating that the equation is in the correct form. The value of ADF −4.68 is less than the critical value of 5% level −2.98, and also less than −3.71 at the 1% significance level, which shows that the tested time series LNNHL is stable under one difference.

The original time series of LNTPL and LNNHL are unstable, but their primary and secondary differential sequences are stationary. So we can't construct the traditional econometrics directly, and we still need to carry on the cointegration test.

64.3.3 Cointegration Test

In this paper, the single equation cointegration test is used to test whether there is cointegration relationship among several variables. The results are shown in Table 64.3

The Prob. value of C and LNNHL is 0 respectively, less than 0.05. So the equation coefficients are all significant. Then *Eviews* is used to make residuals, and then apply ADF unit root tests on the resulting residuals. If the residual sequence is unstable, there is no cointegration relationship between the multivariables. If the residual sequence is stable, there is cointegration relationship between the multivariables. ADF test results are shown in Table 64.4.

In the above results, the corresponding (C, T, K) of Residual error is (N, N, 1), which means that the unit root test does not contain a constant term, a time trend term, and a lag number of 1. The DW value of 1.98 is close to 2, indicating that the equation is in the correct form. The value of ADF-3.50 is less than the critical value of "5% level"-1.95, and it is also less than-2.66 at the level of 1% significance. It is proved that LNTPL and LNNHL have cointegration relationship.

Table 64.3 Regression function results of independent variable LNTPL and dependent variable LNNHL

Variable	Prob
C	0.000
LNNHL	0.000

Table 64.4 Test results of the stationarity of residual sequence

Variable	Residual error
Difference number	1
(C, T, K)	(n, n, 1)
DW	1.98
ADF	−3.5
5% threshold	−1.95
1% threshold	−2.66
Result	I(1)***

*Represents $P \leq 0.05$; ** Represents $P \leq 0.01$; *** Represents $P \leq 0.001$

Table 64.5 LNTPL and LNNHL granger causality results

Null hypothesis	Prob
LNTPL does not granger cause LNNHL	0.3427
LNNHL does not granger cause LNTPL	0.0127

64.3.4 Granger Causality Test

After establishing the long-term co-integration relationship between carbon dioxide emissions and renewable energy consumption, Granger causality test was used to objectively test the causality among these variables. The results are shown in Table 64.5 below:

The Prob. of the first line, LNTPL does not Granger Cause LNNHL is 0.3427, greater than 0.05,which is at a significant level of 5% the original hypothesis holds, indicating that LNTPL is not the Cause of Percy Grainger in LNNHL. The second line, LNNHL does not Granger Cause LNTPL, has an associated probability of 0.0127, which indicates that LNNHL is the Cause of Percy Grainger of LNTPL.

64.4 Conclusion

In this paper, the data of non-renewable energy sources such as oil, natural gas, coal, renewable energy sources such as hydropower, nuclear power and wind power in 21 years from 1990 to 2018 are selected from China Statistical Yearbook, the formula is used to calculate the time series of China's total carbon emissions and renewable energy consumption, and the logarithmic form is given respectively. Then we use econometrics software eviews to test the unit root, cointegration and Percy Grainger test of the time series, and draw the following conclusions: Renewable energy consumption and total carbon emissions there is a long-term cointegration relationship; The Granger causality test shows that only renewable energy consumption is the Percy Grainger's cause of carbon emissions, and that increased renewable energy consumption reduces carbon emissions, and that carbon emissions are not the Percy Grainger's cause of renewable energy consumption, that is, the impact of carbon emissions on renewable energy is not significant. They only have a single causal relationship between them, and not as many scholars have previously studied, is a double causal relationship.

References

An F, Zhou D (2021) Carbon emission reduction potential and its influencing factors in China's coal-fired power industry: a cost optimization and decomposition analysis. Environ Dev Sustain 24:3619–3639

Ang JB (2007) CO_2 emissions, energy consumption, and output in France [J]. Energy Policy 35:4772–4778

Bikse V, Grinevica L, Rivza B et al (2022) Consequences and challenges of the fourth industrial revolution and the impact on the development of employability skills [J]. Sustainability 14(2):6970–6970

Chen A, Huang K (2016) Time series analysis of renewable energy consumption and carbon emission intensity in China [J]. Technol Ind 16(02):139–145

Halicioglu F (2009) An econometric study of CO_2 emissions, energy consumption. Income and foreign trade in Turke [J]. Energy Policy 37:1156–1164

Niu S, Li X, Luo G (2010) An empirical study on the relationship among energy consumption, economic growth and carbon emission—based on panel data of eight Asia-Pacific countries [J]. Chin Soft Sci 15(5):12–19

Peng J, Huang X, Chung T, Zhao Y (2011) Decoupling of China's economic growth and energy carbon emissions [J]. Resour Sci 33(04):626–633

Wang L, Zhao T (2013) Dynamic relationship among renewable energy consumption, economic growth and carbon emission in China [J]. Tech Econ 11:99–104

Zhao J, Fan J (2007) An empirical study on the relationship between economic growth and energy consumption [J]. Econ Res 08:31–42

Zhao G, Chen J, Cui H et al (2010) 1992–2007 study on carbon emissions from energy consumption in Guangzhou [J]. Resour Ind 12(6):180–184

Chapter 65
Influence of Dry Distillation on Spontaneous Combustion Characteristics of Oil Shale

Jun Fang, Dawei Wang, and Xinkuan Cao

Abstract In order to study the effect of dry distillation on the spontaneous combustion characteristics of oil shale, samples collected from the Fushun West Open Pit before and after dry distillation were taken as the experimental research objects. And the laws of oxygen consumption, gas release and heat changes were measured by the high-temperature-programmed experimental system and thermal analyzer. The experimental results show that dry distillation has a great influence on the spontaneous combustion characteristics of oil shale. Due to the removal of some kerogen by dry distillation, the combustion efficiency is increased, but the calorific value of oil shale is reduced. The concentration of carbon and oxygen gases precipitated by spontaneous combustion is increased by dry distillation, and the concentration of hydrocarbon gases is reduced.

Keywords Dry distillation · Oil shale · Spontaneous combustion characteristics

65.1 Introduction

Oil shale, also known as kerogen shale, is mostly associated with coal and has high mineral content, and is a beneficial mineral resource (Ping et al. 2005; Xue 2016). The geological reserves of oil shale in China are very rich. The proven reserves are second only to the United States, Brazil and Russia. Most of the reserves are located in the northeast, and it is the most abundant in Fushun area (Hongtao 2011). In the past, oil shale was often used as a stripping material of open-pit coal mines to fill underground mines, or was directly discharged to dumps and turned into industrial waste together with coal gangue (Xue 2016). The stacking of oil shale gangue occupies a large area of urban land, and the harmful gases released by the spontaneous combustion of oil shale gangue will pollute the surrounding environment and endanger human health (Lanyun et al. 2005). In recent years, Fushun Mining Group has carried out various

J. Fang · D. Wang (✉) · X. Cao
CCTEG Ecological Environment Technology Co., Ltd., Beijing, China
e-mail: 454445817@qq.com

J. Zhang et al. (eds.), *Environmental Pollution Governance and Ecological Remediation Technology*, Environmental Science and Engineering,
https://doi.org/10.1007/978-3-031-25284-6_65

explorations in the comprehensive development and utilization of oil shale, especially for the rich ore with an oil content more than 4.7%, which can be extracted after low temperature dry distillation treatment (Hongtao 2011; Wan 2016; Yang et al. 2015). The waste residue after oil refining can be used as raw material for burning brick factory or cement factory (Hongtao 2011), which makes it feasible to transform the oil shale from waste to treasure. At present, the dry distillation technology of oil shale is constantly maturing and developing. Through dry distillation, the physical and chemical properties of oil shale have been greatly changed, and its spontaneous combustion characteristics have also undergone great changes (Aiwu et al. 2015; Baolin 2013; Chen 2013; Guochao and Jun 2013; Shijie et al. 2013; Wan 2016; Wang and Shuai 2021; Yang et al. 2015).

Aiming at the change of retorting process on the spontaneous combustion characteristics of oil shale, this paper selects oil shale samples from Fushun West open-pit mine, and combines the experimental data to analyze the law of oxygen consumption, gas release and heat change of oil shale samples before and after dry distillation. The research results can provide technical support for comprehensive development and utilization of oil shale.

65.2 Experimental Method

65.2.1 Experimental Device

65.2.1.1 High Temperature Programmed Experimental System

In this study, the XKGW-1 high temperature programmed experimental system is used. The experimental system mainly includes gas source, high-temperature reaction furnace, sample tank, and gas chromatograph, etc.

In order to ensure the smooth of air flow and reduce ventilation resistance, the air path of the experimental device adopts φ8 copper pipe, and the air source, connected to the rotameter through a tee, adopts SPB-3 automatic air pump and high-pressure gas cylinder to supply air. The shape of the high-temperature reaction furnace is box-type. The furnace is 65 cm long × 45 cm wide × 40 cm high, with the inner layer made of ceramic fiber and the outer layer carbon steel. Manual sampling is used for gas collection. According to the specific requirements of the experiment, when the temperature of the sample increases by 15 °C, a disposable syringe is used to connect the air outlet reserved on the gas outlet pipe. And the gas in the reaction sample tank was collected slowly and evenly. Gas chromatographs are used to analyze gas components.

Table 65.1 Temperature-programmed experimental conditions for spontaneous combustion

Granularity (mm)	Weight (g)	Air flow (ml/min)	Temperature range(°C)	Heating rate (°C/min)
Mixed particle size	5000	120	30–600	1

65.2.1.2 Thermal Analyzer

The TA-SDTQ600 thermal analyzer from the United States was selected for the experiment with dual functions of differential scanning calorimetry (DSC) and thermogravimetric analysis (TGA), that is, simultaneous DSC and TGA measurement of the sample are made at the same time. The test temperature range is from room temperature to 1500 °C. The measurable parameters include the phase transition within the material, the associated heat flow and weight change. The information provided is divided into endothermic and exothermic events without weight change and endothermic and exothermic events with weight change.

65.2.2 Experimental Conditions

65.2.2.1 Temperature-Programmed Experimental Conditions for Spontaneous Combustion

The test samples are oil shale and the one after dry distillation. The samples were crushed in the air and sieved into 5 different particle sizes: 0–0.9 mm, 0.9–3 mm, 3–5 mm, 5–7 mm and 7–10 mm, and 1 kg of each mixed particle size sample was taken for temperature-programmed experiments. The experimental conditions are shown in Table 65.1.

65.2.2.2 Experimental Conditions for Thermal Analysis

The sample was crushed to a particle size of 80–120 mesh, and a sample with a weight of 2 mg was weighed. The heating rate was set to 10 °C/min, and the gas distribution was 79% nitrogen +21% oxygen. The temperature was raised from room temperature to 1000 °C. In summary, the heat flow curves of two different samples are obtained.

65.3 Experimental Data Analysis

65.3.1 Analysis of Oxygen Consumption Law

Oxygen consumption law is one of the important macroscopic parameters which can show the spontaneous combustion severity. During the high temperature oxidation process, the oxygen consumption and temperature variation of the above two samples are shown in Fig. 65.1.

In the early stage of combustion, the oxidation reaction was relatively gentle. A large number of active functional groups were not activated. The oxygen consumption was small, and a certain amount of oxygen was adsorbed in the pores of the samples, so the oxygen concentration of the two samples decreased very slowly in the early stage of combustion. Between 150 and 200 °C, the oxygen concentration decline rate increases sharply, and the oxygen concentration of the oil shale reaches the lowest point at about 300 °C after dry distillation. Heated up continuously to 350–400 °C, the sample has just reached the ignition point, and various functional groups in the mineral molecules begin to decompose in large quantities. Most of the heat generated by the initial combustion is consumed by the decomposition of functional groups, resulting in a slight decrease in the combustion effect of the sample, so the oxygen concentration slightly increases. Continued heating up to about 450 °C, the combustion continues, the energy generated also increases, and the oxygen concentration gradually begins to decrease until the end of the experiment.

The overall decreasing trend of oxygen concentration in the two rock samples is similar, but there are still differences. The variation range of oxygen concentration of the dry-distilled kerogen decreases during the spontaneous combustion process, and the oxygen concentration in the later stage of the reaction is smaller than that of the oil shale without dry distillation, indicating that the dry distillation has a great

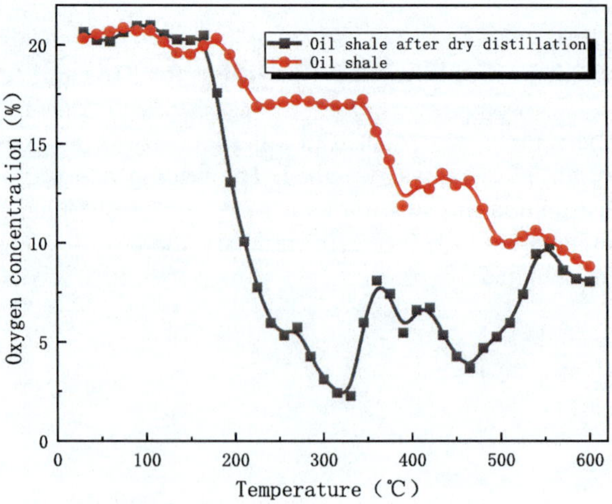

Fig. 65.1 Variation trend of oxygen concentration with temperature during the spontaneous combustion of the sample

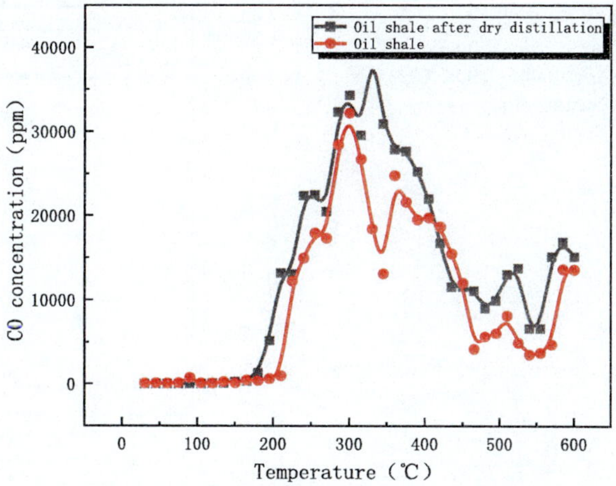

Fig. 65.2 The temperature-influenced variation of sample CO concentration

influence on the oxidative spontaneous combustion reaction of the kerogen. The analysis shows that the oil shale is rich in mica (containing aliphatic hydrocarbon structure), water and minerals, and the mineral content is higher than that of organic matter, making it more difficult to burn, and its burning efficiency can be effectively increased by dry distillation.

65.3.2 *Analysis of Gas Release Law*

65.3.2.1 Carbon Oxide Release Properties

Co

CO gas could be detected in both samples at the beginning of the experiment, and the CO concentration increased parabolically with the increase of temperature. The CO precipitation concentration of the oil shale after dry distillation increased slowly before 165 °C, and the growth rate increased rapidly after 165 °C. The concentration first decreased and then increased, fluctuating slightly between 300 and 315 °C, and reached a peak of 40720 ppm at 330 °C, then beginning to decline. Until the end of the experiment, the CO concentration was basically maintained at about 15000 ppm. The oil shale begins to show a rapid increase in CO concentration after 210 °C. The temperature when the concentration reaches the peak is lower than that of the oil shale after dry distillation, reaching a peak value of 32150 ppm at 300 °C, and then rapidly dropping to 13060 ppm within 300–345 °C. At 360 °C, it increased to 24740 ppm, and then gradually began to decrease, until the end of the experiment, basically maintaining between 5000 and 10000 ppm (Fig. 65.2).

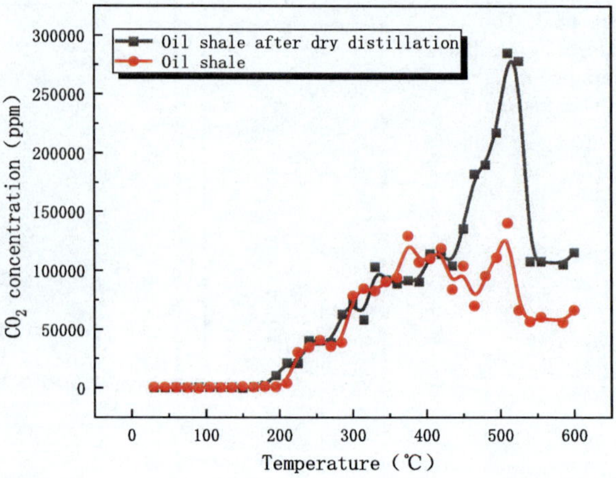

Fig. 65.3 The temperature-influenced variation of sample CO_2 concentration

The analysis shows that dry distillation has advanced the temperature value of the rapid increase of CO concentration during the spontaneous combustion of the rock sample, but has little effect on the temperature at which the CO concentration peaks. It shows that the kerogen composition affects the production of CO gas in the low temperature stage, and the dry distillation can effectively remove the kerogen composition, which makes the rapid increase of CO concentration about 50 °C earlier. In addition, the CO concentration of oil shale after dry distillation is higher than that of oil shale.

Co_2

At the beginning of the experiment, CO_2 gas can be detected in both samples, and the concentration is more than 2 times than that of CO. In the low-temperature oxidation stage, a large amount of adsorbed CO_2 is gradually released due to the van der Waals forces between the desorption and the sample macromolecules, and a small part of the water-oxygen complexes are converted into CO and CO_2 due to their reaction. As the temperature increases, the sample molecules undergo complete oxidation reaction with oxygen, releasing CO_2. The amount of gas increases continuously until it reaches a peak, and then gradually decreases (Fig. 65.3).

The CO_2 concentration released by oil shale began to accelerate at 200 °C, and reached a peak value of 140,400 ppm at 510 °C. During this period, there were large concentration fluctuations, and then the CO_2 concentration decreased rapidly. Dry distillation has advanced the temperature at which the CO_2 concentration of oil shale gradually increases, and the growth rate continues to accelerate before 435 °C, showing a quasi-uniform upward trend. The growth rate increases rapidly after 435 °C, reaching a peak of 285200 ppm at 510 °C, and then begins to decline. Compared with the two samples, the CO_2 precipitation concentration of oil shale is

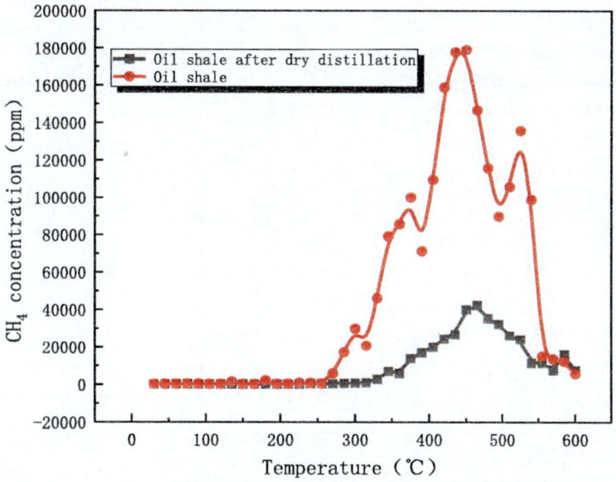

Fig. 65.4 The temperature-influenced variation of sample CH_4 concentration

much lower than that of oil shale after dry distillation, and the main gas product of spontaneous combustion of oil shale after dry distillation is CO_2.

65.3.2.2 Hydrocarbon Release Characteristics

CH_4

CH_4 is the most easily detected hydrocarbon gas and is also the main gaseous product of the oxidative pyrolysis process throughout the sample. The variation curve of its concentration with temperature is shown in Fig. 65.4.

Through comparative analysis, the change trend of CH_4 gas precipitation concentration during spontaneous combustion of the two samples is similar, and the amount of CH_4 precipitation in the low temperature oxidation stage is extremely low. At 285 °C, the oil shale first began to precipitate a large amount of CH_4. After dry distillation, the amount of CH_4 precipitation in the oil shale was small, reaching a peak of only 42,320 ppm at 465 °C, and the oil shale reached a peak at 450 °C with a concentration of 178,964 ppm.

After dry distillation of oil shale, part of kerogen is removed, and kerogen is rich in aliphatic hydrocarbon structure, so the CH_4 concentration of spontaneous combustion of oil shale after dry distillation is lower than that of oil shale.

C_2H_6

It can be seen from Fig. 65.5 that the change trend of the precipitation concentration of C_2H_6 gas during the spontaneous combustion of the two samples is similar, and the precipitation of C_2H_6 begins in the low-temperature oxidation stage, but the content is extremely low. After entering the high temperature combustion stage, the

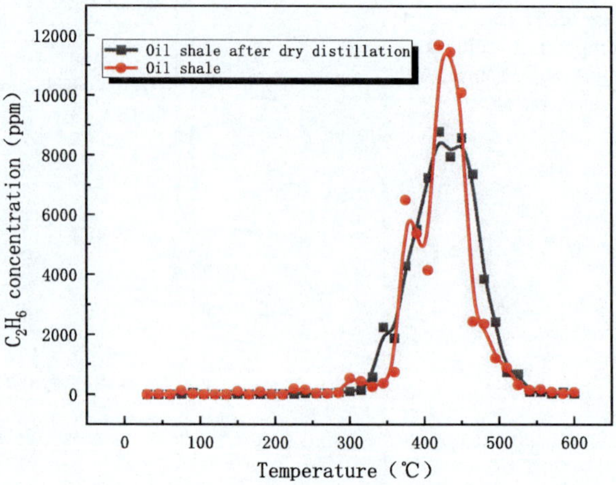

Fig. 65.5 The temperature-influenced variation of sample C_2H_6 concentration

production of C_2H_6 gas increases rapidly, and the gas concentration decreases rapidly after reaching the maximum value. The curve is parabolic.

In general, the precipitation of C_2H_6 gas from the two types of rock samples is extremely low, and both reach their peaks at 420 °C. After dry distillation, the peak concentration of C_2H_6 released from oil shale is 8800 ppm, and the peak concentration of C_2H_6 released from oil shale is 11710 ppm. Consistent with the law of producing CH_4 gas, the C_2H_6 precipitation concentration of oil shale after dry distillation is lower than that of oil shale.

C_2H_4

As shown in Fig. 65.6, the change trend of C_2H_4 is similar to that of C_2H_6, both in parabolic form, and the precipitation of C_2H_4 is extremely low in the early stage. After entering the high temperature combustion stage, a large amount of C_2H_4 begins to precipitate. Its peak concentration time is the same as that of C_2H_6, but its peak concentration is much lower than that of C_2H_6. In comparison, the amount of C_2H_4 gas released from oil shale is greater than that of oil shale after dry distillation.

C_3H_8

As shown in Fig. 65.7, C_3H_8 was produced between 150 and 250 °C in both rock samples, and the precipitation increased after 300 °C. But the overall production was less, reaching a peak between 350 and 450 °C, and then gradually decreased. By comparison, it can be seen that the C_3H_8 gas produced by spontaneous combustion of oil shale after dry distillation is lower than that of oil shale, and the precipitation temperature is delayed.

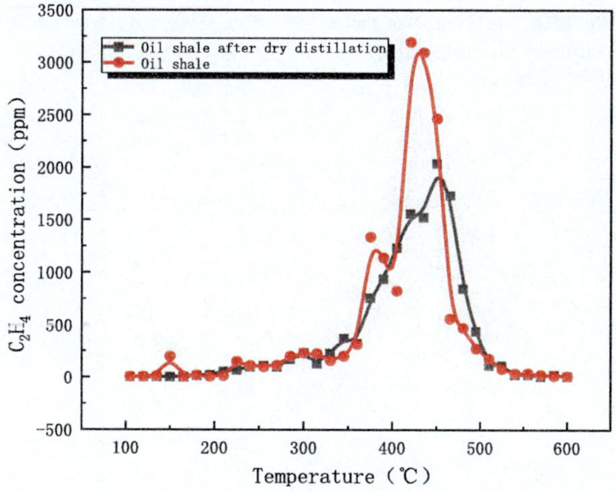

Fig. 65.6 The temperature-influenced variation of sample C_2H_4 concentration

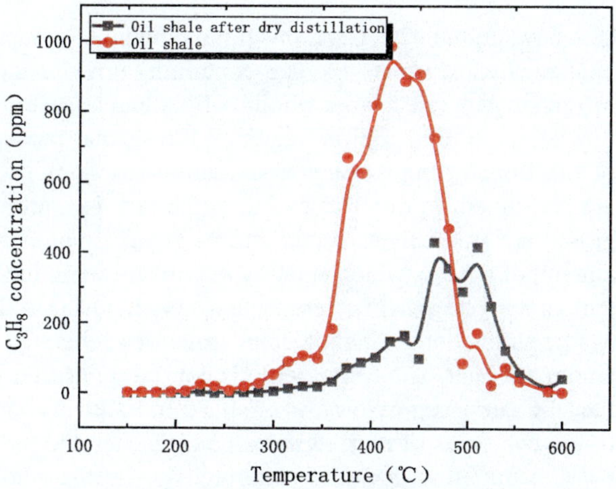

Fig. 65.7 The temperature-influenced variation of sample C_3H_8 concentration

65.3.3 *Analysis of Heat Variation Law in Oxidation and Heating Process*

65.3.3.1 Analysis of Changing Law of Heat Flow Curve in Oxidation Heating

According to the experimental data, the DSC curve is drawn as shown in Fig. 65.8.

In the DSC curve of this experiment, the heat flow values of the two samples have basically similar trends as the temperature increases. In the initial stage of heating, the exothermic value of the rock sample is less than the absorption value. It is in an endothermic process, and the heat flow value decreases. Then the DSC curve keeps rising with the increase of temperature, and reaches the first peak of

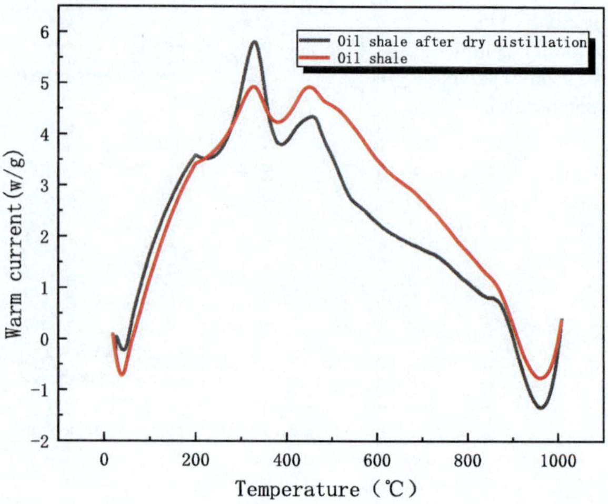

Fig. 65.8 DSC curve of the sample at a heating rate of 10 °C/min

heat flow around 320 °C. At this time, functional groups with higher activity, such as aliphatic hydrocarbons, oxygen-containing functional groups, etc., first reacted with oxygen to generate a large amount of gas and release a lot of heat. And then the heat flow starts to drop. Before reaching the second peak of heat flow, a large amount of functional groups prone to oxidation reactions such as aliphatic hydrocarbons are consumed. In order to maintain the reaction, aromatic hydrocarbon functional groups are gradually activated, but their participation in the reaction requires a large amount of energy, which shows a downward trend in the solid heat flow curve. The temperature is further increased, and oxygen reacts with combustible substances and a large number of small molecular structures generated by cracking, releasing a large amount of heat. The heat release is constantly greater than the heat absorption, and the DSC curve begins to show a rising trend. And at about 460 °C, the heat flow values of the two types of rock samples both reached the second peak. After reaching the peak, it started to decrease continuously. On the whole, the heat flow value of oil shale is greater than that of oil shale after dry distillation.

65.3.3.2 Analysis of Total Heat Release

According to the heat flow curve of the sample, the total heat release of the sample reaction can be obtained by integrating it through the origin software, as shown in Table 65.2.

Table 65.2 Total heat release of the sample

Sample	Oil shale after dry distillation	Oil shale
Exothermic heat (J/g)	2195.078	2573.635

By comparison, it can be seen that the heat release of the oil shale is greater than that of the oil shale after dry distillation, indicating that the oil shale reduces the calorific value of the oil shale. It also shows that the heat released by the generation of hydrocarbon gases is greater than that of carbon and oxygen gases.

65.4 Conclusions

(1) Dry distillation can reduce the oxygen concentration variation range of oil shale in the process of spontaneous combustion, and the oxygen concentration of oil shale after dry distillation is lower than that of oil shale in the later stage of the reaction. When the oil shale spontaneously ignites, the oxygen concentration decreases slowly and the decrease range is small, making it more difficult to burn. Dry distillation can effectively increase its combustion efficiency.

(2) Dry distillation can effectively remove kerosene components, thereby advancing the temperature value of the rapid increase of CO and CO_2 concentrations during the spontaneous combustion process. In addition, the CO and CO_2 concentrations of the oil shale after dry distillation are higher than those of the oil shale, and the main gas product of spontaneous combustion of the kerogen after dry distillation is CO_2.

(3) After dry distillation of oil shale, part of kerogen is removed, and kerogen is rich in aliphatic hydrocarbon structure, so the concentration of hydrocarbon gas precipitated by spontaneous combustion of oil shale after dry distillation is lower than that of oil shale.

(4) Dry distillation reduces the calorific value of oil shale, so that the heat release of oil shale after dry distillation is smaller than that of oil shale, which also shows that the heat released by generating hydrocarbon gas is greater than that of carbon and oxygen gas.

Acknowledagment This work was supported by Special Project of Science and Technology Innovation and Entrepreneurship Fund of China Coal Science and Industry Group Co., Ltd. (2019-ZD004).

References

Aiwu G, Lixin S, Weiwei S et al (2015) Research on low temperature dry distillation process of Suiling oil shale. Contemp Chem Ind 44(12):2856–2858

Baolin P (2013) Experimental study on low temperature pyrolysis of oil shale. Taiyuan University of Technology, Taiyuan

Guochao C (2013) Research on optimization of oil shale dry distillation process. North China University of Technology, Tangshan

Guochao C, Jun G (2013) Research on the transformation of traditional shale low temperature dry distillation technology. Contemp Chem Ind 42(05):608–610

Hongtao C (2011) Comprehensive utilization development and status quo of Fushun oil parent shale. Surf Min Technol 3:76–79

Lanyun C, Weimin Z, Zhenshan Z (2005) Experiment and application of grouting water injection to treat spontaneous combustion of oil mother shale gangue foundation. Railway Constr 06:62–64

Ping X, Shouliang X, Xin Y (2005) Research and evaluation of industrial properties of parent shale in Fushun west open-pit mine. Surf Min Technol S1:84–86

Shijie J, Lei Z, Hao L (2013) Application of low temperature dry distillation process in the oil shale circular economy industry of Longkuang group. Coal Qual Technol 06:69–72

Wan W (2016) Research on the properties of oil shale pyrolysis intermediates and products. Ma'anshan, Anhui University of Technology

Wang Y, Shuai L (2021) Research on new technology of low temperature dry distillation of regenerative external fuel oil parent shale. Contemp Chem Ind Res 02:140–141

Xue G (2016) Research on geochemical reference materials of oil shale. Beijing, China University of Geosciences (Beijing)

Yang Z, Song C, Xiaodong W (2015) Analysis of influencing factors of low temperature dry distillation of oil shale in Fushun furnace. Heilongjiang Sci 6(09):13–15

Chapter 66
Research on Emission Reduction of Container Liners Decision Making Considering Multiple Carbon Trading Products

Jie Wang, Yunpeng Xie, and Meijie Jia

Abstract The bi-level programming model was constructed with the goal of optimal allocation of carbon emission reduction subsidies and carbon emission quota, simulating the interactive decision-making between the government and the liner company and the independent decision-making between enterprises in different carbon emission trading situations. Then, the effect of carbon emission trading and carbon emission reduction subsidies on the promotion of carbon emission reduction of container liner were analyzed. The upper model takes the government as the decision maker to pursue the minimization of the total social cost, and the lower model pursues the lowest cost for the liner company. Liner companies make emission reduction decisions based on the different usage scenarios of the two carbon trading products, the carbon emission quota and the Chinese Certified Emission Reduction (CCER). The relevant data of carbon trading in a certain place were selected for example analysis. The optimal allocation of carbon emission allowances and unit carbon emission reduction subsidies were found. By comparing the single policy with the joint policy, it is concluded that the joint policy can better promote the emission reduction of container liner companies.

Keywords Bi-level programming · Container liner emission reduction · Carbon emission trading · Carbon emission reduction subsidy

J. Wang (✉) · Y. Xie · M. Jia
Dalian Maritime University, Dalian, China
e-mail: dlwjie@163.com

Y. Xie
e-mail: xieyunpeng321@163.com

M. Jia
e-mail: jiaxibei_101@163.com

J. Zhang et al. (eds.), *Environmental Pollution Governance and Ecological Remediation Technology*, Environmental Science and Engineering,
https://doi.org/10.1007/978-3-031-25284-6_66

66.1 Introduction

China proposes new climate goals of peaking carbon dioxide emissions by 2030 and achieving carbon neutrality by 2060. For this reason, policies such as carbon emission trading and carbon emission reduction subsidies have become important means of emission reduction. According to statistics, the global merchant fleet produces about 1 billion tons of carbon dioxide emissions every year, while China's merchant fleet ranks fourth in the world, and the seaborne import and export trade volume accounts for 30% of the global seaborne total.

On the issues related to the emission reduction policies of the shipping industry, the domestic and foreign research focuses on carbon emission trading and carbon tax. Wang et al. (2015, 2021) studied the actual impact of the Emissions Trading System (ETS) on the emission reduction of the shipping industry; Chang and Huang (2019) discuss the allocation of carbon emission allowances whether ships comply with EEDI guidelines; Chen et al. (2020) compared the optimal decision-making of remanufacturing enterprises when they invest in emission reduction and non-abatement in the carbon trading environment; You-dong et al. (2020) analyzed the decision-making of low-carbon operation of an emission-dependent supply chain under the constraints of product carbon emissions and government carbon quotas; Shan-shan and Chuan-xu (2015; Changsong et al. 2015; Jing and Ruiyu 2013) discussed the emission reduction of enterprises under carbon tax policy, carbon quota policy and other policies; Li (2018) discussed the green development path of China's shipping industry in combination with the EU's introduction of a marine carbon tax; Gaofeng and Yanxiang (2019) discussed the emission reduction decision-making of port and shipping enterprises under the background of global carbon tax policy and green preferences of customer groups. Some studies have been carried out from the perspectives of fuel tax, trade policy, incentive and regulatory strategies, Vasileios Kosmas and Acciaro (2017) analyzed the impact of specific and ad valorem fuel taxes on shipping costs; Sturla-Zeren et al. (2020) quantified the changes in the greenhouse gas emissions of the Chilean shipping industry under different trade policies; Linder (2017) discussed the promotion of emissions reductions in the maritime industry compared with government regulation and financial incentives; Limin et al. (2015) analyzed the actual impact of three emission reduction strategies under technical, operational and market measures on the carbon emissions of liner companies.

The above research does not involve the emission reduction subsidies in the shipping industry, nor does it involve the impact of the joint strategy of carbon emissions trading and carbon emission reduction subsidies on the carbon emission reduction; in addition, the proportion of CCER has a certain floating space, and the relevant research does not involve. Therefore, the bi-level programming model was constructed for above problems, simulating the interactive decision-making between the government and the liner company and the independent decision-making between enterprises in different carbon trading situations. In this way, the optimal allocation of carbon emission reduction subsidies and carbon emission quotas are determined, and the actual impact of the joint policy on the emission reduction is analyzed.

66.2 Model Building

66.2.1 Problem Description

Governments and liner companies are at different decision-making levels in shipping carbon emissions, government departments, as the upper management decision-makers, mainly consider the reasonable allocation of carbon emission quotas and carbon emission reduction amounts; as the lower-level decision-makers, liner companies need to formulate carbon emission reduction plans suitable for their own development according to the decisions of government departments.

66.2.2 Hypotheses

J1 Ships are rated for emissions and have the same carbon emissions for the same container ship type.

J2 Container ships only consider carbon emissions from fuel consumption, and only the impact of carbon dioxide.

J3 Liner companies follow the local carbon emission reduction policies and carbon trading policies.

J4 Carbon emission trading market includes two products: carbon emission allowances and CCER. The latter is cheaper, so companies have priority to buy it.

J5 Carbon emission trading fee is borne by the enterprise, and the impact of carbon trading price fluctuations is not considered for the time being.

66.2.3 Notations

See Table 66.1.

66.2.4 Model Formulation

Constructing bi-level programming model for the above problems, the upper-level model seeks to minimize the total social cost, which is mainly composed of the related costs of carbon emissions. The lower model pursues the minimum total cost of the enterprise.

Table 66.1 Notations description

Set	
n	$= \{0,1,2,3...N\}$ is a collection of liner companies
i	$= \{0,1,2,3...I\}$ is a collection of ships owned by a liner company
Constants	
α	Environmental cost factor for carbon emissions
S	Government subsidy limit for carbon reduction
β_1	The price of CCER
β_2	The price of carbon emission quota
p_f	Ship fuel price
L	Liner company carbon emission limits
γ	Fuel consumption coefficient
M_i	Ship decarbonization retrofit costs
J_n	Carbon reduction limit
b%	CCER usage limit
g	Government management fee
u_i^n	The original carbon emissions of the ship of company
Variables	
w_n	Companies carbon emission quota
z	Unit emission reduction subsidies
q_i^n	Carbon emissions after the company's ship carbon policy
d_i^n	The emission reduction of the company's 1st ship
w_i	0–1 variable, whether the ship is decarbonized or not

Specifications of the upper level decision model

The upper-level model pursues the lowest total social cost and consists of three parts, the total cost of government investment, the total cost of carbon emission reduction for liner companies, and the total environmental cost. S_P is the total social cost, G_P is the total cost to the government, C_P is the total cost of the company, C_n is the total cost for a single company, E_P is the total environmental cost, as shown in Eq. (66.1).

$$\min S_P = G_P + C_P + E_P \tag{66.1}$$

The implementation of the government's carbon emission reduction policy requires investment in the total cost of the government's corporate emission reduction subsidies and government management costs, as shown in Eq. (66.2).

$$G_P = \sum_{n=1}^{N}\sum_{i=1}^{I} d_i^n \times z + g \tag{66.2}$$

According to the literature (Zhang et al. 2019), the total environmental cost is proportional to carbon emissions, as shown in Eq. (66.3).

$$\mathrm{E}_P = \alpha \sum_{\mathrm{n}=1}^{N} w_n \tag{66.3}$$

The total business cost is the sum of the costs of each liner company in the region, as shown in Eq. (66.4).

$$C_p = \sum_{n=1}^{N} C_n \tag{66.4}$$

Subject to:

$$0 \le \sum_{n=1}^{N} w_n \le \mathrm{L} \tag{66.5}$$

$$0 \le \sum_{n=1}^{N} \sum_{i=1}^{I} d_i^n \times z \le \mathrm{S} \tag{66.6}$$

Equation (66.5) indicates that the carbon emission quota allocated by the government to each liner company is within a certain limit, Eq. (66.6) indicates that the amount of carbon emission reduction subsidies given by the government to liner companies should not exceed the government's special financial subsidy budget.

Specifications of the lower-level decision model

The lower model aims to minimize the total cost of carbon emission reduction of enterprises, which is mainly composed of production costs of company and the cost of carbon emission reduction. H_n is the production costs of company, T_n is the cost of carbon emission reduction, as shown in Eq. (66.7).

$$\min \mathrm{c_n} = \mathrm{H_n} + \mathrm{T_n} \tag{66.7}$$

The production costs only consider the fuel cost of the ship. Since the carbon emission is equal to the product of the fuel consumption and the emission factor, the fuel consumption can be obtained by multiplying the carbon emission by the inverse of the emission factor, as shown in Eq. (66.8).

$$\mathrm{H}_n = \mathrm{p_f} \sum_{i=1}^{I} \gamma \cdot q_i^n \tag{66.8}$$

In the carbon emission trading market, there are two products, carbon emission allowances and CCER, but local governments have different regulations on the proportion of CCER used. Therefore, the liner company chooses whether to buy or sell the above two products through carbon trading, and completes the carbon emission allowance payment work of the companies for the year. After implementing the carbon emission reduction policy, it can be divided into the following three situations.

Companies do not need to use CCER and purchase carbon emission quota

In this situation, the carbon emission of the companies does not exceed the rated carbon emission quota, and the companies can sell its own CCER and additional carbon emission quota to obtain benefits to reduce the cost of decarbonization and transformation, as shown in Eq. (66.9). Equation (66.10) indicates that the carbon emission after emission reduction is less than the carbon emission quota.

$$\mathrm{T}_n = \sum_{i=1}^{I} \left(\mathrm{M_i} - d_i^n \cdot \beta_1 - d_i^n \times z\right) w_i - \left(w_n - \sum_{i=1}^{I} q_i^n\right) \beta_2 \quad (66.9)$$

$$w_n > \sum_{i=1}^{I} q_i^n \quad (66.10)$$

Companies need to use CCER additionally

In this situation, companies can use CCER to complete, and the CCER usage rate is within the specified limits (b%), do not need to purchase additional carbon emission quota. If the CCER of the company is used to complete the payment and there is a surplus, you can choose to sell it to obtain profits and reduce costs; if the CCER of the company is not enough to complete the payment, the cost of purchasing CCER will be invested, as shown in Eq. (66.11). Equation (66.12) It means that the emission reduction of the liner company is greater than the carbon emission quota, but only CCER can complete the payment.

$$\mathrm{T_n} = \sum_{\mathrm{i}=1}^{I} \left(\mathrm{M_i} - d_i^n \times z\right) w_i - \left[\sum_{i=1}^{I} d_i^n - \left(\sum_{i=1}^{I} q_i^n - w_n\right)\right] \beta_1 \quad (66.11)$$

$$\left[\left(\sum_{i=1}^{I} q_i^n - w_n\right) < \mathrm{b}\% \sum_{i=1}^{I} q_i^n\right] \cap \left[w_n \le \sum_{i=1}^{I} q_i^n\right] \quad (66.12)$$

Equation (66.13) can be obtained after Eq. (66.12) arranging:

$$(1 - \mathrm{b}\%) \sum_{i=1}^{I} q_i^n < w_n \le \sum_{i=1}^{I} q_i^n \quad (66.13)$$

Companies needs to purchase additional carbon emission quotas

In this situation, liner companies after using government-mandated maximum percentage of CCER (b%), the remaining excess is completed by purchasing carbon emission quotas, as shown in Eq. (66.14). Equation (66.15) indicates that the liner companies' carbon emissions exceed the sum of the company's free carbon emission allowance and the maximum use ratio of CCER.

$$\mathrm{T}_n = \sum_{i=1}^{I}(\mathrm{M_i} - d_i^n \times z)w_i - \left[\sum_{i=1}^{I} d_i^n - \mathrm{b}\% \sum_{i=1}^{I} q_i^n\right]\beta_1 + \left(\sum_{i=1}^{I} q_i^n - w_n - \mathrm{b}\% \sum_{i=1}^{I} q_i^n\right)\beta_2 \tag{66.14}$$

$$\left(\sum_{i=1}^{I} q_i^n - w_n\right) \geq \mathrm{b}\% \sum_{i=1}^{I} q_i^n \tag{66.15}$$

Equation (66.16) can be obtained after Eq. (66.15) arranging:

$$w_n \leq (1 - \mathrm{b}\%)\sum_{i=1}^{I} q_i^n \tag{66.16}$$

Subject to:

$$\mathrm{d_i^n} = u_i^n - q_i^n \tag{66.17}$$

$$0 \leq \sum_{i=1}^{I} d_i^n \leq \mathrm{J_n} \tag{66.18}$$

$$0 \leq \mathrm{d_i^n} < u_i^n \tag{66.19}$$

$$0 \leq \sum_{\mathrm{n}=1}^{N}\sum_{i=1}^{I} q_i^n \leq \mathrm{L} \tag{66.20}$$

Equation (66.17) indicates that the carbon emission reduction is equal to the difference between the original carbon emissions and the emissions after implementing the policy. Equation (66.18) is the corporate emission reduction limit, preventing companies from vicious emission reduction in order to obtain additional subsidies. Equation (66.19) indicates the reasonable range of carbon emission reduction for a single ship. Equation (66.20) indicates that the companies' carbon emissions are within a certain limit.

Table 66.2 Raw carbon emissions from ships of three container liner companies

The ship number of company "a"	Annual emissions (kt/year)	The ship number of company "b"	Annual emissions (kt/year)	The ship number of company "c"	Annual emissions (kt/year)
1	30	1	50	1	40
2	40	2	30	2	50
3	30	3	40	3	30
4	50	4	30	4	30
5	80	5	50	5	10
6	40	6	50	6	20
7	20	–	–	7	40
8	30	–	–	–	–
Total emissions	320		230		220

66.3 Case Study

66.3.1 Introduction of Case

Shanghai already has a relatively complete management system. Therefore, the carbon trading data in Shanghai is selected for the case analysis. The CCER price adopts the average value of 13.7 yuan/ton in 2020[1]; the price of carbon allowances is 33.8 yuan/ton. Shanghai stipulates that the use limit of CCER is 3%. According to the 2020 annual report of COSCO SHIPPING Holdings Co., Ltd., the annual carbon dioxide emission of each container ship is about 40,000 tons. Based on this, the annual CO_2 emissions per ship of the three container liner companies are simulated, as shown in Table 66.2; environmental cost factor for carbon emissions is 700 yuan/ton. In 2020, Shanghai's special fund is 868.26 million yuan,[2] and combined with the number of carbon emission companies, it is estimated that the carbon emission reduction subsidy limit of the three liner companies is 8 million yuan. Ship fuel price is 3491 yuan/ton.[3] The carbon emission limit of the three liner companies is 800,000 tons. The heavy fuel oil emission factor is 3.114, so the set fuel consumption coefficient is 0.32. Ship decarbonization retrofit cost amortized to $150,000 per year.[4] Carbon reduction limit is set as the carbon emissions of the liner company in that year.

[1] Cneeex.com.

[2] http://fgw.sh.gov.cn/zgjjl/20201218/a6ac6949e7574f95ba981b5697cc090b.html.

[3] https://www.sci99.com/monitor-429-0.html.

[4] https://www.cnss.com.cn/.

66.3.2 Result Analysis

Solving this bi-level programming problem can be transformed into the mutual iteration of two PSO algorithms (Zhigang et al. 2013). The algorithm parameters are set as follows: $iter_{\max} = 200$, $N = 200$, $W_{\max} = 1$, $W_{\min} = 0$, $c_1 = 0.3$, $c_2 = 0.2$, $v_{\max} = 1$, $v_{\min} = 0.2$, The algorithm is implemented using Matlab2016a.

The solution iteration process is shown in Fig. 66.1, and the optimal allocation of carbon emission quotas and carbon emission reduction subsidies under the single and joint policies are shown in Tables 66.3 and 66.4.

From Tables 66.3 and 66.4, the implementation of individual and joint policies will reduce the carbon emissions of liner companies. However, the emission reduction effect after the implementation of the joint policy is slightly better than that of the carbon trading and carbon emission reduction subsidy policies only. The total social cost after the implementation of the joint policy is obviously better than that of the individual policies. Therefore, it is reasonable for the government to adopt the joint policy, which is conducive to promoting the low-carbon transformation of

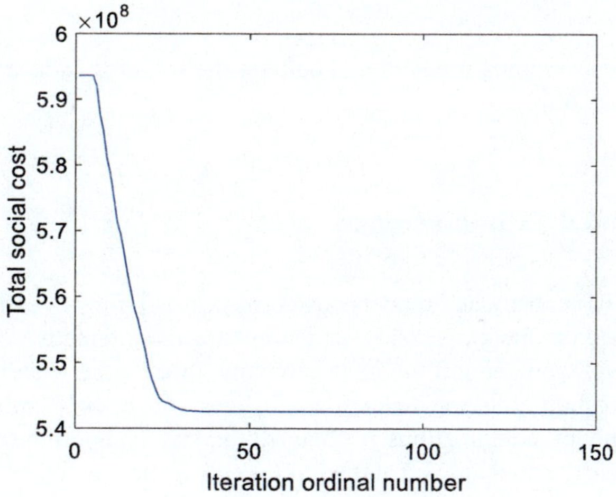

Fig. 66.1 Solution convergence process

Table 66.3 The Data after implementation of different policies

Type	Joint policy	Carbon emissions trading	Carbon reduction subsidies
Total carbon allowance	604,450.8 ton	602,703.8 ton	
Total government subsidy	682,921.6 yuan		446,995.1 yuan
Unit subsidies	6.03 yuan/ton		4.2 yuan/ton
Total social cost	542.562 million yuan	601.562 million yuan	587.466 million yuan

Table 66.4 Emissions of three companies after the implementation of different policies

Companies / Type		a	b	c	Total
Carbon emission quota (Ton)	Joint policy	233,732.6	209,672.8	161,045.4	604,450.8
	Carbon emissions trading	235,239.2	207,269.3	160,194.4	602,703.8
Total carbon emissions (Ton)	Raw emissions	320,000.0	230,000.0	220,000.0	770,000.0
	Joint policy	265,956.2	205,045.0	185,792.8	656,745.3
	Carbon emissions trading	265,956.2	205,044.9	185,792.8	656,793.9
	Carbon reduction subsidies	267,577.6	207,540.4	188,484.6	663,602.5
Total carbon emission reduction (Ton)	Joint policy	54,043.8	24,955.1	34,255.9	113,254.8
	Carbon emissions trading	54,043.8	24,955.1	34,255.9	113,206.1
	Carbon reduction subsidies	52,422.5	22,459.6	31,515.4	106,397.5

the shipping industry and helping the shipping industry achieve the goal of carbon neutrality.

66.4 Conclusions

The constructed model considers the joint policy of carbon trading and carbon emission reduction subsidy, and simulates the interactive decision-making between the government and the liner company. In the case analysis, the optimal allocation of carbon emission reduction subsidies and carbon emission quotas under different policy combinations is obtained, and by comparing the implementation effects, it can be concluded that the joint policy is more suitable for the emission reduction of liner companies than the single policy. The government often adopts a fixed ratio for the allocation of carbon emission allowances, however, the model can set different allocation proportions according to the annual carbon dioxide emissions of the liner companies, which is more in line with the actual emission reduction of the liner companies, and can provide a reference for the carbon emission reduction of the shipping industry.

References

Chang C-C, Huang P-C (2019) Carbon allowance allocation in the shipping industry under EEDI and non-EEDI [J]. Sci Total Environ 678

Changsong M, Chen X, Zhenyu L, Tian L (2015) Production strategy of considering low carbon emission policies regulation under stochastic demand [J]. Control Decis 30(06):969–976

Gaofeng S, Yanxiang M (2019) Game analysis of port and ship decisions on emission reduction under carbon tax policy [J]. Sci Technol Manage Res 39(21):227–235

Jing L, Ruiyu L (2013) Research on the impact of carbon tax on my country's shipping industry and its countermeasures [J]. Marine Technol (05):67–69

Kosmas V, Acciaro M (2017) Bunker levy schemes for greenhouse gas (GHG) emission reduction in international shipping [J]. Transp Res Part D Transp Environ 57:195–206

Li Z (2018) Shipping carbon tax and green road of Chinese shipping industry [J]. Navig China 41(01):113–116

Limin T, Ying C, Huishao L, Zhaokun C (2015) System dynamics model of low-carbon development of Chinese shipping industry under international carbon emission market mechanism [J]. J Shanghai Marit Univ 36(02):25–30

Linder A (2017) Explaining shipping company participation in voluntary vessel emission reduction programs [J]. Transp Res Part D Transp Environ 61(PT.B):234–245

Shan-shan Y, Chuan-xu W (2015) Speed optimization under different carbon emission control policy [J]. J Dalian Marit Univ 41(03):45–50

Sturla-Zerene G, Eugenio FB, Sturla M (2020) Reducing GHG global emissions from copper refining and sea shipping of Chile's mining exports: a world win-win policy [J]. Resour Policy 65

Wang K, Fu X, Luo M (2015) Modeling the impacts of alternative emission trading schemes on international shipping [J]. Transp Res Part A 77:35–49

Wang S, Zhen L, Psaraftis HN et al (2021) Implications of the EU's inclusion of maritime transport in the emissions trading system for shipping companies [J]. Engineering (10)

You-dong L, Xin-peng X, Feng-zheng W, Jing-feng W (2020) Emission dependent supply chain low-carbonization based behavior of enterprise operation decision under cap-and-trade environment [J]. Control Decis 35(09):2236–2244

Yu-yu C, Bang-yi L, Bai Qing-guo X, Jian-teng WZ (2020) Research on production and emission reduction investment decisions of the remanufacturing enterprise under carbon trading environment [J]. Control Decis 35(03):695–703

Zhang S, Chen N, Song X, Yang J (2019) Optimizing decision-making of regional cold chain logistics system in view of low-carbon economy [J]. Transp Res Part A 130

Zhigang Z, Weiqian W, Shuyun H (2013) Bi-level programming problem based on improved particle swarm algorithm [J]. Comput Sci 40(S2):115–119

Chapter 67
Water Ecological Environment Protection and Restoration of Chaohu Lake Under the "Double Carbon" Goal

Zhihua Huang, Leixiang Wu, Yongsheng Zhang, Shi Wu, Zheng Zhou, Xiaohui Chen, and Zhiyuan Tong

Abstract Accomplishing carbon peaking in 2030 and carbon neutrality in 2060 (the "double carbon" goal for short) is one of the strategic goals of China in realizing the sustainable development under the background of global warming. In fact, the accomplishment of the "double carbon" goal covers every aspect, i.e. it is closely related to eco-environment protection and restoration of rivers and lakes. In fact, when the water ecological function of rivers and lakes was restored, the water ecological carrying capacity and the carbon fixation capacity of water bodies can enhance, which is of great significance to creating a green, low-carbon and circular development pattern of river basins. In this paper, we discussed the effective strategies of water eco-environment protection and restoration in Chaohu Lake under the "double carbon" goal, so as to provide a reference.

Keywords Carbon peaking · Carbon neutrality · "Double carbon" goal · Water eco-environment · Protection and restoration · Chaohu lake

67.1 Introduction

Accomplishing carbon peaking in 2030 and carbon neutrality in 2060 (the "double carbon" goal for short) is one of the strategic goals of China. In the 2021 Chinese

Z. Huang (✉) · L. Wu · Y. Zhang
China Institute of Water Resources and Hydropower Research, Beijing 100038, China
e-mail: huangzh@iwhr.com

S. Wu
Anhui Survey and Design Institute of Water Resources and Hydropower Co., Ltd., Hefei 230088, China

Z. Zhou · X. Chen
Anhui Provincial Group Limited for Yangtze-to-Huaihe Water Diversion, Hefei 230000, China

Z. Tong
Anhui Provincial Ecological Development Co., Ltd for Yangtze-to-Huaihe Water Diversion, Hefei 230000, China

J. Zhang et al. (eds.), *Environmental Pollution Governance and Ecological Remediation Technology*, Environmental Science and Engineering,
https://doi.org/10.1007/978-3-031-25284-6_67

government work report, "Making Solid Progress in Carbon Peaking and Carbon Neutrality" was listed as one of the key tasks in 2021. It marks that China has bid farewell to the high resource input mode and shift to a new development mode with high-quality development of the economic society and comprehensive modernization, based on technological progress, innovation-driven and institutional reform (Zhenghao et al. 2022).

As we all know that, water eco-environment construction of rivers and lakes is not only related to people's well-being, but also to the future of the nation. Chaohu Lake is one of the five major freshwater lakes located in the middle and lower reaches of the Yangtze River, Anhui Province. In the recent thirty years, deteriorating water eco-environment has become a key issue restricting the sustainable development of Chaohu Lake Basin. According to the newly revised *Regulations on the Prevention and Control of Water Pollution in Chaohu Lake Basin* in December 2019, the key point of water pollution control in Chaohu Lake basin is to strengthen ecological management. It means that the development of Chaohu Lake under the "dual carbon" goal requires more concept on green development, i.e., tracing the source and system management, accelerating the recovery of water eco-environment and promoting fundamental improvement of water quality, so as to realize the sustainable utilization of water ecological function in Chaohu Lake.

67.2 Current Situation of Chaohu Lake

67.2.1 Water Level

Chaohu Lake is a typical shallow lake. The lake bottom elevation is 3.1–4.1 m (1985 National datum elevation, the same below). The dead water level is 5.1 m, and the normal storage level is about 6.1 m. After the Chaohu Lake Gate was put into operation in 1963, the water level was mainly controlled by the gate. Generally, the mean water level in the flood season from June to September is 6.1–7.1 m, and in the dry season from November to the next March is about 6.1 m. According to the research results of Xiaoyuan et al. in (2018), the annual average water level of Chaohu Lake in 1963–1980, 1981–2000, and 2001–2015 was 6.29 m, 6.65 m and 7.05 m respectively. It can be seen that after the completion of the Chaohu Lake Gate, the water level showed a significant increase. In order to meet the needs of flood control, the controlled water level of the Chaohu Lake Gate is limited to 6.1 m from June to August, 6.6 m in May and September, and 6.6–7.1 m during non-flood season.

67.2.2 Water Quality

According to water quality monitoring data of national control section, in the recent two years, Chaohu Lake was slightly polluted and the main pollution indexes were total phosphorus (TP) and total nitrogen (TN), where water quality of the eastern part was Grade IV and water quality of the western part was Grade V of GB 3838–2002. Among the rivers around Chaohu Lake, water quality of Hangbu River, Baishitian River, Zhaohe River Yuxi River, Shuangqiao River and Zhegao River located in the southwest, south and northeast of the lake was better, belonging to Grade II to Grade III. While, Paihe River and Nanfei River located in the northwest of Chaohu Lake was polluted, and the water quality of Paihe River was maintained at Grade IV, the water quality of Nanfei River was improved from inferior Grade V to Grade V, and the main pollution indexes were ammonia nitrogen (NH_3–N), total phosphorus (TP) and Chemical oxygen demand (COD). As shown in Table 67.1.

From the nutritional status of the lake, the eastern and western parts of the lake were slightly eutrophic, and cyanobacteria blooms occurred frequently. According to the remote sensing data released by the Satellite Environmental Application Center of the Ministry of Ecology and Environment, in 2020, a large-scale water bloom occurred in Chaohu Lake at 11:15 on September 19, with an area of 189.25 km^2, mainly concentrated in the northwest, southeast and northeast areas of the lake. Therefore, Chaohu Lake is faced with a serious situation in eutrophication risk.

Table 67.1 Present water quality of Chaohu Lake

Lake and river	Water quality grade of 2019	Water quality grade of 2020	Main pollution indexes
Eastern part of Chaohu lake	IV	IV	TP, TN
Western part of Chaohu lake	V	V	TP, TN
Hangbu river	III	III	
Baishitian river	II	III	
Zhaohe river	III	III	
Yuxi river	III	III	
Shuangqiao river	II	II	
Zhegao river	II	III	
Shiwuli river	III	III	
Paihe river	IV	IV	NH_3–N
Nanfei river	< V	V	NH_3–N, TP, COD

67.2.3 *Pollution Characteristics*

Urban industrial wastewater and domestic sewage

Although urbanization and the development of industrial and mining enterprises have promoted the economic growth of the surrounding areas of Chaohu Lake Basin, the annual discharge of urban industrial wastewater and domestic sewage is up to hundreds of millions of tons. In addition to a small number of wastewater can meet the emission standards after treatment, most are directly discharged into the lake through the channels around the lake, resulting in increased pollution load and serious water pollution (Hongling and Cheng 2018; Xiaoyuan et al. 2018). With the development of urban economy and population growth in Chaohu Lake basin, the amount of domestic sewage discharge will increase year by year, which is also one of the main pollution sources in Chaohu Lake basin in the future.

Agricultural non-point source pollution

Chaohu Lake is surrounded by farmland. According to statistics, the amount of fertilizer applied in the farmland of the lake basin is about 500,000 tons per year, showing an increasing trend year by year, and the annual loss of fertilizer is about 200,000 to 250,000 tons per year. These unused fertilizers directly enter the lake under the scour of storm runoff, which seriously affect the water quality of the lake. In addition, the random discharge of rural domestic sewage and the outdoor accumulation of domestic garbage around the lake are also the main causes of water pollution in Chaohu Lake.

Endogenous pollution

Endogenous pollution refers to the pollution caused by deposition of nutrients in the lake to the surface of the lake floor under the action of physics, chemistry and biology. Generally, concentration of pollutants in sediments is usually several orders of magnitude higher than that in overlying water (Junwei et al. 2018). When nutrients in the sediment accumulated to a certain amount, it will be released to the water body. The release process is quite complex, which is affected by temperature, wind, wave disturbance, pH and biological activity. Chaohu Lake is a typical shallow sedimentary lake with a long history of sediment deposition. Under the wind and wave disturbance, nutrient such as nitrogen and phosphorus in sediment will continually release to the water, which increased nutrient concentration in water, bring secondary pollution risk to the lake and greatly hindered ecological restoration of lakes.

In conclusion, urban industrial wastewater and domestic sewage are the main sources of point source pollution in Chaohu Lake, and farmland drainage and rural domestic sewage are the main sources of non-point source pollution. In addition, the endogenous pollution caused by sediment release cannot be ignored.

67.3 Water Eco-Environment Problems of Chaohu Lake

In conclusion, the water eco-environment problems of Chaohu Lake are as follows:

67.3.1 Serious Eutrophication and Water Pollution

Unfortunately, after years of renovation, the eutrophication of Chaohu Lake has not been effectively controlled. The whole lake, the eastern and western parts of the lake are slightly eutrophic, and cyanobacteria blooms appeared many times in the western parts of the lake, affecting a large water area. The pollution load caused by urbanization and population growth around the lake will continue to aggravate the water pollution of Chaohu Lake, so the water pollution is still facing a serious situation.

67.3.2 Substandard Water Quality and Water Supply Risk

According to the Water Function Zoning of Anhui Province, the water quality management goal of Chaohu Lake is Grade III. The newly revised Water Pollution Control Regulations for Chaohu Lake Basin in 2019 required that the water quality of Chaohu Lake and some inflow rivers such as Fengle River, Hangbu River, Baishitian River, Zhao River, Zhegao River, Yuxi River and Pai River should be protected as the Chinese surface water quality standards of Grade III, while the water quality of Nanfei River and Shiwuli River should be protected as Grade IV of GB 3838–2002. It can be seen that although the water quality of Chaohu Lake has improved significantly, it was still slightly polluted. Among the rivers around the lake, the water quality of Paihe River and Nanfei River has not met the requirements of Grade III and Grade IV despite some improvement, which would inevitably bring some hidden dangers to water sources and irrigation areas in Chaohu Lake basin.

67.4 Water Eco-Environment Protection and Restoration of Chaohu Lake Under the "Double Carbon" Goal

The realization of the "double carbon" goal involves multi-disciplinary, multi-level and multi-dimensional research and improvement, and the improvement of water ecological carbon sequestration capacity is the most direct response to the realization of the "double carbon" goal (Qiting et al. 2021). In the past 100 years, with the strong intervention of human activities, the water eco-environment of Chaohu Lake has gradually deteriorated, and water pollution and eutrophication have become increasingly serious. In order to restore the water eco-environment of Chaohu Lake, we should adhere to the work strategy of "focusing on protection and governance", strengthen the protection of water quality, strictly control the pollution discharge to rivers and lakes, and enforce treatment of all kinds of pollution sources, so as to improve the water ecological carrying capacity and increase the carbon sink capacity of Chaohu Lake.

67.4.1 Controlling Industrial Pollution and Improving Urban Sewage Treatment Capacity

In Chaohu Lake basin, industrial pollutants must meet the national emission standard. We should carry out special remediation for heavy-pollution industrial enterprises, eliminate those with outdated technology, serious pollution and excessive emission standards, shut down, merge or transfer those has seriously affected lake environment, and encourage those industrial enterprises to implement deep treatment and cleaner production on the basis of stable emission standards to reduce carbon sources.

Meanwhile, we should continue to improve the construction of urban sewage treatment facilities around Chaohu Lake Basin, expand sewage treatment scale, improve sewage treatment process and capacity, upgrade sewage treatment facilities, and ensure that the effluent quality can reach the first class discharge standard of GB8978-1996 and strive to approach the Chinese surface water quality standards of Grade IV, so as to realize the utilization of sewage resources and reduce carbon emission.

67.4.2 Promoting Agricultural Non-point Source Pollution Remediation

According to the distribution of population and cultivated land around Chaohu Lake, We should guide farmers to implement returning farmland to forest or grassland for scattered farmland, and carry out scientific planting in large-scale planting areas, so as to reduce agricultural non-point source pollution caused by excessive use of chemical fertilizers and pesticides.

In addition, we should continue to improve the construction of domestic sewage and garbage treatment facilities in the rural areas around Chaohu Lake basin. The domestic sewage and garbage should be collected and treated uniformly and meet the emission standards, so as to avoid pollution risks to Chaohu Lake caused by arbitrary discharge of sewage and garbage stacking in the open and promote the restoration of water eco-environment.

67.4.3 Developing Lake Ecological Engineering Construction

Sediment dredging

The sediment dredging technology can be applied in the areas with serious sediment pollution, less distribution of aquatic plants, insufficient aquatic biodiversity and frequent cyanobacterial bloom. Chaohu Lake has a long history of sedimentation, a large amount of sediment has been deposited at the bottom of the lake. The implementation of sediment dredging technology can not only remove the contaminants from sediment and clean the internal pollution of water body, but also reduce the release of the sediment pollutants to the water body and create a healthy environment for the restoration of Chaohu Lake ecosystem.

Ecological water supplement

Ecological water supplement usually refers to reducing the nutrient concentration of water body by supplementing clean water to "clean the intestine", so as to improve the water quality. The implementation of ecological water supplement in Chaohu Lake can not only improve the hydrodynamic conditions, accelerate the water flow, but also increase the water environment capacity and improve the self-purification capacity of water body. It is of great significance to create a variety of water eco-environment of Chaohu Lake, improve water quality, and promote the lake recuperation.

67.4.4 Strengthening Watershed Administrative Supervision and Management Capacity

Strengthening the supervision of various pollution sources and potential pollution hazards

The relevant departments should strengthen the supervision of various pollution sources and potential pollution hazards in Chaohu Lake basin, increase the frequency of inspections, and ensure the precise and effective implementation of treatment and emission reduction measures. In addition, it should focus on the supervision and inspection of industrial enterprises around Chaohu Lake, stop in time the projects that may cause serious pollution to the water body, and order the project leaders to rectify the problems within a time limit. In addition, it is necessary to strengthen the inspection of all kinds of water source protection areas within the lake basin. Once any potential pollution risks are found, they should be timely reported to the superior departments and given rectification suggestions.

Enhancing the public awareness of environment protection

The eco-environment protection of Chaohu Lake is not only the task of someone or a department, but the common responsibility of all people. The relevant departments should enhance the education and publicity on the eco-environment protection of Chaohu Lake, improve the environment protection awareness of the residents around the lake, give full play to the subjective initiative of public involvement, and encourage the public to consciously participate in the protection work. In this way, we will make the concept of eco-environment protection deeply ingrained in everyone's consciousness, and call on the whole society to carry out the eco-environment protection of Chaohu Lake.

67.5 Conclusions

As we all know that, the accomplishment of "double carbon" goal is never easy to reach, but require joint efforts from our government department and everyone. It is noted that carbon emission reduction does not intend to compromise economic development, instead, it aims to accomplish high-quality development. For Chaohu Lake, the development under the "dual carbon" goal requires preventing various pollution sources and increasing carbon absorption. The main ways include improving the eco-environment, increasing the water ecological capacity, enhancing the carbon sequestration capacity, and supporting the construction of a green and low-carbon pattern in Chaohu Lake basin.

Acknowledgements Funding from the National Natural Science Foundation of China (Grant No. 51879279) and engineering consulting program on integrated operation of water quality and quantity and risk prevention and control in water transfer project for Yangtze-to-Huaihe are gratefully acknowledged.

References

Hongling J, Cheng Z (2018) Exploration of water pollution causes and treatment measures of Chaohu lake. Acad J Anhui Water Conservancy Tech Coll 18(1):39–413

Junwei W, Junsong Z, Xianjin Z et al (2018) Study on sediment release characteristics of Chaohu lake. J Xinyu Univ 23(4):15–18

Qiting Z, Xi Q, Tao Z (2021) A new journey of water conservancy development of China under the "double carbon" goal. China Water Resour (22):29–33

Xiaoyuan W, Bo J, Mengfei Y et al (2018) Analysis of ecological status and protection measures of Chaohu lake. Yangtze River 49(17):24–30

Zhenghao D, Shuaizheng L, Chengming D et al (2022) Reflection on the construction of new smart cities under the "dual-carbon" strategy. Ind Policy 1:57–63

Chapter 68
Preparation and Performance of $BiVO_4$/CMS Composite Photocatalytic Material

Longmei Zhao, Xiaoxia Ou, Cuixin Ma, and Chunyang Liang

Abstract To improve the performance of bismuth vanadate ($BiVO_4$) photocatalyst, $BiVO_4$/CMS composite photocatalytic material was prepared by hydrothermal method using $CaCO_3$-based natural photocatalysts (CMS) as the carrier. The crystal structure, elemental composition and surface morphology of the composites were characterized and analyzed by XRD, SEM, IR and DRS. The photocatalytic activity of $BiVO_4$/CMS was tested on the removal efficiency of crystal violet (CV). Compared with pure $BiVO_4$, $BiVO_4$/CMS exhibited the enhanced photocatalytic activity to degrade refractory organic pollutant. The spatial structure of $BiVO_4$ was significantly changed after compositing with CMS, and the main components were $BiVO_4$ and $CaCO_3$. The band gap energy of $BiVO_4$, 1.5-BCMS, 1.75-BCMS and 2-BCMS obtained by the plots between $(ahv)^2$ and energy (hv) were about 2.76 eV, 2.97 eV, 2.37 eV and 2.91 eV, respectively. Among the composites with various $BiVO_4$/CMS loading ratios, the prepared 1.75-BCMS photocatalyst presents the best performance with a CV (10 mg/L) removal efficiency of 97.71% within 120 min, Which was 1.84 times of pure $BiVO_4$ (52.97%).

Keywords $BiVO_4$/CMS · Photocatalytic degradation · Crystal violet

68.1 Introduction

With the accelerating industrialization, it has caused serious pollution of the water environment in nature, especially the treatment of dye wastewater is the most concern.

L. Zhao · X. Ou (✉) · C. Ma · C. Liang
College of Environment and Resources, Dalian Minzu University, Dalian 116600, China
e-mail: ouxiaoxia@dlnu.edu.cn

X. Ou
Liaoning Academy of Landfill Leachate Treatment Engineering, Dalian, China

Dalian Carbon Neutral Technology Innovation Center for Sewage Treatment, Dalian, China

J. Zhang et al. (eds.), *Environmental Pollution Governance and Ecological Remediation Technology*, Environmental Science and Engineering,
https://doi.org/10.1007/978-3-031-25284-6_68

$BiVO_4$ is a typical n-type semiconductor material with suitable energy band structure, high response to visible light and favorable narrow band gap (2.4 eV), which become a promising material for degrading organic dye wastewater (Kai et al. 2022).

$BiVO_4$/CMS composite photocatalysts with good stability and catalytic activity were prepared by hydrothermal synthesis, and their loading ratios were adjusted to obtain $BiVO_4$/CMS composite photocatalysts with different loading amounts. The catalytic performance of the composite photocatalysts was tested with CV as the target degradation product.

68.2 Experimental

68.2.1 Materials

Bismuth nitrate ($Bi(NO_3)_3.5H_2O$),ammonium metavanadate (NH_4NO_3) and anhydrous ethanol (CH_3CH_2OH). The reagents were purchased from Comio Chemical Reagent Co., Ltd in Tianjin, China. The shells were purchased from the farmers' market in Jinma Mall, Dalian, China. The purity of all reagents was analytical grade and the reagents were used directly without any further purification.

68.2.2 Preparation of CMS

The collected sand clam shells were brushed and dried, and, were churned with a powdering machine for three minutes after drying. The powdered sand clam shells in the muffle furnace calcination was maintained two hours at 900 °C. Shell powder was added distilled water for centrifugal stirring three times, and was dryed at 105 °C for 24 h until completely dry. Then, the shell powder was ground with agate, sifted with a 100 mesh screen, and loaded into a sealed bag.

68.2.3 Preparation of $BiVO_4$/CMS

Different mass ratios of $BiVO_4$/CMS composites were synthesized by hydrothermal route. Take 1.5-BCMS as an example, 0.4424 g $BiVO_4$ was added to 7 mL ethanol accompanied by ultrasonic mixing for 30 min and magnetic stirring for 30 min to obtain solution C. Simultaneously, 0.2949 g shell powder was dissolved in 20 mL ethanol under magnetic stirring for 30 min to obtain solution D. The solution C was slowly added into D of stirring for 12 h. The suspensions were henceforth transferred into a 50 ml Teflon-lined steel autoclave and subjected to a 12-h hydrothermal treatment at 160 °C. Afterward, the samples were cooled to room temperature and repeatedly washed with ethanol and distilled water until the solids were separated. The obtained products were dried and ground at 60 °C to finally obtain 1.5-$BiVO_4$/CMS composites. According to this process, $BiVO_4$/ CMS were prepared with various molar ratios of $BiVO_4$ and CMS and denoted the composites as 1.5-BCMS, 1.75-BCMS, and 2-BCMS of which the molar ratios of $BiVO_4$ and CMS were 1.5:1, 1.75:1, and 2:1, respectively.

68.2.4 Photocatalytic Degradation Tests

50 mL CV (10 mg/L) and 30 mg catalyst were added to the beaker. The catalyst was uniformly distributed in the CV solution for 10 min ultrasound and the adsorption–desorption equilibrium was obtained by 20 min of stirring. The visible light source was a 500 W xenon lamp ($\lambda > 420$ nm), and the distance between the light source and the liquid level was 14 cm. Under the condition of magnetic stirring, 3 mL of reaction solution was collected and labeled at certain time intervals, and subsequently was centrifuged and filtered through millipore filter. The absorbance was measured at 590 nm by UV–visible spectrophotometer. The removal rate of CV (η) was calculated using the equation method.

$$\text{Removal rate } \eta\% = (1 - A_t/A_0) * 100\% \tag{68.1}$$

In formula (68.1), A_0 is the initial absorbance and A_t is the absorbance of the supernatant at t minute.

68.2.5 Characterization

The crystall ographic structures of the as-prepared samples were conducted by a powder X-ray diffractometer (XRD-600, Shimadzu, Japan) employing CuKα radiation in the 2θ ranging from 10 to 80° with a scan rate of 4° min^{-1}. The microstructure morphology of the sample surface was analyzed by scanning electron microscopy (S-4800, Shimadzu, Japan). The characteristics of the functional groups involved in the sample were tested and characterized by Fourier infrared spectroscopy (IR). The spectral characteristics of the catalyst were characterized by the DRS (Hitachi type-7000 spectrophotometer, Japan).

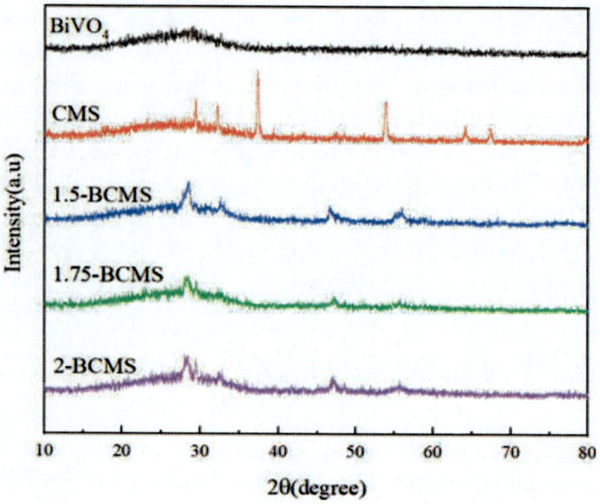

Fig. 68.1 XRD patterns of CMS, $BiVO_4$ and $BiVO_4$/CMS samples

68.3 Results and Discussion

68.3.1 XRD Analysis

The XRD of $BiVO_4$, CMS and $BiVO_4$/CMS composite photocatalysts with different loadings are shown in Fig. 68.1. The characteristic peak of CMS and the weaker characteristic peak of $BiVO_4$ ($2\theta = 28°$) indicated that the composite photocatalytic material was comprised of $BiVO_4$ and CMS. The characteristic diffraction peaks of pure $BiVO_4$ corresponded to a diffraction angle of 28°, and the diffraction peaks of CMS corresponded to diffraction angles of 29°, 34°, 37°, 55°, 65°, and 68°. When the loading was increased from 1.5 to 2, the characteristic diffraction peaks of $BiVO_4$/CMS were all consistent with no impurity peaks. It can be inferred that the addition of CMS did not have any effects on the crystal structure of $BiVO_4$/CMS. Compared with pure $BiVO_4$, no obvious peaks corresponding to $BiVO_4$ were observed in $BiVO_4$/CMS, which may be due to the fact that $BiVO_4$ is too dispersed (Qiang et al. 2020).

68.3.2 SEM Analysis

Figure 68.2 shows the SEM images of pure $BiVO_4$ and $BiVO_4$/CMS composite photocatalysts. It can be seen from Fig. 68.2a that the pure phase of $BiVO_4$ is flower-like nanoparticles with different particle size. Comparing the composite photocatalytic materials with three different loading ratios, it was found that when $m(BiVO_4):m(CMS) = 1.5$, the composites mainly appear in the form of flakes and a few are in the form of particles. A small amount of $BiVO_4$ particles are attached to

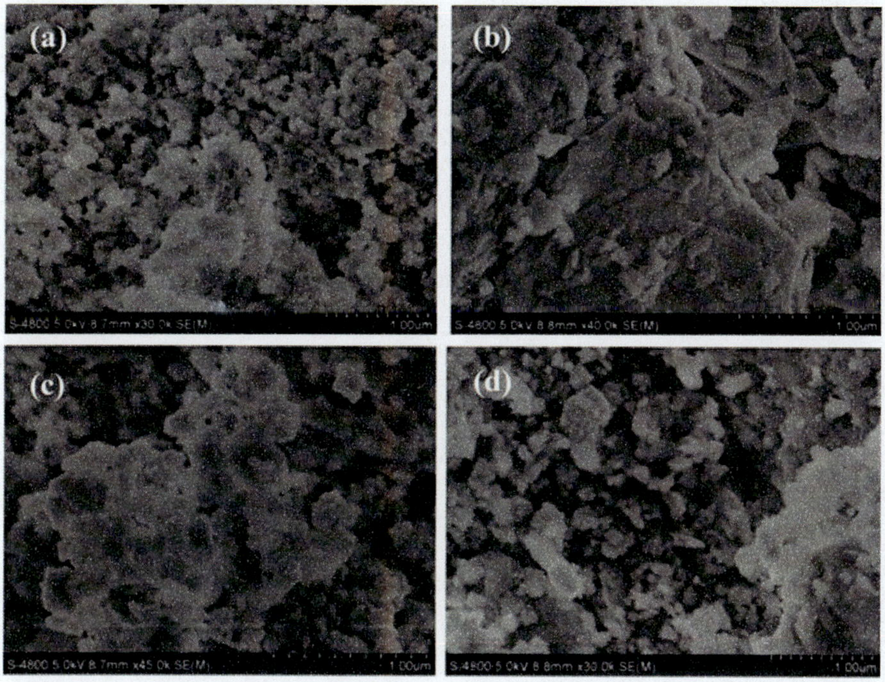

Fig. 68.2 SEM images of pure $BiVO_4$ **a** 1.5-BCMS **b** 1.75-BCMS **c** and 2-BCMS **d**

CMS indicating that the materials are successfully prepared. With the increase of the loading ratio, the microscopic morphology of the composites changed from flake to granular, and the inter-particle dispersion was good without bonding, which increased the contact area between the composite catalyst and the pollutants in solution and the photocatalytic efficiency was improved.

68.3.3 DRS Analysis

All the samples had a certain light absorption in the visible light range ($\lambda >$ 400 nm). As can be seen from Fig. 68.3a, the DRS spectra of pure $BiVO_4$, CMS and $BiVO_4$/CMS composites with different CMS contents, revealing that the pure $BiVO_4$, CMS and $BiVO_4$/CMS samples exhibit a broad absorption in the visible light region. With the increasing amount of CMS, the absorption edge of $BiVO_4$/CMS red shifted to longer wavelength within the range of visible-light, with the spectral response range of CMS being extended. The $BiVO_4$ sample showed an absorption edge around 550 nm, which could be responsible for the visible-light induced photocatalytic activity (Baowei and Jianhong 2013). Comparing to the pure $BiVO_4$, the addition of CMS did not improve the visible light absorption range. The optical band gap energies (Eg) of $BiVO_4$, 1.5-BCMS, 1.75-BCMS and 2-BCMS can be calculated by the following formula (68.2), where a, h, v, A and Eg were the absorption coefficient, Planck constant, light frequency, proportionality constant, and the band gap energy, respectively.

$$(ahv)^2 = A(hv - Eg) \tag{68.2}$$

The Eg of $BiVO_4$, 1.5-BCMS, 1.75-BCMS and 2-BCMS obtained by the plots between $(ahv)^2$ and energy (hv) were about 2.76eV, 2.97 eV, 2.37 eV and 2.91 eV in Fig. 68.3b. So the optimal loading ratio of $BiVO_4$ and CMS is 1.75.

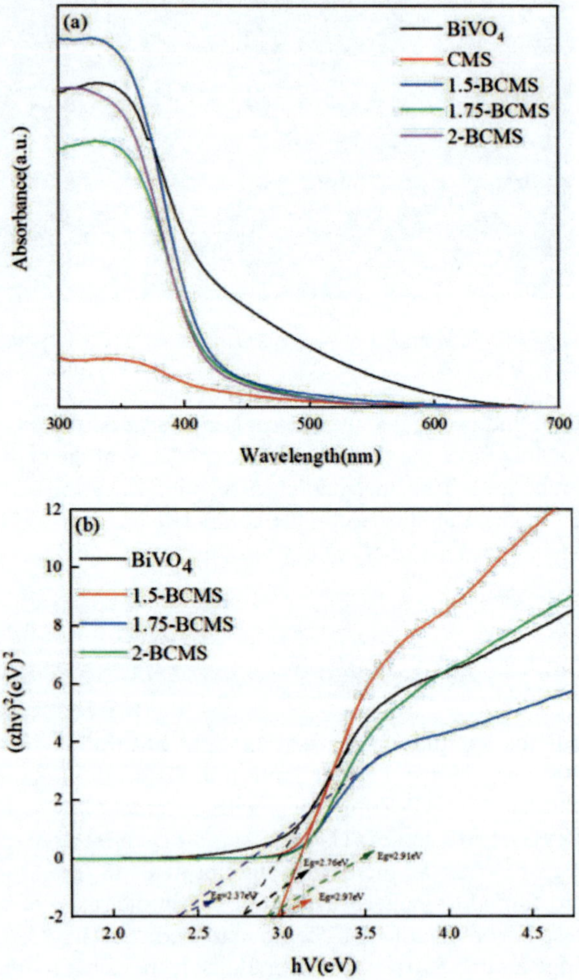

Fig. 68.3 **a** UV–Vis DRS spectra and **b** band gaps of the samples

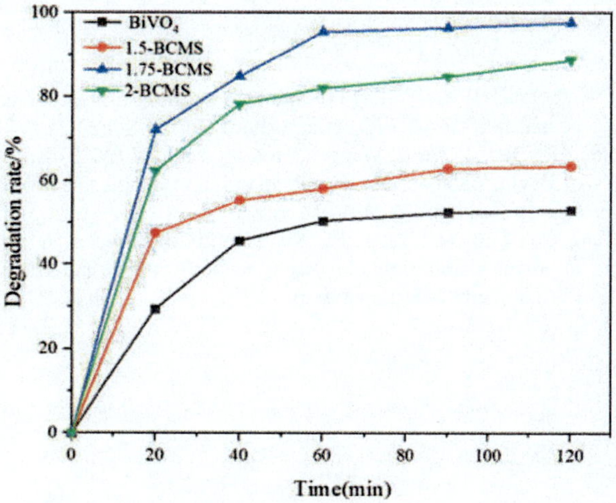

Fig. 68.4 Photocatalytic activity of $BiVO_4$ and $BiVO_4$/CMS for degradation of CV

68.3.4 Photocatalytic Activity Analysis

The photocatalytic activities of $BiVO_4$ and $BiVO_4$/CMS composites were evaluated through the CV degradation under visible light irradiation and the results are graphically illustrated in Fig. 68.4. It can be seen that all the $BiVO_4$/CMS composites possess better photocatalytic activity of CV degradation than pure $BiVO_4$. After 120 min visible light irradiation, the degradation rates of CV over $BiVO_4$ is 52.96%. With an increase of CMS content (1.5-BCMS < 2-BCMS < 1.75-BCM), the CV degradation rate is increased from 52.96 to 97.71%. However, with further increasing CMS content, the photocatalytic activity decrease that indicate that excessive CMS will aggregate, lead to the reduction of active sites and the decrease of photocatalytic efficiency. Among them, 1.75-BCMS exhibits the best photocatalytic performance with almost all CV decomposed within 120 min. From the experimental result, it also can be seen that the photocatalytic activity of $BiVO_4$ was greatly improved after the introduction of CMS. This is due to the fact that the addition of CMS can promote the absorption of visible light by the catalyst and can effectively accelerate the separation of electrons and holes.

68.4 Conclusions

In summary, we successfully prepared $BiVO_4$/CMS composite material by solvothermal method. The structure of the synthesized catalysts was characterized. The introduction of CMS changed the crystal phase structure of $BiVO_4$, and $BiVO_4$/CMS photocatalysts displayed higher photocatalytic activity. When m($BiVO_4$):m(CMS) = 1.75, the $BiVO_4$/CMS samples performed well and could degrade 98% of CV(10 mg/L) within 120 min irradiation.

References

Cao B, Peng J, Xu Y (2013) Simulated sunlight-driven degradation of rhodamine B by porous peanut-like $TiO_2/BiVO_4$ composite. J Cluster Sci 24(3):771–785

Song K, He F, Zhou E, Wang L, Hou H, Yang W (2022) Boosting solar water oxidation activity of $BiVO_4$ photoanode through an efficient in situ selective surface cation exchange strategy. J Energy Chem 68(05):49–59

Tang QY, Luo XL, Yang SY, Xu YH (2020) Novel Z-scheme $In_2S_3/BiVO_4$ composites with improved visible-light photocatalytic performance and stability for glyphosate degradation. Sep Purif Technol 248(prepublish)

Chapter 69
Structure, Conducts, and Performance in Payments for Ecosystem Services: A Case Study in Lishui City of Zhejiang Province, China

Yingzhi Yuan and Youling Sun

Abstract Payment for Ecosystem Service (PES) has become the research focus in various countries. Prior research has quantitatively evaluated the feasibility, influencing variables, and performance of PES, presenting fewer details on the logical structure and practice mechanisms of PES in specific contexts. Based on a revised Structure-Conducts-Performance framework, this paper digs into the market structure, contract and policy content, and performance of innovative PES in Lishui City, Zhejiang Province, China. The study finds that Lishui City redefines Ecosystem Service (ES) and PES based on its circumstances and policy needs, and the various properties of ES construct a pluralistic ESM. Service providers and users interact and establish diverse contracts, cooperation, and policies, making the market not only provide economic advantages but also contribute to societal values.

Keywords Payment for ecosystem service (PES) · Environmental policy · China's pilot · Structure-conduct-performance

69.1 Introduction

In 2001, the UN established The Millennium Ecosystem Evaluation to encourage ecosystem assessment and protection. Since then, Ecosystem Services (ES) have been a worldwide emphasis. Payment for Ecosystem Service (PES) is an emerging natural resource management strategy (Muradian et al. 2010) that financially incentivizes ES providers to make eco-friendly actions (Salzman et al. 2018). PES schemes like US carbon trading (Solomon and Lee 2000), Costa Rica's Pago por Servicios Ambientales (Pagiola 2008), and REDD+ (Pistorius et al. 2012) are significant.

China's PES exploration is notable. In 2006, Xi Jinping stated, "lucid waters and lush mountains are invaluable assets/gold mountains" (Jinping 2015) (Two mountain theory, TMT). He believes that economic progress shouldn't be at the expense

Y. Yuan (✉) · Y. Sun
Hunan University, Hunan, People's Republic of China
e-mail: yuanyz776@163.com

J. Zhang et al. (eds.), *Environmental Pollution Governance and Ecological Remediation Technology*, Environmental Science and Engineering,
https://doi.org/10.1007/978-3-031-25284-6_69

of the environment but rather add value to natural capital to safeguard ES and realize ecological resources' economic and social value. This approach aligns with the UN's Economics of Ecosystems and Biodiversity (TEEB) (The Economics of Ecosystems and Biodiversity 2022). Since then, the Chinese government has pushed for environmental management reform, notably an innovative PES scheme. China's political framework, collective property rights regime, and low per capita resource endowment have allowed it to construct a unique PES mechanism based on experience from other nations (Adhikari and Agrawal 2013). 2021, the official document "Opinions on establishing and improving The Value Realization Mechanism of Ecological Products (VRMEP)" advises China to construct a sustainable, high-cost–benefit, and resource-efficient ecological products (EP) market as soon as feasible (Opinions on establishing the value realization mechanism of ecological products 2022). VRMEP has become the official name of China's localized and innovative PES, and a fresh round of pilots has begun under government support. Lishui City, Zhejiang Province, the birthplace of TMT, explored and practiced the PES mechanism before other cities, becoming the first pilot city of VRMEP. Lishui City's innovation shows China's distinctive ES and market knowledge, varied collaboration methods, and rich PES forms. Its ecological preservation and regional development success represent China's efforts to build viable PES schemes. This paper aims to understand the adaptive reconfiguration, policy choices, and outcomes of the PES mechanism in China through the case of the innovative PES mechanism and the construction of VRMEP in Lishui City and to discuss its implications for academic research, policy formulation, and practice.

69.2 Literature Review

ES is the focus of environmental policy and includes the direct and indirect sustaining, regulating, provisioning, and cultural services and goods (Millennium ecosystem assessment 2005) humans derive from ecosystems (Daily 1997). ES has "fit-for-purpose" categories for different objects and contexts (Brendan et al. 2009). Therefore, there are various classifications based on ES functions (Wallace 2007).

PES has been proposed as a new financing mechanism and management framework to prevent the degradation of ES and maintain ongoing supply. With very dynamic management decisions and practices, PES's definitions, inherent logic, and terminologies (Sven 2015) are developing to reflect its nature and innovations. As the "government-financed" PES variant becomes mainstream (Engel et al. 2008), past Coasean PES definitions are excessively restricted. Wunder (2015) proposes a new normative perception: PES is voluntary transactions between service users and providers that are conditional on agreed rules of natural resource management for generating offsite services (Sven 2015). Some scholars have focused on PES's structure, including institutional work and actors' perspectives. Institutional work focuses on the influence of power relations and institutional design. For example, power relations influence knowledge, behavior, habits, and norms (Gomez-Baggethun and

Muradian 2015) and determine access, control, governance, equity, and efficiency through formal and informal institutions (He 2020). The actors' perspective provides context-sensitive, nuanced, and dynamic elucidation of actors' behaviors in micro scenarios. For example, communities identify the beneficiary groups and determine the norms of benefits distribution (Santiago Izquierdo 2022). Fieldworkers exercise discretion (Silva-Muller 2022) to increase stakeholders' environmental awareness and acceptance and even change the formal system to gain legal and discursive support (Benjamin and Harris 2021). Most studies analyzed PES's impact variables and benefits quantitatively. They emphasize the impact of government trustworthiness (White et al. 2022), nature connection (Esther et al. 2021), and societal norms (Gi et al. 2021) on PES and utilize a matching approach (Zhou 2022) or meta coupling framework to assess PES projects' viability (Robert 2021) and economic (Xutong 2021), ecological (Zhenmin and Shunbo 2021), and social (Rosa 2022) benefits. Other scholars have used qualitative research methodologies, such as case studies and in-depth interviews, based on neoliberal (Pamela 2014) and stakeholder theories (Stosch et al. 2019) to describe PES implementation and the mechanism of action.

Existing studies enhance our knowledge of PES in crucial ways, reveal us to focus on the developmental definition of ES and PES, and comprehend them in local constraints. However, qualitative studies on PES policy arrangements, practical progress, and empirical results of PES in different contexts are not presented and analyzed in detail. Therefore, this paper seeks to comprehend and clarify how and what sort of mechanism is built locally to fit local institutions and culture and generate advantages and dig and characterize the essential elements involved in PES of Lishui City.

69.3 Methodology and Case Selection

69.3.1 Conceptual Approach: A Revised SCP Framework

This paper introduces the Structure-conducts-performance (SCP) paradigm, which was used to analyze the behavior of industrial organizations (Tesfaye and Abdurezak 2018), as an analytical framework. SCP argues that the desired market performance can be achieved through public policies that adjust the market structure and regulate market behavior. This approach is usually used to examine industries and financial and securities markets. Thus, its application's applicability in public administration must be shown. First, the SCP paradigm contains three significant components that drive the PES in Lishui City, and their composition and influence mechanisms may expose distinctive practices. Second, this paper takes the ecosystem service market (ESM) as a study object rather than ecological governance behaviors and is, therefore, interactive with the framework. These two arguments show that employing SCP is sensible.

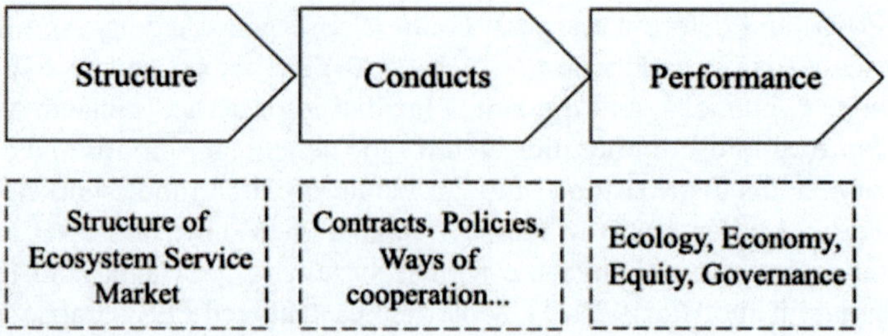

Fig. 69.1 Revised structure-conducts-performance framework based on J. S. Bain (1959) (Joe 1959)

However, the ESM and SCP frameworks' past analysis items are different. ESM is an imperfectly competitive, non-neoliberal market in China and most nations. This is because the externalities and publicness of ES influence the degree of government intervention and performance objectives. Market-driven production and consumption evaluate cost–benefit more than ecological value, which easily leads to resource waste and pollution. At the same time, China's natural resources are publicly owned, and property rights are unclearly defined, leading to inefficient market operations. Thus, the Chinese government employs public institutions, administrative power, and fiscal transfers (Xingliang et al. 2017) to facilitate ES transactions, resulting in imperfectly competitive markets and contracts. In addition to production efficiency, ESM considers ecological conservation and poverty alleviation as performance indicators. Therefore, the previous SCP framework needs to be revised to be more utility. First, using ES typology to analyze ESM structure and to reflect its effects on transaction types and cooperation patterns. Second, using contract and relational contract theories to analyze market conduct (MacNeil 1982) to introduce varied ES transaction methods. Finally, Performance evaluation also considers ecological indicators and fairness. Then, the revised SCP framework is under the impact of ESM structure, service providers and users interact and establish diverse contracts, cooperation, and policies, making the market not only provide economic advantages but also contribute to societal values such as ecological benefits (see Fig. 69.1).

69.3.2 Methods and Study Area

This paper takes a case study as a method to detail PES practices in the Chinese context. Case studies can demonstrate causal mechanisms and event development processes (Yin 1994). This study aims to identify and understand parallels and contrasts between PES in Lishui City and other nations' practices, and to define its features, logic, and performance. Therefore, an exploratory and qualitative case study fits the purpose of the research. In August 2021 and January 2022, the authors and research team visited Lishui City and its subordinate counties and townships

Table 69.1 Summary of data collected during research

Method	Category	Number	Timing	Data
Individual interviews	Public official	4	60 min	More than 30 copies of first-hand interview materials and more than 1,000 min of audio recordings
	Enterprise	7	30 min	
	Professionals	3	20 min	
	Villagers	5	50 min	
Focus groups	Public official	5	50 min	
	Villagers	6	30 min	
Participatory observation	Governments' daily work	5	/	Writing memos of PES mechanism, attitudes and participation of various subjects
	Enterprise conference	3	/	
	Homestays' operation	2	/	
	Villagers' life	3		
Ethnographic materials	Public documents	32	/	Used for data triangulation
	Policy Brochure	9	/	
	News report	56	/	

to conduct field research and interview local party and government leaders, business leaders, citizens, and related people (see Table 69.1). The paper collects data on ecological conservation and PES mechanism development in Lishui City from 1978 to the present. Lishui was chosen for two reasons. First, Lishui is China's first VRMEP pilot city. This signifies advancement and representativeness of Lishui's actions are acknowledged by the country. Second, Lishui is not a special municipality, provincial capital, or mega-city; thus, its experience is applicable to similar environmentally rich but impoverished locations.

Lishui is a mountainous prefecture city in Zhejiang Province. Lishui has natural ecological advantages. Lishui's 81.7% forest coverage rate is 20 points greater than the province and second-ranked in China (Lishui statistical yearbook 2021). Biological species are diverse, accounting for about three-quarters of the province's total number. Lishui has a well-developed river system and high-quality water environment, and the complete index of urban surface water quality ranks first in the province (Bulletin 2021). Lishui was an environmentally rich and undeveloped area at the outset of reforming and opening, and its resources were mainly idle or inefficiently employed, limiting local economic growth. Since the millennium, Lishui City has used ecological conservation, PES, and VRMEP to boost its economy and eliminate poverty based on environmental resources. From 1978 to the present, the policy development history encompasses ecological devastation, ecological restoration, industrial ecology, PES, and VRMEP (see Table 69.2).

Table 69.2 Ecological policy history of Lishui city

Stage	Year	Policy document/Event	Policy goals
Ecological destruction (1978–200)	1992	Decision on accelerating reform and opening-up policy and accelerating economic development	Developed along the river, broadened markets along the road, and developed industries in market towns
	1995	The seminar to emancipate the mind of Lishui Municipal Committee of the CCP	Agriculture should be developed on a larger scale. The added value of agriculture from 1995 to 2000 should reach 6 billion yuan
Ecological restoration & industry (2000–2006)	2000	The first communist party congress in Lishui city	Develop eco-efficient economy. Adjust industrial structure, guide enterprises to upgrade technology and control pollution. Public welfare forest construction
	2005	Lishui city eco-construction and pollution control plan	Improve the environment of the Ou River water system and reservoirs; rectify and close enterprises with excessive emissions; increase environmental quality and pollution source monitoring equipment
PES learn and experiment (2006–2017)	2008	Lishui city ecological civilization construction outline	Improve environmental quality and form a consumption and industrial structure that conserves resources and protects the environment
	2015	Establish regional public brand *Lishui Shangeng*	Though government 's investment to build brands, reduce transaction costs, and enhance brand credibility and premium
	2016	Lishui as pilot area for collective forestry reform	Collective forest land easement system, forest land management property rights, cultivating new forestry business entities, forestry financial mortgage loans, etc.
VRMEP innovate & improve (2017-now)	2017	Suggestions on improving strategy and system of Eco-functional zones	Lishui has been identified as a pilot city of VRMEP in Zhejiang province

(continued)

Table 69.2 (continued)

Stage	Year	Policy document/Event	Policy goals
	2020	Technical specification for accounting gross ecosystem product (GEP)–terrestrial ecosystems	Calculate the final goods, services and economic value provided by ecosystems for human well-being, economic and social development services in given area
	2020	Suggestions on implementing financial incentive mechanism for eco-construction	Based on the absolute value and growth rate of GEP, the funds for incentives and subsidies to local governments are calculated

69.4 PES in China: Based on a Revised Framework

69.4.1 Structure

The market structure emphasizes market features that affect corporate actions and outcomes. This research contends that the characteristics most closely associated with ESM in China and have the most impact on market behavior are ES categorization, property, price, and market type.

To define ES and its categorization in China's ESM, it should be noted that Chinese policy papers use Ecological Products (EP) as the official title for a similar concept to ES, demonstrating that their interpretation of ES is different in some dimensions. (1) Definition. ES refers to all the benefits humans gain from ecosystems that are not affected by human activities (Wallace 2017). EP is the final product or service that ecosystems produce via biological production and joint action with human labor to meet human needs (Technical specification for accounting gross ecosystem product (GEP)—Terrestrial ecosystems 2022), which highlights that both people and nature produce EP in Social-Ecological Systems. (2) Scope. EP is a marketable final service and product, and its scope is more minor than ES. (3) Function and application context. ES focuses on services provided exclusively by natural ecosystems, which can be applied in academic research and raise awareness of the ecosystems' importance. Because increasing economic growth and harmonizing regional development is VRMEP's strategic goal (Opinions on establishing and improving the mechanism for realizing the value of ecological products 2022), EP is more often applied in policy and focuses on the value production by human labor and technical elements, markets' consumption, and exchange. (4) Property. ES tends to operate as public goods offered by ecosystems, expressing tangible information flows. EP has more commodity features and focuses on the ecological transformation of public EP by people and private products created mainly through human participation.

In summary, EP in China emphasizes producers, property, and marketability. EP can be classified as purely public EP, including water, soil, species conservation, flood, and climate regulation. Quasi-public EP includes air and water purification, carbon fixation, public forest, grass, etc. Private EP includes agricultural products, cultural education, health rehabilitation, etc. This categorization reflects the above three aspects and forms diverse ESM structures with property rights and pricing. When EPs are provided more by ecosystems, product property is tilted toward the public, property rights belong to the state or the collective, and governments act as agents or commissioning agencies for exclusive production, creating a monopoly market. Because measuring and establishing property rights is difficult, prices are often negotiated. When quasi-EP and private EP are predominantly produced by humans, they have well-defined private property rights that can minimize transaction costs by identifying benefit streams. This sort of EP has numerous buyers and sellers, and the perfectly competitive market determines the price.

69.4.2 Conducts

China's ESM structure affects the conduct of market stakeholders, including contractual and policy actions. Contractual conduct involves clarifying reciprocal rights, obligations, and signing agreements between two trading partners to manage trader behavior and gain desired advantages. Chinese ESM generates relational (RC) and transactional contracts (TC) based on EP properties (see Fig. 69.2).

RC is utilized in emergent, sophisticated, and specialized trading scenarios where the value of trading items is difficult to assess in monetary terms. And the transaction duration and traders' rights and duties cannot be rationally predicted (MacNeil 1982). Most pure EP and quasi-EP exchange socially or metaphorically in this way. ES

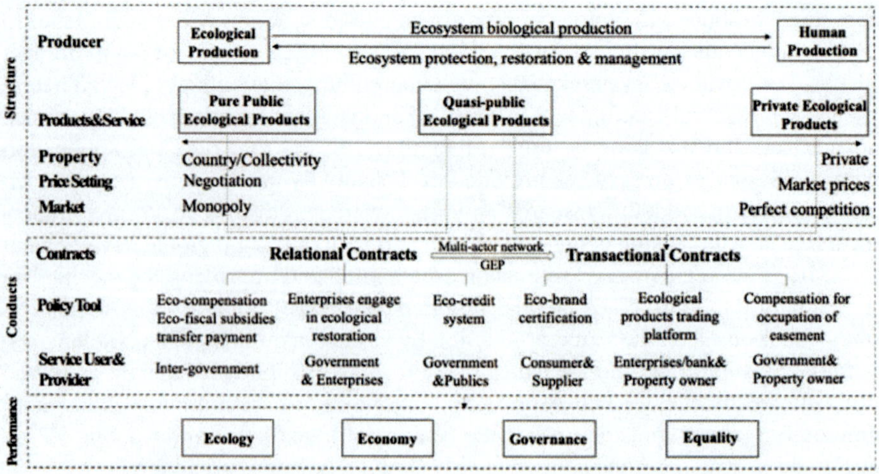

Fig. 69.2 Analysis of Chinese ecological products market with a revised structure-conducts-performance framework

traders agree on a processual, adaptable contractual framework for cooperation. The contract's content constantly evolves and finally attains stable transactions in repeated games by preserving a stable division of labor and collaboration (Argyres et al. 2020). TC demands a precise explanation of the potential states associated with the transaction and the rights and duties of the trader. It applies to quasi-EP and private EP where transaction ownership is evident, and market value may be recognized.

Three PES routes in Lishui City used relational contracts. (1) Inter-governmental eco-compensation. The central or ecologically profitable government pays or offers projects or intellectual assistance to the government that supplies ES. Vertical ecological compensation is a unique upward centralized redistribution strategy in China between the central and local governments. The county and district governments and corporations around Yuxi and Huangcun reservoirs earned 8.55-million-yuan transfer payments for protecting the water ecosystem. Horizontal ecological compensation comprises inter-government, eco-function areas, and watershed payments. Six cities and counties in the Oujiang River watershed agreed to donate 5 million yuan each for ecological compensation that year. Downstream rewards upstream's efforts to maintain the water ecosystem and has claims for water quality deterioration and excessive water consumption. Water conservation is purely public EP with inaccurate transaction circumstances and compensation requirements. It's an RC founded on inter-governmental cooperation, policy incentives, and common interests. (2) Government-private RC. The government invited a big agricultural firm to help rehabilitate abandoned mines and pledged the use right of treated mine land to the company, and it could utilize 17 acres of mine land for ecological-oriented construction, which eventually built an agro-ecological park with high production value. The company solves its land constraints, the government saves fiscal expenditure and attracts huge tax revenue. Farmers and villagers gain employment opportunities, which balances the regional economy and ecological protection simultaneously. The transaction did not assess the economic value of the mine and the land but completed a symbolic RC by the government exercising its agency to manage the land, exchanging land rights, potential economic benefits, and government-enterprise partnership for ecological treatment funds from the enterprise, and building an ecologically oriented industry as an additional condition. The transaction did not assess the mine and land's economic value but was completed with symbolic RC by giving land rights, potential benefits, and a government-enterprise partnership. (3) Government-citizen RC. The Lishui government linked ecological credit behavior with financial credit preferences by establishing personal ecological credit files. Financial institutions use this to rate the grade farmers' ecological behavior and give unsecured loan support to farmers who meet the ecological credit rating. The government employs ambiguous credit ratings as an evaluation standard to construct RC that uses loan equity to gain positive ecological behavior, which solves the high cost of ES value assessment and considers ecological preservation and farmers' personal development.

However, conflicts of interest in RC may cause contract dissolution. RC ensures the actions and performance of traders through resource dependency and reputation mechanism, which leads to the dominant subject in the social and power relationship having more authority over the contracts' contents, resulting in injustice, inefficiency, and poor sustainability. Lishui City solves this challenge by integrating heterogeneous actors and employing coordination, voluntary combination, and mutual restrictions to construct a dynamic multi-actor network. Through supervision to ensure the fairness of contract objectives and benefit distribution, PES can continuously serve the interests of multiple stakeholders. At the same time, Lishui City encourages changing RC into TC to boost transaction efficiency. Lishui City uses Gross ecosystem products (GEP) (Zhiyun et al. 2020), an ecological resource quantification technology to count the sum of EP values, improve the measurability and accuracy of EP, and combine with the cloud data monitoring platform to determine and register ambiguous resource property rights so that more EPs can enter the efficient market. GEP may be utilized to account for the added value of EP and services created by a specific ecological behavior for the three forms of relational contracts listed above, making the contracts fairer and more efficient.

Lishui City has developed three high-efficiency TC based on multi-actor networks and quantitative techniques. (1) "Lishui Shangeng" public agricultural brand. Utilizing ecological certification labels or trademarks provided by the public sector or third-party agencies in marketing to attract EP-sensitive consumers and earn premium revenue. The Lishui Municipal Government commissioned professional institutions to build a regional public brand, "Lishui Shangeng" in 2017 and collaborated with companies to operate. Using GEP and agricultural goods traceability and supervision systems to make product value and ecological process transparent to justify the ecological premium and enhance agricultural product prices. Brand recognition and quality certification lower company sales costs. By the end of 2018, "Lishui Shangeng" products had sold 10.158 billion yuan, with an average price premium of over 30% (Shangeng 2022). The brand has contributed to the flourishing of local ecological agriculture, realized EP value, and improved farmers' income. Lishui City has created seven other industrial brands, including ecotourism, to stimulate whole industries' growth. (2) Ecological Bank (EB). The collective economy in rural China puts ecological resources under the village's community ownership and contracts to individuals for their use, which is unclear and fragmented. Following the principle of people's voluntariness, Lishui City employs the EB as a "middleman" to collect and manage the property rights of arable land, forest land, and other ecological resources through surveys, GEP valuation, and registration of rights to encourage ecological tenure transactions. Villagers and village cooperatives can also get dividend revenue, which boosts the village collective economy and benefits individuals. EB is like the US wetland mitigation bank, but its mission highlights rural construction and economic development. (3) Baishanzu National Park (BNP) easement reform. The easement reform takes the state-owned forest land in BNP as the demand land, collective forest land as the supply land, and the BNP management agency as the easement owner. It establishes compensation and collaborative governance mechanism based on not changing the ownership of collective forest land by signing easement contracts with village committees and villagers. Through mobilization, GEP survey, valuation, registration, and archiving, the borders between state-owned and community forest area in BNP are demarcated. Four hundred forty-five easement registration contracts are made between BNP management agencies and more than 6000 villages, and 270 million yuan in compensation monies are provided yearly.

69.4.3 Performance

Ecological indicators and VRMEP's policy goals, including regional economics, equity, and governance, should be used to analyze Chinese PES performance.

Ecology

Lishui City's general ecological state is favorable, and its water, ecological, and forest assets have increased after exploring the creative and effective PES mechanism. (1) Overall condition. 2021, Lishui City's ecological index (EI) has been the first in the province for 18 years. (2) Water resources. Relying on a horizontal ecological compensation fund in the Oujiang River basin, Lishui City has built more than 4000 km of greenway network in the Oujiang River to purify water quality, and 99% of the city's surface water cross-section water reaches high-quality standards. (3) Bioresources. By implementing easement reform and systematic protection for BNP and other parks, habitats of rare plants and animals such as the Baishanzu fir have been restored. Seven new global and 64 new recorded species in China were discovered in Lishui City. (4) Forest resources. Lishui City has relied on ecological finance loans, BNP, and other schemes to carry out scaled management, maintenance, and forest restoration. 2020, Lishui City reforested 37,785 ha (see Fig. 69.3). The scope of timber extraction has been restricted. The area of forestation was more than 290,000 hectares every year. The amount of timber cut has been significantly reduced since 2010 (see Fig. 69.4). The forest cover in Lishui has been maintained at 81.7% for many years, and the forest reserve in Lishui reaches a new high of 988,583 million steres in 2020. The Ecological Environment Research Center of the Chinese Academy of Sciences measured the GEP and its changes in Lishui City in 2020. The results show that the GEP of Lishui City increased by 212,746 billion yuan from 2006 to 2018, which is strong proof of the efforts made by Lishui City in the field of PES, VRMEP, and ecosystem protection.

Economy

The increase in EP marketization has enabled Lishui City to form a modern ecological economy system and improve the quality and efficiency of industrial structure, gross regional product, and ecological industry development. First, the industrial structure of Lishui City has changed from a small peasant economy led by the primary industry to a high-quality green economy led by the tertiary industry such as ecotourism and finance (see Fig. 69.5). Second, Lishui has ushered in a new growth point. Since 2015, the regional GDP in Lishui has been 100 billion yuan and keeps steadily rising (see Fig. 69.6).

In terms of industrial development, VRMEP has supported ecological agriculture and ecotourism through ecological certification and other measures. First,

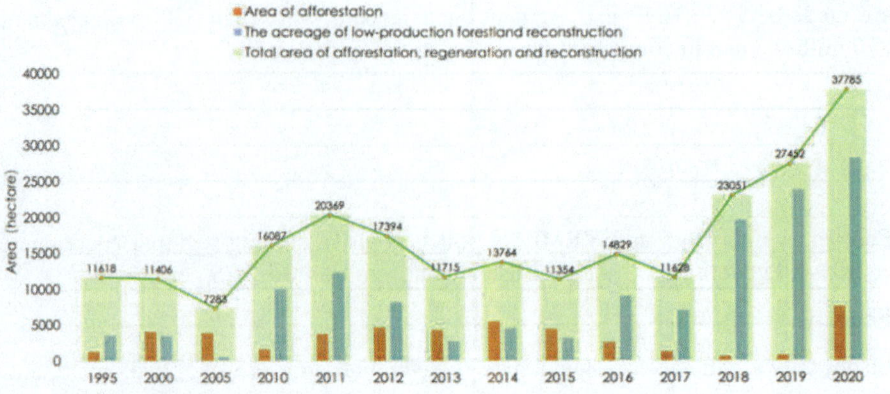

Fig. 69.3 Total area of afforestation, regeneration, and reconstruction of Lishui city from 1995 to 2020 based on the 2021 Lishui statistical yearbook

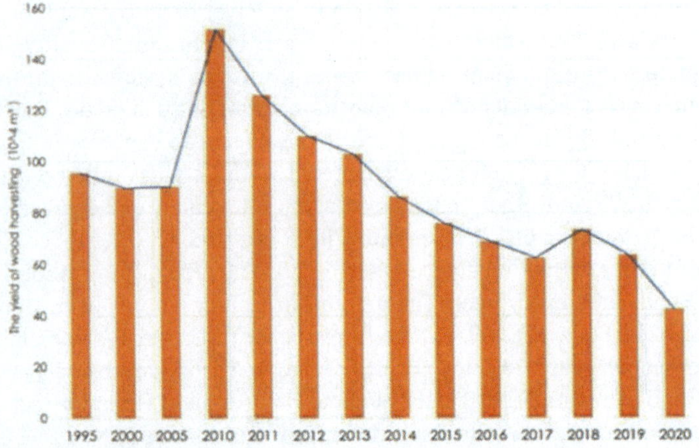

Fig. 69.4 The yield of wood harvesting in Lishui city from 1995 to 2020 based on the 2021 Lishui statistical yearbook

the total agricultural output value of Lishui City grew from 369 million yuan in 1978 to 15,274 billion yuan in 2017. The structure of agricultural production has been continuously adjusted and optimized, changing from solving the subsistence problem to developing boutique eco-agriculture. As of 2017, 1.69 million acres of eco-agricultural product bases have been built. Among them, 695 agricultural products have been registered under "Lishui Shangeng," with a total business income of 15.37 billion yuan. Secondly, eco-tourism is booming, with a total tourism revenue of 65.5 billion yuan in 2020 (see Fig. 69.7). The total revenue has remained high in the face of the impact of the New Crown epidemic.

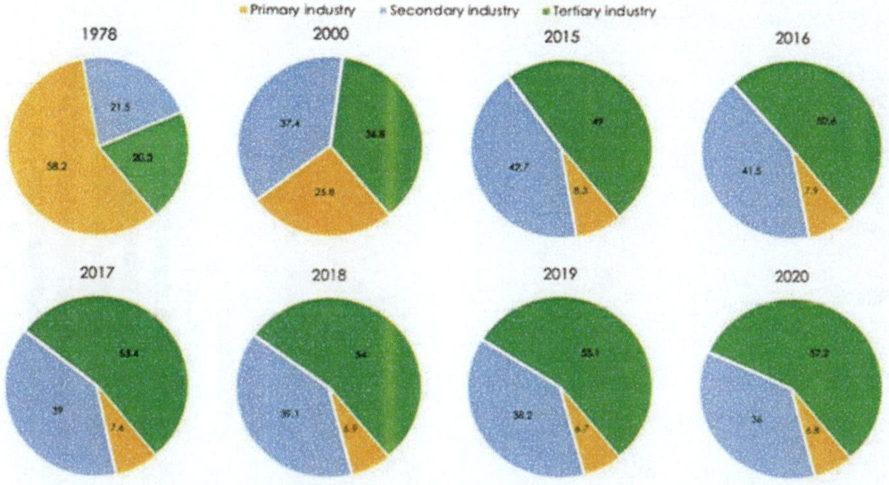

Fig. 69.5 Lishui's industrial structure changes from 1978 to 2020

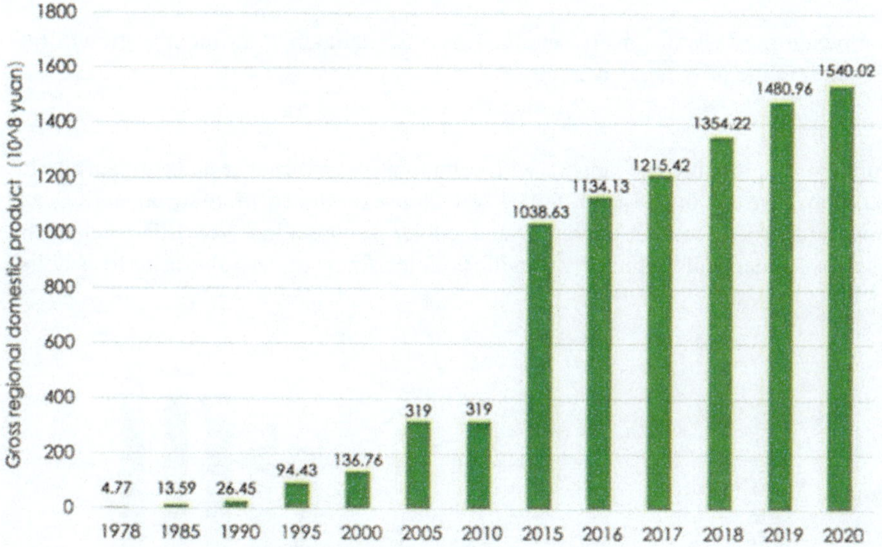

Fig. 69.6 Lishui's gross regional domestic product from 1978 to 2020

Equity

VRMEP pays attention to the poor and disadvantaged groups, giving them policy subsidies and technical support, and implementing industrialized poverty alleviation, effectively promoting low-income farmers to become prosperous. In 2020, rural permanent inhabitants in Lishui City had a per capita disposable income of 23,637 yuan (see Fig. 69.8), and low-income farmers had 12,394 yuan, which not only

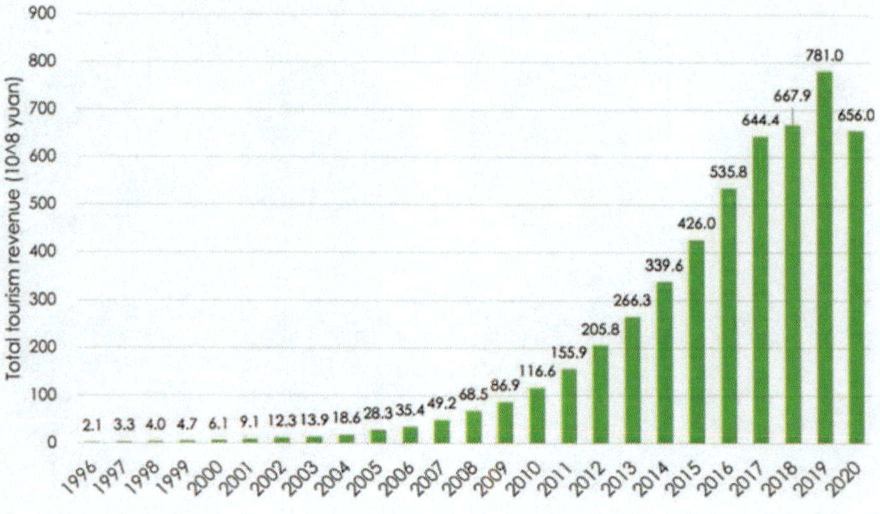

Fig. 69.7 The total tourism revenue of Lishui from 1996 to 2020

completed the task of poverty eradication and maintained an income growth rate higher than the provincial average.

Governance

Policy structures and mechanisms with collaborative governance and efficient administration have become Lishui City's valuable experience of innovative PES. (1) Government-led, market-operated, and public-participation VRMEP mechanism. The governments of Lishui City provide policies, finances, and platforms to establish

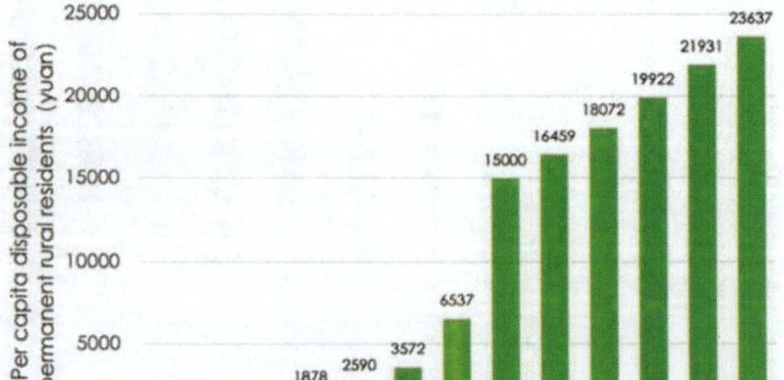

Fig. 69.8 Per capita disposable income of permanent rural residents in Lishui from 2008 to 2020

EP management and trading systems. Social investors cooperate with industrial operators to develop related projects, and enterprises, village collectives, and villagers share the benefits. (2) PES development was the center of an effective administrative management system, which made a comprehensive management department the lead unit and multiple departments participating. Governments used this system to unify VRMEP-related projects, authority, and responsibility. Reduce information barriers and consultation costs. (3) Reflect EP value in the government performance appraisal system. The increase or decrease of EP value is an essential indicator for evaluating and appointing cadres to urge public servants to focus both on economy and ecology.

69.5 Conclusion

This research suggests that three lessons may be learned from China's experimentation with PES and VRMEP processes that can inform the future of PES and ES governance worldwide. (1) The policy needs to be a priority when defining ES. Different countries and programs can modify their understanding and classification of ES to better match their specific contexts, market structures, PES goals, and policy needs. (3) Transforming relational contracts to transactional contracts should be promoted. RCs are necessary for the initial stage of ESM. Still, mature ESM should take the lead in pursuing transparent rights and duties as well as uniform payment standards to maximize efficiency and promote equity. (3) Make economic development and ecological protection as win–win goals. In developing nations, PES is not yet disentangled from prioritizing human rights and basic needs; instead, it emphasizes economic growth predicated on environmental preservation so that those living in poverty can benefit from conservation efforts. For them, the best path forward is to adopt a PES that considers the interconnectedness of natural ecosystems and human social progress.

China's actions to serve the goals of environmental protection and poverty alleviation through innovative PES mechanisms not only reflect China's concern for global environmental issues as a responsible country and generate positive externalities for global ecosystem and economic development but also provide practical, effective, and diverse PES schemes by providing adaptive adjustments and practices.

References

Adhikari B, Agrawal A (2013) Understanding the social and ecological outcomes of PES projects: a review and an analysis. Conserv Soc 11(4):359–374

Argyres N, Bercovitz J, Zanarone G et al (2020) The role of relationship scope in sustaining relational contracts in interfirm networks. Strateg Manag J 41(2):222–245

Brendan FR, Kerry T, Paul M (2009) Defining and classifying ecosystem services for decision making. Ecol Econ 68(3)

Daily C (1997) Introduction: what are ecosystem services? Nature's services: societal dependence on natural ecosystems. Island Press, Washington, D.C, pp 1–10

Engel S, Pagiola S, Wunder S (2008) Designing payments for environmental services in theory and practice: an overview of the issues. Ecol Econ 65(4):663–675

Esther W, Robert M, Josiah A (2021) Determinants of forest dependent household's participation in payment for ecosystem services: evidence from plantation establishment livelihood improvement scheme (PELIS) in Kenya. Glob Ecol Conserv 26

Gi M, An L, Xiaodong C (2021) Normative, livelihood, and demographic influences on enrollment in a payment for ecosystem services. Land Use Policy 108

Gomez-Baggethun E, Muradian R (2015) In markets we trust? Setting the boundaries of market-based instruments in ecosystem services governance. Ecol Econ 117:217–224

He J (2020) Institutional bricolage in payment for ecosystem services: insights from the sloping land conversion programme in upland communities, Southwest China. Asia Pac Viewp 62(2):164–178

Jinping X, Zhejiang (2015) China: a new vision for development. Zhejiang People's Publishing House, Hangzhou

Joe B (1959) Industrial organization. Wiley, New York

Lishui statistical yearbook (2021) http://tjj.lishui.gov.cn/art/2021/11/12/art_1229215928_58834210.html. Accessed 11 Aug 2022

MacNeil R (1982) The new social contract: an inquiry into modern contractual relations. Ethics 93(1):168

Millennium ecosystem assessment (2005) Ecosystems and human well-being: synthesis. Island Press, Washington DC

Muradian R, Corbera E, Pascual U (2010) Reconciling theory, and practice: an alternative conceptual framework for understanding payments for environmental services. Ecol Econ 69(6):1202–1208

Opinions on establishing and improving the mechanism for realizing the value of ecological products, https://www.mee.gov.cn/zcwj/zyygwj/202104/t20210430_831529.shtml. Accessed 21 Aug 2022

Opinions on establishing the value realization mechanism of ecological products, http://www.gov.cn/zhengce/2021-04/26/content_5602763.htm. Accessed 08 Sept 2022

Pagiola S (2008) Payments for environmental services in Costa Rica. Ecol Econ 65(4):712–724

Pamela M et al (2014) Payments for environmental services and contested neoliberalisation in developing countries: a case study from Vietnam. J Rural Stud 36

Pistorius T, Schaich H, Winkel G et al (2012) Lessons for REDD plus: a comparative analysis of the German discourse on forest functions and the global ecosystem services debate. For Policy Econ 18:4–12

Robert WD et al (2021) Protecting habitats in low-intensity tropical farmland using carbon-based payments for ecosystem services. Environ Res Lett 16(11)

Rosa MD et al (2022) Payment for ecosystem services in Peru: assessing the socio-ecological dimension of water services in the upper Santa river. Ecosyst Serv 56

Salzman J, Bennet G, Carol N et al (2018) The global status and trends of payments for ecosystem services. Nat Sustain 1:136–144

Santiago Izquierdo T et al (2022) Contradictory distributive principles and land tenure govern benefit-sharing of payments for ecosystem services (PES) in Chiapas, Mexico. Environ Res Lett 17(5)

Shangeng L. An effective mechanism under the brand premium, https://www.ndrc.gov.cn/xwdt/ztzl/jljqstcpjzsxjz/jdal/202108/t20210827_1294852.html?code=&state=123. Accessed 25 Aug 2022

Silva-Muller L (2022) Payment for ecosystem services and the practices of environmental fieldworkers in policy implementation: the case of Bolsa. Land Use Policy 120

Solomon D, Lee R (2000) Emissions trading systems and environmental justice. Taylor J 42:32–45

Stosch KC et al (2019) Quantifying stakeholder understanding of an ecosystem service trade-off. Sci Total Environ 651(2):2524–2534

Sven W (2015) Revisiting the concept of payments for environmental services. Ecol Econ 117:234–243

Technical specification for accounting gross ecosystem product (GEP)—Terrestrial ecosystems, http://zjamr.zj.gov.cn/art/2020/9/29/art_1229047334_58814039.html. Accessed 21 Aug 2022

Tesfaye B, Abdurezak K (2018) The structure conduct performance model and competing hypothesis: a review of literature. J Financ 8(2):11–25

The Economics of Ecosystems and Biodiversity, http://teebweb.org. Accessed 09 Aug 2022

Thompson BS, Harris JL (2021) Changing environment and development institutions to enable payments for ecosystem services. Role Glob Environ Change 67

Wallace KJ (2007) Classification of ecosystem services: problems and solutions. Biol Cons 139(3/4):235–246

Wallace KJ (2017) Classification of ecosystem services: problems and solutions. Biol Cons 139(3/4):235–246

White AC et al (2022) How can you put a price on the environment? Farmer perspectives on stewardship and payment for ecosystem services. J Soil Water Conserv 77(3):270–283

Xingliang P et al (2017) Payments for ecosystem services in China: policy, practice, and progress. J Clean Prod 158:200–208

Xutong W et al (2021) Integrating multiple influencing factors in evaluating the socioeconomic effects of payments for ecosystem services. Ecosyst Serv 51

Yin RK (1994) Case study research: design and methods. Sage Publications, CA

Zhejiang Provincial Ecological Environment Status Bulletin (2021) http://sthjt.zj.gov.cn/art/2022/6/6/art_1201912_58931709.html. Accessed 11 Aug 2022

Zhenmin D, Shunbo Y (2021) Ecological effectiveness of payment for ecosystem services to identify incentive priority areas: sloping land. Land Use Policy 104

Zhiyun O et al (2020) Using gross ecosystem product (GEP) to value nature in decision making. Proc National Acad Sci USA 117(25):14593–14601

Zhou T et al (2022) Impact evaluation of a payments for ecosystem services program on vegetation quantity and quality restoration in inner Mongolia. J Environ Manage 303

Chapter 70
The Impact of Environmental Regulation on Green Investment: The Regulatory Role of Internal Control

Jinghan Li and Wei Liu

Abstract This paper takes the heavily polluted listed companies in Shanghai and Shenzhen A-shares in China from 2010 to 2018 as the research sample, and uses stata software to empirically test the impact of different dimensions of environmental regulation tools on corporate green investment, and further explores the regulatory role of internal control. The research shows that the relationship between administrative order type environmental regulation and public participation type environmental regulation and green investment of enterprises is inverted U-shaped, while the relationship between market incentive type environmental regulation and green investment of enterprises is positive U-shaped; Market incentive environmental regulation and public participation environmental regulation can strengthen the role of administrative order environmental regulation on green investment; Internal control plays a regulatory role in administrative order type environmental regulation and market incentive type environmental regulation. The conclusions of the study provide useful enlightenment for the government to make reasonable planning for different dimensions of environmental regulation, promote internal control of enterprises and increase green investment of enterprises.

Keywords Environmental regulation · Enterprise green investment · Internal controls

70.1 Introduction

At present, China's environmental protection issues are becoming more and more prominent. Therefore, in order to promote enterprise environmental protection and improve environmental performance, China has made many efforts in enterprise environmental regulation, such as the current research hot spot "dual carbon" policy. Many studies have shown that green investment can better show how enterprises fulfill

J. Li (✉) · W. Liu
Chongqing Technology and Business University, Chongqing, China
e-mail: 1584008142@qq.com

J. Zhang et al. (eds.), *Environmental Pollution Governance and Ecological Remediation Technology*, Environmental Science and Engineering,
https://doi.org/10.1007/978-3-031-25284-6_70

their environmental protection responsibilities. How should government regulations correctly affect green investment? At the same time, the connotation of environmental regulation is gradually expanding. Do different types of environmental regulations have differences in green investment in enterprises?

However, the green investment behavior of enterprises is also affected by many other factors. Internal control is an internal mechanism for enterprises to conduct self-regulation and self-restraint in their daily production and operation activities. Good internal control can effectively restrict various behaviors of enterprises from inside to outside, and help enterprises to truly implement their environmental protection responsibilities into action. To fulfill social responsibility and promote green investment, enterprises need both internal and external forces.

70.2 Hypothesis Presentation

Please Based on the research of existing relevant literature and the theory of information asymmetry, principal-agent theory and Porter's hypothesis, the following theories are proposed in this paper:

Assumption 1. A Positive U—shaped Relationship between Administrative Order Environmental Regulation and Corporate Green Investment.

Assumption 2. An Inverted U—shaped Relationship between Market Incentive Environmental Regulation and Corporate Green Investment.

Assumption 3. An Inverted U—shaped Relationship between Public Participation Environmental Regulation and Corporate Green Investment.

Assumption 4. Public participation environmental regulation can enhance the effect of Administrative Order Environmental Regulation on Corporate Green Investment to some extent.

Assumption 5. Market Incentive Environmental Regulation can enhance the effect of Administrative Order Environmental Regulation on Corporate Green Investment to some extent.

Assumption 6. Internal control plays a moderating role in the impact of Administrative Order Environmental Regulation on Corporate Green Investment.

Assumption 7. Internal control plays a moderating role in the impact of Market Incentive Environmental Regulation on Corporate Green Investment.

Assumption 8. Internal control plays a moderating role in the impact of Public Participation Environmental Regulation on Corporate Green Investment.

70.3 Empirical Design and Model Construction

This article refers to the "Guidelines for Industry Classification of Listed Companies" revised by the China Securities Regulatory Commission in 2012 to subdivide heavily polluting industries into 16 industries. At the same time, the listed companies in the heavily polluting industries whose A-shares disclosed annual environmental protection investment in Shanghai and Shenzhen stock exchanges from 2010 to 2018 were selected as research samples.

The specific variables are defined as follows (Table 70.1).

Through the theoretical analysis and research assumptions above, the econometric model of this paper is set up. To test Hypotheses 1–3, a specific model (1) was designed; to test Hypotheses 4–5, a model (2) was constructed; to test Hypotheses 6–8, a model (3) was constructed as follows, where ER includes ER1, ER2 and IER:

Table 70.1 Variable definitions

Variable name	Variable symbol	Variable explanation
Corporate green investment	EGI	Amount of environmental protection investment collected manually from corporate annual reports
Administrative order environmental regulation	ER1	The total number of local environmental protection laws and regulations promulgated in the year
Market incentive environmental regulation	ER2	Enterprise sewage charges
Public participation environmental regulation	IER	Telephone/internet complaints + total number of letters + total number of visits
Internal control	ICI	Dibo listed company internal control index
Financial leverage	LEV	Total liabilities/total assets
Enterprise size	SIZE	Natural logarithm of average total assets
Year of establishment	AGE	Year since the establishment of the company
Tobin's Q	Q	Enterprise market value/total assets
Board size	BOD	Board of directors
Ownership concentration	FRIST	Shareholding ratio of the largest shareholder
Corporate profit	ROA	Net profit/total assets
Two jobs in one	DUAL	Whether the CEO of the company also serves as the chairman of the board of directors, yes 1 no 0
Executive compensation	EC	The logarithm of the total annual salary of executives
Equity nature	Owner	1 country 0 no
Year	Year	Dummy variable

$$EGI = \alpha_0 + \alpha_1 ER_i + \alpha_2 ER_i^2 + \alpha_3 size + \alpha_4 age + \alpha_5 opp + \alpha_6 dual + \alpha_7 lev + \alpha_8 bod + \alpha_9 first + \alpha_{10} roa + \sum Year + \varepsilon \tag{70.1}$$

$$EGI = b_0 + b_1 ER_i + b_2 ER_i^2 + b_3 ER_i^2 \times ier + b_4 size + b_5 age + b_6 opp + b_7 dual + b_8 lev + b_9 bod + b_{10} first + b_{11} roa + \sum Year + \varepsilon \tag{70.2}$$

$$EGI = c_0 + c_1 ici + c_2 ER_i + c_3 ER_i \times ici + c_4 size + c_5 age + c_6 opp + c_7 dual + c_8 lev + c_9 bod + c_{10} first + c_{11} roa + \sum Year + \varepsilon \tag{70.3}$$

70.4 Results

The second column of Table 70.2 shows that the coefficient of ER1 is −0.497, which is significantly negatively correlated at the level of 1%, and the coefficient of ER1 is 0.001, which is significant at the level of 1%. There is a positive "U"-shaped relationship between them, and the results support Hypothesis 1. The third column of Table 70.2 shows that the result shows that the coefficient of ER2 is 1.44, which is significantly positive at the 1% level, and the coefficient of ER2 is −1.31, which is significantly negatively correlated at the 1% level. This shows that the impact of market incentive-based environmental regulation on corporate green investment is in an inverted "U"-shape, and Hypothesis 2 holds. The fourth column of Table 70.2 verifies the regression relationship between public participation in environmental regulation and corporate green investment. The results showed that the IER was significantly negative at the 1% level and significantly positively correlated at the 10% level. It shows that there is an inverted "U"-shaped relationship between public participation-based environmental regulation and corporate green investment, and Hypothesis 3 is verified.

The fifth column of Table 70.2 shows that the coefficient of, is 0.001, which is significant at the 95% confidence level, and the coefficient of the multiplication term with IER is 0.994, which is significant at the 95% confidence level. This means that public participation-based environmental regulation can positively strengthen the impact of executive-ordered environmental regulation on corporate green investment, and Hypothesis 4 holds. It can be seen from the sixth column of Table 3 that the coefficient of ER2 is 3.98, which is significant at the 99% confidence level, and the coefficient of the multiplication term with ER2 is 2.37, which is significant at the 95% confidence level. This shows that with the intensity of market-incentivized

Table 70.2 Regression results

Variable	Hypothesis 1	Hypothesis 2	Hypothesis 3	Hypothesis 4	Hypothesis 5	Hypothesis 6	Hypothesis 7	Hypothesis 8
ER1	−0.497***			−0.537**	−0.249	−0.436***		
	(−5.37)			(−2.22)	(−1.24)	(−4.61)		
	0.001***			0.001**	0.001	0.001***		
	(5.42)			(2.11)	(1.17)	(3.37)		
ER2		1.44***			3.98***		1.44***	
		(26.21)			(9.13)		(25.99)	
		−1.31***					−4.52	
		(−18.64)					(−0.37)	
IER			0.538***	−6.18				0.005
			(2.86)	(0.48)				(0.95)
				0.994**				3.94**
				(2.02)				(2.19)
×IER			−0.301*					
			(−1.76)					
×ER2					2.37**			
					(2.55)			
ICI						−0.053	0.101	0.277
						(−0.18)	(0.61)	(1.71)
×ICI						0.003**		
						(2.21)		
×ICI							−1.3	

(continued)

Table 70.2 (continued)

Variable	Hypothesis 1	Hypothesis 2	Hypothesis 3	Hypothesis 4	Hypothesis 5	Hypothesis 6	Hypothesis 7	Hypothesis 8
							(−0.71)	
×ICI								−6.14**
								(−2.19)
lev	0.14	0.015	0.255	0.275	0.203	0.23	0.025	0.286
	(0.53)	(0.08)	(0.97)	(0.59)	(0.49)	(0.88)	(0.13)	(0.93)
Size	0.64***	0.169***	0.646***	0.445***	0.294**	0.628***	0.167***	0.435***
	(13.1)	(4.32)	(13.24)	(2.92)	(2.24)	(12.78)	(4.24)	(4.48)
Age	0.013	0.003	0.012	0.039	−0.028	0.013	0.003	−0.102***
	(1.25)	(0.35)	(1.15)	(0.3)	(−0.26)	(1.24)	(0.37)	(−5.32)
Q	−0.167***	−0.128***	−0.145***	0.077	0.105**	−0.148***	−0.124***	0.076
	(−3.5)	(−3.65)	(−3.01)	(1.3)	(1.99)	(−3.08)	(−3.53)	(1.71)
Bod	0.008	−0.006	0.015	0.008	−0.004	−0.005	−0.007	0.003
	(0.37)	(−0.39)	(0.66)	(0.18)	(−0.12)	(0.22)	(−0.44)	(0.11)
Frist	0.016***	0.006***	0.016***	0.011	0.015**	0.015***	0.006***	0.012
	(5.42)	(2.91)	(5.54)	(1.52)	(2.33)	(5.3)	(2.88)	(2.5)
Roa	1.991**	0.798	1.991**	0.821	0.718	1.712**	0.709	0.83
	(2.53)	(1.39)	(2.54)	(0.98)	(0.93)	(2.15)	(1.22)	(1.38)
Dual	0.349***	0.075	0.317***	0.038	−0.014	0.33***	0.077	0.037
	(3.14)	(0.92)	(2.83)	(0.24)	(−0.09)	(2.97)	(0.95)	(0.35)
EC	0.22***	0.078	0.233***	0.099	0.072	0.209***	0.077	0.09
	(3.37)	(1.62)	(3.58)	(0.92)	(0.79)	(3.21)	(1.6)	(1.16)

(continued)

Table 70.2 (continued)

Variable	Hypothesis 1	Hypothesis 2	Hypothesis 3	Hypothesis 4	Hypothesis 5	Hypothesis 6	Hypothesis 7	Hypothesis 8
_Cons	25.992***	−5.053***	−18.995**	35.15*	14.336	21.377***	−5.648***	−12.491***
	(3.05)	(−5.43)	(−16.84)	(1.96)	(0.95)	(2.41)	(−4.18)	(−5.28)
Year	Control	Control	Control	Control	Control	Control	Control	Control
N	1066	1066	1066	1066	1066	1066	1066	1066
R-squared	0.438	0.404	0.44	0.388	0.331	0.442	0.438	0.41

Note ***, **, and * indicate significance at the 1%, 5%, and 10% levels, respectively, and t values are in brackets environmental regulation, the impact of executive-ordered environmental regulation on corporate green investment has been enhanced to a certain extent, and Hypothesis 5 has been verified.

The seventh column of Table 70.2 shows that it is significantly positive at the 1% level, and the regression coefficient of the interaction term with internal control (ICI) is 0.003 and is significant at the 5% level, which indicates that internal control plays an important role in the administrative command-type environmental regulation. It plays a positive role in regulating the green investment of enterprises, and verifies Hypothesis 6. The eighth column of Table 70.2 shows that the regression coefficient of the interaction term with ICI is not significant, which means that internal control does not play a moderating effect on the impact of market incentive environmental regulation on corporate green investment, and hypothesis 7 does not hold. The ninth column of Table 70.2 shows that there is a significant positive correlation at the level of 5%, and the regression coefficient of the interaction term with ICI is significant at the level of 5%, which means that internal control has a significant impact on public participation-based environmental regulation and corporate green investment. The relationship has an inhibitory effect, and Hypothesis 8 holds.

70.5 Conclusion

70.5.1 Government Level

Reasonable and comprehensive utilization of Environmental Regulation tools (Bin and Xihao 2019). Administrative order type environmental regulation is mandatory, universal and basic, and it is also the most mature environmental regulation tool in China (Control 2019). Both market incentive environmental regulation and public participation environmental regulation have positive feedback in the relationship between administrative order environmental regulation and green investment of enterprises (Zhenxing et al. 2019). Therefore, the government should reasonably use all kinds of environmental regulation tools, give full play to the coordinated and efficient allocation of the three types of regulation tools, and play the largest role in the linkage of multiple subjects to promote green investment of enterprises (Hong and Mafei 2015).

70.5.2 Enterprise Level

At the enterprise level, the enterprise should actively improve the internal control system, improve the requirements of the management, and constrain the self-interest and short-term initiatives of senior executives (Maochu et al. 2015). Each department is clearly divided and the division of labor is clear to improve the internal operation efficiency (Zeng and Hu 2015). We will appropriately regulate the green investment behavior of enterprises and increase investment in environmental protection (Yinhua and Xuan 2015). Standardize the corporate governance mechanism, abide by the government's environmental protection regulations, give play to the subjective initiative, and reduce the environmental pollution caused by enterprise production in a planned way. The management personnel of enterprises should establish the awareness of environmental protection, infiltrate and drive the corporate culture, adhere to the green concept in places invisible to the government and the public, and strengthen the operation and management of environmental protection (Lijuan and Xiaolan 2016). Improve the quality of internal control, improve the level of green investment, and then enhance the core competitiveness of the enterprise, establish a good corporate image, and realize the long-term development of the enterprise (Zhonggao and Maohuan 2017).

References

Bin L, Xihao W (2019) Can internal control evaluation improve accounting conservatism? [J]. Nankai Econ Res (06):158–180

Control ZHI (2019) Environmental policy and enterprise environmental protection investment—empirical evidence from listed companies in heavy pollution industries [J]. Finan Account Commun 06:88–91

Hong L, Mafei T (2015) Correlation between internal control, media function, legal environment and accounting information value [J]. Acc Res (06):64–71+97

Lijuan C, Xiaolan J (2016) Effectiveness of internal control, marketization process and environmental information disclosure [J]. J xi'an Univ Financ Econ 29(02):101–107

Maochu Z, Mengjie L, Jian D (2015) Can environmental regulation force the adjustment of industrial structure—an empirical test based on China's inter provincial panel data [J]. China's Popul, Resour Environ 25(08):107–115

Yinhua Q, Xuan Y (2015) Empirical study on the relationship between the effectiveness of internal control and the quality of environmental information disclosure [J]. Statist Decis Making 23:166–169

Zeng J, Hu J (2015) Research on impact factors of public environmental participation in China—empirical analysis based on China's provincial panel data [J]. China's Popul, Resour Environ 25(12):62–69

Zhenxing H, Xiaoyue G, Hao W (2019) Internet finance development, internal control quality and SME financing constraints [J]. Finan Account Commun (30):120–124

Zhonggao L, Maohuan D (2017) The impact of internal control defects and their repair on corporate debt financing costs—an empirical study based on the perspective of internal control regulatory system change [J]. Acc Res (04):73–80+96

Chapter 71
Research on the Impact on ESG Performance on Green Innovation Under the Background of Carbon Emissions Peak-Based on the Data of Listed Companies in Shandong Province

Cui Guo and Guomin Song

Abstract With the intensification of the greenhouse effect, the country has put forward the carbon emissions peak, committed to reducing carbon dioxide emissions and improving the ecological environment, which poses a great challenge to enterprises. On the basis of improving business performance and corporate governance level, enterprises also need to pay attention to environmental protection and the implementation of social responsibility. On the one hand, they can directly save energy and reduce carbon emissions. On the other hand, it can improve social relations, expand capital, and improve the capacity of carbon reduction and green innovation. Therefore, this paper introduces ESG to study the impact of ESG performance on enterprise green innovation. The results show that ESG performance in each aspect and the comprehensive performance will improve the level of green innovation of enterprises.

Keywords Carbon emissions peak · ESG performance · Green innovation

71.1 Introduction

Economic development is bound to consume energy, and in the process of consuming energy, carbon dioxide will be emitted. Excessive emissions of carbon dioxide will lead to climate warming and threaten the living environment of human beings, which has become a global problem. All countries have taken the necessary measures to reduce carbon dioxide emissions. China's carbon emissions are about one fifth of the world's. In this context, the Chinese government has also actively proposed "carbon emissions peak", strive for a downward trend of carbon dioxide emissions after 2030. The country's low-carbon policy should ultimately be led by enterprises to complete,

C. Guo (✉) · G. Song
Shandong Women's University, 2399 Daxue Road, Jinan, Shandong, China
e-mail: 648473792@qq.com

J. Zhang et al. (eds.), *Environmental Pollution Governance and Ecological Remediation Technology*, Environmental Science and Engineering,
https://doi.org/10.1007/978-3-031-25284-6_71

so low-carbon has become a big test faced by enterprises. In order to meet the national policy, enterprises actively adjust the industrial structure, increase innovation inputs, introduce advanced technology, committed to energy saving, emission reduction, clean and low-carbon.

At present, the country attaches great importance to green and low-carbon cycle development, and the concepts of green investment and green production emerge as the times require, which poses great challenges for enterprises. In order to continue to achieve sustainable development under the background of carbon emissions peak, enterprises need to pay attention to the coordination and unification of environment, social responsibility and corporate governance, that is, to practice ESG. E refers to the combination of self-development and carbon emissions peak, and the goal of carbon emissions peak throughout. S means that enterprises need to pay attention to social responsibility and the interests of all stakeholders, and strengthen social responsibility disclosure. G refers to the need for enterprises to optimize corporate governance structure, enhance their own strength, avoid financial fraud and other problems, and ensure the realization of E and S. In short, the purpose of ESG is responsible operation and management, responsible scientific and technological innovation, so as to achieve green and low-carbon development and sustainable development.

The existing literature has studied the impact of environment, social responsibility, corporate governance on green innovation respectively. Wang (2018) believes that both environmental regulation and board governance are conducive to enterprises' green innovation (Wang and Chen (2018). Xiao and Pan (2022) believes that corporate social responsibility is conducive to green innovation (Xiao and Pan 2022). There are few studies about the impact of ESG comprehensive score on green innovation. ESG aims to achieve carbon emissions peak and find a model suitable for their development, which is conducive to green innovation. Based on the background of carbon emissions peak, the paper studies the impact of ESG performance on green innovation.

71.2 Research Design

71.2.1 Research Hypothesis

The concept of carbon emissions peak guides enterprises to actively seek change, change the previous behavior of ignoring ecological environment and blindly pursuing profit maximization, strengthen environmental governance, and fully integrate the concept of carbon emissions peak into every stage of enterprise operation. Enterprises increase green R&D investment to improve the production process, purchase environmental protection equipment, introduce environment-friendly products, to achieve cleaner production, improve production efficiency, so that enterprises can realize economic benefits and environmental benefits at the same time, thus realizing green innovation. Therefore, this paper puts forward hypothesis 1: Environmental performance is conducive to green innovation.

Based on stakeholder theory and social responsibility theory, enterprises should not only pay attention to their own production and operation, but also pay attention to the various effects of enterprise operation on stakeholders. On the one hand, enterprises can reduce conflicts with stakeholders, reduce agency costs, make enterprises and stakeholders get along harmoniously, and promote sustainable and healthy development of enterprises. On the other hand, it can improve the reputation of enterprises, enhance soft power, so that enterprises with a good social image attract investment, healthy development, maintain long-term competitive advantages. In a word, it is easier for enterprises to obtain more resources after undertaking social responsibilities, such as the knowledge, experience, contacts, technical personnel, environmental protection equipment and R&D funds of stakeholders, which are conducive to the development of green innovation activities. Therefore, this paper puts forward hypothesis 2: The implementation of social responsibility is conducive to promoting the green innovation of enterprises.

Green innovation activities belong to the key issues of strategic decision-making of enterprises and cannot be separated from the judgment of senior managers. Based on the principal-agent theory, enterprises must adjust and optimize the corporate governance structure, reduce agency costs, and achieve good governance effects by optimizing the governance structure, improving the incentive mechanism and improving the system of responsibility and rights, so as to improve production efficiency and enterprise value, and commit themselves to the goal of carbon emissions peak and green innovation. Therefore, this paper puts forward hypothesis 3: Good corporate governance is conducive to promoting green innovation.

To sum up, hypothesis 4 is put forward in this paper: ESG performance of enterprises is conducive to promoting green innovation.

71.2.2 Sample Selection and Data Sources

In order to enhance the pertinence of the research, this paper takes the data of A-share listed companies in Shandong Province as the research sample, screens out ST companies, companies with incomplete data and abnormal data, and obtains a total of 59 listed companies. Although Shandong Province has not been included in the carbon emission trading pilot, it has firmly implemented the requirement of carbon emissions peak, formulated the action plan for prevention and control of pollution sources from 2018 to 2020, committed to energy conservation, carbon reduction and structural adjustment, and achieved certain results. Therefore, this paper selected the data of listed companies in Shandong Province from 2018 to 2020. The green innovation data are from CNRDS platform, the ESG data are from the score published by Bloomberg, and the data of other control variables are from CSMAR database.

71.2.3 Variable Setting

Dependent Variable. Existing literature mainly measures innovation by R&D investment, number of technical personnel, number of patents, etc. Considering that patents of invention and utility new-style can better reflect the substantive innovation of enterprises, and patent application means that enterprises have entered the actual innovation stage, and the whole application process will have an impact on enterprise operation. Based on the practice of Zhang (2022), this paper measures green innovation by the number of green invention and green utility new-style patents independently and jointly applied by enterprises. Considering that there may be enterprises with 0 green patent applications, add 1 to the number of applications and take the logarithm.

Independent Variable. There are different measurement methods for ESG in existing literature. Qiu and Yin (2019) measures the quality of ESG with the help of the ESG score of Bloomberg (Qiu and Yin 2019). Zhang and Zhao (2019) uses ESG rating developed by SynTao Green Finance (Zhang and Zhao 2019). Xu et al. (2022) uses ESG rating developed by Huazheng (Xu et al. 2022). This paper refers to the practice of Qiu and Yin (2019) (Zhang 2022) and measures the ESG performance of enterprises with the ESG single and comprehensive scores published by Bloomberg.

Control Variable. The profitability of the enterprise is directly related to the economic strength of carrying out green innovation activities. This paper chooses the return on equity to measure profitability. The higher the return on equity is, the more sufficient capital an enterprise has to carry out green innovation. The debt paying ability of enterprises will also affect the capital required for green innovation activities. This paper chooses the asset liability ratio to measure the solvency. The higher the asset liability ratio is, the more able the enterprises are to obtain debt capital, so as to guarantee the development of green innovation activities. The establishment time of the enterprise will affect many factors, such as innovation enthusiasm and innovation experience, thus affecting the development of green innovation activities. To sum up, the variable definitions are shown in Table 71.1.

71.2.4 Model Constructing

In order to verify the impact of ESG performance on green innovation, this paper constructs the following four models.

$$Patent = \alpha_0 + \alpha_1 E + \alpha_2 ROE + \alpha_3 Debt + \alpha_4 Age + \varepsilon \tag{71.1}$$

$$Patent = \alpha_0 + \alpha_1 S + \alpha_2 ROE + \alpha_3 Debt + \alpha_4 Age + \varepsilon \tag{71.2}$$

Table 71.1 Table of variable definitions

Dependent variable	Green innovation (Patent)	Ln (Number of green patent applications + 1)
Independent variables	Environmental performance (E)	Environmental score published by Bloomberg
	Social performance (S)	Social score published by Bloomberg
	Corporate governance performance (G)	Corporate governance score published by Bloomberg
	ESG performance (ESG)	ESG comprehensive score published by Bloomberg
Control variables	The return on equity (ROE)	Net profit/owner's equity
	The asset liability ratio (debt)	Liability/asset
	Years of establishment (age)	(Target time-established time)/365

$$Patent = \alpha_0 + \alpha_1 G + \alpha_2 ROE + \alpha_3 Debt + \alpha_4 Age + \varepsilon \tag{71.3}$$

$$Patent = \alpha_0 + \alpha_1 ESG + \alpha_2 ROE + \alpha_3 Debt + \alpha_4 Age + \varepsilon \tag{71.4}$$

Among them, α_0 represents intercept, $\alpha_{1\text{-}4}$ represents coefficient, and ε represents residuals.

71.3 Empirical Analysis

71.3.1 Descriptive Statistics

In order to understand the basic situation of ESG performance and green innovation of listed companies in Shandong Province, this paper uses SPSS to conduct descriptive statistical analysis on variable data collected and sorted, as shown in Table 71.2. The minimum value of green patent is 0, and the maximum value is 6.6846. It can be seen that some listed companies have not carried out green innovation, and there is a certain gap between them. The standard deviation of environmental performance is 11.0417, indicating that there are large differences among companies. Different companies pay different attention to climate change, air quality, ecological environment, energy utilization, pollutant emission and other issues. The minimum value is 0, indicating that some companies have not considered environmental issues in the process of operation. The standard deviation of social performance is 7.4428, indicating that there are differences in the importance of social responsibility among companies, but the difference is smaller than that of environmental performance. The

Table 71.2 Table of descriptive statistics

Index	Sample number	Minimum	Maximum	Mean	Standard deviation
Patent	177	0.0000	6.6846	1.6873	1.6784
E	177	0.0000	43.4109	13.1405	11.0417
S	177	7.0175	43.8596	23.5801	7.4428
G	177	33.9286	62.5000	44.7236	5.7038
ESG	177	12.8099	44.6281	22.8315	7.9045
ROE	177	−1.6948	0.6043	0.0735	0.1698
Debt	177	0.0000	0.8621	0.4595	0.1982
Age	177	14	32	22.36	4.216

standard deviation of corporate governance performance is 5.7038, and the degree of difference between companies is less than that of social performance. Considering the differences of the three individual indicators, the overall performance of ESG is also different. The minimum value of ROE is −1.6948, and the maximum value is 0.6043, indicating that some companies have incurred losses. The minimum value of Debt is 0, the maximum value is 0.8621, and the average value is 0.4595, indicating that some companies do not use debt financing, while some companies' debt level is higher than 85%. The company should fully consider leverage benefits and financial risks, and reasonably determine the debt level. The minimum value of Age is 14 years, and the maximum value is 32 years, indicating that the sample companies have been established for more than 10 years, and the differences are quite large.

71.3.2 Regression Analysis

This paper uses the four models to conduct multiple linear regression analysis, and the analysis results are shown in Table 71.3. The VIF values of all variables in Table 71.3 are lower than 2, and the tolerance is higher than 0.1, so there is no multicollinearity problem among all variables. The regression results show that the regression coefficient of environmental performance is 0.030, which is significant at the level of 1%. Hypothesis 1 holds. The regression coefficient of social performance is 0.051, which passes the significance test, and hypothesis 2 holds. The regression coefficient of corporate governance performance is 0.060, and the significance level is less than 5%. Hypothesis 3 holds. The regression coefficient of ESG performance is 0.052, and the sig value is less than 1%, so hypothesis 4 holds. The regression coefficients of ROE in the four models are all positive and have passed the significance test, indicating that the higher the ROE, the stronger the capital used for green patent application and the higher the level of green innovation. The regression coefficient of Debt in the four models is positive, except for model 3, other sig values are significant, indicating that the higher the asset liability ratio, the greater the possibility of listed

Table 71.3 Table of regression results

Index	Model 1	Model 2	Model 3	Model 4
E	0.030[b] (2.661)			
S		0.051[b] (3.138)		
G			0.060[a] (2.545)	
ESG				0.052[b] (3.274)
ROE	1.485[a] (2.075)	1.631[a] (2.313)	1.615[a] (2.267)	1.482[a] (2.098)
Debt	1.576[a] (2.520)	2.091[b] (3.468)	1.298 (1.954)	1.506[a] (2.438)
Age	−0.063[a] (−2.177)	−0.065[a] (−2.272)	−0.065[a] (−2.207)	−0.070[a] (−2.418)

Note [a] indicates significance at 5% level and [b] indicates significance at 1% level

companies to obtain capital through debt, and the greater the possibility of green innovation. The regression coefficients of Age in the four models are all negative and significant at the level of 5%, indicating that the longer a listed company is established, the less enthusiastic it is to try green innovation.

71.4 Conclusion and Suggestion

71.4.1 Conclusion

Listed companies in Shandong Province have made some improvements in fulfilling ESG obligations. Corporate governance performance is the best, followed by social responsibility performance, and finally the environment performance. In a word, ESG has been paid more attention to. Based on the good performance of ESG from a unilateral and comprehensive perspective, the green innovation level of listed companies in Shandong Province has been improved.

71.4.2 Suggestion

Implement the concept of carbon emissions peak and pay attention to environmental protection. Under the background of carbon emissions peak, the company is faced with the pressure of energy saving and carbon reduction. Therefore, the company must follow closely the national policy, implement the concept of carbon emissions peak, carry out low-carbon environmental protection actions in the supply chain field, and promote the green and sustainable development of the company.

Improve the awareness of social responsibility and fulfill social responsibility. Listed companies must deal with the relationship with shareholders, creditors, suppliers, customers and other stakeholders, improve the awareness of social responsibility, enhance the reputation and image of the company, so as to obtain more support and provide guarantee for green innovation.

Improve corporate governance and enhance the comprehensive strength of enterprises. Good corporate governance is the guarantee of the company's operation and development and the basis of green innovation activities. Therefore, the company must improve the governance institutions such as the general meeting of shareholders, the board of directors and the board of supervisors, and strengthen the restraint and supervision mechanism, so as to guarantee the smooth operation of the company.

Acknowledgements This research was supported by Shandong Provincial Key Research and Development Program (Soft Science Project) issues "Research on the Impact of Social Responsibility Fulfillment on the Innovation Performance of Shandong Enterprises under the Background of Carbon Emissions Peak" (Grant No. 2021RKY03049).

References

Qiu, M, Yin H (2019) ESG performance and financing cost of enterprises in the context of ecological civilization construction. J Quant Tech Econ 36(3):108–123

Wang F, Chen F (2018) Board governance, environmental regulation and green technology innovation-based on the empirical test of listed companies in heavy pollution industries in China. Studies Sci 36(2):361–369

Xiao X, Pan Y (2022) Director and executive liability insurance and corporate green innovation: incentive tool or self-interest method. Sci Techn Progress Pol 39(13):94–104

Xu G, Zhuo Y, Zhang Y, Zhang J (2020) Does ESG disclosure increase enterprise value. Comm Fin Account 4:33–37

Zhang L, Zhao H (2019) Does ESG performance affect enterprise value?—An empirical study based on A-share listed companies. Wuhan Fin 10:36–43

Zhang A (2020) Environmental tax collection, social responsibility and enterprise green innovation. Econ Theory Bus Manag 42(1):67–85

Chapter 72
A Pilot Experiment for Phytoremediation of Petroleum Contaminated Soil by *Trifolium repens* L. with Weathering in an Arid and Semi-Arid Region in China

Yuanyuan Shen, Wenke Wang, Haijuan Li, Minjuan Zhao, Bin Yue, Guohua Chang, Zhongyun Li, and Tianpeng Gao

Abstract Phytoremediation can provide an efficient and economical method for the degradation of petroleum in contaminated soil. The present study analyzed *Trifolium repens* L. as a reclamation plant in both non-petroleum soil and weathered petroleum contaminated soil for 120 days. The effects of reclamation were determined using the degradation rates of total petroleum hydrocarbons (TPHs), the chemical of soil and the bioassay of *T. repens* in the four culture stages (35d, 60d, 90d,120d). The bioassay included measurements of height, foot length, fresh weight and root fresh weight. The results showed that the presence of *T. repens* could increase degradation rates of TPHs in soil by up to 51.56% after 120 days. *Trifolium repens* can tolerate a moderate level of petroleum contamination and could effectively degrade TPHs at contamination levels of less than 10,000 mg/kg soil. Obviously, *T. repens* can be used in the phytoremediation of petroleum contaminated soil in arid and semi-arid regions of China.

Keywords Petroleum contaminated soil · Phytoremediation · TPH degradation · *Trifolium repens* L

Y. Shen (✉) · H. Li · M. Zhao · Z. Li · T. Gao
College of Biological and Environmental Engineering, Xi'an University, Xi'an 710065, China
e-mail: shenyuan1982@xawl.edu.cn

W. Wang
School of Water and Environment, Chang'an University, Xi'an 710054, China

Y. Shen · H. Li · M. Zhao · T. Gao
Academy of Eco Xi'an, Xi'an University, Xi'an 710065, China

B. Yue · G. Chang · T. Gao (✉)
College of Urban Environment, The Engineering Center of Mining Pollution Treatment and Ecological Restoration of Gansu Province, Lanzhou City University, Lanzhou 730070, China
e-mail: zkgtp@163.com

J. Zhang et al. (eds.), *Environmental Pollution Governance and Ecological Remediation Technology*, Environmental Science and Engineering,
https://doi.org/10.1007/978-3-031-25284-6_72

72.1 Introduction

Petroleum serves as an important energy source and as a feedstock for the synthesis of organic chemicals (Kalita et al. 2022; Das and Chandran 2011). However, petroleum hydrocarbons are commonly accidentally released into the environment during the process of transport, storage, refinement, and use in a way that often results in widespread industrial pollution (Daâssi and Qabil Almaghribi 2022). Petroleum-based hydrocarbon contamination poses a significant threat to local ecosystems and especially to soil, where polluted soil often creates hazards to human health (Mafiana et al. 2021). In China, approximately 1×10^8 ha of petroleum polluted land has been influenced by oil released by the oil industry (Yuan et al. 2014).

Some approaches including physical and/or chemical remediation along with bioremediation of petroleum contaminated soil, methods that have been reported in the framework of environmental remediation. Bioremediation, including both microbial remediation and phytoremediation, has been widely used, with previous reports finding these methods to be efficient and economical (Peng et al. 2019). However, abiotic stress has a negative effect on microbial communities and their structure. Drought and extreme temperature stress also do not help microorganisms to thrive (Dubrovskaya et al. 2022). Thus, the use of microbial remediation is not an obvious solution to pollution in arid and semi-arid regions because prolonged drought, high temperatures, and high soil erosion rates, etc. do not favor microbes. In addition, plants used in soil reclamation change the chemical, physical, and biological properties in the rhizosphere, so the advantages of microbial remediation are not well known (Zhou et al. 2016). Few studies have been published related to the use of *Trifolium repens* L. (*T. repens*) during bioremediation of petroleum contaminated soil. The use of jasmine, petunia, sunflower, cotton, gaodan grass, and *T. repens* have been studied related to the degradation rates of oil in an area of polluted land in the arid and semi-arid regions in China, but the use of *T. repens* has not attracted much attention among all these plants used during reclamation (Nazarov et al. 2017). Wang et al. reported that the fibrous roots of *T. repens* could not only become well developed, but could also fix nitrogen, a feature that it is very important for plant growth (Wang et al. 2016).

This study used a phytoremediation technique to explore effective treatment options with *T. repens* for four different concentrations of petroleum in arid and semi-arid regions. Specifically, a comparison was made of the degradation rates at sites with and without plants using three concentrations of petroleum-contaminated soil (0 mg/kg, 3000 mg/kg, 7000 mg/kg, and 10,000 mg/kg) after 120 days. Then we detected petroleum products in *T. repens* using a bioassay in areas with four different concentrations of petroleum during the experiment, and made measurements of stem height, root length, fresh weight and root fresh weight. We aimed to compare the response of plants to phytoremediation and measured the efficiency of natural petroleum degradation. In order to study the effect of plants on soil properties during phytoremediation of soil petroleum pollution, the pH, electrical conductivity

(EC) of different treatments in different periods were tested. The results could provide insight into the study of phytoremediation and microbial populations in soils.

72.2 Materials and Methods

72.2.1 Experimental Materials

Seeds of *T. repens* were obtained from the seedlings field of Shuyang Green Flowers (Jiangsu, China). Appropriate amounts of Yanchang petroleum were applied to contaminate the substrate as described below. Uncontaminated silt loam soil samples were collected from the Nie Feng of Gaoling district, Xi′an, Shannxi Province, China, at depths of 0 to 50 cm. Soil was passed through a 2.0 mm sieve and was fully mixed to ensure the soil homogeneity. These samples had chemical properties of ($\mu g \bullet g^{-1}$) 0.31 NO_3^-, 207.55 SO_4^{2-}, 0.39 available P, 24.47 available N, 14.3 available K, 37.42 K, 56.10 Ca, 8.37 Mg, and 73.10 Na. The electrical conductivity was 202 $\mu S \cdot cm^{-1}$, pH 7.24, with 9.2% silt and 90.8% clay based on textural analysis.

72.2.2 Experimental Design and Sample Collection

The experiment was performed at the Water and Environment in Situ Experimental Area, Weishui Campus, Chang'an University, Xi'an, China. Experimental treatments were designed as Table 72.1. Soil for each experimental unit was placed in a square concrete tank (100 cm × 100 cm × 90 cm deep) containing 1.1 ton of experimental soil. 20–30 g of seeds were sown in each reclamation unit. Soil moisture were maintained at 60% of the daily field storage water capacity of the soil during the experiment. Soil samples were collected from each units at 0, 35, 60, 90 and 125 days to test the effects of the experimental treatments. All samples were taken from each unit at 9 stochasticnine random locations in the middle soil layer. Composite samples from each unit were separated into three smaller parts to allow for triplicate analysis. Plant samples were collected from the above plants units at 35, 60, 90 and 120 days to test the plant growth condition.

72.2.3 Soil Chemical Property and Plant Biomass Measurements

The TPHs of soils were based on US Environmental Protection Agency standard 3550 (U.S. EPA Method 2010). Soil pH values were measured by a pH meter (PHS-3E) (w/v ratio of soil and water was set at 1:2.5). Electrical Conductivity (EC) was

Table 72.1 Experiment details with and without plants *T. repens* with different levels of petroleum-contaminated soil

Samples	0 mg/kg TPH	3000 mg/kg TPH	7000 mg/kg TPH	10,000 mg/kg TPH
Soils with *T. repens* B	CK	PA	PB	PC
Soils without *T. repens*	CK1	A	B	C

measured by a conductivity meter (DDS-307A) (w/v ratio of soil and water was set at 1:5).

The biomass of *T. repens* includes stem height, root length, fresh weight and root weight. Ten plants were acquired from per soil sample unit. The roots of *T. repens* were slightly shaken to remove soil adhering to the roots, and then roots were washed with tap water. Then, all plant root surfaces were rinsed using deionized water to assure that all soil had been removed. Finally, stem height, fresh weight and root fresh weight of each plant had been measured at 35, 60, 90, 120 days. The root length had been measured at 60 and 120 days.

72.2.4 Soil Microorganisms Population

The population of microorganisms was collected from rhizosphere soil at 4 °C, analyzed, and conducted by the serial dilution method (Li et al. 1996). The number of microorganism colonies was determined and expressed as colony forming units (CFUs) g^{-1} dry soil.

72.2.5 Statistical Analysis

Data were expressed as means ± SE, which were calculated in all treatments using IBM SPSS Statistics 21. Each treatment is reported as the mean of three replicates. Significant differences between means were analyzed by Duncan's multiple range test ($p < 0.05$). Within each treatment, the values followed by the same letter are not significantly different ($p < 0.05$) according to Duncan's multiple range test.

72.3 Results and Discussion

72.3.1 Chemical Properties of the Soil Samples

The degradation rates of total petroleum hydrocarbons in petroleum-contaminated soil with, and without, *T. repens* are given in Table 72.2. The results showed that the chemical, physical, and biological action caused by degradation of petroleum in contaminated soil without remediation from plants was also weak as a result of only natural attenuation. Conversely, *T. repens* could promote the removal of TPHs during the process of growth. Table 72.2 shows that for all degradation resulting from phytoremediation in petroleum-contaminated soils up to 35.09–51.56% of all TPHs were degraded in reclamation soil, while only 19.46–29.90% of TPHs were degraded in petroleum-contaminated soils without plants. In addition, the degradation of all soil with *T. repens* was significantly ($p < 0.05$) higher than soils without plants at all three tested levels of pollution. The degradation rate of soil reclaimed by *T. repens* was as high as 51.56%, which is higher than that in similar published studies (Wang et al. 2016). Liu et al. analyzed the phytoremediation of petroleum contaminated soil by *bermudagrass, sunflower, cotton, and gaodan grass* over 120 days, the petroleum degradation rates reached 38.23%, 36.57%, 40.67%, and 38.67%, respectively (Liu et al. 2009). Our results were consistent in that *T. repens* was the most tolerant to petroleum pollution when five species were compared: *alfalfa, T. repens, tall fescue, Pennisetum, Ryegrass* (Peng et al. 2009). However, testing for the dissipation or degradation of petroleum contaminants in soil in situ is very difficult. In our study, we applied in situ remediation using *T. repens* to remove TPHs. In recent years, *T. repens* has been found to provide gradual but effective phytoremediation of petroleum-contaminated soil. Sun found that a low concentration of petroleum contamination in soil (0.5%) could promote seed germination. In contrast, a higher concentration (1–3%) had inhibitory effects on both seed germination and plant growth (Shan et al. 2014).

Figure 72.1 indicate the changes in the soil pH, EC by *T. repens* phytoremediation and control treatments, respectively, in the five culture stages under the conditions of three TPHs concentrations. In the phytoremediation groups, under different petroleum pollution treatments, pH gradually increased from (6.83–7.60)

Table 72.2 Degradation of TPHs in soils with and without *T. repens*

Soil treatment (mg/kg)	Degradation of petroleum in soils with *T. repens* (%)	Degradation of petroleum in soils without *T. repens* (%)
3000	51.56 ± 3.24^{a}	29.90 ± 2.12^{a}
7000	40.66 ± 3.07^{b}	21.45 ± 5.24^{b}
10,000	35.09 ± 3.54^{b}	19.46 ± 3.95^{b}

Different letters (a,b) significant differences ($p<0.05$) among treatments by two-way analysis of variance (ANOVA) followed by Newman-keuls multiple comparison test at the initial stage of restoration to the highest (8.33–8.55) after 60 days of restoration in Fig. 72.1a. With the increase of restoration time, pH gradually decreased, slowed down after 90 days and tended to be stable. In the control treatments without vegetation, the pH showed a trend of gradual increase. It showed a downward trend at 60d and reached the maximum at 120d (8.47–8.75). After 60d in the vegetation restoration treatments, the pH remained relatively low. The pH of different concentrations in petroleum-contaminated soils without plants treatments showed a trend of gradually increasing. 35–60d belong to the rapid growth period of plants, regardless of plant height or fresh weight.

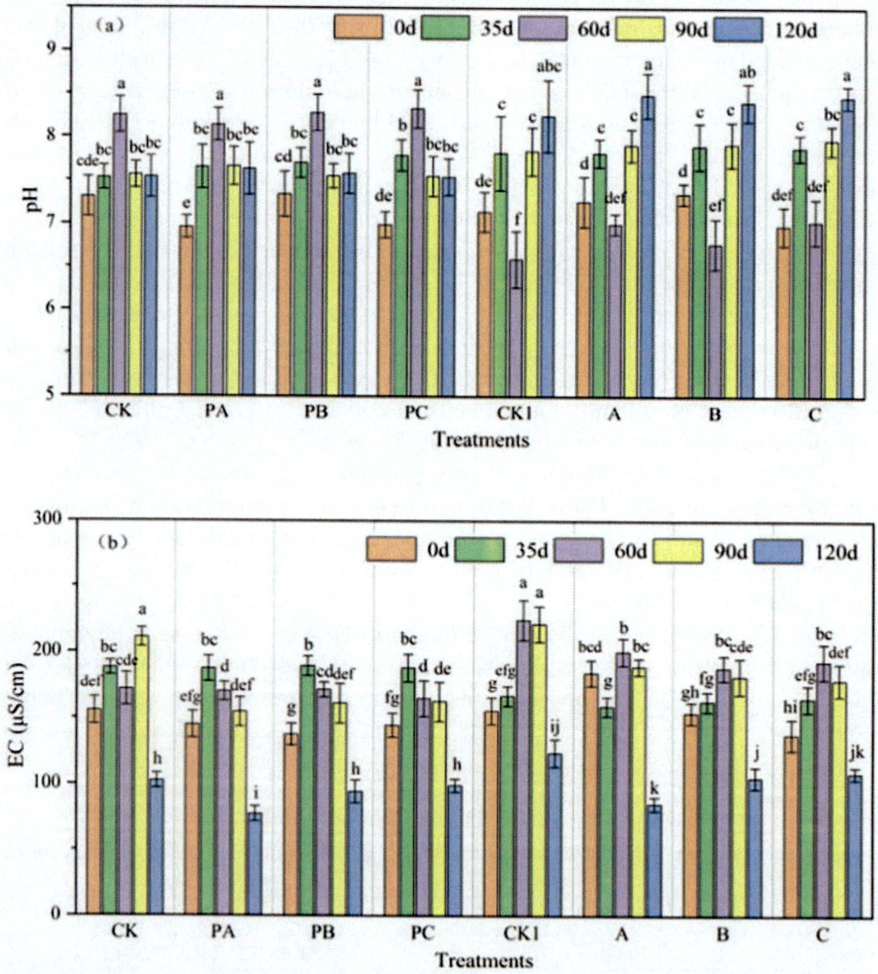

Fig. 72.1 Changes of soil pH (**a**), EC (**b**) in different treatment groups. Correlation analysis was performed separately in phytoremediation groups (CK, PA, PB, PC) and control groups (CK1, A, B, C)

The increase of plant biomass will make the nutrients and water in the soil move to the plants through the roots, and at the same time, it will also secrete a large amount of root exudates, changing the physical and chemical properties of the soil. Huang et al. showed that four plants: *Vigna radiata* L., *Medicago sativa* L., *Lolium perenne* L. and *Arachis hypogaea* L. all increased the pH of pyrene contaminated soil and increased the activity of soil dehydrogenase after 60 days of remediation (Sun et al. 2011). The pH in petroleum-contaminated soils without plants treatments showed an increasing trend in the later stage, while the pH in the phytoremediation treatments showed a decreasing trend due to the presence of small molecular organic acids in the rhizosphere exudates of plants (Kalita et al. 2022). The reason for the increase of soil pH may be due to the accumulation of hydrocarbons in bicarbonate caused by biodegradation (Huang et al. 2019).

It can be seen from Fig. 72.1b, the maximum EC of phytoremediation treatments for petroleum pollution occurred at 35d. After 35d remediation, it showed a decreasing trend. The maximum of EC in the petroleum-contaminated soils without plants treatments appeared at 60d and gradually decreased after 60d. The EC of petroleum-contaminated soils without plants treatments were higher than that of phytoremediation treatments. The EC of the polluted soil treatments is lower than that of the non-petroleum-contaminated soil treatments. It is mainly due to the increase of soil coarse aggregates caused by petroleum pollution (Lu et al. 2014). fHydrophobicity of petroleum will lead to potential drought of soil layer, intensify soil salinization, and then increase the pH of soil (Wang et al. 2019). The increase of pH will affect the availability of soil nutrients and reduce their solubility. It will reduce the concentration of exchangeable base (Ca^{2+}, Na^{+}, Mg^{2+}, K^{+} etc.) and lead to the decrease of EC (Njoku et al. 2009).

72.3.2 *Biomass of the Plant Samples*

The responses of growth of *T. repens* in petroleum contaminated soil are presented in Figs. 72.2 and 72.3. Figure 72.2a is a comparison photograph of different concentrations in the phytoremediation treatments at 35d. It can be seen from the figure that the height of plants has an obvious trend of gradually decreasing with the increase of concentration. Figure 72.2b shows the 35d plant sampling photos. Data on the fresh weight and root length of *T. repens* in all four treatments are presented in Fig. 72.2c, d. The height of *T. repens* plants with different concentrations in the different culture stages is significantly different ($p < 0.05$) (Fig. 72.2c). Similarly, as shown in Fig. 72.3a, there were significant differences between the fresh weight in petroleum-contaminated phytoremediation treatments (PA, PB, PC) and CK in all periods except for the 35d culture stages ($p < 0.05$). As the concentration of petroleum increased, *T. repens* had a significantly shorter height, root length and fresh weight especially in the PB and PC treatments (Fig. 72.2c, d). After the 120 days of phytoremediation, the height of *T. repens* grown in petroleum contaminated soil was 28.61%, 47.59%, and 76.49% shorter than the growth in CK for treatments PA, PB, and PC, respectively. The root length of *T. repens* grown in petroleum contaminated soil was 6.74%, 16.87%, and 35.33% shorter than the growth in CK for treatments PA, PB, and PC. Both parts of Fig. 72.2d.

Figure 72.3b show that the biomass of *T. repens* continued on a downward but insignificant ($p < 0.05$) trend as the concentration of petroleum contamination increased in the tested soil. After the 120 days of phytoremediation, the fresh weight of *T. repens* grown in petroleum contaminated soil was 64.34%, 75.52%, and 96.85% less than the growth in CK for treatments PA, PB, and PC, respectively (Fig. 72.3a). The root fresh weight of *T. repens* grown in petroleum contaminated soil was 26.67%, 26.67%, and 60.00% less than the growth in CK for treatments PA, PB, and PC, respectively (Fig. 72.3b). The degree of inhibition of plant biomass is generally expressed as fresh weight > plant height > root fresh > root length. Overall, although the stem height, root length, fresh weight, root weight of the *T. repens* were affected by the gradient of the petroleum concentration, *T. repens* obviously tolerated lower levels of petroleum contamination in soil (<10,000 mg/kg).

72.3.3 Treatment Effect on the Numbers of Soil Microorganisms

Microorganisms, including bacteria, actinomycetes, and fungi, were not prevalent in severely contaminated soil in the PC treatment (Fig. 72.4). The population density of bacteria, actinomycetes, and fungi decreased by 57.14%, 28.57%, and 45.83%, respectively, in this treatment. In the clean control soils, the populations of bacteria, actinomycetes, and fungi were 1.40×10^8 CFU^{-1} g dry soil, 7.08×10^5 CFU^{-1} g dry soil, and 2.47×10^6 CFU^{-1} g dry soil, respectively. The amount of bacteria obviously decreased gradually. The population of actinomycetes was greatest in soil with the 7000 mg/kg of petroleum hydrocarbons. However, all populations of microorganisms were lowest in the PC treatment soil.

The three types of microorganisms play an irreplaceable role except by volatilization, natural attenuation, eluviations, and photolysis (Ekundayo and Oluekwe 1997). We were surprised to find that the PB treatment contributed to an increase in the populations of fungi and actinomycetes. The degradation of TPHs had a positive effect on the population of microorganisms (Steliga 2012). Plants may modify soil conditions in a way that provides a suitable environment for microorganisms such as bacteria, actinomycetes, and fungi. The main reason was that *T. repens* not only had actively developed a fibrous roots system in the PB treatment, but also grew rhizomatous and creeping roots that helped to increase the root surface area (Galitskaya et al. 2021). Importantly, the rhizospheric environment of plants grown in petroleum contaminated soil had changed during phytoremediation, and the rate of petroleum degradation had obviously improved (Alavi et al. 2017).

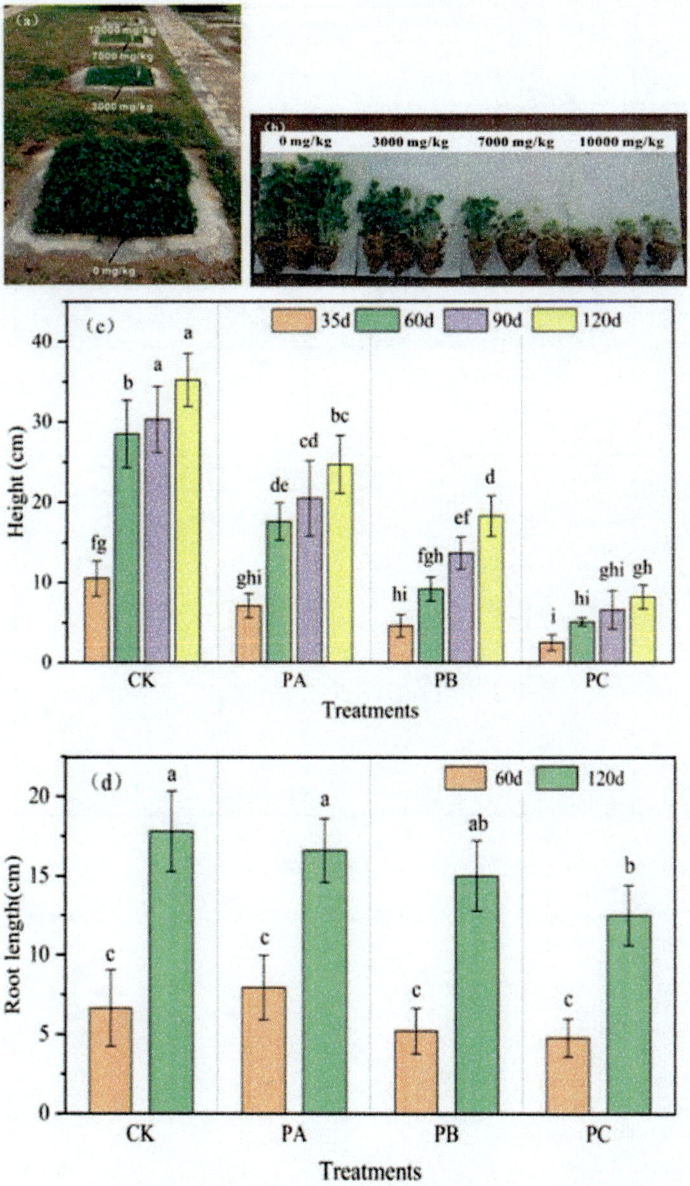

Fig. 72.2 Effects of petroleum contamination on the height and root length of *T. repens*. **a** real view of the restoration area of *T. repens* **b** *T. repens* samples **c** height of *T. repens*, **d** root length of *T. repens*

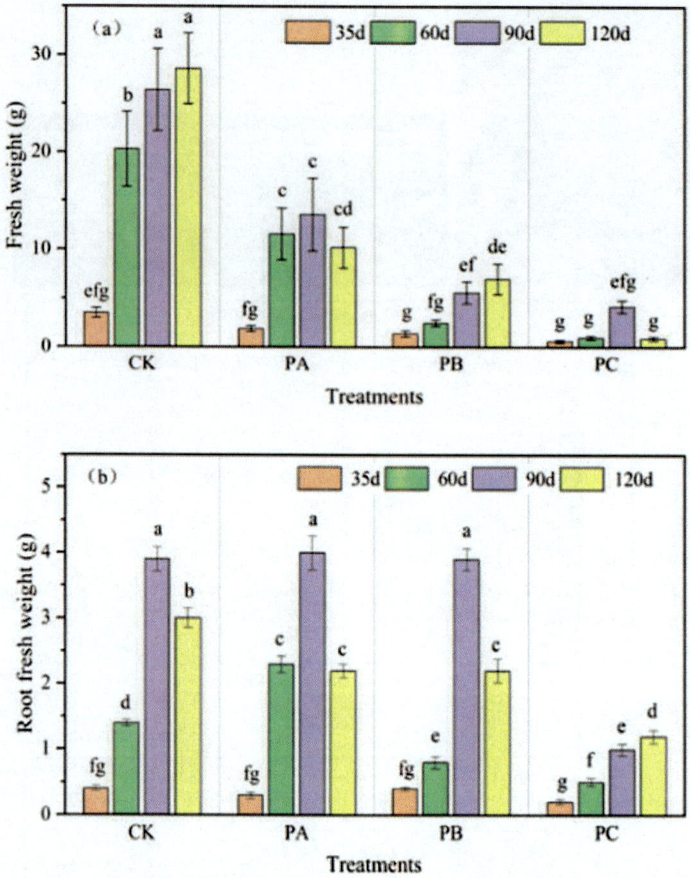

Fig. 72.3 Growth responses in fresh weight (**a**) and root fresh weight (**b**) of *T. repens* grown in the petroleum contaminated soil

Fig. 72.4 Effects of petroleum contamination on the populations of bacteria, fungi, and actinomycetes in soil with *T. repens*. Note: CFU, colony forming units

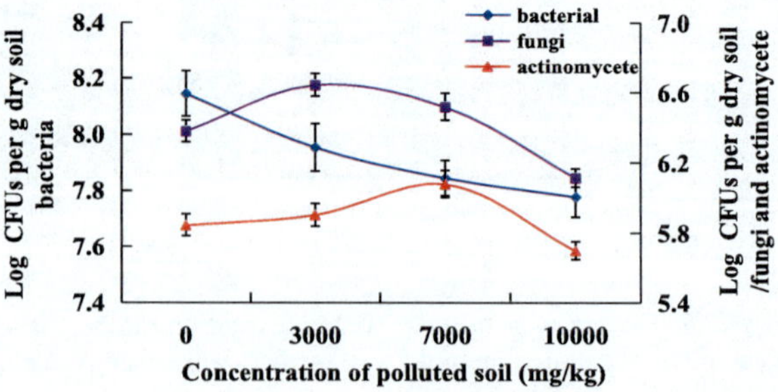

72.4 Conclusions

The data from this study provides useful information on the impact of *T. repens* phytoremediation on the petroleum contamination soil ecosystem in Arid and Semi-arid Region in China. Overall, the legume *T. repens* can degrade petroleum in contaminated soil at amounts of less than 10,000 mg/kg in a semi-arid region. *T. repens* phytoremediation can change the soil pH and EC. Furthermore, pH and EC affected soil ion content and microbial community structure. Although the biomass of *T. repens* was affected based on plant height, root length, fresh weight and root fresh weight, *T. repens* showed a strong ability to endure and grow. In addition, bacteria, actinomycetes, and fungi play a major role in the process of degradation of petroleum in contaminated soil. Also, contaminated petroleum soil had the negative effect on populations of bacteria, actinomycetes, and fungi which were more prevalent with lower levels of petroleum contamination in soil (<10,000 mg/kg). A relatively high petroleum concentration appeared to limit the populations of microbes.

Funding The authors thank the National Natural Science Foundation of China (31860176), Key Research and Development Program of Shaanxi (2020ZDLSF06-06), Natural Science Basic Research Program of Shaanxi (2021JQ-791, 2021JM-510), Science and Technology Program of Xi'an (21NYYF0038,20NYYF0013), Dean's Fund (YZJJ202110), "Space Breeding" program (HTYZ2106). The authors also thank Key Laboratory for ecological restoration and high quality development of Qinling Mountains in the upper and middle reaches of the Yellow River, Xi'an Key Laboratory of plant stress physiology and ecological restoration technology, Engineering Research Center for Groundwater and Ecoenvironment of Shaanxi Province.

References

Alavi N, Parsehn I, Ahmadi M, Jafarzadeh N, Yari AR, Chehrazi M, Chorom M (2017) Phytoremediation of total petroleum hydrocarbons from highly aaline and clay soil using *Sorghum halepense* (L.) *Pers.* and *Aeluropus littoralis* (Guna) *Parl.* Soil Sediment Contam 26(1):127–140

Daâssi D, Almaghribi QF (2022) Petroleum-contaminated soil: environmental occurrence and remediation strategies. 3 Biotech 12:139

Das N, Chandran P (2011) Microbial degradation of petroleum hydrocarbon contaminants: an overview. Biotechnol Res Int 10:6787–6790

Dubrovskaya E, Golubev S, Muratova A et al (2022) Effect of remediation techniques on petroleum removal from and on biological activity of a drought-stressed Kastanozem soil. Environ Sci Pollut Res

Ekundayo EO, Oluekwe CO (1997) Effects of an oil spillon soil physio-chemical properties of a spill site in a Tlypic Paleudult of Midwesterm Nigeria. Environ Monit Assess 45(2):209–221

Galitskaya P, Biktasheva L, Kuryntseva P et al (2021) Response of soil bacterial communities to high petroleum content in the absence of remediation procedures. Environ Sci Pollut Res 28:9610–9627

Huang H, Zhang CL, Zhou YX, Xie T, Liao CJ (2019) Effect of rhizobium-plant remediation on Pyrene contaminated soil. Ecol Environ Sci 28(7):1466–1472

Kalita M, Chakravarty P, Deka H (2022) Understanding biochemical defense and phytoremediation potential of Leucas aspera in crude oil polluted soil. Environ Sci Pollut Res

Li FD, Yu ZN, He SJ (1996) Experimental techniques of agricultural microbiology. China Agricultural Press, Beijin

Liu JZ, Cui S, Zhang YP (2009) Effects of plants and microorganisms on bioremediation of petroleum contaminated soil. J Ecol Rural Environ 25:80–83

Lu L, Tyler H, Song J, Yi Z, Zhiyong JR (2014) Microbial metabolism and community structure in response to bioelectrochemically enhanced remediation of petroleum hydrocarbon-contaminated soil. Environ Sci Technol 48(7):4021.

Mafiana MO, Kang XH, Leng Y, He LF, Li SW (2021) Petroleum contamination significantly changes soil microbial communities in three oilfield locations in Delta State, Nigeria. Environ Sci Pollut Res 28:31447–31461

Nazarov AV, Shestakova EA, Anan'yina LN (2017) Effect of red clover on the microbial transformation of phenanthrene and octadecane in the soil. Eurasian Soil Sci 50(8):971–976

Njoku KL, Akinola MO, Oboh BO (2009) Phytoremediation of crude oil contaminated soil: the effect of growth of *Glycine max* on the physico-chemistry and crude oil contents of soil. Nat Sci 7(10):79–87

Peng SW, Zhou QX, Cai Z, Zhang ZN (2009) Phytoremediation of petroleum contaminated soils by Mirabilis JalapaL. in a greenhouse plot experiment. J Hazard Mater 168:1490–1496

Peng Z, Guo ZH, Xiao XY, Peng C (2019) Dynamic response of enzymatic activity and microbial community structure in metal (loid)-contaminated soil with tree-herb intercropping. Chemo 220:237–248

Shan BQ, Zhang YT, Cao QQ, Kang ZY (2014) Growth responses of six leguminous plants adaptable in Northern Shaanxi to petroleum contaminated soil. Environ Sci 35(3):1125

Steliga T (2012) Role of fungi in biodegradation of petroleum hydrocarbons in drill waste. Pol J Environ Stud 21:471–479

Sun ZL, Xiao X, Zhang S, Wu GL et al (2011) Study on the tolerance of different plants to oil pollution. Environ Sci Manage 36:130–132

U.S. EPA Method (2010) Oil and grease analysis in wastewater. Available online: http://www.epa.gov/waterscience/ methods/method/oil/ (accessed on 23 April 2018)

Wang J, Qi C, Feng Q, Zhu RJ, Liu H, He JJ (2016) The soil and grass in Loess Plateau oil recovery and soil CAT activity. Shaanxi J Agr Sci 62:6–11

Wang WL, Yuan QY, Dong BC, Gao JQ, Han GX, Yu FH (2019) Effects of clonal integration and crude oil pollution on soil physicochemical properties in a *Phragmites australis* wetland. Acta Ecol Sin 39(24):9179–9187

Yuan JM, Yang DM, Xia HC (2014) Bioremediation of petroleum contaminated soil. Environ Eng 32:797–800

Zhou W, Yin XB, Wang L (2016) The effect of plant on the restoration of oil pollution in upper soil. Res Environ Sci 10:182–183

Chapter 73
Research on PRB Technology for Groundwater Remediation

Lixin Zhao, Xiaodong Wang, and Zhiwu Liu

Abstract In groundwater remediation, the permeable reactive barrier (PRB) technology is increasingly being studied and applied as an economical, simple and efficient in situ remediation technique. The key technology is the study of reaction media. This paper reviews the types of PRBs and several media materials and their mechanisms of action, and also describes and outlines the use of electric in-situ remediation techniques in conjunction with PRBs.

Keywords Groundwater remediation · PRB · Medium material

73.1 Introduction

In recent years, as groundwater pollution continues to intensify, remediation techniques have received widespread attention, with both ex-situ and in-situ remediation techniques gradually emerging (Zhang et al. 2012). Among them, in-situ remediation techniques can be divided into bioremediation techniques, groundwater recycling well remediation techniques, electro-remediation techniques, chemical reaction remediation techniques, and permeable reactive barriers (PRB) (Xie et al. 2016).

PRB technology is one of the main research priorities for groundwater remediation due to its simple installation, economy and efficiency, It mainly uses the permeability of the reaction wall to make pollutants flow through the reaction medium through natural hydraulic transmission and transform into environmentally acceptable forms under the action of the reaction medium, in order to purify and intercept pollutants (Jacobs and Rstner 1999). The radioactive substances, heavy metals, inorganic and organic pollutants dissolved in groundwater can be effectively controlled and treated by osmosis reaction wall technology (Guerin et al. 2001). After a lot of research on PRB technology, we have not only studied a variety of PRB types, but also developed

L. Zhao · X. Wang (✉) · Z. Liu
School of Civil Engineering and Architecture, University of Jinan, Jinan 250022, China
e-mail: 13853189137@163.com

J. Zhang et al. (eds.), *Environmental Pollution Governance and Ecological Remediation Technology*, Environmental Science and Engineering,
https://doi.org/10.1007/978-3-031-25284-6_73

a variety of PRB medium materials, and further enhanced the practicability of PRB technology in the treatment of groundwater pollutants.

73.2 Main Type and Reaction Mechanism of PRB

73.2.1 The Main Type of PRB

There are three types of PRB structures common according to the structure type, namely, continuous wall type structure, water partition funnel-water guide gate type and wall curtain type PRB structure. When the groundwater pollution is shallow, the continuous wall structure can be adopted. When the polluted area is large or the aquifer is thick, considering the treatment effect and economic cost, the water partition funnel-water guide gate structure can be adopted in the polluted area to collect the contaminated groundwater body into a smaller range, and then place the reaction wall to conduct centralized treatment of the contaminated groundwater. The wall curtain PRB structure is mainly used for shallow polluted sites (Jeen and Gillham 2001; Statham et al. 2016).

73.2.2 The Main Mechanism of the PRB

Pollutant removal mechanism according to the treated pollutants and reaction medium are mainly different in the physical (physical adsorption), chemical (chemical precipitation, chemical adsorption, redox) and biological (biodegradation) remediation mechanism.

73.3 PRB Media Material and Its Mechanism of Action

From the original iron-based PRB, the variety of PRB media is increasing. For different types of pollutants, all kinds of active materials have corresponding treatment forms. According to its treatment mechanism, the PRB reaction medium can be divided into adsorption medium, precipitation medium, reduction medium, degradation medium and ion exchange medium.

73.3.1 *Adsorbing Medium*

The adsorption medium has a large specific surface area, providing a large number of adsorption sites; under the adsorption, the contaminated water adsorption separation, so as to achieve the purpose of removal, in the groundwater flow conditions, adsorption medium through physical adsorption, chemical adsorption, surface complexation and other adsorption mechanism, the pollutants on the medium surface and pores. There are many kinds of adsorption media, including zeolite, clay minerals, activated carbon, biochar, hydroxyapatite, graphene, metal oxides, fly ash, steel slag, waste rubber particles and other natural minerals, synthetic materials and industrial wastes. Adsorption medium is widely used, for heavy metals, organic matter, and other groundwater pollutants have a good treatment effect. Yang et al. (2013) chose natural river sand, zeolite and sponge iron as the reaction media, and the simulated restoration of groundwater with PRB technology contaminated with ammonia nitrogen showed that ammonia nitrogen was almost completely removed in the zeolite layer.

73.3.2 *Chemical Precipitation Medium*

Precipitation medium is to form the precipitation with the pollutants when repairing the groundwater pollution, and to intercept the pollutants in the PRB, so as to remove the pollutants. To this, requires the sediment reaction of difficult solvent, pollution is not easy to be precipitated, reaction soluble without secondary pollution (Wang et al. 2016), common limestone, apatite, etc. Wilopo et al. (2008) of PRB technology treatment by As, Mn pollution groundwater active filler made test analysis, found that As and Mn is easy to compost and zero price metal adsorption and common precipitation, to be removed.

73.3.3 *Chemical Precipitation Medium*

Redox media are mainly composed of zero-valent iron (ZVI), nano-zero-valent iron (nZVI), zero-valent aluminum, pyrite and other metal materials. In the treatment mode, organic matter is dehalogenated or removed; or redox reaction is produced with heavy metal in the form of simple or insoluble compounds (Fan et al. 2019). At present, the zero-valent iron PRB has been successfully used for the in-situ restoration of the Cr (VI) -contaminated groundwater, and the reaction equation as follows:

$$CrO_7^{2-} + 2FeO + 7H_2O \rightarrow 2Cr^{3+} + 14OH^- + 2Fe^{3+}$$

$$Cr^{3+} + 3OH^- \rightarrow Cr(OH)_3$$

$$2Cr^{3+} + 6OH^- \rightarrow Cr_2O_3 + 3H_2O$$

$$Fe^{3+} + 3OH^- \rightarrow Fe(OH)_3$$

$$2Fe^{3+} + 6OH^- \rightarrow Fe_2O_3 + 3H_2O$$

$$(1 - x)Fe^{3+} + xCr^{3+} + 3OH^- \rightarrow (Cr_xFe_{1-x})(OH)_3$$

Ponder et al. (2020) studied the reduction and removal of hexavalent chromium and divalent lead in water bodies using nanoscale nivalent iron, and showed that hexavalent chromium was reduced to trivalent chromium, and divalent lead was mainly reduced to nivalent lead.

73.3.4 Biodegradation Media

Biodegradation media is a process in which microorganisms attach to the carrier (such as biomass straw) under sufficient electron donors and carbon sources. Degradation media mainly play the role of providing electron donors, carbon sources and microbial carriers. Zhang et al. (2017) Used kinds of agricultural wastes to remove nitrate in water, they found that bagasse, corn stalks, rice husks, wheat straw, corn cobs and sawdust were ideal carbon sources and biofilm carriers of denitrification bacteria, and the removal rate of nitrate was more than 80%.

73.3.5 Ion Exchange Medium

The medium is mainly divided into ion exchange resin and chelation resin. Ion exchange resin uses the difference of the ion hydration radius to make the ion exchange action of the target ions in the solution with the cations or anions on the resin, thus removing the heavy metal pollutants in the groundwater (Chen et al. 2017). Chelating resin uses a chelating group on the chelate resin to selectively chelate with the heavy metal ions in an aqueous solution, and then to remove the heavy metal ions (Sharbatmaleki et al. 2015).

73.4 PRB Technology with Other Technologies (EK-PRBs)

73.4.1 Technical Principles of EK-PRBs

The principle of EK-PRBs is applied on the basis of the hydraulic gradient of PRB to form a power gradient to move the pollutants to the electrodes on both sides under the combined action of electricity and hydropower, and to fully react the pollutants with the fillers in the wall, and to remove or reduce the toxicity through adsorption, degradation or precipitation (Xu et al. 2020). The PRB technology allows primary pollutants to degrade into secondary materials that can be remediated by the electrokinetic technology, while the electric field provided in the electrokinetic remediation technology speeds up the passage of pollutants through the wall on the one hand, and allows pollutants to react more fully within the wall reaction zone on the other, making the two technologies much more efficient when used in combination than alone. Li et al. (2022) used waste cotton to synthesise a new β CD/hydrothermal biochar (KCB). The KCB was used as a reaction material for an electro-osmotic reaction barrier (EK-PRB) to explore the removal characteristics of heavy metals. The final removal efficiency of Pb^{2+} and Cd^{2+} from soil groundwater reached 92.87% and 86.19% respectively. This finding demonstrates that KCB/ EK-PRB can be used as an inexpensive green process to effectively remediate heavy metal contaminated soil and groundwater.

73.4.2 Prospect of the EK-PRBs Technologies

Although this technology is effective, has a wide range of applications and low secondary contamination, the technology has a long treatment time and the consumption of fillers within the permeable reaction wall is all but unavoidable. What is known is that the pollutants will move under the action of electricity, but it is not known whether adsorption or oxidation and decomposition are dominant in the reaction and removal in the wall, if we can study this issue in depth, it will be more helpful to the design and use of fillers. It would also be useful to find better, cheaper and more useful fillers.

73.5 Conclusion and Further Works

PRB has significant advantages over traditional groundwater remediation technologies in terms of its effectiveness, simplicity of placement, low operating costs and improved efficiency in groundwater treatment. Due to the wide choice of media materials available for PRB, this enables it to deal with groundwater from a wide range of contaminants, and in combination with electrokinetic technology it can treat a wide range of contaminants that are difficult to remove by a single removal method, making this technology a promising application and development in the remediation of contaminated groundwater. At the same time, we are aware that the material of the reaction media of PRB is the key to its treatment capacity, so the research and development of composite media and new reaction media and PRB coupling technology will remain the focus of future research on PRB technology.

References

Chen LF, Chen YL, Wang XP et al (2017) A novel silica-based anion exchange resin used for removing uranium from drinking water. J Radioanal Nucl Chem 314:2569–2578

Fan ZX, Zhang Q, Gao B et al (2019) Removal of hexavalent chromium by biochar supported nZVI composite: batch and fixed-bed column evaluations, mechanisms, and secondary contamination prevention. Chemosphere 217:85–94

Guerin TF, McGoverm T, Horner S (2001) A funnel and gate system for remediation of dissovlved phase petroleum hydrocarbons in groundwater. Land Contam Reclam 9(2):209–224

Jacobs PH, Rstner UF (1999) Concept of subaqueous capping of contaminated sediments with active barrier system (ABS) using natural and modified zeolite. Water Res 33:2083–2087

Jeen SW, Gillham RW, Przepiora A (2001) Predictions of long-term performance of granular iron permeable reactive barriers: field-scale evaluation. J Contam Hydrol 123:12350–12364

Li YL et al (2022) Enhanced remediation of heavy metals contaminated soils with EK-PRB using β-CD/hydrothermal biochar by waste cotton as reactive barrier. Chem 286(P1):131470–131470

Ponder SM, Arabj G, Malloukt E (2020) Remediation of Cr(VI) and Pb(II) aqueous solutions using supported, nanoseale zero-valent iron. Environ Sci Techno l34(12):2564–2569

Sharbatmaleki M, Unz RF, Batista JR (2015) Potential mechanisms for bioregeneration of perchlorate-containing ion-exchange resin. Water Res 75:1–10

Statham TM, Stark SC, Snape I et al (2016) A permeable reactive barrier (PRB) media sequence for the remediation of heavy metal and hydrocarbon contaminated water: a field assessment at Casey Station, Antarctica. Chemos 147:368–375

Wang Y, Pleasant S, Jain P et al (2016) Calcium carbonate-based permeable reactive barriers for iron and manganese groundwater remediation at landfills. Waste Manage 53:128–135

Wilopo W, Sasaki K, Hirajima T, et al (2008) Immobilization of arsenic and manganese in contaminated groundwater by permeable reactive barriers using zero valent iron and sheep manure. Mat Trans 49(10):2265–227449

Xie YF, Li XW, Liu XJ et al (2016) Performance evaluation of remediation scenarios for DNAPL contaminated groundwater using analytical models and probabilistic methods. Procedia Environ Sci 31:264–273

Xu JH, Zhou HD, Lv SF, et al (2020) Effect of EK-PRB on Pb (II) contaminated soil. J Univ Shanghai Sci Tech 42(4):368–403

Yang YZ, Liu F, Kong XK, Li SP, Ma JF (2013) Transformation and presence morphology of ammonia nitrogen in multi-media osmosis reaction grille. J Environ Eng 08:2931–2936

Zhang W, Zhang YP, Yin L et al (2017) Experimental research on denitrification carbon source properties of groundwater with 10 agricultural wastes as base materials. J Environ Sci 37(5):1787–1797

Zhang ZL, Wei YF, Guo et al (2012) Review of permeable reaction gate wall technology. Water Res Protect 28(6):70–80

Chapter 74
Research on the Carbon Storage and Economic Value of Carbon Sequestration in Chizhou Forests

Dufang Shi, Jie Bao, Xinwang Xu, and Haozhen Chen

Abstract Forests are carbon storage warehouses in terrestrial ecosystems, and increasing forest carbon sinks is an important way to achieve carbon neutrality in the region. Past studies have focused on national and provincial scales. Based on the forest resources survey data of Chizhou, the forest stock volume method and market value method were used to calculate the carbon storage and value. We can find that, in 2020, the forest carbon storage of Chizhou is 37.1705 million tons, among which the carbon storage of arbor forest is 35.2617 million tons at the most. From 2015 to 2020, its carbon storage has increased by 3.9624 million tons. The per-unit carbon storage is 0.46 to 0.55 million t/km^2 in the southeast (Shitai county, Qingyang County),which is higher than 0.33 to 0.41 million t/km^2 in the northwest (Guichi District, Dongzhi County). In 2020, the total value of forest carbon storage in Chizhou is 2.2 billion yuan. This paper puts forward some countermeasures to increase forest carbon sink by optimizing tree species allocation, increasing investment in science and technology and strengthening forest management.

Keywords Chizhou · Carbon storage · Value of carbon sequestration

D. Shi
Department of Humanities, Anhui Vocational College of Arts, Hefei 230601, China

J. Bao
School of Information Engineering, Guangxi University for Nationalities, Nanning 530006, China

X. Xu (✉)
Department of Garden, Chizhou Vocational and Technical College, Chizhou 247100, China
e-mail: xuxinwang@163.com

X. Xu · H. Chen
Department of Geography and Planning, Chizhou College, Chizhou 247100, China

J. Zhang et al. (eds.), *Environmental Pollution Governance and Ecological Remediation Technology*, Environmental Science and Engineering,
https://doi.org/10.1007/978-3-031-25284-6_74

74.1 Introduction

Forests have the ability to absorb carbon dioxide and fix carbon elements. 80% of the above-ground carbon storage exists in the forest, and 40% of the underground carbon storage also exists in the underground part of the forest system. It can be said that forest is the carbon repository on the earth. Forest carbon is changing faster than other ecosystems. My country is one of the countries with the most abundant forest resources in the world. It has good opportunities to play the role of forest carbon sequestration and participate in regulating the global climate, but also faces huge challenges. China has made plans for changes in carbon sinks. It is expected that carbon will peak in 2030, and we will strive to achieve carbon neutrality by 2060. Interventions in forest carbon sinks are a better approach than reducing CO_2 emissions through industrial production.

The research on forest biomass is the earliest in foreign countries. After the 1950s, the research on biomass and carbon storage shifted from developed countries to the world. The International Biology Program (IBP), as a leader, called on countries to conduct research and count forest organisms worldwide quantity. The research on biomass in China started relatively late. Weicun et al. (1978) studied the biomass of Chinese fir plantation as a symbol of the beginning of forest biomass research in my country (Weicun et al. 1978); Xudong et al. (1992) carried out the masson pine plantation. The biomass research survey shows that my country's forest biomass research has entered a stage of rapid development in the 1980s (Xudong et al. 1992). In the 1990s, the team of Chinese ecologist Fang Jingyun estimated the carbon content of China's forests (Jingyun et al. 1996; Guohua et al. 2000). In recent years, from the provincial scale, there have been studies on forest carbon storage, influencing factors and carbon sequestration potential in Fujian, Anhui, Beijing and other provinces and cities, and fruitful results have been achieved (Ying and Xiaoge 2022; Qingbo et al. 2017; Aiqin et al. 2019). The research at the regional level is relatively insufficient, while the selection of forest species and the layout adjustment at the regional level are more targeted. More accurate selection of suitable plant species for planting (Jingyun 2021). The importance of this study lies in revealing the total value of forest carbon storage and the distribution of carbon pools in different forests in Chizhou City, and further estimating the changes in different stands of arbor forests. This study enriches the forest carbon research cases at the regional level, and combines the analysis of forest carbon storage with economic value for the first time in Anhui Province, in order to provide a reference for the formulation of prefecture-level green development strategies and the selection of local carbon emission reduction strategies.

74.2 Data Sources and Research Methods

74.2.1 Data Sources

The data in this paper are mainly from field investigation, and from the data in the "Chizhou Statistical Yearbook", "Chizhou City Forest Resources Inventory" and published literature. The specific data used are the area and stock volume of different forest species in Chizhou City in recent years.

74.2.2 Research Methods

Estimation method of forest carbon storage. There are many methods for calculating forest carbon storage. The calculation of carbon sink is also based on carbon storage data. The methods often used at home and abroad include the average biomass expansion factor method, the stock volume method, and the carbon density method. Some of these methods are relatively mature and well-established, and are widely used today. For example, the average biomass method was the main method used during the International Biological Program (IBP). The algorithms of these research methods are relatively scientific and rigorous, so the actual calculation is relatively cumbersome. Since the study is on carbon storage in the microscopic field, the scale of the study is small and the calculation process is too complicated, so there are many limitations in operation. In addition to these methods, with the development of science and technology, the model simulation method of the new technology of remote sensing is also gradually developed. The advantage of not destroying forest vegetation in the research process is more prominent, but there is a lack of data for estimation, and parameters can only be set under ideal conditions, and the research results deviate from the actual situation. Although remote sensing technology is relatively mature at present, the combination of remote sensing technology and model method needs further exploration.

In this study, the forest stock volume method was mainly used to measure carbon storage and carbon sink, and the results obtained by this method were more accurate (Nuyun 2016). The carbon storage of forest resources in arbor forest, economic forest, bamboo forest, sparse forest and four-sided forest in Chizhou City was selected for research.

The specific calculation formula of forest stock expansion method is:

$$TC_F = S_i \times C_i + \alpha \times S_i \times C_i + \beta \times S_i \times C_i \quad (74.1)$$

$$C_i = V_i \times \delta \times \rho \times \gamma \quad (74.2)$$

where

TC_F—carbon storage of forest resources.
Si—the area of the i-th forest type.
Ci—the carbon density of the i-th forest type.
Vi—the stock volume per unit area of the i-th type of forest.
α—understory plant carbon conversion coefficient, take 0.195.
β—Forest land carbon conversion coefficient, take 1.244.
δ—biomass expansion coefficient, the value is 1.90.
ρ—bulk density, which is 0.45–0.5t/m^3, which is taken as 0.5 in this study.
γ—Carbon content, take 0.5
Note: According to the default parameter values of the Intergovernmental Panel on Climate Change (IPCC).

The carbon storage accounting of moso bamboo in this study consists of three parts: the carbon storage in the arbor layer of moso bamboo, the carbon storage in the understory of moso bamboo, and the carbon storage in the litter of moso bamboo (Qingbo et al. 2017).

The relevant calculation formulas are as follows:

$$TC_B = \omega \times S_B \times \varepsilon + \theta \times S_B + \mu \times S_B \tag{74.3}$$

TC_B—The carbon storage of bamboo (including the carbon storage of the bamboo arbor layer, the carbon storage of the bamboo forest, and the carbon storage of the bamboo litter).
ε—Carbon content, take 0.5 as the calculated value.
ω—the biomass per unit area, which is 37.6 t•hm^{-2}
S_B—Moso bamboo area.
θ—carbon density per unit area, the carbon density per unit area of understory vegetation is taken as 0.161 t•hm^{-2}
μ—carbon density per unit area, the carbon density per unit area of litter is 0.594 t•hm^{-2}

Estimation of economic value of carbon sequestration. Common methods for estimating the economic value of forest carbon sequestration mainly include carbon sequestration cost method, afforestation cost method, carbon tax method and willingness to pay method. The first three are the same in nature, only the design principle of willingness to pay is more in line with the price formation of the market economy. This study uses the market price method to estimate. The optimal price of China's forest carbon sink researched by Chinese scholars is 0.11–15.17 US dollars / t, and the price of each province varies in different years. This study uses the latest national carbon market daily data released by my country's carbon emission trading website in April 2021. The transaction data is realistic. The carbon sink transaction value in April was 58.5–60 yuan/t. In this study, the median value was 59.25 yuan/t to calculate the carbon storage value of forest resources in Chizhou City.

Formula:

$$V = k * TC \tag{74.4}$$

In the formula: V is the value of carbon sequestration, K is the carbon transaction price per ton, and TC is the carbon storage.

74.3 Results and Analysis

74.3.1 Forest Carbon Storage

The forest carbon storage in Chizhou City was calculated using Formula (74.1) (74.2) (74.3), as shown in Table 74.1, Fig. 74.1.

Overall, the forest carbon storage in Chizhou is increasing, from 33,208,100 t in 2015 to 37,170,500 t in 2020, an increase of 3,962,400 t, and a compound annual growth rate of 11%. In 2015, the carbon storage of arbor forests was 31.3144 million t, which will increase to 35.2617 million t in 2020, with an average annual compound growth rate of 12%. The proportion of arbor forest carbon storage in Chizhou's forest carbon storage is relatively stable, 94% in 2015 as well as in 2020. Although

Table 74.1 The carbon storage of forest resources in Chizhou City by tree structure (10,000 t) (self made)

	2020	2019	2018	2017	2016	2015
Arbor	3526.17	3435.60	3342.32	3270.88	3219.97	3131.44
Economic forest	0.90	0.80	0.39	0.38	0.33	0.33
Bamboo forest	73.30	75.16	67.35	75.14	74.94	73.60
Sparse forest	1.82	1.69	2.39	2.29	2.26	0.68
Si Pang Lin	99.09	87.25	107.18	105.04	101.42	102.08
Scattered forest	15.77	15.25	14.58	14.26	13.80	12.68

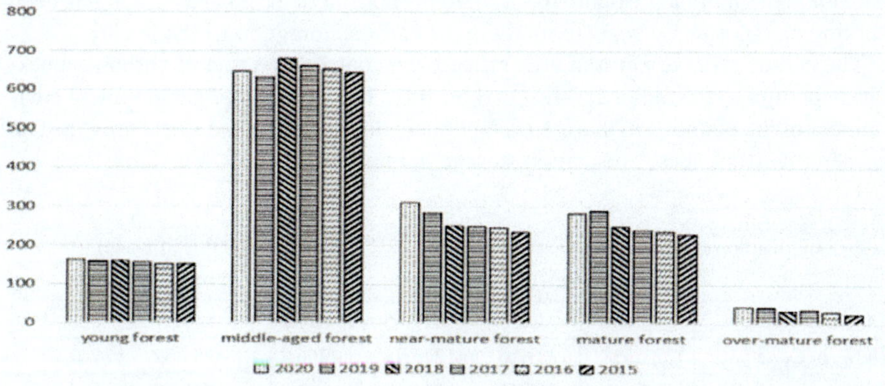

Fig. 74.1 Carbon sinks of arbor forests by age (10,000 t) (self made)

the proportion in the middle five years fluctuated, it did not change much. From the perspective of economic forests, the carbon storage of economic forests has increased in a large area since 2018, and the increase reached as much as 105.12% from 2018 to 2019. It can be seen that Chizhou City has begun to attach importance to the planting of economic forests in 2018, and the economic benefits brought by trees are also increased. From the perspective of the surrounding trees, with the construction and development of the city, the carbon storage of the surrounding forests has decreased in the past two years, and the greening construction of the surrounding trees should be strengthened.

According to the carbon storage of arbor forests according to different forest ages (as shown in Fig. 74.1), among the forests of different ages, the carbon storage of middle-aged forests in each year is relatively large, and the carbon storage of over-mature forests and young forests is small; Judging from the average annual growth rate of carbon sinks, the average annual growth rate of carbon sinks in over-mature forests is faster, reaching 14.92%, while the average annual growth rate of carbon sinks in young forests is slower, only 1.48%.

74.3.2 Regional Differences in Carbon Storage

Using the forest resources survey data of each county and district in Chizhou City, the calculation results of carbon storage in each district are shown in Table 74.2.

From the comparison of carbon storage in different districts, Dongzhi County accounts for the largest proportion of carbon storage, which can reach 35.91% of the total in Chizhou City. It can be seen that Dongzhi County has a wide forest area and a large amount of forest storage. Qingyang County has the smallest proportion of carbon storage at 16.97%. From the perspective of changes in carbon storage in each district, the overall trend is increasing. In terms of the increase rate, Qingyang County has the fastest growth rate, with an average rate of 3.61%. It can be seen that although Qingyang County has a small proportion of carbon storage, it has been working hard for many years to increase the carbon storage in Chizhou City.

The carbon storage per unit area reflects the contribution rate of carbon sequestration per mu to the regional forest carbon sink. The natural and geographical environment of the counties in this area is obviously different, and the forest growth state has certain differences. The carbon storage per unit area is shown in Table 74.3.

Table 74.2 Carbon storage in counties and districts (2015–2020) (10,000 t) (self made)

	2020	2019	2018	2017	2016	2015
Guichi District	949.80	920.13	891.65	870.98	848.87	828.41
Dongzhi County	1321.35	1287.50	1275.76	1266.48	1255.62	1241.62
Shitai County	782.84	766.63	752.67	741.48	730.37	724.55
Qingyang County	624.62	603.36	584.49	551.70	540.59	523.28

Table 74.3 Carbon storage per unit area (10,000 t/km^2) in each district (2015–2020) (self made)

	2020	2019	2018	2016	2015
Guichi District	0.38	0.37	0.35	0.34	0.33
Dongzhi County	0.41	0.39	0.39	0.39	0.38
Shitai County	0.55	0.54	0.53	0.52	0.51
Qingyang County	0.55	0.53	0.52	0.48	0.46

The carbon storage per unit area of Shitai County and Qingyang County is relatively large. It can be seen that the coverage rate and density of forest vegetation in Shitai County and Qingyang County are high, which is the result of long-term protection of virgin forest and increased artificial afforestation. The landforms of Guichi district include mountains, hills, polders, water areas, etc., and the carbon storage per unit area is relatively high.

74.3.3 Carbon Sequestration Value of Forest Resources

According to formula (74.4), the carbon storage value of forest resources in Chizhou City in each year was calculated, and the results are shown in Table 74.4.

Chizhou's forest carbon storage is 2.2 billion, of which arbor forest accounts for the largest proportion, reaching 95%. The carbon sequestration value of trees increased from 1,855,378,200 yuan in 2015 to 2,089,255,700 yuan in 2020, an average annual increase of 46,775,500 yuan, and a compound annual growth rate of 2.52%. From the changes in the value of forest carbon storage in the past five years, it can be seen that except the value of bamboo forests, sparse forests and four-sided forests, all others increased.

Table 74.4 The value of forest carbon in Chizhou City (10,000 yuan) (self made)

	2020	2019	2018	2017	2016	2015
Economic forest	53.33	47.4	23.11	22.52	19.55	19.55
Bamboo forest	4343.03	4453.23	3990.49	4452.05	4440.20	4360.80
Sparse forest	107.84	100.13	141.61	135.68	133.91	40.29
Si Pang Lin	5871.08	5169.56	6350.42	6223.62	6009.14	6048.24
Scattered forest	934.37	903.56	863.87	844.91	817.65	751.29
Arbor	208,925.57	203,559.30	198,032.46	193,799.64	190,783.22	185,537.82
Total	220,234.60	214,232.59	209,402.56	205,478.43	202,203.67	196,758

74.4 Conclusions and Recommendations

Overall, the forest carbon sink in Chizhou is relatively stable and increases slowly, but it has a great growth potential. The important conclusions of this study are as follows: First, the forest carbon storage in Chizhou City will reach 37.1705 million t in 2020, ranking at the forefront of the city at the provincial level. Arbor forests account for more than 94% of the carbon storage. The second is based on 2014, with a carbon sink of 3.9624 million tons, with a compound annual growth rate of 11%, exceeding the provincial average. Third, the total value of forest carbon storage is 2.2 billion yuan, which is equivalent to 34.26% of the GDP of that year. It can be seen that increasing the level of forest carbon sinks in Chizhou will be beneficial to economic and social development. The specific measures suggested include.

74.4.1 Optimize Tree Species Configuration

The forest carbon sink is based on the forest ecosystem. To promote the development of the forest carbon sink, it is necessary to promote the rapid growth of forest resources. The arbor forest in this area has the largest carbon content. Among the trees, Chinese fir, Metasequoia, Masson pine, Paulownia and other tree species should be selected first. Moso bamboo has high carbon density and short forest formation period, so it is the first choice for carbon sink forest construction. The tree species in the surrounding forest are different due to the local microenvironment, and the tree species that can be selected are aspen, willow, camphor and so on.

74.4.2 Increase Investment in Science and Technology

The implementation of forest carbon sink projects requires high technical requirements, so increase investment in scientific research, accelerate the research on basic forestry science projects, encourage universities and research institutes to actively participate in forestry carbon sink projects, promote the transformation of forestry scientific and technological achievements, and improve the level of forestry science and technology services. It is also necessary to improve the scientific and technological level of forestry practitioners, master forestry carbon sink management technology, and enhance the forest carbon sequestration capacity.

74.4.3 Strengthen Forest Management

The government implements the reform of forest tenure system, change collective forests to individual forest farmers, clarify forest farmers' ownership of forest land, encourage forest farmers to actively engage in forest carbon sink forest production activities, define forest carbon sink property rights, and increase forest farmers' income. The government guides forest carbon sink trading, and converts part of the ecological value of the forest into economic value through the market mechanism to achieve compensation for the ecological benefits of the forest. Transform forest carbon sinks from ecological value to market value.

74.4.4 Increase Publicity Efforts

All walks of life have increased the publicity of forest carbon sinks to give full play to the rich advantages of forestry carbon sinks, but the education level of forest households is generally not high. Carbon sinks are a new concept. Carbon sink is a brand new concept, and the Chizhou Municipal Government has guided them to actively participate in the activities of increasing forest carbon sink. In terms of publicity methods, the combination of mobile phones, TV, Weibo and other channels is used to make people truly understand the huge economic value brought by forest carbon sinks, enhance people's awareness of forest carbon sinks, and advocate people's low-carbon life. All of this will increase awareness of forest carbon sinks.

Acknowledgements National Natural Science Foundation of China General Program National Natural Science Foundation of China General Program (41071337); Anhui Province Natural Science Key Project (KJ2016520, KJ2021A1415, KJ2021A1490, 2021gkszgg 049)

References

Aiqin L, Huirong W, Jingjing W, Zhen T, Xiaoniu X (2019) Carbon storage, carbon density dynamics and carbon sequestration potential of forest vegetation in Anhui Province. J Jiangxi Agricult Univ 41(5):953–962

Guohua L, Bojie F, Jingyun F (2000) China's forest carbon dynamics and its contribution to the global carbon balance. Chinese J Ecol 5:733–740

Jingyun F (2021) An ecological perspective of carbon neutrality. Chinese J Plant Ecol 45(11):1173–1176

Jingyun F, Guohua L, Songling X (1966) Biomass and net production of forest vegetation in my country. Chinese J Ecol 5:497–508

Nuyun L (2016) China's forestry carbon sink, Revised. China Forestry Press, Beijing

Weicun P, Licun L, Zhengheng G, Xiangqiong Z, Dongyuan T (1978) Research on the biological yield and productivity of Chinese fir plantation ecosystem. Hunan Forestry Sci Techn 05:1–12

Xudong Z, Xue Minghua X, Jun WT, Yingfa C, Guangping Z, Chao S (1992) Study on the biomass structure of Masson pine plantation. J Anhui Agricult Univ 04:268–273

Ying Z, Xiaoge L (2022) Analysis on the potential of Beijing's forest carbon sink under the carbon neutralization target. Res Indust 24(1):15–25

Zhu Qingbo S, Alan LY, Chengzhen W (2017) Carbon storage and influencing factors of bamboo forests in urban areas of Fujian Province. Guangxi Plants 37(07):891–900

Chapter 75
Application of New Compound Conditioning Agent in Activated Sludge Dewatering

Xiaomeng Zhang, Peijing Kuang, Guohao Gong, Wanjun Zhang, and Yubo Cui

Abstract To overcome the problem of weak sludge dewatering performance in Jinshitan wastewater treatment plant, three organic polymer flocculants and two inorganic polymer flocculants were selected to remove metal ions in water through chelation. Particles in water were coalesced into larger flocs through electrical neutralization, hydrogen bonding and bridging. Meanwhile, economical and environmental friendly sludge flocculants with low specific resistance were determined using sludge specific resistance experiment. The optimal flocculant dosage and mixing time were used to further explore the relationship between different flocculants and sludge dewatering performance. Results suggested that adding 0.34% PDMDAAC + PAFC compound flocculant and mixing sludge for 120 s could obtain the lowest sludge specific resistance and the best sludge dewatering performance.

Keywords Sludge dewatering · Sludge specific resistance · Compound flocculant

75.1 Introduction

Sludge is an inevitable product of sewage treatment plants and sewage treatment. When sludge without proper treatment and disposal enters the environment, it will directly bring secondary pollution to water bodies and the atmosphere, posing a serious threat to the ecological environment and human activities (Long and Najafi 2022). The sludge produced by sewage treatment has high water content, sludge reduction is the key and focus of sludge treatment, sludge reduction has a total of two ways to reduce the content of organic matter in sludge and reduce the water content of sludge, from the technical and financial aspects, reducing the water content of sludge is simple and easy to achieve (Meyer et al. 2020). The common dewatering method used in wastewater plants today is mechanical dewatering, but the current water content of sludge after all types of mechanical dewatering is still very high

X. Zhang · P. Kuang (✉) · G. Gong · W. Zhang · Y. Cui
College of Environment and Resources, Dalian Minzu University, Dalian 116600, China
e-mail: kuangpeijing@dlnu.edu.cn

J. Zhang et al. (eds.), *Environmental Pollution Governance and Ecological Remediation Technology*, Environmental Science and Engineering,
https://doi.org/10.1007/978-3-031-25284-6_75

and does not meet the standards for treatment. In order to perform effective deep dewatering of sludge, sludge needs to be pretreated before mechanical dewatering, usually by using chemical conditioning.

A single sludge conditioning agent is not ideal for improving sludge dewatering performance, for example, the most common organic polymer conditioning agent polyacrylamide (PAM) series of organic synthetic polymer conditioning agent (Feng et al. 2019), the price of such conditioning agent is more expensive, although its dosage is less, but converted to the cost of conditioning per ton of sludge, the cost of using organic synthetic polymer conditioning agent is still higher. Although inorganic polymer conditioning agent is inexpensive, but the amount of injection is large, not enough environmental protection (Zhao et al. 2015). So, the research and development of efficient and inexpensive conditioning agents is a major problem that must be overcome in the field of wastewater treatment.

This study is to improve the problem of activated sludge dewatering faced by Dalian Detai Water Company—Jinshitan Wastewater Treatment Plant in China, by adding polymer flocculants and inorganic flocculants to the activated sludge, removing metal ions in the water through chelation, removing particles in the water through electrical neutralization, hydrogen bonding and bridging. The most economical flocculant with good treatment effect is selected, and the most effective flocculant dosage and mixing time are determined by sludge specific resistance experiment to improve sludge dewatering efficiency and reduce sludge volume.

75.2 Materials and Methods

75.2.1 Experimental Instruments and Reagents

Activated sludge samples were collected from Jinshitan sewage treatment Plant in Dalian, China. The plant mainly treats the production and domestic sewage in the Jinshitan tourist resort.

The main reagents are polydimethyldiallylammonium chloride (PDMDAAC), polyacrylamide (PAM (3395) and PAM (4460) from different manufacturers), polyaluminum chloride (PAC) and polyaluminum ferric chloride (PAFC). The main instruments used are six mixers, experimental equipment for measuring specific resistance of sludge, oven, thermometer, measuring cylinder, beaker, filter paper and glass rod.

75.2.2 Flocculant Configuration

The configured concentrations of PDMDAAC solution and PAM solution are both 0.2%. PAC and PAFC need to be mixed with PDMDAAC and PAM. The compound conditioners of PAM and PAC, PAM and PAFC, PDMDAAC and PAC, PDMDAAC and PAFC are all prepared according to the ratio of 10 g inorganic conditioner, 1 g organic conditioner and 110 g water.

75.2.3 Experimental Methods

The water content of the sludge was measured to determine the nature of the sludge; filter paper moistened with water was placed on the cloth funnel and pressed against the circumference (Liu et al. 2022); the vacuum pump was turned on and the vacuum pressure was adjusted to 1/3 of the experimental pressure and the vacuum pump was turned off (the experimental pressure used in this experiment was 0.05 MPa); the corresponding amount of filtrate in the metering tube was recorded every 5 s and filtered until the vacuum was destroyed. The flocculant dosage was divided into 4 gradients, which were 0.1%, 0.2%, 0.3% and 0.4% of the dry weight of the sludge. The stirring time was set to 5 gradients of 30 s, 60 s, 120 s, 180 s and 300 s, respectively. 1000 mL of sludge sample was measured in a cup with a 1000 mL measuring cylinder, and 5 groups were taken and added according to the set gradient of flocculant dosage. The sludge was stirred rapidly for 10 s at 300 rpm to mix the conditioning agent with the activated sludge, and then the speed was set at 100 rpm to keep the sludge in a dough without being broken up. The sludge specific resistance is calculated as follows (Guoxiang and Ruan 2001).

$$\alpha = \frac{2pF^2}{\mu} \cdot \frac{b}{C}$$

where p is the pressure drop during filtration, g/cm^2; F is the filtration area, cm^2; b is the slope of the filtration line V–t/V, N/m^2; μ is the kinetic viscosity coefficient of the filtrate, g · s/cm^2; C is the amount of solids per unit area of filtrate, g/cm^3.

75.3 Experimental Results

75.3.1 Selection of Flocculant

Sludge specific resistance (SRF) value is a comprehensive index indicating the filterability of sludge, and the smaller the SRF value, the better the dewatering performance of sludge (Xia et al. 2022).

From Fig. 75.1, the sludge specific resistance of activated sludge with compound flocculants is lower than that of organic polymer flocculants alone, which indicates that inorganic flocculants neutralize and compress the double electric layer of colloids with opposite charges in water, thus destabilizing and coagulating the colloids and producing better sludge coagulation effect in combination with organic polymer

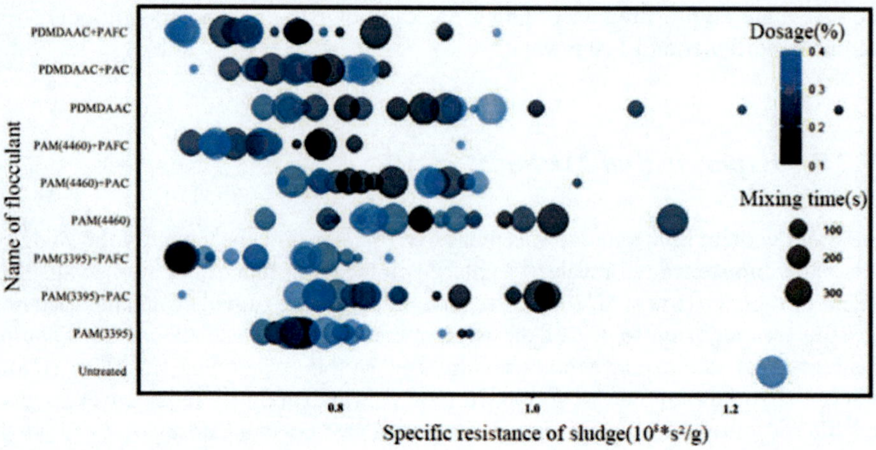

Fig. 75.1 Experimental bubble diagram of sludge specific resistance

flocculants. And three organic polymer flocculants and inorganic flocculants PAFC compound effect are better than inorganic flocculants PAC, because PAFC it is a collection of aluminum and iron salts of their respective advantages, aluminum ions and iron ions have significantly improved the form, the degree of polymerization is greatly improved, the purification of high turbidity water and low temperature low turbidity water treatment effect is particularly obvious. Compared with polymerized aluminum chloride (PAC), polymerized aluminum chloride iron (PAFC) not only has the excellent flocculation performance of PAC and strong electric neutralization effect also has polymerized iron chloride (PFC) adsorption strong precipitation speed characteristics. It can also be seen from the graph that PDMDAAC + PAFC, PAM(4460) + PAFC, PAM(3395) + PAFC three compound flocculants are used in activated sludge in sewage plants with low sludge specific resistance and good sludge dewatering effect.

75.3.2 *Amount of Compound Flocculant Dosing*

In addition to the choice of flocculant type, there is also the influence of dosing amount. The PAM organic polymer flocculant with PAFC inorganic flocculant selected after the test in 3.1 was selected for the sludge specific resistance experiment, and the experimental conditions were improved to improve the dosing amount to 0.2–0.4%. The experimental results are shown in Fig. 75.2, which shows that the sludge specific resistance decreased continuously with the increase of the compound flocculant dosage; it indicates that the sludge specific resistance is positively correlated with the flocculant dosage. Compared with PAM, PDMDAAC has strong electrical neutralization ability in addition to the bridging and sweeping functions of general

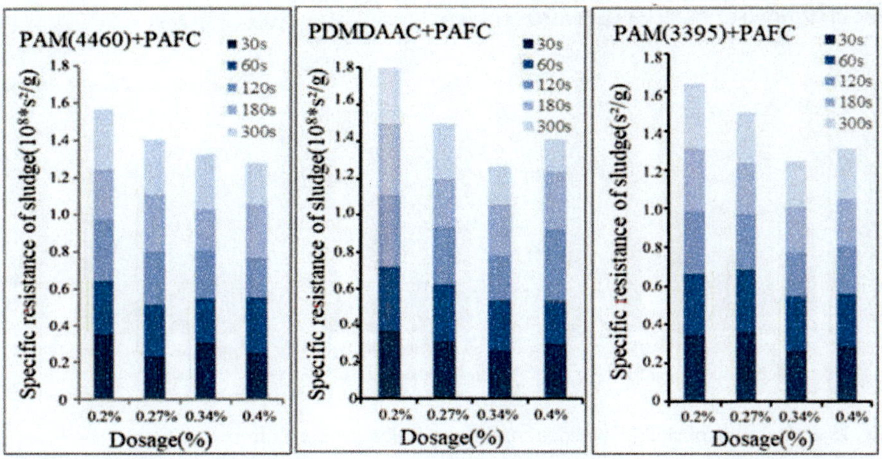

Fig. 75.2 Statistics of sludge specific resistance with dosing amount

polymer flocculants, so that the lowest sludge specific resistance of 1.27*10^8 s^2/g was achieved when 0.34% PDMDAAC + PAFC flocculant was added to the activated sludge.

75.3.3 Determination of Sludge Stirring Time

The sludge stirring time also affects the dewatering of sludge, too short stirring time will lead to the activated sludge not being fully flocculated, which is not conducive to sludge dewatering; too long stirring time will lead to the activated sludge just flocculated to be broken up again. According to the experiment in 3.1, the mixing time was improved to 60 s, 90 s, 120 s, 150 s, 180 s. The experimental results are shown in Fig. 75.3. From Fig. 75.3, it can be seen that the sludge specific resistance mainly shows a trend of decreasing first and then increasing with the increase of agitated activated sludge time, which is consistent with the previous theory. When the activated sludge stirring time is 120 s, the sludge specific resistance is the lowest when 0.34% PDMDAAC flocculant is added, and the sludge dewatering effect is the best.

75.4 Conclusion

The metal ions in the sludge were removed by the chelation of organic polymer flocculant PDMDAAC and inorganic polymer flocculant PAFC, and the particles in the sludge were condensed into larger flocs by electric neutralization, hydrogen

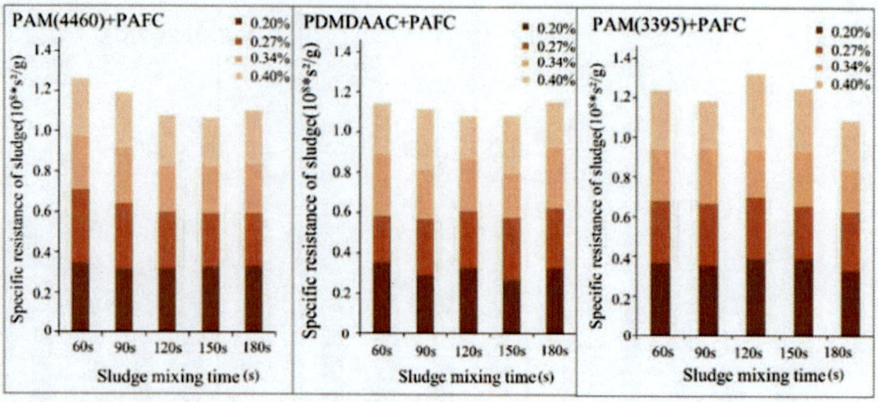

Fig. 75.3 Variation of sludge specific resistance with sludge mixing time

bonding and bridging, which significantly improved the dewatering performance of the sludge. When the dosage is 0.34% and the sludge mixing time is 120 s, the sludge specific resistance of activated sludge is the lowest and the sludge dewatering performance is the best.

Acknowledgements This work was supported by the National Natural Science Foundation of China (No.41402208) and Graduate Innovation Project of Dalian Minzu University in 2022.

References

Feng G, Hu Z, Ma H, Bai T, Guo Y, Hao Y (2019) Semi-solid rheology characterization of sludge conditioned with inorganic coagulants. Wat Sci Technol 80(11):2158-2168

Liu M, Yuan C, Ru S, Li j, Lei Z, Zhang Z, Shimizu K, Yuan T, Li F (2022) Combined organic reagents for co-conditioning of sewage sludge: High performance in deep dewatering and significant contribution to the floc property. J Water Process Eng 48:102855

Long L, Najafi SE (2022) Adaptive fuzzy control of sludge conditioning and pressing. Adv Math Phys, 1–8

Meyer T, Barquin A, Gionest P, Tran H (2020) Multivariate data analysis of full-scale sludge dewatering. J Environ Chem Eng 8(2):103648

Xia X, Chen W, Wang L (2022) Effect of compound conditioner on dewatering performance of abandoned building mud. J Environ Eng 16(04):1313–1322

Xu G, Ruan F (2001) Performance analysis of iron- and aluminum-based inorganic flocculants. Chongqing Environ Sci 3:52–55+72

Zhao K, Zhang YT, Zhang Y, Yang MQ (2015) Effects of EPS on settlement and dewatering properties of activated sludge. Adv Mat Res 1092–1093:848–851. Trans Tech Publications, Ltd.

Chapter 76
The Significance and Path of Promoting Green Finance Under the Concept of Green Development

Zhao Liu

Abstract Green finance a financial development model under the sustainable development is the product of the inter-coordination and mutual development of economy, society and environment. In today's era, we need to adopt a development mode of green finance to move forward the green development and facilitate the development of social stability. At present, economy and environment have attracted great attention to the globe with the ecological protection especially valued by the whole world. Today, the initiation of building a low-carbon society contributes to the green finance the focus by most countries. This article makes an analysis of the significance and development path targeting at the green finance.

Keywords Green finance · Green development · Social economy

76.1 Introduction

Green finance is a financial development model under the sustainable development, and it is the product of the coordinated and common development of economy, society and environment. At present, in order to promote green development and to promote stable social development, we need to adopt a green finance development model. At present, economy and environment have become two topics of wide attention to all countries in the world, especially ecological protection is valued by the world and advocates low-carbon society, so green finance has become the focus of most countries.

Z. Liu (✉)
Henan Mechanical and Electrical Vocational College, Zhengzhou, Henan 451191, China
e-mail: 755463163@qq.com

J. Zhang et al. (eds.), *Environmental Pollution Governance and Ecological Remediation Technology*, Environmental Science and Engineering,
https://doi.org/10.1007/978-3-031-25284-6_76

76.2 The Major Significance of Green Finance for Green Development

76.2.1 Boost Green Economic Growth

The so-called green finance refers to the ability to make full use of the limited resources, and to combine them with environmental protection while developing the economy, so as to maximize the economic interests. The development philosophy of green finance not only takes into account the benefits of enterprises, but also plays a leading role in social capital resources and environmental protection. The development mode of green finance enables enterprises to strike the balance between economic development and environmental protection, so as to effectively solve the contradiction caused by economic development and environmental protection. The adoption of the development concept of green finance ensures the strengthening of environmental protection as well as the realization of economic development thus effectively promoting economic sustainability, driving a virtuous cycle of the economy and society and the environment, so as to promote the green development.

76.2.2 Promoting the Green Industry

The implementation of the development concept of green finance will not only help to achieve green economic development, but also promote the greening of the industry. The adoption of the development concept of green finance plays a vital role in the industrial upgrading and the sustainable cycle of the industry. In order to promote the greening of the industry, it is necessary to strengthen the attention to the specific flow of funds, and adjust the industrial structure based on the capital flow, so that the industrial structure can be the best optimized. In the process of developing the concept of green finance, we should strengthen the pooling role of funds, and promote the green industrial trend of energy conservation and environmental protection while promoting green development.

76.2.3 Promoting Social Green Development

In order to promote the green economic development, the green development of the industry, and the green development of society, it is more necessary to carry out the intervention of green finance. In reality, paying attention to the development concept of green finance can constantly improve people's awareness of environmental protection. While realizing the economic development, environmental problems can also be taken into account. Implementing the development concept of green finance is a necessary measure to conform to the trend of the times. Promoting the development concept of green finance requires enterprises to bravely shoulder their own responsibilities and strengthen environmental protection while developing the economy, so that more and more people can also strengthen their awareness of environmental protection and promote the green development process of the whole society. In addition, enterprises should also create a good employment environment, and make greater benefits for society and enterprises through improving the employment environment. The development model of green finance can not only bring greater profits to enterprises, but also reduce the costs needed by enterprises, so as to realize the green development of enterprises.

76.3 The Path Selection of the Services from Green Finance to Green Growth

76.3.1 Improvement of the Law and Regulatory System of Green Finance

A rigorous and scientific legal and regulatory system is the institutional premise for the development of things, and green finance is no exception. With a certain legal force of constraint, the ability of green financial services for green development can be given full play. One of the important problems to be solved in the legislation on green finance is to coordinate and deal with the relationship and contradiction between environmental problems and corporate profits. One of the main implementing parties of green finance is financial institutions, but financial institutions basically focus on profit and pursue maximum economic benefits according to market operation rules, which is fundamentally inevitable to have conflicts with green activities such as energy conservation, environmental protection and emission reduction regarding social interests. Therefore, in order to realize the practical operability of the legislation, it is advised to establish the relevant laws and regulations of green finance on the basis of financial institutions maintaining basic profits, and to clarify their responsibilities and obligations in serving green development. For example, financial institutions could be reflected in the Bill when providing financial services at the same time. Risks and hazards caused by the environment, energy consumption, emissions and other factors, and possible returns and benefits are included in financial decision-making activities, making financial institutions have a clear list of guidance, and restrain its blindly pursuing benefit maximization but ignoring the behavior of green development.

Specifically, green financial products generally include: green credit, green securities, green fund, green insurance, carbon emissions, etc. In the laws and regulations of green finance development, the corresponding norms should be conducted in line with different products' features.

(1) Green credit aspect. At present, most banks have begun to show the development situation of green finance in their social responsibility reports, normally in the form of green credit balance. However, in addition to the 21 major domestic banks, there are many banks without publishing continuous data year by year. Because green credit can reflect the loan structure of financial institutions, it also guides the capital flow of the financial market in a certain sense. Therefore, if the guidance-based provisions are formulated in legislation for the amount and conditions for the completion of green credit, the "two high and one remaining industries" is a downward compression, which can make financial resources be more reasonably allocated. In terms of supervision, the supervision mechanism should be built more comprehensively and in place, design a more reasonable and strict project environmental risk assessment mechanism, a stricter index evaluation mechanism, and follow-up supervision in the process of project implementation. If the behaviors of operation avoidance occur, the corresponding legal responsibilities should be undertaken.
(2) Green securities and funds. On the basis of the present relevant laws, the environmental supervision system of listed companies should be improved and established. And the provisions of listed companies on green environmental protection should be increased in the promulgated laws and regulations of the current laws and regulations such as the Securities Law and the Company Law. In addition, strict environmental inspections will be conducted on upcoming listed companies and green environmental files of the listed enterprises should be established, and their environmental information ought to be brought in light. At the government level, the government should also incorporate the green environmental protection of listed companies into the performance assessment of economic growth and strengthen the awareness of green responsibility.
(3) Green insurance. On the basis of the real condition of China, it is suggested to increase the relevant provisions of the green insurance system in the Basic Law of Environmental Protection, establish a green insurance system of "compulsory green insurance first, arbitrary liability insurance second", clarify the environmental liability insurance for compensation for environmental damage, and refine the specific content of green environmental protection.
(4) Carbon emission trading. Compared with developed countries, China's legislation on carbon emission trading is still relatively weak, and the national norms on carbon emission trading have not yet been introduced. It is suggested to give certain norms for the trading subject, trading object, trading rules, trading procedures and the verification method of defining environmental capacity.

76.3.2 Adjusting the Structure of Green Financial Products and Enriching the Product Types

At present, the main way to realize green finance is still through a series of financial products such as green credit, green bonds, green insurance and green fund, which are also developed and circulated according to the characteristics of green finance. Among them, the most well-known and the highest promotion rate is green credit products. Due to the habit of traditional financial adaptability, green credit is still the main product to support the development of green finance whose application is much wider than other green financial products, but the simple product structure is easy to lead to the accumulation of green financial resources in banking financial institutions. Some invested enterprises rarely make effective investment to products such as green bonds, green fund products for effective investment. The extreme capital allocation may cause it difficult for China's green finance to develop orderly and stably for a long term, and also wear down the enthusiasm of non-bank financial institutions to invest in green financial products, which is not conducive to the development of green finance itself. The development of green finance urgently needs more novel development measures to change the traditional development mode. The common development of various and diversified products, going forward together, is the best way to develop green finance.

76.3.3 Promoting Green Growth with the Carbon Financial System

With the carbon financial system making inflexible rules on carbon emissions and effect made on the development of related industries from resource-dependent regions through market intervention, it stimulates the transformation of its industrial structure, and realize high efficiency of resource utilization, so as to promote the development of green economy. First, relevant government agencies need to create a policy environment, build a regulatory system and establish a legal framework. Grasp the trend of carbon finance at a macro level and formulate supporting policies on investment and financing, credit scale and tax according to the principle of green and sustainable development, thus opening green channels for clean development mechanism (CDM) project re-approval, tax revenue, investment and financing, also fostering a good policy environment for the development of the carbon trading market. Second, financial supervision institutions draw lessons from advanced international experience and make detailed assessment of the risks of carbon finance business so as to set up control index basis to guide financial institutions to carry out carbon finance business. In this connection, they can have the rules and laws to follow. Third, we should take full account of China's actual situation and innovate the operation model from the perspective of economic structure and participants. Establish a unified and efficient operation mechanism, meanwhile considering the financial institutions in different regions and the differences between resources and talents in order to make the relevant mechanism and rules. Fourth, the diversified development of carbon financial products. China's current carbon financial market business mainly focusing on some scattered carbon tradings, such as CDM projects and environmental

Table 76.1 Carbon peak and Carbon neutral schedule for some countries

Regions	Peak carbon dioxide emissions (year)	Carbon dioxide peaking (year)	Remarks
America	2007	2050	
China	2030	2060	
Japan	2020	2050	After 2035, wifuel vehicles will be prohibited
EU	1990	2050	
UK	1991	2050	
Canada	2007	2050	
Korea		2050	Decarbonize the economy and end coal financing

trading. Banking financial institutions and other financial institutions should cooperate to provide services for the carbon trading market, and vigorously research and develop diversified carbon financial products. Financial institutions can also develop targeted financial products related to carbon trading, and raise funds for carbon financial derivative industries and energy conservation and emission reduction projects. Plus, we must give full play to the service ability of financial intermediaries, innovate carbon financial intermediaries, and provide the parties with professional third-party intermediary financial services such as financial and leasing.

At present, especially, mainly developed economies have made efforts for carbon neutral, and developing countries such as India, Russia in the "carbon dioxide peaking, carbon neutral" goals are hesitant, which because that carbon neutral means that the industrialization of the economy will be greatly affected. Therefore, improve the carbon financial system is still a long way to go (Table 76.1).

76.3.4 Strengthening Publicity and Guidance to the Public and Advocating the Awareness of Green and Environmental Protection

Although green development in the academic circle and relevant government departments have reached a certain consensus, still in the public life, especially in the structure of market main body such as consumer market, business owners and financial practitioners, people have not formed a more profound and rational understanding of the green finance and green development, and have no idea about the urgency and necessity of the green development. However, as participants and even decision makers, they are the key to whether green finance can develop smoothly and can play a better service role in green development. Xie (2018) also pointed out that the lack of adequate understanding of green development often leads to the lack of green environmental awareness in production and life or business decisions, thus restricting the service of green finance to green development.

Therefore, this requires the relevant departments to increase the publicity of the "green development awareness" in the whole society, and cultivate all the economic and social participants to form a "green awareness". First, make the use of school education to publicize green development for students at all stages and strengthen the awareness of green environmental protection. Second, it is to use the form of official open classes to publicize social responsibility to entrepreneurs and strengthen their sense of it so that they can realize the responsibility of enterprises for green development. Third, use the training to publicize the necessity of green development to financial practitioners, and strengthen the ability of financial institutions to develop green finance and the practitioners' awareness of green environmental protection.

In addition, vigorously propagandize the financial institutions or enterprises making a major contribution to development of green financial and set a good example for the industry. Meanwhile, carefully summarize the practical experience and form the development experience of directive function, operability, reproducibility and propaganda, making other industry participants realize the necessity of green development. At the same time, we also need to change the traditional financial development thinking mode of the financial industry. This requires the intervention of relevant government institutions and relevant experts and scholars, drawing on the experience of developed countries, and step up the efforts to the explain and guide the business strategy model of financial enterprises to develop green finance, so that the green development of green finance services can follow the rules and abide by the laws.

76.4 Conlusion

To sum up, in the context of economic globalization, major enterprises should not only pay attention to the short-term economic development, but to pursue sustainable economic development and handle the relationship of economy and environment. During the economic development, environmental protection needs to be strengthened and the traditional economic management mode needs to be broken with adopting green management mode to realize the effective combination of economy and environmental protection. In the pursuit of the economic interests, we also need to strengthen environmental protection to bring greater benefits to the whole society.

References

Wang X, Liu N (2016) Long-term mechanism exploration of the sustainable development of green finance in China. J Theo Explor 32(4):99–105

Xie Y (2018) Key to the development of green finance in China. J Int Fin 9(4):45–49

Xing Y (2017) Theoretical structure of responsible investment, international trend and chinese countermeasures of responsible investment. J Econo 9(9):44–54
Xu Y, Qi Y (2017) Reassessment and mechanism inspection of environmental regulation on enterprise productivity. J Fin Trade Econ 38(06):147–161
Zou X, Hu J, Yao N (2019) Green Securities investment fund financial performance, environmental performance and investor selection. J Shanghai Econ Res 375(12):33–34

Chapter 77
A Review and Prospect of Landscape Resilience Research in China Based on Bibliometric Analysis

Zhiyuan Liang and Lu Feng

Abstract In recent years, the impact of natural disasters on urban development has been deepening, and issues related to resilient cities have gradually become the focus of academic attention. Based on the CNKI database, this paper uses CiteSpace visualization software to conduct a bibliometric analysis from the perspective of resilience. The paper analyzes the current research status, hot topics, and research evolution of landscape resilience. Then discusses the future research trends and possible applications. This paper grasps the research situation of landscape resilience in China as a whole, and lays the foundation for subsequent research as a literature review. The results are as follows: (1) The related research shows a continuous upward trend in general, and has maintained a high degree of popularity in recent years; (2) The research history is mainly divided into three stages; (3) There is not much cooperation among scholars and institutions; (4) The hotspots are gradually changing to climate change, urban resilience, and urban form. Through this study, the enlightenment for the related research on resilient landscape in my country is as follows: strengthen exchanges and cooperation between scholars and institutions; carry out interdisciplinary cooperation and expand research fields; research focus on quantitative analysis, scientific management model and construction of evaluation system and other aspects are extended; strengthen the emphasis on rural landscape resilience research.

Keywords Resilient cities · Bibliometric analysis · Literature review

Z. Liang · L. Feng (✉)
School of Art, Design and Media, East China University of Science and Technology, Shanghai, China
e-mail: candy1702023@163.com

J. Zhang et al. (eds.), *Environmental Pollution Governance and Ecological Remediation Technology*, Environmental Science and Engineering,
https://doi.org/10.1007/978-3-031-25284-6_77

77.1 Introduction

77.1.1 Background

The frequent occurrence of catastrophic events around the world nowadays maps the lack of urban emergency response capacity. Disasters have great negative effects on human survival, social stability and social economy (Minxing et al. 2016). Such as global warming, virus invasion, extreme weather occurrences (drought, high temperature, extreme precipitation, strong typhoon, etc.), and major natural disasters (earthquake, tsunami, etc.) Traditional passive disaster mitigation strategies can no longer meet the requirements of sustainable urban development and resilient cities (Xinyu and Highest 2019). Resilient cities refer to cities that are able to withstand disasters and mitigate the losses incurred in disasters while quickly recovering to their pre-disaster state, and able to learn from past disasters to improve the resilience (Yiwen and Jiang 2015).

77.1.2 Significance

Resilient landscapes are important adjustable green spaces in urban environment. they play an important role in alleviating urban flood disasters, restoring urban habitats, mitigating heat island effects, responding to sudden health crises, and reducing social pressure (Barker and Stockman 2018). Resilient landscapes are the "skeleton" of cities in coping with natural and social risks.

On this basis, this paper chooses to take the research literature related to resilient landscape as the database and uses the data visualization software CiteSpace as the research tool to conduct an bibliometric analysis. The database includes in the core journals of China Knowledge Network (CNKI) and doctoral dissertations. Then this paper draws a series of knowledge maps in order to grasp the dynamics of resilient landscape research more intuitively and accurately.

77.2 Research Methodology and Data Sources

77.2.1 Research Methodology

CiteSpace, as an analytical tool for data visualization, has been gradually used by scholars at home and abroad for hotspot research and trend analysis in various fields. For this software, data visualization analysis can be conducted through the keywords, authors, institutions, and other information of the literature. The software can also help construct the distribution law of the literature to discover the research situation of a certain discipline or related fields in a specific period of time. In this study, the author organized the retrieved literature data and imported them into CiteSpace 5.8 R3 to draw keyword co-occurrence map, author co-occurrence network map, keyword emergence map, etc. through the parameter setting so as to carry out the analysis.

77.2.2 Data Sources

All the data in this study were obtained from the China Knowledge Network (CNKI) database. The key words of the searched subject were "resilience" (or "elasticity"), "landscape" (or "park", "green space", "green infrastructure", "ecological infrastructure", etc.). The retrieval time is June 10, 2022, as a result, a total of 246 journal articles, and 136 doctoral dissertations were obtained from the period from 1996 to 2022. Then the results were exported to CiteSpace software in Refworks format for processing.

77.3 Analysis of Research Status

Bibliometric visualization analysis reveals that since 2000, the overall research literature on resilient landscape in China has shown an upward trend, the overall distribution of the literature is widely distributed across disciplines. Also, more and more scholars and institutions are devoting themselves to related research fields.

77.3.1 Trends in the Number of Studies

The annual distribution and changing trend of the number of research literature can reflect the importance of the research field and the degree of attention it receives. In terms of the number of publications and trends, the total number of literature in this field has been increasing. Although it started late, it has developed rapidly in the past 20 years, and the number of papers rose significantly after 2010 and maintained a steady growth within a decade. After the national strategy of resilient cities was proposed, a new wave of significant growth was ushered in 2020. Generally speaking, with the development of urbanization, domestic scholars have started to think about future development and improvement of landscape from the perspective of resilience.

77.3.2 Distribution of Research Subject Areas

From the retrieved literature, it can be seen that the research discipline with the highest number of publications is building science and engineering, indicating that this discipline occupies an important position in domestic research in this field. Then followed by environmental science and resource utilization. Domestic research disciplines are widely distributed but mostly concentrated in the natural sciences and engineering sciences, lacking a cross-collaboration with the humanities and the social sciences.

77.3.3 Analysis of Authors and Institutions

Through the analysis of the author's profile (Fig. 77.1) and the institution's profile (Fig. 77.2), it is found that, from the cooperation network of the authors, there are sparse links in general; the cooperation among institutions in the research field of resilient landscape is not obvious, and the institutions are mainly colleges of architecture, urban planning and landscape architecture.

From Fig. 77.2, we can see that the College of Design and Architecture of Zhejiang University of Technology has issued the most articles, followed by the College of Landscape Architecture of Beijing Forestry University and the College of Architecture and Urban Planning of Tongji University. From the overall viewpoint, there is little cooperation among research institutions in this field in China; basically, they are all independent individuals, and there is little cross-research. In the future, we should

Fig. 77.1 Author collaboration network map

Fig. 77.2 Institutional cooperation network map

strengthen academic exchanges and cooperation, break through geographical and institutional restrictions, and build cross-regional and interdisciplinary collaborative research mechanisms and platforms.

77.4 Research Hotspots and Frontier Analysis

77.4.1 Analysis of Research Hotspots

The keywords is usually an accurate expression of the subject and content of literature and is a very important part of academic research. Keyword co-occurrence analysis can reflect the hotness and trend of research in a certain field and its inner connection. By setting the network node as "keyword" and the threshold value (5, 6, 200), a keyword network map (Fig. 77.3) and a keyword frequency table (Table 77.1) are obtained. From the figure, we can see that there are 325 nodes and 493 lines, and the high-frequency keywords are "resilience", "landscape architecture", "planning research", "urban resilience", etc. As can be concluded from Fig. 77.3 and in conjunction with the relevant literature, the main research hotspots in this area focus on resilient facility planning, landscape resilience system building, climate adaptation and ecosystem services, etc.

Based on the keyword co-occurrence network map, a further cluster analysis map of keywords was conducted (Fig. 77.4), and the following eight groups of major cluster labels were obtained: landscape gardening, resilience, urban resilience, urban design, land use, ecological safety, seismic events, and ecological restoration.

Through the keyword co-occurrence network anayzing and the relevant literature summarizing, it can be concluded that resilient landscape research mainly includes 3 aspects. (1) Based on the planning and design of urban resilience-related facilities,

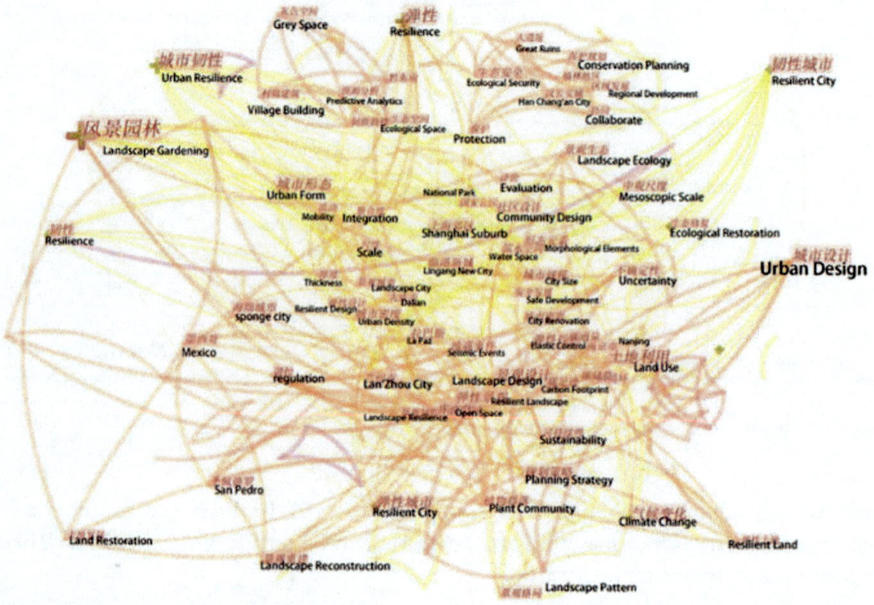

Fig. 77.3 Keyword co-occurrence map

Table 77.1 Keyword frequency table

Serial number	Keywords	Frequency
1	Resilience	34
2	Landscape architecture	25
3	Planning studies	21
4	Urban resilience	11
5	Resilience	11
6	Climate change	9
7	Urban design	8
8	Land use	7
9	Resilient cities	6
10	Elastic cities	6

resilient landscape planning and construction is an important part of the construction of resilient cities. Such as, Zhang (2020) and others based on the concept of resilience, focusing on the improvement of community green infrastructure functions; Zhao (2021) and others take the U.S. city of Norfolk as an example and introduce a resilience-oriented green infrastructure planning approach with risk response mechanism as the main line. (2) Landscape planning and construction strategies based on ecological safety and restoration, the current domestic development environment is becoming increasingly complex, and ecological safety and ecological restoration

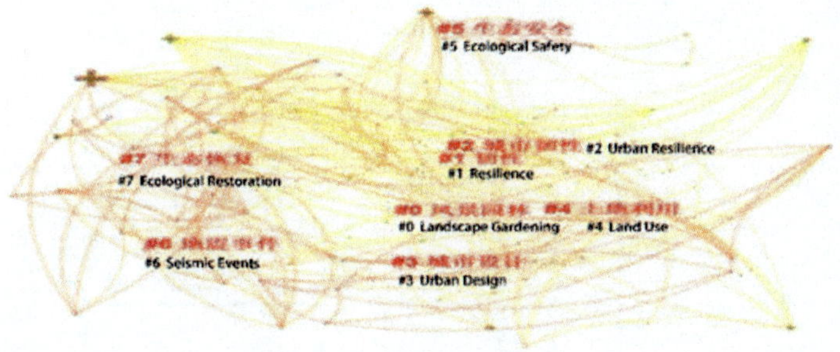

Fig. 77.4 Keyword clustering map

are gradually attracting attention as well as discussion. For example, Yang (2022) and others summarized the research and practical achievements in ecological restoration research, planning and construction of waterfront space in 2021. And on this basis, proposed that the research on this aspect should be further strengthened in future urban planning and construction research. Liu (2020) and others chose to target the construction of dynamic ecological security pattern. On the basis of establishing a model for predicting urban spatial development and evaluating the security of ecological space pattern, the regional ecological space resilient planning approach of "dynamic multi-scenario simulation prediction—ecological space security evaluation—program coordination resilient response" is summarized. Li (2017) discusses the new ideas of urban ecological restoration from two aspects: the path of urban ecological restoration and the value of restoring urban ecological functions, in order to provide a reference for related planning. (3) Reflecting on the resilience in the face of unexpected events such as coronavirus pandemic in 2020. Baoxing (2018) believes that the new sources of uncertainty in cities can be summarized as four aspects, such as extreme climate, technological revolution, coronavirus, sudden attacks, etc. In addition to the pandemic, extreme weather events are also having a significant impact on landscape infrastructure. Rui (2020) believes that resilient design is a method of landscape and urban design in the face of the flood; Yuhong (2022) and others proposed a framework for building a flood resilient city system by promoting the integration of blue-green-grey infrastructure based on the concept of sponge cities, and emphasized the establishment of risk adaptation, risk prevention, and disaster emergency implementation strategies and measures for rain flood resilient cities, so as to enhance the resilience of the whole city in the face of emergencies.

77.4.2 Research Frontier and Trend Analysis

Since keywords are important for literature analysis, they are also used to draw the keyword timeline maps (Fig. 77.5) and emergency word analysis maps (Fig. 77.6), in addition to their concurrent network and cluster analysis maps.

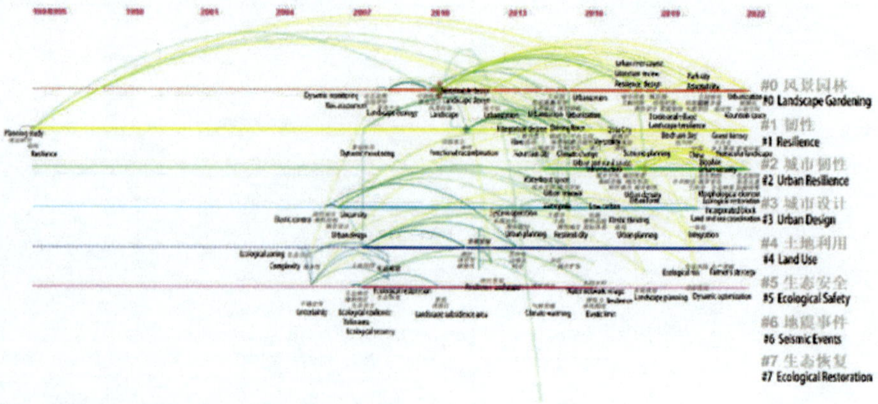

Fig. 77.5 Keyword timeline chart

Top 18 Institutions with the Strongest Citation Bursts

Institutions		Year	Strength	Begin	End	1994 - 2022
规划研究	Planning Research	1994	8.11	**1994**	2009	
韧性	Resilience	1994	6.07	**1994**	2009	
不确定性	Uncertainty	1994	1.12	**2005**	2008	
复杂性	Complexity	1994	1.12	**2005**	2008	
生态安全	Ecological Security	1994	1.76	**2007**	2010	
景观生态	Landscape Ecology	1994	1.22	**2008**	2010	
健康评价	Health Evaluation	1994	1.13	**2010**	2012	
评价	Evaluation	1994	1.78	**2011**	2013	
弹性景观	Resilient Landscape	1994	1.19	**2012**	2013	
弹性	Resilience	1994	3.99	**2014**	2017	
城市设计	Urban Design	1994	1.73	**2015**	2017	
规划策略	Planning Strategy	1994	1.34	**2015**	2019	
多尺度	Multiscale	1994	1.2	**2015**	2018	
城市	City	1994	1.2	**2017**	2020	
风景园林	Landscape Gardening	1994	5.03	**2018**	2022	
城市韧性	Urban Resilience	1994	3.31	**2018**	2022	
城市形态	Urban Form	1994	1.18	**2018**	2022	
气候变化	Climate Change	1994	1.87	**2020**	2022	

Fig. 77.6 Emergent word analysis chart

The keyword timeline diagram (Fig. 77.5) can be combined with the years, which can more intuitively see the research dynamics and hotspots in each year and time period; from the emergent word diagram, we can clearly see when each hot keyword appears and when it ends, so as to judge the duration of the hotspot. Combining the literature and the figure below, the study of resilient landscape can be divided into the following phases: (1) The first phase, from 1994 to 2010, the research of resilient landscape was relatively few and unconcentrated. Resiliency was an unclear concept with a greater use as an indicator in various evaluations to pursue landscape ecological safety. For example, Wang (2005) and others select the Dong village buildings in Qiandongnan, Guizhou Province, as a base to conduct spatial prediction analysis of the built ecological environment with the resiliency theory. (2) The second stage is from 2011 to 2017, which is a period of rapid development. And compared with the previous stage, the research outcomes in this stage are relatively more, and the focus of the research is also shifted from the theoretical discussion to the practical application, mainly around the evaluation of landscape ecosystem as well as resilient landscape design and urban design. Jun (2016) proposed how to cope with climate change under the perspective of resilience and initially explored the resilient organizational structure of cities in response to climate change; Yu (2015) and others carried out the resilient design of Yanweizhou Park in Jinhua, which is one of the important cases representing resilient landscape design; Du (2014) and others proposed the idea of using the concept of resilience to guide the research and practice of rural reshaping in China in the face of the decline of the countryside. They considered that the concept of resilience is an open loop system consisting of "thinking resilience—technical resilience—management resilience". (3) The third stage is from 2018 to 2022, and the research outcomes in this stage are much improved in quantity and quality. In terms of quality, the literature in this period includes the design of resilient landscape, the construction of landscape resilience systems, the assessment of ecological resilience, the construction of landscape infrastructure, etc. The research ranged from landscape design to the construction of health evaluation systems, which also indicates that the research is deepening, and scholars are no longer limited to single-dimensional assessment and concept definition, but extend it to a deeper level such as multi-dimensional assessment and system construction. For example, Wang (2022) and others proposed a planning strategy of the countryside water network in southern Jiangsu based on the characteristics of resilience on the basis of ecological risk evaluation; Wenbin (2022) and others took Moganshan as an example, and supported the construction of the resilient landscape system based on the comprehensive ecological security pattern; Dai (2021) and others discussed the construction of a new strategy and method of building a new water city model. From the analysis of keywords, it can also be intuitively seen that due to the recent impact of climate change, many scholars have taken this as an entry point and conducted corresponding research on this. For example, Yu (2021) believes that creating a green infrastructure that is adaptable and resilient to climate change has become a major challenge and opportunity for landscape architects. Deng (2015) argues that the negative impacts of climate change on public health have inspired localized design practices that emphasize program flexibility; Wangxin (2021) and others argued that blue-green infrastructure is extremely important for coping with future climate change, regulating urban climate and enhance urban resilience, and analyzed the cooling effect of infrastructure from the perspective of the landscape.

Through the analysis of emergent words and the summary of related literature, the author believes that the future research trend will shift the focus from planning and the assessment of resilience to the construction of a scientific management mode and the construction of a complete evaluation system. Secondly, the research will not only limited to urban landscape design, but will also focuses on the rural landscape, through which the research can help the resilience of rural area further improved. Also, the further transformation of research methods will continue to progress, strengthening theoretical construction while focusing on the application of quantitative methods in order to improve the scientific, normative, and reliability of the research. At present, it is mostly a combination of quantitative and qualitative research, with survey data as the main focus and less use of big data and other methods, so we can focus more on the change of research methods in the future. Finally, with the help of technological means, and the use of ecological wisdom, we can explore the optimization of landscape conditions, improve the accuracy of landscape evaluation.

77.5 Conclusion

This study used CiteSpace 5.8 R3 to summarize the hot topics of relevant research on resilient landscape from author and institution analysis, keyword co-occurrence. Thenthe author proposed frontier trends at each stage between 1994 and 2022 based on keyword timeline maps and emergent maps.

77.5.1 Problems

China's resilient landscape research started late. As to the research field, the research is still mainly concentrated in urban planning, landscape ecology, and environmental science, mostly dominated by natural science and engineering disciplines, humanities are rarely involved. And the overall research direction is onefold, as a result, the quantitative and modeling research are lack of technical support from relevant disciplines. As a composite carrier of functions, landscape involves all aspects of social development and has not only ecological value but also economic and social value. In addition to relatively more abundant research on hydrological and climate regulation, there is less research on other ecological aspects of landscape. Secondly, through the analysis of the distribution of scholars and institutions, there is less communication and cooperation between scholars and institutions, the overall academic cooperation still need to be developed. In addition, whether the research method is scientifically applicable is an important way to measure the research maturity (Zhenming 2010). At this stage, China's resilient landscape research is mostly based on macro countermeasures and theoretical studies, and the conclusions are mostly universal and untargeted. And the research is mostly based on static evaluation, and there is a lack of research on the dynamic process and action mechanism of resilient landscape. And there is a lack of sufficient data and case support. Finally, through the keywords analysis in the previous article, it is easy to find that resilient landscape related design is mostly developed in urban spaces and less for rural areas.

77.5.2 Research Suggestions and Future Prospects

First, the exchange and cooperation of domestic research scholars and the cooperation between institutions is insufficient, which is not conducive to the future research development of China's resilient cities. There should be more cooperation which will help to improve the overall level of research. Second, expand the research field and strengthen interdisciplinary cooperation (Yanping and Jixin 2021). The research is mostly concentrated in engineering science and natural science, mainly distributed in the disciplines of landscape architecture and urban planning, and lacks cross-research cooperation with humanities and social science fields, especially the combination with economics and sociology. With the increasing demand of residents for a human living environment, public health has received more attention, so scholars should extend their research about resilient landscape to human physiological and psychological health aspects. Third, the research focus should be extended from planning and design to quantitative analysis, the construction of scientific management models and evaluation systems. By making full use of big data, GIS, and other quantitative research means, we can provide the scientific basis for its planning and construction, and continuously improve the evaluation system of landscape infrastructure. Also, combined with the different developing level of regions in China, to explore a suitable management model. In the later stage of construction, the legal and restraint mechanism should be gradually improved to play a protective role in the management model. Fourth, strengthen the attention to the rural landscape resilience study, because the city infrastructure is more complete than the countryside. So for now, the focus of research is basically locked in the urban aspect. In the subsequent study, we should focus more on the countryside, which is also an essential carrier of resilient landscape development.

China's resilient landscape research started late but is developing rapidly and has a vast development space. Using CiteSpace to visualize and analyze the data of domestic research of resilient landscape can provide a scientific basis and theoretical support for the future study.

Acknowledgements This work is supported by Shanghai Philosophy and Social Science Planning Project (2021ECK002).

References

Baoxing Q (2018) Methods and principles of resilient city design based on complex adaptive systems theory. Urban Develop Res 25(10):1–3

Barker K, Stockman A (2018) Resilient design: reconnecting people and the environment. Lands Arch 6(04):14–31

Bo Y, Tian C, Xiaoyan L, Yangli L, Yanjun L, Yu L, Lei Z, Zhen H, Yu D, Baojun L, Jiguang G (2022) Review of hotspots in waterfront ecological restoration research and planning and construction in 2021. Sci Technol Rev 40(01):204–214

Deng X, Averica JK (2015) Resilient LeChing: climate change, public health and landscape infrastructure. Lands Arch 3(01):70–83

Jun Z (2016) Resilience as an organizational structure for cities to cope with climate change: an example of the winning proposal of the post-Hurricane Sandy reconstruction competition in the United States. Urban Plan 40(08):9–15

Jun L (2017) Ecological node governance and ecological element restoration in urban ecological restoration. Planner 33(S2):10–14

Kongjian Y (2021) Climate adaptation and resilience. Lands Arch (in English and Chinese) 9(06):5–7+4

Kongjian Y, Hongqian SY, Shuiming Z (2015) Resilient landscape—design of Yanweizhou Park in Jinhua. J Arch 04:68–70

Minxing Y, Bo H, Chong C, Zuopeng X (2016) Review and prospect of disaster prevention and control based on resilient city theory. Urban Plan J 01:48–55

Pengfei D, Zeyu S, Jun Z (2021) A preliminary investigation of water urban model based on landscape infrastructure: the case of Shengze, Wujiang, Suzhou. Urban Plan 45(08):96–106

Qinyuan W, Jinhua D (2022) Resilience planning of water network villages based on landscape ecological risk assessment–an example of the Changbaidang area. Southern Arch 05:10–17

Quan Z, Xiaoxue P, Dongmei B (2020) Research on the function improvement strategy of community green infrastructure under the concept of resilience. Constr Econ 41(S1):262–265

Rui N, Yankun S, Lijie W (2020) Resilient design—a proactive approach to building, landscape and urban design in response to flooding—A review of Flood-oriented design: architecture, landscape and urban design for rapid recovery from flooding and climate change. Ind Arch 50(11):214

Song L, Lei L (2020) Research on regional ecological spatial resilience planning based on ecological security—Tengzhou City, Shandong Province as an example. China Garden 36(02):11–16

Wang H, Liu JD, Li J (2005) Spatial prediction analysis of built ecological environment using elasticity theory: an example of Dong village architecture in Qiandongnan, Guizhou Province. Urban Plan 12:93–96

Wangxin S, Qing C, Xiao L, Liukuan Z (2021) A review of the cooling effect of urban blue-green infrastructure. J Ecol 41(07):2902–2917

Wenbin N, Yan S, Fan Y, Xingu N, Zhiyi B (2022) Research on the construction of mountain landscape resilience system based on comprehensive ecological security pattern: the case of Moganshan, Zhejiang. China Garden 38(02):127–132

Wenwu D, Jianlin Z, Cong T (2014) The concept of resilience, thinking about landscape gardening in the reshaping of the countryside. China Garden 30(10):102–106

Xinyu Z, Highest W (2019) The conceptual evolution, research content and development trend of urban resilience. Sci Technol Herald 37(22):94–104

Yanping L, Jixin H (2021) Current status of domestic and international research on facility resilience assessment and future research directions. Dis Sci 36(3):153–159

Yiwen S, Jiang X (2015) Urban Resilience: concept analysis based on international literature review. Int Urban Plan 30(02):48–54

Yuhong L, Liang Z, Li L, Zheyan S, Jing Z (2022) Promoting sponge city construction system to enhance urban rainwater resilience. Western J Habitat 37(01):22–26

Zhao J, Xu X, Tang M (2021) Resilience-oriented research on "norfolk city green infrastructure planning. Int Urban Plan 36(04):148–153

Zhenming C (2010) The disciplinary orientation and knowledge growth of public management. Admin Forum 17(4):17–20

Chapter 78
Utilizing Pyrolytic Char from Waste Drill Cuttings in Shale Gas Industry as a Potential Adsorbent for Heavy Metal Removal

Huan Ye, Qianxia Xu, Yongqiang Liu, Bin Luo, Guihong Lan, Bo Xu, and Haiyan Qiu

Abstract The management of hazardous waste drilling cuttings waste (OBDC) from the shale gas industry is challenging. In this study, OBDC from a shale gas block was pyrolyzed at different temperatures such as 300, 400, 500, and 600 °C to generate activated carbon, which was used to adsorb Cu(II), Zn(II), Pb(II) and Cr (III) in aqueous solution. It was found that the adsorption capacity of OBDC char reached the highest at pyrolysis temperature of 500 °C for 30 min. The pseudo-second-order-kinetics and Langmuir model describe the heavy metal adsorption process well. According to Langmuir isotherm calculation, the maximum adsorption capacities of OBDC-500 for Cu(II), Zn(II), Pb(II) and Cr (III) at 500 °C are 83.2 mg/g, 38.29 mg/g, 143.95 mg/g and 104.1 mg/g, respectively. OBDC-500 has a simple and environmentally friendly preparation process and excellent adsorption performance. It has great potential for heavy metal removal in the aspect of water treatment.

Keywords Oil-based drill cuttings · Pyrolysis temperature · Heavy metals · Adsorption characteristic

---

H. Ye · G. Lan (✉) · B. Xu · H. Qiu
College of Chemistry and Chemical Engineering, Southwest Petroleum University, Chengdu 610500, P.R. China
e-mail: guihonglan416@sina.com

Q. Xu
Sichuan Lejian Dreamer Technology Co., Ltd., Chengdu 610000, China

Y. Liu
Faculty of Engineering and Physical Sciences, University of Southampton, Southampton SO17 1BJ, UK

B. Luo
Sichuan Academy of Eco-Environmental Sciences, Chengdu 610000, China

J. Zhang et al. (eds.), *Environmental Pollution Governance and Ecological Remediation Technology*, Environmental Science and Engineering,
https://doi.org/10.1007/978-3-031-25284-6_78

78.1 Introduction

Copper, zinc, lead and chromium are recognized as the four most toxic metals in common heavy metal wastewater. In order to prevent such wastewater from polluting natural water bodies, these heavy metal ions must be removed before wastewater discharge (Elmac et al. 2009). Heavy metal removal has been intensively studies (Ihsanullah et al. 2016). Among them, the adsorption method has become the most widely used wastewater treatment technology due to its good performance and easy operation. The high cost of adsorbent, however, impedes its wide application (Ali 2012). One of the potential approaches is to utilize natural or industrial waste as adsorbents, which could not only significantly reduce adsorbent cost, but also convert wastes into resources to deal with both problems from waste management and resource depletion.

Drill cuttings are considered as hazardous waste, which is very complex and difficult to be managed. Pyrolysis is a mature technology, which is one of the most effective and relatively cost-effective ways to make OBDC harmless. In addition, it was found that OBDC char has fine and porous structure with significant amount of hydroxyl on the surface, which shows similar compositions and structure properties to those of red mud (Wu et al. 2017) and steel powder in steel works (Bouabidi et al. 2018). Both red mud and steel powder are potential adsorbents for heavy metal adsorption, indicating that OBDC char has similar adsorption capacity.

The specific purposes of this study are as follows: (1) investigating the relationship between the pyrolysis temperature and the adsorption performance of OBDC char; (2) evaluating the effect of adsorption conditions on the adsorption of OBDC coke on single metals system; (3) studying the isothermal and kinetic models of metal adsorption by OBDC char.

78.2 Materials and Methods

78.2.1 Preparation of OBDC Char

OBDC from Shale Gas Blocks (Weiyuan, Sichuan, China) was pyrolyzed and the time was 30 min in a tube furnace. The effects of different pyrolysis temperatures on the adsorption performance of OBDC char were studied.

78.2.2 Characterization of OBDC Char

The morphology and elemental analysis of OBDC were performed by scanning electron microscopy (SEM, ZEISS EVO MA15, US). The crystal structure of OBDC char was detected by diffractometer (XRD, X'Pert PRO, Netherlands). The surface area (BET) analysis was measured by Brunauer-Elmet-Teller nitrogen adsorption (QuadraSorb Station 4, US). Through infrared spectroscopy (FTIR, WQF-520 FT-IR, China) analysis, understand the change of material structure after pyrolysis. The contact Angle of the substance was determined by the contact Angle tester (Dataphysics OCA20, Germany).

78.2.3 Heavy Metal Adsorption Experiments

In each adsorption experiment, 1 g/Ladsorbent was mixed with 50 mL of Cu^{2+}, Zn^{2+}, Pb^{2+} and Cr^{3+}solutions with different initial concentrations such as 30, 60, 100, 260 (mg/L), respectively, in 100 mL Erlenmeyer flasks. The change of adsorption capacity of OBDC char was studied by adjust solution pH to 2, 3, 4, 5 and compared with the unadjusted pH. During the whole experiment, the stirring speed was 200 rpm and the stirring time was 24 h. The adsorption capacity was calculated as follows:

$$q_e = \frac{(C_0 - C_e) \times V}{M} \tag{78.1}$$

where q_e is the adsorption capacity of OBDC (mg/g), C_0 and C_e are the unresponsive and post-adsorption concentrations (mg/L), V is the wastewater volume (L), and M(g) is the amount of OBDC char.

78.2.4 Adsorption Kinetics

In the study of adsorption kinetics, four kinds of common adsorption model fitting analysis were used to see which model could best represent OBDC char. The pseudo-first order [Eq. (78.2)] is the most common one in kinetic studies (Moghaddam et al. 2013). The pseudo-second-order [Eq. (78.3)] proposed by Makey (2000) corresponds to most adsorption process.

$$\ln(q_e - q_t) = \ln q_e - k_1 \times t \tag{78.2}$$

$$\frac{t}{q_t} = \frac{1}{k_2 \times q_e^2} + \frac{t}{q_e} \tag{78.3}$$

where qt is the adsorption capacity with reaction time t (min), k_1 is rate constant in pseudo-first-order; k_2 is rate constant in pseudo-second-order.

78.2.5 Adsorption Isotherms

The Langmuir isotherm [Eq. (78.6)] quantitatively describes the adsorption of monolayer adsorbent on the adsorbent surface, resulting in no further adsorption (Azizian et al. 2018). Another model commonly used to describe single-solute systems is the Freundlich isotherm [Eq. (78.7)].

$$q_e = \frac{q_m k_L C_e}{1 + k_L C_e} \tag{78.6}$$

$$q_e = k_F C_e^{\frac{1}{n}} \tag{78.7}$$

where k_L is the rate constants in Langmuir equations, k_F is the rate constants in Freundlich, q_m (mg/g) is maximum adsorption capacity at equilibrium. 1/n is to represent the bond distribution.

78.3 Results and Discussion

78.3.1 Selection of Pyrolysis Temperature

From Fig. 78.1, it can be seen that OBDC chars produced from different pyrolysis temperature demonstrated different heavy metal adsorption capacity. Generally, the adsorption capacity of OBDC char produced from higher temperature and longer pyrolysis time is higher. When the pyrolysis temperature is above 500 °C, the increase in the adsorption capacity is not obvious. According to this study, the adsorption capacity of OBDC char generated at 500 °C pyrolysis temperature is the largest, these pyrolysis conditions were used for further study.

78.3.2 Characterization of OBDC and OBDC Char

Table 78.1 listed the surface properties of OBDC and OBDC char from pyrolysis at different temperatures. It can be seen that pyrolysis changed the properties of material and pyrolysis temperature caused significant difference of materials properties. Meanwhile, the pore size decreased, resulting in an increase in the S_{BET} from 5.699 to 8.435 m^2/g. When the temperature continues to rise to 600 °C, S_{BET} only slightly increases.

As shown in Fig. 78.2a that the main crystals in the raw OBDC are calcite, quartz and hematite. According to Chaoqiang (2017), the char of OBDC from shale gas exploration mainly contains SiO_2, Al_2O_3, CaO, SO_3, MgO, and Fe_2O_3. Meanwhile, it shows highly stable amorphous structure. The studies by Wu (2017) on red mud

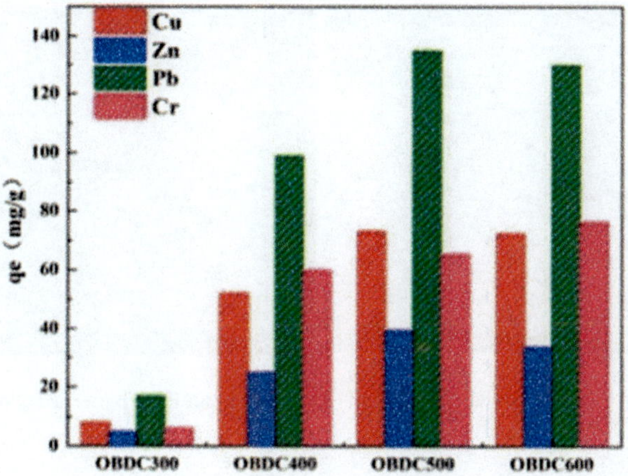

Fig. 78.1 Adsorption capacity at different pyrolysis temperatures

Table 78.1 Comparison of surface properties of OBDC and OBDC char

Adsorbent	OBDC	OBDC300	OBDC400	OBDC500	OBDC600
S_{BET} (m^2/g)	5.075	5.699	6.856	8.435	8.188
Micropore volume (cm^3/g)	0.033	0.038	0.039	0.041	0.040
Pore size (nm)	26.14	24.54	21.74	19.55	18.78

showed similar results. Therefore, like other types of char from different wastes, OBDC char has a strong potential to adsorb heavy metals. It was found from XRD spectra shown in Fig. 78.2a that When the temperature exceeds 500 °C, the diffraction peak of kaolinite ($2\theta = 12.28°$) disappears, indicating that kaolinite has been transformed into metaolinite. Metakaolin was further decomposed to aluminum silicate ($2\theta = 22.491°$).

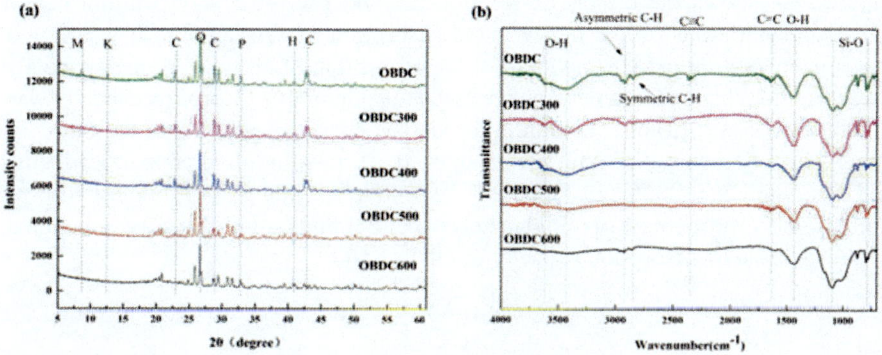

Fig. 78.2 XRD (**a**): C—Calcite; Q—Quartz; M—Muscovite; K—Kaolinite; P—Pyrite. FT-IR (**b**)

Fig. 78.3 SEM of OBDC (**a**) and OBDC-500 (**b**) samples and EDS patterns (**c**) of the raw OBDC

As seen from the infrared image in Fig. 78.2b, symmetric and asymmetric C–H bonds, carbon–carbon double and triple bonds in OBDC char decreased with the increase in pyrolysis temperature. At temperature above 500 °C, C=C, C ≡ C peak almost disappeared, leaving only the O–H bond and Si–O. This indicates that hydrocarbons are completely removed from OBDC char. The absorption peaks appeared around 1465 cm^{-1}, indicating the presence of short-chain –OH and –COOH. Therefore, OBDC char from pyrolysis temperature 500 °C and above without the concern of hydrocarbon leaching, is suitable as adsorbent for heavy metal adoption.

It can be seen in Fig. 78.3 that OBDC-500 has a more rough surface with small particles randomly accumulated while the raw OBDC's surface is more smooth with a lamellar structure. Figure 78.3b shows the appearance of mineral particles becomes clearly visible and well-defined, indicating that pyrolysis has decomposed the base oil and reduced the oil content in the OBDC.

78.3.3 The Adsorption of Heavy Metals by OBDC-500

Figure 78.4 shows the effects of solution pH and Fig. 78.5 shows initial heavy metal concentrations on the heavy metal adsorption by OBDC-500. It was found that higher solution pH is beneficial for the adsorption of all types of heavy metals studied although for each type of heavy metal the maximum adsorption capacity was different. At lower pHs, the alkalinity of OBDC char was neutralized, resulting in lower adsorption capacity. With pH unadjusted, pH of solution of copper, lead, zinc and chromium was 5.3, 5.8, 5.1, and 5.5, respectively, leading to the highest adsorption capacity. The increase of initial metal concentration in solution was accompanied by the increase of adsorption capacity of OBDC-500.

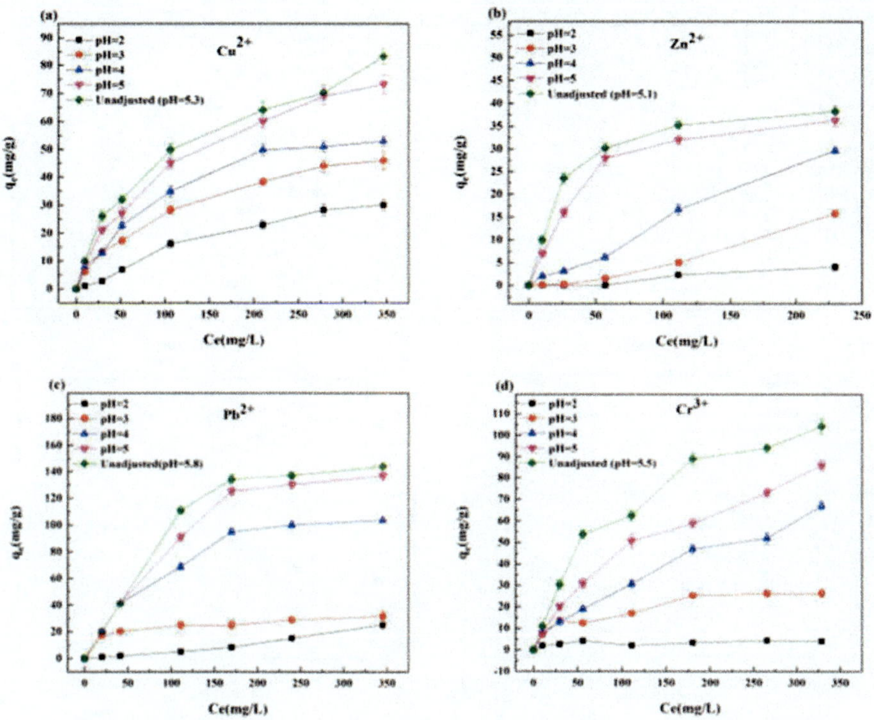

Fig. 78.4 Effect of pH value and concentration on the adsorption capacities by OBDC 500

78.3.4 Adsorption Kinetics of Heavy Metals by OBDC-500 Char

Contact times correspond to the effect of adsorption of copper, zinc, lead and chromium (at optimum pH), respectively, was studied (Fig. 78.5). In the first 2 h, the metal adsorption rate is very fast. From 2 to 10 h, the adsorption continued with a reduced rate and after 10 h, the adsorption reached equilibrium. The adsorption kinetic curves are presented in Fig. 78.6.

As shown in Fig. 78.6b and Table 78.2, it is the relation curve drawn between t/qt and time according to the linear pseudo-second-order model. Found that the size of the adsorption capacity of the model is little difference between OBDC-500 and experimental values, and R^2 is above 0.99, indicating that the adsorption kinetics of OBDC-500 on metal conforms to the second-order model. means that the adsorption type of OBDC-500 to 4 heavy metal ions is mainly chemical adsorption.

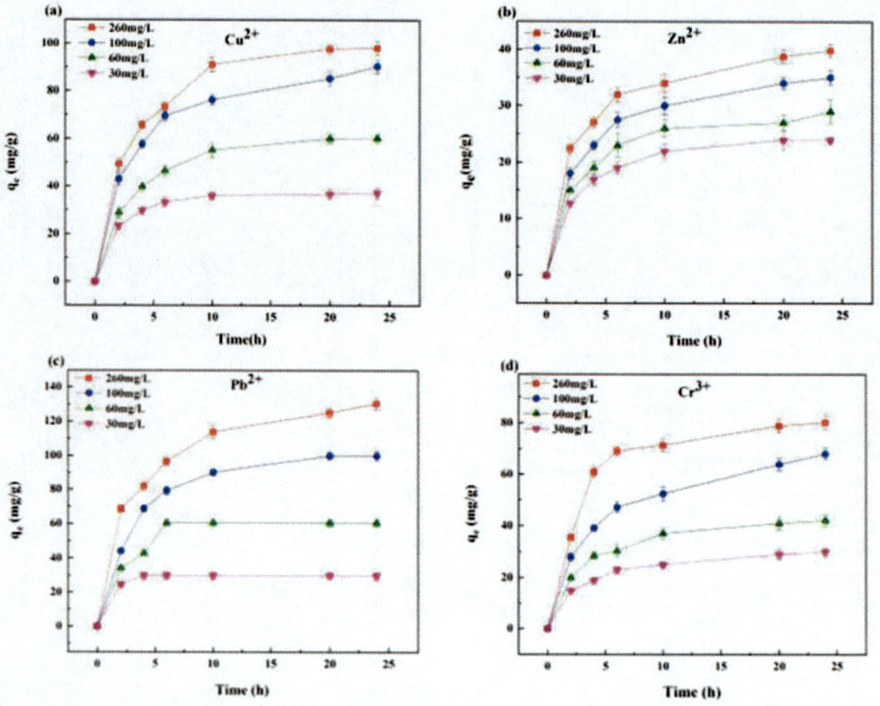

Fig. 78.5 Effect of adsorption time on adsorption capacities of OBDC char at optimum pH

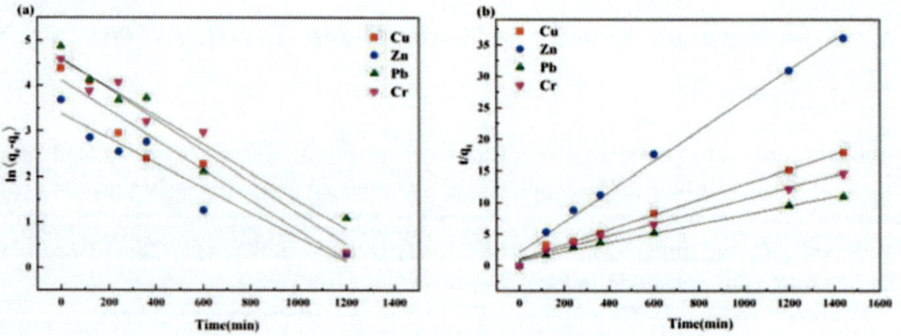

Fig. 78.6 The fitted curves of First-order kinetics (**a**), second-order kinetics (**b**) of OBDC-500 adsorbed heavy metals. The test conditions: concentration are 260 mg/L

Table 78.2 kinetic parameters for copper, zinc, lead and chromium

Metal	Pseudo-first order			Pseudo-second order		
	R^2	k_1	q_e	R^2	k_2	q_e
Cu	0.9383	2.90E−03	51.65	0.9922	1.39E−04	84.03
Zn	0.9247	2.40E−03	25.61	0.9932	2.77E−04	41.49
Pb	0.9381	3.00E−03	105.45	0.9903	6.95E−05	136.99
Cr	0.9639	3.60E−03	78.23	0.9913	9.72E−05	104.17

Table 78.3 Langmuir and Freundlich isotherm parameters of OBDC-500

Metal	Langmuir			Freundlich		
	R^2	k_L	q_m	R^2	k_F	1/n
Cu	0.9853	0.1582	83.20	0.8942	28.98	0.2036
Zn	0.9841	0.1508	38.29	0.9792	18.27	0.2471
Pb	0.9674	0.2475	143.95	0.8407	48.83	0.1472
Cr	0.9836	0.4298	104.10	0.8026	52.00	0.1488

78.3.5 *Adsorption Isotherms of Heavy Metals by OBDC-500 Char*

Langmuir and Freundlich model parameters for 4 heavy metal adsorption are shown in Table 78.3. R^2 proved that the Langmuir model was more consistent than the Freundlich model for all four metals, suggesting a strong surface interaction between heavy metals and OBDC-500. The Langmuir model represents that the adsorption process is homogeneous adsorption by chemisorption (Li et al. 2019). Moreover, Freundlich parameter (1/n) for metal adsorbed was all below 1.The smaller 1/n in Freundlich model means an easier adsorption (Fig. 78.7).

78.4 Conclusions

Hazardous waste OBDC can be converted into non-hazardous materials by pyrolysis for application as adsorbents. Pyrolysis temperature plays an important role to determine the physical structure, physical properties, chemicals and functional groups of OBDC char, thereby affecting its adsorption capacity for heavy metals. The adsorption capacity of OBDC char from the pyrolysis at 500 °C for 30 min reached 83.2 mg/g for Cu^{2+}, 38.29 mg/g for Zn^{2+}, 143.95 mg/g for Pb^{2+} and 104.1 mg/g for Cr^{3+}. The mechanism study shows that the adsorption process of metal ions by OBDC char conforms to Langmuir model. Further studies on adsorption kinetics show that the

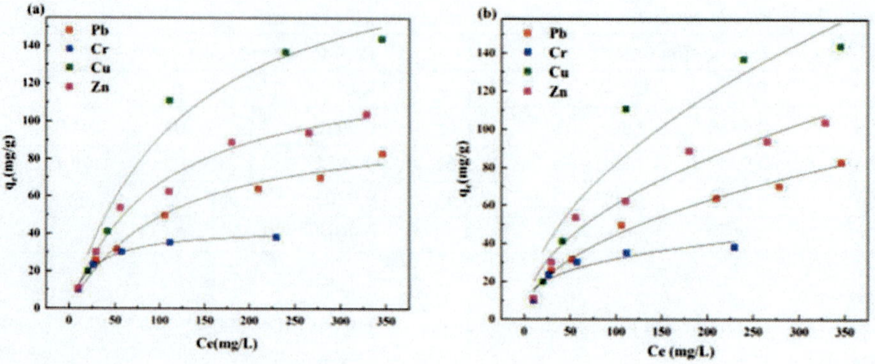

Fig. 78.7 The fitted curves of Langmuir model (**a**) and Freundlich model (**b**) of OBDC-500 adsorbed heavy metals

quasi-second order adsorption model could better represent the adsorption of OBDC-500 on four heavy metal ions. In general, OBDC char could be an cost-effective and efficient alternative to adsorb metal ions from aqueous solution.

Acknowledgements This work was supported by the Open Fund of Oil and Gas Field Applied Chemistry Key Laboratory of Sichuan Province (YQKF202008).

References

Ali I (2012) New generation adsorbents for water treatment. Chem Rev 112(10):5073–5091

Azizian S, Eris S, Wilson LD (2018) Re-evaluation of the century-old Langmuir isotherm for modeling adsorption phenomena in solution. Chem Phys 513:99–104

Bouabidi ZB, El-Naas MH, Cortes D, McKay G (2018) Steel-making dust as a potential adsorbent for the removal of lead (II) from an aqueous solution. Chem Eng J 334:837–844

Elmac A, Zengin N, Yonar T (2009) Removal of chromium(III), copper(II), lead(II) and zinc(II) using Lemna minor L. Fresen Environ Bull 18(5):538–542

Ho YS, McKay G (2000) The kinetics of sorption of divalent metal ions onto sphagnum moss peat. Water Res 34(3):735–742

Ihsanullah N, Abbas A, Al-Amer AM, Laoui T, Al-Marri MJ, Nasser MS, Khraisheh M, Atieh MA (2016) Sep Purif Technol 157:141–161

Li YLY, Taggart MTMA, McKenzie CMC, Zhang ZZZ, Lu YLY, Pap SPS, Gibb SGS (2019) Utilizing low-cost natural waste for the removal of pharmaceuticals from water: Mechanisms, isotherms and kinetics at low concentrations. J Clean Prod 88–97

Moghaddam MR, Fatemi S, Keshtkar A (2013) Adsorption of lead (Pb2+) and uranium cations by brown algae; experimental and thermodynamic modeling. Chem Eng J 231:294–303

Wang C, Lin X, He M, Wang D, Zhang S (2017) Environmental performance, mechanical and microstructure analysis of concrete containing oil-based drilling cuttings pyrolysis residues of shale gas. J Hazard Mater 338:410–427

Wu C, Huang L, Xue S, Huang Y, Hartley W, Cui M, Wong M (2017) Arsenic sorption by red mud-modified biochar produced from rice straw. Environ Sci Pollut R 24(22):18168–18178

Chapter 79
Low-Temperature Synthesis of the β-SiC Particle by Microwave Heating of Solar Waste Silicon Powder

Haifeng Zhao, Dandan Xie, Weiquan Shao, Guoben Zhu, Jun Liu, Aizhen Ma, Baorong Zhang, Ankang Song, Wenwen Song, Zhaoqing Fang, and Lining Han

Abstract The solar waste silicon powder generated by the cutting of crystalline silicon was used as the raw materials, and silicon carbon was synthesized by the microwave heating method. The temperature-raising characteristics of silicon powder by microwave heating and the effects of temperature and holding time on the formation of β-SiC particles were investigated. The formation mechanism of the β-SiC particle was further studied. Results showed that β-SiC particle with diameters of 1–3 μm was formed in the products, and the yield of β-SiC in the products was more than 95% at 800 °C for a holding time of 30 min. The microwave heating decreases the reaction temperature and greatly shortens the reaction time. The local microwave plasma activated by the microwave electromagnetic field promotes the nucleation and growth of β-SiC particles. It realizes the recovery and utilization of industrial solar waste silicon powder, and the microwave heating technology effectively saves energy consumption.

Keywords Solar waste silicon powder · Microwave heating · β-SiC particle · Microwave plasma

79.1 Introduction

Due to the advantages of cleanliness, environmental protection, and abundant resources, the utilization and development of solar energy have attracted more and more attention (Möller 2004; Mathur et al. 2020). Crystalline silicon solar cells play an important role in the photovoltaic industry. During the cutting process of crystalline silicon, nearly half of the ultra-fine silicon powder enters the cutting liquid

H. Zhao (✉) · D. Xie · G. Zhu · J. Liu · A. Ma · B. Zhang · A. Song · W. Song · Z. Fang · L. Han
Qingdao Hiworld New Materials Co., Ltd., 1 Zhengwang Rd., Qingdao, China
e-mail: zhaohiworld@163.com

W. Shao
College of Physics, Qingdao University, 308 Ningxia Rd., Qingdao, China

J. Zhang et al. (eds.), *Environmental Pollution Governance and Ecological Remediation Technology*, Environmental Science and Engineering,
https://doi.org/10.1007/978-3-031-25284-6_79

to form a large amount of cutting waste, resulting in the waste of silicon resources and environmental pollution (Yang et al. 2019; Schwinde et al. 2015). Therefore, the recycling of crystalline silicon cutting waste has become an urgent problem to be solved.

As an important inorganic non-metallic material, β-SiC is widely used in high-temperature structural ceramics, electronics, photocatalysis, and aerospace due to its excellent thermal stability, high thermal conductivity, high mechanical strength and hardness, high electron mobility, wide bandgap, and high-frequency absorbent (Xu et al. 2021; Zeraati et al. 2021; Schneider et al. 2021; Qi et al. 2020; Lin et al. 2022). At present, most of the industrial SiC is manufactured using the Acheson process through the carbon thermal reduction method (Boudard et al. 2014). The drawbacks of this industrial production process lie in high energy consumption, serious pollution, and low added value of products. Recently, a great variety of alternative methods like sol-gel (Kettner et al. 2018), chemical vapor deposition (CVD) (Liu et al. 2022), and plasma (Wang et al. 2022) have been reported in the literature for the synthesis of β-SiC. However, most of the above preparation methods have disadvantages such as low yield, complex synthesis steps, low efficiency, and high cost, which are not suitable for mass industrial production.

Microwave heating utilizes the coupling effect of the microwave electromagnetic field and the material medium. The material itself absorbs microwave energy through the dielectric loss and converts it into heat. This heating mechanism does not depends on heat conduction and has the advantages of rapid, uniform, selective heating and dramatically reduces the reaction time (Bergese 2006; Liao et al. 2019). In recent years, many researchers have successfully synthesized β-SiC by the microwave heating method (Wei et al. 2019; Qin et al. 2022).

In this paper, the preparation of β-SiC particle from the cutting solar waste silicon powder by microwave heating method was studied. The effects of microwave heating temperature and holding time on the formation and the yield of β-SiC were investigated. The generation mechanism of low-temperature synthesis of β-SiC was also explained.

79.2 Experimental

The solar waste crystalline silicon powder (10–30 μm, Qingdao Huike Microelectronics Co., Ltd.) mixed with the cured phenolic resin powder (~150 μm, Jinan Shengquan Group Share-Holding Co., Ltd.) at the molar ratio of 1:1.5 and the mixed powders were pressed into a cylinder sample. Then the sample was heated to the target temperature by microwave heating. Finally, the as-obtained products were calcined in air at 600 °C for 3 h to remove excess carbon.

The phase composition of the product was analyzed by powder X-ray diffraction, and the morphology and size of the product were characterized by scanning electron microscopy. The relative yield of each phase in the product was calculated according to the adiabatic principle method reported by Chung (1974). If there are N phases in a system, the mass fraction of the x phase is (Eq. 79.1):

$$W = \frac{I_x}{K_{\mathrm{A}}^{x} \sum_{i=1}^{N} \frac{I_x}{K_A^x}} \tag{79.1}$$

In this equation, I_x is the integral intensity of the strongest peak of x phase, and K_{A}^{x} is the relative RIR value to the selected standard A phase.

79.3 Results and Discussion

XRD patterns and the yield of each phase of products at different temperatures for 30 min are shown in Fig. 79.1. Diffraction peaks centered at 2θ of 35.65°, 41.40°, 59.98°, 71.77°, and 75.50° corresponding to (111), (200), (220), (311), and (222) planes of cubic β-SiC phase (JCPDS 74–2307) are detected at 600 °C, which is much lower than other microwave syntheses of β-SiC (Wei et al. 2019; Qin et al. 2022). With the temperature rising, the diffraction peak intensity of β-SiC increases, and the yield of β-SiC in the product is 95.31% and 97.22% at 800 °C and 1000 °C, respectively. In addition, a significant stacking fault peak at 2θ of about 33.6° marked with SF is observed due to defects such as twins and stacking faults in β-SiC (Pujar and Cawley 1995).

The morphology of the products at different temperatures for 30 min is shown in Fig. 79.2. It can be seen that the β-SiC particles have a regular cubic structure. When the temperature is 600 °C, the size of β-SiC particles is ranging from 150 to 300 nm (Fig. 79.2a). As the reaction temperature increases to 800 and 1000 °C, the maximum grain size of the β-SiC particles increases to 2.5 μm (Fig. 79.2b) and 3 μm (Fig. 79.2c), while the morphology of β-SiC particles has no obvious change.

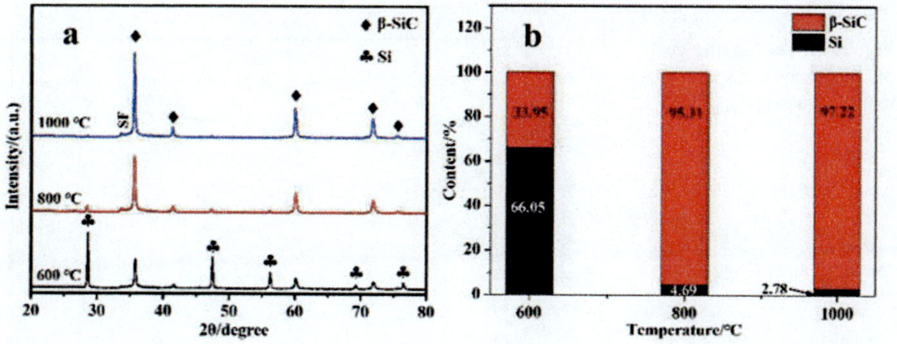

Fig. 79.1 **a** XRD patterns of the products; **b** the yield of each phase in the products at different temperatures for 30 min

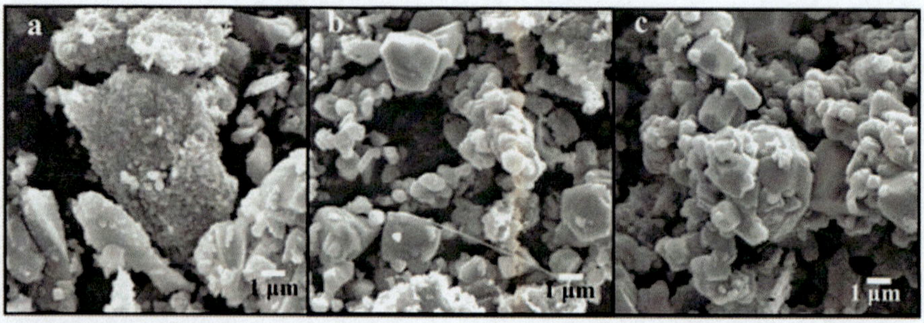

Fig. 79.2 SEM images of β-SiC particles at different temperatures for 30 min: **a** 600 °C; **b** 800 °C; **c** 1000 °C

To study the microwave heating behavior of the sample during the microwave heating process, the silicon powder and the cured phenolic resin powder were respectively pressed into cylinder samples under the same experimental conditions and heated by microwave method, and the heating curve of silicon powder is shown in Fig. 79.3. The temperature of the silicon powder rapidly rises to 660 °C within 2.5 min, indicating the excellent microwave absorbability of silicon powder, and it could be rapidly heated by microwaves (Luo et al. 2020). However, for the cured phenolic resin powder, the temperature remains below 300 °C after heating for 30 min, showing its poor microwave absorbability. It could be concluded that the silicon powder in the sample acts as the thermal source to heat the cured phenolic resin powder around it at the initial microwave heating stage.

Figure 79.4 shows the schematic of the nucleation and growth of β-SiC particle by microwave heating. As the temperature of the sample reaches ~350 °C (Fig. 79.4b), the cured phenolic resin powders pyrolyze and release the carbon-containing gases (CxHyOz) such as methane, ethylene, phenol, methyl phenol, and phenol alkyl derivatives and then convert to amorphous pyrolytic carbon (Zhao et al. 2020). The

Fig. 79.3 The heating curve of silicon powder by microwave heating

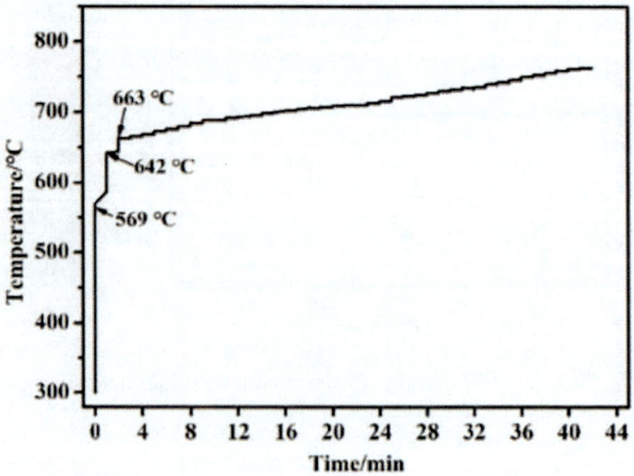

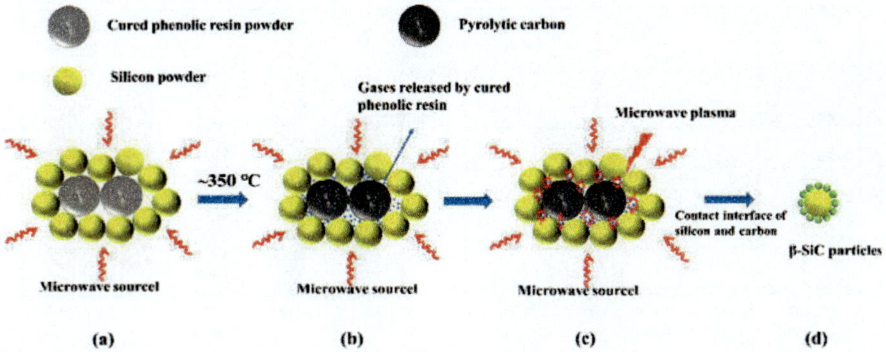

Fig. 79.4 The schematic of the nucleation and growth of β-SiC particle by microwave heating

carbon-containing gases and the air filled in the inter pores of the sample could be activated by microwave and generate a strong local microwave plasma when the temperature rises (Fig. 79.4c; Zhang et al. 2018). This process generates a large amount of energy, resulting in the instantaneous high temperature of the reaction interface, thus promoting the nucleation and growth of primary β-SiC particles.

With continuous microwave heating, the unreacted silicon powder, the primary β-SiC, and pyrolytic carbon all could absorb the microwave to increase the temperature of the sample. As the microwave temperature and holding time increase, the diffusion rate of silicon and carbon atoms of silicon and pyrolyzed carbon gradually increases, promoting the further growth of β-SiC particles.

XRD patterns and the yield of each phase in the products at 800 °C for different holding times are shown in Fig. 79.5. The characteristic diffraction peak intensity of β-SiC gradually increases with the holding time increasing from 5 to 60 min. When the holding time is 30 and 60 min, the yield of β-SiC increases to 95.31 and 97.49%. It can be seen that the increasing time has little effect on the morphology and the yield of β-SiC when the holding time exceeds 30 min. After the reaction cost consideration, the appropriate holding time is 30 min.

Figure 79.6 shows the SEM images of the products at 800 °C for different holding times. At the initial stage of the reaction, nanometer β-SiC particles are formed on the surface of the reactant particles (Fig. 79.6a). With the holding time increasing, β-SiC particles gradually grow up, while the reactant particles are gradually consumed and broken up (Fig. 79.6b, c). When the holding time is 30 min and 60 min, the maximum grain size of β-SiC particles is 2.5 μm and 3 μm, respectively, indicating that the prolonging holding time could promote the growth of SiC particles to a certain extent, which would be attributed to the kinetic process.

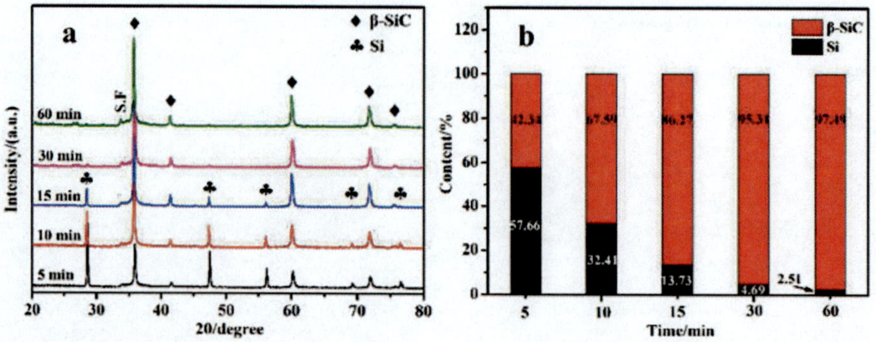

Fig. 79.5 **a** XRD patterns of the products; **b** the yield of each phase in the products at 800 °C for different holding time

Fig. 79.6 SEM images of β-SiC particles at 800 °C for different holding times: **a** 5 min; **b** 10 min; **c** 15 min; **d** 30 min; **e** 60 min

79.4 Conclusions

In summary, in order to realize the preparation of silicon carbide (β-SiC) from the industrial waste solar silicon powder, the microwave heating method was applied in this study. Microwave heating of silicon powder has a higher heating rate, which significantly shortens the reaction time, and greatly reduces the energy consumption. As the temperature raised, the carbon-containing gases released by the cured phenolic resin and the air filled in the inter pores of the sample occur to activate a local microwave plasma by the microwave electromagnetic field, resulting in the instantaneous high temperature of the reaction interface, and thus promoting the nucleation and growth of β-SiC particles. With the continuous heating of the microwave, the size of β-SiC particles increases. It can realize more than 95% β-SiC yield at 800 °C for the holding time of 30 min by microwave heating.

Acknowledgements This work was financially supported by Qingdao Hiworld New Materials Co., Ltd.

References

Bergese P (2006) Specific heat, polarization and heat conduction in microwave heating systems: a nonequilibrium thermodynamic point of view. Acta Mater 54(7):1843–1849

Boudard D, Forest V, Pourchez J et al (2014) In vitro cellular responses to silicon carbide particles manufactured through the Acheson process: impact of physico-chemical features on pro-inflammatory and pro-oxidative effects. Toxicol In Vitro 28(5):856–865

Chung FH (1974) Quantitative interpretation of X-ray diffraction patterns of mixtures. II. Adiabatic principle of X-ray diffraction analysis of mixtures. J Appl Crystallogr 7(6):526–531

Kettner O, Simic S, Kunert B et al (2018) Characterization of surface and structure of in-situ doped sol-gel-derived silicon carbide. Adv Eng Mater 1701067

Liao Y, Wu Z, Feng X (2019) Characteristics of heating process in microwave applicators with elements in periodic motion. Int J RF Microwave Comp Aid Eng 29(3):e21641

Lin LH, Li CS, Qin JY et al (2022) A SiC/C foam microwave absorbing material for PIM darkroom box in simulating space environment. Ceram Int 48(10):14626–14632

Liu JB, Chen ZF, Chai P et al (2022) The effect of deposition temperature on microstructure and mechanical properties of SiC coatings on graphite. J Aust Ceram Soc 58(2):557–562

Luo T, Xu L, Peng JH et al (2020) Efficient preparation of Si_3N_4 by microwave treatment of solar-grade waste silicon powder. ACS Omega 5(11):5834–5843

Mathur N, Singh S, Sutherland JW (2020) Promoting a circular economy in the solar photovoltaic industry using life cycle symbiosis. Resour Conserv Recycl 155:104649

Möller HJ (2004) Basic mechanisms and models of multi-wire sawing. Adv Eng Mater 6(7):501–513

Pujar VV, Cawley JD (1995) Effect of stacking faults on the X-ray diffraction profiles of β-SiC powders. J Am Ceram Soc 78(3):774–782

Qi W, Li MJ, Zhao L (2020) One-step fabrication of photoluminescent SiC quantum dots through a radiation technique. New J Chem 44(31):13301–13307

Qin Q, Chen J, Song M et al (2021) Preparation of SiC nanowires based on graphene as the template by microwave sintering. J Alloys Comp 910:164746

Schneider M, Šimůnková L, Michaelis A et al (2021) Study of anodic oxide films formed on solid-state sintered SiC-ceramic at high anodic potentials. Ceram Int 47(11):15010–15016

Schwinde S, Berg M, Kunert M et al (2015) New potential for reduction of kerf loss and wire consumption in multi-wire sawing. Sol Energy Mater Sol Cells 136:44–47

Wang C, Huang ZZ, Zhou JW et al (2022) One-step synthesis of SiC/C nanocomposites by atmospheric thermal plasmas for efficient microwave absorption. Ceram Int 48(8):10391–10402

Wei S, Guan L, Song B et al (2019) Seeds-induced synthesis of SiC by microwave heating. Ceram Int 45(8):9771–9775

Xu M, Girish YR, Rakesh KP et al (2021) Recent advances and challenges in silicon carbide (SiC) ceramic nanoarchitectures and their applications. Mater Today Commun 28:102533

Yang HL, Liu IT, Liu CE et al (2019) Recycling and reuse of kerf-loss silicon from diamond wire sawing for photovoltaic industry. Waste Manage 84:204–210

Zeraati M, Alizadeh V, Sargazi G et al (2021) Sol-gel synthesis of silicon carbide on silicon pyramids: a promising candidate for supercapacitor electrodes. J Mater Sci Mater Electron 32:22319–22329

Zhang Y, Song BZ, Zhang YP et al (2018) Influences of pre-forming on preparation of SiC by microwave heating. Ceram Int 44(17):21309–21313

Zhao HF, Xie DD, Zhang S et al (2020) Study on improving the high-temperature oxidation resistance of pyrolytic carbons of phenolic resin binder by in-situ formation of carbon nanotubes. React Func Polym 157:104772

Chapter 80
CFD Simulation of Flare Soot Emission Based on Mechanism of Soot Formation

Xiaoya Zhao, Yue Liu, Yunnan Jiao, and Zongjie Han

Abstract As the last guarantee for safe production of the plant, industrial flare is also an essential environmental protection measure to reduce environmental pollution. According to the mechanism of soot formation, the main limiting factors in the process of soot generation and oxidation are identified. The CFD software is used as the solver to optimize the ground flare structure and reduce the concentration of soot emissions. We chose PDF model to calculate turbulent combustion and Moss-Brookes model to predict soot formation. Using C_3H_6 as fuel and air as oxidant, we calculated the temperature and soot volume fraction (SVF) of air assisted ground flare under different combustion hole sizes. Combined with soot formation mechanism, we find that on the premise of ensuring the fixed air flow, properly reducing the size of combustion supporting holes can accelerate the flame propagation speed, shorten the flame height, and promote the formation and oxidation of soot. Accurate air distribution can prevent soot accumulation near the flare burner, thus extending the service life.

Keywords Ground flare · Soot · Turbulent combustion · Numerical simulation

80.1 Introduction

Nowadays more attention has been paid to the environmental protection. In many engineering fields, reasonable reduction of soot emission is an important indicator requirement, and the industrial flare industry is one of them.

In order to achieve efficient combustion and smokeless emission, it is necessary to clarify the combustion mechanism and soot generation mechanism. Currently, the most commonly used combustion reaction mechanism of C_1–C_4 hydrocarbons is LU

X. Zhao
Beijing Aerospace Propulsion Institute, Beijing 100076, China

X. Zhao · Y. Liu (✉) · Y. Jiao · Z. Han
Beijing Aerospace Petrochemical EC and EP Technology Co., Ltd., Beijing 100076, China
e-mail: 15318385331@163.com

J. Zhang et al. (eds.), *Environmental Pollution Governance and Ecological Remediation Technology*, Environmental Science and Engineering,
https://doi.org/10.1007/978-3-031-25284-6_80

1.1. Now LU 1.1 is applied in the flare field. In 2011, Allen and Torres (2012) used LU 1.1 to simulate flare test of John Zink. In 2014, Devesh Singh et al. (2014) used LU 1.1 to study the parameters of air and steam assisted ethylene flare. At present, there is relatively little research on the mechanism of soot formation in the flare field. In 2016, Wang et al. (2016) developed a new combustion mechanism (V_{soot}) for predicting the soot emission of the air-assisted flare.

Flare combustion is widely used in energy, oil refining and chemical processing industries to relieve pressure, discharge harmful gases (Wu et al. 2019). Nowadays the flare can be divided into air and steam assisted. Steam-assisted flare is the main way to eliminate soot. Air-assisted flare has great advantages in cold areas, insufficient steam supply and high burnout rate. At present, the air supply way of air-assisted flare mostly belongs to regional air supply. This method can slightly increase the air distribution of the combustion chamber, but the fuel and air still show obvious nonuniformity, and it is difficult to make full use of the combustion air. Therefore, it is necessary to carry out the research on the flare burner with accurate air distribution.

80.2 Mechanism of Soot Formation

The mechanism of soot formation and oxidation is closely related to fuel mode and flame type, and are closely coupled with fluid dynamics and heat transfer. It is generally believed that polycyclic aromatic hydrocarbons (PAHs) are the precursors of soot. The formation of soot mainly includes the following steps:

- Nucleation of soot. The nucleation of soot refers to the process of homogeneous nucleation of gaseous polycyclic aromatic hydrocarbons through physical and chemical coalescence to form primary soot particles.
- Surface growth of soot. According to the soot surface growth scheme proposed by Henning et al. (2013), the hydrogen free radicals firstly react with the soot surface, and then form active free radical sites there. In the presence of acetylene, the active sites react with them, resulting in an increase in the carbon content of particles. This reaction sequence is generally called the HACA surface growth of soot.
- Agglomeration of soot particles (D'Anna et al. 2010). After the primary soot particles grow to a certain extent, the surface growth rate decreases, and then the soot particles coalesce and agglomerate.
- Oxidation of soot. Jerez et al. (2018) studied the influence of oxygen on soot generation in ethylene diffusion flame based on PAH cross-sectional particle dynamics soot model. It was found that increasing oxygen flow rate will increase soot precursor generation. Among the surface oxidation factors of soot, O_2 is dominant in the lean combustion state at the top of the flame and OH radicals are dominant in the rich combustion state at the middle height of the flame.

The schematic diagram of soot formation process is shown in Fig. 80.1.

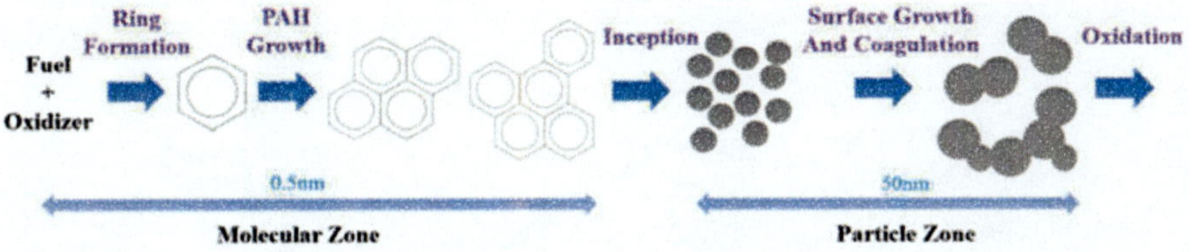

Fig. 80.1 The process of soot formation

Experimental methods for accurate measurement of soot need precise instruments and harsh experimental conditions, which are difficult to apply in engineering. CFD software can be used to simulate the smoke emission of the flare. Fluent provides four models to predict soot emissions. The first is a one-step Khan and Greeves model (Snelling et al. 2012), which predicts the soot formation rate according to a simple empirical formula. The second is a two-step Tesner model (Snegirev 2004), which takes into account the formation of nuclear particles and the formation of soot on the nucleus. The third is Moss-Brookes model (Zucca et al. 2006), which usually uses C_2H_2 as the precursor of soot generation to solve the transport equation of normalized free radical nuclear concentration and soot mass fraction. The fourth is Moss-Brooks Hall model, which has wider applicability.

80.3 CFD Model

The CFD code used to do this work was Ansys Fluent version 2020r2. A three-dimensional pressure-based solver with a coupled algorithm was used in these simulations. For spatial discretization, the PRESTO! scheme was used. A second-order upwind differencing scheme was used for spatial discretization of momentum. For simplicity, this work assumed steady-state conditions. Turbulent mixing was simulated using the k-ε realizable model.

80.3.1 Physical Model

The flare burner used in this paper is a star structure, with 8 branches in total, and each branch contains 12 circular holes. The hole diameter of the fuel hole a near the center is small, 2.6 mm, and the spacing is 25 mm, which is perpendicular to the branch plane; The diameter of fuel hole B is large, 6.2 mm, and the spacing is 14 mm, which is vertical, as shown in Fig. 80.2.

For the structure of the precise air distribution, single side air distribution is selected in this paper. There are eight air combustion supporting holes, each with a length of 120 mm, 12° from the horizontal (see Fig. 80.3).

Fig. 80.2 Burner structure diagram: **a** abbreviated drawing, **b** front view, and **c** side view

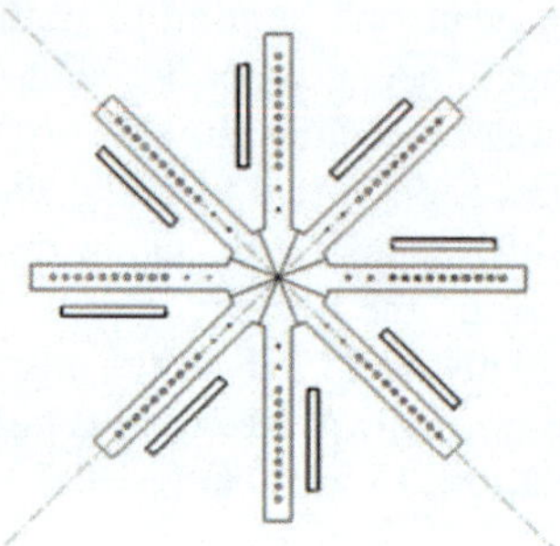

Fig. 80.3 Distribution of combustion supporting holes

In order to make the calculation more accurate, the actual height of the flare barrel is about 20–30 m, and the interval between burners is about 1 m. Therefore, this paper only considers the combustion of a single flare burner. In order to observe the full development of the flame, the size of the combustion area is determined to be 2 × 2 × 15 m.

80.3.2 Grid Division

Use ICEM to divide unstructured grids (see Fig. 80.4). Locally densify the fuel holes and combustion supporting holes. The boundary layer grid is adopted to densify the main core flow field. The total number of grids is about 2.5 million. The grid quality is greater than 0.3, which can be used for subsequent calculation.

80.3.3 Boundary Condition

The fuel is 100% propylene (C_3H_6) and the oxidant is air. The air inlet is the velocity inlet, the fuel inlet is the mass flow inlet, and the outlet is the pressure outlet. The fuel inlet and the combustion supporting air inlet can be simplified as thin plate nozzles.

Fig. 80.4 Mesh view of computational model

According to the known gas temperature and pre injection pressure, combined with the nozzle calculation principle, the basic state parameters, flow rate and turbulence intensity of the gas at the nozzle outlet can be calculated. The wind speed of natural entrainment wind is determined according to the wind speed of natural wind entrainment below the actual flare barrel. We set the air volume of accurate air distribution as 8% of the theoretical air volume. Table 80.1 is a summary of the boundary condition setting parameters.

Table 80.1 Summary of boundary conditions

Name	Type	Turbulence intensity, I (%)	Hydraulic diameter, D (mm)
Natural entrainment air inlet	Velocity inlet, 4 m/s	3.05	2000.0
Combustion supporting air inlet	Velocity inlet, 49.3 m/s	3.93	21.7
Fuel inlet	Mass flow inlet, 0.361 kg/s	3.59	6.0
Outlet	Pressure outlet	3.05	2000.0

Table 80.2 Boundary conditions corresponding to different combustion supporting hole sizes

Width (mm)	The measure of area (dm^2)	Pressure, P (kPa)	Velocity, v (m/s)	Turbulence intensity, I (%)	Hydraulic diameter, D (mm)
9.5	0.9120	2.4	60.22	3.93	17.6
11.9	1.1424	1.6	49.29	3.93	21.7
14.1	1.3536	1.1	40.93	3.95	25.2
19.0	1.8240	0.6	30.28	3.96	32.8

Control the combustion air quantity to keep 8%, change the area of the combustion supporting hole by changing the width. Table 80.2 shows the boundary conditions corresponding to different combustion supporting hole sizes. It can be seen that with the increase of the combustion supporting hole area, the pre injection pressure, air velocity and Reynolds number decrease, the turbulence intensity of the combustion air increases.

80.3.4 *Mathematical Model*

The standard k-ε model is solved simultaneously by using the turbulent kinetic energy equation and the turbulent dissipation rate equation.

$$\frac{\partial}{\partial t}(\rho k) + \frac{\partial}{\partial x_i}(\rho k u_i) = \frac{\partial}{\partial x_j}\left[\left(\mu + \frac{\mu_t}{\sigma_k}\right)\frac{\partial k}{\partial x_j}\right] + G_k + G_b - \rho\varepsilon - Y_M + S_k \tag{80.1}$$

$$\frac{\partial}{\partial t}(\rho \varepsilon) + \frac{\partial}{\partial x_i}(\rho \varepsilon u_i) = \frac{\partial}{\partial x_j}\left[\left(\mu + \frac{\mu_t}{\sigma_k}\right)\frac{\partial \varepsilon}{\partial x_j}\right] + C_{1\varepsilon}\frac{\varepsilon}{k}(G_k + C_{3\varepsilon}G_b) - C_{2\varepsilon}\rho\frac{\varepsilon^2}{k} + S_\varepsilon \tag{80.2}$$

where ρ is the fluid density, k is turbulent kinetic energy, ε is the dissipation rate, μ_i is the velocity vector, μ_t is turbulent viscosity, G_k, G_b are turbulent kinetic energy, σ_k, σ_ε are Prandtl number.

For Moss-Brookes soot model, C_2H_2 is selected as the precursor of soot, and Fenimore Jones is selected as the soot oxidation model. The model predicts soot formation in hydrocarbon flame by solving the transport equation of normalized free radical nuclear concentration and soot mass fraction.

$$\frac{\partial(\rho Y_{soot})}{\partial t} + \frac{\partial}{\partial x_i}(\rho u Y_{soot}) = \frac{\partial}{\partial x_i}\left(\frac{u_t}{\sigma_{soot}}\frac{\partial Y_{soot}}{\partial x_i}\right) + \frac{dM}{dt} \tag{80.3}$$

$$\frac{\partial\left(\rho b_{nuc}^{*}\right)}{\partial t}+\frac{\partial}{\partial x_{i}}\left(\rho u_{i} b_{nuc}^{*}\right)=\frac{\partial}{\partial x_{i}}\left(\frac{u_{t}}{\sigma_{nuc}} \frac{\partial b_{nuc}^{*}}{\partial x_{i}}\right)+\frac{1}{N_{norm}} \frac{dN}{dt} \tag{80.4}$$

where Y_{soot} is the mass fraction of soot, M is the mass concentration of soot, b_{nuc}^{*} is the concentration of nuclei, N is the number density of soot particles.

80.4 Result and Discussion

Accurate air distribution has an important impact on the soot emission of the flare. Here, the combustion state and soot distribution of the flare are calculated under different air distribution hole sizes.

80.4.1 *Verification of Soot Formation Mechanism*

Taking a model with a combustion supporting hole width of 11.9 mm as an example. According to the mechanism of soot formation, the surface growth rate of soot is mainly related to the concentration of H radicals, and the oxidation rate of soot is mainly related to the concentration of OH radicals and O_2.

Figure 80.5 shows the changes of soot surface growth rate and H mole fraction on the axial centerline. The change trend of two curves are similar. According to HACA principle, at the beginning, the increase of H concentration accelerates the breaking of bond and promotes the formation of soot. Although there is still H at higher position, with the increase of soot size, the number of active sites where surface reactions decreases, the specific surface area of soot molecules with large mass decreases, and surface reactions become less important. Figure 80.6 shows the effect of OH and O_2 concentrations on soot oxidation rate. It can be seen that soot oxidation occurs at a higher position than surface growth. The change trend of OH concentration is more similar to the trend of soot oxidation rate than O_2 concentration, so that OH radicals play a major role in oxidation in this project.

80.4.2 *Temperature Distribution*

Figure 80.7 shows the temperature distribution on the XY plane under different combustion supporting hole sizes. Figure 80.8 shows the temperature distribution on the axis with different combustion supporting hole sizes. No accurate air distribution can be regarded as the case of infinite combustion supporting hole. With the increase of the combustion supporting hole size, the maximum temperature of the flame decreases slightly, but always in the range of 2000–2100 K. The reduction of the size

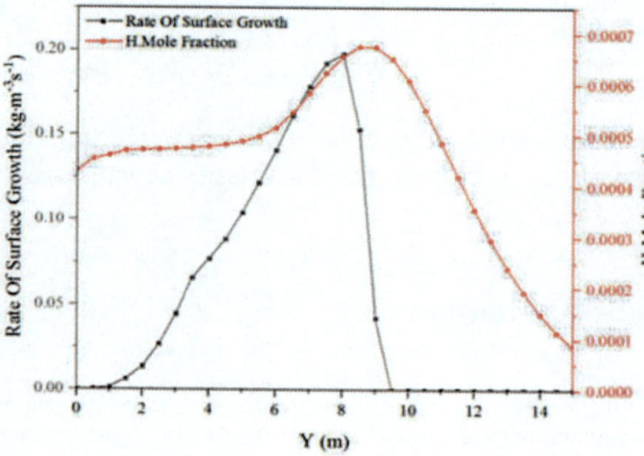

Fig. 80.5 Comparison of soot surface growth rate and H mole fraction distribution on the axial centerline

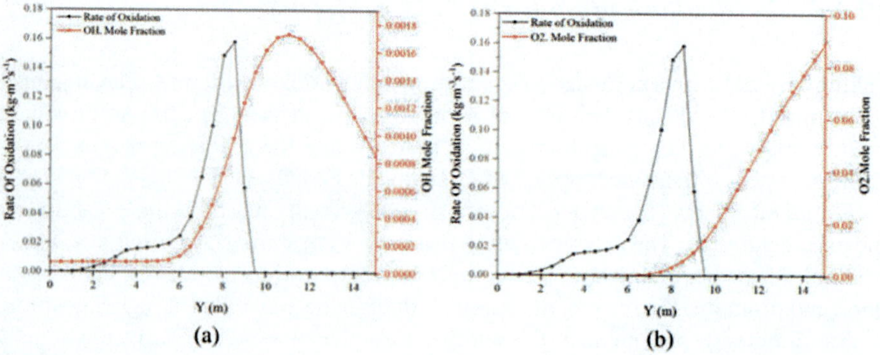

Fig. 80.6 Comparison of soot oxidation rate and **a** OH, **b** O_2 mole fraction distribution on the axial centerline

of the combustion supporting hole accelerates the flame propagation speed, promotes the mixing of fuel and oxygen, and shortens the flame height.

80.4.3 *Soot Volume Fraction Distribution*

Figure 80.9 shows the SVF distribution on the XY plane under different combustion supporting hole sizes. Figure 80.10 shows the SVF distribution on the axis with different combustion supporting hole sizes. With the increase of combustion supporting hole size, the maximum of SVF decreases, and the location of soot concentration distribution area becomes higher. The soot elimination effect of the flare is mainly characterized by the content of soot in the gas discharged from the flare

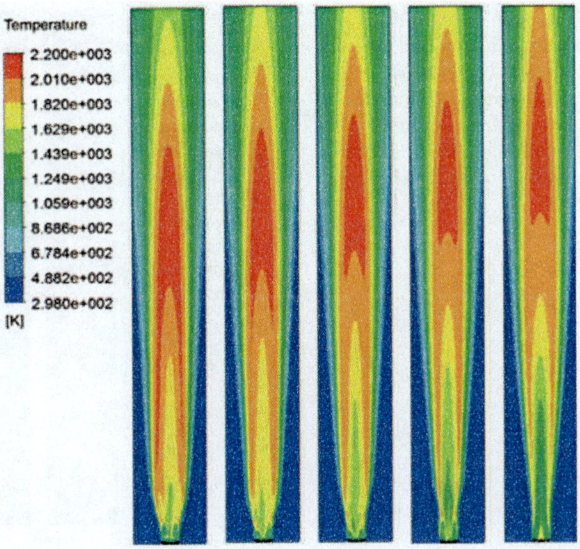

Fig. 80.7 Temperature distribution under different combustion supporting hole widths: **a** 9.5 mm, **b** 11.9 mm, **c** 14.1 mm, **d** 19.0 mm, **e** no combustion supporting hole

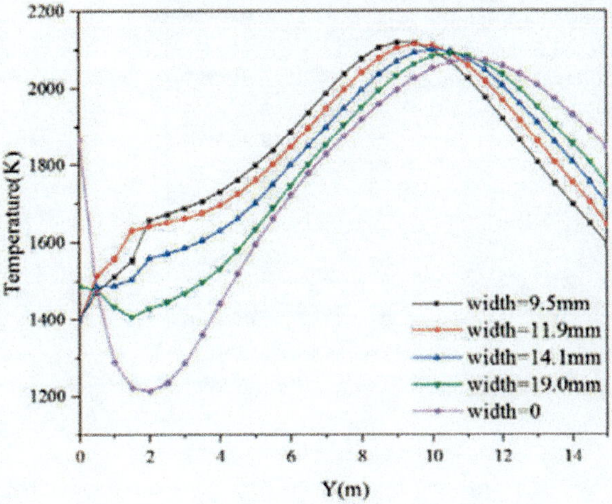

Fig. 80.8 Temperature distribution along the axial centerline under different combustion supporting hole sizes

barrel. Therefore, the smaller the combustion supporting hole, the better the soot elimination effect.

Compared with the temperature distribution, it can be seen that the distribution trends of soot and temperature are similar, but the concentrated distribution position of soot is lower than the core area of temperature. This indicates that the increase of flame temperature leads to the acceleration of soot growth rate and nucleation rate.

It can be found from Fig. 80.10 that the soot distribution corresponding to the combustion supporting holes with widths of 9.5 and 11.9 mm is very similar. In

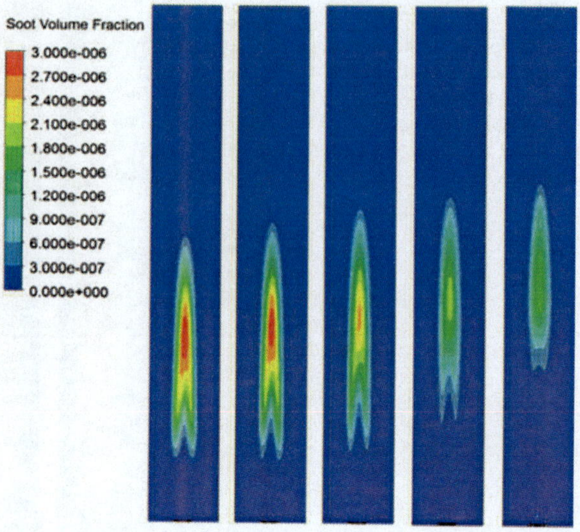

Fig. 80.9 Soot volume fraction distribution under different combustion supporting hole widths: **a** 9.5 mm, **b** 11.9 mm, **c** 14.1 mm, **d** 19.0 mm, **e** no combustion supporting hole

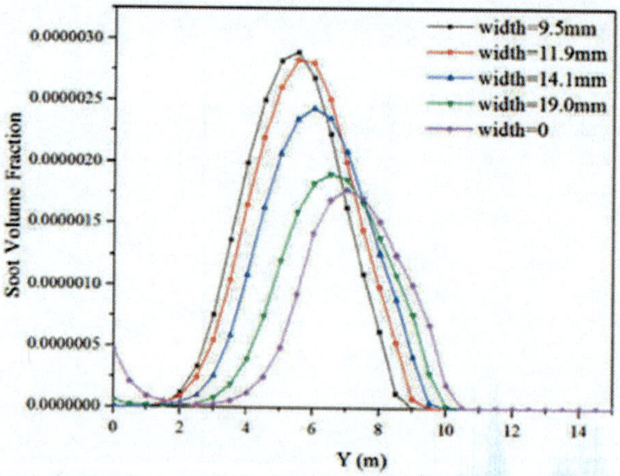

Fig. 80.10 Soot volume fraction distribution along the axial centerline under different combustion supporting hole sizes

order to reduce the load of the blower, it is better to set the combustion supporting hole width to 11.9 mm.

In addition, we can see that there will be a small amount of soot near the burner when there is no accurate air distribution. Soot near the burner may block the burner, so accurate air distribution helps to improve the service life of the burner.

The concentration of H and OH radicals on the axis under different combustion supporting hole sizes is shown in Fig. 80.11. It can be seen that with the increase of combustion supporting hole size, the position of the maximum value of H and OH become higher. That means the surface growth and oxidation of soot appear eariler, promoting the formation of soot. As can be clearly seen from Fig. 80.11a, when there is an accurate air distribution, the H concentration first increases and then decreases.

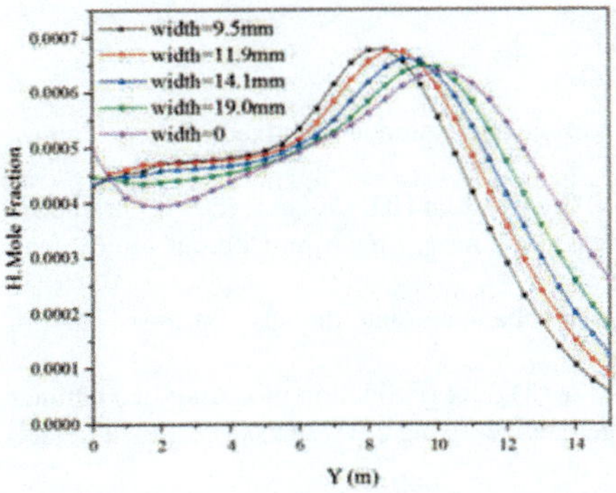

(a)

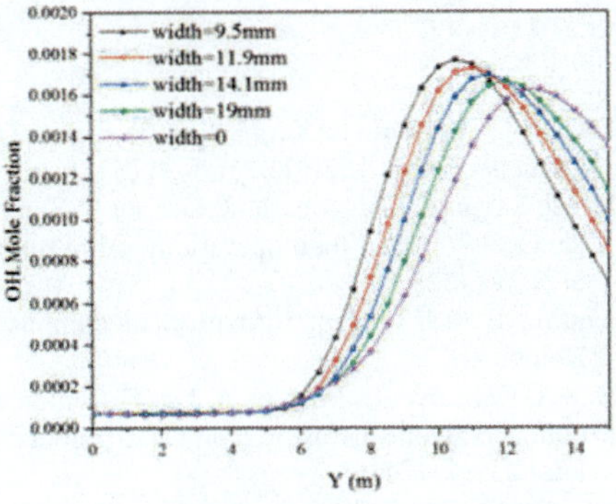

(b)

Fig. 80.11 **a** H and **b** OH mole fraction distribution along the axial centerline under different combustion supporting hole sizes

When there is no accurate air distribution, the H concentration first decreases, then increases, and then decreases. This means that the accurate air distribution near the burner affects the concentration of H, and the generation of soot at the bottom of the burner is suppressed by suppressing the surface growth of soot.

80.5 Conclusions

In this paper, the temperature and soot distribution of the flare burner under different air distribution structure sizes are studied by using computational fluid dynamics in combination with soot formation mechanism.

The results verified that H and OH play an important role in the formation of soot. Confirmed that along the central axis, the change laws of temperature and soot distribution are similar, but the concentrated distribution position of soot is lower than the core area of temperature. The reduction of the combustion supporting hole size accelerates the flame propagation speed, promotes the mixing of fuel and oxygen, shortens the flame height, and promotes the generation and oxidation of soot. However, the size of the combustion supporting hole is not the smaller the better. When the combustion supporting hole is too small, the load of the blower increases and the cost increases. Furthermore, accurate air distribution can prevent soot accumulation near the flare burner, thus increasing the service life of the flare burner.

References

D'Anna A, Sirignano R, Kent R (2010) A model of particle nucleation in premixed ethylene flames. Combust Flame 157(11):2106–2115

Devesh Singh K, Gangadharan P, Dabade T, Shinde V, Chen D, Lou HH, Li X et al (2014) Parametric study of ethylene flare operations using numerical simulation. Eng Appl Comput Fluid Mech 8(2):211–228

Henning B (2013) Soot formation in combustion: mechanisms and models. Springer, Berlin, Heidelberg

Jerez A, Consalvi JL, Fuentes A, Liu F, Demarco R (2018) Soot production modeling in a laminar coflow ethylene diffusion flame at different oxygen indices using a PAH-based sectional model. Fuel 231:404–416

Snegirev AY (2004) Statistical modeling of thermal radiation transfer in buoyant turbulent diffusion flames. Combust Flame 136(1/2):51–71

Snelling DR, Thomson KA, Smallwood GJ, Oslash LGU, lder et al (2012) Spectrally resolved measurement of flame radiation to determine soot temperature and concentration. Aiaa J 40(9):1789–1795

Torres VM, Herndon S, Allen DT (2012) Industrial flare performance at low flow conditions. steam- and air-assisted flares. Indust Eng Chem Res 51(39):12569–12576

Wang A, Lou HH, Chen D, Yu A, Dang W, Li X et al (2016) Combustion mechanism development and CFD simulation for the prediction of soot emission during flaring. Front Chem Sci Eng 10(4):1–13

Wu AT, Wei BX, Gao CX, Han DJ, Huang EW (2019) Study on the risk analysis and system safety integrity of enclosed ground flare. Therm Sci Eng Progr

Zucca A, Marchisio DL, Barresi AA, Fox RO (2006) Implementation of the population balance equation in CFD codes for modelling soot formation in turbulent flames. Chem Eng Sci 61(1):87–95

Chapter 81
Aviation Carbon Verification Method Based on Energy Efficiency

Jiaxue Liu, Jiamin Zhang, and Jingjie Chen

Abstract With the increasingly serious problem of climate change, aviation carbon emissions have received international attention. Aviation carbon verification is an important part of the aviation carbon emission monitoring system. In order to verify the rationality of the reported carbon emission data, an aviation carbon verification method based on energy efficiency is constructed in view of the limitations of single factor carbon verification and the applicability of the verification method. From the perspective of energy efficiency, the method takes ton kilometers as the basis for verification. Firstly, based on the fuel consumption analysis of each flight phase, the energy efficiency double exponential decay trend is obtained. The parameter estimation method is used to establish the energy efficiency double exponential decay model based on ton kilometers, and the energy efficiency estimation interval with a certain confidence level is obtained. Then, based on the energy efficiency estimation interval, an aviation carbon verification method based on energy efficiency is constructed. Finally, an example is given to verify the effectiveness of the model and the applicability of the method, which provides a reference for verifying the carbon emission data reported by aircraft operators.

Keywords Climate change · Carbon emission of civil aviation · Carbon verification · Double exponential model · Parameter estimation

81.1 Carbon Emissions and Aviation Carbon Verification

With the development of the global economy, the carbon dioxide emitted by human production activities has led to the increasingly serious problem of climate warming. The carbon emission of air transport industry accounts for 2% of the total carbon emission of human production activities, which has attracted the attention of the international community. Aviation carbon verification verifies the rationality of fuel

J. Liu (✉) · J. Zhang · J. Chen
Civil Aviation University of China, Tianjin 300300, China
e-mail: 936601270@qq.com

J. Zhang et al. (eds.), *Environmental Pollution Governance and Ecological Remediation Technology*, Environmental Science and Engineering,
https://doi.org/10.1007/978-3-031-25284-6_81

consumption data reported by aircraft operators by verifying relevant fuel consumption data (Certification and Accreditation Technology Institute of the National Certification and Accreditation Administration 2014) and ensures the credibility of carbon emission data (Yue 2017). At present, aviation carbon verification is only based on a single factor, that is, flight fuel consumption is verified based on flight distance or flight hours. Among them, the International Civil Aviation Organization (ICAO) adopts the ICAO carbon emission calculator based on the great circle distance, and only uses the information of takeoff and landing Airports and aircraft types to estimate the fuel consumption of aircraft segments (ICAO). The Committee on Aviation Environmental Protection (CEAP) has launched an aircraft carbon emission estimation and reporting tool, which is used to estimate carbon emissions. It is a scaling factor method based on the linear estimation of fuel consumption based on the standard great circle distance or flight hours of the flight segment under the condition of a fixed aircraft type (ICAO). Yao et al. (2021) based on the single factor of flight hours and combined with the seasonal characteristics of flight hours on the route, established an interval estimation model of aircraft fuel consumption with a certain confidence level, and proposed an aviation carbon verification method based on flight hours, which provides a reference method for the verification organization. Wang (2021) considered the density and deviation degree of fuel consumption data distribution under the single factor of flight range, established the interval estimation model of aircraft fuel consumption based on flight range, and obtained the interval estimation results of aircraft fuel consumption with a certain confidence level. Aghayeva (2017) based on the single factor of the range, aiming at the flight distance of the A320 in the three stages of climb, cruise and landing under the medium range, the mathematical model of fuel consumption in each flight stage is established to estimate the fuel consumption of the segment. Sun (2014) established an estimation model of the average fuel consumption of civil airliner routes based on the model type level through regression analysis of the relationship between route fuel consumption and segment distance. The average relative error of the estimation model is small and the effect is good. In order to improve the prediction results of the fuel consumption interval of the flight segment under the given confidence level, literature (Chen and Zhao 2022) proposed a range fuel consumption interval prediction algorithm based on deep learning, which is used as the basis of a carbon verification auxiliary tool to judge the rationality of the fuel consumption data reported by the aviation company. Because there are many factors that affect aircraft fuel consumption, including range or flight hours, total weight, flight altitude and route meteorology, and there is a complex nonlinear relationship between fuel consumption in the segment and these factors, it is limited to estimate fuel consumption only based on flight hours or flight distance. In this paper, from the perspective of energy efficiency, considering the two factors of flight distance and total weight or operating load at the same time, an energy efficiency interval estimation model with a certain confidence level is established based on ton kilometers, and an aviation carbon verification method based on ton kilometers is constructed, which not only includes energy consumption and converted carbon emission level, but also further enriches the physical basis of carbon verification.

81.2 Energy Efficiency Estimation Model

81.2.1 Definition of Energy Efficiency

From the perspective of airline operating revenue, energy efficiency refers to the fuel consumption per unit turnover. From the perspective of aircraft performance, the total takeoff weight of the aircraft should be used to replace the load in the transportation turnover. These two kinds are called operating energy efficiency (OE) and performance energy efficiency (PE), and their expressions are respectively:

$$EE_{OE} = \frac{W_f}{W_L \times D_s} \tag{81.1}$$

$$EE_{PE} = \frac{W_f}{W_G \times D_s} \tag{81.2}$$

EE refers to the average fuel consumption per ton km, W_f is the total fuel consumption of the aircraft segment, D_S refers to the standard great circle distance, W_L, W_G is the flight load and the total takeoff weight.

81.2.2 Double Exponential Decay Model of Energy Efficiency

Figure 81.1 shows the distribution trend of the operating energy efficiency of each flight with the change of ton kilometers for a total of 553 flights on 32 routes operated by b738. It can be observed from the figure that with the increase of ton kilometers, the overall fuel consumption of the aircraft per ton kilometers shows an exponential decline and has a steady-state trend.

From the perspective of the proportion of flight hours in each flight phase of the mission, with the increase of flight distance and block flight hours, the cruise time and cruise distance of the aircraft increase, and the proportion of flight hours in each phase will also change. The proportion distribution of flight hours in each stage of ground taxiing, takeoff and climb, descent and approach and cruise of each flight in the total flight hours of the voyage is shown in Fig. 81.2.

It can be seen from the time proportion change trend chart of each flight phase in Fig. 81.2 that with the increase of flight distance and required flight time, the proportion of flight hours in the aircraft taxiing phase, climbing phase and descent phase gradually decreases, and cruise gradually occupies the main process of flight mission, and the proportion continues to rise. As the proportion of flight hours in each flight phase changes with the block flight hours, the proportion of fuel consumption in each flight phase will also change with the block flight hours. Figure 81.3 shows

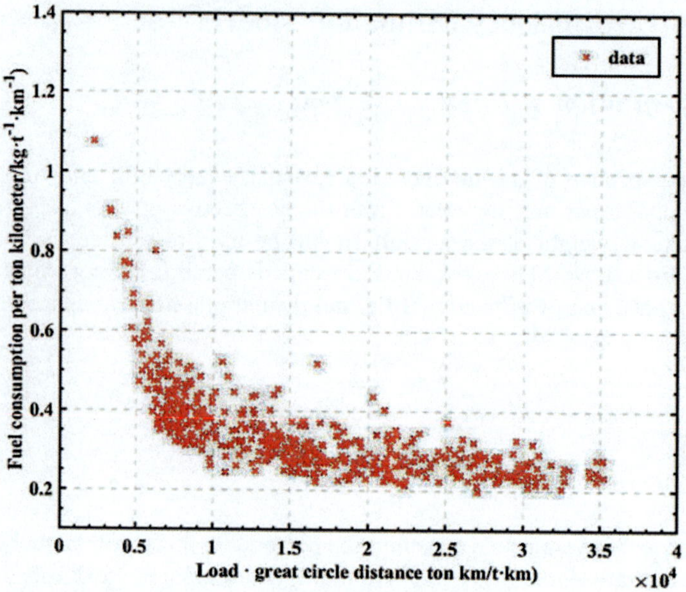

Fig. 81.1 Scatter diagram of fuel consumption per ton-kilometer

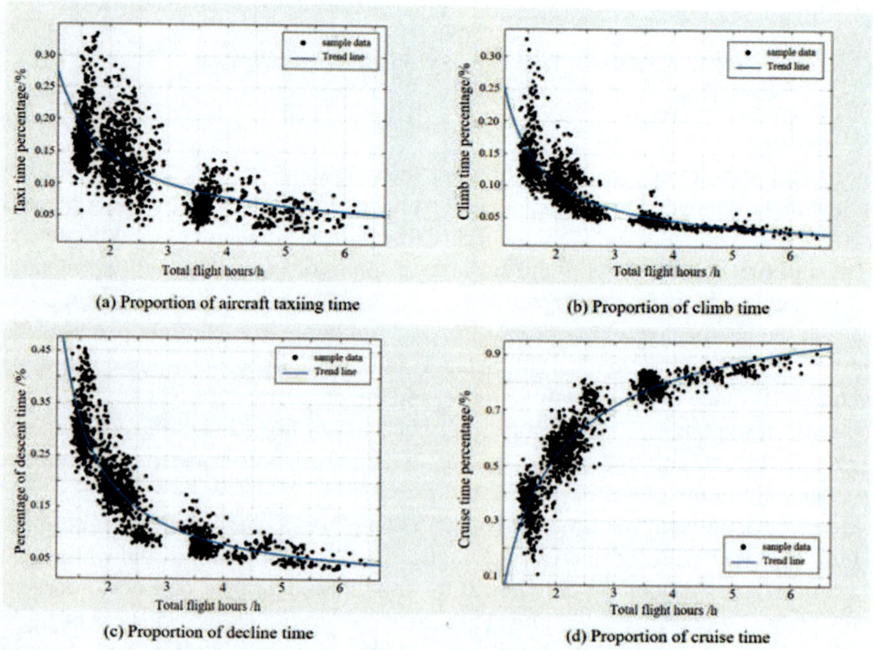

Fig. 81.2 Trend chart of the proportion of each flight phase time to the total flight time

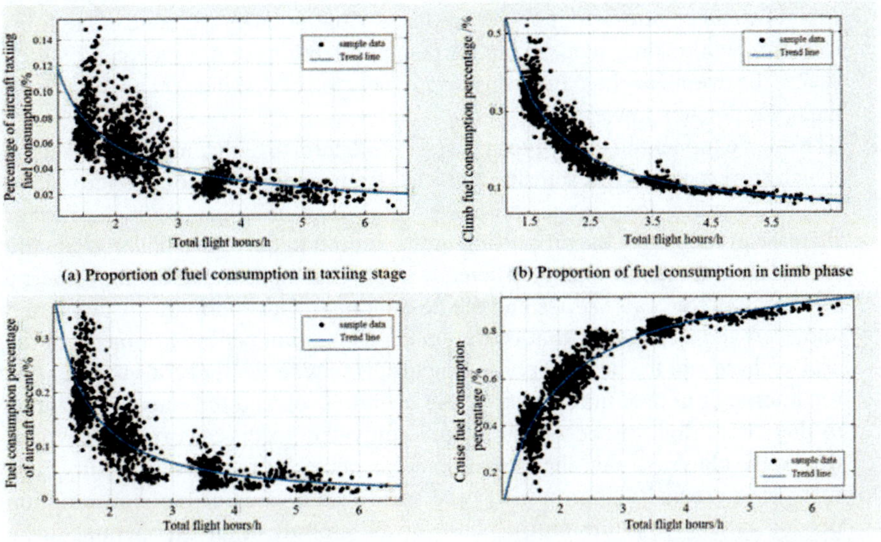

(a) Proportion of fuel consumption in taxiing stage

(b) Proportion of fuel consumption in climb phase

(c) Proportion of fuel consumption in descending stage

(d) Proportion of fuel consumption in cruise phase

Fig. 81.3 Trend chart of fuel consumption in each flight phase to total flight fuel consumption

the trend of the proportion of fuel consumption in each flight phase as the block flight hours increase.

It can be seen from Fig. 81.3 that with the increase of range and flight hours, the proportion of climb phase and descent phase in the fuel consumption of the segment continues to decrease, while the proportion of fuel consumption in the cruise phase continues to increase. In the climb phase, the kinetic energy of the aircraft increases continuously and obtains greater potential energy, consumes more energy, the fuel flow and engine output power of the aircraft are large, the potential energy in the cruise phase is almost unchanged, the fuel flow and engine output power of the aircraft are lower than that in the climb phase, the potential energy in the descent phase is continuously converted into kinetic energy, and the fuel flow and engine output power of the aircraft are smaller. As the total takeoff weight or operating load remains unchanged, the greater the engine output power, the lower the energy efficiency. According to the two different situations of flight time or flight distance, the proportion of fuel consumption in each flight stage is different and the proportion of fuel consumption per unit time changes. Based on this basis and combined with the perspective of work, the energy efficiency is analyzed, and the energy efficiency double exponential attenuation model based on ton kilometers is constructed:

(1) An exponential decay type is used for the transition process from low to medium ton kilometers, including short range, low or moderate load, and moderate range, low or no load. The take-off and climb section greatly affects the overall fuel consumption index per ton kilometers of the flight. The climb section continuously obtains greater potential energy in the vertical direction, and the aircraft's

fuel flow and engine output power are large, Continuously generate enough lift to climb, while doing work in the horizontal direction for displacement, which makes the overall average fuel consumption per ton of KM higher and the overall energy efficiency lower.

(2) The other exponential decay type is used for medium to high tonnage kilometers, which corresponds to the transition process from medium and long range to long range, including long range, normal or high load missions. With the further increase of the range, the proportion of the aircraft in the cruise phase gradually increases, and the low energy efficiency in the climb phase and the high energy efficiency in the cruise phase and the descent phase are continuously converted and averaged as a whole, that is, the fuel consumption per ton kilometer in the cruise phase and the descent phase is neutralized with the fuel consumption per ton kilometer in the climb phase on the overall level, and the energy efficiency in the whole flight process is gradually equivalent to the energy efficiency in the cruise phase, so that the overall energy efficiency returns to stability, The regression gradually approximates the steady-state value of fuel consumption per ton kilometer in the cruise phase of the aircraft in an exponential form. Therefore, the energy efficiency model based on ton kilometers can be expressed as:

$$f(x) = ae^{bx} + ce^{dx} + g + e_n \tag{81.3}$$

$f(x)$ stands for energy efficiency, a, c and g are constants greater than zero, b and d are constants less than zero, e_n represents the model noise with the mean value of zero.

81.2.3 Model Parameter Estimation

The sample is from the flight data of 553 groups of b738 under the whole range. The energy efficiency adopts performance energy efficiency. Because the objective fitting function is nonlinear and complex, it is difficult to obtain a set of optimal solutions by traditional fitting methods. In this paper, particle swarm optimization algorithm is used to solve the parameters to be estimated, in which each particle represents a set of solutions. Here, the dimension of each particle corresponding to the five parameters to be estimated is 5 dimensions, and the particle swarm size is set to 50. In particle swarm optimization, the position of the ith particle distributed in h-dimensional space is $X_i = (x_{i1}, x_{i2}, \cdots, x_{ih})$. The flight speed is $V_i = (v_{i1}, v_{i2}, \cdots, v_{ih})$, x_{ih} and v_{ih} is the component of particle i in the h dimension; $P_i = (p_{i1}, p_{i2}, \cdots, p_{ih})$ is the optimal position searched by particle I so far, that is, the individual extreme value P_{best}; $g_i = (g_{i1}, g_{i2}, \cdots, g_{ih})$ is the optimal position searched by particle swarm at present, that is, the global extremum g_{best}. In the iteration process, the particle updates its own flight speed and position through individual extreme value and global extreme value. The expression (Jiang et al. 2021) is:

$$v_{ih}^{t+1} = wv_{ih}^{t} + c_1 r_1 (P_i^t - x_{ih}^t) + c_2 r_2 (P_g^t - x_{ih}^t) \quad (81.4)$$

$$x_{ih}^{t+1} = x_{ih}^{t} + v_{ih}^{t+1} \quad (81.5)$$

In the formula, w represents the weight coefficient, which is set to 0.73; c_1 and c_2 denotes a learning factor, and two learning factors c_1 and c_2 are set to 2; t is the current iteration number, with a total of 300 iterations; v_{ih}^{t+1} and v_{ih}^{t} respectively represent the velocity of particle i in the h-th dimension at iteration $t+1$ and t times; P_i^t is the individual extreme value of particle i at the t-th iteration; P_g^t is the global extreme value when iterating to the t-th time, r_1 and r_2 represent random numbers between (0,1) and are used to increase search randomness. The objective is to find the optimal function parameters so that the sum of the residual squares of the function expression and the discrete data is the smallest. In combination with the needs of aviation carbon verification, more than 90% of the fuel consumption data of most flights are recognized through rapid and one-time verification, that is, they fall within the estimated confidence interval, while a small number of flight data that do not fall within the interval range need to be independently verified to meet the requirements of aviation carbon verification. Therefore, based on the fitting function and the 90% confidence interval of sample statistics, See Table 81.1 for the parameter results of the double exponential model.

The estimation interval of fuel consumption per unit gross weight Round ton kilometer of double exponential decay type is shown in Fig. 81.4.

The root mean square error (RMSE) is used to evaluate the fitting effect. The smaller the root mean square error is, the better the fitting effect is. The formula is as follows.

$$RMSE = \sqrt{\frac{1}{n}\sum_{i=1}^{n}(y_i - \overline{y_i})^2} \quad (81.6)$$

In the formula, n is the number of samples, y_i is the actual value of the ith sample, and $\overline{y_i}$ is the estimated value of the ith sample. After the parameter is brought into the function expression, the RMSE of the discrete data with the sample is 0.001397. The small RMSE indicates that the overall parameter estimation effect is good. There are

Table 81.1 Double exponential model parameters

Parameter	Objective function	Confidence interval
a	0.1724	(0.1536, 0.184)
b	-7.199×10^{-5}	$(-7.145 \times 10^{-5}, -5.719 \times 10^{-5})$
c	0.04035	(0.02743, 0.04404)
d	-1.784×10^{-5}	$(-1.95 \times 10^{-5}, -1.524 \times 10^{-5})$
e	0.0525	(0.0508, 0.0552)

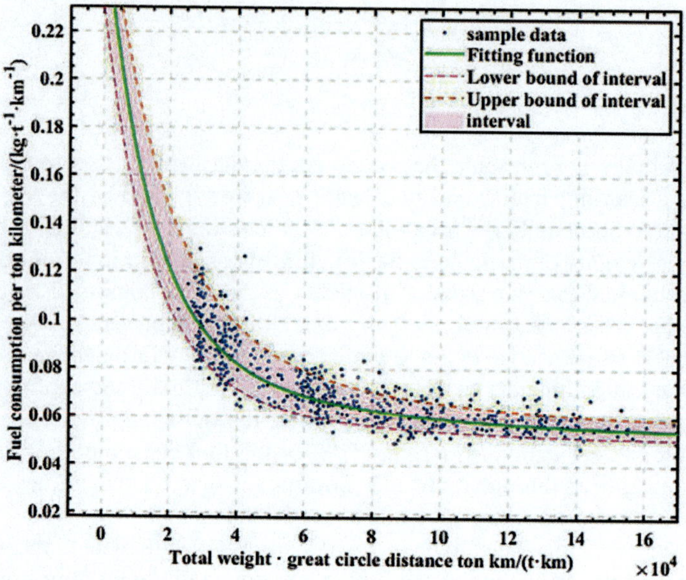

Fig. 81.4 Estimation interval

553 groups of data in total. The relative error between the actual value and the fitting value of each group of data is shown in Fig. 81.5.

It can be seen from the relative error scatter diagram in Fig. 81.5 that the error is low, and the average relative error between the fitting value and the actual value is

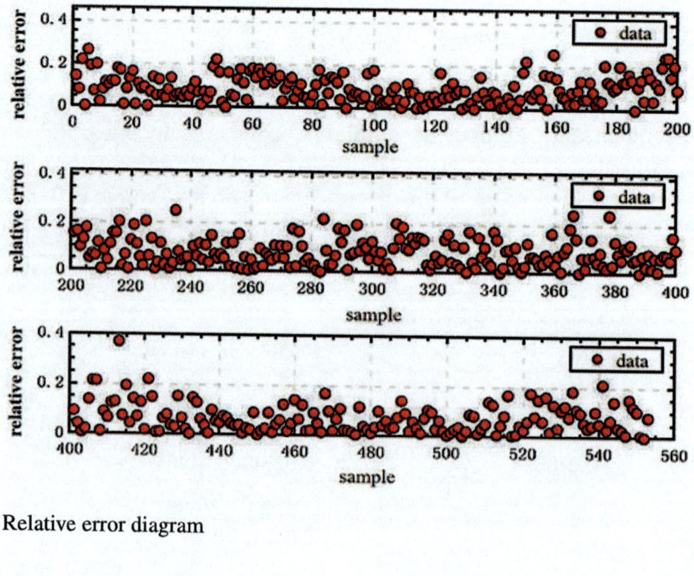

Fig. 81.5 Relative error diagram

0.079. The small relative error of each group of data indicates that the parameter estimation effect is good, so that the model takes into account the accuracy of scientific calculation and the rationality of certification practice.

81.3 Case Analysis

Based on the double exponential decay energy efficiency interval estimation model, from the perspective of energy efficiency, the flight energy efficiency is recognized, and the approved flight fuel consumption data is indirectly equivalent, so as to verify the fuel consumption under the ton kilometer. It can also be used as a reference for airlines to understand their own operational energy efficiency and formulate energy efficiency targets, and also provide a reference standard for the evaluation of transport aircraft energy efficiency.

The data is from QAR data of b738 model of an aviation company. 73 groups of historical operation data of b738 are randomly selected for example verification and analysis.

According to the double exponential attenuation type energy efficiency estimation interval in Sect. 81.2.3, the sample data to be verified is added for example verification analysis. The frequency distribution histogram of the selected sample data is shown in Fig. 81.6, and the example verification results are shown in Fig. 81.7.

It can be seen from Fig. 81.7 that through the example analysis of historical operation data, 66 groups of data accounting for more than 90.4% of 73 groups of samples are within the interval range, which can verify the authenticity of the data within the interval. However, a small number of flight data outside the estimated interval need to be independently verified in the verification procedure to verify

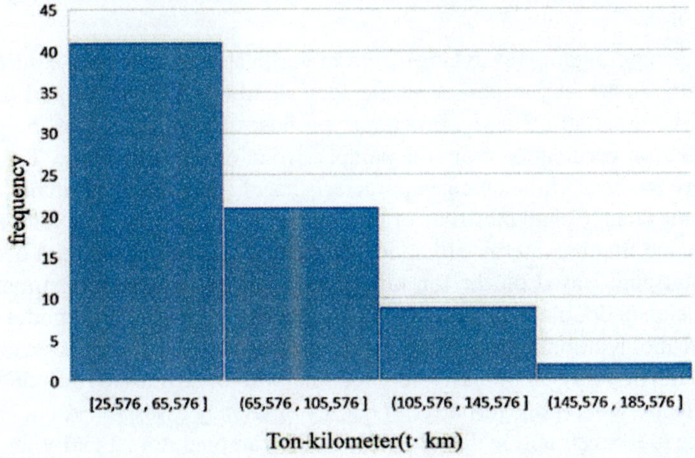

Fig. 81.6 Sample frequency distribution

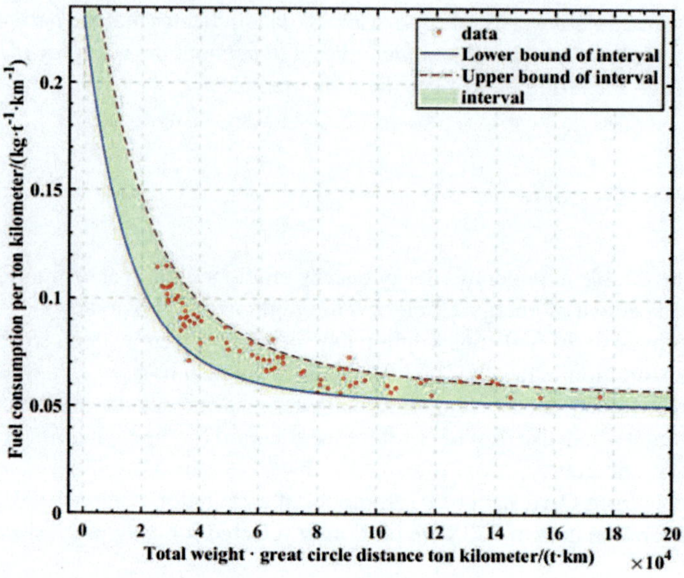

Fig. 81.7 Case verification results

the authenticity and rationality of the data reported by the aircraft operator. The effectiveness and applicability of the method are verified by an example, which can be used to verify the authenticity and rationality of the data reported by the aircraft operator.

81.4 Conclusion

Based on the combination of aviation carbon verification and carbon control objectives, aiming at the limitations of single factor carbon verification and the applicability of verification methods, this paper proposes ton kilometer as the basis for carbon emission verification from the perspective of energy efficiency. Through the comparative analysis of the change characteristics of fuel consumption per ton kilometer and the change characteristics of hourly fuel consumption in each flight phase, it is concluded that the energy efficiency presents a double exponential decay type distribution trend. Based on the ton kilometer nonlinear parameter estimation, the energy efficiency double exponential decay type interval estimation model is established, and the aviation carbon verification method based on the ton kilometer is constructed. It not only provides a reference standard for airlines to formulate carbon control objectives and transport aircraft energy efficiency assessment, but also helps to verify the fuel consumption data reported by aircraft operators. Finally, an example is analyzed to verify the validity of the model and the applicability of the method, which provides a reference for verifying the rationality of the fuel consumption data reported by airlines.

References

Certification and Accreditation Technology Institute of the National Certification and Accreditation Administration (2014) Carbon emission and carbon emission reduction certification and accreditation strategy. China Standards Press, Beijing

Chen J, Wang X (2021) Interval estimation of aircraft fuel consumption based on US-D-DD. Comput Eng Des 42(07):2037–2043

Chen J, Zhao H (2022) Segment fuel consumption depth learning high quality interval prediction algorithm. Comput Eng Des 43(02):450–457

ICAO. ICAO carbon emissions calculator methodology version 11 [EB/OL]. https://www.icao.int/environmental-protection/CarbonOffset/Pages/default.aspx

ICAO. Committee on aviation environmental protection (CAEP) [EB/OL]. (2020-04-05) [2020.11.01]. https://www.icao.int/environmental-protection/CORSIA/Pages/SARPs-Annex-16-Volume-IV.aspx

Jiang H, Guo T, Yang Z (2021) Wheat storage quality prediction model based on improved particle swarm optimization back propagation neural network algorithm. Sci Technol Eng 21(21):8951–8956

Sun Y (2017) Research on carbon verification methods in China. Beijing University of chemical technology, Beijing

Şahin Ö, Aghayeva C (2017) Modeling of fuel consumption for a medium range aircraft. Ömer Halisdemir Üniversitesi Mühendislik Bilimleri Dergisi 2017(2)

Sun H, Huang G, Wang X (2014) An average fuel consumption evaluation model for civil airliner routes. Ind Eng 17(05):41–45

Yao R, Liu J (2021) Aviation carbon verification method based on flight hours. Sci Technol Eng 21(20):8713–8718

Chapter 82
How Does Global Value Chain Embedding Affect Corporate Green Technological Innovation

Zhiyu Li, Tianzun Wang, and Pengjuan Lv

Abstract This article uses Chinese provincial panel data from 2000 to 2019, and uses empirical methods to investigate the factors affecting corporate green technology innovation. The study concluded that: under the background of the division of labor in the global value chain, corporate green technology innovation is facing path dependence and low-end lock-in dilemmas. High energy consumption mode and resource mismatch are important factors that restrict the development of corporate green transformation. At the same time, the concentration of innovative elements and the strengthening of environmental regulations can effectively improve the quality of green technology innovation. Further analysis of heterogeneity shows that the rise in the ranking of industrial structure and value chain is helpful to increase the level of green development in participating in international value chain division and cooperation.

Keywords Global value chain division of labor · Green technology · Innovation

82.1 Introduction

At present, global carbon peaking and carbon neutral actions will bring a series of changes to the technological systems and production methods of enterprises, and will surely lead to a new round of global value chain reconstruction, which will have a major impact on the green transformation of enterprises. The export-oriented growth model has made China as the "world factory". In March 2020, the United Nations released a report that pointed out that about 20% of the world's manufacturing trade in intermediate goods came from China, but Chinese companies' participation in the

Z. Li (✉) · T. Wang
Institute of Economic Research, Hebei University of Economics and Business, Shijiazhuang 050061, China
e-mail: lizhiyu100@heuet.edu.cn

P. Lv
School of Accounting, Hebei University of Economics and Business, Shijiazhuang 050061, China

J. Zhang et al. (eds.), *Environmental Pollution Governance and Ecological Remediation Technology*, Environmental Science and Engineering,
https://doi.org/10.1007/978-3-031-25284-6_82

global division of labor is mainly based on abundant labor, Relying on resources and low costs, it integrates into the global value chain division of labor system through the development of processing trade and other methods, and by built on the basis of high material consumption, high energy consumption, high pollution, and low added value. In the traditional global industrial value chain, the majority of Chinese enterprises are in the most recessed areas of the chain, and the development model that relies on demographic dividends, policy dividends, and environmental pollution dividends will be unsustainable. The restructuring of the value chain will lead to changes in the added value of each link in the value chain, and knowledge creation has become the core driving force in the maturity and the transfer phase of the manufacturing green innovation system. To solve environmental problems, enterprises need to carry out green and low-carbon technological innovation (Jabbour et al. 2008). The realization of green technological innovation in the manufacturing industry is full of uncertainties and needs to rely on technological innovation (Sun et al. 2020). It is important to realize the dual effects of participating in global technological progress and energy saving and emission reduction (Wangab et al. 2021). With the increasingly close division of labor in the global value chain, Chinese companies use green technological reforms to promote the continuous improvement of green competitive advantages and seek to reconstruct the path of the global value chain, which has important practical significance for the green transformation.

Pertinent foreign scholars have conducted rich research on the relationship between value chain embedding and green technological innovation from different perspectives (Lv et al. 2018), the study found that the division of labor in the global value chain led most manufacturing companies to enter the low value-added production link, and the embedding of the company in the value chain did not produce the expected technological upgrading effect. It is necessary to strengthen independent innovation and introduce and absorb innovation Combining to improve the level of technological innovation (Liu and Zhang 2009) and others believe that the construction of the value chain may be the only way for developing countries to achieve high-end upgrades and ultimately obtain international competitive advantages. According to Hashim et al. (2015), enterprise Continuously improve green competitiveness through inter-enterprise cooperation mechanisms (Hashim et al. 2015). The study found that participation in the GVC division of labor has both promotion mechanism and restraint mechanism for the technological upgrading of the manufacturing industry in developing countries (Hui et al. 2020). Qian and Zhou (2014) found that there is a benign positive feedback transaction mechanism between technological progress and the upgrading of foreign trade industries. The research found that green technology innovation and the manufacturing global value chain upgrade are U-shaped, that is, green technology innovation can only promote the upgrade of the manufacturing global value chain by crossing a certain threshold (Han and Yan 2018). Environmental regulations and standards are important factors in promoting international green technology transfer (Qu et al. 1962). Malin et al. (2017) research results show that the deeper the integration into the GVC, the looser the financing environment would be, and the stronger the green innovation abilities of the enterprises. Pan (2019) believes that the high-end embedded path of the global value chain generated by the learning-by-doing effect that can effectively drive the overall transformation and upgrading of the equipment manufacturing industry. By increasing the rank of GVC, Can significantly promote the green growth of China's manufacturing industry (Qu et al. 1962).

Although the existing literature has conducted fruitful research on global value chain embedding and green technology innovation, there are still different opinions and controversies, and the research on the influence mechanism is relatively lacking. This paper attempts to construct a theoretical analysis framework for the impact of green technology progress on China's manufacturing upgrades from the perspective of global value chain division of labor, in order to enrich the theory of green upgrading of foreign trade; empirically test the internal mechanism of enterprises participating in international value chain division of labor affecting green technology progress, to provide a theoretical reference for better promoting the green competitiveness.

82.2 Model Setting and Variable Descriptions

82.2.1 Model Setting

Based on the influence mechanism of traditional manufacturing green transformation from the perspective of international division of labor, this paper takes into account the two factors of manufacturing enterprises' participation in international division of labor and the realization of structural upgrading in the empirical model, and comprehensively considers the influence factors of manufacturing green transformation to construct a panel data measurement model. Model 1 mainly examines the impact of manufacturing companies' participation in international division of labor on green technology innovation; Model 2 examines the impact of manufacturing companies' export product structure upgrades on green technology innovation; Model 3 comprehensively examines the impact of the two and analyzes the context of value chain reconstruction Factors affecting green technology innovation in manufacturing. Among them, the number of green patent applications (GT) is used to measure the green technological innovation of the manufacturing industry; VSS represents the degree of participation of manufacturing enterprises in global sub-projects; GE represents the proportion of high-tech products exports; Control represents the controlled variable, and the subscripts i and t represent region and year respectively. It is a random disturbance item.

$$GTit = \beta 0 + \beta 1 VSSit + \gamma CONTROLit + \varepsilon it \tag{82.1}$$

$$GTit - \beta 0 + \beta 1 TEit + \gamma CONTROLit + \varepsilon it \tag{82.2}$$

$$GTit = \beta 0 + \beta 1 VSSit + \beta 2 TEit + \gamma CONTROLit + \varepsilon it \tag{82.3}$$

82.2.2 Variable Descriptions

Enterprise Green Technology Innovation (GT). Considering that it is difficult to directly measure green technological innovation, the current measurement of green technological innovation mainly focuses on two methods: one is to use green total factor productivity decomposition variable considering environmental pollution to measure green technological progress; the other is to use R&D Investment or technology patents represent the level of green technology innovation. Since there is a high degree of correlation between the number of green patent applications and innovation, it is less affected by external factors, which can reflect the green innovation capabilities of enterprises in various regions. This article uses the proportion of green patent applications in total patent applications to measure the level of green technological innovation of enterprises.

Global Value Chain Sub-Engineering Degree (VSS). China's manufacturing industry participates in the division of labor in the global value chain based on its comparative advantages. It can improve the quality of export products through the import of new intermediate inputs and the technology diffusion effect, and the "learning by doing" effect formed by the division of labor in production processes. This article refers to the method of (Yin 2018), and uses the ratio of the total processing trade to the total import and export in each region to measure the extent to which China's manufacturing industry participates in the division of labor in the global value chain.

Upgrade of foreign trade structure (TE). In the traditional division of labor in international trade, my country's manufacturing industry mainly uses processing trade, relying on high energy consumption and low cost to participate in international trade, and most of the products have low technical content. This article measures the degree of upgrading of foreign trade structure based on the proportion of high-tech products exports in total international trade.

Energy efficiency (ECE). As energy consumption, especially in the industrial sector, is the main source of pollution emissions, adjusting the energy structure and improving energy efficiency in the production process is important means to promote the green transformation of the manufacturing industry. This paper utilizes the rate of decrease in energy consumption per unit of GDP to measure energy consumption efficiency.

Other variables as follows: Environmental regulation (ER) is expressed by the proportion of environmental pollution control investment in GDP; human resources (HR) is calculated by the average number of years of education of employees in each province; Financial development (FD) is expressed by the proportion of provincial loan balances in GDP; marketization index (MI) is expressed by Fan Gang marketization index; foreign direct investment (FDI) is expressed by provincial foreign direct investment data. Through the collection of inter-provincial panel data from 2000 to 2019 in 30 provinces (except Tibet), 600 observations are obtained. The descriptive statistics of the data are shown in Table 82.1.

Table 82.1 Descriptive statistics of panel data

Variable	Mean	Standard deviation	Minimum	Max
GT	1.528	0.739	0.051	2.980
VSS	0.222	0.169	0.001	0.706
TE	0.160	0.173	0.000	0.791
ECE	1.201	0.775	0.226	5.140
ER	1.424	0.805	0.300	5.690
HR	8.564	1.105	5.440	12.783
MI	6.105	2.168	0.320	11.400
FDI	0.367	0.355	0.003	2.394

82.3 Empirical Analysis

82.3.1 Benchmark Regression Results

In order to select the most appropriate panel data measurement model, this paper firstly judges by F test and LSDV test. The results show that the fixed effect model is better than the mixed regression model, and then the Hausman test is used to determine that the fixed effect is better than the random effect. However, the Friedman test, Modified Wald test and Wooldridge test results show that the model has error structures such as cross-sectional correlation, heteroscedasticity, and serial correlation. Therefore, generalized least squares (FGLS) are used for regression analysis. Table 82.2 reports the FGLS estimation results.

From the analysis results of Model 1 and Model 2, the participation of enterprises in the international division of labor has a negative impact on the performance of green technology innovation, and the increase in the proportion of high-tech products exports has a positive impact on green technology innovation, indicating that the higher the degree of participation in international division, the more unfavorable to green technology innovation. Technological innovation, and the increase in the proportion of high-tech products exports has a positive effect on green technological innovation. It may be explained by the fact that most of the companies participating in the division of labor in the global value chain are manufacturing companies that undertake processing tasks. They have low technical content and at the cost of energy consumption, which not only causes more environmental problems, but also due to insufficient innovation capabilities, many manufacturing companies would rather bear the costs brought about by environmental regulations than invest more in research and development funds to carry out green technological innovation. Generally speaking, green technology research and development and application are mainly introduced by external and internal research and development. Most companies need to solve environmental problems by purchasing environmental protection equipment, and choose more economical environmental protection management methods according to the requirements of environmental regulations.

Table 82.2 Benchmark regression results

Explanatory variables	Model 1	Model 2	Model 3
VSS	– 0.0346*** (– 5.22)		– 0.0568*** (– 5.31)
TE		0.0168* (– 2.45)	0.0286*** (2.63)
ECE	– 0.0101*** (– 5.48)	– 0.0096*** – 5.09	– 0.0098*** (– 5.34)
ER	0.0055*** (3.87)	0.0057*** (3.90)	0.0051*** (3.64)
HR	0.0112*** (9.33)	0.0122*** (9.86)	0.0099*** (7.71)
MI	– 0.0043*** (– 6.47)	– 0.0050*** (– 7.48)	– 0.0042*** (– 6.40)
FDI	– 0.0085*** (– 2.99)	– 0.0102*** (– 3.57)	– 0.0075*** (– 2.66)
Constant	0.0142 (1.35)	0.0042 (0.39)	0.0249* (-1.38)

Note The value in parentheses below the regression coefficient is the T value of the coefficient
*, ** and ***indicate significance at the 10%, 5%, and 1% significance levels, respectively

From the regression results of all variables in Model 3, human resources and environmental regulations have a positive effect on green technological innovation and are important driving factors for promoting green technological innovation. The decline in energy consumption, the degree of marketization, and the three indicators of foreign direct investment have a negative impact on the performance of green technology innovation. It shows that the agglomeration of high-end talents is a key factor in advancing technological innovation, and the promotion of green technological innovation requires the support of talents. Environmental regulation is manifested as the government's investment in green development. Although at a certain stage, it will have a crowding-out effect on the company's green technological innovation, but from the regression results, environmental regulation has shown a positive role in guiding and promoting green technological innovation. The reduction of energy consumption efficiency indicates the improvement of production efficiency and environmental efficiency, and promotes the growth of green technology. The negative effects of marketization index and green technology innovation may be due to the problem of resource misallocation. At the same time, companies' green technology innovation is affected by multiple factors such as capital, technology, and information. It is generally difficult for a single company to integrate all the elements restricts the pace of green technology innovation for enterprises.

82.3.2 Robustness Test

Considering that there may be a two-way causal relationship between international value chain embedding and green technology innovation, the application system GMM solves the system to solve possible endogenous problems. The regression results show that the significance and sign of the green technology innovation coefficient of global value chain embedding have not changed. This proves that the regression results are robust. Unresolved estimation errors caused by missing variables, this paper uses the synergy of independent variables and dependent variables to solve the endogenous problem caused by insufficient control variables. considering that the synergy term may lead to a new two-way causality in the model, all explanatory variables are regarded as endogenous variables. At the same time, the lag and difference terms are introduced for the system GMM regression. The estimated results are basically consistent with the benchmark regression, indicating that the regression results have Robustness. At the same time, by replacing the explanatory variable green technology with the proportion of green technology patents in the total number of patents, other variables remain unchanged, and the regression analysis is performed again. The results obtained do not show significant differences, indicating that the regression settings and results are robust.

82.3.3 Mechanism Test

The previous article found that the division of labor in the global value chain of enterprises has a negative effect on green technology innovation. Based on literature review and empirical analysis, the important impacts between the two include three aspects: the upgrading of foreign trade structure, energy consumption efficiency and environmental regulation. Participate in the division of labor in the global value chain and introduce an analysis model to further analyze the mechanism of influence. As shown in Table 82.3, the three factors and the interaction items of the division of labor participating in the global value chain all have a positive effect on green technology innovation. Among them, the improvement of energy consumption efficiency is significant. Strengthening the combination of these factors and participating in the international value chain can help Better promote the green upgrade of enterprises.

82.3.4 Sub-Sample Regression Analyses

Owing to different regions and different time periods, various factors may have different effects on green technology innovation. Therefore, the samples are grouped into eastern, central, and western regions to analyze the heterogeneity characteristics of different regions. At the same time, because the 2008 global financial crisis had an

Table 82.3 Mechanism test results

Explanatory variables	Model1	Model2	Model 3	Model 4	Model5	Model 3
VSS	0.2833*** (0.15)	0.6415*** (0.22)	0.2725*** (0.12)	– 0.0875* (0.19)	0.2661*** (0.12)	0.1937 (0.19)
TE		– 0.0724 (0.24)				
TE × VSS		– 0.5294 (0.50)				
ECE			– 0.0391 (0.04)	– 0.0761 (0.04)		
ECE × VSS				0.3358*** (0.14)		
ER					0.0323* (0.02)	0.0228 (0.03)
ER × VSS						0.0441 (0.09)
Observations	600	600	600	600	600	600
Control variable	Yes	Yes	Yes	Yes	Yes	Yes
Fixed item	Yes	Yes	Yes	Yes	Yes	Yes

Note The value in parentheses below the regression coefficient is the T value of the coefficient
*, ** and ***indicate significance at the 10%, 5%, and 1% significance levels, respectively

important impact on the macro economy and had an important impact on the global value chain, the sample data was divided into two time periods before and after 2008 for comparative analysis.

From the results of sub regional analysis, the eastern region's participation in the international trade division of labor has a positive impact on green technology innovation with a coefficient of 0.211, while the central and western regions' participation in international trade division has a negative impact on green technology innovation. It may be because the eastern region has a relatively high proportion of high-tech industries and a relatively high level of green, but the central and western regions may have a relatively high proportion of low-end manufacturing, relatively high energy consumption and environmental pollution, and low green growth efficiency. The impact of overseas trade product structure on green technological innovation is also different. The eastern region is under a negative growth effect, while the central and western regions have a negative growth effect. This may be due to the rapid growth of green technology innovation in the eastern and western regions, while the high-tech export structure is relatively in a stable development stage, but the proportion of high-tech products exports in the western region is relatively low, and the demand for upgrading the foreign trade structure drives the growth of green technology innovation. The positive impact of human resources and environmental regulations in the eastern region of green technological innovation is also higher than that in the central and western regions.

Table 82.4 Analysis of heterogeneity

Explanatory variables	Regression results by region			Regression results by time period	
	East	Middle	West	2000–2008	2009–2019
VSS	0.211 (1.35)	– 0.050** (– 2.81)	– 0.084*** (– 3.42)	– 0.034*** (– 3.35)	– 0.073 *** (– 2.94)
TE	– 0.093*** (– 3.79)	0.050*** (2.99)	0.579** (2.46)	0.031** (2.15)	0.048* (1.64)
ECE	– 0.187*** (– 3.79)	– 0.03 (– 0.80)	– 0.003*** (– 2.94)	0.001*** (0.24)	0.001 (0.01)
ER	0.008*** (2.77)	0.002*** (1.44)	0.003 (1.25)	-0.001*** (-0.45)	0.004 (1.33)
HR	0.021*** (10.7)	0.018 (7.48)	0.006* (1.81)	0.010*** (7.71)	0.013*** (3.50)
MI	– 0.007*** (– 5.82)	– 0.001 (– 0.18)	– 0.003* (– 1.36)	– 0.003*** (– 3.79)	– 0.001 (– 0.17)
FDI	– 0.001 (– 0.12)	– 0.003 (– 0.12)	– 0.012 (– 0.64)	– 0.010*** (– 3.84)	– 0.004 (– 1.03)
Constant	– 0.051* (– 2.46)	– 0.080* (– 3.22)	– 0.058* (2.27)	0.001* (0.05)	– 0.036 (– 0.83)

Note The value in parentheses below the regression coefficient is the T value of the coefficient
*, ** and ***indicate significance at the 10%, 5%, and 1% significance levels, respectively

From the regression results by time period, the impact of various indicators on green technology innovation is not obvious: First, the two indicators of corporate participation in the international division of labor and the upgrading of foreign trade structure have more significant effects on green technology innovation after 2008. Second, the impact of environmental regulations on green technology innovation is different. Before 2008, environmental regulation had a negative effect on green technological innovation, and after 2008, it had a positive effect, indicating that environmental regulation factors have shown a "U"-shaped effect on green technological innovation, indicating that environmental regulation has entered the stage of promoting green development (Table 82.4).

82.4 Conclusions

Based on the perspective of participating in the global value chain division of labor, this paper uses the provincial panel data model and uses the FGLS model to analyze the participation in global value chain division of labor in green technology innovation. Influencing factors. The research found that: Firstly, participating in the division of labor in the global value chain by processing trade has a negative effect on the performance of green technology innovation, and the upgrading of foreign trade product structure can promote the improvement of the level of green technology innovation. Secondly, the improvement of human resources and environmental regulations are conducive to improving the level of green technology innovation of foreign trade companies, while the level of energy consumption, marketization and foreign direct investment have a negative effect on the performance of green technology innovation. Thirdly, due to the high level of development of high-tech industries in eastern China, participation in the value chain division of labor may drive green technology innovation, while the participation of the central and western regions in the global value chain division of labor has a negative effect on green technology innovation. Fourth, green R&D investment may have a crowding-out effect on the normal production activities of enterprises. It is necessary to further promote the upgrading of foreign trade products, the level of energy consumption, and the strengthening of environmental regulations to be more closely integrated with participation in the division of labor in the global value chain, so as to better promote enterprises to enhance green competitiveness.

Acknowledgements The authors gratefully acknowledge the financial support from the Hebei Social Science Fund Project "Research on the High-quality Economic Growth Path of Hebei Province under Carbon Emission Constraints" (HB20GL013); Hebei University of Economics and Trade research Project "Multi-Scenario Simulation and Path Optimization Research on the Development Potential of Hebei Green Economy Transformation" (2020YB06); Hebei Provincial Science and Technology Program Soft Science Research Special Fund (215576131D).

References

Hashim R, Bock AJ, Cooper S (2015) The relationship between absorptive capacity and green innovation. World Acad Sci Eng Technol 9:1040–1047

Han J, Yan S (2018) The impact of green technology innovation capability on the climbing of manufacturing value chain. Sci Technol Manage Res 38(24):177–182

Jabbour CJC, Santos FCA, Nagano MS (2008) Environmental management system and human resource practices: is there a link between them in four Brazilian companies? J Clean Prod 16(17):1922–1925

Liu Z, Zhang J (2009) Building a national value chain to break through from the low end of the locked global value chain. J Chinese Acad Soc Sci 004

Pan Q (2019) Research on the impact of global value chain embedding on the transformation and upgrading of China's equipment manufacturing. World Econ Res 9:78–96

Pan H, Tang H, Zhang H (2020) How does participation in the division of labor in the global value chain affect the technological upgrading of the manufacturing industry? Based on the theoretical analysis of the mechanism of the GVC division of labor on the technological upgrading of the manufacturing industry. Econ Syst Reform 06:115–122

Qu C, Shao J, Cheng Z (1962) Can embedding in global value chain drive green growth in China's manufacturing industry? J Clean Prod:121962

Qian S, Zhou Y (2014) PVAR analysis of financial development, technological progress and upgrading of China's foreign trade industry. Zhejiang Acad J 2:172–178

Song ML, Wang SH (2017) Participation in global value chain and green technology progress: evidence from big data of Chinese enterprises. Environ Sci Pollut Res 24(2):1648–1661

Sun Y, Bi K, Yin S (2020) Measuring and integrating risk management into green innovation practices for green manufacturing under the global value chain. Sustainability 12(2):545
Yin B, Liu Y (2018) The impact of green technological progress under GVC on the upgrading of manufacturing in China. Progr Sci Technol Countermeas 35(04):56–62
Yue L, Shuai C, Bin S (2018) Will embedding in the global value chain lead to "low-end lock-in" made in China? Manage World 34(08):11–29
Wangab S, Hea Y, Songc M (2021) Global value chains, technological progress, and environmental pollution: inequality towards developing countries. J Environ Manage 2021:110999

Chapter 83
Influence of Environmentally Friendly Snow Melt Agent on Road Ecological Environment

Hongjun Li, Tao Liu, and Kun Zhao

Abstract In order to explore the impact of environment-friendly snow melt agent on road soil and vegetation, snow melt experiments were carried out on the Jilin-Changchun section of G12 Huan-Wu Expressway, and the effects of environment-friendly snow melt agent and traditional industrial salts on road soil and plant growth were compared and analyzed. It was found that compared with the industrial salt plots, the soil physical and chemical properties were more stable, the soil organic matter content was higher, and the plant growth was better in the areas where the environmentally friendly snow melt agent was used, indicating that it had less impact on the ecological environment.

Keywords Environmentally friendly snow melt agent · Expressway · Ecological environment · Soil · Vegetation

83.1 The Introduction

The winter in northern China is long, and the snowfall is frequent and heavy, which seriously affects the safe operation of roads and easily causes traffic accidents. Studies have shown that the traffic accidents rate on snowy days is 12 times higher than that on non-snowy days (Dong et al. 2020). In order to remove snow quickly and efficiently, a variety of snowmelt technologies have emerged. Snowmelt agent is widely used in winter road snow removal because of its advantages such as simple operation, high efficiency and low cost (Shi et al. 2009). Snowmelt agents can be divided into three types: chlorine salts, non-chlorine salts and mixed types. Due to the price advantage, chlorine snowmelt agent has been widely used for a long time, which is easy to cause metal corrosion (Ii and Scherer 2007), soil compaction (Yan et al. 2008), water pollution (Ramakrishna and Viraraghavan 2005), vegetation damage (Viskari and

H. Li (✉) · T. Liu · K. Zhao
China Academy of Transportation Sciences, 240 Huixinli, Chaoyang District, Beijing, China
e-mail: 196962282@qq.com

J. Zhang et al. (eds.), *Environmental Pollution Governance and Ecological Remediation Technology*, Environmental Science and Engineering,
https://doi.org/10.1007/978-3-031-25284-6_83

Karenlampi 2000) and other adverse effects. The environmental problems caused by snowmelt materials have gradually aroused widespread concerns.

At present, most studies focus on the effects of snowmelt agents on soil environment (Azovtseva and Smagin 2018), groundwater (Kelly et al. 2019) and plant growth (Laffray et al. 2018). Studies have shown that in Europe and the United States and other countries, due to the long-term use of snowmelt agents, the concentration of soil metal ions increases, the content of soil water and organic matter decreases, the pH value of soil increases, and soil conductivity increases and there exists a series of negative effects on soil (Galuszka et al. 2011). Therefore, based on the Jilin-Changchun section of the Hun-Wu Expressway, this study carried out the ice-melting test of the environmentally friendly snow melt agent, and analyzes the comprehensive impact of the environmentally friendly snow melt agent on the soil and vegetation.

83.2 The Research Methods

83.2.1 Research Area

The test site is located at the Jilin-Changchun section of the G12 Hunchun-Wulanhaote Expressway, in Jiutai District, Changchun City, in the transition zone between Changbai Mountain and Songliao Plain, with temperate continental monsoon climate, annual average temperature of 5.8 °C and annual precipitation of 560.8 mm. The average annual snowfall is 19.3 mm. In winter, it is necessary to melt and de-ice the road surface, and promote environmentally friendly snow melt agent in K450 + 000 ~ K480 + 000.

83.2.2 Materials and Methods

Materials

The environmentally friendly snow melt agent consists of anhydrous calcium chloride, sodium chloride, soil amendment, humate, indoleacetic acid and other substances. It can effectively reduce the corrosion of metal and concrete structures and improve the antioxidant capacity of seeds, plants and soil under the premise of ensuring the efficiency of snowmelt. It can also be used at the temperature of – 25 °C.

Method

From the winter of 2013–2015, test materials were spread on the left half of the road and conventional snowmelt agent or industrial salt on the right half of the road in the K450 + 000 to K480 + 000 section of the G12 Hunwu Highay. Mechanical spreading

Table 83.1 The use of test materials

Number	Snowfall time	Snow removal (m^3)	Materials (t)	Snowfall time	Snow removal (m^3)	Materials (t)
1	2013/11/15	6774	0	2014/12/10	24,710	15.6
2	2013/11/17	25,645	21.38	2014/12/15	10,129	15.6
3	2013/11/18	38,710	12.38	2014/12/20	4847	8.78
4	2013/11/19	8387	16.88	2014/12/29	21,484	20.48
5	2013/11/20	6774	12.38	2015/1/19	19,637	20.48
6	2013/11/24	7677	15.75	2015/1/26	21,000	20.48
7	2013/12/8	18,806	31.50	2015/2/4	8139	9.75
8	2013/12/10	5806	9.00	2015/2/11	27,234	7.8
9	2013/12/12	7419	20.25	2015/2/16	56,548	47.78
10	2013/12/18	2581	9.00	2015/2/20	57,097	30.23
11	2013/12/24	3274	11.25	2015/2/22	18,823	11.7
12	2014/1/19	15,226	30.38	2015/2/28	16,097	21.45
13	2014/11/30	23,452	19.50	2015/3/8	15,298	20.48
14	2014/12/1	4911	8.78	2015/3/9	5766	9.75

was the main method, and manual spreading was adopted in some sections according to the snow and road conditions. The specific implementation scale is shown in Table 83.1. No snowmelt material was applied to the central divider.

Data collection and processing

In K455 + 200, K463 + 500 and K478 + 625, three tracking and monitoring areas for ecological environmental impact of road domain were set up (Table 83.2). Seven soil sample sampling points were set in each monitoring area, and the sampling points were located on both sides of the subgrade slope and the central divider. The sampling points on the slope are 1.0 m, 2.0 m and 5.0 m away from the shoulder of the dirt road, respectively. A, B, C, D, E, F and G were used to represent each sampling point, and the specific Settings were shown in Fig. 83.1. Soil samples were collected in two layers of 0–20 and 20–40 cm, including 3 undisturbed soil samples and 1 mixed soil sample. The soil samples were sealed and brought back to the laboratory for testing, and the soil density, non-capillary porosity, mechanical composition (represented by the ratio of powder to viscosity), electrical conductivity (represented by TDS), pH value, organic matter content and other related indexes were determined.

Soil samples were collected in mid-September 2013 (before the snow season), early May 2014 (at the end of the snow season), mid-October 2014 (at the end of the rainy season), and early May 2015 (at the end of the snow season).

The 1 m × 1 m herb quadrat and the 2 m × 2 m shrub quadrat were set up on the side slopes and the central divider of the three monitoring areas, respectively. Five

Table 83.2 The basic situation of plants in the monitoring area

Station	Slope	Central
K455 + 200	*Sabina chinensis, Pinus sylvestnis* var, *Acer ginnala, Prunus padus* L., *Lolium perenne* L	*Sabina chinensis*
K463 + 500	*Sophora japonica, Sabina chinensis, Pinus sylvestnis* var, *Amygdalus triloba, Sabina procumbens, Lolium perenne* L., *Ixeris chinensis*	*Sophora japonica Sabina chinensis*
K478 + 625	*Amorpha fruticose, Salix babylonica, Lolium perenne* L., Herba Taraxaci	Anti-glare board

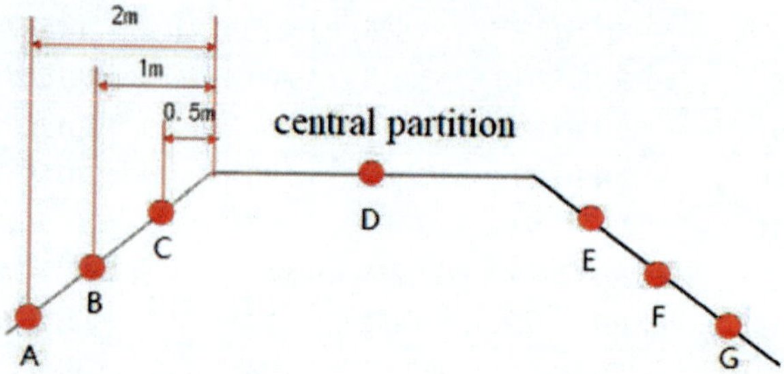

Fig. 83.1 The sampling points

shrubs were selected and marked in the shrub quadrat, and the growth of the plants was observed regularly.

The soil bulk density was determined by the ring knife method, the soil conductivity was determined by the conductivity meter, the pH value was determined by the pH meter, and the soil organic matter content was determined by the potassium dichromate method.

83.3 Data Processing Analysis

83.3.1 Soil Physical Properties

Soil bulk density

It can be seen from Fig. 83.2 that the soil bulk density on the slope was higher than the central partition zone, and it showed a trend that the farther away from the road, the greater the soil bulk density. In the same snow season, the increase of soil bulk density of the slope with test materials was significantly lower than that of the area with industrial salt, and the change of soil bulk density of 20–40 cm soil was slightly lower than that of 0–20 cm soil. The slope with test materials at 1 and 2 m away from the shoulder even showed a decrease in soil bulk density. During the same period (before snow than before, and after snow than after snow), the soil bulk

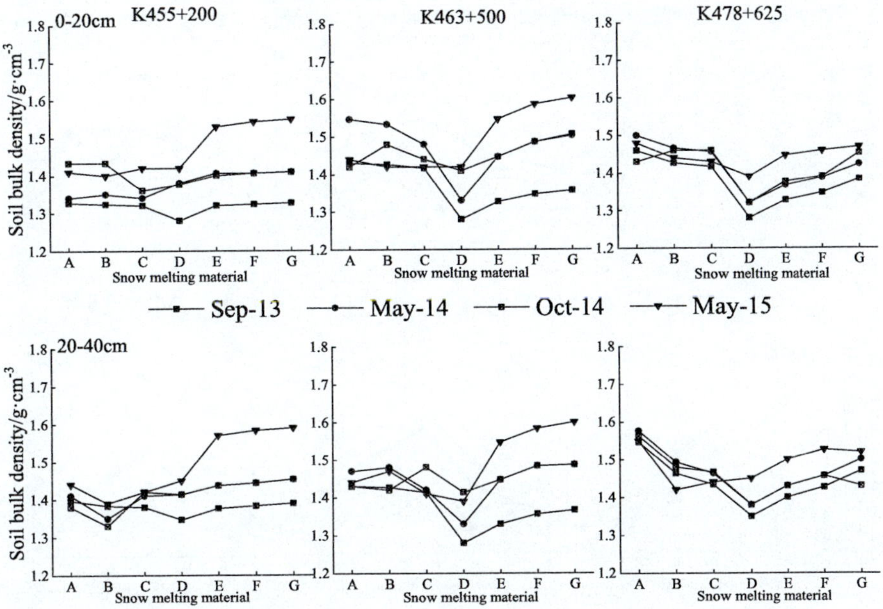

Fig. 83.2 The soil bulk density

density increased, indicating that after natural degradation, the influence of the snow melting agent could not be completely eliminated in short term. The increase of soil bulk density of the slope with test material was relatively small and stable, and there was no sudden increase. The maximum increase of soil bulk density of the slope with industrial salt slope was 17.1%.

Soil non-capillary porosity

According to Fig. 83.3, the non-capillary porosity of soil in the central partition zone was always above 15%, which was the largest in the same period. The farther away from the road surface, the smaller was the soil non-capillary porosity value. After two years of follow-up observation, the decrease of non-capillary porosity of the soil sprayed with industrial salt was larger than that of the experimental material, and the maximum decrease of non-capillary porosity of the soil using industrial salt was 36.54%. The non-capillary porosity of 0–20 cm soil was more regular than that of 20–40 cm soil.

Compared with May 2014 and October 2014, there was no significant difference in soil non-capillary porosity in the industrial salt area. At K455 + 200 and K463 + 500, the non-capillary porosity of soil showed no change, indicating that the influence of industrial salt on soil was difficult to be eliminated in a short period of time. However, the non-capillary porosity of soil increased slightly in October in the area where the test material was used, and its influence on soil could be eliminated

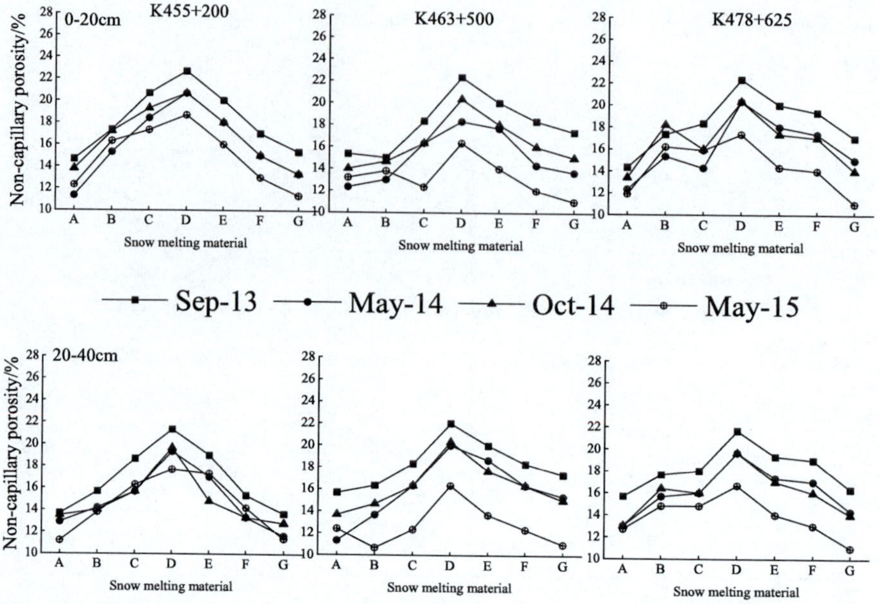

Fig. 83.3 The soil non-capillary porosity

to a certain extent in the short term due to climate changes such as high temperature and rainfall in summer.

Soil pH

It can be seen from Fig. 83.4 that the pH value of the native soil in the monitoring area was 6.5–7.0, which was a weakly acidic soil. After two years of using industrial salt, the soil pH value ranged from 7.0 to 8.4, indicating that the soil was salinized to a certain extent, and the pH value increased by about 5% compared with the initial monitoring period. The change of soil pH value in the central partition zone and the test material was relatively small, indicating that the adsorption components in the test material had a certain inhibitory effect on salt ion diffusion.

Soil conductivity

The electrical conductivity of soil can directly reflect the content of mixed salt, and its value has a linear positive correlation with the total soluble solid content (TDS) of soil.

It can be seen from Fig. 83.5 that the soil conductivity of the central partition zone was significantly lower than that of the slope on both sides of the road. The use of snowmelt materials increased the soil conductivity to a certain extent, and the soil conductivity of the experimental materials at the same time was lower than that of the slope with industrial salt. After two years of observation, the increase of soil conductivity of the test material was between 0 and 14%, and that of the industrial salt area was between 0 and 27%. The increase of soil conductivity of K455 +

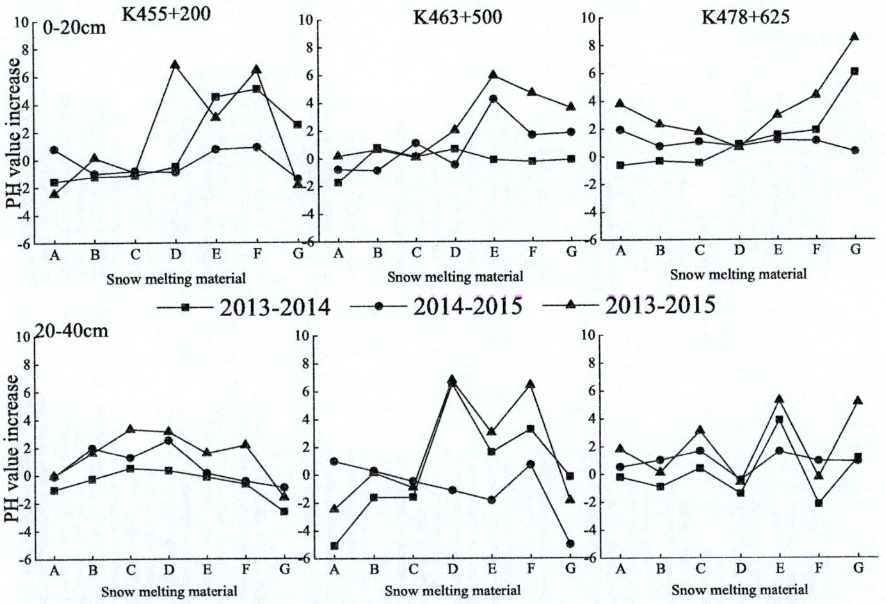

Fig. 83.4 The changes in soil pH

200 and K463 + 500 with scattered drainage was significantly higher than that of K478 + 625 with centralized drainage, and there was no significant difference in soil conductivity on both sides of K478 + 625, indicating that the drainage method had a great influence on soil salinity input. The increase of soil conductivity of 0–20 cm in the central isolation zone was higher than that of 20–40 cm, and there was no obvious rule of soil conductivity of both sides of the slope.

83.3.2 Soil Organic Matter

As can be seen from Fig. 83.6, the content of organic matter in 0–20 cm soil was between 8 and 80 g/kg, and that in 20–40 cm soil was between 10 and 30 g/kg. Among them, the content of organic matter in 0–20 cm soil in K455 + 200 monitoring area was significantly higher than that in the other two monitoring areas. The soil organic matter content in the 0–20 cm central partition zone was significantly higher than that in the two sides of the slope, while the soil organic matter content in the 20–40 cm zone was not significantly different. During the 3-year monitoring period, the soil organic matter content showed a decreasing trend after the use of snowmelt materials. Among them, the organic matter content in the test material area decreased by 0–14%, and the industrial salt decreased by 4–26%. The soil organic matter content in the industrial salt area of K478 + 625, 5 m away from the road shoulder, was

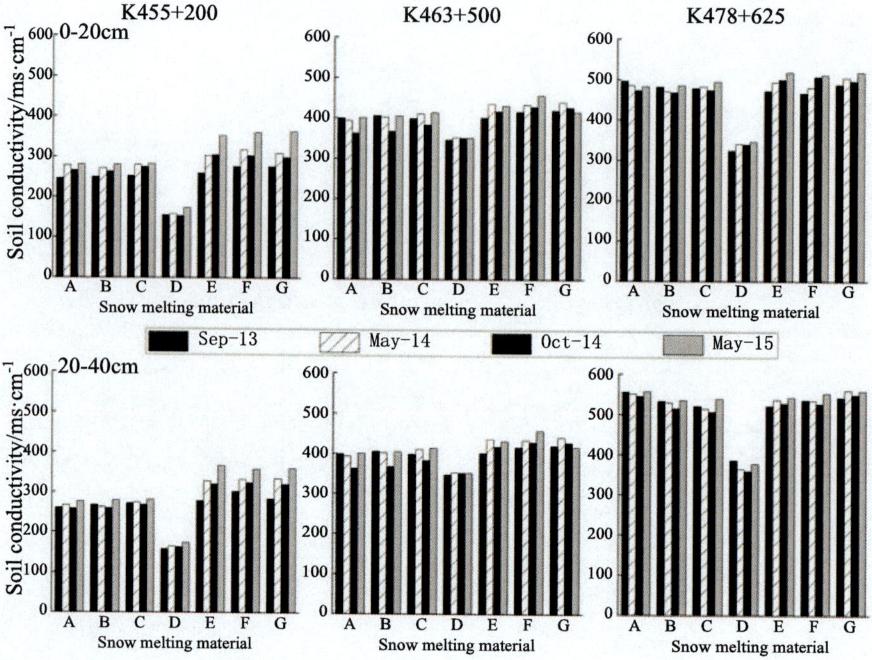

Fig. 83.5 The soil conductivity

significantly higher in 0–20 cm. It's due to a partial accumulation of litter. Among them, the organic matter content in the area with test materials decreased by 0–14%, and that in the area with industrial salt decreased by 4–26%. The soil organic matter content of 0–20 cm in the area with industrial salt 5 m away from the road shoulder was significantly higher, which was caused by the accumulation of some litter.

83.3.3 Plant Growth

Judgment level

The growth of plants is divided into four grades: dead, yellow, damaged and normal (Fig. 83.7). The determination methods of each grade are shown in Table 83.3.

Growth of plants

The growth status of trees and shrubs in the monitoring area was investigated in September 2013 and May 2015, and the results are shown in Tables 83.4 and 83.5.

It can be seen from Table 83.4 that the growth status of trees and shrubs in the area where industrial salt was used decreased compared with that two years ago. Some damaged trees had poor growth status after two years of industrial salt snow removal,

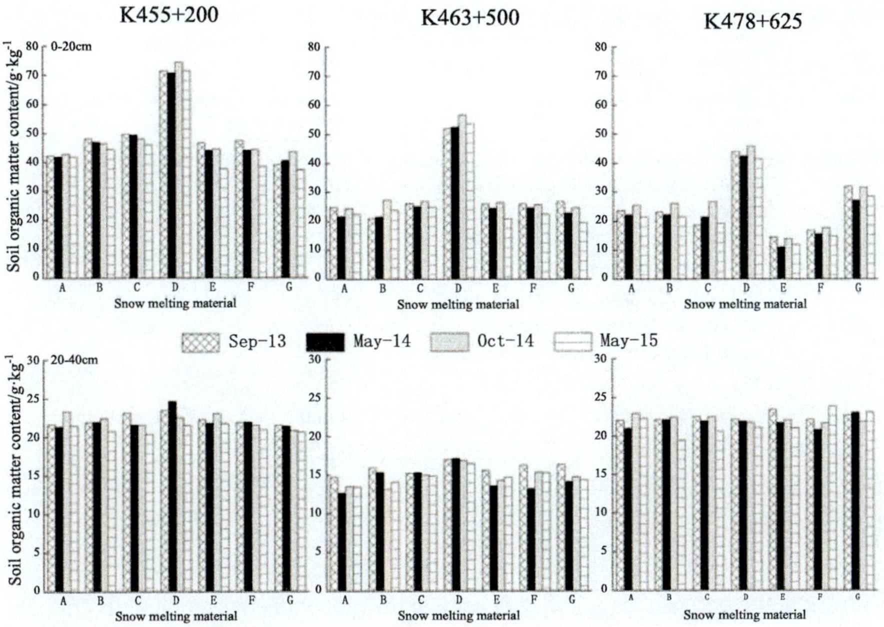

Fig. 83.6 The soil organic matter content

Fig. 83.7 The on-site investigation

mainly *Pinus sylvestris* and Cypress. The growth conditions of trees and shrubs in the central partition zone and the experimental material area were relatively stable. Dead cypress trees were found in the central partition zone of K463 + 500, which may be caused by the lack of water in spring and autumn. The growth status of the damaged Amorpha and willow seedlings in K478 + 625 tended to be good at the end of the monitoring period.

It can be seen from Table 83.5 that the growth status of herbs in the industrial salt plots decreased compared with that two years ago, and ryegrass was yellow and died in a small amount. The growth of herbaceous plants in the central partition zone and the experimental material area was relatively stable and tended to be good at the end of the monitoring period.

Table 83.3 The determination method of plant growth status

Level	Trees and shrubs	Herb
Dead	The plants are dead	Plants die in patches with less than 30% coverage
Yellow	The plant is alive, the growth of branches and leaves is poor, the plant vigor is low, a large number of leaves are yellow and withered, and some branches are dead	The plants are alive and yellow in a large area, with a coverage of 30–50%
Damaged	The plant is alive, the growth of branches and leaves is basically normal, the plant activity is affected to some extent, and some leaves are yellow and withered	The plants are alive, the leaves are yellowed and focal. The vegetation coverage is less than 80%
Normal	The plants are alive, the growth of branches and leaves is normal, and the leaves are not yellowing	The Plant growth is normal, leaves without yellowing, vegetation coverage greater than 90%

83.4 Conclusions

The results showed that the soil bulk density of the plot with environmentally friendly snow melt agent was relatively stable without significant fluctuation. The soil non-capillary porosity of the plots in the central partition zone without snowmelt materials was maintained above 15%. The further away from the road shoulder, the smaller was the soil non-capillary porosity. The decrease of soil non-capillary porosity in the plots using environmentally friendly snow melt agent was smaller than that in the plots using industrial salt. During the monitoring period, the changes in soil pH in plots using environmentally friendly snow melt agent were relatively small, and the increase in soil electrical conductivity was 0–14%, almost half of that in plots using industrial salt. The drainage method of highway slope has great influence on soil salt ion content, and the soil electrical conductivity of the slope with centralized drainage was obviously lower. After the use of snowmelt materials, the soil organic matter content showed a downward trend, and the soil organic matter content in the plot with environmentally friendly snow melt agent decreased slightly, and the plant growth situation was generally stable without significant change. In general, the soil quality and plant growth in the communities with environmentally friendly snow melt agent were relatively stable, but soil nutrients were still lost to a certain extent. Therefore, soil nutrients could be added in the subsequent research and development of environmentally friendly snow melt agent.

Table 83.4 The growth of trees and shrubs

Time	Station	Test materials			Central partition	Industrial salt		
2013.9	K455 + 200	Normal	–	–	Damaged	–	Normal	Damaged
	K463 + 500	Damaged	Normal	–	Normal/damaged	Damaged	Normal	Normal
	K478 + 625	Normal	–	Damaged	–	Normal	Damaged	Normal
2015.5	K455 + 200	Normal	–	–	Damaged	–	Normal	Yellow
	K463 + 500	Damaged	Normal	–	Normal/dead	Yellow	Damaged	Normal
	K478 + 625	Normal	–	Normal	–	Normal	Damaged	Normal

Table 83.5 The growth of herbaceous

Time	Station	Test materials			Central partition	Industrial salt		
2013.9	K455 + 200	Normal	–	–	Damaged	–	Normal	Damaged
	K463 + 500	Damaged	Normal	–	Normal/damaged	Damaged	Normal	Normal
	K478 + 625	Normal	–	Damaged	–	Normal	Damaged	Normal
2015.5	K455 + 200	Normal	–	–	Damaged	–	Normal	Yellow
	K463 + 500	Damaged	Normal	–	Normal/dead	Yellow	Damaged	Normal
	K478 + 625	Normal	–	Normal	–	Normal	Damaged	Normal

Acknowledgements This study was supported by the Scientific and Technological Achievements Promotion Project of the Ministry of Transport (No. 2014 316 T10 010). We would like to thank the staff of the Liaoning Expressway Operation and Management Co. LTD. and Inner Mongolia Shenheng Environmental Protection Technology Co. LTD.

References

Azovtseva NA, Smagin AV (2018) Dynamics of physical and physicochemical properties of urban soils under the effect of ice-melting salts. Eurasian Soil Sci 51(1):120–129

Dong W, Fuqiang Z, Zhongnan T et al (2020) Preparation and effect of an environment-friendly snow-melting agent. J Chongqing Jiaotong Univ Nat Sci Edn 39(6):92–98

Galuszka A, Migaszewski ZM, Podlaski R et al (2011) The influence of chloride deicers on mineral nutrition and the health status of roadside trees in the city of Kielce, Poland. Environ Monit Assess 176(1/2/3/4):451–464

Ii JJV, Scherer GW (2007) A review of salt scaling: I. Phenomenology. Cem Concr Res 37(7):1007–1021

Kelly VR, Findlay SE, Hamilton SK et al (2019) Seasonal and long-term dynamics in stream water sodium chloride concentrations and the effectiveness of road salt best management practices. Water Air Soil Pollut 230:13

Laffray X, Alaoui-Sehmer L, Bourioug M et al (2018) Effects of Sodium chloride salinity on ecophysiological and biochemical parameters of oak seedlings (*Quercus robur* L.) from use of deicing salts for winter road maintenance. Environ Monitoring Assess 190(5):266

Ramakrishna DM, Viraraghavan T (2005) Environmental impact of chemical deicers—a review. Water Air Soil Pollut 166(1/2/3/4):49–63

Shi X, Fay L, Yang Z et al (2009) Corrosion of deicers to metals in transportation infrastructure: introduction and recent developments. Corros Rev 27(1–2):23–52

Viskari EL, Karenlampi L (2000) Roadside scots pine as an indicator of deicing salt use—a comparative study from two consecutive winters. Water Air Soil Pollut 122(3/4):405–419

Yan X, Li F, Liu T et al (2008) Effects of deicing chemicals on ecological environment. Chin J Ecol 27(12):2209–2214

Chapter 84
Talk About the Status and Development of Algae Removal Technology

Genyuan Zhang and Ying Fu

Abstract With the increasing frequency and scale of water bloom outbreaks, the safety of water sources, human health and sustainable development of society are under serious threat. As a large country with scarce water resources, how to deal with the management of water pollution problems caused by algal pollution, the choice of practical new algae removal technology for algal bloom removal and eutrophication of water treatment, to protect human health and ecological environment has become an urgent issue. In this paper, the existing algae removal technologies are sorted out and summarized, and the current situation, advantages and disadvantages of physical, chemical and biological methods are reviewed, as well as the development of new algae removal technologies, in order to provide some reference for the future research and development direction of algae removal methods.

Keywords Eutrophication of water bodies · Algae removal technology · Research progress

84.1 Introduction

In recent years, due to the rapid development of social economy and the acceleration of industrial and agricultural modernization, a large amount of domestic sewage and industrial and agricultural wastewater containing nitrogen and phosphorus are discharged into surface water without adequate treatment, which not only causes physical and chemical changes in natural water bodies and deteriorates the water quality, but also further leads to the occurrence of eutrophication. This has led to the flooding of "water wars" and "red tides", where algae grow and multiply wildly in these water bodies. For example, cyanobacteria outbreaks in Dianchi (Zhang 2020) and algae outbreaks in Taihu Lake (Zhu et al. 2020) have occurred in recent

G. Zhang · Y. Fu (✉)
School of Civil Engineering and Architecture, University of Jinan, Jinan 250022, Shandong, China
e-mail: cea_fuy@ujn.edu.cn

J. Zhang et al. (eds.), *Environmental Pollution Governance and Ecological Remediation Technology*, Environmental Science and Engineering,
https://doi.org/10.1007/978-3-031-25284-6_84

years. Therefore, how to choose practical algae removal technology has become the urgent focus and difficulty to be solved. The current algae removal technologies are mainly divided into chemical, physical and biological methods. This study reviews the current status, advantages and disadvantages of various algae removal technologies, proposes several new technologies for algae treatment, and provides an outlook on the future development direction of algae removal technologies, in order to provide relevant references for the subsequent research of algae removal technologies.

84.2 Physical Method Algae Removal Technology

84.2.1 Ultrasonic Algae Removal Method

The ultrasonic algae removal method can cause localized high temperature and pressure in the water column, cavitation bubble resonance, free radical oxidation and microjet shear effects to damage the structure and function of algal cells (Güyer and Kabashi 2017). Fan et al. (2014) studied the effect of power density and frequency parameters on the removal of *Microcystis aeruginosa*, and the results showed that the removal rate of *Microcystis aeruginosa* reached 99.8% within 30 min at an ultrasound frequency of 580 kHz and an ultrasound intensity of 0.089 W/mL. When the cavitation bubble effect occurs, fewer cavitation bubbles are formed at low frequency ultrasound, the cavitation bubbles grow sufficiently and collapse violently when they collapse, and higher frequency ultrasound requires a greater energy output to achieve the same effect. Srisuksomwong et al. (2011) used 29, 43, 108, 200 and 1000 kHz ultrasound to irradiate fish pond water containing *Microcystis aeruginosa* and found that 200 kHz ultrasound had the greatest effect on deposited *Microcystis aeruginosa* and was effective in reducing Microcystis toxin concentration and moldy odor. Since the energy of ultrasonic waves decays rapidly during propagation, it is difficult to affect algae cells at long distances, and because of concerns about water ecological safety, there are currently few cases of practical application in large lakes.

84.2.2 Air Flotation and Algae Removal Method

The air flotation process can suspend a large number of bubbles in the water, and the impurity particles in the water will adhere to the surface of the bubbles to form bubble flocs to speed up the separation rate and eventually form scum on the water surface, and the air flotation method has a better treatment effect for low turbidity high algae water. Gao et al. (2014) used high-efficiency shallow air flotation process for the treatment of low turbidity high algae water, in the process parameters are: PAC, PAM dosage of 3 and 0.1 mg/L, dissolved air tank inlet pressure of 0.35–0.4 MPa, reflux ratio ≥ 15%, hydraulic load ≤ 6.5 $m^3/(m^2\ h)$ when algae removal was > 80%. Sun (2022) used ozonation microbubble enhanced air flotation for the removal of *Microcystis aeruginosa* and showed that ozone and microbubbles effectively destroyed the cell integrity based on ensuring efficient algal cell removal Dissolved gas pressure was 0.4 MPa, reflux ratio was 20%, PAC and cetyltrimethylammonium bromide (CTAB) were dosed at 130 and 1.0 mg/L, and ozone dosing was 20 mg/L, the removal rate of algal cells reached 94.6%. Nguyen et al. (2019) installed a low-pressure air flotation system on a ferry to remove algae from the Han River and showed that the removal of chlorophyll a was 99% using a low pressure of 1.5 kgf/cm^2, a PAC dosing rate of 30 mg/L, a water circulation rate of 10%, and a hydraulic retention time of 30 min.

84.3 Chemical Method

84.3.1 Oxidation Method

Chemical pre-oxidation is an effective means of remove algae (Li et al. 2015), and the oxidation methods currently used include pre-chlorination, ozone pre-oxidation, and potassium permanganate pre-oxidation. Alireza Khataee et al. (2018) used ozone oxidation to remove oocysts and drum algae, and the results showed that ozone molecules, hydroxyl radicals and superoxide radicals in water destroy algal cells simultaneously, and the removal rates of oocysts and drum algae were 85.18% and 58% at an ozone flow rate of 3 L/h and an oxidation time of 20 min, respectively. Wang et al. (2021a) showed that the effect of potassium permanganate pre-oxidation on algae removal was better than that of ozone pre-oxidation in a pilot study. Wang et al. (2021b) investigated the treatment of *P. aeruginosa* with Fe(II)/peroxymonosulfate pre-oxidation, and the results showed that the pre-oxidized algal cells could remain relatively intact during the removal, and when the PMS dosing rate was 20 mg/L and Fe(II) dosing rate was Yang et al. (2022) combined peroxymonosulfate (PMS) with iron ions for the removal of *Microcystis aeruginosa*, and the results showed that peroxymonosulfate (PMS) could oxidize the extracellular organic matter of algal cells and destabilize the algal cells without causing their large-scale lysis, while it could effectively degrade the algal toxins and PMS will cause the formation of redox cycle of iron ions, resulting in 97.5% removal of algal cells. Zhang et al. (2020) showed that H_2O_2 at a dosage of 80 μmol/L can result in the removal of *Microcystis aeruginosa* at more than 95%, and the oxidation of hydroxyl radicals can effectively control the release of algal toxins.

84.3.2 Coagulant

Luo (2017, 2018) used ferric chloride to remove *Microcystis aeruginosa* at different pH, and the results showed that the removal rate of *Microcystis aeruginosa* at pH 6 was 83%, which was 33% and 37% higher than that at pH 7 and 8, respectively, and it was also found that the extracellular organic matter (EOM) of *Microcystis aeruginosa* would chelate with metal ions and inhibit the removal effect of iron salts on algae. Yu et al. (2017) studied the removal effect of PAC on high algal waters at different alkalinity levels, and the results showed that the best removal effect on high algal waters was achieved at an alkalinity of 2.0 and a dosing rate of 5 mg/L.

84.4 Biological Method

84.4.1 Aquatic Animal Method

Aquatic biology method is also divided into classical biomanipulation and non-classical biomanipulation classical biomanipulation refers to changing the species composition or multiplicity of predators (fish) to manipulate the structure of phytoplankton communities, but there are some problems with the classical biomanipulation theory, which is more successful in shallow lakes with light eutrophication, but difficult to implement in deep lakes with severe eutrophication (Gu et al. 2022). In China, Liu et al. (1999) proposed a non-classical biological manipulation theory to control algae directly by using phytoplankton-feeding fish and mollusks to control lake eutrophication, and successfully applied it in Wuhan East Lake. Yang et al. (2019) conducted in situ fencing silver and bighead carp algae control experiments in Hongshan Bay, Erhai Sea, and the results showed that in, silver and bighead carp could reduce chlorophyll a concentration and had a significant reduction effect on phytoplankton composed of *Microcystis aeruginosa* as the dominant species, and did not lead to eutrophication of water bodies.

84.4.2 Aquatic Plant Method

Nakai et al. (2000) showed that the removal rate of *P. aeruginosa* at 20 °C was 1.9 times higher than that at 30 °C, the removal rate of *P. aeruginosa* at 25 μmol/(m^2 s) was 1.5 times higher than that at 75 μmol/(m^2 s), and the release of chemosensitive substances by *P. foxglove* could effectively Rojo et al. (2013) found that phenolic compounds produced by macrophyte communities could produce a synergistic effect and have a stronger chemosensory inhibition effect on cyanobacteria and diatoms.

Table 84.1 Comparison of physical, chemical and biological methods

Alga-killing techniques	Advantage	Disadvantages
Physical method	Simple operation, long-term effect, little impact on the environment, no secondary pollution	The root cause cannot be fundamentally eliminated, high operation cost
Chemical method	Low use cost, simple operation, quick and good effect	Drug resistance, cause secondary pollution, with great negative effect
Biological method	Remove odor, purify water quality, no toxic substances, no secondary pollution, little impact on the surrounding environment	Difficult to implement, slow effect, possible to introduce alien organisms

84.4.3 Microbiological Method

With the increasing understanding of aquatic ecosystems, microbial algae control techniques will receive more and more attention. Yu et al. (2019) showed that Streptomyces lysogenicus HG-16 could bind to *Microcystis aeruginosa* to form flocs and kill cyanobacterial cells. The algal toxin synthesis of treated *Microcystis aeruginosa* was reduced by 91.2%, and it also released anti-acid, alkaline and proteinase K active substances, which helped to alleviate cyanobacterial blooms. Li et al. (2013) recombined and transferred the genes encoding endopeptidase and perforin 123–124 L from *M. sansevii* into the cells, and the growth of cyanobacteria was significantly slowed down and the cell wall gradually dissolved to disappear (Table 84.1).

84.5 New Technology Research Progress

84.5.1 Advanced Oxidation Method

Chen et al. (2022) loaded bimetallic oxide materials onto the surface layer of granular activated carbon (AC) to produce Cu–FeOx@AC composite catalyst for activation of PMS, the composite catalyst can reduce a portion of algal cells in solution by adsorbing them and also activate PMS to reduce HSO_5^-, promote Fe(III) and Fe(III), Cu(II) and Cu(I) The mutual transformation between Fe(III) and Fe(III), Cu(II) and Cu(I), generating a variety of oxidation groups to achieve the destruction and removal of algal cells. The degradation rate of the system for *Microcystis aeruginosa* was up to 97.25% at 90 min using catalyst 0.5 g/L and PMS 0.2 g/L.

84.5.2 *Photocatalytic Materials*

Wu et al. (2016) used montmorillonite loaded with Cu_2O nanoparticles as adsorbent, flocculant and photocatalyst for the removal of Cu green microcystis, and the results showed that the nano-Cu_2O modified montmorillonite could destroy the surface structure of algal cells and degrade the intracellular components, and the algae removal rate could reach 90.4% after 3 h of visible light irradiation. Wang et al. (2020) successfully produced Z-crystalline g-C_3N_4-MoO_3 (Mo-CN) composite photocatalyst to treat algal pollution, and the best photocatalyst (15Mo-CN) was used to remove 97% of *Microcystis aeruginosa* under visible light irradiation for 3 h. Due to the synergistic effect of h^+ and·OH radicals on the photocatalytic deactivation process of algal cells in the photocatalytic process, the prepared photocatalyst had a degradation effect on the organic matter released from the algae while killing them. The prepared photocatalysts have some degradation effect on the organic matter released from the algae while killing them.

84.5.3 *Floating Algae Removal Technology*

Floating algae removal technology is a new idea of algae removal proposed in recent years, using the combination of algae removal substances and floating materials to prepare floating algae removal substances, which in turn move algae out of the water column. Fan et al. (2021) prepared a new self-floating Ag/AgCl@$LaFeO_3$ (ALFO) photocatalytic hydrogel for the removal of *Microcystis aeruginosa*, and the results showed that the ALFO hydrogel had excellent photocatalytic activity, with 99.4% removal of chlorophyll a within 4 h of photocatalysis, and more than 95% of chlorophyll a could still be removed after six consecutive cycles.

84.6 Outlook

The frequent outbreak of cyanobacterial blooms has brought great trouble to people's production and life. Although single control methods can play a certain effect, but at the same time there are more limitations and drawbacks, in recent years domestic and foreign scholars began to use the combination of different single methods, while the restoration of biological communities to improve the resistance of aquatic ecosystem, in order to achieve better treatment effect. In addition, the further improvement and development of green, efficient, long-life and low-cost new algae removal technologies will become the key research trend of future cyanobacterial water bloom treatment technology.

References

Chen HT, Xiang P, Jiang CW et al (2022) Efficiency and mechanism of Cu-FeOx@AC activated peroxymonosulfate system for algae removal. J Chongqing Univ:1–13

Fan GD, Liu DM, Zhu GC (2014) Influence factors in kinetics during removal of harmful algae by ultrasonic irradiation process. Desalin Water Treat 52(37–39):7317–7322

Fan C, Chen Z, Cu S et al (2021) Self-floating photocatalytic hydrogel for efficient removal of *Microcystis aeruginosa* and degradation of microcystins-LR. Chemosphere 284:131283

Gao X, Fu L, Chen CA et al (2014) Study on algae removal efficiency of efficient shallow air flotation process. China Water Wastewater 30(22):18–22

Güyer GT, Kabashi F (2017) Removal of microcystins in water by ultrasound coupled advanced oxidation techniques. Ultrason Sonochem 30(2):1–22

Gu J, Cui XH, Liu M et al (2022) Review on the applications and problems of biomanipulation in ecological restoration of eutrophic lakes. Ecol Sci 41:237–248

Khataee A, Fazli A, Fathinia M et al (2018) Simultaneous elimination of two species of algae from a contaminated water through ozonation process: mechanism and destruction intermediates. Ozone Sci Eng 41(1):35–45

Li SH, Gao WB, Ou T et al (2013) Cloning and expression analysis of the major capsid proteins, perforin, and endopeptidase genes in the algae-phage Pav-Ld. Acta Hydrobiol Sin 37:252–260

Li J, Wang YJ, Gao P (2015) Damaging effects of hydrogen peroxide on *Microcystis aeruginosa.* Acta Sci Circum 35(04):1183–1189

Liu JK, Xie P (1999) Uncover the mystery of the disappearance of cyanobacteria. Blooms in Wuhan east lake. Resour Environ Yangtze Basin 03:85–92

Luo LN (2017) Study on removal of *Microcystis aeruginosa* from water by coagulation efficiency. Hebei University of Technology

Luo LN, Wang LJ, Yang M et al (2018) Effect and mechanism of extracellular substances on removal of *Microcystis aeruginosa* by ferric salt coagulation. Acta Sci Circum 38(2):437–442

Nakai S, Inoue Y, Hosomi M et al (2000) Myriophyllum spicatum-released allelopathic polyphenols inhibiting growth of blue-green algae *Microcystis aeruginosa.* Water Res 34:3026–3032

Nguyen HV, Kim JK, Chang SW (2019) A case study of low pressure air flotation ferryboat for algae removal in korean rivers and lakes. J Ind Eng Chem 69:32–38

Rojo C, Segura M, Rodrigo MA (2013) The Allelopathic capacity of submerged macrophytes shapes the microalgal assemblages from a recently restored coastal wetland. Ecol Eng 58:149–155

Srisuksomwong P, Whangchai N, Yagita Y et al (2011) Effects of ultrasonic irradiation on degradation of microcystin infish ponds. Int J Agric Biol 13(1):67–70

Sun WT (2022) Study on the efficiency and mechanism of aeroremoval. Shandong Jianzhu University

Wang D, Ao Y, Wang P (2020) Effective inactivation of *Microcystis aeruginosa* by a novel Z-scheme composite photocatalyst. Visible Light Irradiation 746:141–149

Wang XY, Jiang ZW, Fu AM (2021a) Effect of raw water hardness on algae removal by ozone or potassium permanganate preoxidation. China Water Wastewater 37(01):46–50

Wang BJ, Zhang Y, Qin YY et al (2021b) Removal of *Microcystis aeruginosa* and control of algal organic matter by Fe(II)/peroxymonosulfate pre-oxidation enhanced coagulation. Chem Eng J 403:126381

Wu YX, Wang KT, Li H et al (2016) Montmorillonite-supportedwith Cu_2O nanoparticles for damage and removal of *Microcystis aeruginosa.* Visible Light (132/133):79–89

Yang X, Yao L, Wang Y et al (2022) Simultaneous removal of algae, microcystins and disinfection byproduct precursors by peroxymonosulfate (PMS)-enhanced Fe(III) coagulation. Chem Eng J 445:136689

Yu M, Song XQ, Yang P, Zhang WJ et al (2017) Study on the coagulation removal effect of polymer aluminum chloride (PAC) coagulant on algae-rich water. Proc Sci Technol Annual Conf Chinese Soc Environ Sci 2:796–806

Yang JJ, Guo LG, Yin CJ et al (2019) Preliminary evaluation of ecological effects of silver and bighead carps to control cyanobacterial blooms in the early eutrophication lakes. J Lake Sci 31:386–396

Yu Y, Zeng YD, Li J et al (2019) An algicidal streptomyces amritsarensis strain against *Microcystis aeruginosa* strongly inhibits microcystin synthesis simultaneously. Sci Total Environ 650:34–43

Zhang SW (2020) Causes and treatment measures of cyanobacteria outbreak in Dianchi Lake. Technol Innov Appl 33:116–117

Zhu X, L GB, Wang SR (2020) Treatment of a cyanobacteria outbreak in Taihu Lake. Water Resour Prot 36(06):106–111

Zhang XF, Ma YX, Tang TT et al (2020) Removal of cyanobacteria and control of algal organic matter by simultaneous oxidation and coagulation—comparing the H_2O_2/Fe(II) and H_2O_2/Fe(III) processes. Sci Total Environ 720:137653

Chapter 85
Research on Innovative Technology for Cyanide Regeneration from Thiocyanate

Chao Xu, Guangsheng Li, Xingfu Zhu, Tengyue Gao, Yanbo Chen, and Mingming Cai

Abstract The cyanide leaching process produces large amounts of SCN- ions. Ultraviolet light can decompose thiocyanate into highly toxic cyanide, posing a hazard to aquatic organisms. However, the complete decomposition of thiocyanate technology is costly, and its industrial application imposes a heavy burden on enterprises. In this study, the technology of ozone oxidation of thiocyanate to regenerate cyanide is developed through experimental equipment and condition control. By controlling the concentration of metal ions, the initial pH value, and the dosage of the agent (ozone), the regeneration rate can reach over 85%, which achieves the oxidative regeneration of thiocyanate and provides direction and data support for cyanide and the realization of resource reuse.

Keywords Ozone · Thiocyanate · Cyanate · Cyanide · Regeneration

85.1 Introduction

During the cyanidation leaching process of gold smelting, cyanide ions or their metal complexes react with sulfur atoms generated by alkali abrasion reactive sulfides (such as pyrrhotite), and subsequently form thiocyanate (SCN^-). And ultraviolet light can decompose thiocyanate into highly toxic cyanide, which is harmful to aquatic organisms. Therefore, the thiocyanate waste liquid after cyanidation leaching needs further treatment to reduce the thiocyanate concentration to meet the standard before discharge (Soto et al. 2000; Bakir Ogutveren et al. 1999; Dictor et al. 1997; Rice 1997).

Most domestic mainstream gold smelters use flotation gold concentrate as the raw material for cyanidation smelting, and a large amount of SCN^- ions will be produced during cyanidation leaching (Polish Standards for Drinking Water 2002). After years

C. Xu (✉) · G. Li · X. Zhu · T. Gao · Y. Chen · M. Cai
Metallurgical Laboratory Branch of Shandong Gold Mining Technology Co., Ltd., Yantai 261441, Shandong, China
e-mail: xuchao84_08@163.com

J. Zhang et al. (eds.), *Environmental Pollution Governance and Ecological Remediation Technology*, Environmental Science and Engineering,
https://doi.org/10.1007/978-3-031-25284-6_85

of recycling production water, the SCN^- concentration has accumulated to a higher concentration and maintains an upward trend (Kruithof and Kamp 1998).

The cyanidation process for extracting gold or silver usually involves the treatment of sulfide ores, which react with cyanide to form thiocyanate. The formation of thiocyanate is an undesired side reaction, and the process will typically consume excessive amounts of cyanide. There is an increase in the international research on inhibiting thiocyanate formation and treating thiocyanate solution at this stage. So far, there is no commercial application for cyanide regeneration from thiocyanate, but there are several applications to recover cyanide from thiocyanate. Cyanide technology can reduce the toxicity of tailings and reuse thiocyanate to form cyanide to minimize cyanide purchases (Duguet and Wabler 1994).

In order to decompose thiocyanate, it is generally necessary to use reagents with strong oxidizing properties, such as hydrogen peroxide, hydrogen peroxide, chlorine, peroxymonosulfuric acid. (Glaze et al. 1987).

(1) Oxygen: Oxygen is relatively inexpensive in the atmosphere, and using oxygen to regenerate cyanide from thiocyanate has obvious advantages. Related work has shown that, in practice, the oxygen oxidation rate is very slow, and researchers are evaluating the possibility of using oxygen for this reaction in the presence of solid and aqueous catalysts.
(2) Chlorine: Chlorine has been widely used for the degradation of cyanide, which is usually carried out at pH values above about 10.0 and results in complete oxidation of thiocyanate and/or cyanide to form cyanate. At low pH, chlorine may partially oxidize thiocyanate to cyanide in lower yields.
(3) Hydrogen peroxide: Similar to chlorine, hydrogen peroxide has been widely used for cyanide destruction and has been proved to affect cyanide regeneration by thiocyanate at low pH conditions. Related studies have found that hydrogen peroxide accelerates the oxidation of cyanide to cyanate ester, where soluble copper serves as a catalyst. Even at low pH, the presence of dissolved copper leads to the rapid catalytic oxidation of cyanide to cyanate, thereby destroying any regenerated cyanide. The study also found that hydrogen peroxide is weakly oxidizing, which hinders the rate of thiocyanate oxidation to cyanide.
(4) Peroxymonosulfuric acid: It has been found that at low pH conditions, peroxymonosulfuric acid can regenerate cyanide from thiocyanate.
(5) Sulfur dioxide: The most widely used method for cyanide destruction is the Inco method, which can effectively destroy thiocyanate under the condition of manganese-nickel catalyst, and oxidize cyanide to cyanate through reaction. The reaction of thiocyanate to regenerate cyanide may occur at acidic pH, but such a method using sulfur dioxide and oxygen has not been thoroughly studied.

The complete decomposition of thiocyanate technology is costly, and its industrial application imposes a heavy burden on enterprises. Oxidation and regeneration of thiocyanate into cyanide for resource reuse has become an explorable way to solve existing problems.

85.2 The Principle of Reaction

This study mainly aims to regenerate cyanide by ozone oxidation of thiocyanate. The reaction between thiocyanate and ozone requires a two-step process. The first step is the oxidation of thiocyanate to cyanide by ozone, and the second step is cyanide reaction with hydrogen peroxide and ozone to further convert into cyanate, and the reaction equation is as follows:

$$SCN^- + O_3 + H_2O \rightarrow SO_4^{2-} + HCN + H^+ \tag{85.1}$$

$$CN^- + O_3 \rightarrow CNO^- + O_2 \tag{85.2}$$

The cyanate salt generated by the second step reaction is only a reaction intermediate product that meets the waste liquid discharge standard and has no actual economic value. The technical difficulties of cyanide regeneration from thiocyanate oxide lie in the test device and the inhibition of the second-step reaction.

85.3 Test Facility Instruments

In order to overcome the technical difficulties, it is necessary to make a device for preparing cyanide by ozone oxidation of thiocyanate, which has the following characteristics: suppression of the second oxidation step by aeration and other means; regeneration rate of cyanide; adjustment of pH value during the reaction to ensure the blowing off of HCN; completely sealed setting to avoid potential safety hazards caused by HCN overflow; addition of an ozone destruction device to prevent excessive ozone from entering the absorption device; sing alkaline solution to recover cyanide with economic value after thiocyanate oxidation, and improve the recycling value of thiocyanate waste liquid. The specific test device is shown in Fig. 85.1.

Sulfuric acid (initially adjusted pH value) is added to the multi-port reactor with a constant pressure liquid separator; ozone is prepared in an ozone generator with air or oxygen as the production source and fed into the multi-port reactor through positive pressure. The first step of thiocyanate oxidation takes place in a multi-port reactor, where the raw HCN gas is blown out by aeration and positive pressure, and then absorbed with alkali to cyanide as a product.

85.4 Experimental Study on Regeneration of Cyanide by Ozone Oxidation of Thiocyanate

The reaction device is used for ozone oxidation of thiocyanate to regenerate cyanide, and the SCN^- content in the test solution is 5 g/L. During the test, the oxidation rate

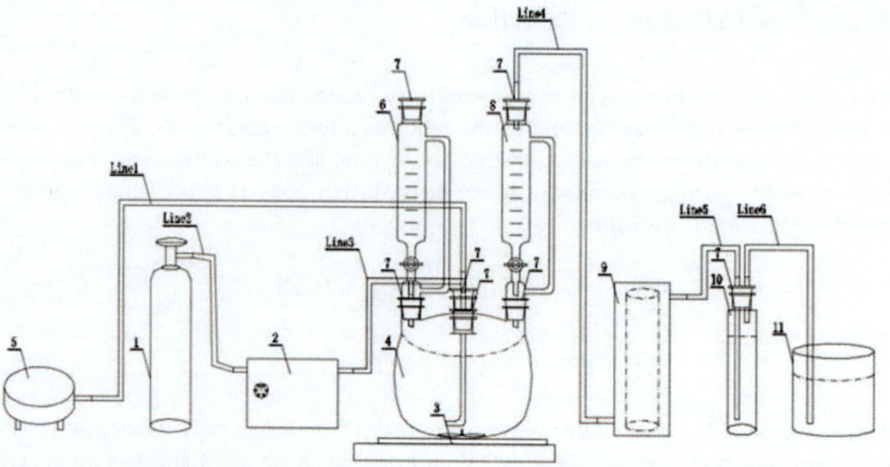

Fig. 85.1 Test device for regeneration of cyanide by ozone oxidation of thiocyanate (1—oxygen tank; 2—ozone generator; 3—magnetic stirring device; 4—multi-port glass reactor; 5—air pump; 6—constant pressure liquid separation device (alkaline liquid); 8—constant pressure liquid separation device (acidic liquid); 9—ozone destruction device; 10—absorption device; 11—tail gas collecting tank)

of thiocyanate is calculated by detecting the SCN^- content in the original solution, the cyanide production is calculated by detecting CN^- in the absorption solution, and the cyanide regeneration rate and conversion rate by ozone oxidation of thiocyanate are calculated by the molar ratio relationship between the reaction original solution and the product. The research finds that the concentration of metal ions, the initial pH value, and the dosage of the agent (ozone) are the main factors affecting the cyanide regeneration rate.

85.4.1 Effect of Metal Ions (Cu, Fe, and Zn) Concentration on Cyanide Regeneration Rate

Cyanide regeneration tests are carried out when Cu, Fe, and Zn contents are 0, 5, 10, 20, 30, 50, and 100 mg/L to determine how the metal ion concentration affects the results and the concentration of metal ions (Cu, Fe, and Zn) are plotted schematically based on the test results. Figure 85.2 shows the regeneration curve of cyanide by ozone oxidation of thiocyanate.

After analysis, we can obtain the following conclusions

(1) During the oxidation of thiocyanate to cyanide, the pH value decreases continuously and finally becomes a strong acid (pH = 2 or so) as H^+ is generated during the reaction;

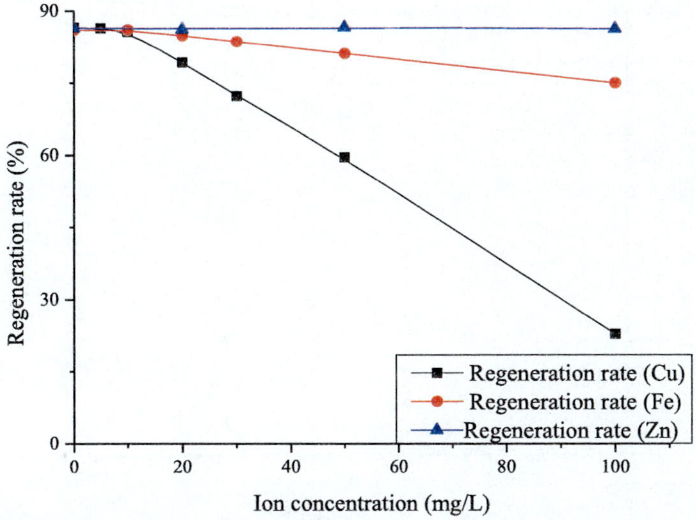

Fig. 85.2 Effect curve of metal ions (Cu, Fe, Zn) concentration on regeneration of cyanide by ozone oxidation of thiocyanate

(2) The concentration of Zn ions influences the process the least, that of Cu ions affects the process the most, and that of Fe ions affects the process intermediately;
(3) Some of the cyanide produced by the oxidation of thiocyanate by iron ions will be complexed by iron ions to form ferricyanide complexes, so the increase of Fe ions will affect the regeneration rate;
(4) Cu ions play a catalytic role in the second step of thiocyanate oxidation and promote the generation of cyanide (CNO^-), which affects the regeneration rate. When the copper ion concentration is less than or equal to 10 mg/L, the catalytic effect is weakened and has little effect on the regeneration rate.

85.4.2 Effect of Initial pH Value on Cyanide Regeneration Rate

Cyanide regeneration tests have been conducted at initial pH values of 10, 8, 6, 5, 4, and 3 to determine their effects. According to the test results, the curves of initial pH on cyanide regeneration by thiocyanate ozonation are drawn, as shown in Fig. 85.3.

According to the results, the low H^+ in the solution is not beneficial for HCN stripping, and SCN^- is continuously oxidized to produce the reactant (2) CNO^-. Therefore, the regeneration rate of cyanide is low when the initial pH is neutral or alkaline. As the initial pH decreases, the yield of cyanide regeneration by ozone-oxidized thiocyanate keeps increasing, and the cyanide regeneration rate reaches the highest value when the initial pH value is less than or equal to 4.

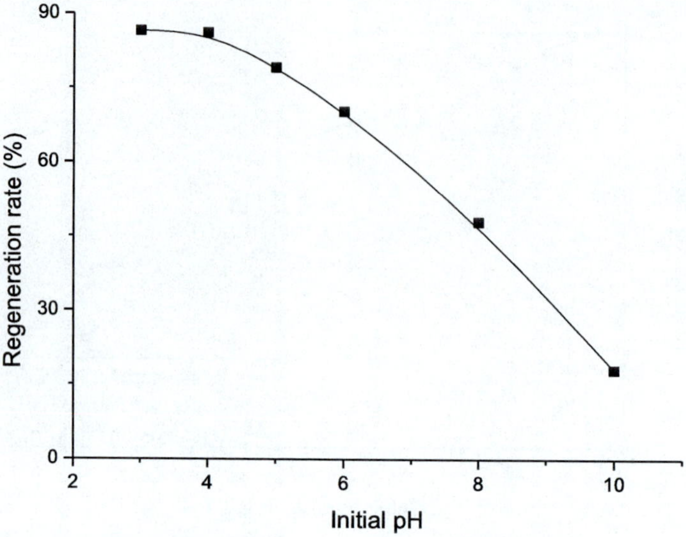

Fig. 85.3 Effect curve of initial pH value on the regeneration of cyanide by hydrogen peroxide oxidation of thiocyanate

85.4.3 *Effect of Ozone Dosage on Cyanide Regeneration Rate*

Cyanide regeneration tests were carried out under the conditions of mol(O_3): mol(SCN) = 0.4, 0.6, 0.8, 1.0, 1.2, 1.4, and 1.6 to determine their effects and the effects of ozone dosage on ozone oxidation thiocyanate regeneration were plotted according to the test results. The cyanide curve is shown in Fig. 85.4.

The research results show that: with the continuous increase of ozone, the SCN oxidation rate and the cyanide regeneration rate both show an increasing trend, and the conversion rate remains above 85%. The optimal dosage of ozone-oxidizing thiocyanate to regenerate cyanide is mol(O_3): mol(SCN) = 1.4.

85.5 Conclusion

Reasonable test device configuration can inhibit the direct oxidation of thiocyanate (CNO^-) by ozone to generate intermediate product CN^-, and combine the volatilization and lye absorption characteristics of HCN to achieve product recovery and resource utilization. By controlling the concentration of metal ions, the initial pH value, and the dosage of chemical (ozone), the regeneration of cyanide from ozone-oxidized thiocyanate can be achieved, and the regeneration rate can reach more than 85%.

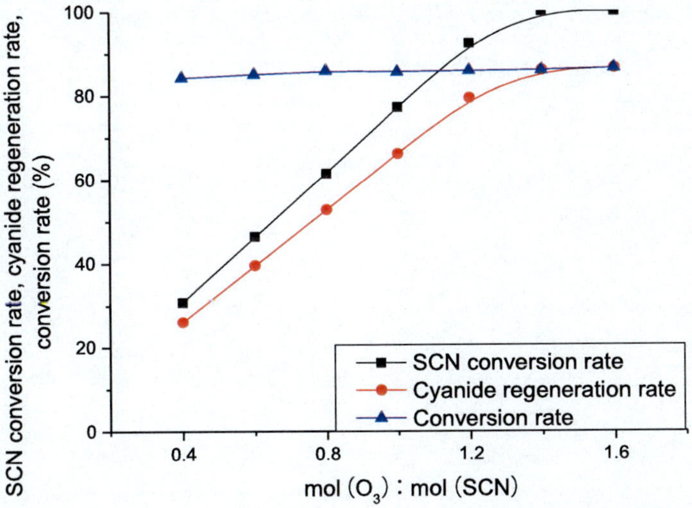

Fig. 85.4 The influence curve of ozone amount on the regeneration of cyanide by oxidizing thiocyanate

References

Bakir Ogutveren U, Toru E, Koparal S (1999) Removal of cyanide by anodic oxidation for wastewater treatment. Water Res 33(8):1851–1856

Dictor MC, Battaglia-Brunet F, Morin D, Bories A, Clarens M (1997) Biological treatment of gold ore cyanidation wastewater in fixed bed reactors. Environ Poll 97(3):287–294

Duguet JP, Wabler O (1994) How to optimize the ozone and ozone/hydrogen peroxide process in order to comply with regulations on micropollutants disinfection and oxidation by-products. In: Proceedings of the international ozone symposium: application of ozone in water and wastewater treatment, Warsaw, Poland, pp 140–148

Glaze WH, Kang JW, Chapin DH (1987) The chemistry of water treatment processes involving ozone, hydrogen peroxide and ultraviolet radiation. Ozone Sci Eng 9:335–352

Kruithof JC, Kamp PC (1998) The dilemma of pesticide control and by-product formation in the Heemsterk drinking water treatment plant design. In: Proceedings EAG ozone generation and application to water and wastewater treatment, Moskwa, pp 331–347

Polish Standards for Drinking Water (2002) Dz. U. Nr 203, poz. 1718

Rice RG (1997) Application of ozone for industrial wastewater treatment—a review. Ozone Sci Eng Int Ozone Assoc 1997:477–515

Soto H, Nava F, Leal J, Jara J (2000) Regeneration of cyanide by ozone oxidation of thiocyanate in cyanidation tailings. Min Eng 8:273–281

Chapter 86
Environmental Regulation, Technological Progress and Chinese Green Industrial Agglomeration

Hong Li

Abstract Industrial agglomeration and pollution prevention and control are two major themes to promote the overall green transformation of China's economic and social development. This paper brings technological progress into the analysis framework of environmental regulation and green industry agglomeration, and uses dynamic panel model and mediating effect model to determine the mechanism of environmental regulation's impact on green industry agglomeration. The two main findings are: (1) The impact of environmental regulation on green industry agglomeration shows a trend of inhibition before promotion; (2) Environmental regulation promotes green industry agglomeration through technological progress. According to the empirical evidence, this paper proposes some policy implications for China to accelerate green industry agglomeration.

Keywords Environmental regulation · Technological progress · Green industrial agglomeration · Mediating effect

86.1 Introduction

Since the reform and opening-up, China has rapidly promoted the process of urbanization and industrialization through the concentration of factors such as labour, capital and technology, and achieved a remarkable development miracle. However, the traditional extensive development mode characterized by the increase of resource input and the expansion of economic scale has also led to serious resource and environmental problems. According to Yale University's 2020 Environmental Performance Index Report, China's environmental performance index is only 37.3, ranking 120th in the world, far behind developed countries such as the United Kingdom (80.3) and France (80.0), and even emerging countries such as Brazil (51.2) and Russia (50.5); Among them, the air quality index score is lower, only 27.1, ranking 137th

H. Li (✉)
School of Digital Economics, Sichuan University Jinjiang College, Sichuan 620860, China
e-mail: xylihong@163.com

J. Zhang et al. (eds.), *Environmental Pollution Governance and Ecological Remediation Technology*, Environmental Science and Engineering,
https://doi.org/10.1007/978-3-031-25284-6_86

in the world. (These data have been taken from *Environmental Performance Index 2020*).

To reduce the environmental impact of the economic development, the central and local governments have issued different types of environmental regulation policy. By rapidly gathering industrial capital elements in a certain space, industrial agglomeration strengthens technological exchange, speeds up technological diffusion and optimizes resource allocation, thus helping to improve production efficiency and achieve energy conservation and emission reduction. Existing studies on the relationship between environmental regulation and industrial agglomeration generally have three different views. First, environmental regulation increases the level of industrial agglomeration. The level of environmental regulation may increase the production cost of enterprises. On one hand, enterprises may improve production and environmental efficiency through technological innovation and other ways to promote clean production and industrial agglomeration. On the other hand, polluting enterprises may be stimulated to further expand production to compensate for investment in environmental governance, leading to an increase in industrial agglomeration (Ding et al. 2020; Du and Chen 2021). Second, environmental regulation decreases the level of industrial agglomeration. By raising the environmental standards, environmental regulation leads to the relocation or withdrawal of polluting enterprises, which is not conducive to industrial agglomeration (Efthymia and Anastasios 2013). Third, the relationship is non-linear (Liu and Ran 2015).

However, the energy needs and pollution emission levels of different industries differ significantly, and they also respond differently to environmental regulatory policies. The goal of promoting green, low-carbon and high-quality economic development can only be achieved if environmental regulation policies are helpful to promoting the agglomeration of industries with "low energy consumption, low pollution and high quality". Most studies consider the overall industrial agglomeration, but do not consider the green industrial agglomeration and environmental regulation mechanism of green industrial agglomeration. Considering the heterogeneity of industrial agglomeration, this paper constructs the green industry agglomeration index and discusses the direct impact of environmental regulation on green industry agglomeration. In this study, technological progress is incorporated into the analysis framework of environmental regulation and green industry agglomeration, and dynamic panel model and mediating effect model are used to determine the mechanism of the impact of environmental regulation on green industry agglomeration. Finally, based on the research results, we put forward some policy suggestions to promote the green industrial agglomeration and realize the coordination in China between industrial agglomeration and environmental improvement.

86.2 Methodology and Variables

86.2.1 Research Model

Due to different capital structures, different enterprises respond differently to environmental regulatory policies. Generally speaking, for industries with "low energy consumption, low pollution and high quality", environmental regulation can improve the comparative advantage of clean industry in market competition and promote clean production and clean industry agglomeration through subsidy effect and technological innovation effect. Xiong and Deng empirically confirm that the industrial agglomeration effect of environmental regulation has industrial heterogeneity (Xiong and Deng 2017).

Environmental regulation affects industrial agglomeration through technological progress. When faced with certain environmental regulations, in order to control their pollution emission level within the scope of environmental regulations, manufacturers can control the pollution level by increasing certain pollution control expenditures, or reduce the pollution level through technological innovation to improve production efficiency. Deng and Wang et al. take China's Ten Thousand Enterprise Energy-saving and Low-carbon Program implemented in 2011 as a "quasi-natural experiment" to investigate the environmental regulation effect on firms' green innovation capability. Empirical results show that this energy-saving and low-carbon policy significantly improves firms' green innovation capability (Deng et al. 2021). Niu and Liu take the adjustment of the SO2 pollutant discharge fee standard in all provincial governments after the promulgation of the document Guofa 2007(15) as a "quasi-natural experiment". The results show that the relationship between environmental protection tax and technological innovation of enterprises is non-linear (Niu and Liu 2021).

Based on the research model, the basic model is presented as model (86.1):

$$GIA_{i,t} = \alpha_0 + \alpha_1 ER_{i,t} + \alpha_2 ER_{i,t}^2 + \alpha_3 fdi_{i,t} + \alpha_4 open_{i,t} + \alpha_5 isu_{i,t} + \alpha_6 ur_{i,t} + \varepsilon_{i,t} \tag{86.1}$$

where *i* refers to the 30 provinces in China, and *t* refers to the years from 2005 to 2018. *GIA* denotes green industrial agglomeration. *ER* indicates environmental regulation. $\alpha_i, i = 0, 1...6$ are the regression coefficients, ε is the random error term. *fdi, open, isu,* and *ur* indicate foreign direct investment, trade openness, industrial structure, and the urbanization level, respectively.

We also consider the fact that there may be strong inertia and continuity effects in the explained variable, the green industrial agglomeration. Therefore, this paper includes $GIA_{i,t-1}$ into the basic model as model (86.2):

$$\begin{aligned} GIA_{i,t} &= \alpha_0 + \delta_1 GIA_{i,t-1} + \alpha_1 ER_{i,t} + \alpha_2 ER_{i,t}^2 \\ &\quad + \alpha_3 fdi_{i,t} + \alpha_4 open_{i,t} + \alpha_5 isu_{i,t} + \alpha_6 ur_{i,t} + \varepsilon_{i,t} \end{aligned} \tag{86.2}$$

Mediating effect tests the indirect effect mechanism through which explanatory variables indirectly affect the explained variables (Mackinnon et al. 2000). This paper uses stepwise regression method to confirm whether there is an indirect effect of technological progress (Baron and Kenny 1986).

To investigate the environmental regulation's technological effect, we construct the mediating effect model as model (86.3) and model (86.4). If θ_1 and γ are significant, $\theta_1 * \gamma$ is the indirect effect of technological progress.

$$\begin{aligned} TFP_{i,t} &= \theta_0 + \eta TFP_{i,t-1} + \theta_1 ER_{i,t} + \theta_2 fdi_{i,t} \\ &+ \theta_3 open_{i,t} + \theta_4 isu_{i,t} + \theta_5 ur_{i,t} + \varepsilon_{i,t} \end{aligned} \tag{86.3}$$

$$\begin{aligned} GIA_{i,t} &= \rho_0 + \eta_1 GIA_{i,t-1} + \rho_1 ER_{i,t} + \rho_2 ER_{i,t}^2 + \rho_3 fdi_{i,t} \\ &+ \rho_4 open_{i,t} + \rho_5 isu_{i,t} + \rho_6 ur_{i,t} + \gamma TFP_{i,t} + \varepsilon_{i,t} \end{aligned} \tag{86.4}$$

86.2.2 Variables

This study uses a panel dataset of 30 Chinese provinces (Tibet, Macao, Taiwan, and Hongkong are not included), for 2005–2018. According to the research model, the variables are green industrial agglomeration, environmental regulation, technological progress, foreign direct investment, trade openness, industrial structure, and the urbanization level.

Green industrial agglomeration (*GIA*) is the location quotient of the province n (n = 1, 2, …, N), which is defined as the proportion of green industry GDP in province n divided by the proportion of green industry GDP in the whole country. The classification of green industries is based on the calculation of comprehensive pollution emission scores of all industries over the years. The industries whose average annual pollution emission score is less than the median of each industry is classified as green industries.

As an important measure to reduce environmental pollution caused by market failures, environmental regulation (*ER*) is conducive to reduce pollution emissions and achieve the goal of environmental quality improvement through environmental constraints imposed to the enterprise. At present, many scholars estimate environmental regulation mainly from the perspectives of environmental investment, environmental effects, and per capita income etc. (Zhang et al. 2011). From the perspective of pollution control and emission reduction performance, this paper uses the comprehensive utilization rate of solid waste as the level of environmental regulation. The higher the comprehensive utilization rate of solid waste in a region, the higher the level of environmental regulation.

Table 86.1 Descriptive statistics of the variables used in this paper

Symbol	Mean	Std.	Min	Max
GIA	1.183	0.345	0.197	2.528
ER	4.139	0.326	2.939	4.603
TFP	1.604	0.770	0.070	2.980
fdi	2.283	1.774	0.000	8.190
open	2.895	0.981	0.520	5.148
isu	3.811	0.214	2.923	4.119
ur	3.948	0.251	3.291	4.495

This paper uses total factor productivity (*TFP*) as technological progress. To measure technological progress with DEA-Malmquist non-parametric method, this paper selects capital and labor as input and real GDP as output.

Foreign direct investment (*fdi*) is expressed as foreign direct investment divided by gdp. Trade openness (*open*) is defined as the total imports and exports divided by GDP. Industrial structure (*isu*) is defined as the economic output share of the secondary industry in GDP. The urbanisation level (*ur*) is defined as the ratio of the urban population in the total population.

86.2.3 Data

The corresponding data on the provincial industry's waste gas, wastewater, solid waste pollution, discharge (or production) as well as the environmental governance investment mainly come from China's Environmental Statistics Yearbook and China Environment Yearbook, etc. The provincial industrial value added of industry data mainly come from China's Industrial Statistical Yearbook. The provincial foreign direct investment, GDP, added value of the secondary industry, urban population, total population can be collected from the China Provincial Statistical Yearbooks. Table 86.1 is the descriptive statistical table of variables.

86.3 Results and Discussion

86.3.1 Results for Non-linear Model

Considering there is a lagged term of the explained variables in the model, the estimation by ordinary least squares method will lead to biased estimation results, so GMM estimation is adopted (Hao et al. 2020). Therefore, this paper uses the dynamic panel method based on SYS-GMM. The corresponding estimated results

Table 86.2 Estimation results for the non-linear model and mediating effect model

Variables	(1) Direct effect	(2) Technological effect	(3) Total effect
L.*GIA*	0.967***		0.994***
L.*TFP*		0.535**	
TFP			– 0.00792***
ER	– 0.763***	– 1.387***	– 0.413**
ER * *ER*	0.100***		0.0531**
fdi	0.0013	– 0.0525***	0.00342
open	– 0.0243***	– 0.0427	– 0.0209***
isu	– 0.0297**	0.199	– 0.0216**
ur	0.0276*	1.884***	0.0615***
Constant	1.548**	– 1.512*	0.709*
N	390	390	390
AR(1)[b]	– 1.698**	– 2.865***	– 1.697*
AR(2)	– 0.00802	– 1.777*	0.223
Hansen (*p*-value)[c]	28.12 (0.709)	29.91 (0.668)	24.36 (0.831)

Notes

[a] ***, ** and *denote significance at the 1%, 5% and 10% levels, respectively

[b] The AR(1) and AR(2) are the first and second-order autocorrelations tests respectively

[c] The *p*-value for Hansen test is given in parentheses

of empirical model (2) and mediating effect model (3) (4) are presented in Table 86.2.

According to column (1) in Table 86.2, the lag period of the explained variable (*GIA*) is significant, the Hansen over-identification test passes. In addition, there is only first-order autocorrelation. The above tests indicate the estimation results of SYS-GMM are reliable (Cheng et al. 2018; Arellano and Bond 1991).

In Table 86.2, Column (1) shows the direct effect of the environment regulation on green industrial agglomeration. The coefficients of environmental regulation are – 0.763 and 0.100, respectively, revealing there is a U curve between explanatory variable and environmental regulation. As discussed above, with the improvement of environmental regulation, enterprises may expand output to compensate for the increase of pollution control cost, which is not conducive to green industrial agglomeration. Then, enterprises will be encouraged to carry out technological innovation and improve production efficiency, thus promoting green industrial agglomeration. We can calculate that, the inflection point is 3.815 [= – (– 0.763)/(0.100 * 2)], and the mean of environmental regulation in 2018 is 3.983, which reveals that environmental regulation is helpful to improve the level of green industrial agglomeration.

As for the control variables, foreign direct investment *(fdi)* has an insignificant positive effect on green industry agglomeration. The coefficient on open is negative at 1% significance level. The likely reason is that China relies on extensive economic growth to expand imports and exports, which is conducive to the agglomeration of polluting industries, but not helpful to the agglomeration of green industries. The industrial structure has a significant negative effect. The urbanisation promotes the agglomeration of green industries.

86.3.2 *Results for the Mediating Effect Model*

Concerning the mechanism of environmental regulation on green industrial agglomeration from the perspective of technological progress, this paper constructs a mediating effect model. As Column (2) in Table 86.2 shows, environmental regulation has a significant negative impact on technological progress. The possible reason lies in the higher level of environmental regulation, which leads some enterprises to push out of the market or shift their production to other regions with lower level of environmental regulation, rather than improving technological progress. As Column (3) shows, the technological progress has a negative effect on green industrial agglomeration. The partial mediating effect is 0.01098504 [0.01098504 = (− 1.387) * (− 0.00792)]. The regression results verify that environmental regulation indirectly promotes green industry agglomeration through technological progress.

86.4 Conclusions and Policy Implications

This paper explores the impact of environmental regulation on green industry agglomeration using Chinese provincial datasets from 2005 to 2018. The paper discusses the mechanism of environmental regulation of green industry agglomeration in China by putting technical progress into the framework of environmental supervision and green industry agglomeration analysis. The results of nonlinear regression analysis show that there is a U-shaped relationship between environmental regulation and green industry agglomeration. Nowadays, the level of environmental regulation in China helps to promote green industry agglomeration. Based on the test of mediating effect, environmental regulation promotes green industry agglomeration through technological progress.

Due to the characteristics of technological innovation, such as large capital demand, high cost of trial and error, high risk of R&D and positive externalities, enterprises lack resources and internal incentives for independent innovation. Therefore, in order to stimulate the innovation momentum of enterprises and promote the agglomeration of green industries, it is suggested to adopt the following policies: First, enhance the universality and equalization of digital finance, optimize the allocation of financial resources, and guide financial resources to green and efficient enterprises; Second, fully seize the opportunities of the new round of scientific and technological revolution and industrial revolution, strengthen cross-regional and cross-organizational technological cooperation, realize the complementarity of resource advantages of all regions, strengthen technological exchanges and cooperation, promote knowledge and technology spillovers, and raise innovation capacity and level; Third, innovation system design, comprehensive use of a number of environmental regulation policies, forced enterprises to speed up the green transformation. Through perfecting the system design, combination of incentives and reversed transmission mechanism, guide the social resources from traditional extensive industry department to "low energy consumption, low pollution, high efficiency" sector, and provide all kinds of policies, resources support for the enterprise technology innovation, improve the level of technology, promote the polluting enterprises, thereby promoting green industry agglomeration.

However, there may be some main limitations in this paper. First, this paper does not include the latest data. Second, environmental regulation may also have spatial spillover effects.

Acknowledgements This work was financially supported by Sichuan University Jinjiang College 2021 Young Teachers' Scientific Research Fund Project (QNJJ-2021-B06).

References

Arellano M, Bond S (1991) Some tests of specification for panel data: Monte Carlo evidence and an application to employment equations. Rev Econ Stud 58(2):277–297

Baron RM, Kenny DA (1986) The moderator-mediator variable distinction in social psychological research. J Pers Soc Psychol 51:1173–1182

Cheng Z, Li L, Liu J (2018) Industrial structure, technical progress, and carbon intensity in China's provinces. Renew Sustain Energy Rev 81:2935–2946

Deng YP, Wang L, Zhou WJ (2021) Does environmental regulation promote green innovation capability? Evidence from China. Stat Res 38(07):76–86

Ding F, Zhuang GY, Liu D (2020) Environmental regulation, industrial agglomeration and urban carbon emission intensity: empirical analysis based on panel data of 282 prefecture-level cities in China. J China Univ Geosci (soc Sci) 3:90–104

Du WC, Chen B (2021) Environmental regulation, industry concentration and environmental pollution. J xi'an Jiaotong Univ (soc Sci) 1:69–77

Efthymia K, Anastasios X (2013) Environmental policy, first nature advantage and the emergence of economic clusters. Reg Sci Urban Econ 43:101–116

Hao Y, Zheng S, Zhao M et al (2020) Re-examining the relationships among urbanization, industrial structure, and environmental pollution in China—new evidence using the dynamic threshold panel model. Energy Rep 6(28):28–39

Liu JL, Ran MS (2015) Environmental regulation, industry heterogeneity and regional industry agglomerations—based on provincial level dynamic panel data model with GMM method. Collected Essays Fin Econ 1:16–23

Mackinnon DP, Krull JL, Lockwood CM (2000) Equivalence of the mediation, confounding and suppression Effect. Prev Sci 1(4):173–181

Niu MC, Liu Y (2021) Will higher sewage charges promote enterprise innovation? With implications for the environmental protection tax in China. Stat Res 38(07):87–99

Xiong HH, Deng WT (2017) Empirical analysis on relationship between environmental regulation, industrial agglomeration, and energy efficiency. Stat Decis 21:117–121

Zhang C, Lu Y, Guo L, Yu TS (2011) The intensity of environmental regulation and technological progress of production. Econ Res 2:113–124

Chapter 87
Rivers and Runoff Transferred Cu to the Jiaozhou Bay

Dongfang Yang and Zaibin Chang

Abstract In terms of the data investigated in May and September as well as October of 1994, the paper determined the horizontal distribution of Cu and its content in the Jiaozhou Bay. The data we got showed that the Cu content in the Jiaozhou Bay of May, September and October changed between the range from 0.72 μg/L to 2.19 μg/L, meeting the seawater standard in quality at Scale I (5.00 μg/L). We could see that the Jiaozhou Bay was immune from Cu pollution in May, September and October, in terms of Cu content. In May, Cu content in the Jiaozhou Bay varied from 0.87 to 1.78 μg/L. In the entry of the Licun River, the Cu content become relatively high of 1.78 μg/L. In northern Jiaozhou Bay, in its coastal areas, the Cu content was 1.66 μg/L. In September, Cu content in the Jiaozhou Bay ranged 1.11–2.19 μg/L. In the northern offshore areas in the Jiaozhou Bay, the Cu content was 2.19 μg/L, a quiet high level. In the entering part of the Licun River, the value of Cu content was 2.14 μg/L. In October, the Cu content was between 0.72 and 1.60 μg/L. In the start point of the Haibo River, the Cu was at a high level, 1.60 μg/L. In addition, the Cu content reached 1.45 μg/L in Licun River's estuary. Speaking of Cu, the water there was of good quality, which means that the water was clean and had no pollution of Cu. The Cuin the Jiaozhou Bay were mainly from two channels: rivers and runoff. The Cu content in rivers was 1.45–2.14 μg/L, and the Cu transported from surface runoff was 1.66–2.19 μg/L. The Cu carried by surface runoff, Licun River and Haibo River conformed to the state standard of seawater at Scale I, 5.00 μg/L, which meant that the surface runoff, the Licun River and the Haibo River did not be polluted by Cu. From May to October, Licun River transported Cu content to Jiaozhou Bay all the way. From May to September, that is, starting from the rainy season to the end, the land surface runoff had always transported Cu content to Jiaozhou Bay. In different months, the sources of Cu content were the rivers on land and surface runoff. Moreover, regardless of time changes, there was only the transportation of Cu content from land but not the transportation from atmosphere and ocean. During

D. Yang (✉) · Z. Chang
Xi'an Traffic Engineering Institute, Xi'an, China
e-mail: dfyang_dfyang@126.com

D. Yang
North China Sea Environmental Monitoring Center, SOA, Qingdao, China

J. Zhang et al. (eds.), *Environmental Pollution Governance and Ecological Remediation Technology*, Environmental Science and Engineering,
https://doi.org/10.1007/978-3-031-25284-6_87

the migration of Cu, the Cu in the land surface runoff was at a high level with the continuous precipitation of Cu. However, as the Cu content carried by surface runoff merged into rivers, the Cu involved in the river was reduced.

Keywords Cu content · Horizontal distribution · Transportation of rivers and land surface runoff · Pollution degree · Jiaozhou Bay

87.1 Introduction

Cu is widely distributed in nature. Human activities emit Cu to land and then surface runoff and rivers put the Cu to the seawater from land (Yang et al. 2017a, b, 2018a, b, c). Therefore, humans are actively paying attention on the migration of Cu in the ocean. Taking the data obtained in 1994 as the background, the paper determined the content of Cu and horizontal distribution as well as the sources, obtained the water quality, the sources of Cu and the sourced quantity, and presented the transportation channels, transported quantity and transported distance, providing tools of science and theory for the sake of research on both sources and pathways of Cu content and its migration in the Jiaozhou Bay.

87.2 Investigated Waters and Involved Materials and Methods

87.2.1 The Jiaozhou Bay's Natural Environment

Nestling in the southern part of the Shandong Peninsula, the Jiaozhou Bay's longitude and latitude is 120° 04′–120° 23′ E and 35° 58′–36° 18′ N, where the bay's boundary is draw up by Tuan Island and Xuejia Island. Near to the Yellow Sea, the Jiaozhou Bay, which is typical of its semi-enclosed form, has an area of about 446 km^2 and its average water depth reaches about 7 m. There are several rivers going to the sea through the Jiaozhou Bay. They are the Yang River, the Haibo River and the Dagu River and the Licun River as well as the Loushan River, which is located in Qingdao with huge flows and a high level of sand content. They all have obvious seasonal characteristics (Yang et al. 2018d, e).

87.2.2 Materials and Methods

The North Sea Monitoring Center of the State Oceanic Administration offered the numbers on Cu content in the Jiaozhou Bay investigated in May, September and

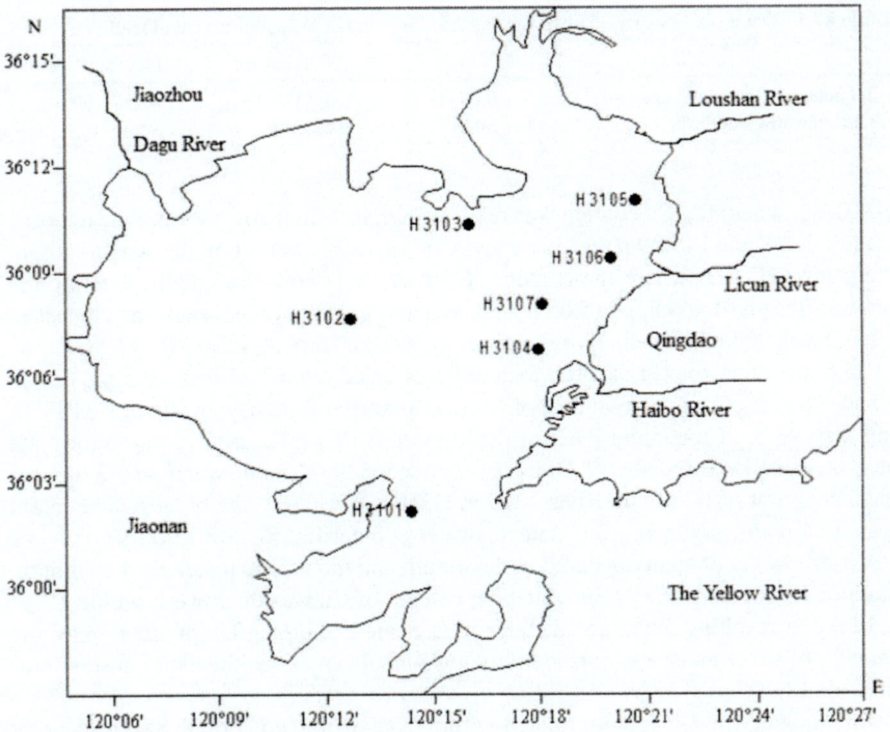

Fig. 87.1 Investigation sites in the Jiaozhou Bay

October of 1994. In May and September as well as October of 1994, 7 sites were sampled, as shown in Fig. 87.1. The samples were taken three times in terms of the water depth (Samples would be taken in surface and bottom waters when it came to waters deeper than 10 m, and samples would be taken in the surface only when the depth is lower than 10 m). The research on Cu was implemented in terms of the state seawater standard. This approach was written in the national "Marine Monitoring Code" (State Oceanic Administration 1991).

87.3 Results

87.3.1 Cu Content

In May, the Cu ranged within 0.87–1.78 μg/L (Table 87.1). The first area with huge Cu appeared in Licun River's sea-entering point, site H3106, where the Cu content was the highest, 1.78 μg/L, meeting the national standard at Scale I (5.00 μg/L). The second high value area showed up in north of the Jiaozhou Bay, in the site

Table 87.1 Water quality in surface of the Jiaozhou bay in May, September and October

	May	September	October
Cu content in the seawater/μg·L^{-1}	0.87–1.78	1.11–2.19	0.72–1.60
State seawater standard	Scale I	Scale I	Scale I

H3103, in which the Cu content was relatively high, 1.66 μg/L, meeting the national standard at Scale I (5.00 μg/L). The low value area showed up in the bay mouth, in the station H3101, where the content of Cu was very low, 0.87 μg/L, meeting the national standard at Scale I (5.00 μg/L). When it came to other areas, the Cu was at a low level, and also reached the state seawater standards at Scale I (5.00 μg/L).

In September, the Cu of the Jiaozhou Bay ranged from 1.11 to 2.19 μg/L (Table 87.1). The water with a great deal of Cu was in north of the bay, in site of H3103. In this station, the Cu content reached the highest, 2.19 μg/L, conforming to the state seawater standard at Scale I (5.00 μg/L). The second high value water was in the sea-entering point of the Licun River, station H3107. In H3107, the highest Cu content was 2.14 μg/L, achieving the state seawater standard at Scale I (5.00 μg/L). The water of low Cu content appeared in the mouth and the waters inside the bay mouth, stations H3101 and H3104, in which the content of Cu was the lowest, within 1.11–1.17 μg/L, reaching the national standard at Scale I (5.00 μg/L). In other areas, the number of Cu content was considerably low and also reached the national standards at Scale I (5.00 μg/L).

In October, the Cu content ranged within 0.72–1.60 μg/L (Table 87.1). The first waters of high value showed up at the Haibo River's estuary, which was station H3104. In the site, the valure of Cu content was considerably high, reaching 1.60 μg/L, meeting the national standard at Scale I (5.00 μg/L). The second area with a high Cu level appeared in the estuary of Licun River, site H3106, in which the Cu content reached its highest, 1.45 μg/L, meeting the national standard of Scale I (5.00 μg/L). The area of low Cu content was in the bay's eastern part, where was the site H3107. In this station, the content of Cu was very low, 0.72 μg/L, meeting the national standard at Scale I (5.00 μg/L). In other waters, the value of Cu content was low and also met the national standards at Scale I (5.00 μg/L).

Therefore, Cu content ranged from 0.72 to 2.19 μg/L in May and September as well as October, which met the seawater standards at Scale I. It showed that the Cu content was less than 3.00 μg/L, far less than the national standards of Scale I. The water was totally clean and had no pollutant of Cu (Table 87.1).

87.3.2 Horizontal Distribution in the Surface Waters

In the estuary point of the Licun River of May, in site H3106, the Cu content was considerably high, at 1.78 μg/L. A area with high Cu content was showed up centered on the estuary point of Licun River, causing the appearance of a set of semicircles in varied gradients. The Cu content become less and less from the high number of 1.78 μg/L in the core to 0.87 μg/L in the mouth of the bay along the gradient, (Fig. 87.2). In the bay's northern areas, in site H3103, the degree of the Cu content was the highest, 1.66 μg/L. A water with high Cu content came to existence, taking the northern coastal waters as the core, a set of parallels of different gradients showed up. The Cu content become less and less from the high content of 1.66 μg/L to the content of 0.87 μg/L in the bay mouth along the gradients (Fig. 87.2).

In September, in the northern offshore regions, in site H3103, the value of Cu content was high, 2.19 μg/L. A high Cu content water showed up centered on the bay's northern offshore waters, coming a set of semicircles in varied gradients to existence. The Cu content reduced from 2.19 μg/L in the center to 1.11 μg/L in the waters inside the mouth of the along the gradients (Fig. 87.3). In Licun River's estuary point, in the site of H3107, the Cu content was comparatively high of 2.14 μg/L. An area of high Cu content was formed, taking the estuary position of Licun River as its core, forming a range of semicircles with varied gradients. The Cu content decreased from 2.14 μg/L in the center to 1.11 μg/L inside the mouth according to the gradient (Fig. 87.3).

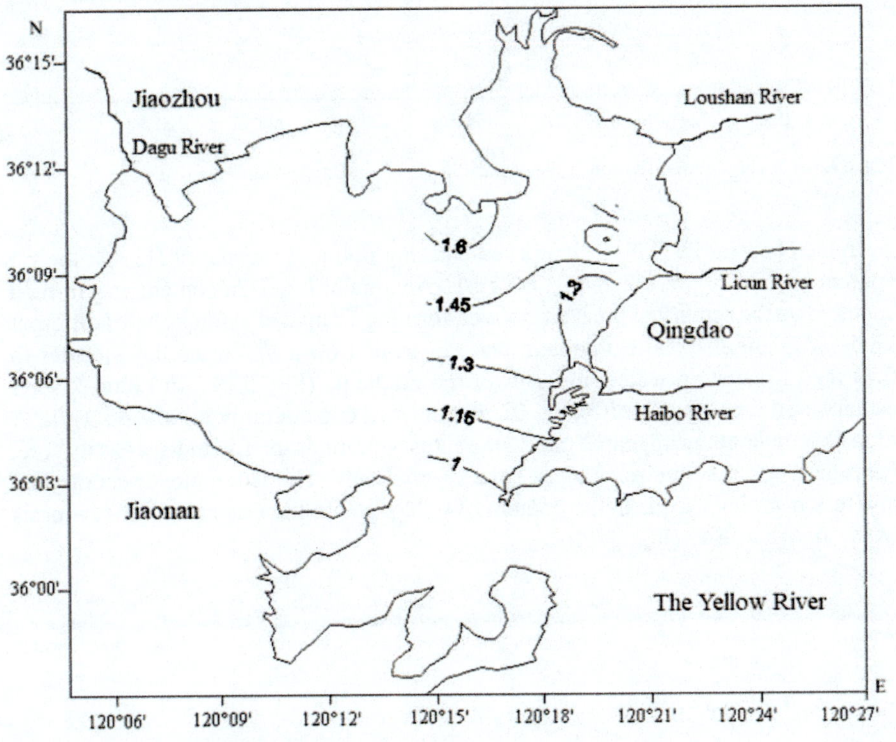

Fig. 87.2 Cu content distribution at the surface in Jiaozhou Bay in May (μg/L)

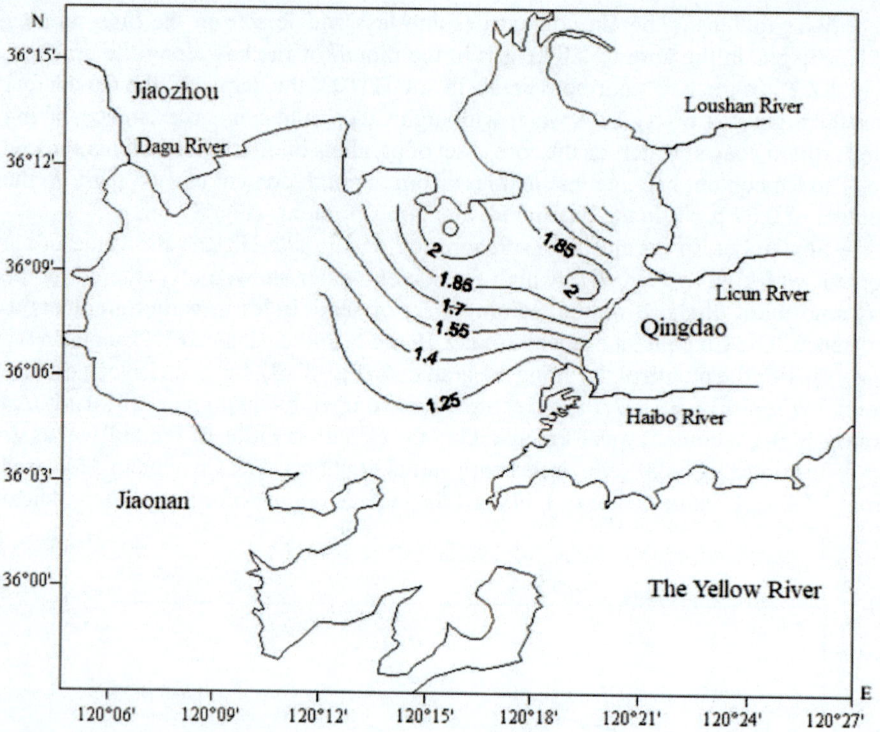

Fig. 87.3 Cu content distribution at the surface in Jiaozhou Bay in September (μg/L)

In October, in the Haibo River's sea-entering point, in the site of H3104, the Cu content was the highest level of 1.60 μg/L. An area of large Cu content was formed centered on the estuary of the Haibo River, forming a range of semicircles of different kinds of gradients. The Cu content was cut from 1.60 μg/L in the bay's center to 0.72 μg/L in eastern waters in terms of the gradients (Fig. 87.4). In Licun River's estuary, at the point of H3106, the Cu content was considerably high, 1.45 μg/L. A high Cu content area appeared taking the estuary point of the Licun River as its core, forming a set of semicircles in all sorts of gradients. The Cu content become less and less from 1.45 μg/L in the content of 0.72 μg/L in the eastern regions in terms of the Jiaozhou Bay (Fig. 87.4).

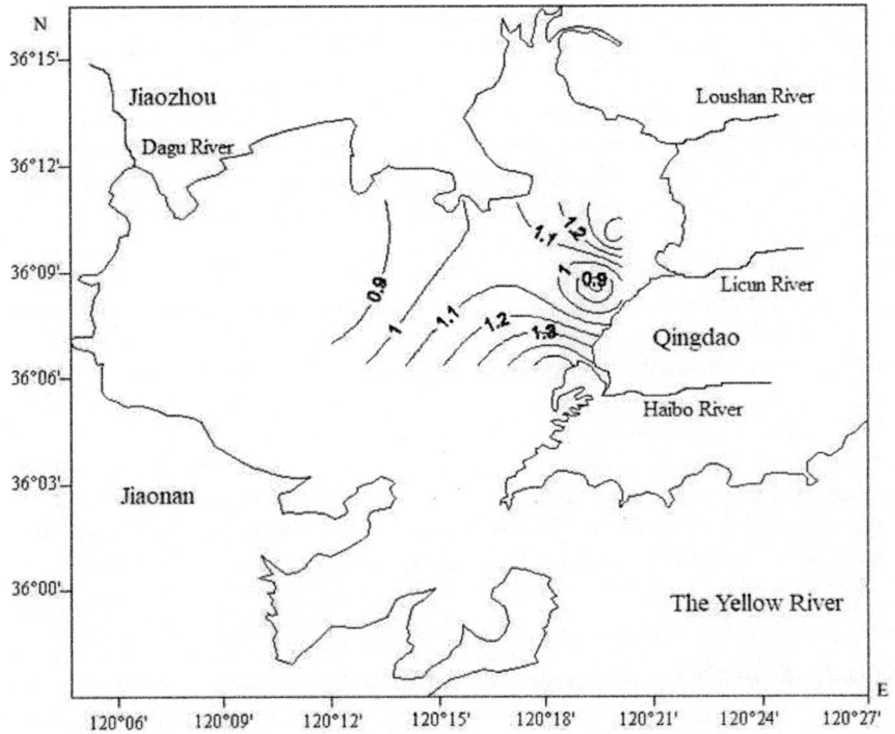

Fig. 87.4 Jiaozhou Bay's Cu content distribution at the surface in October (μg/L)

87.4 Discussion

87.4.1 Water Quality

In May and September as well as October, the value of the Cu content was between 0.72 and 2.19 μg/L, achieving the state seawater standards at Scale I (5.00 μg/L). It showed that speaking of Cu content, in the same three months, the whole Jiaozhou Bay had no Cu pollution.

During May, the changing scale of Cu content was 0.87–1.78 μg/L, suggesting that the bay's water body had no Cu pollution. In the estuary point of the Licun River, the content of Cu became a considerably high value of 1.78 μg/L. In the bay's offshore regions, the level of Cu was 1.66 μg/L, We can see that the water in the estuary point of Licun River and the bay's northern part did not be tainted by Cu. In other areas, the Cu content was at a relative low level, much less than 5.00 μg/L, suggesting that in other regions of the bay, the water there was of high quality and did not get tainted by Cu.

There months later, September, the changing range of Cu content was between 1.11 and 2.19 μg/L, registering that water mentioned above was not tained by Cu at all. In its northern part, the Cu content increased to a considerably high level, 2.19 μg/L, and in the Licun River's estuary point, the Cu content reached relatively high of 2.14 μg/L. This showed that in the Jiaozhou Bay's offshore regions and its north and the estuary waters of the Licun River, the water quality there was not tainted by any Cu. In addition, in other regions, the number of Cu content was comparatively low, much less than 5.00 μg/L. This meant that the water there, in terms of Cu content, had reached a level of high quality, which meant that the water was clean, and it was completely free of any pollution of Cu.

In the next month of September, the changing degree of Cu content was 0.72–1.60 μg/L, registering that the waters were not subject to any pollution of Cu. In Haibo River, the Cu content reached a comparatively high value, 1.60 μg/L, and in the Licun river's estuary point, the Cu content was 1.45 μg/L, in the estuary waters of the Licun River in Jiaozhou Bay, the number of Cu content reached a high value of 1.45 μg/L, which showed that in the estuary points of the Haibo River and the estuary points of the Licun River, the water was not contaminated by any Cu. In addition, in other water areas, the Cu content was comparatively low, much less than 5.00 μg/L, which registered that in other regions of the bay, the water there had high quality with clean water and did not get polluted by Cu in terms of Cu content.

87.4.2 Sources

During the time of May, region with a high content of Cu came to existence in the estuary point of the Licun River, which showed that the Cu carried by the Licun River increased to a considerably high degree of 1.78 μg/L. In the north, a Cu content region appeared, which registered that the Cu content was transferred from the runoff, which was 1.66 μg/L.

Three months later of the same year, areas with high Cu content was formed in the bay's north, indicating that the Cu content came from land surface runoff, and the Cu content was comparably high, 2.19 μg/L. Moreover, in the estuary positions of the Licun River, a region with a great deal of Cu was formed, which indicated that the source channel of Cu was from the Licun River, and its Cu content was 2.14 μg/L, and the content of transport was rather high.

In the next month of September,, areas with high Cu concentration showed up in the entry of the Haibo River, manifesting that the Cu content of a comparatively high content of 1.60 μg/L was from Haibo River. In the estuary points of the Licun River, a region carried high Cu content was formed as well, meaning that the Cu content came from the Licun River with a considerably low quantity of 1.45 μg/L.

Therefore, there were two sources of Cu. One was the transfer by rivers, the other was the transportation by land surface runoff (Table 87.2). The Cu content carried by rivers was 1.45–2.14 μg/L and transferred by land surface runoff was 1.66–2.19 μg/L.

Table 87.2 The Cu content transported from different sources in Jiaozhou bay

Sources	Transportation of rivers	Transportation of runoff
Cu content/μg·L^{-1}	1.45–2.14	1.66–2.19

87.4.3 Change of Cu Content in Sources

During May, September and October, the Cu content in the Jiaozhou Bay sourced from the transfer of rivers and land surface runoff.

In May, Cu content sourced from the Licun River reached 1.78 μg/L.

In May, Cu content sourced from land surface runoff was 1.66 μg/L.

In September, Cu content sourced from land surface runoff was 2.19 μg/L.

In September, Cu content sourced from Licun River was 2.14 μg/L.

In October, Cu content sourced from Haibo River was 1.60 μg/L.

In October, Cu content sourced from Licun River achieved 1.45 μg/L.

In different months, the transportation of Cu content was from rivers and land surface runoff. In addition, regardless of the time changing, the Cu was only transported on the land, and there was no transportation from atmosphere and ocean. Moreover, the Cu content transferred by runoff was slightly higher than that transported by rivers, which revealed that during the migration of Cu content, as the Cu content continued to subside, the Cu content transported from surface runoff was considerably high, and the Cu transported by the runoff merged into rivers, so the Cu from the river decreased.

In May, September and October, Licun River had always transported Cu content to the Jiaozhou Bay all the time, which demonstrated that there was always Cu content in Licun River from May to October.

In May and September, the land surface runoff had always transported Cu content to the Jiaozhou Bay all the time, which demonstrated that there was always Cu content in the land surface runoff from May to September, that is, from the start of rainy season to the end. Cu content had always existed on land.

87.4.4 Pollution Degree in Sources

In May, the Cu from the Licun River was 1.78 μg/L. It was 1.66 μg/L of Cu content transported by land surface runoff, achieving the state seawater standard at Scale I (5.00 μg/L), registering the Licun River and land surface runoff did not get contaminated by Cu (Table 87.2).

In September, the Cu carried by runoff was 2.19 μg/L and was 2.14 μg/L transported by Licun River, achieving the state seawater standard at Scale I (5.00 μg/L), which meant the Licun River and runoff did not be contaminated by Cu (Table 87.2).

In October, the Cu from the Haibo River was 1.60 μg/L and was 1.45 μg/L transported by Licun River, meeting the national standard at Scale I (5.00 μg/L), which manifested that the Licun River and Haibo River had no Cu pollutant (Table 87.2).

Therefore, the Cu carried by runoff, Licun River and Haibo River matched to the state standard at Scale I (5.00 μg/L), which meant that the land surface runoff, the Licun River and the Haibo River did not get polluted by Cu at all.

87.4.5 Pathways Transported Cu

There are two pathways to transport Cu content: rivers and land surface runoff. Therefore, human activities bring Cu to land, and reach the Jiaozhou Bay through these two pathways.

From land, through ocean and then to the Jiaozhou Bay, the migration presented two pathways to transport Cu content. (1) Human activities directly emitted Cu land. Through the rain, the Cu content was transferred to the Jiaozhou Bay through runoff which carried the Cu of 1.66–2.19 μg/L. (2) Human emitted Cu to the land, where the Cu was transferred to the ocean through rivers. The Cu transferred to the Jiaozhou Bay by rivers was 1.45–2.14 μg/L, which revealed that the Cu content from human activities to land arrived the Jiaozhou Bay by two pathways (Fig. 87.5).

The sequence of Cu transferred from different sources to the Jiaozhou Bay in descending is: land surface runoff and rivers. Therefore, the Cu content was brought

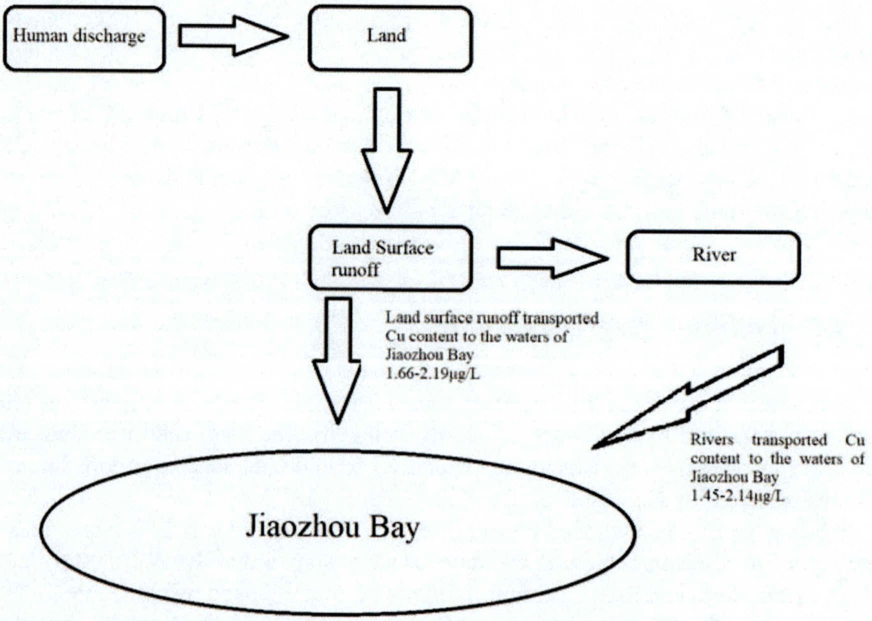

Fig. 87.5 The transportation pathways of Cu contents to the Jiaozhou Bay

to land through human activities and was transferred to the ocean by runoff and rivers. In the aspect of Cu content, the ocean had no Cu pollutant. The environmental protection from humans had shown significant effect.

87.5 Conclusion

In Jiaozhou Bay, the variation range of the value of Cu content in May, September and October was between 0.72 and 2.19 μg/L, reaching the water quality standards for Class I seawater (5.00 μg/L). This shows that in terms of Cu content, in May, September and October, in the entire waters of Jiaozhou Bay, the water quality was not subject to any pollution of Cu content. In May, the range of Cu content's variation in the Jiaozhou Bay was 0.87–1.78 μg/L, meaning that the water body of the Jiaozhou Bay was not subject to any pollution of Cu content. The water quality in Licun River's estuary and northern Jiaozhou Bay had no pollutant of Cu. In other water areas of the bay, the water had high quality with clean water and did not get polluted by Cu in terms of Cu content.

In September, In September, the change range of Cu content in Jiaozhou Bay was from 1.11 to 2.19 μg/L, registering that the Jiaozhou Bay was clean and did not polluted by Cu. In the coastal waters north of Jiaozhou Bay and the estuary waters of the Licun River, the water quality is not subject to any pollution of the Cu content. In addition, in other water areas of the bay, the water had high quality with clean water and did not get polluted by Cu in terms of Cu.

During October, Cu content in the Jiaozhou Bay was between 0.72 and 1.60 μg/L, and the waters of the Jiaozhou Bay was not subject to any pollution of the Cu content.. In the estuary waters of the Haibo River in Jiaozhou Bay and the estuary waters of the Licun River, the water quality is not subject to any pollution of Cu content; In other water areas, the water had high quality with clean water and did not be polluted by Cu.

There are two sources of Cu content in the waters of Jiaozhou Bay, which come from the transport of rivers and surface runoff. The Cu content transported by rivers was 1.45–2.14 μg/L, and the Cu content transported from surface runoff was 1.66–2.19 μg/L. The Cu content in surface runoff, the Licun River and the Haibo River to the Jiaozhou Bay conformed to the National first-class seawater quality standards (5.00 μg/L), which showed that the surface runoff, the Licun River and the Haibo River were clean and not contaminated with any Cu content.

In May, September and October, the Licun River had always transferred Cu to the Jiaozhou Bay all the time. There was always Cu content in the land surface runoff from May to September, that is, from the start of rainy season to the end. Cu content had always existed on land and transported to the Jiaozhou Bay.

In different months, the transportation of Cu was from rivers and land surface runoff. In addition, regardless of the changing time, Cu was only transported on the land, and there was no transportation from atmosphere and ocean. Moreover, the Cu content carried by runoff was slightly higher than that transported by rivers, which revealed that during the migration of Cu content, as the Cu continued to subside, the Cu content transported from the land surface runoff was considerably high, and the Cu content transferred by the runoff merged into the water body of rivers, so the Cu content transported by the river decreased.

References

State Oceanic Administration (1991) The specification for marine monitoring. China Ocean Press, Beijing

Yang D, Zhu S, Chai J, Wu Y, Yang X (2017a) Transport trend of Cu in waters and the mechanism in Jiaozhou Bay. WOP Eng Sci Res 6:148–151

Yang D, Ma W, Wang Z, Tao X, Zhu S (2017b) Mechanism of Cu's high settling process in Jiaozhou Bay. WOP Eng Sci Res 6:152–155

Yang D, Su C, Zhu S, Wu Y, Zhou W (2018a) Theories of transporting processes of Cu in Jiaozhou Bay. Earth Environ Sci 113(012130):1–4

Yang D, Suo H, Zhu S, Wang M, Xiuqin Y (2018b) A comprehensive research on the transporting laws of Cu in Jiaozhou Bay. WOP Eng Sci Res 9:237–240

Yang D, Suo H, Zhu S, Su C, Zhou W (2018c) Quantification of the vertical transporting process of Cu in Jiaozhou Bay. Earth Environ Sci 128(012167):1–6

Yang D, Miao Z, Zhu S, Fengyou W, Xiuqin Y (2018d) Examination of Yang's marine water filter phenomenon. Earth Environ Sci 153(062064):1–7

Yang D, Miao Z, Li H, Zhang X, Wang Q (2018e) The homogeneity changing process of Cu in Jiaozhou bay waters. Earth Environ Sci 153(062066):1–7

